IUPAP C20 Conference on Computational Physics

(CCP 2011)

Journal of Physics: Conference Series Volume 402

Gatlinburg, Tennessee, USA
16-20 October 2011

Editors:

G Malcolm Stocks
M Claudia Troparevsky

ISBN: 978-1-62276-868-4
ISSN: 1742-6588

Printed from e-media with permission by:

Curran Associates, Inc.
57 Morehouse Lane
Red Hook, NY 12571

Some format issues inherent in the e-media version may also appear in this print version.

Printed by Curran Associates, Inc. (2013)

For permission requests, please contact the Institute of Physics
at the address below.

Institute of Physics
Dirac House, Temple Back
Bristol BS1 6BE UK

Phone: 44 1 17 929 7481
Fax: 44 1 17 920 0979

techtracking@iop.org

Additional copies of this publication are available from:

Curran Associates, Inc.
57 Morehouse Lane
Red Hook, NY 12571 USA
Phone: 845-758-0400
Fax: 845-758-2634
Email: curran@proceedings.com
Web: www.proceedings.com

Materials/condensed matter theory and nano science

Strongly correlated systems and quantum phase transitions

Astrophysics

Nuclear and high energy physics

Macroscopic transport, mesoscopic methods and related

Biological physics

PREFACE

Increasingly, computational physics stands alongside experiment and theory as an integral part of the modern approach to solving the great scientific challenges of the day on all scales – from cosmology and astrophysics, through climate science, to materials physics, and the fundamental structure of matter. Computational physics touches aspects of science and technology with direct relevance to our everyday lives, such as communication technologies and securing a clean and efficient energy future.

This volume of *Journal of Physics: Conference Series* contains the proceedings of the scientific contributions presented at the 23rd Conference on Computational Physics held in Gatlinburg, Tennessee, USA, in November 2011. The annual Conferences on Computational Physics (CCP) are dedicated to presenting an overview of the most recent developments and opportunities in computational physics across a broad range of topical areas and from around the world. The CCP series has been in existence for more than 20 years, serving as a lively forum for computational physicists. The topics covered by this conference were: Materials/Condensed Matter Theory and Nanoscience, Strongly Correlated Systems and Quantum Phase Transitions, Quantum Chemistry and Atomic Physics, Quantum Chromodynamics, Astrophysics, Plasma Physics, Nuclear and High Energy Physics, Complex Systems: Chaos and Statistical Physics, Macroscopic Transport and Mesoscopic Methods, Biological Physics and Soft Materials, Supercomputing and Computational Physics Teaching, Computational Physics and Sustainable Energy.

We would like to take this opportunity to thank our sponsors: International Union of Pure and Applied Physics (IUPAP), IUPAP Commission on Computational Physics (C20), American Physical Society Division of Computational Physics (APS-DCOMP), Oak Ridge National Laboratory (ORNL), Center for Defect Physics (CDP), the University of Tennessee (UT)/ORNL Joint Institute for Computational Sciences (JICS) and Cray, Inc. We are grateful to the committees that helped put the conference together, especially the local organizing committee. Particular thanks are also due to a number of ORNL staff who spent long hours with the administrative details. We are pleased to express our thanks to the conference administrator Ann Strange (ORNL/CDP) for her responsive and efficient day-to-day handling of this event, Sherry Samples, Assistant Conference Administrator (ORNL), Angie Beach and the ORNL Conference Office, and Shirley Shugart (ORNL) and Fern Stooksbury (ORNL) who created and maintained the conference website.

Editors:
G Malcolm Stocks (ORNL) and M Claudia Troparevsky (UT)
http://ccp2011.ornl.gov

Chair:
Dr Malcolm Stocks (ORNL)

Vice Chairs:
Adriana Moreo (ORNL/UT)
James Gubernatis (LANL)

Local Program Committee:
Don Batchelor (ORNL)
Jack Dongarra (UTK/ORNL)
James Hack (ORNL)
Robert Harrison (ORNL)
Paul Kent (ORNL)
Anthony Mezzacappa (ORNL)
Adriana Moreo (ORNL)
Witold Nazarewicz (UT)
Loukas Petridis (ORNL)
David Schultz (ORNL)
Bill Shelton (ORNL)
Claudia Troparevsky (ORNL)
Mina Yoon (ORNL)

International Advisory Board Members:
Joan Adler (Israel Institute of Technology, Israel)
Constantia Alexandrou (University of Cyprus, Cyprus)
Claudia Ambrosch-Draxl (University of Leoben, Austria)
Amanda Barnard (CSIRO, Australia)
Peter Borcherds (University of Birmingham, UK)
Klaus Cappelle (UFABC, Brazil)
Giovanni Ciccotti (Università degli Studi di Roma 'La Sapienza', Italy)
Nithaya Chetty (University of Pretoria, South Africa)
Charlotte Froese-Fischer (NIST, US)
Giulia A. Galli (University of California, Davis, US)
Gillian Gehring (University of Sheffield, UK)
Guang-Yu Guo (National Taiwan University, Taiwan)
Sharon Hammes-Schiffer (Penn State, US)
Alex Hansen (Norweigan UST)
Duane D. Johnson (University of Illinois at Urbana-Champaign, US)
David Landau (University of Georgia, US)
Joaquin Marro (University of Granada, Spain)
Richard Martin (UIUC, US)
Todd Martinez (Stanford University, US)
Bill McCurdy (Lawrence Berkeley National Laboratory, US)
Ingrid Mertig (Martin Luther University, Germany)
Alejandro Muramatsu (Universitat Stuttgart, Germany)
Richard Needs (Cavendish Laboratory, UK)
Giuseppina Orlandini (University of Trento, Italy)
Martin Savage (University of Washington, US)
Thomas Schulthess (ETH, Switzerland)

Dzidka Szotek (Daresbury Laboratory, UK)
Hideaki Takabe (Osaka University, Japan)
William M. Tang (Princeton University, US)
James Vary (Iowa State, US)
Enge Wang (Chinese Academy of Science, China)
Jian-Guo Wang (Institute of Applied Physics and Computational Mathematics, China)
Jian-Sheng Wang (National University, Singapore)
Dan Wei (Tsinghua University, China)
Tony Williams (University of Adelaide, Australia)
Rudy Zeller (Julich, Germany)

Conference Administrator:
Ann Strange (ORNL)

IUPAP C20 Conference on Computational Physics (CCP 2011) IOP Publishing
Journal of Physics: Conference Series **402** (2012) 012001 doi:10.1088/1742-6596/402/1/012001

Temperature dependence in interatomic potentials and an improved potential for Ti

G.J.Ackland
School of Physics, JCMB, The King's Buildings, University of Edinburgh, EH9 3JZ.
E-mail: gjackland@ed.ac.uk

Abstract. The process of deriving an interatomic potentials represents an attempt to integrate out the electronic degrees of freedom from the full quantum description of a condensed matter system. In practice it is the derivatives of the interatomic potentials which are used in molecular dynamics, as a model for the forces on a system. These forces should be the derivative of the free energy of the electronic system, which includes contributions from the entropy of the electronic states. This free energy is weakly temperature dependent, and although this can be safely neglected in many cases there are some systems where the electronic entropy plays a significant role. Here a method is proposed to incorporate electronic entropy in the Sommerfeld approximation into empirical potentials. The method is applied as a correction to an existing potential for titanium. Thermal properties of the new model are calculated, and a simple method for fixing the melting point and solid-solid phase transition temperature for existing models fitted to zero temperature data is presented.

1. Introduction

Classical molecular dynamics simulations are widely used in many areas of the physical sciences. Eliminating the explicit treatment of electronic degrees of freedom brings a computational advantage of around six orders of magnitude in the simulation of metallic systems, large enough to tackle systems which would be crippled by finite size effects in an electronic structure calculation. For applications in microstructure, plasticity and radiation damage where one wishes to study the behaviour of the atoms rather than electronic structure, this makes classical MD the method of choice.

The price for this is that the interatomic forces must be represented by the derivative of an interatomic potential, which is necessarily a poor approximation to the electronic structure. Typically this potential involves some functional form fitted to empirical or *ab initio*-calculated data. It is useful to distinguish the two distinct sources of error involved in this process: incorrect functional form and poor fitting.

Many MD simulations are still conducted using a pairwise-additive interatomic potential, such as Lennard Jones or Coulomb charges. Such pairwise potentials constrain possible values of the elastic constants. Most notably, the "Cauchy" relation which relates C_{12} with C_{66}. In a pairwise potential, elastic constants are given by the second derivative of the energy with respect to strain. Regarding the potential as a function of r^2 rather than r, it is easy to show that for *any* pair potential:

$$C_{12} = C_{66} = \frac{2}{\Omega} \sum_{ij} V''(r_{ij}^2) x_{ij}^2 y_{ij}^2$$

Journal of Physics: Conference Series **402** (2012) 012001 doi:10.1088/1742-6596/402/1/012001

where r_{ij}, x_{ij}, y_{ij} are the separation between atoms i and j, and x,y components thereof. where i, j run over all atoms and Ω is the volume of the system. Experimentally, this is true only for strongly ionic materials. Table 1 shows that elasticity in NaCl can be described by pairwise forces, but it is impossible to fit the three independent elastic constants for typical oxides, noble gases, water and metals. No amount of fitting can circumvent this mathematical constraint: the problem is one of of incorrect functional form.

Table 1. Elastic constants for selected materials showing the violation of the cauchy relation $C_{12} = C_{66}$ and by implication the impossibility of describing these materials by a pairwise additive potential. For water the constants are C_{13} and C_{44}, the hexagonal equivalent relation

Material	C_{11}	C_{12}	C_{66}
NaCl	482	128	127
MgO	291	96	152
Argon	233	149	117
SiC	385	135	257
Cu	168	121	75
Fe	237	141	116
Water ice		59	31

There are many similar examples where an incorrect functional form renders the process of fitting impossible. About 25 years ago there was a strong move away from pairwise forces, in particular in metals, in favour of models which describe the local environment such as the Embedded atom method[3], Finnis-Sinclair[2], and Tersoff[4]. Easily justified by elementary consideration of the electronic structure, these functional forms allowed enough freedom in fitting to avoid the Cauchy constraint and similar relations governing surface and vacancy energies. They satisfy the needs of most applications, and even if transferrability is imperfect have remained the workhorse of atomistic simulation ever since.

Some more recent potential developments have simplified the fitting process or made it more intuitive. The Modified Embedded Atom Method[5] maps local coordination onto spherical harmonics and hence establishes a link to electron orbitals. Two band and magnetic potentials[6, 7, 8] explicitly introduce limited electronic degrees of freedom, and then show how to eliminate them analytically, leaving a simple potential of embedded atom form, but justifying more complicated parameterisations.

Ultimately, however, derivation of a functional form for a potential should go back to first principles. The overwhelming success of the Kohn-Sham formulation of the density functional theory suggests that we should start with the Kohn-Sham Hamiltonian[9], which in standard notation is:

$$F(\rho) = T[\rho] + \frac{1}{2}\int \frac{\rho(\mathbf{r})\rho(\mathbf{r}')}{4\pi\epsilon_0|\mathbf{r}-\mathbf{r}'|}d^3\mathbf{r}d^3\mathbf{r}' + E_{xc}[\rho] + \sum_i \int \frac{Z_i e\rho(\mathbf{r}')}{4\pi\epsilon_0|\mathbf{R_i}-\mathbf{r}'|}d^3\mathbf{r}' + \sum_i\sum_j \frac{Z_i Z_j e^2}{4\pi\epsilon_0|\mathbf{R_i}-\mathbf{R_j}|} \tag{1}$$

For molecular dynamics we are interested in forces. These can be calculated using the Hellmann-Feynman theorem[10] which tells us that the forces are simply the *partial* derivative of the Kohn-Sham Hamiltonian with respect to ionic positions R_i. Inspection of eq.1 shows that only the final two, electrostatic, terms contribute to this.

At finite temperature the electron free energy for a metal changes as the Fermi distribution becomes non-singular. The energy increases as states above the Fermi are occupied, while the electronic entropy also increases due to partial occupations. According to Sommerfeld theory[11] there is an additional temperature-dependent contribution to the free energy which depends on temperature and the density of states at the Fermi energy as

$$F_{som}(T) \propto T^2 n(E_F)$$

In many electronic structure packages this effect is exploited to improve the numerical stability of the self consistency loop, by calculating with finite temperature electrons, so-called "Fermi Smearing". Typical effective temperatures can run to thousands of Kelvin, and the result is adjusted back to zero Kelvin. Nevertheless, the Sommerfeld temperature dependence of the electronic free energy is real, and so any potential involving integrating out electronic degrees of freedom should itself be temperature dependent. It is worth recalling at this point that the Hellmann-Feynman theorem is valid only for an electronic energy which is a *variational minimum* with respect to the parameters describing the electronic structure. For the Fermi-smeared system the variational quantity is the electronic *free energy* ($U_{el} - TS_{el}$), so the Hellman-Feynman forces are the derivative of this. The physical assumption underlying this is similar, though not identical, to the Born-Oppenheimer approximation: the ionic motion should be slow enough for the electrons to relax into their equilibrium distribution.

The temperature dependence of the Sommerfeld electronic free energy for Ti is shown in Fig.1 alongside the associated density of states. These figures were calculated using the CASTEP[13] program with settings as given in previous work[1], using Fermi-Dirac smearing with the smearing width corresponding to the quoted temperature. In all cases the atoms were located on their ideal lattice sites. The density of states was calculated using the *castepdos* program, which interpolates and integrates the density of states using the octahedron method between the explicitly calculated k-points.[14] At 0.05eV (about 600K) the Sommerfeld contribution in bcc is 34meV compared with 7meV for hcp. This is about 20% of the free energy difference at 0K.

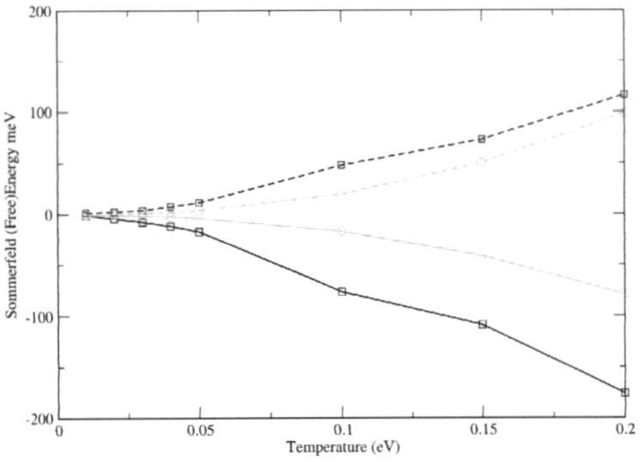

Figure 1. Sommerfeld electronic energy (dashed lines) and free energy (solid lines) for bcc (black, squares) and hcp (green, circles) Ti

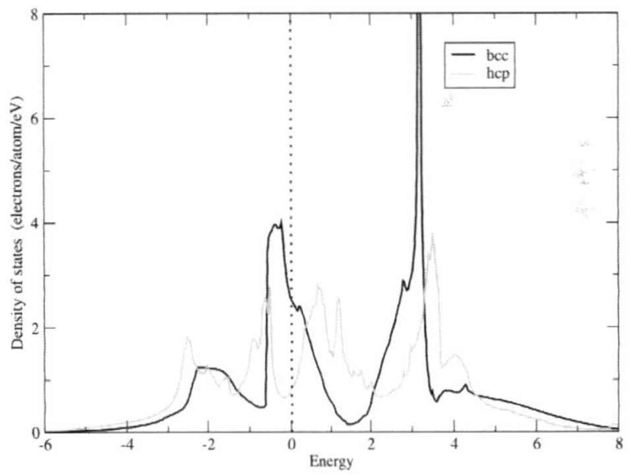

Figure 2. Density of states for bcc (black) and hcp (green) Ti

2. Sommerfeld Potentials

In order to include Sommerfeld effects in molecular dynamics, the potential must be written in the form of a sum of energies per atom j in an arrangement of atoms i in positions $\{\mathbf{r_i}\}$:

$$U(\{\mathbf{r_i}\}) = \sum_j U_0(\mathbf{r_j}, \{\mathbf{r_i}\}) + F_{som}(\mathbf{r_j}, \{\mathbf{r_i}\}) \quad (2)$$

The vast majority of interatomic potentials are fitted to (extrapolated) zero-temperature experimental data and/or *ab initio* data calculated on the Born-Oppenheimer surface. Thus in many cases good parameterisations for U_0 already exist. Here we consider the second term in equation 2, which is proportional to T^2. The proportionality constant can be treated as a fitting parameter (A_T). The challenge is to obtain a sensible measure for the local density of states at the Fermi energy $n_j(E_f)$ in terms of the atomic positions without performing a full electronic structure calculation. The most convenient approach is to write it as a pairwise function:

$$n_j(E_f) = \sum_i f(|\mathbf{r_i} - \mathbf{r_j}|)$$

We could stop at this point, and simply take f as an arbitrary function to be fitted, however it is worth discussing how it might look.

2.1. *Heuristics for f(r) in transition metals*

$f(r)$ attempts to measure the density of states at the Fermi level projected onto an individual atom. This quantity can be readily calculated for various crystal structures (e.g. Fig 2), and may vary quite widely between them. If we consider a canonical d-band model, the density of states is very different between close-packed and bcc structures. The DoS can be defined in terms of moments, in particular the third and fourth moment determine the skew and kurtosis. The moments theorem[16] relates these to a real-space picture involving closed paths of near neighbour hops. This defines the key difference between close packing (many loops of three hops) and body centred cubic (no three-hop loops- many four loops). Open structures seldom feature in transition metals and their alloys, so we can neglect these.

Although counting paths of hops cannot be done directly by a pair potential, it suggests a related local measure. For practical purposes the skew/kurtosis relation can be approximated by using a function which is either sharply peaked around first neighbour (favours three-loop equilateral triangles) or a broader function which favours four-loops. Which is energetically favoured for a given material depends on the band filling.

In most existing EAM and Finnis Sinclair potentials this narrow vs broad minimum in the potential determines the stability of close packed vs bcc materials. In the moments picture, the longer range acts as a proxy for measuring four-membered rings. Use of exponential fitting functions, as in early EAMs, tends to give a narrow minimum; polynomial functions, as used in Finnis Sinclair potentials, give a broader minimum. This, rather than any deep physics, explains their early application to fcc and bcc respectively.

2.2. *Application to Titanium*

For titanium the Fermi level falls at a minimum of the hcp density of states, and a maximium of the bcc density of states. Thus we can expect that electronic entropy contributions will be different and significant in determining the phase transition temperature. Indeed, electronic entropy was shown to provide almost half the excess entropy of bcc Zr[12], a material very similar to Ti. The density of states is for valence electrons, so $f(r)$ should have significant value only outside the repulsive core. It is also desirable that its derivatives be continuous and the number of adjustable parameters is minimised. Consequently we write the function as:

$$f(r) = X^2(1-X)^2 \qquad 0 < X < 1; \qquad X = (r - r_o)/d$$

$$F_{som} = \sum_{ij} A_T T^2 X^2 (1-X)^2$$

this gives a three parameter model, defining the range of the interaction from r_0 to $r_o + d$ and the strength A_T

Most published potentials are fitted to zero temperature data, so this Sommerfeld term can simply be added. Moreover, the parameterisation is relatively straightforward: we have calculated the temperature-dependent contribution to the free energy for fcc and hcp Ti. These two pieces of data are sufficient to parameterise the model.

For Ti we consider a potential introduced in 1992[21] which is known to have too low a basal stacking fault but otherwise reproduces the behaviour of hcp titanium reasonably well. From Fig. 1 above we see that at low energies (up to 1000K) the Sommerfeld contribution to bcc is about 4.7 times that of hcp. This suggests an approximate solution $r_o = 2.84\AA$, $d = 1.46\AA$, $A_T = -7.5 \times 10^{-7} meV/K^2$.

2.3. *Which temperature to use?*

In order to use the temperature-dependent potential, it is necessary to specify the electronic temperature. In practice, most MD simulations are relatively small and one can assume that the electrons are able to reach thermal equilibrium throughout the region.

Thus for an NVT/NPT calculation run using a thermostat, one can simply use the thermostat temperature to fix a constant potential throughout the run. For NVE one can use the instantaneous global temperature of the simulation for all atoms, so although the potential may vary in time, it is constant across space.

3. Tests of the new potentials

To test the thermal effects we have carried out melting point calculations from bcc and hcp Ti, and thermal expansion calculations.

3.1. *Melting Point Calculations*

We determined melting point using the coexistence method[18], using the MOLDY program in the NPE ensemble[17]. By using intermediate sized systems, 16000 and 13824 atoms for hcp and bcc respectively, it is possible to suppress the bcc-hcp transition and calculate melting points from both hcp and bcc phases. The original potential[21] gives a melting point of 1395 ± 10K (hcp) and 1790 ± 10 K (bcc), the difference indicating the stability of bcc at high temperature. This contrasts with the 0K energies of -4.853eV (hcp) and -4.807eV (bcc) which indicate low temperature stability of hcp. The transition at intermediate temperature is via a soft phonon analogous to zirconium[12, 15], and care must be taken to ensure that the solid phase does not transform during the simulation, which we did by monitoring local coordination using the BallViewer code[19].

Experimentally, the melting temperature of Ti is 1941K, and the bcc-hcp transition is at 1155 K. The original potential therefore gives too low a melting temperature. The effect of the Sommerfeld correction on the melting is small (1315K and 1800K), slightly stabilising bcc and destabilising hcp.

Metals melt at high temperature because the liquid phase has higher entropy than the solid. Thus the more of the phase space the system can reach without too high potential energy, the lower will be the melting point. Since a melt samples interatomic distances shorter than typically realised in the solid, softening the potential inside nearest neighbour distances can be expected to favour the liquid. In the present potential, the a_6 spline parameter controls the short range repulsion. The 1992 potential was not fitted to any data for short-ranged interactions, and in Fig. 3 we show the evolution of the melting temperature as a function of the parameter a_6: the short ranged spline starting close to nearest neighbour separations (see Appendix A). The effect on the melting point is very pronounced, while the 0K properties are essentially unaffected.

The original potential had $a_6 = 0.494\text{eV}/\text{Å}^3$: the experimental melting point (1941K) and ab initio hcp-bcc difference (90meV) were not fitted in 1992, and to do so requires a significantly stiffer short-ranged potential.

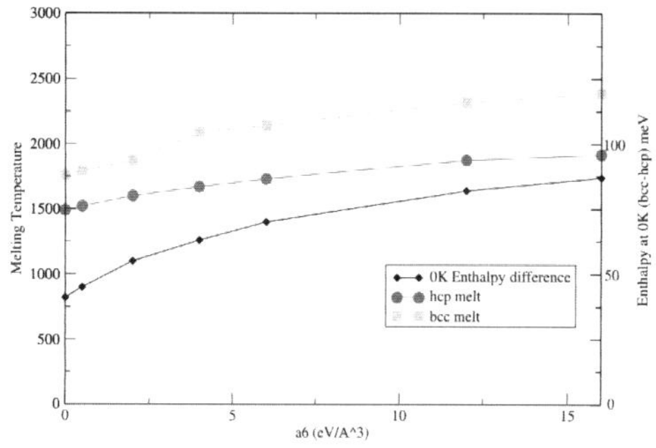

Figure 3. Effect of short ranged repulsion of melting point and zero-temperature hcp-bcc stability. See Appendix A for definition of a_6

3.2. Thermal Expansion

The thermal expansion is not typically fitted in empirical potentials, and in consequence is often significantly wrong. The Sommerfeld correction introduces additional temperature-dependent anharmonicity into the potential, which affects the thermal expansion. For this potential the thermal expansion is close to the experimental data but highly temperature dependent, especially at high temperature where the softening around second neighbours leads to high values.

3.3. Phase Transformation Calculations

Whatever value is used, the Sommerfeld correction will be insufficient to reduce the enthalpy of bcc below hcp: the bcc structure is always stabilised by phonon entropy. The constraints imposed by the crystal symmetry mean that it is not possible to calculate the phase transition temperature between hcp and bcc by coexistence. Rather, it requires a full free energy calculation[20, 24] which can be performed by integrating using the Gibbs-Helmholtz equation:

$$\Delta(\frac{G}{T}) = -\int H/T^2 dT$$

The right hand side of this equation can be calculated from a single-phase NPT molecular dynamics calculation (see e.g. Appendix B), varying the temperature slowly by incrementing the required temperature of a Nose thermostat. In practice we use this equation with the two melting points and assume that H/T^2 varies linearly with T. The stabilising effect on bcc lowers the transition temperature considerably, the actual value depending on the chosen value of a_6.

Figure 4 shows the effect of the Sommerfeld potential in a dynamics calculation of the phase transition. Although this is only a lower bound for the thermodynamic transition temperature, it does illustrate the increased stability range of bcc.

4. Conclusions

The thermal excitation of electronic degrees of freedom introduces a temperature dependence into the electronic free energy. If the electronic degrees of freedom are integrated out to make an interatomic potential, this temperature dependence should remain. Where the excitation arises simply from the Fermi-Dirac distribution via Sommerfeld theory, this temperature dependence scales as T^2. The magnitude of the effect has been calculated for Ti using density functional theory, and found to give a contribution of a few tens of meV to the bcc-hcp free energy difference. This in turn is enough to lower the calculated hcp-bcc phase transition temperature significantly,

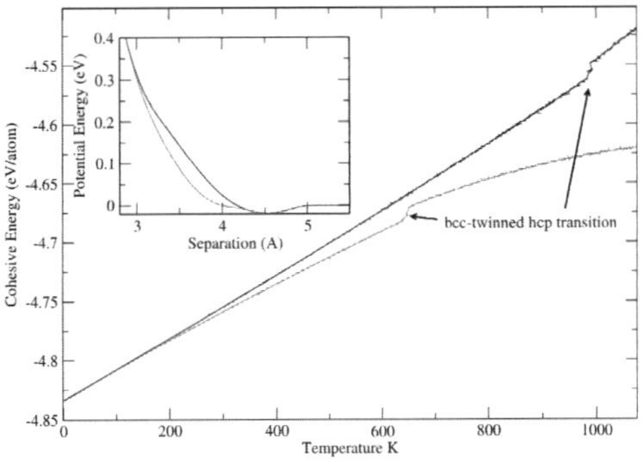

Figure 4. Plot of cohesive energy per atom vs temperature for MD simulation on heating with and without Sommerfeld correction. Main figure shows linear increase in energy with temperature, as expected from virial theorem (original potential, black line). Corrected potential (red) shows deviation from virial behaviour as electronic degrees of freedom are excited at high T. Both cases have a discontinuity as the cooling bcc transforms to twinned hcp, the correction stabilises bcc, dropping the observed transformation from 1000K to 650K. Inset shows pair potential V(r) at 0K and 1250K .

but has less effect on the melting temperature. Neither effect is large compared to the errors of typical EAM-type potentials which do not include melting temperature as a fit parameter.

By contrast, the melting temperature has strong correlation with the short ranged part of the potential, inside the normal separations found and fitted at 0K. By adjusting only the short-range terms it is possible to refit the melting point of a potential fitted to elastic moduli, cohesive energy, vacancy and surface formation etc. without disturbing those properties. This suggests a simple way to improve the high temperature phase performance of existing potentials without major redevelopment[26, 27, 28]. Note that energies of self interstitials[2, 3, 7] and interstitial impurities[14, 29] may be affected.

In addition to Sommerfeld correction, a strong temperature dependence of the electron free energy may be found in magnetic materials such as iron[30, 31, 32], where the fcc phase is stabilised by paramagnetic free energy. This effect could be captured by a temperature-dependent potential, although the excitation of all magnetic modes means the temperature dependence would be more complex, requiring an additional many-body term rather than a pairwise one. This will be the subject of future work.

Appendix A.

The potential has a functional form for the energy of the ith atom

$$U_i = \sum_j \sum_k H(r_k - R_{ij}) a_k (r_k - R_{ij})^3 \tag{A.1}$$

$$\sum_j A_T T^2 X^2 (1-X)^2 H(r_o + d - R_{ij}) H(R_{ij} - r_o) \tag{A.2}$$

$$- \sqrt{\sum_j \sum_k H(R_k - R_{ij}) A_k (R_k - R_{ij})^3} \tag{A.3}$$

with X as defined above, R_{ij} the separation between i^{th} and j^{th} nearest atoms, H the Heaviside step function and the other parameters given in Table A1 below.

k	1	2	3	4	5	6
a_k	-0.785715	1.110966	-0.299450	-0.143061	1.025368	0.494293
r_k	5.09113	5.00767	4.673828	3.964408	3.338449	2.9508
A_k	0.547614	-0.551266				
R_k	5.09113	4.381714				
	$A_T = -7.5\times10^{-7}$	r_o=2.84	d=1.4			

Table A1. Parameters for the Ti potential, from ref.[21], converted into units of eV and Angstroms, and for the temperature-dependent correction. Note that a_6 can be varied to fit the melting temperature.

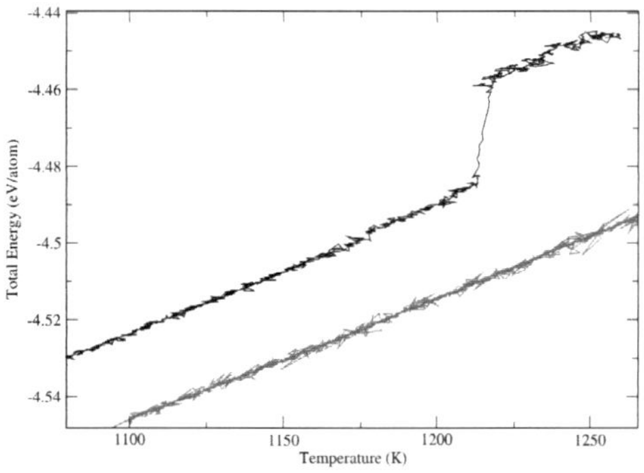

Figure A1. Plot of enthalpy vs temperature showing failure of thermodynamic integration due to phase transition from bcc to deformed hcp on cooling. Upper line (black) is bcc at high temperature, transforming to twinned hcp on cooling. Reheating gave hysteresis in the retransformation. Note the different transuition point compared with fig 4. Lower line (red) is hcp on heating from 0K: recooling produced the same curve. NPT simulations were run with a cooling rate of 1K/ps and a 1fs timestep, with a Nose relaxation time parameter of 100fs. Phases were initialised in a supercell with periodic boundaries compatible with the phase in question: although the transformation could happen in principle in Parrinello-Rahman dynamics, in practice the transformation is slow enough to gather phonon statistics.

Appendix B.

One potential drawback of the thermodynamic integration is that the system will undergo a phase transformation as the temperature is varied, rather than remaining in the metastable phase. When this occurs, the latent heat causes a discontinuity in a graph of enthalpy vs temperature. Fig.A1. shows an example of this in cooling of the bcc phase. Some authors have maintained that if the transformation occurs quickly enough, the temperature of the discontinuity would be the transformation temperature. However, better statistics than are available here are needed to ensure that this is so[25]. Moreover, Fig A1. shows that the transformation is not to the pure hcp phase, and BallViewer analysis[19] shows that the low temperature phase is a twinned hcp.

3. References

[1] B.E.Tegner, L. Zhu, and G.J. Ackland Phys.Rev.B (2012)
[2] Finnis,M.W. and Sinclair,J.E., Phil.Mag.A, 50, 45. (1984)

[3] Daw,M.S. and Baskes,M.I., Phys.Rev.B 29, 6443 (1984)
[4] J.Tersoff, Phys.Rev.Lett. **56**, 632 (1986)
[5] M.I.Baskes, Phys Rev B, 46, 2727-2742 (1992)
[6] G.J.Ackland, S.K.Reed Phys Rev B 67 174108 (2003)
[7] Dudarev SL and Derlet PM J.Phys.CM, 19 239001 2007
[8] G.J.Ackland J.Nucl.Mater. 351, 1-3, 20 (2006)
[9] W.Kohn and L.J.Sham. *Phys.Rev.*, **140** 1133, (1965).
[10] Hellmann H (1937) *Einfuhrung in die Quantumchemie* (Leipzig: Franz Deutsche 1937); R. P Feynman, *Phys Rev* **56**, 340 (1939).
[11] N.W.Ashcroft and N.D.Mermin, Solid State Physics, Chapter 2.
[12] F.Willaime and C.Massobrio Phys. Rev. Lett. 63, 22442247 (1989)
[13] S.J.Clark, M.D.Segall, C.J.Pickard, P.J.Hasnip, M.J.Probert, K. Refson, M.C.Payne Zeitschrift fr Kristallographie 220(5-6) 567 (2005)
[14] G.J.Ackland, Phys.Rev.Letters, 80, 2233 (1998)
[15] U.Pinsook and G.J.Ackland, Phys.Rev.B 59 13642 (1999)
[16] F Cyrot-Lackmann J.Phys.Chem. Solids 29 1235 (1968)
[17] G.J.Ackland, K D'Mellow, S.L.Daraszewicz, D.J.Hepburn, M Uhrin and K.Stratford Comp.Phys.Comm., 182 2587 (2011)
[18] J.R.Morris, C.Z.Wang, K.M.Ho, and C.T.Chan Phys. Rev. B 49, 31093115 (1994)
[19] G.J.Ackland, and A.P.Jones Phys Rev B 73 054104 (2006)
[20] G.J.Ackland J.Phys.CM 14 2975 (2002)
[21] G.J.Ackland, Phil. Mag. A 66, 917-932 (1992).
[22] `http://www2.ph.ed.ac.uk/~gja/dos/dos_manual`
[23] X.G. Lu M. Selleby, B. Sundman Calphad, 29 6889 (2005)
[24] E. M. Lopasso, M. Caro, A. Caro, and P. E. A. Turchi Phys. Rev. B 68, 214205 (2003)
[25] D. Alfe, C. Cazorla, and M. J. Gillan J. Chem. Phys. 135, 024102 (2011);
[26] R.C. Pasianot and L. Malerba J.Nucl.Mater. 360 118 (2007)
[27] Bonny G., Pasianot R. and Malerba L. Phil.Mag, 89 3451 (2009)
[28] J M Winey, A Kubota and Y M Gupta Modelling Simul. Mater. Sci. Eng. 17 055004 (2009)
[29] D.J.Hepburn and G.J.Ackland Physical Review B, 78, 16, 165115 (2008)
[30] H. Hasegawa and D. G. Pettifor Phys. Rev. Lett. 50, 130133 (1983)
[31] Hasegawa H, Finnis MW, Pettifor DG,J Phys F Met Phys, 17 2049 1987.
[32] P.-W. Ma, C.H. Woo, S.L. Dudarev, Phys. Rev. B78 024434 (2008)

Simulations of nanosensors based on single walled carbon nanotubes

Polina Pine[1], Yuval E. Yaish[2] and Joan Adler[3]

[1]RBNI, Technion-Israel Institute of Technology, 32000, Haifa, Israel
[2]Department of Electrical Engineering, Technion-Israel Institute of Technology, 32000, Haifa, Israel
[3]Department of Physics, Technion-Israel Institute of Technology, 32000, Haifa, Israel

E-mail: pine@tx.technion.ac.il

Abstract. The potential of single-walled carbon nanotubes as mass sensors is examined. The change in mass leads to proportional changes in the nanotube vibrational frequencies, which are monitored during atomistic simulations. We observed a frequency shift as a result of replacement of carbon C_{12} with its isotope C_{13}. For a zigzag (12,0) nanotube of about 10 nm length, we found zeptogram sensitivity.

1. Introduction

There is a growing interest in biological application of carbon nanotubes (CNTs) [1, 2, 3, 4] especially for medical technology [5] and sensors [6, 7]. These can be broadly classified into two categories: chemical sensors [8] and biosensors [9, 10, 11] and include ultrasensitive nano-bio sensors [3], electroanalytical nanotube devices [4] and electromechanical actuators for artificial muscles [12]. The development of nano-bio sensors [13] and nanoscale bioreactor systems based on CNTs has been driven by the experimental evidence that biological entities such as proteins, enzymes, bacteria can be immobilized either in the hollow cavity or on the surface of carbon nanotubes [1, 2]. In the light of successful fabrication of various nanotube devices, prepared as single walled carbon nanotube (SWCNT) transistors, significant success has been achieved in attempts to use CNTs as superior biosensor materials. They have shown promising sensitivities required for such applications as antigen recognition [14], enzyme-catalyzed reactions [15] and DNA hybridizations [16].

Resonance based sensors offer the deeper potential of achieving the high-fidelity requirement of many sensing applications. The key issue of mass detection is in quantifying the change in the resonant frequency due to the added mass. There are some early studies [17, 18, 19] in this young and rapidly developing field, but there is a lack of systematic basic research, and especially of atomistic studies. These enable modeling of such systems that are as close as possible to the real atomistic behavior.

2. Motivation and previous studies

Chowdhury *et al* [17] examined the potential of single-walled carbon nanotubes (SWCNTs) as mass sensors, using a continuum mechanics based approach based on the Euler-Bernoulli(EB) beam theory [20] and Finite Element Analysis (FEA). The carbon nanotube resonators were

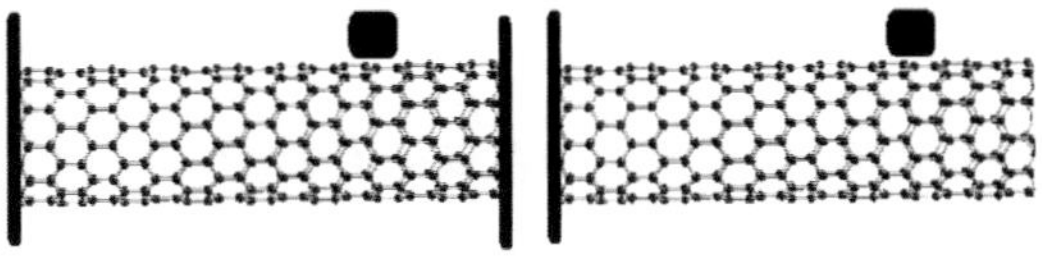

Figure 1. SWCNT with attached mass in a bridged configuration at left and a cantilevered configuration at right.

assumed to be either in bridged (doubly clamped) or in cantilevered (clamped at one end) configurations with an attached mass in the atomic systems as depicted schematically in Fig. 1. The length of the CNTs studied in [17] was 8 nm and the length of the biological attached mass varied between 0.5 - 3.5 nm. It was observed that the proposed sensor-equations work reasonably well when the length of the bacteria is more than 1 nm for both cantilevered and bridged configurations. The numerical results indicated that the mass sensitivity of carbon nanotube-based nanobalances can reach up to 10^{-24} kg.

Joshi *et al* [18] presented a FEA simulation of the mechanical responses of individual carbon nanotubes treated as thin shells with a thickness based on Chowdhury *et al* [17]. The resonant frequencies of the cantilevered and the bridged SWCNT were investigated. This analysis explores the resonant frequency shift of SWCNTs caused by the changes in the size of the CNT as a function of length as well of mass. They found sensitivity of the SWCNTs to different masses and different lengths. The mass sensitivity could reach 10^{-24} kg and it increases when smaller size nanotube resonators are used in mass sensors as predicted by EB beam theory.

The study of Aydogdu *et al* [19] is the most detailed until now. They studied axial vibrational behavior of single-walled carbon nanotube-based mass sensors using nonlocal elasticity theory. Carbon nanotubes with different lengths, attached masses and their positions on the CNTs, as well as different boundary conditions (cantilevered or bridged) were considered. They showed that for the cantilevered CNTs the frequency decreased with increase of the mass distance from the point of CNT clamping, however for the bridged ones the frequency change symmetrically with respect to the midpoint of the SWCNT where the minimum frequency is obtained. They found a sensitivity of nanotube-based mass sensors in the zeptogram range.

In one of our earlier papers, denoted below as I, [21] we observed some problems with EB theory for doubly clamped CNTs for certain length and radius choices including some of the values in the above studies. In another, paper II, [22] we showed that while choice of nanotube symmetry is not a relevant parameter if the CNT is doubly clamped, under partial clamping, (paper III [23]) we showed that vibrational symmetry is broken. Thus the continuum models described above do not tell the full story and an atomistic approach is needed. In the present study, because the mass distribution is not uniform in all cases this is even more crucial.

3. Details of the simulation method

In our atomistic approach [21, 22, 23] we calculate the thermal vibrations of SWCNTs using molecular dynamics (MD). In each case we generate high quality data with MD simulations using the Brenner [24] potential on carefully equilibrated clamped tubes. The thermal vibrations were decomposed into different modes. In order to keep temperature constant a Berendsen thermostat [25] was applied. To ensure stable nanotube structure and eliminate intrinsic tension, we collected data after a period of slow initial thermalization to 300K (periodic boundary conditions with no frozen edges) waiting until the length of the equilibrated nanotube remained constant up to insignificant fluctuations. This relaxed nanotube has no tension but in a study currently underway we are exploring the implications of introducing tension into the tube. In order to

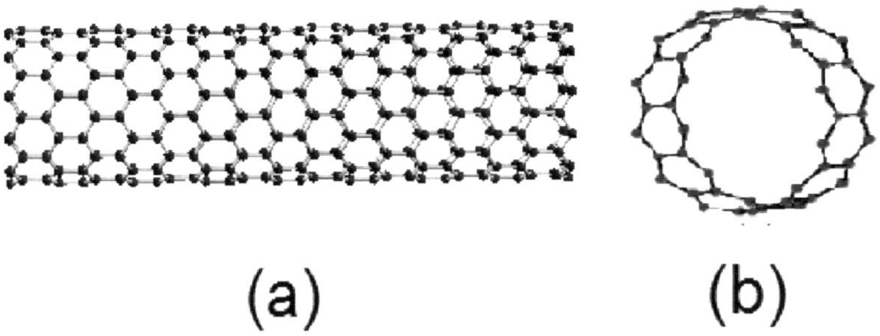

Figure 2. (a) (12,0) zigzag SWCNT; (b) a single period.

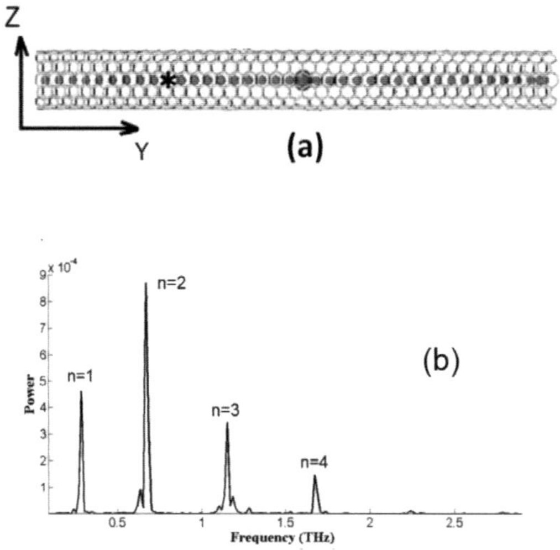

Figure 3. (a) Single Walled Carbon Nanotube (green), with the CA made up of the centers of mass of each period of the nanotube as (red) spheres inside the nanotube, and a (blue) large atom to indicate its CM. The $*$ marks one place where all the first 4 modes have finite amplitude; (b) FFT analysis of the first four thermal vibrational modes (n=1-4) in the Z direction at this point at 300K, under doubly clamped conditions.

obtain adequate statistics for all vibrational modes we let every nanotube vibrate 1000 times more than the period of its lowest frequency, using a MD timestep of 0.5 fsec. In order to reduce the amount of data needed for the vibrational analysis we selected some special points where data was recorded. The centers of mass (spheres inside the nanotube in Fig.3(a)) for each period were identified, making up a line of points which we call the centered axis (CA). The center of mass (CM) of the entire nanotube (large sphere in Fig.3(a)) falls in the middle of this line. We then applied a fast fourier transform (FFT) analysis to the data from the selected points, (including one near the center of mass) to calculate the power density as a function of frequency (see Fig.3(b)).

In paper I [21] we reported carefully equilibrated MD simulations of doubly clamped armchair SWCNTs including a precise analysis of the four lowest modes of vibrations. We provided clear evidence for the failure of simple analytic models such as Euler-Bernoulli to accurately extract resonance frequencies as the ratio, (R/L), between the tube radius (R) and the length (L) increases. Our results were in excellent agreement with the Timoshenko beam [26] model, which includes the effect of both rotary inertia and of shearing deformation. Invoking the EB model implies that the bending rigidity (which includes the product of the Young's modulus, E, and the nanotube wall thickness) is constant. We showed that if λ_n is the wavelength of the nth mode, for larger values of R/λ_n this is most definitely not true. Yakobson's paradox [27], relates to a scatter of between 1 and 5 TPa in the Young's modulus from atomistic simulations. We shed light on this by giving an upper cutoff estimate for the effective SWCNT thickness, and showed that in the Timoshenko model, there are two different sources for the nanotube thickness. The issue of nanotube thickness, was also addressed by [28] from a different viewpoint and our E and thickness results are in good agreement.

In an attempt to determine whether the nanotube type affects vibration we reported in II [22] on the vibrational behavior of four different types of SWCNTs: armchair, zigzag and two different chiral ones. All were fully clamped at both ends as in our previous study I [21]. Comparison between the vibrational behavior of these four types of nanotubes gave the result that the SWCNT structure does not affect the vibrational frequencies under these conditions.

Doubly clamped nanotubes of all three types exhibit the same vibrational modes, and their vibrations in the directions perpendicular to the nanotube axis are degenerate. In nanotube applications such as sensors based on nano-electro-mechanical systems their ends are not fully clamped, thus their vibrational behavior could differ. In paper III [23] we studied SWCNTs with boundary conditions which imitate the partly clamped experimental conditions. Our results demonstrate that armchair, zigzag and chiral nanotubes indeed vibrate differently. The symmetry between the two perpendicular directions is broken, and the SWCNT type does then influence the vibrational modes.

4. Simulation model for mass change

Using the approach described above, we have now carried out atomistic simulations of a vibrating nanotube suitable for use in sensors based on (12,0) 110.76 $\AA$ zigzag carbon nanotube (see Fig. 2(a)). This CNT consists of 24 periods (see Fig. 2(b)). Two first/last periods were frozen in order to retain bridged boundary conditions, giving a 91.96 $\AA$ vibrating length. We first calculated the frequencies of first four modes of the unloaded nanotube. This result was published in I.

Our carbon atoms in the previous studies were assumed to be the usual C_{12}. Since the concept of the nanosensor application is based on the possibility of measuring adsorbed masses, the first step towards the proof of concept, is showing that additional masses can be accurately measured. In our simulations based on classical MD potentials, introducing adsorbed atoms or molecules of different types would require careful evaluation of additional parameters of interactions. Hence we chose to use a different isotope where to a good approximation the interatomic potential would only need to be adjusted for the different nuclear mass. Thus in order to simulate attached mass we decided to change some carbon atoms to the isotope C_{13}, changing the mass of segments of the SWCNT and calculating vibrational frequencies of nanotube with some periods replaced with C_{13}.

The weight of a single C_{12} is 12 amu and of C_{13} is 13 amu which is equivalent to $1.9926466 \cdot 10^{-23}$ gr and $2.1587005 \cdot 10^{-23}$ gr respectively. Changing the whole vibrational segment (20 periods) to isotope C_{13} would give us a theoretical resolution of $1.5941174 \cdot 10^{-21}$ gr, see Table 1. This zeptogram sensitively is in agreement with that reported by [17, 18, 19].

Table 1. Mass resolution in grams due to replacement of C_{12} with C_{13}.

periods	num. of atoms	Δm
1	48	$7.9705872 \cdot 10^{-23}$
2	96	$1.5941174 \cdot 10^{-22}$
3	144	$2.3911762 \cdot 10^{-22}$
4	192	$3.1882349 \cdot 10^{-22}$
20	960	$1.5941174 \cdot 10^{-21}$
24	1152	$1.9129409 \cdot 10^{-21}$

Table 2. Changes in vibrational frequencies (THz) of first four modes as a result of isotope replacement.

n	C_{12} (all atoms)	C_{13} (4 periods)	C_{13} (all atoms)
1	0.2686	0.2686	0.2604
2	0.6429	0.6429	0.6185
3	1.107	1.107	1.074
4	1.609	1.595	1.546

5. Results and discussion

The road to predicting the actual change of frequencies is somewhat complex, as fitting to the correct Timoshenko model is an iterative process, requiring a set of different lengths/radii, and cannot be done for other than fully clamped boundary conditions anyway. Thus we decided to measure the frequencies directly. The results for three cases are presented in Table 2. When the isotope C_{13} is inserted only in the first four periods it does not noticeably affect the frequencies, although a slight change is seen in the 4th mode. The replacement of all atoms of the nanotube gives a significant effect as can be seen from Table 2. Since a very few atoms gave no clear change above the numerical error it would be desirable to determine what minimal load, or number of substituted atoms is needed in order to obtain a differential response for all four modes.

6. Conclusions

Our current results are only the beginning of a detailed investigation of the mechanism of action of resonant sensors based on SWCNTs. In future we plan not only to change isotopes to those heavier than C_{13} but also to study the vibrational behavior of nanotubes with mismatches, i.e. with defects (without atoms in specific points of nanotube). We also plan to attach different molecules to the nanotube's surface and study vibrational changes and sensing limitations of such devices.

7. Acknowledgements

We thank the organisers of CCP2011 for the opportunity to present these results and the Russell Berrie Nanotechnology Institute for support via the NEVET program, use of the NANCO computer, and a fellowship for P.P. Y.E.Y thanks the ISF (grant number 1334/06) for support. P.P. thanks the Ramon foundation and the Israel Industrial club for financial support.

References

[1] S. C. Tsang, J. J. Davis, M. L. Green, H. A. O. Hill, Y.C. Leung and P. J. Sadler, J. Chem. Soc., Chem. Commun, **17** 1803 (1995)

[2] J. Davis, M.L.H. Greena, H. A. O. Hill, Y. C. Leung, P. J. Sadler, J. Sloan, A.V. Xavier and S. Chi Tsang, Inorganica Chimica Acta **272** 261 (1998)
[3] R. Baughman, C. Cui, A.A. Zakhidov, Z. Iqbal, J.N. Barisci, G.M. Spinks, G.G. Wallace, A. Mazzoldi, D. De Rossi, A.G. Rinzler, O. Jaschinski, S. Roth M. Kertesz Science **284** 1340 (1999)
[4] M. Mattson, R. Haddon and A. Rao, Journal of Molecular Neuroscience **14**,175 (2000)
[5] F. Lu, L. Gu, M. Meziani, X. Wang,P. G. Luo, L. V. Veca, L. Cao and Y-P. Sun, Advanced Materials **21** 139 (2009)
[6] Y. Lin, S. Taylor, H. Li, K. A. S. Fernando, L. Gu, W. wang, L. Gu, B. Zhou and Y-P. Sun, Journal of Materials Chemistry, **14** 527 (2004)
[7] L. Gu, T. Elkin, X. Jiang, H. Li, Y. lin, L. Qu, T-R. J. Tzeng, R. Joseph and Y-P. Sun, Chemical Communications **7** 874 (2005)
[8] J. Wang and M. Musameh, Analytical Chemistry **75** 2075 (2003)
[9] B.L. Allen, P.D. Kichambare and A. Star, Advanced Materials **19** 1439 (2007)
[10] K. Balasubramanian and M. Burghard, Analytical and Bioanalytical Chemistry **385** 452 (2006)
[11] M. Atashbar, D. Rocha, P.M. Sarro, M.J. Vellekoop, (Eds.), Proceedings of the IEEE Sensors, vols.1-3, Vienna, Austria, October 24-27, 1048 (2004)
[12] J. Davis, K.S. Coleman, B.R. Azamian, C.B. Bagshaw, M.L.H. Green, ChemistryA European Journal **9** 3732 (2003)
[13] S. Lee and D.S. Yoon, Biochip Journal **1** 193 (2007)
[14] R. J. Chen, S. Bangsaruntip, K. A. Drouvalakis, N. Wong Shi Kam, M. Shim, Y. Li, W. Kim, P. J. Utz, H. Dai, Proceedings of the National Academy of Sciences of the United States of America **100** 4984 (2003)
[15] J. J. Gooding, R. Wibowo, J. Liu, W. Yang, D. Losic, S. Orbons F. J. Mearns, J. G. Shapter, D. B. Hibbert, Journal of the American Chemical Society **125** 9006 (2003)
[16] J. Li, H.T. Ng, A. Cassell, W. Fan, H. Chen, Q. Ye, J. Koehne, J. Han, M. Meyyappan, Nano Letters **3** 597 (2003)
[17] R. Chowdhury, S. Adhikari, J. Mitchell, Physica E, **42** 104 (2009)
[18] A.Y. Joshi, S.P. Harsha, S.C. Sharma, Physica E, **42** 2115 (2010)
[19] M. Aydogdua, S. Filiz, Physica E, **43** 1229 (2011)
[20] L. D. Landau and E. M. Lifshitz, Elasticity Theory (Pergamon,Oxford, 1986)
[21] P. Pine, Y. E. Yaish, J. Adler, Phys. Rev. B, **83** 155410 (2011)
[22] P. Pine, Y. E. Yaish, J. Adler, Phys. Rev. B, **84** 245409 (2011)
[23] P. Pine, Y. E. Yaish, J. Adler, J. Appl. Phys. **110** 124311 (2011)
[24] D. W. Brenner, Phys. Rev. B **42** 9458 (1990)
[25] H.J.C. Berendsen, J.P.M. Postma, W.F. van Gunsteren, A. DiNola, J.R. Haak, J. Chem. Phys. **81** 3684 (1984)
[26] S. P. Timoshenko and J. N. Goodier, Theory of Elasticity (McGraw Hill, New York, 3d ed., 1987).
[27] O. A. Shenderova, V. V. Zhirnov, and D. W. Brenner, Crit. Rev. Solid State Mater. Sci., **27** 227 (2002)
[28] Y. Huang, J. Wu, and K. W. Hwang, Phys. Rev. B, **74** 245413 (2006)

Monte Carlo Simulation of Secondary Electron Emission from Dielectric Targets

Maurizio Dapor[1,2,3]
[1]Interdisciplinary Laboratory for Computational Science (LISC)
FBK-CMM and University of Trento
38123 Povo, Trento, Italy
[2]Department of Materials Engineering and Industrial Technologies
University of Trento
38123 Trento, Italy
[3]INFN, Trento, Italy

E-mail: dapor@fbk.eu

Abstract. In modern physics we are interested in systems with many degrees of freedom. The Monte Carlo (MC) method gives us a very accurate way to calculate definite integrals of high dimension: it evaluates the integrand at a random sampling of abscissa. MC is also used for evaluating the many physical quantities necessary to the study of the interactions of particle-beams with solid targets. Letting the particles carry out an artificial random walk and taking into account the effect of the single collisions, it is possible to accurately evaluate the diffusion process.

Secondary electron emission is a process where primary incident electrons impinging on a surface induce the emission of secondary electrons. The number of secondary electrons emitted divided by the number of the incident electrons is the so-called *secondary electron emission yield*. The secondary electron emission yield is conventionally measured as the integral of the secondary electron energy distribution in the emitted electron energy range from 0 to 50eV.

The problem of the determination of secondary electron emission from solids irradiated by a particle beam is of crucial importance, especially in connection with the analytical techniques that utilize secondary electrons to investigate chemical and compositional properties of solids in the near surface layers. Secondary electrons are used for imaging in scanning electron microscopes, with applications ranging from secondary electron doping contrast in p-n junctions, line-width measurement in critical-dimension scanning electron microscopy, to the study of biological samples.

In this work, the main mechanisms of scattering and energy loss of electrons scattered in dielectric materials are briefly treated. The present MC scheme takes into account all the single energy losses suffered by each electron in the secondary electron cascade, and is rather accurate for the calculation of the secondary electron yield and energy distribution as well.

1.Introduction

Many features appear in the secondary electron spectra, due to the different collision processes involved prior to low-energy secondary electron emission. When a stream of electrons impinges on a solid target, the incident particles (during their travel within the solid) lose energy and change direction at each collision with the atoms bound in the solid. Nuclear collisions deflect the electrons without relevant kinetic-energy transfer, owing to the large difference between the masses of the electron and the nucleus: the differential elastic scattering cross-section has to be used in order to describe these processes (Mott cross section). On the other hand, excitations and ejections of atomic electrons, and plasmon, phonon, and polaron excitations affect the energy dissipation but only slightly affect the direction of the electron in the solid. These can be described as inelastic processes, governed by the equations for the differential inelastic scattering cross-sections.

Transport MC is a very well-known numerical method which allows to predict the behavior of charged particles traveling through the matter. In particular, it has been widely utilized for the study of the electron-beams interacting with solid targets. This topic has many applications. Among them, on the side of the chemical analysis of the surfaces and the interfaces of materials, let's mention Scanning Electron Microscopy, Transmission Electron Microscopy, Reflection Electron Energy Loss

Published under licence by IOP Publishing Ltd

Spectroscopy, Auger Electron Spectroscopy, X-Ray Photoelectron Spectroscopy. There are many other applications relative to the study of materials. If we limit ourselves to the secondary electron emission, MC simulation can be used for the study of image contrast in silicon p-n junctions, for line scan calculations of resist materials with given geometrical cross sections deposited on silicon substrates, for electric fields induced in insulating materials by electron beam irradiation, and so on.

2.Monte Carlo Modeling

While for electron kinetic energies higher than 10keV, MC simulations provide excellent results by the simple use of the Rutherford differential elastic scattering cross section and of the Bethe-Bloch stopping power formula (or semi-empirical stopping power formulas), when the electron energies of interest are much lower than 5keV – and this is the case of secondary electron emission – this approach can no longer be utilized[1]. There are many reasons, and the most important ones are related to the three following facts: (i) The Rutherford formula is a high energy approximation, being the result of the first Born approximation. (ii) The Bethe-Bloch formula is valid only for quite high energies as well; in particular, the Bethe-Bloch stopping power does not work well when the electron energy E becomes lower than the mean ionization energy J. It reaches a maximum for $E \approx 2.5J$ and then goes to zero for $E = J/1.166$. Below this energy value, the predicted stopping power becomes negative. Thus numerical approaches based on the calculation of the dielectric function–as a function of the energy loss and of the momentum transfer–are necessary to calculate low energy inelastic processes. (iii) The use of the stopping power, and hence of the continuous-slowing-down approximation, for describing energy loss, completely neglects that actually electrons lose their energy in several inelastic collisions. Sometimes an electron can even lose all its energy in a single collision. In other words, any realistic modeling of the electron trajectories should avoid the approximation of continuity in describing the electron energy losses.

A detailed approach able to accurately describe low energy elastic and inelastic scattering and to appropriately take into account the energy straggling (i.e. the different energy losses suffered by each electron of the penetrating beam) is required for the description of secondary electron cascade. The whole cascade must be followed: indeed any truncation, or cut off, underestimates the secondary electron emission yield. What is more, as we will discuss below, for insulating materials the main mechanisms of energy loss cannot be limited to the electron-electron interaction for, when the electron energy becomes very low (let's say lower than 10-20eV), inelastic interactions with other particles or quasi-particles are responsible for electron energy losses. In particular, at very low electron energy, electron-phonon interactions and electron-polaron interactions (polaronic effects) become the main mechanisms for the electron energy loss.

3.Elastic and Inelastic Scattering

The results presented in this paper were obtained with the differential and total elastic scattering cross sections calculated using the Mott theory, i.e. obtained numerically solving the Dirac equation in a central field: this procedure is known as the "relativistic partial wave expansion method" and it has been demonstrated to provide excellent results when compared to experimental data[2-9].

The main mechanism which determines the inelastic scattering cross section and the relative energy losses, for energies higher than 20eV, is the interaction of the incident electrons with the collective excitations of the electron sea, known as plasmons. Such energy loss mechanisms can be described by calculating the so-called energy loss function, i.e. the reciprocal of the imaginary part of the dielectric function. The Ritchie theory[10-20] was used–starting from the knowledge of the dependence of the dielectric function upon both the energy loss and the momentum transfer–to calculate the differential inverse electron inelastic mean free path and the electron inelastic mean free path. When the electron energy is higher than 50eV, both the electron inelastic mean free path and the electron stopping power calculated within the dielectric formalism are in very good agreement with the experiment (and with theoretical data obtained by other investigators). For details of the present calculation of the electron-electron inelastic mean free path λ_{ee} in SiO_2, see Refs.[16,17].

When, on the other hand, the electron energy becomes lower than 50eV, the dielectric formalism alone is no longer able to accurately describe the energy loss phenomena. In fact, as the electron energy decreases, the electron inelastic mean free path calculated using only the electron-electron interaction increases indefinitely (while the stopping power goes quickly to zero). This means that if only electron-electron interactions were active for inelastic scattering, electrons with such a low energy would no longer interact inelastically (i.e. losing energy) with the solid (we are neglecting here, for the sake of simplicity, the little energy losses associated to the electron-atom elastic scattering collisions, which also occur for they actually are quasi-elastic). As a consequence they would travel without any change in their kinetic energy. For a semi-infinite target, this very long travel in the solid would continue forever or until the electron reaches the surface of the material and is able to emerge. Also note that the condition for emerging from the surface is not always satisfied. The interface with the vacuum presents, in fact, a potential barrier, and not all the electrons that reach the surface are able to go beyond it. When the electrons reaching the surface are not able to emerge, they are reflected back in the material.

In general, in order to be able to emerge from the surface and to be detected, an electron must have a residual energy E satisfying the relationship

$$E\,\cos^2\vartheta \geq \chi \quad , \tag{1}$$

where ϑ is the angle of the electron direction with respect to the normal to the surface and χ is the electron affinity, namely the potential barrier between the vacuum level and the minimum of the conduction band[21,22]. Experimental and MC results relative to the slow electrons (secondary electron energy is lower than 50eV) depend on the value of the electron affinity: for SiO_2, χ = 0.9eV.

Thus, if the only mechanism of energy loss was the electron-electron interaction, as the energy of the electrons become lower than approximately 10-20eV, the particles would no longer suffer energy losses, the inelastic mean free path would tend to infinite (and the stopping power would approach zero) so that many electrons could virtually walk in the solid without end. A MC simulation, in such a case, would require so long CPU times to converge as to be, practically, without any utility.

Luckily nature is much cleverer than MC code programmers. Indeed, when the electron energy is so low, electrons lose small and numerous amounts of energies interacting with lattice vibrations, or phonons. In fact, another important mechanism of electron energy loss for electrons traveling in solid targets, particularly effective when the electron energy is low (few eV), is represented by the interaction of the electrons with the optical modes of the lattice vibrations. Notice that this kind of interactions are responsible for electron energy gain as well. In his theory of the electron-phonon interaction, Fröhlich[23] considered, in particular, the interaction of free conduction electrons with the longitudinal optical mode lattice vibrations. The interaction was treated considering both phonon creation and phonon annihilation, corresponding to electron energy loss and to electrons energy gain, respectively. As the phonon generation probability is much higher than the phonon absorption probability, the last is often neglected in MC simulations[24]. Furthermore, since, according to Ganachaud and Mokrani[25], the dispersion relation of the longitudinal phonons can be neglected in the optical branch, one can use a single phonon frequency ω. For the present case we put $\hbar\omega$ = 0.1eV. According to Fröhlich[23], Llacer and Garwin[24], and Ganachaud and Mokrani[25], the inverse inelastic mean free path for the electron-phonon interaction, λ_{eph}^{-1}, can be calculated by

$$\lambda_{eph}^{-1} = \frac{1}{a_0}\left[\frac{n(T)+1}{2}\right]\left[\frac{\varepsilon(0)-\varepsilon(\infty)}{\varepsilon(0)\varepsilon(\infty)}\right]\frac{\hbar\omega}{E}\ln\left\{\frac{[1+(1-\hbar\omega/E)^{1/2}]}{[1-(1-\hbar\omega/E)^{1/2}]}\right\}, \tag{2}$$

where

$$n(T) = \frac{1}{[e^{\hbar\omega/kT}-1]} . \tag{3}$$

Here a_0 is the Bohr radius, k the Boltzmann constant, $\varepsilon(0)$ the static dielectric constant and $\varepsilon(\infty)$ the high frequency dielectric constant respectively. For SiO_2, $\varepsilon(\infty) = 2.13$ and $\varepsilon(0) = 3.9$.
Electron-phonon interaction is not the only mechanism of energy loss for very slow electrons interacting with insulating materials. Indeed a low energy electron moving in an insulating material induces around it a polarization field with a stabilizing (trapping) effect on the electron itself. This is known as the polaronic effect, as the quasi-particle constituted by the slow electron with the polarization field is known as a polaron. According to Ganachaud and Mokrani[25], the inverse inelastic mean free path for electron-polaron interaction, λ_{ep}^{-1} , is given by

$$\lambda_{ep}^{-1}(E) = Ce^{-\gamma E}, \tag{4}$$

where, for the present calculations relative to SiO_2, we put $C = 1\text{nm}^{-1}$, $\gamma = 0.085\text{eV}^{-1}$.
In Fig. 1 the effects the inelastic electron mean free paths in SiO_2 corresponding to the various mechanisms of energy loss (electron-electron, λ_{ee}, electron-phonon, λ_{eph}, electron-polaron, λ_{ep}) are presented as long as the inelastic mean free path λ which takes into account all these inelastic scattering mechanisms and is given by

$$\lambda^{-1} = \lambda_{ee}^{-1} + \lambda_{eph}^{-1} + \lambda_{ep}^{-1} \,. \tag{5}$$

It is clear from the figure that, due to the inclusion of the electron-phonon and of the electron-polaron interactions in the calculation, the electron inelastic mean free path λ approaches zero as the electron energy goes to zero.

4.Results and Discussion

The MC results we are going to present were obtained with a detailed modeling which takes into account all the described mechanisms of energy loss (electron-electron, electron-phonon, electron-polaron) and treats the elastic scattering events using the Mott cross section. Furthermore, the entire cascade of secondary electrons is followed. For a description of the present detailed MC approach, see Refs.[25-28].
The whole MC spectrum representing the energy distribution of electrons emerging from a SiO_2 sample due to a 250eV electron beam irradiation is presented in Fig. 2. Similar spectra have been observed for metals as well[29,30].
For the present calculation the target is considered as semi-infinite. The case of thin films requires special attention, in particular when the samples are thinner than the mean free path. The study of the secondary electron emission from thin films will be the subject of further investigations.
Many electrons of the primary electron beam can be backscattered, after having interacted with the atoms and electrons of the target. A fraction of them conserves the original kinetic energy, having suffered only elastic scattering collisions with the atoms of the target: these electrons constitute the so called elastic peak, or zero loss peak, whose maximum is located at the energy of the primary beam. The plasmon peak in Fig. 2 represents the electrons of the primary electron beam that emerge from the surface after having suffered a single inelastic collision with a plasmon. Multiple collisions with plasmons are also present in the spectrum, but they are of very low intensity so that they are not visible on this scale. Electron-phonon energy losses are also not visible in Fig. 2, for (i) their intensity is much lower than that of the elastic peak and (ii) they are very close to the much more intense elastic peak whose width is of the order of 1eV.
On the other hand the secondary electrons produced by a cascade process are those electrons which have been extracted from the atoms by inelastic electron-electron collisions and are able to emerge from the target surface. Their energy distribution presents a pronounced peak in the very low energy region of the spectrum, typically below 50eV. The secondary electron emission yield is conventionally

measured integrating the area of the spectrum from 0 to 50eV (including, in such a way, also the tail of backscattered electron whose number, on the other hand, in this energy region is negligible – unless the primary energy be very low as well).
The comparison of the MC simulated secondary electron yield from SiO_2 with the Dionne experimental data[31] was shown in Fig. 3, to validate the code (in this case we used three values of γ: $0.080eV^{-1}$, $0.085eV^{-1}$, and $0.090eV^{-1}$).
As the primary energy increases, the yield increases until a maximum is reached. Then the yield decreases by increasing the primary energy. Indeed, at very low primary energy, few secondary electrons are generated, and increasing the primary energy the number of secondary electrons emerging from the surface increases. Furthermore, at very low primary energy, the average depth at which the secondary electrons escaping from the surface are generated also increases as the primary energy increases. When the energy becomes higher than a threshold which depends on the target (the turning point for SiO_2 being approximately at 250eV, according to the mean value between experiment and simulation), the average depth of generation of the secondary electrons becomes so deep that just a small amount of the generated secondary electrons are able to reach the surface satisfying the condition, represented by Eq. (1), necessary to emerge from the sample and to be detected.
The agreement between MC results and experimental data is quite satisfactory. The MC model well predicts, in fact, the shape of the experimental secondary electron yield curve of SiO_2 as a function of the electron primary energy. In particular we observe that the dashed line looks similar in shape, but shifted in energy. In order to compare the shapes of the curves, the Monte Carlo secondary electron yields corresponding to γ=$0.085eV^{-1}$ were horizontally translated (shift in energy was -50 eV). With this shift in energy, the experimental and Monte Carlo data yield a very close match (see Fig. 4).
Adjusting other parameters does not yield a closer match. We conclude that the differences between experimental and Monte Carlo data corresponding the original calculation (i.e. without any energy translation) could be reasonably attributed to the poor approximation adopted to describe the polaronic effect.
Eq. (4), proposed by Ganachaud and Mokrani[25], is based on the principle that the lower the electron energy, the higher the probability for the electron to lose its energy in the generation of a polaron. It is assumed that, once the polaron has been generated, the residual kinetic energy of the electron could be neglected. Therefore, the electron is considered to stay trapped at the interaction site. Of course, this represents a quite rough approximation. In fact, as recognized by Ganachaud and Mokrani[25], trapped electrons can actually hop from one trapping site to the next, due to phonon-induced processes, which cannot be adequately taken into account with the present oversimplified model represented by Eq.(4). Further investigations are thus required to improve the code in order to better understand and describe the polaronic effect, which represents one of the more relevant phenomena in determining the yield of secondary electron emission from insulators.

5.Conclusion

The main mechanisms of scattering and energy loss of electrons scattered in dielectric materials have been briefly treated within the Ganachaud and Mokrani heuristic model[25]. A MC scheme which takes into account all the energy losses suffered by each electron in the secondary electron cascade is used for calculating the secondary electron energy distribution and secondary electron emission yield from SiO_2. Satisfactory agreement has been found between the shapes of the curves corresponding to the MC simulation and the Dionne experimental data[31].

References

[1] Dapor M 1992 *Phys. Rev. B* **46** 618

[2] Mott NF 1929 *Proc. R. Soc. London* Ser. **124** 425

[3] Mott NF, Massey HSW 1965 *The Theory of Atomic Collisions*, Oxford University Press, Oxford

[4] Lin S-R, Sherman N, Percus JK 1963 *Nucl. Phys.* **45** 492

[5] Bunyan PJ, Shonfelder JL 1965 *Proc. Phys. Soc.* **85** 455

[6] Salvat F, Mayol R 1993 *Comput. Phys. Commun.* **74** 358

[7] Dapor M 1996 *J. Appl. Phys.* **79** 8406

[8] Dapor M 2003 *Electron-Beam Interactions with Solids: Applications of the Monte Carlo Method to Electron Scattering Problems*, Springer Tracts in Modern Physics **186**, Springer, Berlin

[9] Jablonski A, Salvat F, Powell CJ 2004 *J. Phys. Chem. Ref. Data* **33** 409

[10] Ritchie RH 1957 *Phys. Rev.* **106** 874

[11] Ritchie RH, Howie A 1977 *Phil. Mag.* **36** 463

[12] Penn DR 1987 *Phys. Rev. B* **35** 482

[13] Ashley JC 1988 *J. Electrons Spectrosc. Relat. Phenom.* **46** 199

[14] Yubero F, Tougaard S 1992 *Phys. Rev. B* **46** 2486

[15] Cohen-Simonsen A, Yubero F, Tougaard S 1997 *Phys. Rev. B* **56** 1612

[16] Dapor M 2006 *J. Electrons Spectrosc. Relat. Phenom.* **151** 182

[17] Dapor M 2006 *Surf. Sci.* **600** 4728

[18] Filippi M, Calliari L, Dapor M 2007 *Phys. Rev. B* **75** 125406

[19] Taioli S, Simonucci S, Calliari L, Filippi M, Dapor M 2009 *Phys. Rev. B* **79** 085432

[20] Taioli S, Simonucci S, Calliari L, Dapor M 2010 *Phys. Rep.* **493** 237

[21] Dapor M, Inkson BJ, Rodenburg C, Rodenburg JM 2008 *EPL* **82** 30006; **82** 49901

[22] Dapor M 2009 *Nucl. Instr. Meth. B* **267** 3055

[23] Fröhlich H 1954 *Adv. Phys.* **3** 325

[24] Llacer J, Garwin EL 1969 *J. Appl. Phys.* **40** 2766

[25] Ganachaud JP, Mokrani A 1995 *Surf. Sci.* **334** 329

[26] Dapor M, Ciappa M, Fichtner W 2010 *J. Micro/Nanolith. MEMS MOEMS* **9** 023001

[27] Koschik A, Ciappa M, Holzer S, Dapor M, Fichtner W 2010 *Proc. of SPIE* **7729** 77290X

[28] Dapor M 2011 *Nucl. Instr. Meth. B* **269** 1668

[29] Cimino R, Collins IR, Furman MA, Pivi M, Ruggiero F, Rumolo G, Zimmermann F 2004 *Phys. Rev. Lett.* **93** 014801

[30] Furman MA, Chaplin VH 2006 *Phys. Rev. Spec. Top. – Accel. Beams* **9** 034403

[31] Dionne GF 1975 *J. Appl. Phys.* **46** 3347

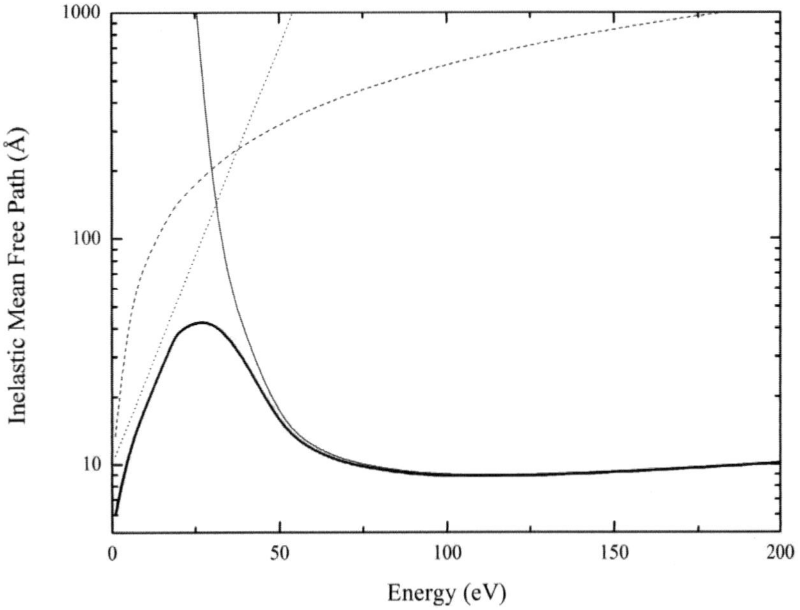

Figure 1. Electron inelastic mean free paths in SiO_2 corresponding to the various mechanisms of energy loss. Electron-electron inelastic mean free path, λ_{ee}, is represented by the solid line. Electron-phonon inelastic mean free path, λ_{eph}, is represented by the dashed line. Electron-polaron inelastic mean free path, λ_{ep}, is represented by the dotted line. The electron inelastic mean free path λ is given by $\lambda^{-1} = \lambda_{ee}^{-1} + \lambda_{eph}^{-1} + \lambda_{ep}^{-1}$ and is represented by the bold solid line. It approaches zero as the electron energy goes to zero. $\hbar\omega$=0.1eV, C=1nm^{-1}, γ=0.085eV^{-1}.

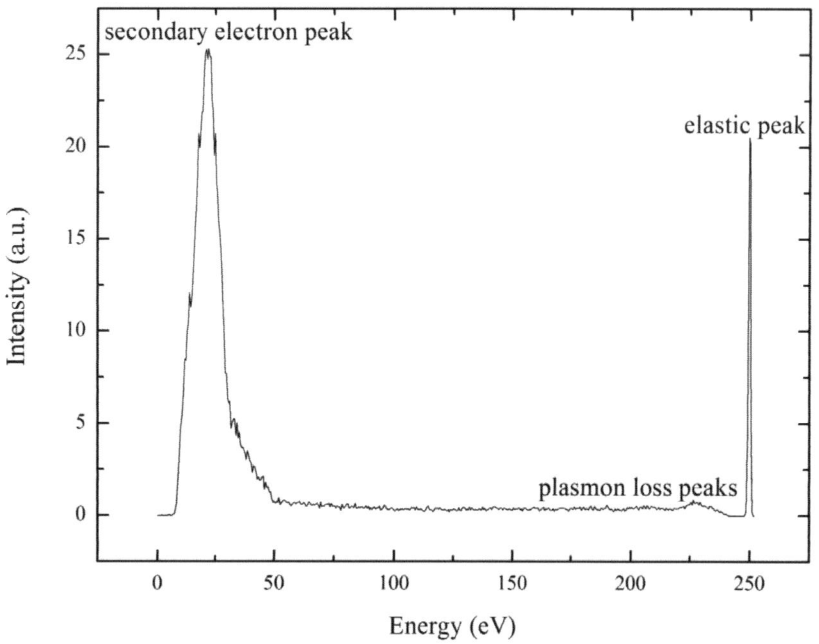

Figure 2. MC spectrum representing the energy distribution of electrons emerging from a SiO_2 sample due to a 250eV electron beam irradiation. The elastic peak, or zero loss peak, whose maximum is located at the energy of the primary beam, represents the electrons which suffered only elastic scattering collisions. The plasmon peak represents the electrons of the primary electron beam that emerge from the surface after having suffered a single inelastic collision with a plasmon. Multiple collisions with plasmons are also present in the spectrum, but they are of very low intensity so that they are not visible on this scale. The secondary electrons energy distribution presents a pronounced peak in the very low energy region of the spectrum, typically below 50eV. χ=0.9eV, $\hbar\omega$=0.1eV, C=1nm^{-1}, γ=0.085eV^{-1}.

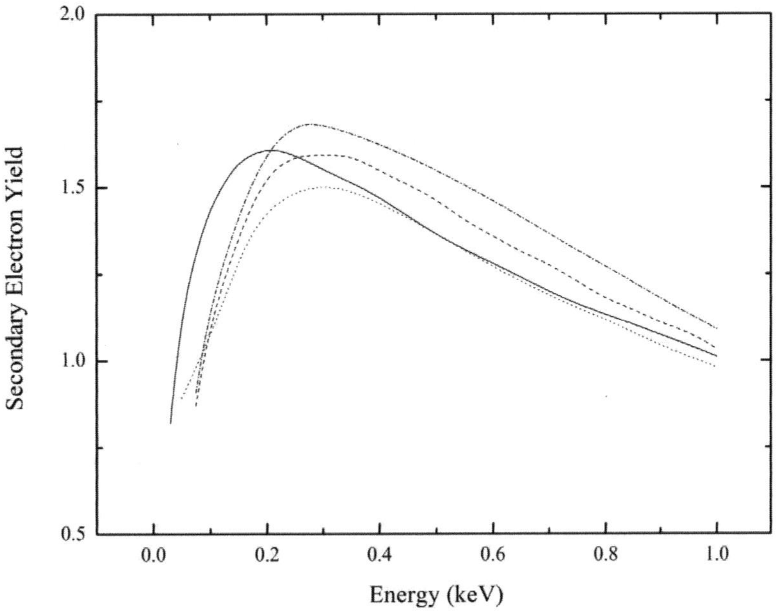

Figure 3. Secondary electron yield of SiO_2. Comparison of the Dionne experimental data[31] (solid line) with the MC simulated data (dotted line: γ=0.080eV^{-1}; dashed line: γ=0.085eV^{-1}, dotted-dashed line: γ=0.090eV^{-1}). χ=0.9eV, $\hbar\omega$=0.1eV, C=1nm^{-1}.

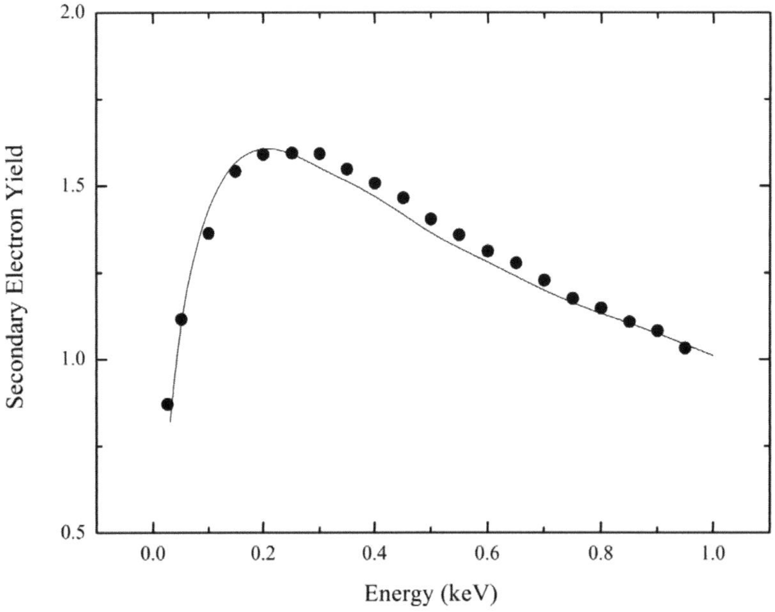

Figure 4. Secondary electron yield of SiO_2. Comparison of the Dionne experimental data[31] (solid line) with the MC simulated data (bullets). γ=0.085eV^{-1}, χ=0.9eV, $\hbar\omega$=0.1eV, C=1nm^{-1}. The Monte Carlo data were horizontally translated (shift in energy was -50 eV) in order to compare the shapes of the curves.

IUPAP C20 Conference on Computational Physics (CCP 2011) IOP Publishing
Journal of Physics: Conference Series **402** (2012) 012004 doi:10.1088/1742-6596/402/1/012004

Electronic properties of rippled graphene

Gui Gui[1], Jianxin Zhong[2,*] and Zhenqiang Ma[1,*]

[1]Department of Electrical and Computer Engineering, University of Wisconsin-Madison, USA
[2]Laboratory for Quantum Engineering and Micro-Nano Energy Technology, Xiangtan University, P. R. China

*Email: mazq@engr.wisc.edu; jxzhong@xtu.edu.cn

Abstract. Short range periodic ripples in graphene have been modeled. The electronic properties of the rippled graphene have been investigated using first-principles calculations. Compared with flat graphene, there is a band gap opening in rippled graphene. Generally, the value of energy gaps increases as the height of ripples increase, but it decreases as the range of ripples enlarges. The maximum value of energy gaps in rippled graphene can reach several hundred meV, which turns rippled graphene into a good semiconductor. As a result, the magnitude of energy gaps can be tuned effectively by controlling the range and height of ripples in graphene.

1. Introduction

Since the successful isolation of graphene by Novoselov et al [1] in 2004, the prospects of this new material have motivated widespread research. Many interesting electronic properties have been investigated in graphene, such as the anomalous quantum-Hall effect [2-5], the quantization minimum conductivity [6], especially zero effective mass and thus extremely high mobility for electrons and holes [7]. It has become one of the most promising nano-scale materials for building the next generation of electronic devices.

As a unique two dimensional material, graphene has a hexagonal ring structure, which results in π bands crossing and a direct zero band gap near the Fermi level [8]. Undoubtedly, absence of band gap sets limitations on its practical applications as a semiconductor. Therefore it is exceedingly important to find methods to effectively open and tune band gaps in graphene.

Researchers have been working on the band gap engineering in graphene for years. Some effective methods were proposed, such as graphene nanoribbons [9] and mesh [10] by making electronic confinements, and graphene oxidation [11] and hydrogenation [12] by chemical treating. However, all above methods introduce additional serious problems, including edge roughness, disorders and impurities, which greatly reduce the carrier mobility in graphene.

Suspended graphene sheets were observed being not perfectly flat showing ripples, with amplitude of about one nanometer [13, 14]. Recent experiments also show that graphene on iridium is slightly rippled [15]. These ripples generate band gaps in graphene by developing spatially varying potentials [16] or effective magnetic fields (also known as gauge fields) [7]. Lau et al [17] demonstrated that periodic ripples can be created and controlled in suspended graphene by thermal treatment, which has been proposed to be a promising way to band gap engineering of graphene by placing it on an especially prepared substrate [18]. Thus understanding the effect of ripples on the electronic properties of graphene is significant for its practical applications in modern nanoelectronics.

In this work, we study the electronic properties of rippled graphene by the first principles method. Our calculation results show band gap opening in the rippled graphene. It is confirmed that electronic transport properties and physical features of graphene can be effectively tuned by controlling the periodic ripple structure in graphene.

2. Simulation methods

We employ the density functional theory (DFT) method implemented in the plane wave basis VASP [19, 20] code that allows the electronic structure of systems with periodic boundary conditions to be calculated.

The projector augmented wave (PAW) [21, 22] potentials and the GGA of Perdew and Wang known as PW91 [23] are adopted. The total energy is converged to within 1 meV with a sufficiently large plane wave cutoff of 500 eV and a Monkhorst-Pack [24] k-point mesh of 15×15×1. The structural is fully relaxed until the interatomic forces are less than 0.01 eV/angstrom. In order to eliminate inter-layer interactions, the vertical distance of graphene layers is taken to be 1.55 nm.

Our simulation proceeds by taking a proper super cell as shown in Fig. 1. Periodic ripples are achieved as follows: firstly, pull some specific atoms in the middle part of a super cell up from their original places and fix them; then, relax the rest atoms with fixing their z-coordinates in plane to get a final equilibrium structure; plus, the in-plane lattice constants were allowed to relax throughout the simulations. The size of a super cell increases as the range of ripples increases. The amplitude of ripples is defined as the height of carbon atoms away from the original plane, shown in Fig.1 (a). Fig. 1 (b) depicts the curvature of the graphene sheet under the case of pulling out 3 atoms in a 2x2 super cell which will be shown in details later in this paper.

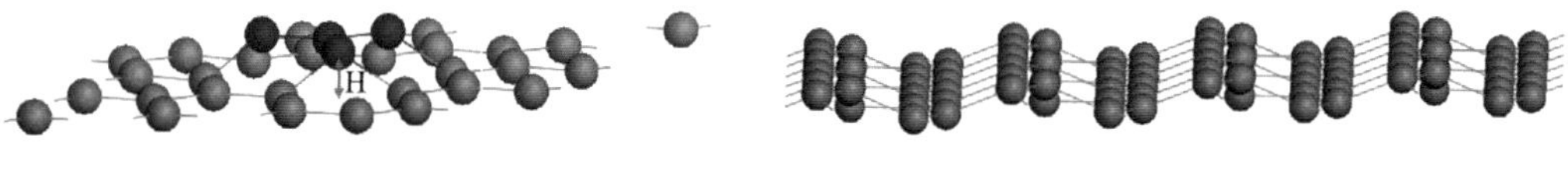

(a) (b)

Figure 1 (a) Rippled graphene structure. H is the height of carbon atoms away from the original plane. (b) The overall curvature of the graphene sheet under the case of pulling out 3 atoms in a 2×2 super cell.

3. Simulation results and discussions

Before doing systematic calculations, we investigate the effect of super cell sizes on the simulation results. Figure 2 shows the value of band gaps as a function of the super cell sizes and the ripple amplitudes, where one carbon atom is pulled up in this case. We can see from Fig. 2 that the value of band gaps increases as the height of out-of-plane atom increases, and decreases not monotonically as the size of super cell increases. It indicates that choosing an appropriate super cell size for the calculations can affect the magnitude of band gaps in rippled graphene. Therefore, it is critical to control the range and amplitude of ripples in graphene for band gap engineering.

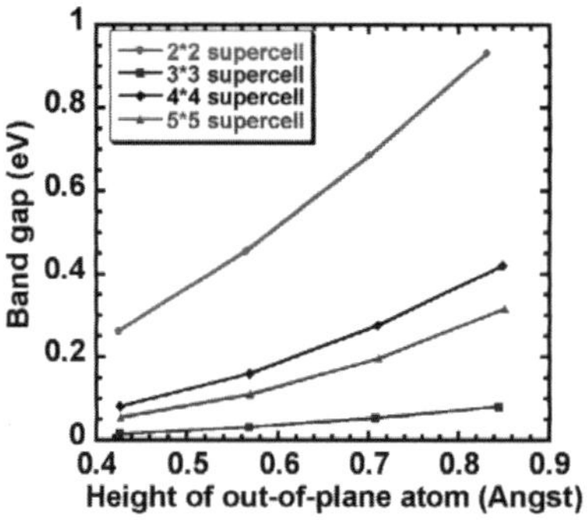

Figure 2 The effect of super cell size and ripple amplitude on the opening of band gaps in graphene.

In the following, we will show the DFT calculations of the electronic structures and band gap opening in the rippled graphene systems with 2×2, 3×3 and 4×4 super cells. A regular k-points mesh of 60 points has been used to calculate the band structure, and extra 40 points were added near the high symmetry point where the band gap opens. The way to define high symmetry point in band structures can refer to our previous work [25].

The simulation results of the rippled graphene with a 2×2 super cell are shown in Fig. 3. Figure 3 (a) is band structures of graphene systems with one rippled atom of 0.56 Å and 0.83 Å heights, which show a significant direct band gap at R point. From Fig. 3 (b), we can find that the value of band gaps increases as the height of out-of-plane atoms increases from 0.28 Å to 0.83 Å, and decreases as the number of rippled atoms increases from one to three. The maximum value of band gaps is about 0.93 eV when the height of one rippled atom reaches 0.83 Å.

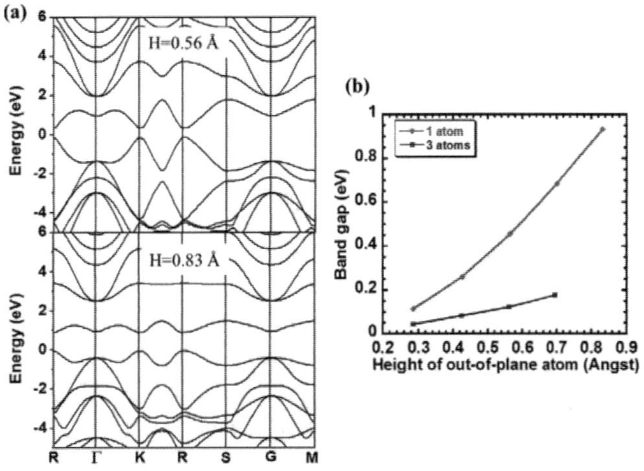

Figure 3 (a) Band structures for rippled graphene with a 2×2 super cell. (b) The value of band gaps as a function of the height of out-of-plane atoms.

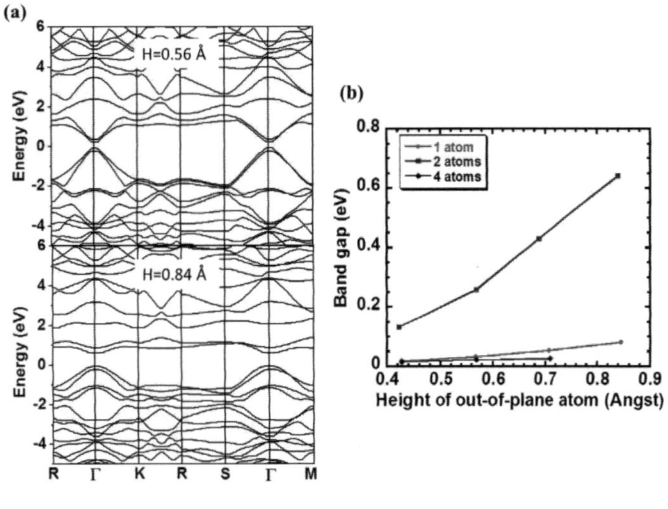

Figure 4 Band structures for rippled graphene with a 3×3 super cell. (b) The value of band gaps as a function of the height of out-of-plane atoms.

Then we increase the super cell size from 2×2 to 3×3 and do simulations for graphene systems with one, two and four rippled atoms. Similarly, the value of band gaps increases as the height of out-of-plane atoms increases from 0.42 Å to 0.84 Å, and first increases and then decreases as the number of rippled atoms increases from one to four, as shown in Fig. 4 (b). Fig. 4 (a) displays band structures of graphene systems with two rippled atoms of 0.56 Å and 0.84 Å heights, which show a significant direct band gap at Γ point. As for this 3×3 case, the maximum value of band gaps is about 0.64 eV when the heights of two rippled atoms reach 0.84 Å.

We further increase the super cell size to 4×4 and do simulations for graphene systems with one, two and four rippled atoms. The calculation results indicate a similar trend as the above two cases. The value of band gaps increase as the height of out-of-plane atoms increases from 0.43 Å to 0.85 Å, and decrease not monotonically as the number of rippled atoms increases from one to four, as shown in Fig. 5 (b). Band structures of graphene systems with four rippled atoms of 0.57 Å and 0.85 Å heights are shown in Fig. 5 (a), which exhibit a significant direct band gap at both K and R points. As for this 4×4 case, the band gap of 0.42 eV is maximal when the height of one rippled atom reaches 0.85 Å.

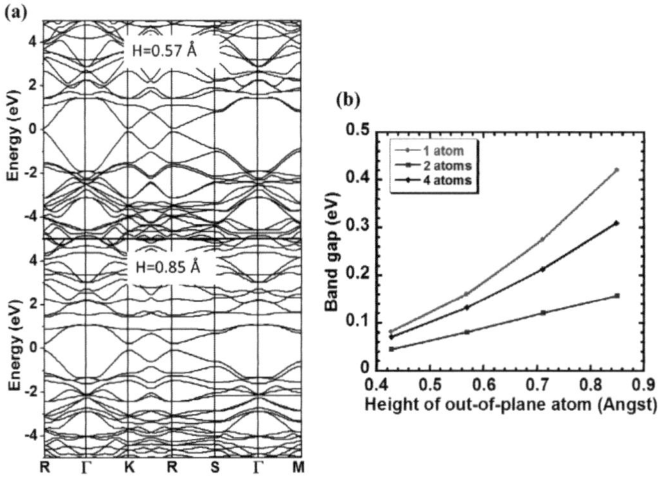

Figure 5 Band structures of rippled graphene with a 4×4 super cell. (b) The value of band gaps as a function of the height of out-of-plane atoms.

4. Conclusions

In summary, we have presented a study on electronic properties of rippled graphene systems by the *ab initio* calculations. Our results show that the maximum value of band gaps decreases as the super cell size increases. Additionally, the band gap increases as the amplitude of ripples increases. A direct band gap of 0.93 eV was achieved, which presents rippled graphene as a good semiconductor. It is consistent with previous experimental work [15, 17] that by controlling the range and amplitude of periodic ripples can effectively tune band gaps in graphene. Although our calculations just modeled a related short range of periodic ripples in graphene, it can be scaled up for the practical prediction.

Acknowledgements
The work was financially supported by the Office of Naval Research (Grant No. 00549970).

References

[1] K. S. Novoselov, A. K. Geim, S. V. Morozov, D. Jiang, Y. Zhang, S. V. Dubonos, I. V. Grigorieva and A. A.Firsov 2004, *Science* **306**, 666.
[2] Y. Zhang, Z. Jiang, P. Small, M. S. Purewal, Y. W. Tan, M. Fazlollahi, J. D. Chudow, J. A. Jaszczak, H. L. Stormer and P. Kim 2006, *Phys. Rev. Lett.* **96**, 136806.
[3] Dmitry A. Abanin, Patrick A. Lee and Leonid S. Levitov 2006, *Phys. Rev. Lett.* **96**, 176803.

[4] C. L. Kane and E. J. Mele 2005, *Phys. Rev. Lett.* **95**, 226801.
[5] Y. Zhang, Y. Tan, H. L. Stormer and P. Kim 2005, *Nature* **438**, 201.
[6] J. Tworzydlo, B. Trauzettel, M. Titov, A. Rycerz and C. W. J. Beenakker 2006, *Phys. Rev. Lett.* **96**, 246802.
[7] A. H. Castro Neto, F. Guinea, N. M. R. Peres, K. S. Novoselov and A. K. Geim 2009, *Rev. Mod. Phys.* **81**, 109.
[8] Y. Zheng and T. Ando 2002, *Phys. Rev. B* **65**, 245420.
[9] X. Wang, Y. Ouyang, X. Li, H. Wang, J. Guo and H. Dai 2008, *Phys. Rev. Lett.* **100,** 206803.
[10] J. Bai, X. Zhong, S. Jiang, Y. Huang and X. Duan 2010, *Nature Nanotech.* **5**, 190.
[11] J. Yan, L. Xian and M. Chou 2009, *Phys. Rev. Lett.* **103,** 086802.
[12] R. Balog, B. Jørgensen, L. Nilsson, M. Andersen, E. Rienks, M. Bianchi, M. Fanetti, E. Lægsgaard, A. Baraldi, S. Lizzit, Z. Sljivancanin, F. Besenbacher, B. Hammer, T. Pedersen, P. Hofmann and L. Hornekær 2010, *Nature Mater.* **9**, 315.
[13] A. Geim and K. Novoselov 2007, *Nature Mater.* **6,** 183.
[14] A. Fasolino, J. Los and M. Katsnelson 2007, *Nature Mater.* **6** 858.
[15] I. Pletikosić, M. Kralj, P. Pervan, R. Brako, J. Coraux, A. N'Diaye, C. Busse and T. Michely 2009 *Phys. Rev. Lett.* **102,** 056808.
[16] C. Park, L. Yang, Y. Son, M. Cohen, and S. Louie 2008, *Nature Phys.* **4**, 213.
[17] W. Bao, F. Miao, Z. Chen, H. Zhang, W. Jang, C. Dames and C. Lau 2009, *Nature Nanotech.* **4**, 562.
[18] R.Miranda and A. Parga 2009, *Nature Nano.* **4**, 549.
[19] G. Kresse and J. Furthmüller 1996, *Comp. Mater. Sci.* **6**, 15.
[20] G. Kresse and J. Furthmüller 1996, *Phys. Rev. B* **54**, 11169.
[21] P. Blöchl 1994, *Phys. Rev. B* **50**, 17953.
[22] G. Kresse and D. Joubert 1999, *Phys. Rev. B* **59**, 1758.
[23] J. P. Perdew and Y. Wang 1992, *Phys. Rev. B* **45**, 13244.
[24] H. J. Monkhorst and J. D. Pack 1976, *Phys. Rev. B* **13**, 5188.
[25] G. Gui, J. Li and J. Zhong 2008, *Phys. Rev. B* **78**, 075435.

Journal of Physics: Conference Series **402** (2012) 012005 doi:10.1088/1742-6596/402/1/012005

Structural search for dense packing of concave and convex shapes in two dimensions

Nabiha T. Elias, Toby S. Hudson

E-mail: `Toby.Hudson@sydney.edu.au`
School of Chemistry, The University of Sydney, NSW 2006, Australia.

Abstract. We describe and implement a structural search methodology which divides the global configuration space into enumerable subspaces, termed isopointal sets, and then searches within them using simulated annealing. This division allows us to follow a natural order for searching the infinite global configuration space, by respecting the observed tendency for molecular or colloidal particles to form crystals with a very limited set of local environments and symmetries. The method also produces near-optimal structural candidates with each symmetry searched, a useful feature for systems exhibiting polymorphic crystallization. We have applied this to dense packing simply-connected hard shapes in two dimensions, but the same method is applicable to optimizing arbitrary Hamiltonians, or *ab-initio* structural searches in higher dimensions.

1. Introduction

When we say that natural materials crystallize, or self-assemble, we are actually summarising the result of a momentous structural search process. There are infinitely many ways of arranging particles into a crystal, and the nucleation of a crystal from a melt or solution involves stochastically testing many different configurations, in search of those that lower the free energy of the system. Throughout the fluid, there are of the order of 10^{23} particles, continually forming and reforming local clusters searching for the rare arrangement of a supercritical nucleus. Even with this inbuilt search-parallelism, there is no guarantee that the globally optimal configuration will be found. Crystallization is an intrinsically kinetic phenomenon. Indeed many molecular crystals exhibit polymorphism, where a number of crystalline arrangements are observed as metastable products of this process. What hope then do we have of predicting these arrangements with the humble computer?

The good news is that in most cases, crystallization is a reliable process, and even if not thermodynamically optimal, the polymorphs are very good, and limited in number. Additionally, the structure formed is usually simple, having a low number of different particle environments (distinct occupied Wyckoff sites). The equivalent observation in the mathematics of tiling is that tile-transitive, or isohedral tilings (where a generating tile can be mapped onto any other tile by the symmetry operations) are quite frequent, but tiles that require anisohedral tiling are rarer. This means that a comprehensive search of structures with a small number of unrelated local environments should usually be sufficient to find the thermodynamically optimal structure.

One conventional option is to approximate the small number of distinct particle environments by a small, flexible unit cell.[1, 2, 3] This is reasonable for crystals with low overall symmetry, but becomes wasteful in highly symmetric situations. For example, the dense packing of binary spheres seen in the super-lattice analogous to the atomic $NaZn_{13}$ structure, formed by some colloidal opals, has only three inequivalent sites (and two of those are on fixed high-symmetry points). Simulating the structure as a single flexible unit cell would require 112 particles, with a corresponding search space with 338 degrees of freedom. Making use of the high symmetry allowed an exhaustive search of the relevent isopointal set, with only three degrees of freedom, obtaining structures with a density higher than that of segregated uniform spheres.[4]

This paper reports the implementation of a proposed methodology which divides the global search domain into enumerated sub-domains, ordered by a gradual increase in the number of distinct particle sites.[5] We obtain three potential advantages over conventional structural search methodologies. The first is that a local optimum is found within each sub-domain searched, thus generating a range of good candidate structures with different symmetries. This is useful for the study of molecules exhibiting polymorphism, because a single exploration of the global search domain can locate multiple discrete polymorphs. The second advantage is the ability to search within only specific sub-domains when there is reason to believe they will contain the solution. Thirdly, because each subdomain is defined by the symmetry of the configurations it explores, each sub-search uses the minimal set of degrees of freedom, obtaining the efficiency advantages discussed in the previous paragraph.

As a proof-of-concept, we apply our search to the problem of packing simply-connected hard shapes in two dimensions (2-d), including non-convex shapes. We have chosen this limited testing domain for a few reasons. There are only seventeen distinct types of crystalline symmetry in 2-d, called the wallpaper groups. The equivalent number of space groups in three-dimensional (3-d) space is 230, which makes the 2-d case considerably cheaper computationally. Many conjectured dense packings and theorems are available for comparison and validation in 2-d, because of the overlap with the mathematical packing and tiling community. A final benefit is that solutions are easily visualized. Despite the limited testing domain used in this paper, the same method is applicable to optimizing arbitrary Hamiltonians or *ab-initio* systems in higher dimensions. We have previously had success when searching specific isopointal sets for binary hard sphere packings in 3-d,[6, 4, 5, 7] and intend to apply the method to other particle shapes and interactions.

The packing of shapes in space is a fundamental constraint and ordering principle with widespread applications throughout the natural sciences. This kind of problem has been studied since the time of Aristotle,[8] but is sufficiently difficult that the best packing of even one of the simplest shapes, identical spheres, was unproven until around 2005.[9] Although the consequences of packing are evident at many size scales, including atomic compounds,[10] precious opals,[11], biological systems, and even everyday fruit packing, one of the most ubiquitous are the self-assembly of molecules or nano-particles into crystalline materials. In fact, the modern mathematical field of packings and coverings had its origins in chemical crystallography.[12]

Even 2-d packing problems have some scientific relevence. Molecular packing appears to be the primary determinant of the structure of monolayers formed at interfaces, across a wide range of molecules. This hypothesis is based on the observation that the symmetry groups which allow an arbitrary shape to contact the largest number of neighbouring objects (6-fold coordination), the wallpaper groups $p2$, $p1$, $p2gg$, and pg, are much more frequently observed in molecular monolayers than other wallpaper groups.[13] A similar observation has been made for 3-d molecular crystal structures.[14] Although the optimality of these actual molecular packings has not been tested, this is convincing evidence of the general importance of packing, even in systems where specific interactions of functional groups also play a role. Despite this understanding, the prediction of crystal structure based on the composition of the individual building blocks is still in its infancy.[13] This is in part due to the algorithmic paralysis caused by the infinite search space which presents itself whenever a large collection of shapes is to be arranged optimally.

To the best of our knowledge, few systematic studies of congruent-shape packing principles have been undertaken, but some results are known or conjectured. Any convex shape can be packed in a dual lattice (with a wallpaper group with at least the symmetries of $p2$) on the plane with a density of at least ≈ 0.866025.[15] Many specific shapes, including parallelograms, hexagons, triangles, and a range of specific irregular pentagons, can tile the plane (achieve a perfect density of 1) using only congruent copies of a single shape.[16] Circles have long been known to have the fairly poor optimal density of $\pi/\sqrt{12} \approx 0.9069$. A handful of other convex shapes are thought to have a lower optimal packing, including regular pentagons ($(5 - \sqrt{5})/3 \approx 0.92131$),[15] regular heptagons (≈ 0.8926),[15] and the smoothed-octagon (≈ 0.902414). The best-known packing of the smoothed octagon has been shown to be optimal amongst a specific subset of packings, but is conjectured to be the worst best-packing of all centrally-symmetric convex shapes.[17]

2. Algorithm design

The key difference between our program and other methods of structural search is that we search only within one isopointal set at a time. Because the structures in an isopointal set must have specific symmetries, the number of free parameters is reduced relative to the general search problem. If a particular Wyckoff site is occupied, it may have zero, one or two free parameters that specify the exact positioning of that site, but all of the particles specified by that site are symmetry-related, and thus their orientations and positions are governed by those same few variables. To complete the global search, the possible isopointal sets must be generated and searched individually. This is possible because isopointal sets are enumerable. The enumeration proceeds as follows.

Firstly, we iterate the maximum number of distinct Wyckoff sites to be occupied. Since most molecular packings on surfaces, and known or conjectured optimal congruent-shape packings have all particles of one type in the same symmetry-orbit, corresponding to a single occupied Wyckoff-site (also called isohedral or tile-transitive), we have concentrated our efforts on searching for structures with only one occupied Wyckoff site. We have also briefly tested our program on some shapes that are known to only pack optimally in structures which are not isohedral, such as the pentagonal shapes discussed by Schatschneider.[16] The combinatorial explosion quickly takes hold, so even a comprehensive study of ternary systems would be computationally expensive.

The second level of the enumeration is to loop through all 17 wallpaper groups. These set the underlying minimum symmetry of the packing. Again arguments from experiments can be made that some of these are much more commonly observed than others. This is rationalized by the fact that these common wallpaper groups are those that allow an arbitrary particle shape to pack such that all particles are in contact with six neighbours. In this paper we have chosen not to limit the search in this way. Instead we will search for optimal configurations within every wallpaper group, to establish the feasability of generating multiple candidates for polymorphism.

Within a wallpaper group, there are a number of different Wyckoff sites. One of these is always the general site, a location with no special symmetry requirements, and as such, any type of particle could occupy this general site. The others are locations on particular symmetry elements of the group, such as mirror planes, or points of rotation. These locations impose requirements about what kind of particle can be placed on such a site, and so, depending on the shape we are studying, they sometimes cannot be occupied. Each allowed combinations of Wyckoff sites defines a so called isopointal set. So, for example, an isopointal set labelled $p3$(ad) would indicate the search space defined by the wallpaper set $p3$ with one particle on Wyckoff site a and another on site d. This is a space of structures rather than a particular structure, because

Wyckoff site d in wallpaper group $p3$ has variable coordinates, x and y, a particle rotation, ϕ, and the (hexagonal) cell parameter can also be varied. In some cases, these parameters can be varied until the structure actually obtains a higher overall symmetry, but for search purposes, we consider all accessible structures to be within the isopointal set.

In principle any ergodic search algorithm could be used to search within the isopointal set. In our case the core algorithm we have chosen is a standard simulated annealing Monte Carlo optimization algorithm, which varies the free parameters associated with the packing structures, including the unit cell angle and side lengths, along with the positions and orientations of the particles. The function it seeks to optimize is simply the density of the structure - the inverse of its packing fraction. This density is determined entirely by the shape and number of the particles, and the cell volume, provided the particles do not overlap. In the case of overlap, the cell is expanded until the overlap is removed.

3. Simulations

We ran structural searches to optimize the packing of shapes with a range of different symmetries, including some known for their packing properties. The shapes we report here are the unit circle, the regular polygons with 5, 6, 7, 9, and 12 sides, the smoothed octagon, a pointy-egg shape with mirror symmetry, and a (non-convex) comma.

The egg and comma are defined parametrically, with an angle-dependent radius of the form given in equations 1 and 2. These specific shapes have no particular importance, they were simply selected to increase the diversity of the test set. Our current implementation allows for the study of any shape that can be specified parametrically as a function of the angle, including non-differentiable or non-convex shapes like these. In the results we report here, the shapes were constructed using 3600 radial spokes, which were tested for intersections to determine overlaps with neighbouring shapes. The area of each shape was also numerically determined from these segments. Thus the calculated packing values have a small numerical error, of the order of ± 0.001. The numerical results of our simulations are shown in Table 1.

$$r_{egg}(\theta) = \frac{1}{2\sin^2((3\pi - \theta)/4)} \tag{1}$$

$$r_{comma}(\theta) = 0.4 + 0.6 \times (2(\sqrt{\theta/2\pi} - 0.5))^4 \tag{2}$$

It is immediately notable that many of the simulations for a given shape have converged on the same dense-packed structure even though they were exploring isopointal sets in different wallpaper groups. Nevertheless, depending on the symmetries of the shape to be packed, simulations in some wallpaper groups are unable to symmetrize like this. Thus there are often a range of diverse dense-packed structures, with different symmetry, which could be used as candidates in free energy calculations to predict potential polymorphism. This is particularly true of the less symmetrical shapes, so for example, Figure 1 shows five different structures all with high packing fractions.

We also note that the wallpaper groups most frequently observed in molecular monolayers, $p2$, $p1$, $p2gg$, and pg,[13], are consistently good performers across the range of shapes we have tested. This suggests that searches limited to these groups may be sufficient in systems or situations where an exhaustive global search is not required or proves too expensive computationally.

Detailed inspection of the structural and numerical outcomes for each of the well-known shapes gives maximum-packing results generally in line with literature expectations, as well as a range of high-density polymorphs. For regular pentagons and heptagons, the densest structures found both have symmetry $p2mg$, and are identical to those produced by Kuperberg's constructive method,[15] which are the densest known packings of these shapes and appear in many different contexts.[18, 19, 20] (Our result in the pg set appears slightly better for the

Table 1: Maximum packing fractions found for searches with a particular wallpaper group. In each isopointal set, only one Wyckoff site was occupied. In cases where more than one isopointal set is possible within the same wallpaper group, the best performing search is reported here. The full structures for all isopointal sets are available in the supplementary material.

Wallpaper group	comma	egg	pentagon	heptagon	smoothed octagon	nonagon	hexagon	dodecagon	circle
$p1$	0.939	0.912	0.819	0.864	0.902	0.892	1.000	0.928	0.907
$p2$	0.950	0.881	0.897	0.890	0.902	0.894	1.000	0.928	0.907
pm	0.881	0.807	0.691	0.741	0.825	0.760	0.751	0.804	0.785
pg	0.941	0.912	0.922	0.893	0.901	0.897	1.000	0.928	0.907
cm	0.853	0.911	0.854	0.862	0.902	0.888	1.000	0.928	0.907
$p2mm$	0.756	0.794	0.691	0.739	0.825	0.757	0.750	0.804	0.785
$p2mg$	0.889	0.919	0.921	0.893	0.902	0.869	1.000	0.928	0.907
$p2gg$	0.923	0.916	0.870	0.885	0.903	0.893	1.000	0.928	0.907
$c2mm$	0.817	0.859	0.854	0.842	0.902	0.827	1.000	0.928	0.907
$p4$	0.818	0.833	0.843	0.844	0.851	0.831	0.841	0.862	0.842
$p4mm$	0.493	0.794	0.674	0.729	0.825	0.776	0.804	0.804	0.785
$p4gm$	0.820	0.816	0.823	0.844	0.850	0.827	0.833	0.862	0.842
$p3$	0.903	0.897	0.859	0.883	0.893	0.895	1.000	0.928	0.907
$p3m$	0.646	0.639	0.651	0.638	0.670	0.884	0.933	0.928	0.907
$p31m$	0.696	0.897	0.871	0.883	0.890	0.888	1.000	0.928	0.907
$p6$	0.689	0.790	0.736	0.761	0.776	0.784	1.000	0.928	0.907
$p6mm$	0.473	0.778	0.706	0.572	0.718	0.734	1.000	0.928	0.907

pentagons, but converges structurally and numerically as the shape resolution is increased.) For smoothed octagons, we have also converged on the known optimal density of ≈ 0.9024,[17] and we realise this with two distinct packings, one in the $c2mm$(a) isopointal set, and one in the $p2mg$(c) set. For the hexagon, dodecagon and circle, the high symmetry of these particles means that the known optimal packings of these shapes can easily be obtained from many different isopointal sets. The one known failure in these simulations occurs for the nonagons, for which a $p2$ structure with a density of 0.9010 was recently reported.[3] Our optimal solution has a very close density of 0.897, but is structurally different. Our corresponding $p2$ search appears to have been trapped in a local optima in the six dimensional search space, but we still consider it beneficial to have reduced it from the nine dimensions required for a global search of the two particle unit cell.

4. Conclusions

We have successfully demonstrated a proof-of-concept of the enumerated isopointal set structural search, treating the example problem of searching for homogeneous packings of congruent hard shapes in 2-d. Overall densest packings were almost always in line with the best known prior results, apart from one case which fell slightly short due to the simulated annealing simulation getting trapped in a local optima. These simulations also reproduced the tendency observed for molecular or colloidal particles to favour crystals with a very limited set of local environments and symmetries. The division of the global search into subspaces ensured diverse competitive structural outcomes, especially in the cases with less symmetric particles. This aspect bodes well for the use of this search methodology in the study of polymorphic systems.

The step up from 2-d to 3-d structure search will result in over an order of magnitude expansion in the number of isopointal sets that must be searched. As noted earlier, there are

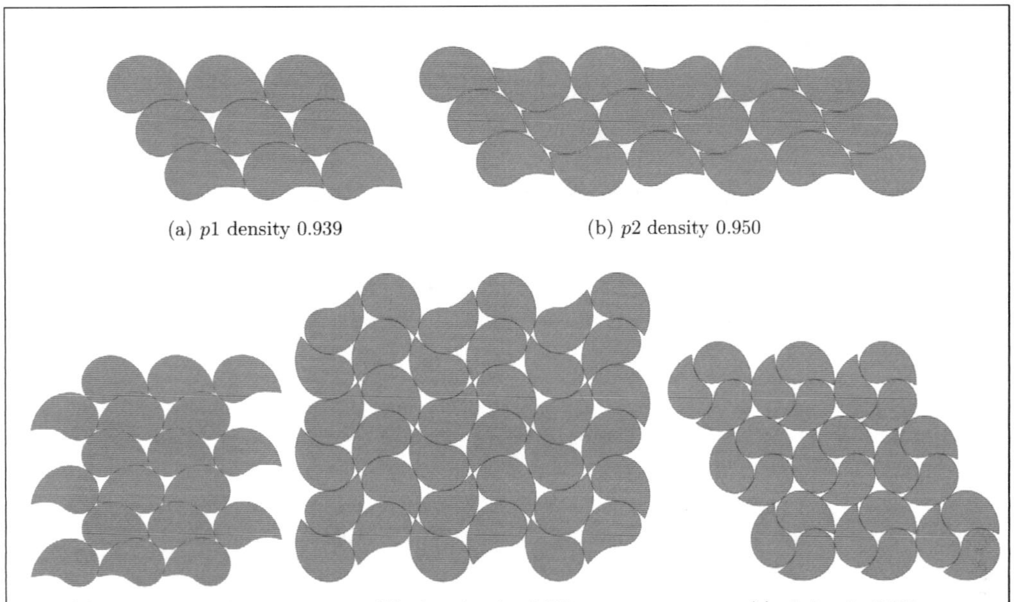

Figure 1: The five densest packings found for the comma shape, each with a different wallpaper group.

230 space groups as compared to 17 wallpaper groups, and space groups often have more Wyckoff sites. Thankfully in many applications this can be scaled back by prioritizing the space groups known to dominate the known crystal structures. But even when an exhaustive search is desired, it will be feasible given that each result reported here required a few hours on a single CPU core, and different isopointal searches can be run on different cores. The speed of an individual test for overlapping particles is directly related to the efficiency. We have not sought to optimize this, but many algorithms are available, due to its importance in computer graphics. Our shapes were defined with 3600 segments, which should be easily sufficient to accurately specify most 3-d shapes.

Acknowledgments
TSH wishes to acknowledge helpful discussions with Prof. Peter Harrowell.

Supporting information
Supporting information is available, which gives optimized packing structures for each isopointal set searched. This material is available under Supplementary Data.

References
[1] Torquato S and Jiao Y 2009 *Nature* **460** 876–879
[2] Filion L, Marechal M, van Oorschot B, Pelt D, Smallenburg F and Dijkstra M 2009 *Physical Review Letters* **103**(18) 188302
[3] de Graaf J, van Roij R and Dijkstra M 2011 *Physical Review Letters* **107**(15) 155501
[4] Hudson T S 2010 *The Journal of Physical Chemistry C* **114** 14013–14017
[5] Hudson T S and Harrowell P 2011 *Journal of Physics: Condensed Matter* **23** 194103
[6] Kummerfeld J K, Hudson T S and Harrowell P 2008 *The Journal of Physical Chemistry B* **112**(35) 10773–10776

[7] O'Toole P I and Hudson T S 2011 *The Journal of Physical Chemistry C* **115** 19037–19040
[8] Aristotle 350 BC *On the Heavens* vol III translated by J. L. Stocks.
[9] Hales T C 2006 *Discrete & Computational Geometry* **36** 5–20
[10] Parthé E 1961 *Zeitschrift für Kristallographie* **115** 52–79
[11] Sanders J V 1980 *Philosophical Magazine A* **42** 705–720
[12] Fejes Tóth G 2004 *Handbook of discrete and computational geometry* ed Goodman J E and O'Rourke J (CRC Press) chap 2
[13] Plass K E, Grzesiak A L and Matzger A J 2007 *Accounts of Chemical Research* **40** 287–293
[14] Mighell A D, Himes V L and Rodgers J R 1983 *Acta Crystallographica Section A* **39** 737–740
[15] Kuperberg G and Kuperberg W 1990 *Discrete & Computational Geometry* **5**(1) 389–397
[16] Schattschneider D 1978 *Mathematics Magazine* **51**(1) 29–44
[17] Hales T C 2011 *ArXiv e-prints* ArXiv:1103.4518v1
[18] Henley C L 1986 *Physical Review B* **34**(2) 797–816
[19] Baker T S, Caspar D L D and Murakami W T 1983 *Nature* **303** 446–448
[20] Schilling T, Pronk S, Mulder B and Frenkel D 2005 *Physical Review E* **71**(3) 036138

Investigating robustness of interatomic potentials with universal interface

Bohumir Jelinek[1], Kiran Solanki[2], John F Peters[4] and Sergio D Felicelli[1,3]

[1] Center for Advanced Vehicular Systems, Mississippi State University, 200 Research Boulevard, Starkville, Mississippi 39759
[2] School for Engineering of Matter, Transport and Energy, Arizona State University, ERC 357, Temple, Arizona 85281
[3] Mechanical Engineering Department, Mississippi State University, 210 Carpenter Building, Mississippi State, Mississippi 39762
[4] U.S. Army Corps of Engineers, Vicksburg, Mississippi 39180

E-mail: bj48@cavs.msstate.edu

Abstract. We present a set of Python routines to perform basic tests of classical atomistic potentials and their example applications. These routines are implemented using universal Atomic Simulation Environment (ASE) and LAMMPS molecular dynamics code. ASE is utilized to create atomic configurations, to write input scripts for LAMMPS, and to read results from output files. Evaluated properties are formation energies and volumes of simple point defects (vacancies, substitutions, and interstitials), formation energies of basic surfaces, heats of formation of simple binary compounds, and elastic constants. The flexibility of LAMMPS allows easy switching between different semi-empirical potentials, while the universality of ASE allows to compare results with a variety of electronic structure codes.

1. Introduction

Molecular dynamics (MD) simulations provide a useful insight into phenomena between the scales of quantum and continuum level simulations. Interatomic potentials used in MD simulations require validation by comparison with available experimental properties, with high accuracy electronic structure calculations, and their quality is also assessed by comparison with other MD potentials. In order to perform validation tests of interatomic potentials, one needs to learn formats of input parameter files, atomic configuration files, and output files of classical MD code implementing MEAM method (LAMMPS [1]) and electronic structure code chosen for validation. Also, atomic configurations for single crystal structures, crystallic compounds, point defects (vacancies, interstitials, substitutions), planar defects (various surfaces and stacking faults), and strained structures needs to be created. Therefore, a tool applicable to quickly evaluate basic properties from classical MD potentials (and electronic structure methods) is desirable. Ideally, a single universal tool should be able to create basic atomic configurations and manipulate them, to serve these atomic configurations as inputs for a variety of methods/simulation codes, and to obtain their energies. In the present effort, we developed Python routines for basic tests of atomistic potentials using Atomic Simulation Environment (ASE) [2]. These routines were constructed to validate the Modified Embedded Atom Method (MEAM) potentials for Al, Si,

Published under licence by IOP Publishing Ltd

Journal of Physics: Conference Series **402** (2012) 012006 doi:10.1088/1742-6596/402/1/012006

Mg, Cu, and Fe alloys. Testing routines, results of MEAM potential tests, and sample tests of other MD potential [3] are available for download from a public repository [7].

2. Atomic Simulation Environment (ASE)

Atomic Simulation Environment (ASE) [2] is a universal object-oriented Python interface to a variety of electronic structure codes (calculators). It creates atomic configurations, input scripts, and parse output files for a variety of electronic structure codes. Recently, an ASE calculator interface to classical MD code LAMMPS [1] was implemented at the Department of Theoretical Chemistry, Technical University of Munich by Jörg Meyer.

ASE provides simple graphical user interface and visualization tools with an extensive documentation and tutorials [4]. It facilitates creation and manipulation of molecules, nanotubes, basic surfaces, and arbitrary crystalline structures utilizing its database of spacegroups. It supports structure optimization, symmetry analysis, dissociation, diffusion, constrains, nudged elastic band (NEB) method, vibration analysis, phonon calculations, infrared intensities, MD with a broad selection of ensembles and dynamics, STM imaging, and electron transport. Electronic structure codes (calculators) working with ASE are listed in table 1.

Table 1. Calculators working with ASE, as listed in ASE online documentation. Broad range of calculator types is supported: density functional theory (DFT), linearized augmented planewave (LAPW), effective medium theory (EMT), Hartree-Fock (HF), and classical molecular dynamics (CMD) calculators.

Code	Description	Type
VASP	Planewave PAW code	DFT
ABINIT	Planewave pseudopotential code	DFT
Castep	Planewave pseudopotential code	DFT, HF
Siesta	Linear combination of atomic orbitals pseudopotential code	DFT
Exciting	Full potential LAPW code	DFT, LAPW
Jacapo	ASE interface to Dacapo, planewave ultrasoft pseudopotentials	DFT
DFTB+	DFT based tight binding code	DFT
TURBOMOLE	Fast atom orbital code	DFT, HF
FHI-aims	Numeric atom-centered basis functions, full potential code	DFT, HF
FLEUR	Full potential LAPW code	DFT, LAPW
EMT	Effective medium theory calculator (written in Python)	EMT
Asap	Highly efficient EMT code (written in C++)	EMT
GPAW	Grid-based real-space PAW code	DFT, HF
Dacapo	Old interface to Dacapo	DFT
LAMMPS	Classical molecular dynamics code	CMD

3. Bulk properties

3.1. Equilibrium atomic volume, energy, and bulk modulus

The equilibrium atomic volume V_0 for a given structure is the volume per atom at which the structure has minimal energy. Equilibrium bulk modulus is defined as

$$B_0 = -V \left. \frac{\partial p}{\partial V} \right|_{V=V_0} \quad (1)$$

ASE provides a command line tool to calculate bulk properties by fitting values of energy at varying volumes to the stabilized jellium equation of state (SJEOS) [5]. Properties obtained from fit are equilibrium atomic volume V_0, corresponding equilibrium energy E_0, and bulk modulus B_0.

4. Point defects

4.1. Monovacancies

The formation energy of a single vacancy E_f^{vac} is defined as the energy cost to create a vacancy:

$$E_f^{vac} = E_{tot}[N-1] - (N-1)\varepsilon_0, \tag{2}$$

where $E_{tot}[N-1]$ is the total energy of a system with $N-1$ atoms containing a vacancy and $\varepsilon_0 = E_0/N_0$ is the equilibrium energy per atom in the bulk (section 3). Python script that calculates monovacancy formation energy takes the atomic specie, crystal structure, and equilibrium lattice constant(s) as input parameters. First, energy without vacancy at equilibrium lattice parameter is calculated. Then, one atom is removed and energy without relaxation is obtained. After relaxation of atomic positions and cell volume optimization, relaxed vacancy formation energy and volume are evaluated.

4.2. Interstitials

The formation energy of an interstitial point defect E_f^{int} is given by

$$E_f^{int} = E_{tot}[N+1] - N\varepsilon_X - \varepsilon_Y \tag{3}$$

where $E_{tot}[N+1]$ is the total energy of a system with N type-X bulk atoms plus one impurity atom of type-Y inserted at the interstitial site, and ε_X (ε_Y) is the total energy per atom of type-X (type-Y) in its most stable bulk structure. The inserted atom Y can be of the same type as the bulk, in which case the point defect is called a self-interstitial defect. Self-interstitial formation energies are calculated at the octahedral, tetrahedral, and dumbbell sites. Dumbbell orientations are [100] for fcc, [0001] for hcp, and [110] for bcc and diamond structures. Systems sizes are 5×5×5 primitive fcc and bcc cells, 3×3×3 primitive diamond cells, and 8×4×4 orthogonal hcp cells. Script that calculates energy of an interstitial takes atomic specie, crystal structure, equilibrium lattice constant, and interstitial type as parameters. Evaluated properties are interstitial formation energy and volume.

4.3. Substitutions

The formation energy of a substitutional point defect E_f^{sub}, in the case of the substitution of a type-X atom of the host with a type-Y atom, is defined by

$$E_f^{sub} = E_{tot}[(N-1)+1] - (N-1)\varepsilon_X - \varepsilon_Y \tag{4}$$

where $E_{tot}[(N-1)+1]$ is the total energy of a system of $N-1$ host type-X atoms and one type-Y atom that replaced type-X atom in the original bulk position, ε_X and ε_Y are the total energies per atom for type-X and type-Y atoms in their ground state bulk structures. Parameters of the script that calculates formation energy of a substitution are host element type, structure, and lattice constant, substitutional element type, and the equilibrium energy ε_Y of a substituting atom. Evaluated properties are energy without substitution, unrelaxed and relaxed energy with substitution, and formation volume of substitution.

5. Planar defects

5.1. Surfaces

The surface formation energy per unit surface area E_{surf} is defined as

$$E_f^{surf} = \frac{E_{tot}[N] - N\varepsilon_0}{A}, \tag{5}$$

where $E_{\rm tot}[N]$ is the total energy of the structure with two surfaces, N is the number of atoms in the structure, ε_0 is the equilibrium energy per atom in the bulk, and A is the total area of both surfaces. Surface formation energy of the fcc (111), (110), (100), diamond (111), (100), hcp (0001), $(10\bar{1}0)$, bcc (111), (110), (100) surfaces are evaluated. Input parameters for the script are atomic specie, crystal structure of the surface and its orientation, lattice parameter, equilibrium energy per atom in the bulk structure, number of repetitions of the base cell in the surface structure, and the vacuum thickness. Surface formation energy is evaluated for both unrelaxed and relaxed surfaces.

6. Heats of formation for binary compounds

The heat of formation per atom, ΔH, is defined as

$$\Delta H = \frac{E_{\rm tot}[N_{\rm X} + N_{\rm Y}] - N_{\rm X}\varepsilon_{\rm X} - N_{\rm Y}\varepsilon_{\rm Y}}{N_{\rm X} + N_{\rm Y}}, \tag{6}$$

$E_{\rm tot}$ is the equilibrium energy of the compound structure, $N_{\rm X}$ and $N_{\rm Y}$ are the numbers of type-X and type-Y atoms, $\varepsilon_{\rm X}$ and $\varepsilon_{\rm Y}$ are the equilibrium energies per atom for type-X and type-Y in their ground state bulk structures, respectively. Equilibrium volume and bulk modulus are obtained by fitting energy to SJEOS with lattice parameter varying by -2%, -1%, 1% and 2% as described in section 3. The heats of formation at equilibrium atomic volumes for B1, B2, B3, C1, C15, $\rm D0_3$, A15, $\rm L1_2$ binary compounds are evaluated following equation (6), along with their equilibrium volumes and bulk moduli.

7. Elastic constants

7.1. Trigonal and tetragonal shear modulus

Starting from an unstrained equilibrium configuration, the change of energy density due to small strain can be expressed as

$$\triangle E_V = \frac{1}{2}\sum_{i=1}^{6}\sum_{j=1}^{6} C_{ij}\epsilon_i\epsilon_j \quad \text{with} \quad C_{ij} = C_{ji}, \tag{7}$$

where ϵ_i are strains in modified Voigt notation. For small deformations of a cubic crystal, the change of energy density due to small strain is

$$\begin{aligned}\triangle E_V &= \frac{1}{2}C_{11}\left(\epsilon_1^2+\epsilon_2^2+\epsilon_3^2\right) + C_{12}\left(\epsilon_1\epsilon_2+\epsilon_2\epsilon_3+\epsilon_3\epsilon_1\right)\\ &+ \frac{1}{2}C_{44}\left(\epsilon_4^2+\epsilon_5^2+\epsilon_6^2\right) + O(\epsilon_i^3).\end{aligned} \tag{8}$$

The trigonal shear modulus C_{44} was determined from rhombohedral deformation given by $\epsilon_1 = \epsilon_2 = \epsilon_3 = 0$ and $\epsilon_4 = \epsilon_5 = \epsilon_6 = \delta$ in equation (8), leading to

$$\triangle E_V(\delta) = \frac{3}{2}C_{44}\delta^2 + O(\delta^3). \tag{9}$$

The tetragonal shear modulus $(C_{11} - C_{12})/2$ was determined from the deformation given by $\epsilon_1 = \delta, \epsilon_2 = \frac{1}{1+\delta} - 1$ in equation (8), leading to

$$\triangle E_V(\delta) = (C_{11} - C_{12})\,\delta^2 + O(\delta^3). \tag{10}$$

Input parameters of the script are element specie(s), crystal structure, and equilibrium lattice parameter. Equilibrium volume is first validated as described in section 3. Evaluated properties are tetragonal and trigonal shear moduli.

8. Conclusions

We developed Python routines for basic tests of classical atomistic potentials using universal Atomic Simulation Environment [2]. These scripts, together with example application to validate MEAM potential for Al, Si, Mg, Cu, and Fe alloys [6], and resulting evaluated properties are available for download from a public Subversion repository [7] under the LGPL license.

References

[1] Plimpton S J 1995 *J. Comput. Phys.* **117** 1–19 URL `http://dx.doi.org/10.1006/jcph.1995.1039`

[2] Bahn S R and Jacobsen K W 2002 *Comput. Sci. Eng.* **4** 56–66 URL `http://dx.doi.org/10.1109/5992.998641`

[3] Mendelev M I, Kramer M J, Becker C A and Asta M 2008 *Philos. Mag.* **88** 1723–1750 URL `http://www.informaworld.com/10.1080/14786430802206482`

[4] 2011 Atomic Simulation Environment, Center for Atomic-scale Materials Design, Danish Technical University URL `https://wiki.fysik.dtu.dk/ase/`

[5] Alchagirov A B, Perdew J P, Boettger J C, Albers R C and Fiolhais C 2001 *Phys. Rev. B* **63**(22) 224115 URL `http://link.aps.org/doi/10.1103/PhysRevB.63.224115`

[6] Jelinek B, Groh S, Moitra A, Horstemeyer M, Houze J, Kim S, Wagner G and Baskes M 2011 *Arxiv preprint arXiv:1107.0544* URL `http://arxiv.org/abs/1107.0544`

[7] Jelinek B 2011 ASE Atomistic Potential Tests URL `http://code.google.com/p/ase-atomistic-potential-tests`

Symbolic programming package *NCoperators* with applications to theoretical atomic spectroscopy

Rytis Juršėnas, Gintaras Merkelis

Institute of Theoretical Physics and Astronomy of Vilnius University, A. Goštauto 12, LT-01108, Vilnius, Lithuania

E-mail: `Rytis.Jursenas@tfai.vu.lt`

Abstract. A symbolic programming package *NCoperators* with applications to atomic physics is introduced. The package runs over *Mathematica* and it implements *NCAlgebra*, the noncommutative algebra package. *NCoperators* features the algebra of irreducible tensor operators, the second quantization representation, the angular momentum theory, and the effective operator approach exploited in many-body perturbation theory, including Wick's theorem. The comprehensiveness is yet another characteristic feature of the present package: The generation of expressions is performed in a way as if it were done by hand. Although the theoretical atomic spectroscopy is a direct target of *NCoperators*, the package, with minor modifications, if any, is believed to appropriate other areas of theoretical physics as well.

1. Introduction

To this day, a number of *Mathematica* sources featuring the properties of creation and annihilation operator products are observed in literature: *Nostromo* [1], *SeQuant* [2], *Wick* [3], *Quantum* [4], *SNEG* [5] etc. Against the variety of areas of theoretical physics the packages supply with, none of them provide an opportunity to manage the Racah algebra [6–8] in a fully compatible fashion; the capability of constructing the irreducible tensor form of complex operators that act on the basis of many-electron open-shell wave functions is advantageous in a special manner. The package *NCoperators* is just what one needs to solve this kind of tasks [9] along with many other ones, eg. the summation of Clebsch–Gordan products to the $3nj$–symbols [10]. Whilst the *Racah* package [11–13], based on *Maple*, is known to handle several similar procedures such as the calculation of transformation matrices and recoupling coefficients, evaluation of many-particle matrix elements etc., the advantages and distinctive features of *NCoperators* are that (i) it manages the reduction of the product of more than two SU(2)–irreducible representations therefore giving an opportunity to calculate the matrix elements on the basis of many-open-shell wave functions (eg. six open shells appear even in the second-order Rayleigh–Schrödinger perturbation theory based on effective operator approach [14]); (ii) it is based on the algebraic manipulations of the quantities considered neither the diagrammatic representations such as the Jucys graphs exploited in the angular momentum theory [15], though the diagrammatic visualization, to some extent, is implemented as well; (iii) the output of expressions, including Clebsch–Gordan coefficients and $3nj$–symbols, obtained within *Mathematica* interface fits the standard text-based form; (iv) *Racah* refers to Varshalovich et al [16], while *NCoperators* implements the sum rules given by Jucys et al [15; 17]; in many

Published under licence by IOP Publishing Ltd

particular cases, the difference is in the phase factor and different notions (eg. the Clebsch–Gordan coefficient versus the Wigner 3j–symbol); (v) at last, but not least, the package *NCoperators* allows one to generate expansion terms of the Rayleigh–Schrödinger perturbation theory up to the third-order [18], and afterward express them in the irreducible tensor form. To calculate energy corrections, such an operator representation puts into direct action the technique that was first proposed by Judd [19] and later extensively developed by Rudzikas et al [20–22].

At the present moment, the package under consideration is used for private research purposes only, though the source files can be sent directly to anyone concerned with the subject.

In this paper, we review the package *NCoperators* from the perspectives of what it is likely for the application of techniques exploited in modern theoretical atomic spectroscopy.

2. Effective operator approach in RSPT

Our proposed formulation of the Rayleigh–Schrödinger perturbation theory (RSPT) relies on the following statement [9, Theorem 2.7]: The nonzero terms of effective Hamiltonian, $\widehat{\mathscr{H}}$, on the model space $\mathcal{P}$ are generated by a maximum of eight types of the n–body parts of wave operator, $\widehat{\Omega}$, with respect to the single-electron states of the set $I_n(\alpha\bar{\beta})$ for all $n \geqq 1$.

Several results immediately proceed from the present statement. First, it allows us to reduce significantly the number of n–body parts of $\widehat{\Omega}$, denoted $\widehat{\Omega}_n$, that are generated by using the generalized Bloch equation [18]. That is to say, only 8 of the 9^n possible distributions of single-electron states (valence, core, excited) have non-zero contribution to the expansion series of $\widehat{\mathscr{H}}$. Second, both approaches the coupled-cluster (CC) and the RSPT are treated similarly by means of the tensor form of expansion terms – it is only the scalar multiplier (also known as the effective matrix element calculated on the basis of single-particle wave functions) that separates CC and RSPT. Therefore, once the irreducible tensor form of $\widehat{\mathscr{H}}$ is constructed, one can easily make the replacements of routine multipliers depending on whether CC or RSPT approach is imposed on. In particular, the package *NCoperators* generates the terms in the RSPT approximation.

Let us clarify the above proposed statement in a more detail. Let α and $\bar{\beta}$ denote the single-electron states the creation operator a_α creates and the annihilation operator $a^\dagger_{\bar{\beta}}$ annihilates. The single-electron states are valence (v), core (c) and excited or virtual (e). Then the model space $\mathcal{P}$ is constructed by declaring that (A) $a_{\rm c}\widehat{P} = 0$, (B) $a^\dagger_{\bar{\rm e}}\widehat{P} = 0$, (C) $a_{\rm v}\widehat{P} \neq 0$, and (D) $a^\dagger_{\bar{\rm v}}\widehat{P} \neq 0$. Here, $\widehat{P}$ denotes the projection operator that projects vectors in $\mathcal{H}$, the N–electron Hilbert space, onto the vectors in $\mathcal{P}$; one assumes that the initially given Hamiltonian $\widehat{H}$ acts on $\mathcal{H}$, whereas the effective Hamiltonian $\widehat{\mathscr{H}}$ acts on $\mathcal{P}$. That is, the effective Hamiltonian reads

$$\widehat{\mathscr{H}} = \widehat{P}\widehat{H}\widehat{P} + \sum_{n=1}^{\infty}\sum_{m=1}^{2}\sum_{\xi=1}^{\min(2m,2n)} :\{\widehat{P}\widehat{V}_m\widehat{\Omega}_n\widehat{P}\}_\xi:, \tag{1}$$

which is obtained by applying the Wick's theorem (:: denotes the normal ordering, ξ denotes the ξ–pair contraction between $\widehat{V}_m$ and $\widehat{\Omega}_n$). The quantities $\widehat{V}_1$ and $\widehat{V}_2$ denote the single-particle and two-particle interaction operators. The wave operator $\widehat{\Omega}_n$ is of the form

$$\widehat{\Omega}_n = \sum_{I_n(\alpha\bar{\beta})} :\widehat{Q}a_{\alpha_1}a_{\alpha_2}\ldots a_{\alpha_{n-1}}a_{\alpha_n}a^\dagger_{\bar{\beta}_n}a^\dagger_{\bar{\beta}_{n-1}}\ldots a^\dagger_{\bar{\beta}_2}a^\dagger_{\bar{\beta}_1}\widehat{P}: \omega_n(\alpha\bar{\beta}), \tag{2}$$

for $\widehat{Q}$ is the orthogonal complement of $\widehat{P}$. Here, $\omega_n(\alpha\bar{\beta})$ is the above mentioned effective n–particle matrix element with the energy denominator included. The set $I_n(\alpha\bar{\beta}) = \{\alpha_1, \alpha_2, \ldots, \alpha_{n-1}, \alpha_n, \bar{\beta}_1, \bar{\beta}_2, \ldots, \bar{\beta}_{n-1}, \bar{\beta}_n\}$. In agreement with items (A)-(D), $\widehat{\Omega}_n$ is nonzero if and only if the α_i states are valence and/or excited, while the $\bar{\beta}_j$ states are valence and/or core. In addition, $\widehat{\Omega}_n$

is zero if and only if the sum of SO(3)–irreducible representations $l_{\mathrm{v}_i} + l_{\bar{\mathrm{v}}_i}$ over all $i = 1, 2, \ldots, n$ is even,

$$\sum_{i=1}^{n} (l_{\mathrm{v}_i} + l_{\bar{\mathrm{v}}_i}) \equiv 0 \mod 2. \tag{3}$$

Eq. (3) may be thought of as an additional parity selection rule that directly follows from items (C)-(D). A subsequent result is the initially proposed statement. In agreement with the parity selection rule, the wave operator $\widehat{\Omega}_n$, eq. (2), that distinguishes between single-, two-, three-, four-particle effects ($n = 1, 2, 3, 4$) reads

$$\widehat{\Omega}_1 = \sum_{I_1^{(1)}} a_{\mathrm{e}} a^{\dagger}_{\bar{\mathrm{v}}} \omega_{\mathrm{e}\bar{\mathrm{v}}} + \sum_{I_1^{(2)}} a_{\mathrm{v}} a^{\dagger}_{\bar{\mathrm{c}}} \omega_{\mathrm{v}\bar{\mathrm{c}}} + \sum_{I_1^{(3)}} a_{\mathrm{e}} a^{\dagger}_{\bar{\mathrm{c}}} \omega_{\mathrm{e}\bar{\mathrm{c}}}, \tag{4a}$$

$$\begin{aligned}\widehat{\Omega}_2 = {} & \sum_{I_2^{(4,5,8)}}{}' a_{\alpha} a_{\alpha'} a^{\dagger}_{\bar{\beta}'} a^{\dagger}_{\bar{\beta}} \omega_{\alpha\alpha'\bar{\beta}\bar{\beta}'} + \sum_{I_2^{(3)}} a_{\mathrm{e}} a_{\mathrm{v}} a^{\dagger}_{\bar{\mathrm{c}}} a^{\dagger}_{\bar{\mathrm{v}}} \omega_{\mathrm{ev}\bar{\mathrm{v}}\bar{\mathrm{c}}} + \sum_{I_2^{(1,6)}}{}' a_{\mathrm{e}} a_{\mathrm{v}} a^{\dagger}_{\bar{\beta}'} a^{\dagger}_{\bar{\beta}} \omega_{\mathrm{ev}\bar{\beta}\bar{\beta}'} \\ & + \sum_{I_2^{(2,7)}}{}' a_{\alpha} a_{\alpha'} a^{\dagger}_{\bar{\mathrm{v}}} a^{\dagger}_{\bar{\mathrm{c}}} \omega_{\alpha\alpha'\bar{\mathrm{c}}\bar{\mathrm{v}}},\end{aligned} \tag{4b}$$

$$\begin{aligned}\widehat{\Omega}_3 = {} & \sum_{I_3^{(1,4)}}{}' a_{\alpha} a_{\alpha'} a_{\mu} a^{\dagger}_{\bar{\mathrm{v}}''} a^{\dagger}_{\bar{\mathrm{v}}'} a^{\dagger}_{\bar{\mathrm{v}}} \omega_{\alpha\alpha'\mu\bar{\mathrm{v}}\bar{\mathrm{v}}'\bar{\mathrm{v}}''} + \sum_{I_3^{(2,5)}}{}' a_{\mathrm{v}} a_{\mathrm{v}'} a_{\mathrm{v}''} a^{\dagger}_{\bar{\mathrm{c}}} a^{\dagger}_{\bar{\beta}'} a^{\dagger}_{\bar{\mathrm{v}}} \omega_{\mathrm{vv}'\mathrm{v}''\bar{\mathrm{v}}\bar{\beta}'\bar{\mathrm{c}}} \\ & + \sum_{I_3^{(3,6,7,8)}}{}' a_{\alpha} a_{\alpha'} a_{\mu} a^{\dagger}_{\bar{\mathrm{c}}} a^{\dagger}_{\bar{\beta}'} a^{\dagger}_{\bar{\mathrm{v}}} \omega_{\alpha\alpha'\mu\bar{\mathrm{v}}\bar{\beta}'\bar{\mathrm{c}}},\end{aligned} \tag{4c}$$

$$\begin{aligned}\widehat{\Omega}_4 = {} & \sum_{I_4^{(1,4)}}{}' a_{\mathrm{e}} a_{\alpha'} a_{\mathrm{v}} a_{\mathrm{v}''} a^{\dagger}_{\bar{\mathrm{v}}'''} a^{\dagger}_{\bar{\mathrm{v}}''} a^{\dagger}_{\bar{\mathrm{v}}'} a^{\dagger}_{\bar{\mathrm{v}}} \omega_{\mathrm{e}\alpha'\mathrm{vv}''\bar{\mathrm{v}}\bar{\mathrm{v}}'\bar{\mathrm{v}}''\bar{\mathrm{v}}'''} + \sum_{I_4^{(2,5)}}{}' a_{\mathrm{v}} a_{\mathrm{v}'} a_{\mathrm{v}''} a_{\mathrm{v}'''} a^{\dagger}_{\bar{\mathrm{c}}} a^{\dagger}_{\bar{\beta}''} a^{\dagger}_{\bar{\mathrm{v}}'} a^{\dagger}_{\bar{\mathrm{v}}} \omega_{\mathrm{vv}'\mathrm{v}''\mathrm{v}'''\bar{\mathrm{v}}\bar{\mathrm{v}}'\bar{\beta}''\bar{\mathrm{c}}} \\ & + \sum_{I_4^{(3,6,7,8)}}{}' a_{\mathrm{e}} a_{\alpha'} a_{\mathrm{v}} a_{\mathrm{v}''} a^{\dagger}_{\bar{\mathrm{c}}} a^{\dagger}_{\bar{\beta}''} a^{\dagger}_{\bar{\mathrm{v}}'} a^{\dagger}_{\bar{\mathrm{v}}} \omega_{\mathrm{e}\alpha'\mathrm{vv}''\bar{\mathrm{v}}\bar{\mathrm{v}}'\bar{\beta}''\bar{\mathrm{c}}}.\end{aligned} \tag{4d}$$

Here and elsewhere, it is assumed that the Greek letters denote all three types of single-electron states. The sums with primes denote the following operations

$$\sum_{I_2^{(4,5,8)}}{}' \equiv \sum_{I_2^{(4)}} \delta_{\alpha\mathrm{e}} \delta_{\beta\mathrm{v}} + \sum_{I_2^{(5)}} \delta_{\alpha\mathrm{v}} \delta_{\beta\mathrm{c}} + \sum_{I_2^{(8)}} \delta_{\alpha\mathrm{e}} \delta_{\beta\mathrm{c}}, \tag{5a}$$

$$\sum_{I_x^{(a,b)}}{}' \equiv \sum_{I_x^{(a)}} \delta_{\beta\mathrm{v}} + \sum_{I_x^{(b)}} \delta_{\beta\mathrm{c}}, \quad \text{for } x = 2, a = 1, b = 6 \text{ and } x = 3, 4, a = 2, b = 5, \tag{5b}$$

$$\sum_{I_x^{(a,b)}}{}' \equiv \sum_{I_x^{(a)}} \delta_{\alpha\mathrm{v}} + \sum_{I_x^{(b)}} \delta_{\alpha\mathrm{e}}, \quad \text{for } x = 2, a = 2, b = 7 \text{ and } x = 4, a = 1, b = 4, \tag{5c}$$

$$\sum_{I_3^{(1,4)}}{}' \equiv \sum_{I_3^{(1)}} \delta_{\alpha\mathrm{v}} \delta_{\mu\mathrm{e}} + \sum_{I_3^{(4)}} \delta_{\alpha\mathrm{e}} \delta_{\mu\mathrm{v}}, \tag{5d}$$

$$\sum_{I_3^{(3,6,7,8)}}{}' \equiv \sum_{I_3^{(3)}} \delta_{\alpha\mathrm{v}} \delta_{\beta\mathrm{v}} \delta_{\mu\mathrm{e}} + \sum_{I_3^{(6)}} \delta_{\alpha\mathrm{v}} \delta_{\beta\mathrm{c}} \delta_{\mu\mathrm{e}} + \sum_{I_3^{(7)}} \delta_{\alpha\mathrm{e}} \delta_{\beta\mathrm{v}} \delta_{\mu\mathrm{v}} + \sum_{I_3^{(8)}} \delta_{\alpha\mathrm{e}} \delta_{\beta\mathrm{c}} \delta_{\mu\mathrm{v}}, \tag{5e}$$

$$\sum_{I_4^{(3,6,7,8)}}' \equiv \sum_{I_4^{(3)}} \delta_{\alpha \mathrm{v}} \delta_{\beta \mathrm{v}} + \sum_{I_4^{(6)}} \delta_{\alpha \mathrm{v}} \delta_{\beta \mathrm{c}} + \sum_{I_4^{(7)}} \delta_{\alpha \mathrm{e}} \delta_{\beta \mathrm{v}} + \sum_{I_4^{(8)}} \delta_{\alpha \mathrm{e}} \delta_{\beta \mathrm{c}}. \tag{5f}$$

As seen from eq. (5), no more than eight types of sets $I_n(\alpha\bar{\beta})$ are observed: Although each $\widehat{\Omega}_n$ contains more terms than those in eq. (4), but only the displayed eight types of sets $I_n(\alpha\bar{\beta})$ have nonzero contribution to the effective Hamiltonian $\widehat{\mathscr{H}}$, eq. (1).

In particular, one may express the third-order effective Hamiltonian $\widehat{\mathscr{H}}^{(3)}$ by

$$\widehat{\mathscr{H}}^{(3)} = \sum_{I_{m+n-\xi}} \sum_{m=1}^{2} \sum_{n=1}^{4} \sum_{\xi=1}^{\min(2m,2n)} :\{\widehat{P}\widehat{V}_m \widehat{\Omega}_n^{(2)} \widehat{P}\}_{\xi}:, \tag{6}$$

where $\widehat{\Omega}_n^{(2)}$ is given by eq. (4) with the effective matrix elements $\omega_n(\alpha\bar{\beta})$ replaced by the second-order effective matrix elements $\omega_n^{(2)}(\alpha\bar{\beta})$. $I_{m+n-\xi}$ denotes the set of all single-particle orbitals.

Even if eq. (3) is exploited, there are hundreds of generated $\widehat{\Omega}_n^{(2)}$ and $\widehat{\mathscr{H}}^{(3)}$ terms. Therefore, the role of *NCoperators* becomes essential.

3. The generation of expansion terms

The package *NCoperators* takes into consideration the anticommutation properties of creation and annihilation operators, $\{a_{\alpha_i}, a^\dagger_{\beta_j}\} = \delta(\alpha_i, \beta_j)$, $\{a_{\alpha_i}, a_{\beta_j}\} = 0$, $\{a^\dagger_{\alpha_i}, a^\dagger_{\beta_j}\} = 0$. The realization is illustrated in Fig. 1.

```
In[58]:= NonCommutOpMulti[4, α, β]
Out[58]= a_α1 ** a_α2 ** a†_β2 ** a†_β1

In[59]:= AntiCommutator2[%, {α, 2}, {β, 2}] //. Trule
Out[59]= -a_α1 ** a†_β2 ** a_α2 ** a†_β1 + a_α1 ** a†_β1 δ_α,β

In[60]:= % /. style
Out[60]= -a_α1 a†_β2 a_α2 a†_β1 + a_α1 a†_β1 δ_α,β

In[61]:= TwoContractions221[4, α, β, 2, μ, ν]
Out[61]= 2 a_α2 ** a†_β2 δ_α1,ν1 δ_β1,μ1 - 2 a_α1 ** a†_β2 δ_α2,ν1 δ_β1,μ1 - 2 a_α2 ** a†_β1 δ_α1,ν1 δ_β2,μ1 + 2 a_α1 ** a†_β1 δ_α2,ν1 δ_β2,μ1

In[62]:= % /. style
Out[62]= 2 a_α2 a†_β2 δ_α1,ν1 δ_β1,μ1 - 2 a_α1 a†_β2 δ_α2,ν1 δ_β1,μ1 - 2 a_α2 a†_β1 δ_α1,ν1 δ_β2,μ1 + 2 a_α1 a†_β1 δ_α2,ν1 δ_β2,μ1
```

Figure 1. Manipulating anticommutation properties of creation and annihilation operators in *NCoperators*. The outputs `Out[60]` and `Out[62]` are fully compatible with LaTeX.

As a representative example, let us consider the single-particle operator $:\{\widehat{P}\widehat{V}_1\widehat{\Omega}_1^{(2)}\widehat{P}\}_1:$. The interaction operator $\widehat{V}_1 = \sum_{\alpha\bar{\beta}} a_\alpha a^\dagger_{\bar{\beta}} v_{\alpha\bar{\beta}}$, where $v_{\alpha\bar{\beta}}$ is the single-particle matrix element. By eq. (4a), the wave operator

$$\widehat{\Omega}_1^{(2)} = \sum_{\mathrm{ev}} a_{\mathrm{e}} a^\dagger_{\bar{\mathrm{v}}} \omega^{(2)}_{\mathrm{e}\bar{\mathrm{v}}} + \sum_{\mathrm{vc}} a_{\mathrm{v}} a^\dagger_{\bar{\mathrm{c}}} \omega^{(2)}_{\mathrm{v}\bar{\mathrm{c}}} + \sum_{\mathrm{ec}} a_{\mathrm{e}} a^\dagger_{\bar{\mathrm{c}}} \omega^{(2)}_{\mathrm{e}\bar{\mathrm{c}}}. \tag{7}$$

The generation of $:\{\widehat{P}\widehat{V}_1\widehat{\Omega}_1^{(2)}\widehat{P}\}_1:$ terms is performed in Fig. 2. In *NCoperators*, the single-electron states are designated by

$$\begin{array}{llll} a_{\rm c}: & {\rm cc1, cc2, etc.;} & a_{\bar{\rm c}}^\dagger: & {\rm ca1, ca2, etc.} \\ a_{\rm v}: & {\rm vc1, vc2, etc.;} & a_{\bar{\rm v}}^\dagger: & {\rm va1, va2, etc.} \\ a_{\rm e}: & {\rm ec1, ec2, etc.;} & a_{\bar{\rm e}}^\dagger: & {\rm ea1, ea2, etc.} \end{array}$$

In a standard output (such as `Out[4]` in Fig. 2), the notation of orbitals is simplified to

$$\begin{array}{llll} {\rm c}: & a_1, b_1, c_1, d_1, e_1, f_1 & \bar{\rm c}: & a_2, b_2, c_2, d_2, e_2, f_2 \\ {\rm v}: & m_1, n_1, p_1, q_1, k_1, l_1 & \bar{\rm v}: & m_2, n_2, p_2, q_2, k_2, l_2 \\ {\rm e}: & r_1, s_1, t_1, u_1, w_1, x_1 & \bar{\rm e}: & r_2, s_2, t_2, u_2, w_2, x_2 \end{array}$$

One can recognize that the function `OneContraction[2,μ,ν,2,vc1,ca1]` corresponds to the term at the sum $\sum_{\rm vc}$ in eq. (7); similarly, the rest two functions `OneContraction[2,μ,ν,2,ec1,ca1]` and `OneContraction[2,μ,ν,2,ec1,va1]` correspond to the terms at $\sum_{\rm ec}$ and $\sum_{\rm ev}$, respectively. In `Out[4]`, the effective matrix element $\langle x|v_2^{eff}|y\rangle$ denotes $\omega_{xy}^{(2)}(\varepsilon_y - \varepsilon_x)$; ε_x is the single-electron energy, also observed in $\widehat{H}_0 = \sum_{\alpha_1} a_{\alpha_1} a_{\bar{\alpha}_1}^\dagger \varepsilon_{\alpha_1}$ (recall that the Hamiltonian $\widehat{H} = \widehat{H}_0 + \widehat{V}_0 + \widehat{V}_1 + \widehat{V}_2$). Other functions such as `KronDelta[]`, `NormalOrder[]`, `MatrixEl[]` should be obvious to identify by their names. Therefore, `Out[4]` shows us that the generated $:\{\widehat{P}\widehat{V}_1\widehat{\Omega}_1^{(2)}\widehat{P}\}_1:$ terms are

```
In[2]:= A111 // Clear
  A111 :=
   Block[{x1, x2, x3, x4, x4a, x5, x6, x7, x8},
    x1 = 1 / 1 * (
        P ** NormalOrder[OneContraction[2, μ, ν, 2, vc1, ca1]] ** P * MatrixEl[μ1, v1, ν1] * MatrixEl[vc11, v2, ca11] /
          (εca11 - εvc11) +
         P ** NormalOrder[OneContraction[2, μ, ν, 2, ec1, ca1]] ** P * MatrixEl[μ1, v1, ν1] *
          MatrixEl[ec11, v2, ca11] /
          (εca11 - εec11) +
         P ** NormalOrder[OneContraction[2, μ, ν, 2, ec1, va1]] ** P * MatrixEl[μ1, v1, ν1] *
          MatrixEl[ec11, v2, va11] /
          (εva11 - εec11)) /. {KronDelta[μ1, vali_] :→ 0, KronDelta[μ2, vali_] :→ 0} //. Trule // Expand;
    x2 = Table[SymbolSum[ x1[[i]], {μ1, ν1}] //. SumSimplify, {i, 1, Length[x1]}] // Total;
    x3 = (x2 /. {μ → vc2, ν → va2}) + (x2 /. {μ → ec2, ν → ea2}) + (x2 /. {μ → cc2, ν → ca2});
    x4 = x3 /. {a___ * SymbolSum[z__, var_List] :→ a * z};
    x5 = x4 //. Elimination;
    x6 = (x5 // NormalOrdering) /. P → 1;
    x7 = x6 /. {v1 → v1, v2 → v2eff} /. style /. style /. style /. style /. style /. style
   ];

In[4]:= A111
```

$$\text{Out[4]=} -\frac{a_{m_1} a_{p_2}^\dagger \langle a_2|v_1|p_2\rangle\langle m_1|v_2^{\rm eff}|a_2\rangle}{\varepsilon_{a_2} - \varepsilon_{m_1}} + \frac{a_{p_1} a_{m_2}^\dagger \langle r_1|v_2^{\rm eff}|m_2\rangle\langle p_1|v_1|r_1\rangle}{\varepsilon_{m_2} - \varepsilon_{r_1}}$$

Figure 2. The generation of $:\{\widehat{P}\widehat{V}_1\widehat{\Omega}_1^{(2)}\widehat{P}\}_1:$ terms with *NCoperators*.

$$:\{\widehat{P}\widehat{V}_1\widehat{\Omega}_1^{(2)}\widehat{P}\}_1:= \sum_{\rm ev} a_{\rm v} a_{\bar{\rm v}}^\dagger v_{\rm ve}\omega_{\rm e\bar{v}}^{(2)} - \sum_{\rm cv} a_{\rm v} a_{\bar{\rm v}}^\dagger v_{\rm c\bar{v}}\omega_{\rm vc}^{(2)}. \tag{8}$$

The next step is to obtain the scalars $\omega_{\alpha\bar{\beta}}^{(2)}$. This is the most time consuming process. In agreement with the generalized Bloch equation, $\omega_{\alpha\bar{\beta}}^{(2)}$ represents the sum of coefficients $\omega_{ij;i+j-1}^{(2)}(\alpha\bar{\beta})$ obtained from the terms $:\{\widehat{R}\widehat{V}_i\widehat{\Omega}_j^{(1)}\widehat{P} - \widehat{R}\widehat{\Omega}_j^{(1)}\widehat{P}\widehat{V}_i\widehat{P}\}_{i+j-1}:$; $\widehat{R}$ denotes the resolvent. Thus,

$$\omega_{\alpha\bar{\beta}}^{(2)} = \omega_{11;1}^{(2)}(\alpha\bar{\beta}) + \omega_{12;2}^{(2)}(\alpha\bar{\beta}) + \omega_{21;2}^{(2)}(\alpha\bar{\beta}) + \omega_{22;3}^{(2)}(\alpha\bar{\beta}).$$

In Figs. 3-4, the algorithms for obtaining the multipliers $\omega^{(2)}_{11;1}$ are displayed. As seen, $\omega^{(2)}_{11;1}$ consists of nine terms in total,

```
In[16]:= RV1Ω1PconnectedOnePair // Clear
         RV1Ω1PconnectedOnePair = Block[{w1, w2, w3, w4, w5, w6, w7, w8, w9, w10, w11, w12, w13, w14, w15},
           w1[α_, β_] := NormalOrder[OneContraction[2, α, β, 2, ec2, ca2]] * MatrixEl[ec2₁, v₂, ca2₁] / Δ[2, ec2, ca2] +
             NormalOrder[OneContraction[2, α, β, 2, ec2, va2]] * MatrixEl[ec2₁, v₂, va2₁] / Δ[2, ec2, va2] +
             NormalOrder[OneContraction[2, α, β, 2, vc2, ca2]] * MatrixEl[vc2₁, v₂, ca2₁] / Δ[2, vc2, ca2];
           w2[α_, β_] := w1[α, β] // Expand;
           w3[α_, β_] := w2[α, β] * MatrixEl[α₁, v₁, β₁] // Expand;
           w4[β_] := (w3[α, β] /. {α -> ec1}) + (w3[α, β] /. {α → vc1}) + (w3[α, β] /. {α → cc1});
           w5 = (w4[β] /. {β → ea1}) + (w4[β] /. {β → va1}) + (w4[β] /. {β → ca1});
           w6 = w5 // Expand;
           w7 = w6 /. {
              KronDelta[Subscript[ec1, k_], Subscript[ea2, l_]] :→ 0,
              KronDelta[Subscript[vc1, k_], Subscript[va2, l_]] :→ 0,
              KronDelta[Subscript[ca1, k_], Subscript[cc2, l_]] :→ 0};
           w8 = Table[Q ** w7[[f]] ** P, {f, 1, Length[w7]}] // Total;
           w9 = (w8 //. MBPTrules) //. PQrules;
           w10 = w9 // NormalOrdering;
           w11 = w10 //. Elimination;
           w12 = (w11 / δε) // Expand;
           w13 = w12 //. denominator11;
           w14 = Table[RenameState[w13[[i]], {cc1₁, ca1₁, ec1₁, ea1₁, vc1₁, va1₁, cc2₁, ca2₁, ec2₁, ea2₁, vc2₁, va2₁}],
              {i, 1, Length[w13]}] // Total;
           w15 = w14 //. WaveOpSimplify //. WaveOpSimplify2

         ];

In[18]:= (SymbolSum[RV1Ω1PconnectedOnePair, {ca2₁, ea1₁, va1₁}] //. SumExpand //. SumSimplify) /. {
                SymbolSum[z___, var_List] :→ z} /. P → 1 /. Q → 1 /. style /. style /. style /. style /. style
```

Out[18]=

$$
-\frac{a_{p_1} a^{\dagger}_{a_2} \langle a_1|v_1|a_2\rangle\langle p_1|v_2|a_1\rangle}{(\varepsilon_{a_1}-\varepsilon_{p_1})(\varepsilon_{a_2}-\varepsilon_{p_1})} + \frac{a_{m_1} a^{\dagger}_{c_2} \langle m_1|v_1|p_1\rangle\langle p_1|v_2|c_2\rangle}{(\varepsilon_{c_2}-\varepsilon_{m_1})(\varepsilon_{c_2}-\varepsilon_{p_1})} + \frac{a_{r_1} a^{\dagger}_{c_2} \langle r_1|v_1|p_1\rangle\langle p_1|v_2|c_2\rangle}{(\varepsilon_{c_2}-\varepsilon_{p_1})(\varepsilon_{c_2}-\varepsilon_{r_1})} - \frac{a_{t_1} a^{\dagger}_{a_2} \langle a_1|v_1|a_2\rangle\langle t_1|v_2|a_1\rangle}{(\varepsilon_{a_1}-\varepsilon_{t_1})(\varepsilon_{a_2}-\varepsilon_{t_1})} + \frac{a_{m_1} a^{\dagger}_{c_2} \langle t_1|v_2|c_2\rangle\langle m_1|v_1|t_1\rangle}{(\varepsilon_{c_2}-\varepsilon_{m_1})(\varepsilon_{c_2}-\varepsilon_{t_1})} + \frac{a_{r_1} a^{\dagger}_{c_2} \langle r_1|v_1|t_1\rangle\langle t_1|v_2|c_2\rangle}{(\varepsilon_{c_2}-\varepsilon_{r_1})(\varepsilon_{c_2}-\varepsilon_{t_1})} - \frac{a_{t_1} a^{\dagger}_{m_2} \langle a_1|v_1|m_2\rangle\langle t_1|v_2|a_1\rangle}{(\varepsilon_{a_1}-\varepsilon_{t_1})(\varepsilon_{m_2}-\varepsilon_{t_1})} + \frac{a_{r_1} a^{\dagger}_{p_2} \langle r_1|v_1|t_1\rangle\langle t_1|v_2|p_2\rangle}{(\varepsilon_{p_2}-\varepsilon_{r_1})(\varepsilon_{p_2}-\varepsilon_{t_1})}
$$

Figure 3. Generating the $:\{\widehat{R}\widehat{V}_1\widehat{\Omega}^{(1)}_1\widehat{P}\}_1$: terms.

$$
\omega^{(2)}_{11;1}(\alpha\bar{\beta})(\varepsilon_{\bar{\beta}}-\varepsilon_{\alpha}) = \sum_{\mu(\alpha\beta)} v_{\alpha\mu}\omega^{(1)}_{\mu\bar{\beta}} - \sum_{\nu(\alpha\beta)} v_{\nu\bar{\beta}}\omega^{(1)}_{\alpha\nu},
$$

where $\omega^{(1)}_{\alpha\bar{\beta}} = v_{\alpha\bar{\beta}}/(\varepsilon_{\bar{\beta}}-\varepsilon_{\alpha})$ is the effective matrix element drawn in $\widehat{\Omega}^{(1)}$. If performing analogous computations for the remaining coefficients $\omega^{(2)}_{ij;i+j-1}$, one finds that

$$
\begin{aligned}
\omega^{(2)}_{12;2}(\alpha\bar{\beta})(\varepsilon_{\bar{\beta}}-\varepsilon_{\alpha}) &= \sum_{\mathrm{c}}\sum_{\mu=\mathrm{v,e}} v_{\mathrm{c}\mu}\big(\omega^{(1)}_{\mu\alpha\mathrm{c}\bar{\beta}} - \omega^{(1)}_{\mu\alpha\bar{\beta}\mathrm{c}}\big),\\
\omega^{(2)}_{21;2}(\alpha\bar{\beta})(\varepsilon_{\bar{\beta}}-\varepsilon_{\alpha}) &= \sum_{\mathrm{c}}\sum_{\mu=\mathrm{v,e}} \big(v_{\mathrm{c}\alpha\mu\bar{\beta}} - v_{\mathrm{c}\alpha\bar{\beta}\mu}\big)\omega^{(1)}_{\mu\mathrm{c}},\\
\omega^{(2)}_{22;3}(\alpha\bar{\beta})(\varepsilon_{\bar{\beta}}-\varepsilon_{\alpha}) &= \sum_{\mathrm{c,v,e}} \big(v_{\mathrm{c}\alpha\mathrm{ve}} - v_{\mathrm{c}\alpha\mathrm{ev}}\big)\big(\omega^{(1)}_{\mathrm{evc}\bar{\beta}} - \omega^{(1)}_{\mathrm{ev}\bar{\beta}\mathrm{c}}\big),
\end{aligned}
$$

where $\omega^{(1)}_{\alpha\beta\bar{\mu}\bar{\nu}} = v_{\alpha\beta\bar{\mu}\bar{\nu}}/(\varepsilon_{\bar{\mu}}+\varepsilon_{\bar{\nu}}-\varepsilon_{\alpha}-\varepsilon_{\beta})$; $v_{\alpha\beta\bar{\mu}\bar{\nu}}$ denotes a two-particle matrix element. Finally, obtained coefficients $\omega^{(2)}_{\alpha\bar{\beta}}$ are substituted in eq. (8). The same procedure should be carried out for the rest $:\{\widehat{P}\widehat{V}_m\widehat{\Omega}^{(2)}_n\widehat{P}\}_{\xi}$: terms in order to establish all terms of $\mathscr{H}^{(3)}$ (see eq. (6)). The last step is to represent obtained expressions in the irreducible tensor form.

```
In[19]:= RΩ1PV1POnePair // Clear
   RΩ1PV1POnePair = Block[{w1, w2, w3, w4, w5, w6, w7, w8, w9, w10, w11, w12, w13, w14, w15},
     w1[μ_, ν_] := NormalOrder[OneContraction[2, ec2, ca2, 2, μ, ν]] * MatrixEl[ec2₁, v₂, ca2₁] / Δ[2, ec2, ca2] +
       NormalOrder[OneContraction[2, ec2, va2, 2, μ, ν]] * MatrixEl[ec2₁, v₂, va2₁] / Δ[2, ec2, va2] +
       NormalOrder[OneContraction[2, vc2, ca2, 2, μ, ν]] * MatrixEl[vc2₁, v₂, ca2₁] / Δ[2, vc2, ca2];
     w2[μ_, ν_] := w1[μ, ν] // Expand;
     w3[μ_, ν_] := w2[μ, ν] * MatrixEl[μ₁, v₁, ν₁] // Expand;
     w4[ν_] := w3[μ, ν] /. {μ → vc1};
     w5 = w4[v] /. {v → va1};
     w6 = w5 // Expand;
     w7 = w6 /. {
        KronDelta[Subscript[ec1, k_], Subscript[ea2, l_]] :→ 0,
        KronDelta[Subscript[vc1, k_], Subscript[va2, l_]] :→ 0,
        KronDelta[Subscript[ca1, k_], Subscript[cc2, l_]] :→ 0,
        KronDelta[Subscript[ec2, k_], Subscript[ea1, l_]] :→ 0,
        KronDelta[Subscript[vc2, k_], Subscript[va1, l_]] :→ 0,
        KronDelta[Subscript[ca2, k_], Subscript[cc1, l_]] :→ 0
       };
     w8 = (Q ** (w7 // NormalOrdering) ** P) //. MBPTrulesExpand;
     w9 = (w8 //. MBPTrules) //. PQrules;
     w10 = w9 //. Elimination;
     w11 = (w10 / δε) // Expand;
     w12 = w11 //. denominator11;
     w13 = RenameState[w12, {cc1₁, ca1₁, ec1₁, ea1₁, vc1₁, va1₁, cc2₁, ca2₁, ec2₁, ea2₁, vc2₁, va2₁}];
     w14 = w13 //. WaveOpSimplify //. WaveOpSimplify2

    ];

In[21]:= (SymbolSum[RΩ1PV1POnePair, {va2₁}] //. SumSimplify) /. P → 1 /. Q → 1 /. style /. style /. style /. style
```

$$\text{Out[21]=}\ \frac{a_{t_1} a^{\dagger}_{m_2} \langle t_1 | v_2 | m_1 \rangle \langle m_1 | v_1 | m_2 \rangle}{(\varepsilon_{m_1} - \varepsilon_{t_1})(\varepsilon_{m_2} - \varepsilon_{t_1})}$$

Figure 4. Generating the $:\{\widehat{R}\widehat{\Omega}_1^{(1)}\widehat{P}\widehat{V}_1\widehat{P}\}_1$: terms.

4. The irreducible tensor form of expansion terms

As first demonstrated by Judd [19], the operator a_α represents the irreducible tensor operator of $\mathrm{SU}^J(2)$ (in jj–representation) or $\mathrm{SO}^L(3)\times\ \mathrm{SU}^S(2)$ (in LS–representation). This operator is denoted by a^{λ_α}, where $\lambda \equiv j$ or $\lambda \equiv l\frac{1}{2}$. Note that in the LS–representation, a^{λ_α} represents a double tensor. According to this, we introduce a general form

$$:\{\widehat{P}\widehat{V}_m\widehat{\Omega}_n^{(2)}\widehat{P}\}_\xi:=\sum_{\Lambda}\sum_{M=-\Lambda}^{+\Lambda}\sum_{\Gamma}\widehat{O}_M^{\Lambda}([w]\varkappa)\, C_{mn;\xi}^{(3)}(\Gamma\Lambda), \tag{9}$$

for $\widehat{O}^{\Lambda}([w]\varkappa)$ is the irreducible tensor operator of rank Λ; this notation fits both representations. The coefficient $C_{mn;\xi}^{(3)}(\Gamma\Lambda)$ is SU(2)–invariant. The letter Γ denotes additional numbers necessary to obtain the above expression. The letter w labels irreducible representations of $S_{2(m+n-\xi)}$, the symmetric group; these are of the type $[w] = \left[2^{h_2}1^{h_1}\right]$; $\varkappa$ denotes additional numbers, if necessary, in order to distinguish different reduction schemes of $\widehat{O}^{\Lambda}([w]\varkappa)$. As one can easily observe, the $S_{2(m+n-\xi)}$–irreducible representations $[w]$ label the irreducible tensor operators $\widehat{O}^{\Lambda}([w]\varkappa)$ associated with the $(m+n-\xi)$–particle operators $:\{\widehat{P}\widehat{V}_m\widehat{\Omega}_n^{(2)}\widehat{P}\}_\xi:$. For example, $\widehat{O}^{\Lambda}([1^2]) = \left[a^{\lambda_{\mathrm{v}}} \times \widetilde{a}^{\lambda_{\bar{\mathrm{v}}}}\right]^{\Lambda}$, where $\widetilde{a}^{\lambda_{\mathrm{v}}}$ denotes the irreducible tensor operator obtained from the transposed annihilation operator. The complete classification of irreducible tensor operators obtained by reducing the products of creation and annihilation operators one can find in [10].

The matrix elements of irreducible tensor operators of rank Λ calculated on the basis of many-electron wave functions can be found, for example, in [20; 22]. For this reason, the SU(2)–invariants $C_{mn;\xi}^{(3)}(\Gamma\Lambda)$ are to be established to complete the task. As it appears from eq. (9), that the structure of $C_{mn;\xi}^{(3)}(\Gamma\Lambda)$ depends on $([w]\varkappa)$, that is to say, the reduction scheme. In turn, the number of possible schemes depends on $m+n-\xi$. Obviously, for a fixed $m+n-\xi$, all reduction schemes are equivalent, and they are related by recoupling coefficients. These coefficients can be found by exploiting *NCoperators*.

```
In[13]:= SymbolSum[CGcoeff[{α12, β12}, {α3, β3}, {α123, β123}] *
           CGcoeff[{α123, β123}, {α456, β456}, {α, β}] *
           CGcoeff[{α56, β56}, {α4, β4}, {α456, β456}] * CGcoeff[{α3, β3}, {α4, β4}, {α34, β34}] *
           CGcoeff[{α12, β12}, {α34, β34}, {α1234, β1234}] * CGcoeff[{α1234, β1234}, {α56, β56}, {α, β}],
         {β3, β4, β123, β12, β456, β56, β456, β34, β1234}]

Out[13]= SymbolSum[[α12 α3 α123 ; β12 β3 β123] [α12 α34 α1234 ; β12 β34 β1234] [α123 α456 α ; β123 β456 β] [α1234 α56 α ; β1234 β56 β]
           [α3 α4 α34 ; β3 β4 β34] [α56 α4 α456 ; β56 β4 β456], {β3, β4, β123, β12, β456, β56, β34, β1234}]

In[14]:= Make[%, {{α12, β12}, {α3, β3}}, {{α34, β34}, {α12, β12}}, {{α123, β123}, {α456, β456}},
           {{α1234, β1234}, {α56, β56}}, {{α4, β4}, {α3, β3}}, {{α56, β56}, {α4, β4}}]

Out[14]= SymbolSum[(-1)^(-α12+α1234-α3-α4) [α12 α3 α123 ; β12 β3 β123] [α123 α456 α ; β123 β456 β] [α1234 α56 α ; β1234 β56 β]
           [α34 α12 α1234 ; β34 β12 β1234] [α4 α3 α34 ; β4 β3 β34] [α56 α4 α456 ; β56 β4 β456], {β3, β4, β123, β12, β456, β56, β34, β1234}]

In[15]:= Recoupling[% //. SumSimplify, {α12, β12}, {α3, β3}] /. {α1234_{α12,α3} → α123p, β1234_{β12,β3} → β123}

Out[15]= (-1)^(-α12+α1234-α3-α4) SymbolSum[[α12 α3 α123 ; β12 β3 β123] [α123 α456 α ; β123 β456 β] [α1234 α56 α ; β1234 β56 β] [α56 α4 α456 ; β56 β4 β456]
           SymbolSum[(-1)^(2 α1234) √(1+2 α123p) √(1+2 α34) [α12 α3 α123p ; β12 β3 β123] [α123p α4 α1234 ; β123 β4 β1234]
           {α12 α3 α123p ; α4 α1234 α34}, {α123p, β123}], {β3, β4, β123, β12, β456, β56, β1234}]

In[16]:= % //. SumCollect //. SumSimplify /. CGCrule //. SumSimplify

Out[16]= (-1)^(-α12+3 α1234-α3-α4) √(1+2 α123) √(1+2 α34) {α12 α3 α123 ; α4 α1234 α34}
           SymbolSum[[α123 α4 α1234 ; β123 β4 β1234] [α123 α456 α ; β123 β456 β] [α1234 α56 α ; β1234 β56 β] [α56 α4 α456 ; β56 β4 β456],
           {β123, β1234, β4, β456, β56}]

In[17]:= Recoupling[%, {α4, β4}, {α56, β56}] /. {α_{α4,α56} → α456p, β_{β4,β56} → β456} //. SumCollect

Out[17]= (-1)^(-α12+3 α1234-α3-α4) √(1+2 α123) √(1+2 α34) {α12 α3 α123 ; α4 α1234 α34}
           SymbolSum[(-1)^(α+α123+α4+α56) √(1+2 α1234) √(1+2 α456p) [α123 α456 α ; β123 β456 β] [α123 α456p α ; β123 β456 β]
           [α4 α56 α456p ; β4 β56 β456] [α56 α4 α456 ; β56 β4 β456] {α123 α4 α1234 ; α56 α α456p}, {α456p, β123, β4, β456, β56}]

In[18]:= Make[%, {{α123, β123}, {α456, β456}}, {{α123, β123}, {α456p, β456}}, {{α4, β4}, {α56, β56}},
           {{α4, β4}, {α56, β56}}]

Out[18]= (-1)^(-α12+3 α1234-α3-α4) √(1+2 α123) √(1+2 α34) {α12 α3 α123 ; α4 α1234 α34}
           SymbolSum[(-1)^(α+α123+α456) √(1+2 α1234) √(1+2 α456p) [α123 α456 α ; β123 β456 β] [α123 α456p α ; β123 β456 β]
           [α4 α56 α456 ; β4 β56 β456] [α4 α56 α456p ; β4 β56 β456] {α123 α4 α1234 ; α56 α α456p}, {α456p, β123, β4, β456, β56}]

In[19]:= % //. SumSimplify /. CGCrule //. SumSimplify

Out[19]= (-1)^(α-α12+α123+3 α1234-α3-α4+α456) √(1+2 α123) √(1+2 α1234)
           √(1+2 α34) √(1+2 α456) {α12 α3 α123 ; α4 α1234 α34} {α123 α4 α1234 ; α56 α α456}
```

Figure 5. The computation of recoupling coefficient with *NCoperators*. The phase factor can be reduced by applying the triangle rule.

For example, the computation of recoupling coefficient

$$(\alpha_1\alpha_2(\alpha_{12})\alpha_3(\alpha_{123}),\alpha_4\alpha_5(\alpha_{45})\alpha_6(\alpha_{456})\alpha|\alpha_1\alpha_2(\alpha_{12}),\alpha_3\alpha_4(\alpha_{34})(\alpha_{1234}),\alpha_5\alpha_6(\alpha_{56})\alpha)$$

is illustrated in Fig. 5. The quantity $\left[\begin{smallmatrix} j_1 & j_2 & j \\ m_1 & m_2 & m \end{smallmatrix}\right]$ denotes the Clebsch–Gordan coefficient for the tensor product $j_1 \times j_2 \to j$. $\left\{\begin{smallmatrix} a & b & e \\ d & c & f \end{smallmatrix}\right\}$ is the $6j$–symbol. In Fig. 5, the obtained coefficient must be multiplied by $(-1)^{2\alpha_4+\alpha_5+\alpha_6+\alpha_{56}}\{(2\alpha_{45}+1)(2\alpha_{56}+1)\}^{1/2}\left\{\begin{smallmatrix} \alpha_4 & \alpha_5 & \alpha_{45} \\ \alpha_6 & \alpha_{456} & \alpha_{56} \end{smallmatrix}\right\}$ due to the recoupling in $\left[\begin{smallmatrix} \alpha_4 & \alpha_5 & \alpha_{45} \\ \beta_4 & \beta_5 & \beta_{45} \end{smallmatrix}\right]\left[\begin{smallmatrix} \alpha_{45} & \alpha_6 & \alpha_{456} \\ \beta_{45} & \beta_6 & \beta_{456} \end{smallmatrix}\right]$.

Let us give a brief examination of the example studied above, in eq. (8). For the irreducible tensor operator whose reduction scheme is labeled by the S_2–irreducible representation $[1^2]$, the associated coefficient $C^{(3)}_{11;1}$ is found by pulling out the SU(2)–invariant parts from the following products of matrix elements (see eq. (8)): $\sum_{\mathrm{e}} v_{\mathrm{ve}}\omega^{(2)\pm}_{\mathrm{e\bar{v}}}$ and $\sum_{\mathrm{c}} v_{\mathrm{c\bar{v}}}\omega^{(2)\pm}_{\mathrm{vc}}$, where the superscript $\pm$ indicates the sign of M in eq. (9). The final result for the plus sign is,

$$\begin{aligned} C^{(3)}_{11;1}(m_0\Lambda) =&(-1)^{\lambda_{\mathrm{v}}-\lambda_{\bar{\mathrm{v}}}}\sqrt{2\tau_0+1}\sum_x\sqrt{2x+1}\begin{bmatrix} \tau_0 & x & \Lambda \\ m_0 & M-m_0 & M \end{bmatrix} \\ &\times\Bigg((-1)^{\Lambda}\sum_{\mathrm{e}} f(\tau_0\lambda_{\mathrm{v}}\lambda_{\mathrm{e}})\Omega^{(2)}_{\mathrm{e\bar{v}}}(x)\begin{Bmatrix} \tau_0 & x & \Lambda \\ \lambda_{\bar{\mathrm{v}}} & \lambda_{\mathrm{v}} & \lambda_{\mathrm{e}} \end{Bmatrix} \\ &-(-1)^{x}\sum_{\mathrm{c}} f(\tau_0\lambda_{\mathrm{c}}\lambda_{\bar{\mathrm{v}}})\Omega^{(2)}_{\mathrm{vc}}(x)\begin{Bmatrix} \tau_0 & x & \Lambda \\ \lambda_{\mathrm{v}} & \lambda_{\bar{\mathrm{v}}} & \lambda_{\mathrm{c}} \end{Bmatrix}\Bigg), \end{aligned}$$

where $f(\tau_i\lambda_\alpha\lambda_{\bar{\beta}}) = -\{(2\lambda_\alpha+1)/(2\lambda_{\bar{\beta}}+1)\}^{1/2}[n_\alpha\lambda_\alpha\|v^{\tau_i}\|n_{\bar{\beta}}\lambda_{\bar{\beta}}]$ is proportional to the reduced matrix element of interaction operator v^{τ_i}, $i=0,1,2$. This matrix element is calculated on the basis of spherical harmonics. In particular, $i=0$ befits the interaction drawn in the second quantized form of $\widehat{V}_1$ in $:\{\widehat{P}\widehat{V}_1\widehat{\Omega}^{(2)}_1\widehat{P}\}_1:$. The coefficients $\Omega^{(2)}_{\alpha\bar{\beta}}(x)$ are found from

$$\omega^{(2)}_{\alpha\bar{\beta}} = (-1)^{\lambda_{\bar{\beta}}+m_{\bar{\beta}}}\sum_x\begin{bmatrix} \lambda_\alpha & \lambda_{\bar{\beta}} & \Lambda \\ m_\alpha & -m_{\bar{\beta}} & M \end{bmatrix}\Omega^{(2)}_{\alpha\bar{\beta}}(x),$$

and they contain 27 Goldstone diagrams in total [9].

5. Conclusion

We have reviewed the symbolic programming package *NCoperators* concentrating on the applications to many-body perturbation theory. We have shown that the package is capable to generate expansion terms up to the third-order RSPT and represent them in the irreducible tensor form by exploiting the angular momentum theory. On the basis that makes *NCoperators* be multifaceted we hope that with minor modifications, if such are necessary at all, the package can be adapted to many other areas that apply the methods briefly considered in the present paper.

References

[1] Bochevarov A D and Sherrill C D 2004 *J. Chem. Phys.* **121** 3374

[2] `http://www.files.chem.vt.edu/chem-dept/valeev/software/sequant/sequant.html`

[3] Derevianko A 2010 *J. Phys. B: At. Mol. Opt. Phys.* **43** 074001

[4] `http://homepage.cem.itesm.mx/lgomez/quantum/`

[5] Žitko R 2011 *Comp. Phys. Commun.* **182** 2259

[6] Racah G 1942 *Phys. Rev.* **61** 186

[7] Racah G 1942 *Phys. Rev.* **62** 438

[8] Fano U and Racah G 1959 *Irreducible Tensorial Sets* vol 4 (New York: Academic Press)

[9] Juršėnas R and Merkelis G 2010 *J. Math. Phys.* **51** 123512

[10] Juršėnas R and Merkelis G 2011 *Cent. Eur. J. Phys.* **9** 751

[11] Fritzsche S 1997 *Comp. Phys. Commun.* **103** 51

[12] Gaigalas G, Fritzsche S and Fricke B 2001 *Comp. Phys. Commun.* **135** 219

[13] Fritzsche S, Inghoff T, Bastug T and Tomaselli M 2001 *Comp. Phys. Commun.* **139** 314

[14] Juršėnas R and Merkelis G 2011 *At. Data Nucl. Data Tables* **97** 23

[15] Jucys A P, Levinson Y B and Vanagas V V 1960 *Mathematical Apparatus of the Theory of Angular Momentum [in Russian]* vol 3 (Gospolitnauchizdat)

[16] Varshalovich D A, Moskalev A N and Khersonskii V K 1975 *Quantum Theory of Angular Momentum [in Russian]* (Leningrad: Nauka)

[17] Jucys A P and Bandzaitis A A 1977 *Theory of Angular Momentum in Quantum Mechanics [in Russian]* (Vilnius: Mokslas)

[18] Lindgren I and Morrison J 1982 *Atomic Many-Body Theory* vol 13 (Springer Series in Chemical Physics)

[19] Judd B R 1963 *Operator Techniques in Atomic Spectroscopy* (New York: McGraw-Hill)

[20] Rudzikas Z and Kaniauskas J 1984 *Quasispin and Isospin in the Theory of Atom [in Russian]* (Vilnius: Mokslas)

[21] Kaniauskas J, Č Šimonis V and Rudzikas Z B 1987 *J. Phys. B: At. Mol. Opt. Phys.* **20** 3267

[22] Rudzikas Z 1997 *Theoretical Atomic Spectroscopy* (Cambridge: Cambridge Univ. Press)

IUPAP C20 Conference on Computational Physics (CCP 2011) IOP Publishing
Journal of Physics: Conference Series **402** (2012) 012008 doi:10.1088/1742-6596/402/1/012008

Hybrid algorithms in quantum Monte Carlo

Jeongnim Kim[1,2], Kenneth P. Esler[1], Jeremy McMinis[3], Miguel A. Morales[4], Bryan K Clark[5], Luke Shulenburger[6] and David M. Ceperley[1,3]

[1] National Center for Supercomputing Applications, University of Illinois at Urbana-Champaign, Urbana, IL, USA
[2] Oak Ridge National Laboratory, Oak Ridge, TN, USA
[3] Department of Physics, University of Illinois at Urbana-Champaign, Urbana, IL, USA
[4] Lawrence Livermore National Laboratory, Livermore, CA, USA
[5] The Princeton Center for Theoretical Science, Princenton University, Princeton, NJ, USA
[6] Sandia National Laboratories, Albuquerque, NM, USA

Abstract. With advances in algorithms and growing computing powers, quantum Monte Carlo (QMC) methods have become a leading contender for high accuracy calculations for the electronic structure of realistic systems. The performance gain on recent HPC systems is largely driven by increasing parallelism: the number of compute cores of a SMP and the number of SMPs have been going up, as the Top500 list attests. However, the available memory as well as the communication and memory bandwidth per element has not kept pace with the increasing parallelism. This severely limits the applicability of QMC and the problem size it can handle. OpenMP/MPI hybrid programming provides applications with simple but effective solutions to overcome efficiency and scalability bottlenecks on large-scale clusters based on multi/many-core SMPs. We discuss the design and implementation of hybrid methods in QMCPACK and analyze its performance on current HPC platforms characterized by various memory and communication hierarchies.

1. Introduction

Continuum quantum Monte Carlo (QMC) methods employ explicitly correlated wave functions to stochastically solve the Schrödinger equation [1]. QMC is one of the most accurate *ab initio* methods to describe the behavior and properties of quantum many-particle systems. It is general and applicable to a wide range of physical and chemical systems in any dimension with arbitrary boundary conditions. The favorable scaling, 3-4 power of the problem size, and ample opportunities for parallelizations, e.g., over multiple Bloch **k**-vectors, make QMC methods uniquely powerful tools to study the electronic structure of realistic systems on large-scale parallel computers.

The objects in QMC algorithms can be divided into two categories: i) *ensemble data* to represent the physical entities like trial wave functions and ii) *walker data* to represent each $3N$ dimensional configuration and its dynamic state. The *ensemble data* are large but are unmodified during a simulation once the initialization step is completed and are shared by all the walkers. On the other hand, the *walker data* evolve as we sample the configurations using various QMC algorithms. The simplest parallelization strategy is distributing walkers among the tasks, while replicating the *ensemble data* on each task. This fits nicely to the distributed memory model of the Message Passing Interface (MPI) library [2].

Until recently, QMC methods have been able to exploit the steady increase in clock rates and the enhancement of performance enabled by advances in CMOS VLSI technology for faster and bigger simulations without any major changes in this parallel algorithm based on MPI. However, this decade has seen some significant changes in the mode of increase of computational power. The performance gain has been largely driven by increasing parallelism in a shared-memory processor (SMP) unit in the form of multi-core processors and GPUs (Graphics Processing Units). A notable feature of multi/many-core architectures is relatively constant memory per core of a few GB. This can limit the problem size QMC can handle if the *ensemble data* is replicated on each core for efficiency. Recognizing that the aggregated memory of each SMP node is steadily growing with the number of cores on a processor, one can easily overcome the memory limitation of multi-core processors by using shared-memory programming models such as OpenMP [3].

In this article, we present hybrid QMC algorithms using OpenMP/MPI. Similar to the standard MPI implementation, we replicate the large but *read-only* ensemble data on each MPI task and distribute the walkers among tasks. OpenMP parallel constructs are used to further divide the walkers among threads. The threads and objects are carefully managed to minimize OpenMP overhead and false sharing, while maximizing data locality and cache reuse. The computational cost per core and the overall parallel efficiency are unaffected by the additional parallelization with OpenMP at a modest core count of a few 1000s. As we increase the number of nodes, the hybrid methods become more effective because of the unfavorable scaling of MPI collectives with respect to the task count and much reduced communication needs for the load balance. We expect that the performance advantage of the hybrid methods will expand with increasing parallelism on newer processors.

The hybrid QMC algorithms are implemented in QMCPACK, an open-source QMC package designed for large-scale parallel computers [4]. It makes extensive use of object-oriented and generic programming and design patterns [5] for reusability and extensibility. High computational efficiency is achieved through inlined specializations of C++ templates and by using SIMD intrinsics for core kernels. It utilizes standard file formats for input and output, using XML standard and HDF5 [6] to streamline the QMC workflow and facilitate data exchange with other applications.

The rest of the paper is organized as follows. First, we introduce QMC methods and core algorithms. We analyze critical computational components and scaling properties with problems size. In section 4, we present hybrid algorithms using OpenMP/MPI at the walker level and detail our efforts to achieve high computational and parallel efficiency on large-scale clusters. Our on-going works to overcome the limitations of current implementations are discussed in section 5 and the conclusions follow.

2. QMC Methods

In quantum mechanics, all physically observable quantities for a system containing N particles can be computed from the $3N$-dimensional *wave function*, $\Psi(\mathbf{r}_1, \ldots, \mathbf{r}_N)$. The exact wave function for the system satisfies the time-independent Schrödinger equation,

$$\hat{H}\Psi = E_0\Psi, \tag{1}$$

where E_0 is the ground-state energy for the system and $\hat{H}$ is the Hamiltonian, or total energy operator. For example, the Hamiltonian for an N-electron system is given as

$$\hat{H} = \sum_{i=1}^{N} -\frac{1}{2}\nabla_i^2 + \sum_{i=1}^{N} V_{\text{ext}}(\mathbf{r}_i) + \sum_{i<j} \frac{1}{|\mathbf{r}_i - \mathbf{r}_j|} \tag{2}$$

where $V_{\text{ext}}(\mathbf{r}_i)$ is the *external* potential generated by the nuclei, applied electric fields, etc. Since the exact solution Ψ is known only for simple few-body systems, QMC methods employ a *trial* wave function Ψ_T, which takes the general form

$$\Psi_T(\mathbf{R}) = \prod_k \psi_k(\{\alpha\}, \mathbf{R}). \tag{3}$$

Here, each component ψ_k is given as an analytic function of a number of optimizable parameters $\{\alpha\}$.

For any $\Psi_T(\mathbf{R})$, we can compute an energy as the expected value of $\hat{H}$,

$$E_T = \frac{\int d^{3N}\mathbf{R}\ \Psi_T^*(\mathbf{R})\hat{H}\Psi_T(\mathbf{R})}{\int d^{3N}\mathbf{R}\ |\Psi_T(\mathbf{R})|^2}, \tag{4}$$

where $\mathbf{R}$ is a $3N$-dimensional vector representing the positions of the N particles. A *variational principle* guarantees that $E_T \geq E_0$ for all Ψ_T, which implies that the lower the energy a given Ψ_T has, the closer it is to the exact solution of the Schrödinger equation. In QMC, E_T is estimated by stochastic sampling as

$$E_T \approx \frac{\sum_i^M w(\mathbf{R}_i)E_L(\mathbf{R}_i)}{\sum_i^M w(\mathbf{R}_i)}, \tag{5}$$

where E_L is a local energy, $E_L(\mathbf{R}) = \hat{H}\Psi_T(\mathbf{R})/\Psi_T(\mathbf{R})$ and w weight at $\mathbf{R}$.

Various algorithms have been developed to find the ground-state solution, i) either by minimizing the energy by varying parameters in a Ψ_T as in the variational Monte Carlo (VMC) method or ii) by projecting out the ground state by repeatedly applying an imaginary time Green's function or *propagator* $\exp(-\tau\hat{H})$ on Ψ_T as in the diffusion Monte Carlo (DMC) method [1]. In actual calculations, the parameters $\{\alpha\}$ are optimized within VMC to obtain improved, yet still compact wave functions. Once a trial wave function is optimized, it is further improved through a stochastic projection by DMC. This typical QMC workflow is illustrated in Fig. 1(a).

For bosons, the DMC method is exact (within statistical errors) if the time-step, τ, is chosen sufficiently small. For fermions, such as electrons, however, the *fixed-node approximation* is introduced to alleviate the *fermion sign problem*. It yields exact results if the nodes of Ψ_T (i.e. the $(3N-1)$-dimensional surface defined by $\Psi_T(\mathbf{R}) = 0$) are exact. With approximate nodes, the error in the energy is quadratic in the error in the nodes. In practice, the error is usually very small if Ψ_T is reasonable. For example, using a trial wave function derived from density functional theory, the accuracy of DMC usually greatly surpasses that of the DFT calculation.

For stochastic sampling methods, the convergence of a simulation is determined by the standard error as

$$\delta = \frac{\sigma}{\sqrt{M}}, \tag{6}$$

for M *uncorrelated* samples. In VMC, all the configurations are equally weighted, *i.e.*, $w = 1$, and the energy variance is

$$\begin{aligned} \sigma^2 &= \frac{\sum_i^M E_L(\mathbf{R}_i)^2}{M} - \left(\frac{\sum_i^M E_L(\mathbf{R}_i)}{M}\right)^2 \\ &= \langle E_T^2\rangle - \langle E_T\rangle^2. \end{aligned} \tag{7}$$

The generalization for $w = \exp(-\tau(E_L - E_T))$ in DMC is straightforward [1]. Finally, we can define the efficiency of a QMC simulation as

$$\kappa = \frac{1}{\sigma^2 T_{\text{corr}} T_{\text{CPU}}}, \tag{8}$$

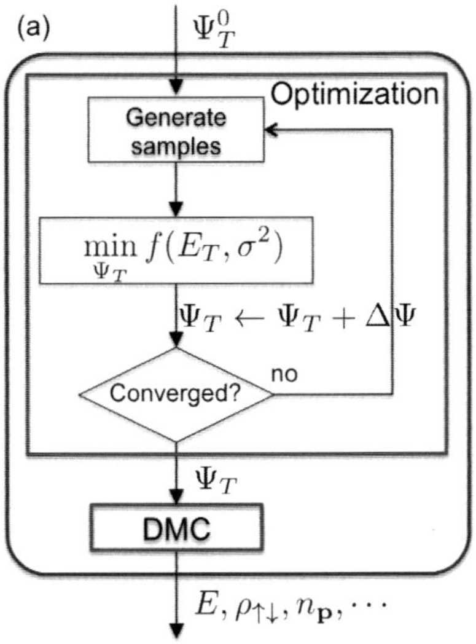

(b) DMC

for generation $= 1 \cdots M$ **do**
  #pragma omp parallel for
  for walker $= 1 \cdots N_w$ **do**
    let $\mathbf{R} = \{\mathbf{r}_1 \ldots \mathbf{r}_N\}$
    for particle $i = 1 \cdots N$ **do**
      set $\mathbf{r}_i^{'} = \mathbf{r}_i + \delta$
      let $\mathbf{R}^{'} = \{\mathbf{r}_1 \ldots \mathbf{r}_i^{'} \ldots \mathbf{r}_N\}$
      ratio $\rho = \Psi_T(\mathbf{R}^{'})/\Psi_T(\mathbf{R})$
      if $\mathbf{r} \to \mathbf{r}'$ is accepted **then**
        update state of a walker
      end if
    end for{particle}
    Compute $E_L = \hat{H}\Psi_T(\mathbf{R})/\Psi_T(\mathbf{R})$
  end for{walker}
  Reweight and branch walkers
  Update E_T
end for{generation}

Figure 1. (a) QMC workflow and (b) DMC algorithm.

where $T_{\rm CPU}$ is the time to achieve a target error and $T_{\rm corr}$ is the auto-correlation time [7]. An accurate Ψ_T with small σ and $T_{\rm corr}$ is essential to achieve high quality results with small statistical error. Choosing a compact representation for Ψ_T with a low computational cost is equally important for efficient QMC calculations. The advantage of increasing parallelism for QMC is apparent, as $T_{\rm CPU} \propto T_{\rm MC}/N_p$, where $T_{\rm MC}$ denotes the computational time to generate a MC sample and N_p, the number of parallel processing units.

3. QMC algorithms

The vast majority of the computational time in an accurate QMC simulation is spent in DMC and therefore minimizing the wall-clock time to obtain a DMC solution within a target error is desired. The optimized computational kernels, physical abstractions and parallelization schemes, however, are also applicable to all other QMC algorithms.

In the DMC algorithm (Fig. 1 b), an ensemble of walkers (*population*) is propagated stochastically from generation to generation, where each walker is represented by $\mathbf{R}$. In each propagation step, the walkers are moved through position space by a *drift-diffusion process.* At the end of each step, each walker may reproduce itself, be killed, or remain unchanged (*branching process*), depending on the value of E_L, the local energy at that point in space. The walkers in an ensemble are tightly coupled through a trial energy E_T and a feedback mechanism to keep the average population at the target $10^3 - 10^5$ walkers. This leads to a fluctuating population (Fig. 2b) and necessitates shuffling of walkers among parallel processing units to maintain good load balance in an appropriate interval.

The direct evaluation of the many-dimensional integrals of Eq. (4) by stochastic sampling enables us to employ highly accurate variational wave functions which can capture crucial many-body effects in an efficient manner. The Slater-Jastrow trial wave functions, commonly used in QMC applications to electronic structure, are sums of Slater determinants of single-particle

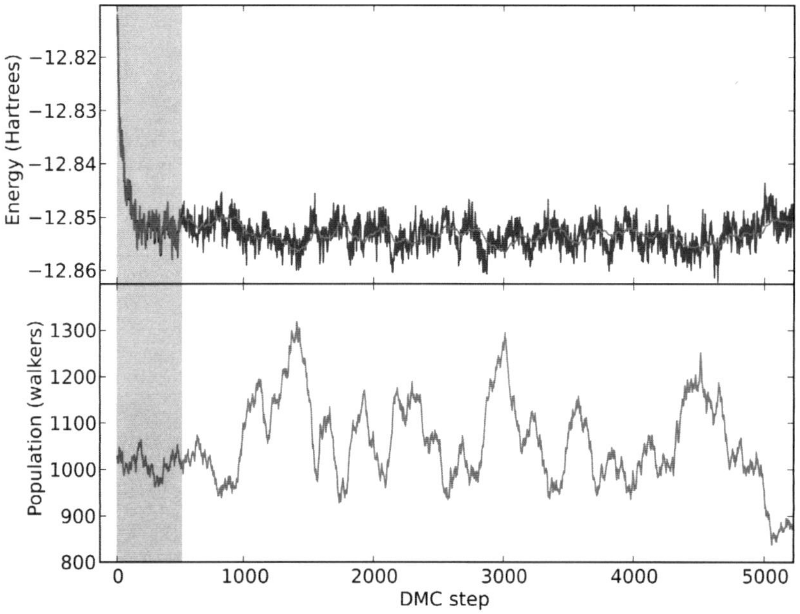

Figure 2. Trace of the DMC energy and walker population during a simulation. The gray shaded region indicates the equilibration period, during which the data should be discarded.

orbitals multiplied by a Jastrow factor:

$$\Psi_T = \exp(J_1)\exp(J_2)\cdots \underbrace{\sum C_i D_i^{\uparrow}(\phi) D_i^{\downarrow}(\phi)}_{\Psi_{AS}}, \tag{9}$$

with $N = N^{\uparrow} + N^{\downarrow}$ for $\uparrow\downarrow$ spins. Here, $\{\phi\}$ denote a set of single-particle orbitals and

$$D_i^{\uparrow} = \det|\mathcal{M}| = \det \begin{vmatrix} \phi_1(\mathbf{r}_1) & \cdots & \phi_1(\mathbf{r}_{N^\uparrow}) \\ \vdots & \ddots & \vdots \\ \phi_{N^\uparrow}(\mathbf{r}_1) & \ldots & \phi_{N^\uparrow}(\mathbf{r}_{N^\uparrow}) \end{vmatrix}, \tag{10}$$

which ensures the antisymmetric property of a Fermionic wave function upon a pair exchange of electrons. The Jastrow factors, which are factorized for computational efficiency, describe the dynamic correlation, whereas static correlation is described by the sum of Slater determinants. The optimization of the many-body trial wave functions is a crucial ingredient for accurate QMC calculations, since accurate trial wave functions reduce statistical and systematic errors. Several new methods for optimization have been developed recently [8, 9]. These new methods enable us to optimize the determinantal coefficients $\{C\}$ in addition to the Jastrow parameters, reducing the dependence of the results on the input trial wave function.

The product form of a trial wave function (Eqs. 3 and 9) is rewritten as

$$\ln \Psi_T(\mathbf{R}) = \sum_k \ln \psi_k(\{\alpha\}, \mathbf{R}), \tag{11}$$

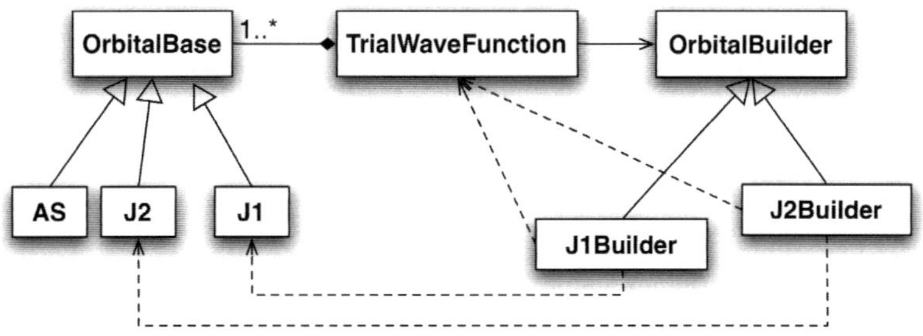

Figure 3. Composite pattern for TrialWaveFunction Ψ_T. It owns a number of `OrbitalBase` objects which are instantiated at runtime as defined in an input file using the builder pattern. The derived classes, denoted as J1, J2, and AS for Ψ_{AS}, implement the virtual functions defined in the `OrbitalBase` class.

for numerical stability of large-scale QMC calculations. The use of the log representation eliminates the need to normalize the trial wave function, which has to be computed with other properties, and to prevent overflow or underflow in using ill-conditioned values, especially the determinants of large matrices. Each component can be defined in a self-contained class derived from an abstract base class. We express the computations associated with Ψ_T as sums of wave function components, e.g.,

$$\frac{\nabla\Psi_T}{\Psi_T} = \nabla \ln \Psi_T = \sum_k \nabla \ln \psi_k, \tag{12}$$

which is needed for E_L. The actual form of the trial wave function is not known at compile time. We apply the builder pattern to construct a trial wave function component by component according to the input parameters as illustrated in Fig. 3 [5]. The many-body Hamiltonian in Eq. (2) follows the same patterns employed for the trial wave function.

The dominant computation for large systems is set by the antisymmetric part Ψ_{AS} of the trial wave function: i) to evaluate the ratios, T_{ratio}; ii) to update Slater determinants, $T_{\mathcal{M}^{-1}}$; and iii) to fill in the Slater determinants with single-particle orbitals, T_{spo}. Since the particle-by-particle moves we employ change only one column of the $\mathcal{M}$ matrices at a time, we employ a rank-1 update of $\mathcal{M}^{-1}$ using the Sherman-Morrison formula. This allows the inverse to be updated in $\mathcal{O}(N^2)$ time rather than $\mathcal{O}(N^3)$ time for a full inversion and the ratios in $\mathcal{O}(N)$.

The single-particle orbitals are a set of three-dimensional functions defined on the simulation domain. They are typically expressed in a basis set. It is possible to optimize $\{\phi\}$ in a given basis set, but, in practice, we use the solution of an approximate method such as Hartree-Fock or density functional theory (DFT). The most commonly used basis sets are plane-wave basis set for solids in periodic boundary conditions and atomic orbitals for molecular systems. The cost to compute the value of a ϕ scales linearly with the number of basis function evaluations which tends to grow with the system size. This amounts to $\mathcal{O}(N^2)$ cost for each particle move and T_{spo} becomes the bottleneck at large N with these basis sets.

For this reason, it is more efficient to use a localized basis with compact support. In particular, 3D tricubic B-splines provide a basis in which only 64 elements are nonzero at any given point in space [10, 11]. The one-dimensional cubic B-spline is given by,

$$f(x) = \sum_{i'=i-1}^{i+2} b^{i',3}(x)\; p_{i'}, \tag{13}$$

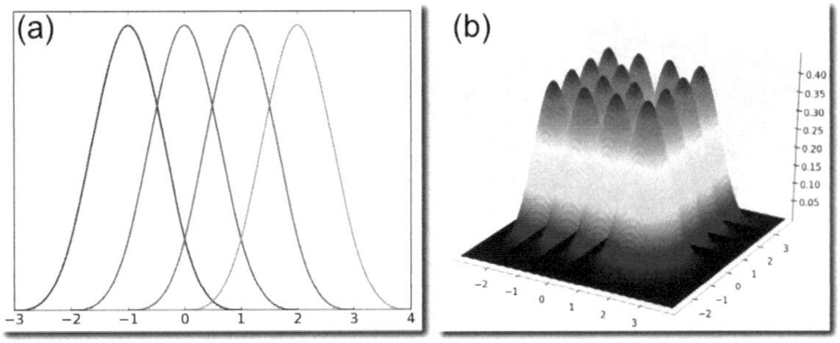

Figure 4. Cubic B-spline basis functions in (a) 1 D and (b) 2 D.

where $b^i(x)$ are the piecewise cubic polynomial basis functions shown in Figure 4, and $i = \text{floor}(\Delta^{-1}x)$ is the index of the first grid point $\leq x$. Constructing a tensor product in each Cartesian direction, we can represent a 3D orbital as

$$\phi_n(x,y,z) = \sum_{i'=i-1}^{i+2} b_x^{i',3}(x) \sum_{j'=j-1}^{j+2} b_y^{j',3}(y) \sum_{k'=k-1}^{k+2} b_z^{k',3}(z)\; p_{i',j',k',n}. \tag{14}$$

This allows the rapid evaluation of each orbital in constant time. Furthermore, they are systematically improvable with a single spacing parameter, so that accuracy is not compromised.

The use of 3D tricubic B-splines greatly improves the computational efficiency. Even for a modest problem size of 32 electrons, the speed up of B-spline interpolations is more than sixfold over an equivalent PW basis set. The gain in computation time of real-space methods becomes increasingly large as the system size grows. On the downside, this computational efficiency comes at the expense of increased memory use.

4. Hybrid QMC implementation

Among the many ways to parallelize the DMC algorithm, the multiplicity of the walkers in an ensemble provides the most natural units for data and task parallelizations. The internal state of each walker, its configuration **R** and other data we wish to avoid recomputing, is encapsulated in the `Walker` class. The operations to propagate each walker during the drift-diffusion process are expressed as a parallel loop over the `Walkers`. Once a generation has evolved, the properties of all the walkers in an ensemble are collected to determine E_T and N_w, the number of walkers of the next generation, which employs global reduction operations among MPI tasks. The redistribution of `Walkers` during the load-balance step can be done efficiently by exchanging a serialized `Walker` object as a large message between paired MPI tasks.

The communication overhead in QMC calculations is quite negligible compared to $T_{\rm MC}$, the computation time. For the reduction operations to perform averages and record results, MPI collectives are found to be the most efficient choice on many platforms. The message size and number of messages for the collectives are independent of N_p, the number of tasks, and the problem size. But, the communication time for the collectives can become a sizable fraction of the total run time and decrease the parallel efficiency at large N_p, as the collectives scale as $\mathcal{O}(\log_2 N_p) - \mathcal{O}(N_p^2)$ depending on the network topology and specific algorithms in use.

There are several different and independent ways to parallelize a QMC calculation beyond the walker level, e.g., over parameter and configuration spaces and over Bloch **k**-vectors. Such high-level parallelism can be easily managed by mapping a DMC ensemble on a MPI group.

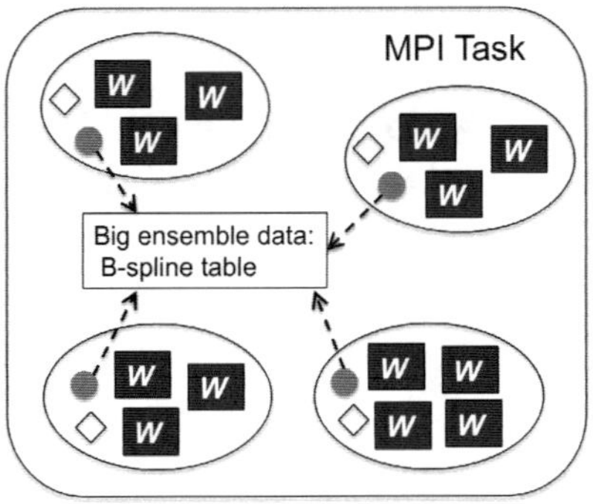

Figure 5. Schematic view of object distribution per MPI task. The ovals denote the OpenMP threads which own `Walker` objects (boxes), and other objects, e.g., Ψ_T (circle) and $\hat{H}$ (diamond).

At present, power and thermal considerations constrain the growth in computational power to come almost exclusively through parallelism in the form of multi-core processors. As we scale out the problem size and use finer grids to achieve higher accuracy, the memory required to store the trial wave function including a large table for B-spline interpolations will become larger than that available on a single core, necessitating the distribution of a single trial wave function data over the cores and adding communications to access the data on remote MPI tasks. This limitation, however, can be easily overcome using shared memory programming models that provide direct access to the entire memory of a node rather than that of a core.

Among various parallel programming models, hybrid programming using OpenMP and MPI provides QMC methods with simple but very effective solutions. We can utilize the large shared-memory address space on a SMP node for the data that can be shared. Solutions based on using an explicit shared memory segment via shmem or mmap or partitioned global address space programming models exist but we choose OpenMP/MPI in QMCPACK: it is the most widely supported programming model on HPC, portable and its performance has been steadily improving by better compilers supports. The loop over the `walkers` can be trivially parallelized using the `parallel-for` construct of OpenMP. However, achieving high performance with OpenMP requires careful data and thread managements.

Our hybrid implementation employs the distributed-memory programming model within OpenMP. Each thread manages a set of `Walkers` using thread-local storage. The objects associated with Ψ_T and $\hat{H}$ are replicated on all the threads except for large, *read-only* objects, such as B-spline tables. They are allocated at the task level and shared among all the threads in a task, significantly reducing the memory footprint of the B-spline representation. Figure 4 illustrates the distribution of `Walkers` and essential objects among OpenMP threads and MPI tasks.

We fully exploit the language features of C++ to manage objects and threads to maximize data locality, eliminate false sharing and minimize the OpenMP overheads. The code modification to implement OpenMP/MPI hybrid algorithms is limited to the high-level QMC drivers: as shown in the DMC pseudocode of Fig. 1(b), complete MC sweeps over a walker are encapsulated within a parallel construct, `parallel for`. This ensures large granularity of computations per thread and eliminates any need for explicit synchronization or memory locking during the sweeps. The methods to migrate walker data sets between nodes for load balancing in DMC and accumulating statistical averages are unaffected by introducing the thread-level parallelism.

This hybrid OpenMP/MPI scheme has several advantages over the standard MPI-only scheme.

- Communication optimized: the communication overhead, especially for the collective

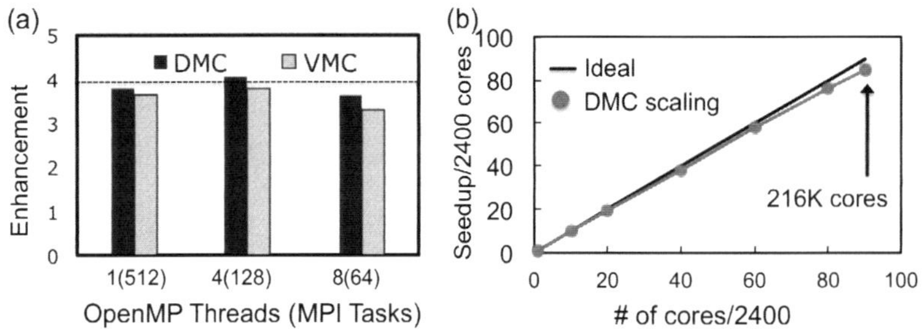

Figure 6. (a) Performance enhancement of OpenMP/MPI hybrid runs on a Cray XT4 when all the cores are utilized. The dotted line denotes the ideal enhancement of 4 on dual quad-core nodes. Values greater then 4 simply reflect statistical noise on our measurement. (b) Strong scaling of DMC on Cray XT5 using 6 OpenMP threads and two MPI tasks per dual hex-core node. All the data are collected on Jaguar at NCCS [12].

operations necessary to determine E_T and measure the properties of an ensemble, is significantly lowered by using fewer MPI tasks.

- Load balanced: `Walkers` in an ensemble are evenly distributed among threads and MPI tasks. The two-level parallelism reduces the population imbalance among MPI tasks and reduces the number of point-to-point communications of large messages (serialized objects) for the `Walker` exchange.
- Memory optimized: large read-only data to store single-particle orbitals and other shared properties to represent the trial wave function and many-body Hamiltonian can be shared among threads, which reduces the memory footprint of a large-scale problem.
- Data-locality optimized: the data associated with an active `Walker` are in cache during the compute-intensive drift-diffusion process and the operations on a `Walker` are optimized for cache reuse. Thread-local objects are used to ensure the data affinity to a thread.

Effectiveness of the hybrid scheme is evident in Fig 6(a). The parallel efficiency remains high for any combination of OpenMP threads and MPI tasks even at this modest scale of 512 cores. In fact, DMC scales nearly perfectly with respect to the number of threads: the additional parallelism from multi threading allows more walkers per MPI task and improves the overall parallel efficiency and load balance among SMP nodes. The VMC efficiency reflects the memory bandwidth-limited nature of QMC algorithms with B-splines when all the computing cores are used. The performance enhancement of less than the ideal of 4 for runs using only a quarter of the physical cores is expected, since resources are shared among the multiple cores. This is not unique to OpenMP, as indicated by the pure MPI run using 512 tasks in Fig. 6(a).

For QMC simulations, the added parallelism with more cores on a SMP node is always beneficial, as $T_{\rm CPU} \propto 1/N_{\rm cores}$. The parallel efficiency of hybrid runs depends on the memory architecture and is subject to the quality of compilers and MPI implementations. In general, the best performance for a wide range of problem sizes is obtained when a MPI task is mapped over a NUMA node as shown in Fig. 6(a) with 4 threads on a quad-core processor.

Hybrid QMC algorithms have allowed us to apply QMC methods to problem sizes at unprecedented scales on large-scale clusters of multi-core SMPs, efficiently utilizing all the available computing resources. At present, we can obtain the energy of a defect in a 64-Si crystalline system (256 electrons) to within 1 mHa in an hour using 4800 cores, completing a QMC workflow of VMC optimizations and DMC projection. The B-spline table for the system with a defect, for which no symmetry can be exploited to reduce the table size, requires more than 4 GB and it is necessary to employ 6 OpenMP threads and 2 MPI tasks on the dual hex-core nodes with 16 GB of memory. Figure 6(b) shows the parallel efficiency of DMC of the same system on Cray XT5 [12]. It shows near 95% of the ideal speedup up to 216K cores, which implies that the same *chemical accuracy* can be achieved in a much shorter wall-clock time.

5. Limitations and works in progress

As previously discussed, QMC computations are dominated by evaluation of Ψ_T and the time to generate a MC sample grows with the problem size N as

$$\begin{aligned} T_{\text{MC}} &= N \times (T_{\mathcal{M}^{-1}} + T_{\text{ratios}} + T_{\text{spo}}) + T_{\text{others}} \\ &= \mathcal{O}(N^3) + \mathcal{O}(N^2) + \mathcal{O}(N^2 N_b), \end{aligned} \tag{15}$$

where N_b=1 with B-splines. Other computations involving Jastrow functions scale at most qudratically to N with much smaller prefactors. The number of samples required to reach the same accuracy weakly depends on the problem size, making the overall scaling of $\mathcal{O}(O^3)-\mathcal{O}(N^4)$. This favorable scaling and near-perfect parallel efficiency make QMC methods uniquely powerful tools to study the electronic structure of realistic systems on the current generation of HPC systems. So far, we could reap increasing computing power – faster CPU clock, more cores on a SMP node and more nodes – to improve the efficiency of QMC calculations through `Walker` parallelization. However, significant changes are needed for QMC methods to play expanded roles on emerging architectures which will have an order of magnitude higher concurrency and a deeper memory and communication hierarchy [13].

Our OpenMP/MPI hybrid methods employ many optimizations to improve computational efficiency at the expense of increased memory use. Each walker needs temporary data for update methods, which scales as $\mathcal{O}(N^2)$. This is in addition to the $\mathcal{O}(N^2)$ memory required for the B-spline table shared by all the walkers. As we scale out the problem size, the required memory will eventually become larger than what is available on a SMP node.

Several methods have been under development to overcome the memory limitation. (i) *Single-precision* can be used to store the B-spline table, halving the memory use and time in evaluating the single-particle orbitals. Our extensive tests have shown that results of the double and mostly single-precision GPU implementations agree within the statistical error we have achieved thus far [14]. (ii) We can use a *mixed-basis* approach. The mesh spacing required to accurately represent the orbitals is determined by their smallest feature size. The shortest wavelength features are concentrated around the atomic cores, while in the area between the atoms, the functions are smooth. For this reason, we divide space into spherical regions called muffin tins surrounding the atoms, and an interstitial region between them. Utilizing this dual basis in place of 3D B-splines alone typically allows the same accuracy to be achieved with 5× to 10× less memory, with about the same performance [15]. (iii) We are exploring *distribution* of the B-spline table over a group of SMP nodes using the Global Arrays framework [16]. It provides a partitioned local view of data in the MPI processes and a global-shared-memory abstraction and allows direct access to portions of gloabal arrays that are stored at a remote node's memory.

These improvements will extend the problems size we can study with QMC on large clusters of SMPs, the most common HPC architecture of today and in a near future. However, the scaling of $T_{\text{MC}} \sim \mathcal{O}(N^{3-4})$ with the problem size will ultimately place limits on QMC's applicability. Again, the increasing parallelism on newer processors and thread-based parallel programming models on a SMP, e.g., OpenMP 3.0 [3] and Intel TBB [17], provide QMC with solutions.

The forms of Ψ_T and $\hat{H}$ as sums of components, lend themselves to task parallelism using well-encapsulated objects. The critical computational step to evaluate the single-particle orbitals and the determinant updates can be carried out in parallel. Furthermore, we can expose the parallel loops over ions and electrons as in the GPU implementation [14]. Preliminary results show promising performance gains with OpenMP task and nested parallelism by defining large chunks of computational units for the tasks and carefully managing objects. The fine-grained task parallelism removes the constraint on the number of `Walkers`, currently at least one `Walker` per core, and enables practical QMC calculations of larger systems with much reduced time-to-solution.

As we improve the computational efficiency and memory use, other elements of the QMC algorithms that have been ignored will become new barriers to overcome. Our scaling studies expose the performance issues with blocking collectives and parallel I/O at large scales. The *true* parallel efficiency of a complete QMC workflow is also limited by the initialization stage. The data collected during the equilibration, shown as the gray region of Fig. 2, are discarded. The equilibration time depends on the total number of walkers and cannot be reduced by just adding more processors. Although the number of steps to reach the equilibration is much smaller than the DMC generations needed for a target accuracy, reducing the equilibration time is essential to attain high efficiency with increasing parallelism. Solutions are being explored including increased intervals between the synchronizations, asynchronous I/O, use of non-blocking collectives [18] and variable time steps for DMC to speed up the equilibration.

6. Conclusions

Many levels of parallelism are afforded by QMC algorithms, including, but not limited to the natural parallelism inherent in Monte Carlo methods, making QMC intrinsically scalable and ideally suited to take advantage of the growth in computational power. We presented OpenMP/MPI hybrid implementations that reduce the memory footprint by selectively replicating data for computational efficiency, while improving the overall parallel efficiency on large-scale clusters of SMPs based on multi-core processors. The walker-parallelization allows QMC methods to handle 1000s of electrons without incurring extra communication overhead and enables high accuracy QMC calculations of realistic systems. We have identified the barriers to overcome, as we further scale out problem size and improve the accuracy of QMC calculations, and offer several solutions to extend QMC's reach on future architectures. With this progress, QMC simulations will be as ubiquitous as DFT calculations are at present and will become the tool of choice for materials design by simulations.

7. Acknowledgments

This work was supported by the U.S. Department of Energy (DOE) under Contract No. DOE-DE-FG05-08OR23336 and by the National Science Foundation (NSF) under No. 0904572. This work was performed in part under the auspices of the US DOE by LLNL under Contract DE-AC52-07NA27344 and by the Laboratory Directed Research and Development Program of Oak Ridge National Laboratory, managed by UT-Battelle, LLC, for the U. S. Department of Energy. Sandia National Laboratories is a multiprogram laboratory managed and operated by Sandia Corporation, a wholly owned subsidiary of Lockheed Martin Corporation, for the U.S. DOE's National Nuclear Security Administration under Contract No. DE-AC04-94AL85000. This research used resources of the National Center for Computational Sciences and the Center for Nanophase Materials Sciences, which are sponsored by the respective facilities divisions of the offices of Advanced Scientific Computing Research and Basic Energy Sciences of DOE under Contract No. DE-AC05-00OR22725 and allocations under TG-MCA93S030 and TG-MCA07S016 of advanced computing resources provided by the NSF.

References

[1] W. M. C. Foulkes, L. Mitas, R. J. Needs, and G. Rajagopal. Quantum monte carlo simulations of solids. *Rev. Mod. Phys.*, 73(1):33–83, Jan 2001.

[2] Message passing library. http://www.mpi-forum.org/.
[3] Openmp application program interface. http://www.openmp.org.
[4] Jeongnim Kim, K. Esler, J. McMinis, B. Clark, J. Gergely, S. Chiesa, K. Delaney, J. Vincent, and D. Ceperley. QMCPACK simulation suite.
[5] Erich Gamma, Richard Helm, Ralph Johnson, and John Vlissides. *Design Patterns: Elements of Reusable Object-Oriented Software*, chapter Methods for Coupled Electronic-Ionic Monte Carlo. Addison-Wesley, 1994.
[6] HDF (hierarchical data format). http://www.hdfgroup.org
[7] G. E. P. Box and G. Jenkins. *Time Series Analysis: Forecasting and Control.* Holden-Day, 1976.
[8] C. J. Umrigar, J. Toulouse, C. Filippi, S. Sorella, and R. G. Hennig. Alleviation of the fermion-sign problem by optimization of many-body wave functions. *Phys. Rev. Lett.*, 98:110201, 2007.
[9] K. P. Esler, J. Kim, D. M. Ceperley, W. Purwanto, E. J. Walter, H. Krakauer, S. Zhang, P. R. C. Kent, R. G. Hennig, C. Umrigar, M. Bajdich, J. Kolorenc, L. Mitas, and A. Srinivasan. Quantum monte carlo algorithms for electronic structure at the petascale; the endstation project. *Journal of Physics: Conference Series*, 125:012057 (15pp), 2008.
[10] D. Alfè and M.J. Gillan. Efficient localized basis set for quantum Monte Carlo calculations on condensed matter. *Phys. Rev. B.*, 70:1661101(R), 11 Oct. 2004.
[11] K. Esler. Einspline B-spline library, http://einspline.sf.net.
[12] Jaguar (Cray XT), ORNL Leadership Computing Facility (OLCF).
[13] J. Dongarra, P. Beckman, P. Aerts, F. Cappello, T. Lippert, S. Matsuoka, P. Messina, T. Moore, R. Stevens, A. Trefethen, and M. Valero. The international exascale software project: a call to cooperative action by the global high-performance community. *Int. J. High Perform. Comput. Appl.*, 23(4):309–322, 2009.
[14] K. P. Esler, J. Kim, L. Shulenburger, and D. M. Ceperley. Fully accelerating quantum monte carlo simulations of real materials on gpu clusters. *Computing in Science and Engineering*, 99:preprint, 2010.
[15] K. P. Esler, R. E. Cohen, B. Militzer, J. Kim, R. J. Needs, and M. D. Towler. Fundamental high-pressure calibration from all-electron quantum monte carlo calculations. *Phys. Rev. Lett.*, 104(18):185702, May 2010.
[16] S. Tirukkovalur, J. Dinan, Q. Niu, P. Sadayappan, J. Kim, A. Srinivasan, S. Kumar, and J. Hammond. A global address space approach to automated data management for parallel quantum monte carlo applications. Submitted to SC11, 2011.
[17] James Reinders. *Intel Thread Building Blocks.* O'Reilly Media, 2007.
[18] M. Gillan, M. D. Towler and D. Alfe, Petascale Computing opens new vistas for qauntum Monte Carlo. Psik Newsletter, 103: 32, February, 2011.

IUPAP C20 Conference on Computational Physics (CCP 2011) IOP Publishing
Journal of Physics: Conference Series **402** (2012) 012009 doi:10.1088/1742-6596/402/1/012009

Dynamics of chemical reacting systems in low dimensional lattices via the Equation Free Method

Giacomo Mazzi

Department of Computer Science, K.U. Leuven, Celestijnenlaan 200A, Heverlee 3001, Belgium

E-mail: giacomo.mazzi@cs.kuleuven

Abstract. In this report we show how it is possible to reconstruct the dynamics of a chemical reacting systems on lattice via an appropriate choice of macroscopic variable set. When chemical reactions fire on a low dimensional setting the usual mean field approximation (MF) may fail to accurately describe the proper dynamics of the system. Consequently it is not possible to use the concentrations of different species as sole quantities to describe macroscopically such systems and more involved approximations are needed. Traditional approaches introduces higher order correlation functions as extra macroscopic variables, however such expansions usually require closure procedures and tend to quickly demand high number of terms even for simple systems. It is also possible to describe the system in a microscopic way via a Kinetic Monte Carlo (KMC) algorithm, but this approach requires much longer computation time. In this report we study such a problem from a different perspective using the Equation Free Method (EFM) framework; with such approach it is possible to "bridge" between the macroscopic dynamics (mean field) and the microscopic one (KMC). We analyze different techniques to move between these two descriptions of the system trying to understand what information needs to be preserved at the macroscopic level in order to recover correctly all the features of the microscopic dynamics.

1. Introduction

The dynamics of chemical reacting systems is usually described via the reaction–diffusion equation for the coverages of the different species present in the compound:

$$\frac{\partial u_i(\vec{x},t)}{\partial t} = f_i\left(\vec{u}(\vec{x},t)\right) + D_i \nabla^2 u_i(\vec{x},t), \tag{1}$$

where $\overline{u}(\vec{x},t) = \{u_1(\vec{x},t),\ldots,u_n(\vec{x},t)\}$ is the vector of concentrations for each species i in $\vec{x}$ at time t, f_i is the mass action laws term for reactions involving species i and D_i is the matrix of diffusion coefficients.

However spatial constraints like surfaces, and low mobility of the particles may render such macroscopic description not accurate enough to capture all the microscopic fluctuations which arise at the microscopic level. In case of nonlinear reactions such fluctuations are enhanced and appear also at the macroscopic level, this renders (1) unable to describe the dynamics of $\overline{u}$ [10, 12, 9].

For these cases it is possible to build a Monte Carlo propagator for the microscopic state; such propagator updates the system via a series of jump process according to given dynamic rules [2]. In this approach the evolution of the system is described by the chemical master equation

[4]

$$\frac{d\mathcal{P}(\sigma,t)}{dt}=\sum_{\sigma'}[W(\sigma'\to\sigma)\mathcal{P}(\sigma',t)-W(\sigma\to\sigma')\mathcal{P}(\sigma,t)], \tag{2}$$

where $\mathcal{P}(\sigma,t)$ is the probability of finding the system in configuration σ at time t and $W(\sigma'\to\sigma)$ is the transition probability per unit time that the system switches from configuration σ' to configuration σ. Due to the fact that the spatial location where the reaction fires is important for such simulation, these Monte Carlo methods are usually variants of traditional Kinetic Monte Carlo (KMC) methods like Stochastic Simulation Algorithm (SSA) [11], where also the reaction site is taken into account. Usual accelerated techniques (like τ leaping) are not easily implemented in lattice KMC, unless other approximation (like some type of coarse graining [13]) are also taken into account. For a general review of such methods see [3].

2. Chemical reactions on low dimensional lattices

Throughout this text we will consider a particular example system in a one–dimensional lattice (such that σ is a vector of length N with periodic boundary condition, we also assume that all the lattice sites are occupied and that there is no diffusion. The reaction laws are:

$$XAX \leftrightarrow XXX. \tag{3}$$

This model contains the simplest mechanism to obtain macroscopic behaviour that non-trivially depends on the local configurations [9]. The reaction laws (3) are also typical part of realistic, more complex reaction schemes [1]. For the system of interest the mean field approximation reads [8]

$$\frac{dx}{dt}=-x^3+ax^2, \tag{4}$$

where a,x are respectively the coverages for the species A,X. The transition probabilities in (2) may be rewritten as lattice sites transitions, i.e. the probability that $A\leftrightarrow X$ in a given lattice site j

$$w(\sigma_j\to-\sigma_j,t)=\frac{1}{4}\left[1-\sigma_{j-1}(t)\right]\left[1-\sigma_{j+1}(t)\right], \tag{5}$$

where we have set $\sigma_j=+1$ if the site j is occupied by a X and $\sigma=-1$ otherwise. Given the simplicity of such model system it is possible to develop a two step Monte Carlo propagator [8] for the microscopic state σ, obeying (3), $\sigma(n+1)=s(\sigma(n))$, as follow :

- pick randomly a lattice site J from a uniform integer distribution with range $[1,N]$;
- if $\sigma_{J+1}+\sigma_{J-1}=-2$ then swap the element :$\sigma_J=-\sigma_J$.

At each step we advance the time by 1, so the total time of the simulation is counted in number of Monte Carlo steps.

It is possible to relate the macroscopic (4) dynamics and the microscopic (2) by expanding the latest in term of concentrations and higher orders particle correlations [9]. The main issue is that the equations for such correlations are not closed, and analytical closures have been developed such as the quasi–chemical approach where pair correlations are also taken into account [8]. However these closures depends strongly on the system considered, and when more than pair correlations are needed the set of equations becomes quickly intractable. Generally one would like to obtain the dynamics for some macroscopic quantities but also study the equilibrium states dependence on the reaction rates via bifurcation analysis, and these results might not be easily obtained with such approximations.

3. The Equation Free Method

A possible alternative is to use equation-free method [5] to develop a numerical closure procedure, this same method may be also used to derive bifurcation diagrams [7]. The basic idea of this approach is to assume that the system may be described by a set of macroscopic quantities, but that the equations of motion for such quantities are not available, for instance they might depend on unknown parameters or, as in this case the set might need a closure procedure. We proceed by building two maps which relate the set of macroscopic variables with the microscopic description and vice–versa (named lifting and restriction respectively). By the use of ad–hoc short microscopic simulations it is possible to obtain a numerical propagator for the macroscopic variables even if no analytic expression is available. Due to the fact that the lifting map (from the macroscopic description to microscopic one) is a one–to–many map we may use a sequence of lift–microscopic run–restrict steps to study the error made by approximating the microscopic system with the macroscopic variables. For a general review on Equation Free Methods (EFM) see[6]. EFM has already been used to get bifurcation results on chemical systems with lateral interactions [7].

A crucial issue in the equation-free approach is the selection of the appropriate macroscopic state variables, and the corresponding lifting operator. Lifting induces a numerical closure, in fact it assumes that all the information lost at the lifting stage, i.e. all those which are not explicitly included in the macroscopic variable set, are depending on it. In other related works [7] this is ensured by assuming that the fast reactions quickly equilibrate the system after any reaction fires, slaving all the spatial correlations to the species coverages. However in our system the lack of diffusion nullifies the slaving and consequently more involved liftings are needed.

We introduce two operators that make the transition between microscopic and macroscopic state variables. We define a *lifting operator*,

$$\mathcal{L} : \mathbf{M} \mapsto \sigma, \tag{6}$$

which maps a macroscopic state $\mathbf{M}$ to a realization of a chain of length N , and the associated *restriction operator*,

$$\mathcal{R} : \sigma \mapsto \mathbf{M} \tag{7}$$

which maps a microscopic realization to the corresponding macroscopic state.

In this work instead of analyzing possible options for $\mathbf{M}$ we focus on the lifting map itself $\mathcal{L}$ while keeping $\mathbf{M}$ fixed, in the next section we show that for the system hereby considered it is enough to use the concentrations as macroscopic variables, so that $\mathbf{M} = a$.

The restriction operator is readily defined using an empirical mean for the coverages:

$$\mathcal{R}(\sigma) = \frac{1}{2}\left(N + \sum_{i=1}^{N} \sigma_i\right). \tag{8}$$

In the lifting step, we need to generate a microscopic realization of the chain, consistent with the macroscopic state $\mathbf{M}(t^*)$ obtained at time t^*.

Let us now detail the complete algorithm. Given an initial condition for the macroscopic state variables $\mathbf{M}(t^*)$ at time t^*, one time step of the coarse time-stepper consists of a three-step procedure:

(i) *Lifting*, *i.e.* the creation of initial conditions

$$\sigma(t^*) = \mathcal{L}(\mathbf{M}(t^*)),$$

for the microscopic model, consistently with the macroscopic state $\mathbf{M}$ at t^*.

Journal of Physics: Conference Series **402** (2012) 012009 doi:10.1088/1742-6596/402/1/012009

(ii) *Simulation*the microscopic state is evolved over a one step time interval $[t^*, t^* + 1]$. Given the microscopic propagator s defined in the previous section we set:

$$\sigma(t^* + 1) = s(\sigma(t^*)).$$

(iii) *Restriction*, *i.e.* the observation (estimation) of the macroscopic state at $t^* + 1$:

$$\mathbf{M}(t^* + 1) = \mathcal{R}(\sigma(t^* + 1)).$$

4. Numerical Results

Given (3) via numerical Monte Carlo simulations it is possible to show that for a 1-D ring starting from a uniform X coverage, $\sigma = \{-1, -1, \ldots, -1\}$, the system reaches a steady state with $a_{\text{eq}} \simeq 0.28$. If instead we look at the solutions of the mean field equations (4) we have a steady state with equal concentrations and $a_{\text{eq}} \simeq= x_{\text{eq}} = 0.5$. The choice of starting from a uniform X coverage is by no mean restricting, in fact if there are some A, it is simply necessary to keep track of any cluster larger than a single A as these clusters will not be affected by the dynamics. If we start from a uniform X coverage then any A appearing in the ring is always surrounded by two X, so any cluster longer than a isolated A is physically impossible.

Therefore we might introduce two different liftings using only the coverage of A as macroscopic quantity. The simplest lifting we develop uses only the coverage $\mathbf{M} = a$ as macro variable and it returns a ring σ with as many A as $N \times a$; the position of these A is completely random, we label this method $\mathbf{lift}_R$. For the second lifting we consider the fact that there are no clustered A, so we proceed as the random lifting, but the position of the A now is picked such that only physically allowed states may be created: specifically we create an initial ring of length $2N \times a$: $\{A, X, A, X \ldots A, X\}$, we then insert randomly as many X are needed,

$$r_X = N(1 - 2a). \tag{9}$$

We call this other lifting $\mathbf{lift}_A$.

From the numerical results we observe that lift_R has $a_{\text{eq}} = 0.5$ as steady state, on the other hand it is enough to take out the non physical impossible states as it is done with lift_A to get already a perfect agreement between the microscopic and the macroscopic dynamics. We underline that there is no difference for these two liftings at the level of macroscopic variables, rather lift_A preserves a specific feature of the microscopic dynamics, while lift_R does not. For lift_R we observe that by allowing the formation of any possible states at the lifting stage of the algorithm, we effectively shuffle the system at each step. This shuffling has the practical effect of stirring the system, so the mean field requirements (well stirred condition) are satisfies.

The reason of the results obtained by lift_A may be explained by looking at the probability that a general element σ_j may swap:

$$\mathcal{P}(A \to X), \qquad \mathcal{P}(X \to A). \tag{10}$$

In the mean field limit any lattice site may swap, so $\mathcal{P}(A \to X) = \mathcal{P}(X \to A)$, for the micro dynamics (3) instead we have that at equilibrium $a < x$, this is due to the fact that not all the X are allowed to swap to A while all the A are free to do so.

For instance, if we assume that on the ring all the A are "isolated", i.e. the distance between two A is never smaller than two sites, we see that for any A there are two neighboring X which if chosen by the random pick will not react, we define such X frozen. The number of frozen X determines the concentrations value at equilibrium. For example if the sole frozen X are those aside A then we have a $1/3$ ratio between A and X; looking at the probabilities we see that:

$$\mathcal{P}(A \to X) \propto \langle A \rangle,$$
$$\mathcal{P}(X \to A) \propto 1 - 3\langle A \rangle.$$

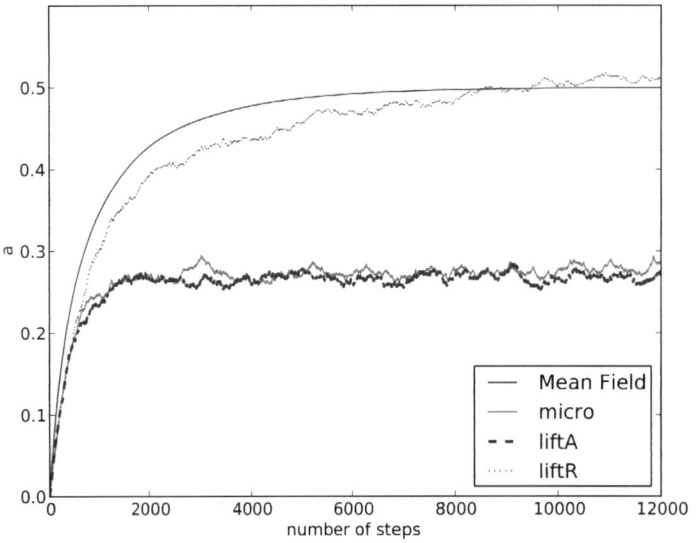

Figure 1. Comparison of micro dynamics with lift_r and lift_A.

In the general case however, the A will not be "isolated", so situations like $\{\ldots A, X, A, X, A\ldots\}$ might happen. In this case the ratio between A and frozen X is not $1/3$ as before but depends on the length of the chain and it is the reason for having $a_{\text{eq}} = 0.28$ which is little less than $1/3$.

In our numerical tests we observed that on average lift_A produced a ring with less frozen X than the corresponding microscopic state coming from the microscopic simulations, while lift_R produced one with more. It is clear that while the errors in lift_A are immediately corrected by the dynamics, those made in lift_R reduce the possible reaction paths and the dynamics is drifted toward the mean field equilibrium value.

5. Conclusions

In this work we have proposed an alternative strategy to obtain a numerical closure for chemical reactions on low dimensional lattices. These systems cannot be described via the mean field approximation, and other approaches are needed using also higher order particle correlations. When no analytic form for the macroscopic variables is available, or when the number of equations become too large, it is possible to use the Equation free method framework to obtain the numerical closure via the inclusion of higher order moments (pair and triplet correlations) in the macroscopic variables set.

Within the Equation Free framework we have shown that in some cases there is no need to introduce other variables if the lifting strategy retains some particular features of the microscopic description, in this way the approach is simpler and easier to apply. By carefully choosing the lifting we show that it is possible to get an equivalent macroscopic dynamics which resembles the data coming from the microscopic run, in this sense the lifting makes the numerical closure.

In more general cases it is possible to decrease the number of macroscopic variables needed by choosing carefully the lifting strategy. The general idea behind this approach trying to preserve the number of "elementary blocks" of the reactions, isolated A and triplets of X for the model presented here. Currently we are working on a generalization of the procedure presented in this report that allows the treatment of more complex systems. We are also implementing a projective integration algorithm [6] for the coarse time stepper in order to accelerate the dynamics.

References

[1] Abad, E., Grosfils, P., and Nicolis, G. Nonlinear reactive systems on a lattice viewed as boolean dynamical systems. *Phys. Rev. E 63* (Mar 2001), 041102.

[2] Albano, E. The monte carlo simulation method: a powerful tool for the study of reaction processes. *Heterogeneous Chemistry Reviews 3* (1996).

[3] Chatterjee, A., and Vlachos, D. An overview of spatial microscopic and accelerated kinetic monte carlo methods. *Journal of Computer-Aided Materials Design 14* (2007), 253–308. 10.1007/s10820-006-9042-9.

[4] Glauber, R. J. Time-dependent statistics of the ising model. *Journal of Mathematical Physics 4*, 2 (1963), 294–307.

[5] Kevrekidis, I. G., Gear, W. C., Hyman, J. M., Kevrekidis, P. G., Runborg, O., and Theodoropoulos, C. Equation-Free Multiscale Computation: enabling microscopic simulators to perform system-level tasks. *Commun. Math. Sci. 1*, 4 (Sept. 2003), 715–762.

[6] Kevrekidis, I. G., and Samaey, G. Equation-free multiscale computation: Algorithms and applications. *Annual Review of Physical Chemistry 60*, 1 (2009), 321–344. PMID: 19335220.

[7] Makeev, A., Maroudas, D., Panagiotopoulos, A. Z., and Kevrekidis, I. G. Coarse bifurcation analysis of kinetic monte carlo simulations: A lattice-gas model with lateral interactions. *J. Chem. Phys. 117*, 18 (2002), 8229–8240.

[8] Prakash, S., and Nicolis, G. Dynamics of fluctuations in a reactive system of low spatial dimension. *Journal of Statistical Physics 82* (1996), 297–322. 10.1007/BF02189232.

[9] Provata, A., Turner, J. W., and Nicolis, G. Nonlinear chemical dynamics in low dimensions: An exactly soluble model. *Journal of Statistical Physics 70* (1993), 1195–1213. 10.1007/BF01049428.

[10] Shabunin, A. V., Baras, F., and Provata, A. Oscillatory reactive dynamics on surfaces: A lattice limit cycle model. *Phys. Rev. E 66*, 3 (Sep 2002), 036219.

[11] T, D., and Gillespie. A general method for numerically simulating the stochastic time evolution of coupled chemical reactions. *Journal of Computational Physics 22*, 4 (1976), 403 – 434.

[12] Velikanov, M. V., and Kapral, R. Perturbation theory for the breakdown of mean-field kinetics in oscillatory reaction-diffusion systems. *The Journal of Chemical Physics 109*, 1 (1998), 281–293.

[13] Vlachos, D. G. Temporal coarse-graining of microscopic-lattice kinetic monte carlo simulations via τ leaping. *Phys. Rev. E 78* (Oct 2008), 046713.

The use of atomic level stress to characterize the structure of irradiated iron

Madhusudan Ojha[1], D.M. Nicholson[2], Kh. Odbadrakh[2], Bala. Radhakrishnan[1], R. E. Stoller[2], and Takeshi Egami[12]

[1]University of Tennessee, Knoxville, TN 37996; [2]Oak Ridge National Laboratory, Oak Ridge, TN 37831

E-mail: nicholsondm@ornl.gov

Abstract. The behaviour of irradiated material near a primary knock on atom immediately after impact is of great importance for designing reactor materials. Currently, molecular dynamics simulations with classical force fields provide the foundation for understanding the resulting cascade. However, modern density functional calculations can now treat large enough numbers of atoms that they can provide additional details of the magnetic and electronic nature of irradiated samples. In this paper we calculate from first principles the atomic level stresses for an instantaneous configuration following the initiation of a low energy cascade in iron.

1. Introduction

Radiation damage occurs when a highly energetic particle incident on a solid transfers its energy to the atoms in the solid displacing atoms from their original crystal lattice positions through absorption of energy in excess of the displacement threshold energy. This displacement of atoms creates vacancy (V) and interstitial (I) defects in pairs, Frenkel pairs. The initially displaced atom is called the primary knock-on atom (PKA); it interacts with other atoms in the crystal displacing them from their lattice sites and thereby generating a displacement cascade. Thus, irradiating a solid with ions or neutrons creates a cascade of Frankel pairs. This process continues until the crystal absorbs all the energy of the incident particle. In the case of iron, the cascade continues to generate defects until no single atom has energy in excess of the displacement threshold energy (40 eV in Fe) to cause further displacements. The remaining kinetic energy is eventually taken away by lattice phonons.

A complex sequence of events is initiated by the irradiation, which produces mainly undesirable changes in the material's properties. The radiation damage caused by fast neutrons to solids is due to the direct transfer of kinetic energy in the range of several tens of keV to the lattice atoms. This transfer happens on a length scale of nanometers and in the time scale of sub-picoseconds (ps) to a few ps. Figure 1 shows the evolution of Frenkel pairs in displacement cascades in alpha iron at 100K for different PKA energies [1].

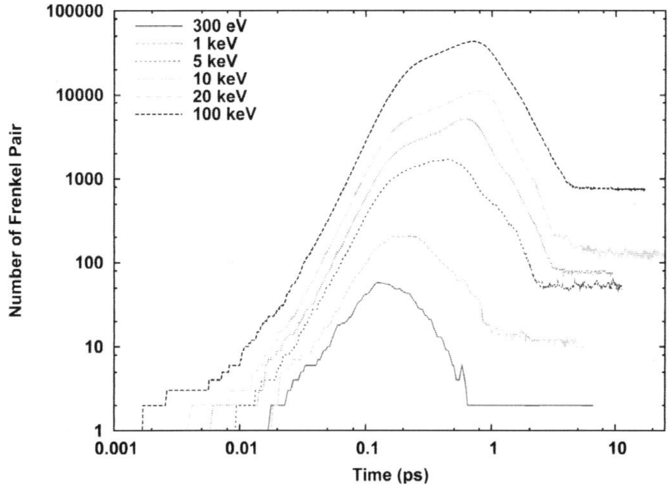

Figure 1. The calculated (Modified Finnis and Sinclair [2] molecular dynamics (MD)) number of Frenkel pairs is shown as a function of time after cascade initiation for a series of PKA energies

A detailed and basic understanding of the initial damage event itself is very important to understand all the later events and the changes caused by radiation damage in microstructure and mechanical properties.

2. Local energy and pressure

A general first principles approach to the problem of defining the local stress [3] is to start by defining a first principles local energy associated with each atom in the system. The local stress can then be determined from the response of the local energy to the application of small affine strains to the system. The local stresses and elastic moduli can be determined by the first and second derivatives of the local energy with respect to strain. In this paper we make a single, simple choice for the definition of local energy and observe the correlation of several local quantities with the local pressure. The local stress has been used for systems that interact through classical potentials to explain glass stability [4] and as the basis for an equipartition theorem [5] for local elastic energy in MD simulations. For classical pair potential systems the local stress has been derived on the basis of pairwise forces without resorting to the definition of a local energy [6]. This result has been extended to the Embedded Atom Method [7], which has force fields that include many-body interactions. For both these types of force fields, expressions can be obtained by first defining a local energy and then evaluating the variation of the local energy with respect to strain, for example an atom can be assigned a local energy given by half the energy of bonding with each atom within the range of the potential.

In the ideal gas limit there are no forces and the only stress is the pressure that results from collision with the walls of the container. In another extreme case, a zero temperature solid, nuclei are stationary and forces are present but sum to zero on each atom. Even though the forces on each atom sum to zero they can contribute to large local stresses, for example: in intermetallic compounds, defects in single component materials, or even in single component perfect crystals if the volume differs from the equilibrium value. Here we will ignore the pressure due to the motion of the nuclei and will evaluate the contribution from the electron bonding. This bonding includes electron but not nuclear kinetic energy.

The energy of the electron-nuclear system is the foundation of Density Functional Theory (DFT) based calculations [8]. The energy is often separated into the kinetic energy, T_S, of an auxiliary system with the same electron density as that of the interacting ground state, the classical electrostatic energy of all charges in the system, U_c, and correction energy, E_{xc}, that accounts for the correlation (including that which results from exchange) between electrons. The correlation affects both kinetic and electrostatic energy. The challenge is to further decompose the energy into local contributions that have the greatest descriptive power.

In the Local Density Approximation (LDA) [9], E_{xc}, is defined locally; this makes the association with an atomic site apparently straightforward. Although we will take local E_{xc} to be given by the integral over the local atomic volume of the product of the electron density and its energy density, we acknowledge that other assignments are possible especially when approximations beyond LDA are considered.

The kinetic energy can be evaluated as the negative of the expectation value of the Laplacian. A natural choice for the local contribution is that obtained by limiting the expectation value integral to the volume of the atomic site. An alternative choice is based on the symmetric form of the kinetic energy, which is the integral of the modulus squared of the gradient of the wavefunction. The two definitions differ by a surface integral that vanishes over the boundary of the supercell or at a boundary at infinity if the system is finite. We argue in favor of the asymmetric form because: 1) it has the form of an expectation value, 2) it conforms to our physical notions, that the kinetic energy contribution is highest at the potential minimum (as opposed to zero for the symmetric form) for a harmonic oscillator in its ground state, and 3) it is more closely associated with the sum of eigenvalues, a quantity that has proved very useful in qualitative descriptions of bonding.

Because the classical electrostatic force is pairwise, the underlying nature of the classical electrostatic energy is that of a pair interaction. It can be recast as a local quantity by considering half the bonding energy of the local electron density with all the system charge (because it is classical it includes the self-interaction of electrons) and half the bonding of the local nucleus with all other charge. The classical electrostatic interactions are hence treated in exactly the same manner as in the classical formula for the stress in a classical pair-potential system. The treatment of the classical electron-electron interaction differs only in that the electron is treated as a continuous distribution; this leads to the inclusion of self-interaction, which is absent in the case of force fields, being excluded by the "i≠j" exclusion on energy sums.

$$E_i^\lambda = \int_{V_i} d\mathbf{r}\left[-\sum_{\varepsilon_i \le \varepsilon_F} \psi_k^{\dagger\lambda}\nabla^2\psi_k^\lambda + n^\lambda\left\{\frac{1}{2}e^2\phi_e\left[n^\lambda\right] - \sum_j \frac{e^2 Z_j}{\left|\mathbf{r} - \lambda\mathbf{R}_j\right|} + \varepsilon_{xc}\left(n^\lambda\right)\right\}\right] + \frac{1}{2}\sum_{j\neq i}^{N}\frac{e^2 Z_i Z_j}{\left|\lambda\mathbf{R}_i - \lambda\mathbf{R}_j\right|}$$

$$n^\lambda = \lambda^{-3} n^\lambda\left(\mathbf{r}/\lambda\right)$$

$$\psi^\lambda = \lambda^{-3/2}\psi^\lambda\left(\mathbf{r}/\lambda\right)$$

$$P_i = -\left.\frac{dE_i^\lambda}{d\lambda}\right|_{\lambda=1}\left(\frac{dV_i}{d\lambda}\right)^{-1};\ \frac{dV_i}{d\lambda} = 3V_i$$

3. Procedure

Calculations were performed with the Locally Self-Consistent Multiple Scattering Method (LSMS) [10] on 9826 cores of Jaguar-PF, a high performance computer of the Oak Ridge Leadership Computing Facility. The electron multiple scattering includes atomic scattering up to *l=3* for atoms within 7.5 a.u. and to *l=2* out to 11.3 a.u. The local Kohn-Sham potential was treated in the Atomic Sphere Approximation [11]. We applied the LSMS to calculate the atomic level pressure of a 9826-

atom sample of alpha Fe after a PKA with a velocity in the [113] direction that corresponds to an impulse delivered by a 1 keV neutron. The sample was initially in equilibrium at a temperature of 100K, periodic boundary conditions were in effect, the pressure was maintained at zero, and the atoms moved according to embedded atom forces integrated with variable time steps to account for initially high atomic velocities [12]. Specifically, the MOLDY-MD code [13] was used with the force fields developed by Finnis and Sinclair [14] as modified by Calder and Bacon [2]. The LSMS calculations correspond to a time 3.6ps after the initial impact.

4. Distribution of volume and pressure

The variation of the local Voronoi polyhedron (VP) volumes is depicted in figure 2; sphere sizes indicate the difference in volume from the average; color also indicates the difference from the average, red is expanded and blue is contracted. Therefore, the sites with volume very close to that of the bulk appear as small white dots. At a particular instant of time the VP-volumes vary from site to site due to the thermal motion of the atoms; there are also significant changes in VP-volume at defects. Furthermore, the impulse from the cascade initiates a pressure wave that reflects from the periodic boundary conditions. Multiple reflections from the boundaries of our small cell result in modulation of the strain field over a length scale of many lattice spacings that results in the extended red and blue regions in the figure. In figure 3 the local pressure at each site is shown. The representation is identical to that of the VP-volumes; sites with pressure near zero appear as small white dots. Comparing figures 2 and 3 a correlation between volume and pressure can be seen.

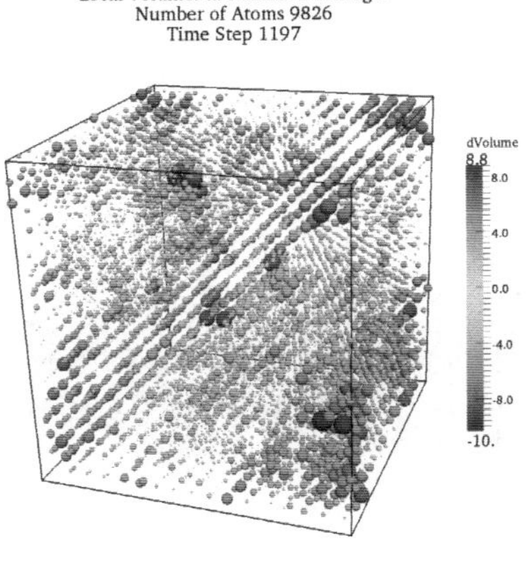

Figure 2. Deviation of VP-Volume from average value (79.5 a.u.).

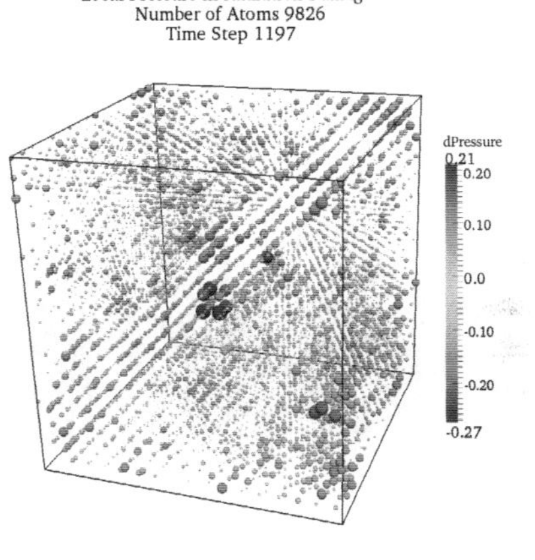

Figure 3. Deviation of local pressure from average value (in a.u.).

In figure 4 the VP-volumes appear correlated with the pressure in natural way, compressive (positive) pressure is associated with lower volume. Although the volumes at a particular pressure have considerable scatter, the maximum volume appears to be clearly demarked by a line, a line of critical volume as a function of pressure. For a given pressure the population of VP-volumes lies overwhelmingly below this line. Its slope defines a pseudo bulk modulus, B_c=-VdP/dV, of 134 GPa. By construction, the average derivative of pressure with respect to volume (multiplied by the negative of the volume) of all local pressures is the actual bulk modulus, B. For comparison the volume versus pressure curve is shown for a perfect crystal of alpha iron at zero temperature calculated to the same level of approximation as the 9826-atom damage-sample. The perfect crystal line passes through the center of the distribution of points, showing that the average distribution of volumes as a function of

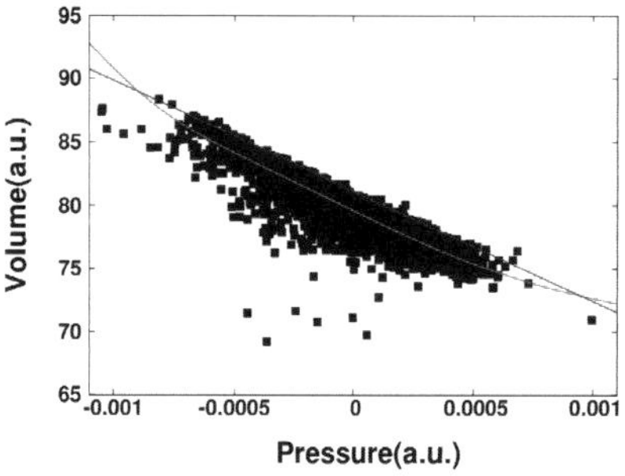

Figure 4. Distribution of VP-volumes as a function of local pressure. The straight-blue line indicates an upper or critical limit on volume as a function of pressure in the damaged-system. The curved-red line is the volume versus pressure curve for perfect crystalline alpha Fe.

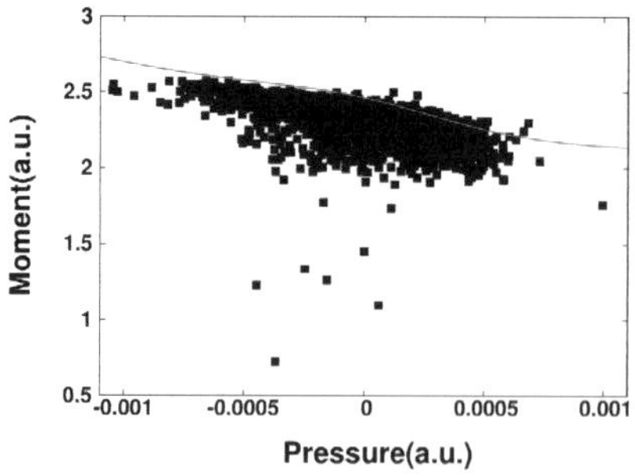

Figure 5. Distribution of local moments as a function of local pressure. The red line indicates the moment versus pressure curve for perfect crystalline alpha Fe.

pressure in the damage-sample is reproduced approximately by volume versus pressure curve of the perfect crystalline-sample for most of the pressure range. The two line cross at low and high volumes. The high volumes on the critical line in the damage-sample have a smaller "compressibility" than the perfect crystal while those at low volume have a greater compressibility. It will be interesting, in further studies, to determine whether or not, under increased hydrostatic pressure, the sample as a whole responds mainly through a reduction of volume of sites at the critical volume.

Magnetic moments also decrease with pressure. This is a correlation that is affected through the volume; the bandwidth increases as the distance to neighbors is reduced, resulting in a smaller exchange splitting. Figure 5 shows the local moments as a function of pressure; for comparison the atomic moment in perfect bulk Fe calculated in exactly the same way is also shown. The bulk moment provides an upper bound to the moment at each pressure. The same relationship is found in plotting the moment versus volume (figure 6). The size of the local moment can be explained in part by the change in volume but other aspects of the local atomic environment are clearly important. We have not attempted to analyze the local moment in terms of local geometry other than the local volume. A clear next step is to correlate the moment with the local value of the overlapped "electron-density" evaluated with an Embedded Atom Method as was proposed by Dudarev [15].

5. Conclusions

We have briefly described first principles local stress and computed the local pressure for a sizable (9826 atom) model of a single time step in the evolution of a radiation cascade. We find the local pressure to be correlated with other local quantities in a way that is reasonable and aids in our understanding of connection between different local environments and local properties. We propose that the sensitivity of the local stress to the local atomic environment will lead to its expanded use in understanding materials behavior at the atom level. In this paper we have made comparisons only to the local pressure in perfect bulk iron. Planned further comparisons include, for example, to other time steps within the cascade and to the same size system but with thermal displacements only. Further comparisons with samples at other temperatures, ambient pressure, and with impurities will clarify the meaning of the local stresses and add to its utility as a descriptive tool.

6. Acknowledgements

This research was supported by the Division of Materials Science and Engineering, Office of Basic Energy Sciences, Department of Energy, and as part of the Center for Defect Physics in Structural Materials (CDP), an Energy Frontier Research Center funded by the U.S. Department of Energy, Office of Science, Office of Basic Energy Sciences. This research used resources of the Oak Ridge Leadership Computing Facility at Oak Ridge National Laboratory.

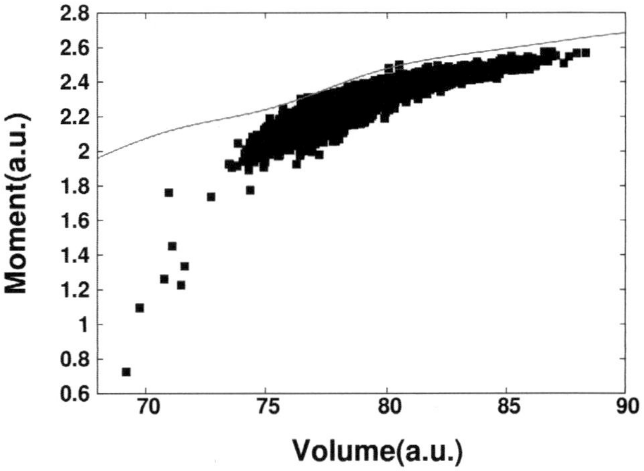

Figure 6. Local moments as a function of VP-volume; red line is the moment per atom in perfect crystal.

References

[1] Stoller R E, 2012 "Primary Radiation Damage Formation," in Comprehensive Nuclear Materials, Rudy Konings, Todd R. Allen, Roger E. Stoller, Shinsuke Yamanaka, Editors, Elsevier Ltd., Amsterdam

[2] Calder A F and Bacon D J 1993 *J. Nucl. Mater.* **207**, 25

[3] Nielson O H and Martin R M 1983 *Physical Review Letters* **50**, 697; Filippetti A and Fiorentini V 2000 *Physical review* B **61**, 8433

[4] Egami T, Poon S J, Zhang Z and Keppens V 2007 *Physical review* B **76**, 024203

[5] Levashov V A, Egami T, Aga R S, et al. 2008 *Physical review* B **78**, 064205

[6] Egami T, Maeda K and Vitek V 1980 *Phil Mag* A**41**, 883

[7] Baskes M I 1999 *Phys. Rev. Lett.* **83**, 2592

[8] Hohenberg P and Kohn W 1964 *Physical review* B **136** 864

[9] Kohn W and L J Sham 1965 *Physical Review* **140** 1133

[10] Wang Y, Stocks G M, Shelton W A, Nicholson D M C, Szotek Z, Temmerman W M 1995 *Physical Review Letters* **75** 2867 and Nicholson D M C, Stocks G M, Wang Y, Shelton W A, Szotek Z, Temmerman W M 1994 *Physical review* B 50 14686

[11] Zhang X G and Nicholson D M C 1999 *Physical review* B **60** 4551-4557

[12] Stoller R E and Calder A F 2000 *J. Nucl. Mater.* **283-287** 746

[13] Finnis M W 1988 *AERE R-13182*, UK AEA Harwell Laboratory

[14] Finnis M W and Sinclair J E 1984 Phil. Mag. A**50**, 45 and Finnis W M and Sinclair J E 1986 *Phil. Mag.* A**53** 161

[15] Dudarev S L and Derlet P M 2005 *J. Phys.: Condens. Matter* **17** 7097

Journal of Physics: Conference Series **402** (2012) 012011 doi:10.1088/1742-6596/402/1/012011

Coarse Grained Approach to First Principles Modeling of Radiation Cascade in Large Fe Supercells

Kh. Odbadrakh[1], D. M. Nicholson[2], A. Rusanu[1], G. D. Samolyuk[1], R. E. Stoller[1], X.-G. Zhang[1], G. M. Stocks[1]

[1] Materials Science and Technology Division, Oak Ridge National Laboratory, Oak Ridge, Tennessee 37830

[2] Computational Science and Mathematics Division, Oak Ridge National Laboratory, Oak Ridge, Tennessee 37830

E-mail: `odbadrakhk@ornl.gov`

Abstract. Classical Molecular Dynamics (MD) simulations characterizing dislocations and radiation damage typically treat 10^5-10^7 atoms. First principles techniques employed to understand systems at an atomistic level are not practical for such large systems consisting of millions of atoms. We present an efficient coarse grained (CG) approach to calculate local electronic and magnetic properties of large MD-generated structures from the first principles.

Local atomic magnetic moments in crystalline Fe are perturbed by the presence of radiation generated vacancies and interstitials. The effects are most pronounced near the defect cores and decay slowly as the strain field of the defects decrease with distance. We develop the CG technique based on the Locally Self-consistent Multiple Scattering (LSMS) method that exploits the near-sightedness of the electron Green function. The atomic positions were determined by MD with an embedded atom force field.

The local moments in the neighborhood of the defect cores are calculated with first-principles based on full local structure information. Atoms in the rest of the system are modeled by representative atoms with approximated properties. The calculations result in local moments near the defect centers with first-principles accuracy, while capturing coarse-grained details of local moments at greater length scales. This CG approach makes these large scale structures amenable to first principles study.

Introduction

Radiation induced damage has substantial impact on the longevity of structural materials used in fission and fusion reactors. The primary radiation damage such as displacement defects initiate a build up of defect aggregates at the micro scale, and eventually lead to the altering of the material's mechanical properties [1, 2]. Fundamental properties of point defects and defect accumulations in bcc-Fe under irradiations have been investigated experimentally [3, 4, 5, 6]. The radiation induced damage in structural metals, and its evolution dynamics are highly complex processes and extremely difficult to investigate in-situ. Therefore, efficient numerical modeling at multiple scales represents a valuable tool in understanding the underlying physical processes associated with defects.

Classical Molecular Dynamics simulations have been the most prolific tool for investigating time evolution of defects within nanosecond range [7, 8, 9]. Recent developments in classical MD simulations of defects utilize classical force fields, e.g. embedded atom method (EAM) [10], and incorporate interatomic potentials computed from first-principles calculations. A magnetic EAM force fields that include the local volume effect on local magnetic moments has been developed by Dudarev et al [11]. Using Kinetic Monte Carlo based technique is an approach to investigate long term temporal and spatial evolution of defects [12, 13].

In recent years, there have been extensive efforts underway in coupling the classical and quantum simulation techniques along with continuum mechanical simulations [16, 14, 15]. The so called multi-scale Coarse-Grained (CG) technique used mostly for biological molecules represents one or more atoms by one CG site and computes interactions between the CG regions from atomistic configurations [16, 17, 18, 19, 20].

These approaches focus on the evolution dynamics of defects or complex molecules. The CG approach presented here looks at radiation damage from a slightly different point of view: that is local electronic and magnetic properties at time snapshots during a classical MD simulation. As it has been shown in MD simulations [21], radiation in the bulk iron produces only a few highly localized defect cascades, and transfers most of its energy as thermal displacements of atoms in the bulk. The thermal displacements change the local electronic and magnetic states, but can still be reasonably averaged in a manner that reflects the local strain.

The CG approach presented here exploits this contrast between the defect cores and rest of the system by treating the the defect core atoms explicitly while representing the rest of the system with CG average parameters. In addition, we employ the LSMS method that exploits the near-sightedness of the electronic Green function. This technique enables us to investigate local properties of radiation damage samples from first principles using very large MD simulation data and is applicable to two broad categories of simulation: relaxation of atomic positions to study the structure of static dislocation cores and molecular dynamics (MD) simulations of the defects.

Computational Details

LSMS

The core of our computational tool is the Locally Self-consistent Multiple Scattering method (LSMS) [22], an all-electron, real space multiple scattering theory based approach within the framework of density functional theory (DFT) with local spin density approximation (LSDA)[23].

Being implicitly parallel, the key feature of the LSMS is to associate a processor with every atom and vacant site. Each processor solves the multiple scattering equations for the Green function of a cluster (local interaction zone), of which it is the center and calculates the electron density (as a function of the maximum occupied energy) in the Voronoi polyhedron directly from the Green functions. The union of the site densities computed up to the Fermi energy gives the electron density of the system, and is set to satisfy global charge neutrality constraint. The potential is reconstructed by solving Poisson's equation and adding the exchange-correlation potential in the usual manner of LSDA and the self-consistent cycle proceeds to convergence in the familiar fashion. The long range electrostatic contribution to the potential is evaluated by the Madelung procedure assuming periodic boundary conditions.

Coarse Grained Approach

There are several issues to be resolved in constructing a local portion of a large MD configuration in first principles calculations. The CG approach takes all of the structural information as input, this includes explicit coordinates and boundary conditions that imply, for example, periodic repetition in some directions and termination in others.

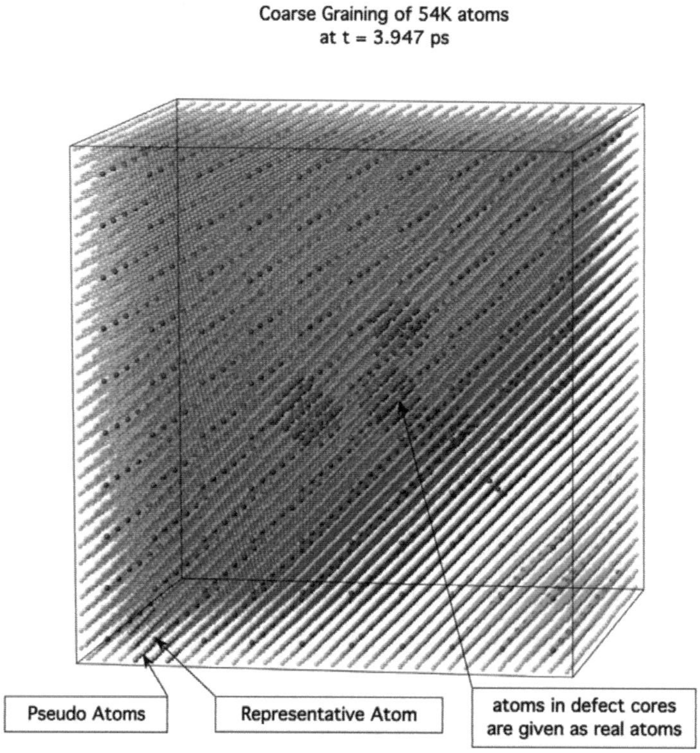

Figure 1. The CG scheme used in iron radiation damage sample. The system is represented by fewer selected atoms (magenta). The defect core atoms are given explicitly (cluster in magenta), and the rest of the system is represented by atoms selected on the nodes of simple mesh. As an example, a set of pseudo-atoms represented by one atom is shown gold colored.

The coordinates of atoms in the desired first principles region are given explicitly. In addition, we choose representative atoms spread throughout the simulation volume by some prescription and give their coordinates explicitly. The multiple scattering equations are solved self consistently on these atoms only. The remaining atoms are assigned charges (higher multipoles in a full potential calculation) and scattering t-matrices from the nearest representative atom. The t-matrices of the periodic image sites are dealt with as in the original LSMS. Given these associations the LSMS calculates self-consistently the density and potential appropriate to the local strain in the regions surrounding each representative atom and transfers this information to its associated pseudo-atoms. This scheme converts most of the original MD structure into spatially coarse grained structure, while leaving the region under investigation accessible from first principles. The schematics of this CG approach is shown in Fig.1.

Results

To validate the CG approach, we first investigated a 9826 atom radiation damaged bulk iron sample using both full atom first principles calculation and the CG calculation using just 629 representative atoms. The primary damage state of the sample and following cascade states

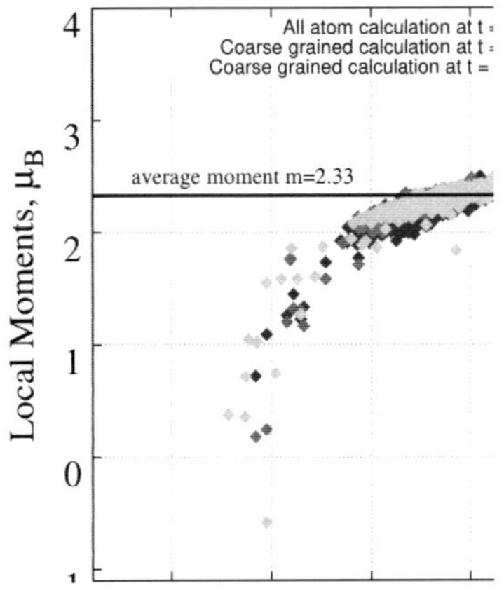

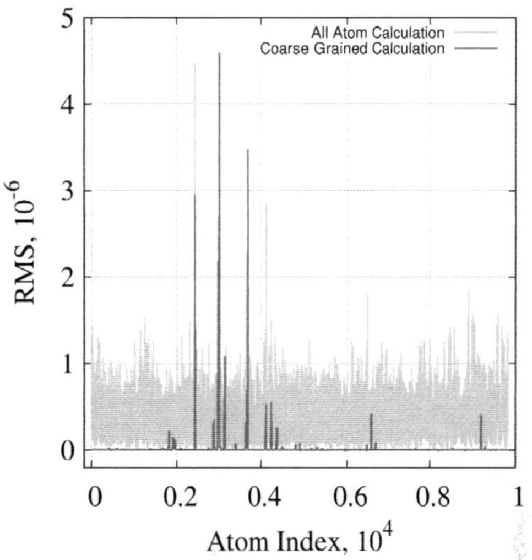

Figure 2. Distribution of local moments vs. local volumes for the 9826 atom radiation damaged bulk iron sample (left). At time step 1197 (t = 3.6 ps), site moments calculated by the CG technique (red) matches the full atom calculation (blue) closely. Earlier time step (1170, t = 3.48 ps) exhibits greater perturbation of local moments (green), resulting in locally anti-ferromagnetic coupled states (negative values). Apart from the few defect atoms, most of the system is perturbed with thermal displacements. The Root Mean Square (RMS) has different convergence behaviors for full atom and CG calculations owing it to vastly different number of representative atoms (left). As a result, the average RMS is achieved at different rates for the two calculations.

were obtained through MD simulations using empirical, embedded-atom inter-atomic potentials [24, 25]. The MD simulation started as a 500 eV radiation event at constant pressure and was terminated when all atomic motions have thermalized at t=3.6 ps.

In the full atom first principles calculation, we treated all 9826 atoms explicitly and converged the LSMS self consistently to obtain moments on each atom. Local volumes for each site have been calculated directly from the structural data. The local volumes change by as much as 12.5% from the average and this change is distributed throughout the simulation volume consistent with a reflected shock wave under periodic boundary conditions. A closer inspection of the structure reveals the compact nature of defects left by the radiation.

In order to investigate the defect regions from the first principles, the atoms at the defect cores have been identified from the structural data. Equipped with this information, atoms in the region encompassing the defects are given explicitly. The rest of the system is modeled by a set of representative atoms that are chosen on the nodes of 8x8x8 mesh in the simulation volume. As a result, the 9826 atom structure at t=3.6 ps is now reduced to 629 atom CG model, in which atoms in the defect region are given explicitly.

The CG calculation captures the spatial distribution of local volumes in the full atom calculation as expected. Atom by atom comparison between the calculations shows that the local moments near the defect cores closely match. For the rest of the system, the CG model captures the thermal fluctuation range of the full atom case (Fig.2, left). At t=3.6 ps, site moments calculated by the CG technique (red) matches the full atom calculation (blue) closely. Despite treating the defect regions explicitly, moments at some of the atoms in these region (red)

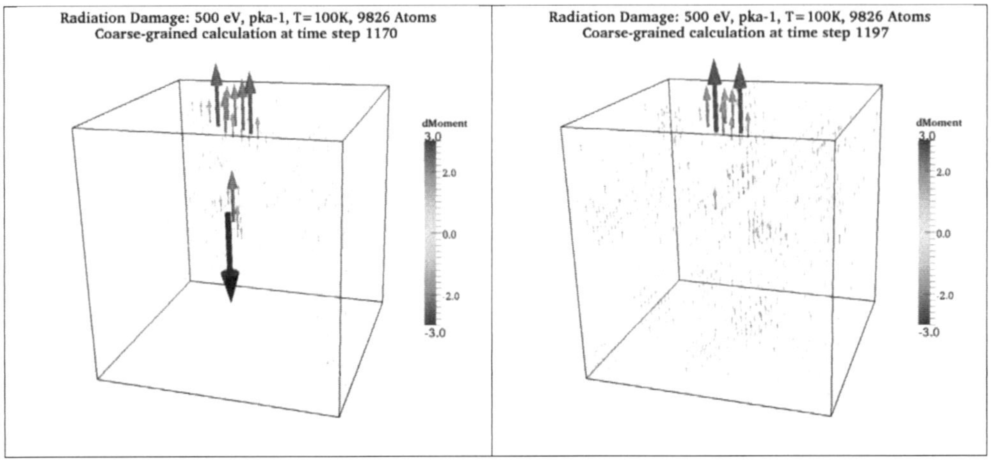

Figure 3. Local moments in the 9826 atom radiation damaged bulk iron sample at different time snapshots. The color and size of the arrows shown here represent deviation from an average value of moments, and their directions indicate local magnetization directions in the system. Blue colors correspond to decreased local moments, and increased moments are shown in the red of the color scale. Earlier time step at 1170 (t=3.48 ps, left) contains more point defects with greater magnetic disturbance than time step 1197 (t=3.6 ps, right). In particular, there is one anti-ferromagnetic coupling formed in the defect region (the largest blue arrow pointing down).

differ from that of full atom calculation (blue). Further analysis revealed that this difference is caused by different convergence behavior of the two calculations. The convergence value set for the Root Mean Square (RMS), (*Definition: square root of average over atomic volume of the square of the difference between input and output electron densities*) is achieved at different rates since the two calculations use vastly different number of atoms (Fig.2, right).

The evolution of radiation cascades can be investigated by comparing snapshots of the system at different time steps. CG model of the system at t=3.48 ps is constructed using the same procedure described above. Being earlier stage in the cascade, the system now contains more atoms in the defect regions, resulting in 657 atoms for the CG model. The calculation reveals more pronounced perturbations in the system than the t=3.6 ps (Fig 3).

After validation on a 9826 atom system, we extended the CG calculation to a larger system consisting of 54K atoms. The classical MD simulations of 1 keV radiation damage has been performed using the same techniques as before. Treating a system of this size from first principles requires significant computational resources and poses challenges associated with numerical convergence. We constructed a CG model of this system using 1241 representative atoms only. Classical MD simulations show that the number of atoms involved in point defects has no direct correlation with the system size and depends only on the energy introduced during the irradiation. Since the number of atoms in the defect regions is comparable to that of the 9826 atom system, increase in the number of representative atoms is due mostly to the number of mesh points that represent the larger simulation volume. Changes in local volumes due to the irradiation is available from the structural data and are closely captured in the CG model (Fig.4, left). The damage is concentrated in a very compact region near the center of the simulation volume, and is mostly compressed local volumes (blue). The rest of the system exhibits atomic

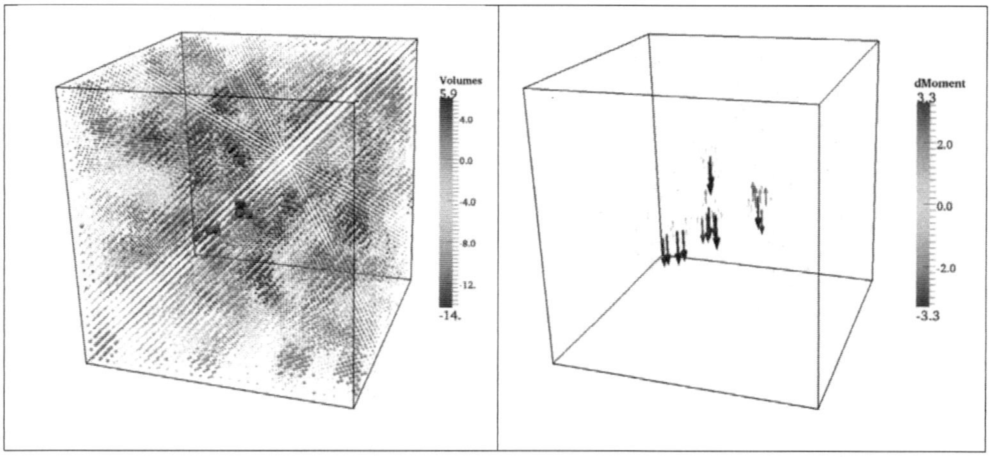

Figure 4. Local volumes (left) and moments (right) for 54K atom radiation damaged bulk iron sample (1 keV, pka-1, T=100 K, t=3.947 ps). The color and size of the arrows shown in the right represent deviation from an average value of moments, and their directions indicate local magnetization directions in the system. Blue colors correspond to decreased local moments, and increased moments are shown in the red of the color scale. The blue arrows pointing down correspond to locally anti-ferromagnetic couplings formed in the defect cores. The calculation is performed atomistically in the defect regions while treating rest of the system using the CG technique.

displacements within thermal fluctuations. These features are well reflected in the local moments obtained by LSMS first-principles calculation using the CG model (Fig.4, right).

It should be noted that there is no comparable first-principles data currently available for systems of this size. The radiation induces compact damage aggregates consisting of point like defects, some of which form local environments that favor an anti-ferromagnetic state. The locally anti-ferromagnetic coupling between iron atoms has been observed before, and are signature of an early stage of radiation damage [21].

Summary

We developed a coarse-grained scheme for solving multiple scattering equations by representing a system with fewer sites. This approach allows us to treat part of a system from the first principles, while substituting the rest of the system with CG model. Comparison with the full atom calculation of 9826 atom system shows that the technique calculates magnetic properties of the selected region accurately from the first principles, while capturing CG distribution of the properties for the rest of the system. Calculation of local properties for 54K atom system demonstrates that this CG technique can efficiently handle larger systems. We also observed that the convergence behavior of the self-consistent calculation improves due to the reduced number of sites. This CG technique substantially reduces the computational resources needed.

Acknowledgment

This material is based upon work supported as part of the Center for Defect Physics in Structural Materials (CDP), an Energy Frontier Research Center funded by the U.S. Department of Energy, Office of Science, Office of Basic Energy Sciences. This research used resources of the Oak Ridge

Leadership Computing Facility at Oak Ridge National Laboratory, which is supported by the Office of Science of the Department of Energy under contract DE-AC05-00OR22725.

References

[1] R. Stoller and G. Odette. *Journal of Nuclear Materials*, 154(2-3):286–304, July 1988.
[2] G. Odette and G. Lucas. *Radiation Effects and Defects In Solids*, 144(1-4):189–231, 1998.
[3] B. Eyre and A. Bartlett. *Philosophical Magazine*, 12(116):261–&, 1965.
[4] M. Jenkins, C. English, and B. Eyre. *Philosophical Magazine A-Physics of Condensed Matter Structure Defects and Mechanical Properties*, 38(1):97–114, 1978.
[5] A. Vehanen, P. Hautojarvi, J. Johansson, J. Ylikauppila, and P. Moser. *Physical Review B*, 25(2):762–780, 1982.
[6] S. Takaki, J. Fuss, H. Kugler, U. Dedek, and H. Schultz. *Radiation Effects and Defects In Solids*, 79(1-4):87–122, 1983.
[7] D. Case, T. Cheatham, T. Darden, H. Gohlke, R. Luo, K. Merz, A. Onufriev, C. Simmerling, B. Wang, and R. Woods. *Journal of Computational Chemistry*, 26(16):1668–1688, Dec. 2005.
[8] B. Brooks, R. Bruccoleri, B. Olafson, D. States, S. Swaminathan, and M. Karplus. *Journal of Computational Chemistry*, 4(2):187–217, 1983.
[9] D. Bacon, F. Gao, and Y. Osetsky. *Journal of Nuclear Materials*, 276:1–12, Jan. 2000.
[10] M. Daw and M. Baskes. *Physical Review B*, 29(12):6443–6453, 1984.
[11] S. Dudarev and P. Derlet. *Journal of Physics-Condensed Matter*, 17(44):7097–7118, Nov. 2005.
[12] C. S. Becquart, A. Barbu, J. L. Bocquet, M. J. Caturla, C. Domain, C.-C. Fu, S. I. Golubov, M. Hou, L. Malerba, C. J. Ortiz, A. Souidi, and R. E. Stoller. *Journal of Nuclear Materials*, 406(1):39–54, Nov. 2010.
[13] F. Soisson, C. S. Becquart, N. Castin, C. Domain, L. Malerba, and E. Vincent. *Journal of Nuclear Materials*, 406(1):55–67, Nov. 2010.
[14] E. Tadmor, M. Ortiz, and R. Phillips. *Philosophical Magazine A-Physics of Condensed Matter Structure Defects and Mechanical Properties*, 73(6):1529–1563, June 1996.
[15] Q. Peng, X. Zhang, L. Hung, E. A. Carter, and G. Lu. *Physical Review B*, 78(5):054118, Aug. 2008.
[16] G. Lu, E. Tadmor, and E. Kaxiras. *Physical Review B*, 73(2):024108, Jan. 2006.
[17] L. Lu, S. Izvekov, A. Das, H. C. Andersen, and G. A. Voth. *Journal of Chemical Theory and Computation*, 6(3):954–965, Mar. 2010.
[18] Q. Shi, P. Liu, and G. A. Voth. *Journal of Physical Chemistry B*, 112(50):16230–16237, Dec. 2008.
[19] W. G. Noid, J.-W. Chu, G. S. Ayton, V. Krishna, S. Izvekov, G. A. Voth, A. Das, and H. C. Andersen. *Journal of Chemical Physics*, 128(24):244114, June 2008.
[20] W. G. Noid, P. Liu, Y. Wang, J.-W. Chu, G. S. Ayton, S. Izvekov, H. C. Andersen, and G. A. Voth. *Journal of Chemical Physics*, 128(24):244115, June 2008.
[21] Y. Wang, D. M. C. Nicholson, G. M. Stocks, A. Rusanu, M. Eisenbach, and R. E. Stoller. *Journal of Applied Physics*, 109(7):07E120, Apr. 2011.
[22] Y. Wang, G. Stocks, W. Shelton, D. Nicholson, Z. Szotek, and W. Temmerman. *Physical Review Letters*, 75(15):2867–2870, Oct. 1995.
[23] C. Woodward, D. R. Trinkle, J. Hector, L. G., and D. L. Olmsted. *Physical Review Letters*, 100(4):045507, Feb. 2008.
[24] M. Finnis and J. Sinclair. *Philosophical Magazine A-Physics of Condensed Matter Structure Defects and Mechanical Properties*, 50(1):45–55, 1984.
[25] A. Calder and D. Bacon. *Journal of Nuclear Materials*, 207:25–45, Dec. 1993.

Ab initio molecular dynamics study of diffusion mechanism in liquid B_2O_3 under pressure

Satoshi Ohmura and Fuyuki Shimojo

Department of Physics, Kumamoto University, Kumamoto 860-8555, Japan

E-mail: 095d9003@st.kumamoto-u.ac.jp

Abstract. Computer simulation based on molecular dynamics is powerful approach to clarify the microscopic mechanism of atomic diffusion in liquid state. Here, the diffusion properties and the microscopic diffusion mechanisms in liquid B_2O_3 under pressure are studied by *ab initio* molecular dynamics simulations. It is found that diffusivities of liquid B_2O_3 show anomalous pressure dependence. Diffusion coefficients of liquid materials usually decrease with pressure. However, that of liquid B_2O_3 increases with increasing pressure up to about 10 GPa. Additionally, diffusivity of boron becomes about twice larger than that of oxygen under pressure above 20 GPa. These anomalous pressure dependences are strong related to the microscopic atomic diffusion in liquid B_2O_3.

1. Introduction

In the structures of crystalline and B_2O_3 in ambient pressure, each boron atom is coordinated to three oxygen atoms, forming a triangular BO_3 unit, and each oxygen atom is twofold coordinated to boron atoms [1, 2]. Under compression, crystalline B_2O_3 is transformed from the trigonal to the orthorhombic phase at about 6.5 GPa [3, 4]. The high-pressure structure consists of interconnected tetrahedral BO_4 units. Although each boron atom bonds to four oxygen atoms, two-thirds of the oxygen atoms are threefold coordinated to boron atoms and the remaining maintain twofold coordination.

In liquid state, the local coordination around each atom remains the same upon melting at atmospheric pressure [6], i.e. boron atoms are predominantly threefold coordinated to oxygen atoms, and most of oxygen atoms bridge two adjacent boron atoms. Unlike crystalline state, however, there must take place the rearrangement of the B-O covalent bonds with long-range atomic diffusion in liquid state. In terms of bond switching process of liquid B_2O_3 at ambient pressure has been investigated by first-principle molecular dynamics simulation [7]. The investigation reveals that a non-bridging oxygen double bonded to a twofold-coordinated boron is always involved with atomic diffusion accompanied by rearrangement of the covalent bonds to reduce the formation energy of the overcoordination defects.

The transport properties of liquid B_2O_3 under pressure have been studied by *in situ* viscosity measurements experiments [8] and *ab initio* molecular dynamics simulations [9, 10]. The former has revealed that the viscosity decreases with increasing pressure up to 8 GPa. The latter has shown that the diffusivity of atoms is enhanced by pressure below 10 GPa, as in other covalent liquids, such as SiO_2 and GeO_2 [12, 13]. Additionaly, the simulation has found that the diffusivity of boron becomes about two times larger than that of oxygen under pressures above 20 GPa, while the former is 10−20% smaller than the latter at lower pressures. This is because the number of threefold-coordinated B atoms decreases more rapidly than that of twofold-coordinated O atoms with increasing pressure. Although it is clarified the origin of the diffusion maximam and the anomalous diffusion property under hipressure above 20 GPa, it is unclear why the diffusivity decrease with increasing pressure above 10 GPa.

In this paper, we investigate the dynamic properties of liquid B_2O_3 under pressure by *ab initio* molecular-dynamics (MD) simulations with interatomic forces calculated quantum mechanically. We focus on the diffusion mechanism under pressure including the pressure region from 10 GPa to 20 GPa in which the diffusivity decrease with pressure. The purposes of our study is to discuss the mechanism of atomic diffusion accompanied by the B-O bond exchange under pressure and to explain the pressure dependence of diffusivity in terms of microscopic diffusion mechanisms in all pressure region.

2. Method of Calculation

We use a system of 120 (48B+72O) atoms in a cubic supercell was used under periodic boundary conditions. The equations of motion for atoms were solved via an explicit reversible integrator [14] with a time step of $\Delta t = 1.2$ fs. The atomic forces were obtained from the electronic states calculated using the projector-augmented-wave (PAW) method [15, 16] within the framework of density functional theory (DFT) in which the generalized gradient approximation (GGA) [17] was used for the exchange-correlation energy. The plane-wave cutoff energies are 30 and 200 Ry for the electronic pseudo-wave functions and the pseudo-charge density, respectively. The energy functional was minimized using an iterative scheme. [18, 19] Projector functions are generated for the 2*s* and 2*p* states of B and O. The Γ point was used for Brillouin zone sampling. To obtain a liquid state, we began by carrying out an *ab initio* MD simulation for about 5 ps at a temperature of 5000 K starting from the low-pressure crystalline phase [1]. The temperature is selected to be high enough to make the system reach a completely disordered state without the effects of the initial configuration. Then, we decreased the temperature of the system gradually to a target temperature of 2500 K. The density was taken from the extrapolation of the experimental data [20] obtained up to 1500 K. The temperature of 2500 K was chosen so as not only to be sufficiently high to maintain the liquid state even at the high pressures, but also in order to observe enough number of atomic-diffusion events to analyze the diffusion mechanism in a statistically meaningful way within a limited amount of simulation time. To determine the density of the liquid state under pressure, a constant-pressure MD simulation [21] was performed for 2.4 ps at each given pressure. Using the time-averaged density, the static and diffusion properties were investigated by MD simulations in the canonical ensemble [22, 23]. The time-averaged pressure [24, 25] was calculated at each density, and we obtained the density-pressure relation as shown in Fig. 1. The thermodynamic states investigated in this study cover a density range from 1.69 to 3.99 g/cm^3, and a pressure range from 1.4 to 97.0 GPa. The quantities of interest were obtained by averaging over 14.4 ∼ 21.6 ps to achieve good statistics after the initial equilibration, which takes at least 2.4 ps.

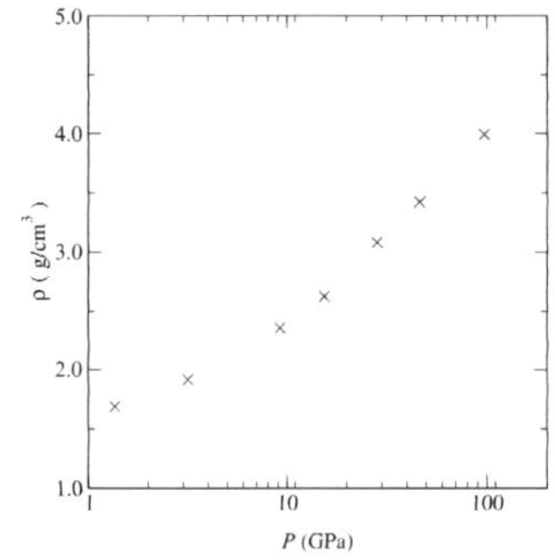

Figure 1. Pressure dependence of the density of liquid B_2O_3 at 2500 K

3. Result and Discussion

3.1. Pressure dependence of diffusivity

Figure 2 shows pressure dependence of diffusion coefficient and coordination number distribution obtained by our simulation. Cleary, the diffusion coefficients of B and O increase with pressure up to 10GPa. This is consistent with the experimental result that the viscosity of liquid B_2O_3 decrease with increasing pressure. Above 10GPa, the diffusivity decreases with increasing pressure, which is not surprising but quite natural. It is, however, unusual that the diffusivity of O atoms is reduced more quickly than that of B atoms with compression above 10 GPa. D_{B} becomes about two times larger than D_{O} when the pressure exceeds 20 GPa. Diffusion properties must be related to the weakening of the covalent interaction between atoms accompanying the increasing in the coordination number. In order to discuss the pressure dependence of the local structure in detail, we show the pressure dependence of the coordination-number distribution $f^{(n)}_{\alpha-\beta}$ in Fig. 2. $f^{(n)}_{\alpha-\beta}$ is the ratio of the number of α-type atoms that are coordinated to n β-type atoms to the total number of α-type atoms. To obtain $f^{(n)}_{\alpha-\beta}$, we count the number of β-type atoms inside a sphere with radius R centered at each α-type atom, where $R = 1.9$ Å is a cutoff distance determined from the first-minimum position of $g_{\mathrm{B-O}}(r)$ at 1.4 GPa. Both $f^{(3)}_{\mathrm{B-O}}$ and $f^{(2)}_{\mathrm{O-B}}$ have high values of about 0.98, i.e., 98% of B and O atoms have threefold and twofold coordination, respectively, under pressure up to about 3 GPa. With increasing pressure, for $P > 3$ GPa, $f^{(3)}_{\mathrm{B-O}}$ and $f^{(2)}_{\mathrm{O-B}}$ decrease, and instead $f^{(4)}_{\mathrm{B-O}}$ and $f^{(3)}_{\mathrm{O-B}}$ increase. While $f^{(3)}_{\mathrm{B-O}}$ and $f^{(4)}_{\mathrm{B-O}}$ are exchanged for each other at about 25 GPa, $f^{(2)}_{\mathrm{O-B}}$ and $f^{(3)}_{\mathrm{O-B}}$ are interchanged at a higher pressure of about 50 GPa because of the composition ratio of B and O atoms. At about 100 GPa, the number of fourfold-coordinated B atoms approaches about 90%, and about 5% of B atoms have fivefold coordination.

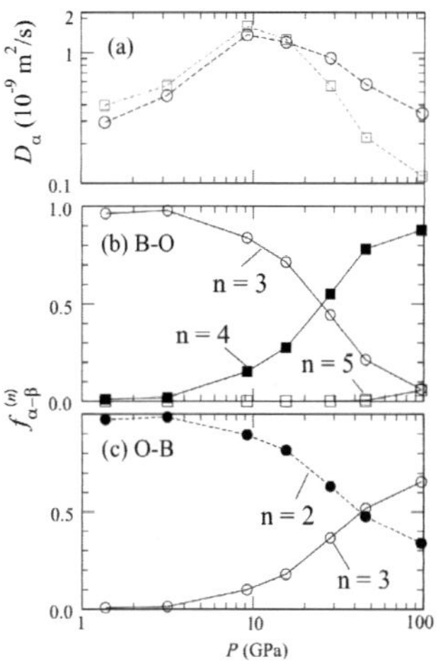

Figure 2. (a) Pressure dependence of the diffusion coefficients D_α for α = B (solid circles) and O (open squares) atoms. Pressure dependence of the coordination-number distribution $f^{(n)}_{\alpha-\beta}$ for α-β = (b) B-O and (c) O-B.

3.2. Diffusion mechanism

In liquid B_2O_3, long-range atomic diffusion occurs by several diffusion mechanisms, depending on pressure. At ambient pressure, one non-bridging O atom double bonded to a twofold-coordinated B atom is always involved with atomic diffusion accompanied by the B-O bond switching. Figure 3 shows a typical example of the generation of a non-bridging oxygen, where snapshots of atomic configurations is displayed with the time evolution of the bond-overlap populations [26, 27] associated with the B and O atoms of interest. The bond-overlap populations give a semiquantitative estimate of the strength of the covalent-like bonding between atoms.

In the atomic configuration at 0.01 ps (the top panel of Fig. 3), all B and O atoms displayed are threefold- and twofold-coordinated, respectively, to hetero atoms, i.e. there is no bond defect. As shown in the bottom panel of Fig. 3, $O_{\mathrm{B2-O2}}(t)$ begins to increase at about 0.04 ps, which means that a covalent bond is formed between B2 and O2 (in the snapshot at 0.07 ps). Due to

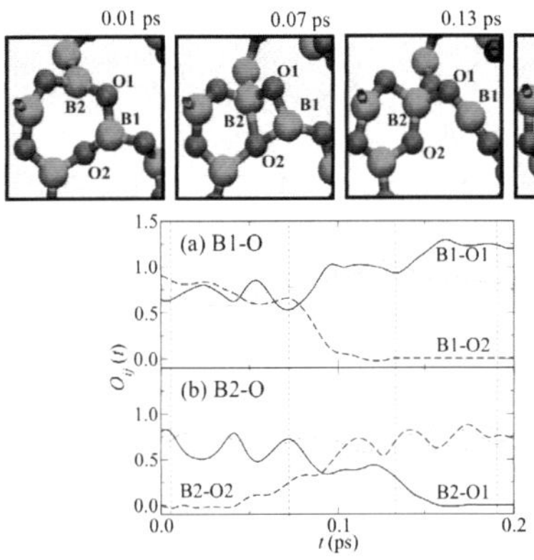

Figure 3. (Top panel) Atomic configurations at t = 0.01, 0.07, 0.13 and 0.19 ps. The large and small spheres show B and O atoms, respectively. (Bottom panel) The time evolution of bond-overlap populations $O_{ij}(t)$ for (a) i = B1, $j \in$ O and (b) i = B2, $j \in$ O in the process of the formation of a non-bridging oxygen. The thick solid and thick dashed lines show $O_{ij}(t)$ associated with the B and O atoms of interest as denoted in the figures. The thin lines show $O_{ij}(t)$ between the B atoms of interest (labeled as 'B1' and 'B2') and their neighboring O atoms except O1 and O2.

the formation of the bond, both B2 and O2 are over-coordinated. Since the over-coordination is energetically unstable, one of the covalent bonds around the threefold-coordinated O2 atom is broken (in the snapshot at 0.13 ps). Note that B1 is coordinated to only two oxygens, while B2 is still over-coordinated. Finally, the covalent bond between B2 and O1 is broken, as $O_{B2-O1}(t)$ is almost zero for $t > 0.16$ ps. While the threefold coordination of B2 is recovered, O1 is coordinated to only one boron B1 (in the snapshot at 0.19 ps). In this way, the non-bridging O atom (O1) is generated with the twofold-coordinated B atom (B1). We see that there is a double bond between B1 and O1 because $O_{B1-O1}(t)$ has higher values for $t > 0.16$ ps after the breaking of the B2-O1 bond. It should be noted that this process takes place in the participation of only two over-coordinated (one fourfold-coordinated B and one threefold-coordinated O) atoms, namely only one BO_4 unit.

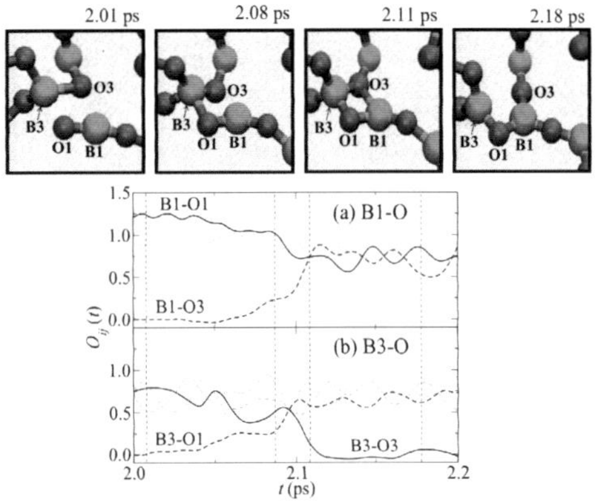

Figure 4. (Top panel)Atomic configurations at t = 2.01, 2.08, 2.11 and 2.18 ps. The large and small spheres show B and O atoms, respectively.(Bottom panel)Same as Fig. 3 but for O3.

After about 2 ps, the double bond between B1 and O1 disappears as shown in Fig. 4. First, the non-bridging O1 approaches to B3 to form a new covalent bond between them. We see that $O_{B3-O1}(t)$ gradually increases for $t > 2.01$ ps as shown in the bottom panel of Fig. 4, and that O1 is bonded to B3 in the atomic configuration at 2.08 ps. It is also seen that $O_{B1-O3}(t)$ increases for $t > 2.07$ ps, which means the formation of a covalent bond between B1 and O3 (in the snapshot at 2.11 ps). In this configuration, O1 and B1 have the proper coordination numbers, and instead B3 and O3 are over-coordinated. As $O_{B3-O3}(t)$ becomes nearly zero at about 2.12 ps, the B3-O3 bond is broken, and an atomic configuration with no bond defect is finally obtained (in the snapshot at 2.18 ps).

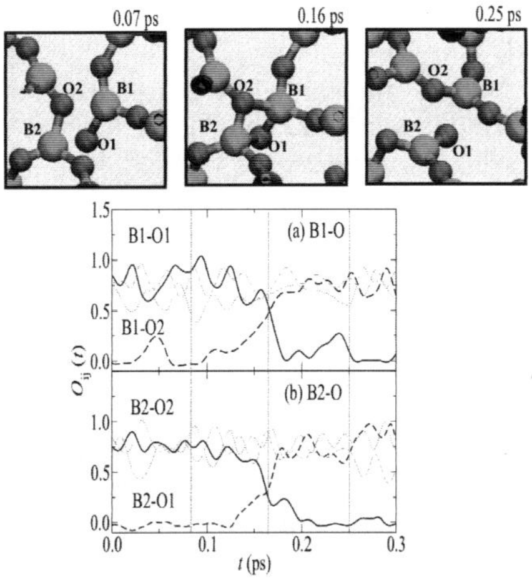

Figure 5. (Top panel)Atomic configurations at t = 0.07, 0.16 and 0.25 ps. The large and small spheres show B and O atoms, respectively.(Bottom panel)Same as Fig. 3.

When pressure increase above 3 GPa, the usual concerted reactions with two BO_4 unit can take place because B atoms are more easily coordinates to four O atoms under high pressure. In Fig. 5, the time evolution of the bond-overlap populations $O_{ij}(t)$ associated with the B and O atoms of interest is displayed with snapshots of atomic configurations. As displayed in the snapshot at 0.09 ps, two BO_4 groups are generated as an intermediate by forming two new B-O bonds between adjoining BO_3 units, in contrast to the fact that only one BO_4 group is required to produce the nonbridging O atoms. The concerted reactions occur more frequently under higher pressures, which will result from the covalent-bond weakening due to the compression Therefore, the diffusion enhancement with pressure is originated from this process. We also observed that the reactions with the nonbridging O atoms decrease with increasing pressure and almost disappear for $P > 10$ GPa.

The diffusivity of B and O atoms decrease with compression above 10 GPa. Since BO_4 group becomes more stable above 10 GPa, it takes longer time to exchange covalent bonds compared to the precess of Fig. 5. A typical bond exchange precess is shown in Fig. 6. In this process, B

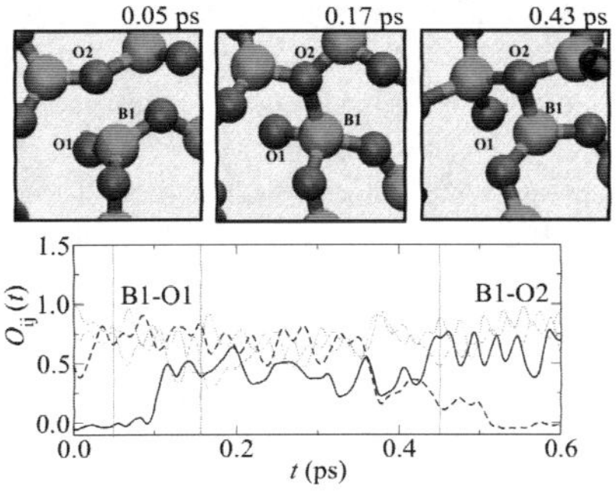

Figure 6. (Top panel)Atomic configurations at t = 0.05, 0.17 and 0.43 ps. The large and small spheres show B and O atoms, respectively.(Bottom panel)Same as Fig. 3 but for B2.

atom labeled "B1" which has threefold coordination exchange O1 for O2 through BO_4 group. Since BO_4 group is more stable in this pressure region, B1 keeps fourfold coordination for about 0.3 ps. After that, B1 recovers to threefold coordination.

This decrease of diffusivity is quite natural in liquid materials. It is, however,unusual that the diffusivity of O atoms is reduced more quickly than that of B atoms. In order to consider this anomalous pressure dipendence of diffusion coefficient, it is worth noting that the number of fourfold-coordinated B atoms is much larger than that of threefold-coordinated B atoms while both twofold- and threefold-coordinated O atoms exist as shown in Fig. 2. We focused on diffusion processes associated with fourfold-coordinated B atoms shown in Fig. 7. In the beginning of this process, the B atom labeled 'B1' has fourfold coordination. The atomic configuration at 0.02 ps (and also at 0.17 ps) shows that fourfold-coordinated B atoms have a tetrahedral arrangement, which demands sp^3 hybridization around them. As displayed in the bottom panel of Fig. 7, $O_{\mathrm{B1-O1}}(t)$ begins to decrease at about 0.08 ps and almost vanishes for $t > 0.11$ ps. This time change means that the covalent bond between B1 and O1 is broken within 0.03 ps. On the other hand, $O_{\mathrm{B1-O2}}(t)$ increases gradually between 0.08 and 0.15 ps, indicating that a covalent bond is formed between B1 and O2, taking 0.07 ps to form. The B1 atom has threefold coordination with a planar arrangement before the formation of the new B1-O2 bond is complete(in the atomic configuration at 0.10 ps). In this way, one of the B-O bonds is broken quickly, and a new B-O bond is formed gradually when B atoms move between the fourfold-coordinated sites.

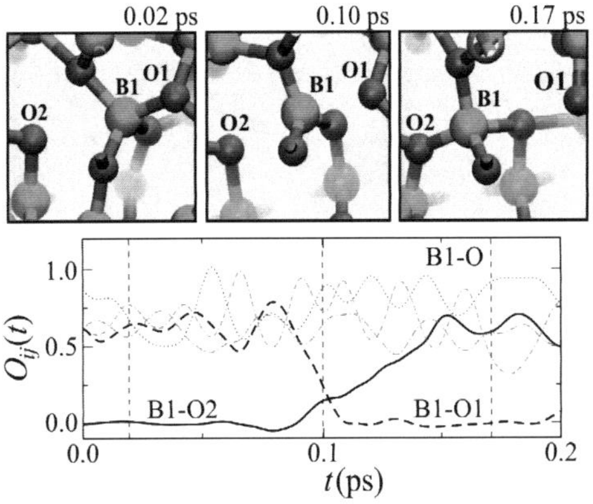

Figure 7. (Top panel)Atomic configurations at $t = 0.02$, 0.10 and 0.17 ps. The large and small spheres show B and O atoms, respectively.(Bottom panel)Same as Fig. 6.

In the diffusion process shown in Fig. 7, twofold-coordinated O atoms, toward which fourfold-coordinated B atoms move, are necessary as B1 moves toward O2. Note that O2 has twofold-coordination before B1 bonds to it. As shown in Fig. 2(b), twofold-coordinated O atoms exist even at high pressures. This fact indicates that this diffusion process occurs rather frequently, and that B atoms can diffuse fairly easily. On the other hand, for the concerted reaction involving the migration of O and B atoms, threefold-coordinated B and twofold-coordinated O atoms are necessary. However, the number of threefold-coordinated B atoms decreases rapidly with increasing pressure for $P > 20$ GPa, which means that the concerted reaction is suppressed at such high pressures. This process is origin of rapid decreasing of D_{O} with increasing pressure.

4. Summary

We have investigated the diffusion properties and the microscopic diffusion mechanisms in liquid B_2O_3 under pressure by *ab initio* molecular dynamics simulation. It is found that long-range atomic diffusion in liquid B_2O_3 occurs by several diffusion mechanisms, depending on pressure.

Anomalous pressure dependence of diffusivity has been explained successfully in terms of these diffusion mechanisms.

Acknowledgments
The present work was supported by KAKENHI (Grant-in-Aid for Scientific Research (B) No. 23340106 and Grant-in-Aid for JSPS Fellows No. 22·1853). The authors acknowledge support by Hamamatsu Photonics K.K., Japan. They furthermore thank the Research Institute for Information Technology, Kyushu University for the use of its facilities. The computations were also performed using the computer facilities at the Supercomputer Center, Institute for Solid State Physics, University of Tokyo.

References

[1] Gurr G E, Montgomery P W, Knutson C D and Gorres B T 1970 Act. Cryst. **B26** 906.
[2] Johnson P A V, Wright A C, and Sinclair R N 1982 J. Non-Cryst. Solids **50** 281.
[3] Prewitt C T and Shannon R D 1968 Act. Cryst. **B24**, 869.
[4] Nieto-Sanz D, Loubeyre P, Crichton W, and Mezouar M 2004 Phys. Rev. B **70**, 214108.
[5] Lee S K, Eng P J, Mao H, Meng Y, Newville M, Hu M Y, and Shu J 2005 Nature Mater. **4**, 851.
[6] Sakowski J and Herms G 2001 J. Non-Cryst. Solids **293-295**, 304.
[7] Ohmura S and Shimojo F, 2008 Phys. Rev. B **78**, 224206.
[8] Brazhkin V V, Farnan I,Funakoshi K, Kanzaki M, Katayama Y Lyapin A G, and Saitoh H, 2010 Phys. Rev. Lett. **105**, 115701.
[9] Ohmura S and Shimojo F, 2009 Phys. Rev. B **80**, 020202(R).
[10] Ohmura S and Shimojo F, 2010 Phys. Rev. B **81**, 014208.
[11] Diefenbacher J and McMillan P F 2001 J. Phys. Chem. A **105**, 7973.
[12] Karki B B, Bhattarai D, and Stixrude L 2007 Phys. Rev. B **76**, 104205.
[13] Hoang V V, Zung H, and Hai T N 2007 J. Phys.: Condens. Matter **19**, 116104.
[14] Tuckerman M, Berne B J and Martyna G J 1992 J. Chem. Phys. **97**, 1990.
[15] Blöchl P E 1994 Phys. Rev. B **50**, 17953.
[16] Kresse G and Joubert D 1999 Phys. Rev. B **59**, 1758.
[17] Perdew J P, Burke K and Ernzerhof M 1996 Phys. Rev. Lett. **77**, 3865.
[18] Kresse G and Hafner J 1994 Phys. Rev. B **49**, 14251.
[19] Shimojo F, Kalia R K, Nakano A and Vashishta P 2001 Comp. Phys. Comm. **140**, 303.
[20] Macedo P B, Capps W, and Litovitz T A 1966 J. Chem. Phys. **44**, 3357.
[21] Martyna G J, Tobias D J, and Klein M L 1994 J. Chem. Phys. **101**, 4177.
[22] Nosé S 1984 Mol. Phys. **52**, 255.
[23] Hoover W G 1985 Phys. Rev. A **31**, 1695.
[24] Nielsen O H and Martin R M 1985 Phys. Rev. B **32**, 3780.
[25] Corso A D and Resta R 1994 Phys. Rev. B **50**, 4327.
[26] Mulliken R S 1955 J. Chem. Phys. **23**, 1841.
[27] Shimojo F, Nakano A, Kalia R. K, and Vashishta P 2008 Phys. Rev. E **77**, 066103.

Journal of Physics: Conference Series **402** (2012) 012014 doi:10.1088/1742-6596/402/1/012014

Irreversible Thermodynamics

David M Rogers[1] **and Susan B Rempe**[2]

Center for Biological and Materials Sciences, MS 0895,
Sandia National Laboratories, Albuquerque, New Mexico 87185, USA

E-mail: dmroge@sandia.gov,[1] slrempe@sandia.gov[2]

Abstract. We present a maximum-entropy theory of mesoscopic kinetics. The theory gives fully nonlinear nonequilibrium thermodynamic relationships and has no explicit requirement for either microscopic bath variables, an equilibrium energy, or an equilibrium partition function. The entropy maximization process is instead carried out over transition probability distributions with constraints on particle position and velocity updates. The Lagrange multipliers for these constraints express the instantaneous temperature and pressure of *external* (or microscopic) thermostatic driving systems, with which the distinguished system may or may not eventually reach equilibrium. We show that the analogues of the Gibbs-Maxwell relations and free energy perturbation techniques carry over to fluctuation-dissipation theorems and nonequilibrium ensemble reweighting techniques as should be expected. The result is a fully time-dependent, non-local description of a nonequilibrium ensemble coupled to reservoirs at possibly time-varying thermostatic or mechanical states. We also show that the thermodynamic entropy production extends the generalized fluctuation theorem through the addition of an instantaneous information entropy term for the end-points, leading to a concise statement of the second law of thermodynamics.

1. Far From Non-Equilibrium

Standard treatments of statistical mechanics are heavily influenced by equilibrium, thermostatic, ideas. This reliance is especially evident when one attempts to formulate a kinetic theory without reference to a potential of mean force. As a case-in-point, Eyring's transition-state theory and Marcus' electron transfer rate equations are fundamentally based on the 'free energy change' in a reversible reaction. Such a reaction has no driving force, does not proceed in the forward direction on average, and dissipates no heat. Although nonequilibrium statistical mechanics is the underlying picture that must be used for analyzing laboratory experiments taking place in open systems, the standard analysis of basic phenomena such as Joule-heating, piezoelectricity, or the Seebeck effect[1] relies on a theory of small, linear, deviations from equilibrium. Further, these deviations assume a spatial dependence that pre-supposes a spatial scale on which the fluid energy function is local. This seems to preclude our discussion of anything but local thermostatic quantities.

On the other hand, a set of fluctuation-dissipation theorems have been derived based on a projection from a fully dynamical system described by a Liouville equation.[2, 3, 4, 5] The projection operation removes the pernicious microscopically reversible information from the phase-space distribution and so allows its information entropy to increase.[6] Driving such a projected system can be modeled by adding a time-dependent Hamiltonian, again with reference to the equilibrium picture. In this case the projection causes dissipation of the added energy into

Journal of Physics: Conference Series **402** (2012) 012014 doi:10.1088/1742-6596/402/1/012014

the unmodeled degrees of freedom. Although this picture gives a complete theory, it introduces unnecessary microscopic variables whose only purpose is to set the rate of heat dissipation from the mesoscopic system. That this rate is linked to the mesoscopically observable fluctuations (the famous fluctuation-dissipation theorem) requires a result from second-order perturbation theory on the system's equilibrium partition function. This state of affairs therefore seems far from nonequilibrium.

2. Maximum Caliber

In the late 1970s, Jaynes[7] introduced an important idea to solve some of the above difficulties. He conceived of an 'ensemble' of trajectories, $x(t)$, in place of the usual ensemble of microstates, x. The probability of each trajectory, $\mathcal{P}(x(t))$, was then to be determined by maximizing a functional of $\mathcal{P}(x(t))$ subject to some constraints on the average values of functionals of $x(t)$. In order to avoid overloading the term entropy, he named the functional to maximize,

$$\sigma = \int -\log \mathcal{P}(x(t))\, \mathcal{P}(x(t))\ \mathcal{D}x, \tag{1}$$

the 'caliber.' Pursuing the analogy further, Jaynes showed[7, 8] that the Lagrange multipliers constraining averages of path functionals correspond to thermodynamic forces, and that the first derivatives of a suitable function of the caliber yield those averages. Further, its second derivative (minus the space and time-correlation functions) gives the "first-order" perturbation to the path averages. A more complete set of the analogies which could be anticipated is shown in Table 1.

Equilibrium	Nonequilibrium
Phase Space	Trajectory Space
Free Energy	Path Free Energy
Entropy	Caliber
Average Value Constraint	Average Flux Constraint
Equilibrium Average	Path Average
Conditional Free Energy (PMF)	Conditional Path log-Probability
Thermodynamic forces (δPMF)	Changes in Path Flux (work)
Heat Capacity	Thermal Conductivity
(none)	Irreversibility and Entropy Production

Table 1. Correspondence between single-time and time-dependent path maximum entropy formulations of statistical mechanics.

However, the published articles of Jaynes showed very few calculations using this method. Instead, in Ref. [9], the maximum entropy formalism is given to justify adding a Lagrange multiplier to the Hamiltonian, and the standard perturbation theory used by, e.g. Chapter 4 of Kubo [10],[1] was employed for carrying out the calculation. The maximum caliber approach has seen a resurgence in recent work[11, 12, 13] outlining its connection to the fluctuation and entropy production theorems.[14] Connections to local equilibrium theories have been strengthened in Refs. [15, 16, 17], and applications to discrete master equations have been explored in Refs. [18, 19], and from another perspective in Ref. [20].

The pervasive reference to perturbation theory in these applications is likely because the argument leading to straightforward maximization of Eq. 1 contains a fatal flaw. To see this, first note that the average of a quantity at time t is dependent on the "control" parameters

[1] It is worthwhile to note that Jaynes gave special attention to perturbation terms of all orders here.

throughout the whole trajectory. This is because logical inference does not contain a preferential time direction. Knowledge that the system has a given property at time $t+\tau$ constrains the state at time t, as it should, but a force λ_k acting at a future time should not be able to influence the past trajectory. Although this could be alleviated by requiring only causal information to enter into the determination of the state at time t, this approach leads to a probability distribution valid only for x_t and not throughout a trajectory.

Here, we show for the first time the benefits of maximizing the entropy of the *transition* probability distribution under suitable constraints. We note here that an argument for using this approach in the long-time limit has been given in Ref. [21], and that several authors have also noted the need for another 'entropy' associated with time-dependent processes.[22, 23, 24, 25]

Limiting the scope of the maximum trajectory entropy procedure in this way automatically corrects a related re-normalization problem. Suppose phase space were to branch at a future time $t+\tau$. In this case, a uniform measure on path space would assign points at time t a different weight depending on future events. A simple example is the Monty Hall problem with a prize assigned to the first door without loss of generality.[26] The contestant's choice of one of three doors plus Monty opening another (not concealing a prize) constitutes four possible paths, and maximum path entropy would give each path an equal weight – an intuitive, but incorrect, solution. The correct probability assignment is a uniform distribution for each *transition*, leading to ending weights of $1/3 \times 1/2 = 1/6$ for the two paths following the selection of the first door[2]. In more abstract terms, the marginal probability at each time should not depend on the future – a concept expressed mathematically by defining a progressively measurable function with respect to the natural filtration of stochastic processes[27].

In practice, this flaw can be avoided by considering only processes where assigning equal *a priori* weight to all paths is equivalent to assigning equal weight to all transitions. By Liouville's theorem, this is obviously true for deterministic processes. More generally, the approaches are equivalent when the number of possible transitions does not depend on the starting point. The path entropy approach of Jaynes is therefore valid in the absence of factors re-normalizing for starting-point dependent differences in the number of possible paths ($Z[\lambda_i, x_i] = Z[\lambda_i]$ in Eq. 6).

3. Maximum Transition Probability Entropy

A realization of a time-dependent process is a history, $\{x_0, x_1, \ldots, x_S\}$, giving the state, x, of the system at times $i = 0, \ldots, S$. In a stochastic process, knowledge of the history, $\{x_0, x_1, \ldots, x_i\} \equiv \Gamma_i$ is not sufficient to determine the future beyond time i. Instead, knowing the process history gives us a probability distribution, $\mathcal{P}(x_{i+1}|\Gamma_i)$, over the state at time $i+1$ *conditional* on the observed history. The conditional probability is formally a function of any predictive information we know, such as the past history, $x_{j\le i}$. It is required to be normalized, $\int \mathcal{P}(x_{i+1}|\Gamma_i)\, dx_{i+1} = 1$. For a Markov process, the conditional probability is fixed and a function of only the previous state, x_i. For a general system with known averages, $\langle f_k(x_{i+1}; \Gamma_i)|\Gamma_i\rangle$ (where f is some function, and the notation '$|\Gamma_i$' is read as conditional on the history up to time i), we introduce one (unknown) Lagrange multiplier, $\lambda_{k,i}$, for each average and carry out the standard constrained maximization of the transition entropy,

$$\delta \mathcal{H}_{i+1|\Gamma} = \delta \sum_{\{x_{i+1}\}} p(x_{i+1}|\Gamma_i)\Big[- \ln \frac{p(x_{i+1}|\Gamma_i)}{p^0(x_{i+1}|\Gamma_i)} \tag{2}$$

$$- (\ln Z[\Gamma_i] - 1) - \sum_{k=1}^{m} \lambda_{k,i} f_k[x_{i+1}, \Gamma_i]\Big] \tag{3}$$

[2] We leave updating the player's state of knowledge about the prize out of the discussion, and reiterate that the full path probabilities are $1/6, 1/6$, (contestant's correct choice plus Monty choosing randomly from the remaining doors), and $1/3, 1/3$ (contestant's two incorrect choices plus Monty's forced choice of an incorrect door).

to yield

$$p(x_{i+1}|\Gamma_i) = p^0(x_{i+1}|\Gamma_i)e^{-\eta_i[x(t)]}/Z[\lambda_i, \Gamma_i]. \tag{4}$$

The terms in the exponential make a quantity like a Boltzmann weight,

$$\eta_i[x(t)] \equiv \sum_{k=1}^{m} \lambda_{k,i} f_k[x_{i+1}, \Gamma_i], \tag{5}$$

and there is a quantity corresponding to a partition function

$$Z[\lambda_i, \Gamma_i] = \sum_{\{x_{i+1}\}} p^0(x_{i+1}|\Gamma_i)e^{-\eta_i[x(t)]}, \tag{6}$$

which is in fact a functional of the instantaneous Lagrange multipliers and the system history. The transition probability entropy can be re-written in terms of Z and the averages as in the equilibrium theory,

$$\mathcal{H}_{i+1|\Gamma} = \ln Z[\lambda_i, \Gamma_i] + \langle \eta_i | \Gamma_i \rangle \tag{7}$$

$$d\mathcal{H}_{i+1|\Gamma} = \sum_{k=1}^{m} \lambda_{k,i} d\langle f_k | \Gamma_i \rangle. \tag{8}$$

These are conditional averages, meaning that they are summed only over the possible values of x_{i+1} and are still functions of the history, Γ_i. We can also take Legendre transforms to yield quantities corresponding to state-functions.

$$\Psi_{i+1|\Gamma} \equiv \sum_{k=1}^{m'} \lambda_{k,i} \langle f_k | \Gamma_i \rangle - \mathcal{H}_{i+1|\Gamma} \tag{9}$$

$$d\Psi_{i+1|\Gamma} = \sum_{k=1}^{m'} \langle f_k | \Gamma_i \rangle d\lambda_k - \sum_{k=m'+1}^{m} \lambda_{k,i} d\langle f_k \rangle. \tag{10}$$

Of course here the functions Ψ are functionals of the system history and of λ_i.[28] We will see that the λ_i control rates of change, e.g. using constraints of the form $f = x_{i+1} - x_i$, so that the equilibrium notion of a state function is replaced here with a single-time, transition state function. That is, the Lagrange multipliers specify the state of the external, driving system between times i and $i+1$. In the continuous limit, the multipliers λ_i exist between times i and $i+1$, and the sum over x_{i+1} should be viewed as a Stratonovich path integral between these times. This distinction is important when λ are functions of Γ_i. We have also found that differentiation of the transition partition function, Z, gives results identical to the stochastic integration of quantities such as instantaneous work and heat.

In the deterministic case, we know that x_{i+1} can be related to x_i, x_{i-1} through a discrete action,[29] such as

$$\mathcal{A} = \tfrac{(x_{i+1}-x_i)^2}{2\epsilon^2} + \tfrac{(x_i - x_{i-1})^2}{2\epsilon^2} - V(x_i) + \text{const.}(\{\ldots, x_{i-1}, x_{i+1}, \ldots\}). \tag{11}$$

Here ϵ represents the length of a small time-step. Differentiation of this equation with respect to x_i must yield zero, so the process is deterministic with the usual Verlet integration,

$x_{i+1} = x_i + (x_i - x_{i-1}) - \epsilon^2 \frac{\partial V}{\partial x_i}$. To model an external environment, we may instead assume that $\frac{\partial \mathcal{A}}{\partial x}$ is only approximately zero, and construct a maximum entropy distribution with a constrained average, $\left\langle \frac{\delta \mathcal{A}}{\delta x_i}^2 \right\rangle$. Without other constraints, allowing this squared-deviation implies interaction with a completely chaotic (infinite temperature) bath. It will not conserve, for example, energy and momentum except in the deterministic limit.

When two systems are coupled, the combined system should obey a set of conservation laws. As formalized by Noether's theorem,[30] such conservation laws can be derived for a single system directly from the action formulation by considering continuous symmetry transformations of the trajectory $x(t) \to x(t) + q(t, a)$ in a region around $a = 0$, where $q(t, 0) = 0$. Because the action is stationary with respect to small perturbations in $x(t)$, there exists a vanishing quantity,

$$\begin{aligned} \frac{dI(t)}{dt} &\equiv - \left. \frac{\partial \mathcal{A}}{\partial a} \right|_{a=0} \\ &= - \left. \frac{\partial q(t,a)^T}{\partial a} \right|_{a=0} \frac{\delta \mathcal{A}}{\delta x(t)}, \end{aligned} \tag{12}$$

(with T reserved for denoting transposition) and it is possible to define an invariant (using $\left. \frac{\partial q(t,a)}{\partial a} \right|_{a=0} \equiv y(t)$),

$$I(t) - I(\infty) = \int_{-\infty}^{t} -y^T(t') \frac{\delta \mathcal{A}}{\delta x(t')} dt'.$$

If the action functional is invariant to this transformation ($\mathcal{A}[x(t)] \to \mathcal{A}[x(t) + q(t, a)]$), then the corresponding invariant is a conventional conserved quantity. The substitution $x(t) \to x(t) + a$ ($y = \mathbf{1}$, the ones-vector) generates the momentum, $x(t) \to x(t+a)$ ($y = \dot{x}$) generates the energy, and so on.

It is instructive to consider a case where the action is not invariant to the transformation. For example, momentum is not conserved in the simple harmonic oscillator ($L = m\dot{x}^2/2 - kx^2/2 \neq m\dot{x}^2/2 - k(x+a)^2/2$). However, if we assume the existence of a generalized momentum, $\hat{p}$, that is nonetheless an invariant of some 'complete' system, we can define (using Eq. 12)

$$\frac{d\hat{p}(t)}{dt} = -\mathbf{1}^T (F - \dot{p}) = m\ddot{x} + kx$$

as a momentum *exchange*. If the net momentum of the observed system changes by more than its internal force, it implies that an external system has lost exactly this amount of momentum. In the presence of noise, this quantity may not be conserved ($d\hat{p}$ is stochastic). Thus one effect of the noise is to re-distribute $\hat{p}$ over the system and the bath.

In general, we can define a set of constraints on $\{\langle dI_j(t)/dt \rangle\}_{j=0}^{m}$ using an $N \times m+1$ matrix Y, whose columns correspond to the constraint directions $\left. \frac{\partial q_j(t,a)}{\partial a} \right|_{a=0}$ and whose leftmost column is reserved for the energy $Y_{*,0} = \dot{x}$. In this case, the vector of exchanges for a given trajectory over a given time interval, ϵ, is

$$dI(t) \equiv -Y(t)^T \frac{\delta \mathcal{A}}{\delta x(t)} \epsilon. \tag{13}$$

In this equation, the presence of ϵ is used to denote the Stratonovich integral implicitly. The selection of constraints must be made based on the allowed interactions of the system with its reservoir.

These considerations have shown a simple method for including the influence of an external system on the dynamics. If we assume the existence of some "total" invariant between the system and the bath, then it makes sense to enforce a *constraint on the average change*, $\langle dI(t) \rangle$.

We should note our fixed sign convention, where I is always taken to be a quantity belonging to the central system under consideration. Before moving on to discuss the obvious connection of these changes to the thermodynamic work, we shall first consider the significance of the Lagrange multipliers in this formalism.

Writing Eq. 4 with $f_{-1} = \epsilon \frac{\delta \mathcal{A}}{\delta x(t)}^2$, $\lambda_{-1} = (2C)^{-1}$, and a vector of constraints, dI (Eq. 13), enforced using a vector of Lagrange multipliers, $\lambda = \beta/2$, we find a Gaussian-distributed probability for

$$\frac{\delta \mathcal{A}}{\delta x(t)}\epsilon = CY(t)\beta\epsilon/2 + (C\epsilon)^{1/2} z(t),$$

where z denotes a standard normal random variable. To constrain only the energy exchange (the leftmost column of Y), we use $f_0 = -\epsilon \dot{x}^T \frac{\delta \mathcal{A}}{\delta x(t)}$(=$dH$, the change in the Hamiltonian) and $\lambda_0 = \beta_0/2$, so

$$\dot{p}\epsilon = F(x(t))\epsilon - C\left(\dot{x}\beta_0\epsilon/2\right) + C^{1/2} d\mathcal{W}(t). \tag{14}$$

Here the centered (Stratonovich) Wiener process increment, $d\mathcal{W}(t) = \epsilon^{1/2}z(t)$, has been substituted for $z(t)$.

This equation is exactly the Langevin equation. Note that our method of derivation generalizes trivially to include imposed time-correlations between $\frac{\delta \mathcal{A}}{\delta x(t)}$ and $\frac{\delta \mathcal{A}}{\delta x(t')}$ as well as arbitrary external forces. Note especially that the correspondence between a time-correlation function and a time-dependent damping (dissipation) function are automatic in the Gaussian theory.

It is well-known from the Fokker-Planck equation[31],

$$\frac{\partial \rho(x,p)}{\partial t} = -(M^{-1}p)^T \nabla_x \rho - \nabla_p^T \left[F\rho - C\dot{x}\beta_0\rho/2 - C\nabla_p\rho/2\right],$$

that the solution to Eq. 14 is the canonical distribution with temperature $\beta_0^{-1} = k_B T$, where k_B is the Boltzmann constant. See also Ref. 32 for the formulation in the presence of memory terms. This solution is independent of C, suggesting that a natural parametrization for the Langevin equation is in terms of the temperature and $C^{1/2}$, which is related to the thermal conductivity or rate of temperature equilibration with the unmodeled external system.

Generally, if an invariant, I, can be expressed as a function of x and p, then a similar result should hold for other common equilibrium thermodynamic ensembles as well, such as the N, P, T ensemble where $\langle dV \rangle$ constitutes an additional dynamic variable and constraint.[33] This establishes the *physical interpretation of the Lagrange multipliers as the thermostatic variables of the bath* that dictate the eventual equilibrium of the system. Note that increasing β tends to decrease $\langle dI \rangle$, for example increasing pressure will drive the volume downward.

4. Irreversible Thermodynamics

Having firmly established the connection between the constrained dynamical equations and the equilibrium distribution, we next construct a thermodynamic view of any process that employs a series of transition probabilities to effect a change in the state of the system. This construction will lead naturally to a view of a (general, nonequilibrium) thermodynamic process in terms of a thermodynamic path transforming one type of energy into another with a concomitant irreversible entropy production. In this section, as in the above, we do not rely on results from the equilibrium theory.

To begin, we exactly define a system state, A, as any information known about a system that is sufficient to construct a probability distribution for its variables, $\mathcal{P}(x|A)$. The machinery of statistical mechanics can then be used to propagate this information to system states at other time points and under alternate possible processes.

The work of Joule and Thomson showed that there exists a series of mechanical operations that can be performed to effect a transition between any two thermostatic states;[34] however, this transition can only take place in the direction of increasing entropy. The entropy increase comes about because of experimental inability to control the detailed motions of all particles, and is therefore zero in the case of completely controllable mechanical work. Thus, it is important to define a mechanical, adiabatic process in which all work is completely controlled by letting $C \to 0$ with constant external force experienced by the system $F^{\text{sys}}_{\text{ext},j} \equiv -\frac{C}{2} y_j \beta_j$. At a finite temperature, there is some amount of uncertainty about the exact state of I on short time scales. This leads to a discrepancy between the force exerted by the external system, $F_{\text{ext},j}$, and its "long time scale" counterpart experienced by the system, $F^{\text{sys}}_{\text{ext},j} = \frac{\partial x(t)}{\partial I_j(t)} F_{\text{ext},j}(t)$. The mechanical, adiabatic process is formally a zero-temperature, continuous-time limit. In general, only a subset of work values can be controlled. Before proceeding it will be necessary to solidify the concept of controllable work.

If this work is to be delivered by an external thermostatic system, for example an adiabatically coupled piston, then the first law of thermostatics gives $F_{\text{ext},j} = -dU_{\text{ext}}/dI_j = dU_{\text{ext}}/dI_{\text{ext},j} = -\beta_j/\beta_0$ if the force can be assumed constant over a sufficiently short time-step. The β variables are the Lagrange multipliers (2λ in Eqns. 2-10) controlling the rates of change of I between the two systems. Mechanically, this force corresponds to the force on a wall exerted by a spring placed externally to it. The total force on the wall is, of course,

$$F_{\text{tot},j} = -\frac{dU_{\text{int}} + dU_{\text{ext}}}{dI_j} \equiv F_{\text{int},j} + F_{\text{ext},j}.$$

This implies that if a two spring system were disconnected after a change $d\tilde{I} \equiv [dI_1, \ldots, dI_m]^T$, their internal potential energies would have changed by an amount

$$\begin{aligned} dU_{\text{int}} &= -d\tilde{I}^T F_{\text{int}} \\ dU_{\text{ext}} &= -d\tilde{I}^T F_{\text{ext}} = d\tilde{I}^T \tilde{\beta}/\beta_0 \\ \tilde{\beta} &\equiv [\beta_1, \ldots, \beta_m]^T. \end{aligned} \tag{15}$$

The sum of these two energy changes is not necessarily zero due to the possibility of momentum change. Using the known energy change of the system, it should then be possible to solve for the change in kinetic energy of the constraint. For the system, the *total* energy change is given by

$$\begin{aligned} dE &= -\frac{\delta A}{\delta x}^T \dot{x}\epsilon = dI_0 \\ &= dW + dQ \\ dW_j &\equiv -\frac{\delta A}{\delta x}^T y_j \dot{a}_j \epsilon \\ dW &= d\tilde{I}^T \dot{a} \end{aligned} \tag{16}$$

$$dQ = -\left(\dot{x} - \tilde{Y}\dot{a}\right)^T \frac{\delta A}{\delta x}\epsilon, \tag{17}$$

where we have used Eq. 12 for $\frac{\partial A}{\partial a}$, and multiplied by da/dt to define work values and a corresponding "non-mechanical" energy transfer, dQ. In the adiabatic limit, all energy transferred to the system by external forces should be reflected by known mechanical changes (related to $\{I\}$), which is precisely what $\dot{a}$ allows us to do. Note also that all quantities refer to the central system unless specifically denoted 'ext.' Thus dW means the work done on the system.

In the mechanical limit, $-\frac{\delta \mathcal{A}}{\delta x(t)}$ is just $\dot{p} - F_{\text{int}} = F_{\text{ext}}^{\text{sys}}$, the external force experienced by the system in the absence of the thermostatting random noise, and $Y_{*,0} = \dot{x}$. Substituting this quantity from the Langevin equation (14), an adiabatic, mechanical system must satisfy

$$\begin{aligned} dQ &= -(\dot{x} - \tilde{Y}\dot{a})^T \tfrac{\delta \mathcal{A}}{\delta x} \epsilon \\ &= -\frac{1}{2}(\dot{x} - \tilde{Y}\dot{a})^T C \tilde{Y} \tilde{\beta} \epsilon = 0. \end{aligned} \tag{18}$$

The last section gave some physical insight into the quantities $\dot{a}$. A mathematical consideration of the previous equation shows that $\tilde{Y}\dot{a} \equiv \sum_{j=1}^{m} y_j \dot{a}_j$ ($\tilde{Y}$ being identical to Y with the first column removed) can be understood as a projection, removing components of $\dot{x}$ parallel to $\delta\mathcal{A}/\delta x$. If we therefore define (writing the Moore-Penrose pseudoinverse of A as A^+)

$$\dot{a} \equiv (C^{-1/2}\tilde{Y})^{+} C^{-1/2} \dot{x}, \tag{19}$$

then $dQ = 0$ identically. For an example application, $y = \mathbf{1}$ (ones-vector) is associated with the system momentum, and Eq. 19 generates the average velocity $\dot{a} = \mathbf{1}^T \dot{x}/N$ when $C = cI$. Eq. 19 is invariant to multiplication of C by a constant, and so persists in the deterministic limit. The work done on the system (Eq. 16) is

$$dW = -\frac{\delta \mathcal{A}}{\delta x}^T \tilde{Y} (C^{-1/2}\tilde{Y})^{+} C^{-1/2} \dot{x} \epsilon.$$

Note the information-theoretic quality of the work defined by the equation above. If separate reservoirs existed that were able to influence the motion of each particle in the system independently, then $\tilde{Y}$ would become an identity matrix, and dW would equal the total energy added to the system, dE, identically. In the presence of noise, the work taken from the thermostatic reservoirs does not necessarily equal the work done on the system. In the $y = \mathbf{1}$ example above, the momentum work done on the system is

$$\tfrac{1}{N}\Big(\sum_a \dot{p}_a - F_a\Big)\Big(\sum_a \dot{x}_a\Big)\epsilon. \tag{20}$$

Here "$\sum_a$" denotes a sum over N (one-dimensional) particles, with obvious extension to multiple dimensions. Because this interaction controls only the total system momentum, the work is computed using the average velocity change. The expectation of this stochastic integral is $\langle dW_p \rangle/\epsilon = \frac{\mathbf{1}^T \dot{x}}{N} \mathbf{1}^T C(\lambda' \mathbf{1} - \beta_0 \dot{x})/2$, while $dU_{\text{ext},p}/\epsilon = \frac{\lambda'}{\beta_0} \mathbf{1}^T C(\lambda' \mathbf{1} - \beta_0 \dot{x})/2$. Similarly, an electric field can couple only to the net dipole moment of a system. This implies the transformation of applied energy to heat if it cannot be manipulated to perfectly match fluctuations in the driven variable.

The kinetic energy change of I, ascribed to the reservoir, can be determined in the mechanical limit from

$$\begin{aligned} 0 &= dU_{\text{ext}} + dK - dW \\ \Rightarrow dK &= -dI^T(\tilde{\beta}/\beta_0 - \dot{a}), \end{aligned} \tag{21}$$

using Eqns. 15 and 16[3]. The equations above thus completely describe any exchange of mechanical energy between deterministic systems exerting known forces. If the two-spring

[3] Note the similarity of this form to $\int \dot{p} - F\, dt$, describing a possible parallel to the flux connection, or integrated applied potential.[28]

system considered above were disconnected at time S, with an outside observer absorbing the kinetic energy of the wall (Eq. 21), the energy change of the reservoir would reduce to the usual thermostatic potential change $\int_0^S dU_{\text{ext}}(t)$.

A well known consequence of Liouville's theorem is that the entropy change is zero in a completely deterministic process.[35, 16] Using adiabatic processes, then, it is possible to propagate a starting state, A, to any state with constant entropy. If, however, phase space volume were not preserved, it would amount to discarding information on the state of the system at a given time (for example, by integrating the probability over short time intervals). Then the amount of work that can be recovered from the system will become less than the amount input. In an extreme case, all information about the system may have been lost, flushing the corresponding information content to zero. Starting from this unknown state, p^0, the probability of a given frequency distribution, p, is approached by $P(p) \propto e^{\mathcal{H}[p]}$, with $\mathcal{H}[p]$ the familiar information entropy functional,

$$\mathcal{H}[p] = -\sum_i p_i \ln p_i / p_i^0. \tag{22}$$

Further, if exchanges of conserved quantities during some process $A \to B$ are known, then the set $I(B)$ are also known from A. This information can be usefully employed to increase the amount of work that can be recovered, showing the entropy as a measure of "lost work." Consequently, any adiabatic process should be described by not only the mechanical work values, but also the change in information entropy due to information loss.

Next, consider allowing heat exchange in addition to controllable work. Assuming an infinitely large reservoir (or a short enough time-step), added heat will cause a negligible change in reservoir temperature. Because we have assumed the work done on each reservoir (Eq. 15) can be reversibly stored, these are not associated with an entropy change. We therefore introduce the physical entropy change in the reservoir as due only to exchange of heat, or "non-work" energy, $dS_{\text{ext}} = \beta_0 dQ_{\text{ext}}$, with $\beta_0^{-1} = k_B T_{\text{ext}}$. In this article, different notations are used for the information entropy, $\mathcal{H}$, and the physical entropy, S. Both of these quantities are defined to be unitless and have some relation to the information entropy of Eq. 22. The energy rejected to the reservoir as heat, dQ_{ext}, originates from heat removed from the system plus a kinetic energy, dK, assumed to be recoverable only as heat.

$$\begin{aligned} dQ_{\text{ext}} &= -dQ + dK = -dE - dU_{\text{ext}} \\ &= \frac{\delta \mathcal{A}}{\delta x(t)}^T \dot{x}\epsilon - d\tilde{I}^T \tilde{\beta}/\beta_0 = -dI^T \beta/\beta_0 \end{aligned} \tag{23}$$

This heat can alternately be understood as the total energy dumped to the environment minus the energy removed "reversibly" (dU_{ext}). This interpretation shows that if some of the changes in the environment were re-classified as irreversibly stored, so that the information $\beta_j dI_j$ becomes lost, then this is equivalent to adding that energy to the total dQ_{ext}. The total energy rejected to the environment is recoverable (making a given step mechanical as considered above) if and only if $dQ_{\text{ext}} = 0$, implying $dE = dI_0 = -d\tilde{I}^T \beta/\beta_0$. These considerations again highlight the subjective nature inherent in the definition of irreversibility.

Connecting back to the usual thermostatics, U_{ext} plays the role of an energy for the reservoir.

$$\begin{aligned} dU_{\text{ext}} &= dE_{\text{ext}} - \beta_0^{-1} dS_{\text{ext}} \\ &= d\tilde{I}^T \tilde{\beta}/\beta_0 \end{aligned}$$

To use this relation, we must provide an experimental justification for the ability to use or store the energy terms appearing in the last equation.

For any transformation, $A \to B$, the total entropy change deriving from information loss and interactions with the environment is given by the change in information entropy (Eq. 22) plus the heat exchange term (Eq. 23). This can be integrated over the time in which the transformation takes place as

$$\Delta S_{\text{inf,tot}} = \mathcal{H}[P(x|B)] - \mathcal{H}[P(x|A)] + \beta_0 Q_{\text{ext}}. \tag{24}$$

It should be noted that this formula is still not complete if there is a change in the phase space between A and B, for example if particles are added/removed, or if the state space is uniformly dilated. In this case, we have extra information on the region of phase space occupied after a transition. In general, if the state at time i is known to be x_i, then the size of the region of configuration space accessible at time $i+1$ is $Z_i(\beta, x_i)$ (Eq. 6). This reduces the entropy at $i+1$ to $S_{i+1} - \ln Z_i[\beta, x_i]$. Writing this down for each transition,

$$\Delta S_{\text{tot}} = \left\langle \sum_{i=0}^{S-1} dS_i \right\rangle$$

$$dS_i \equiv -\ln \frac{\mathcal{P}_{i+1}(x_{i+1}) Z_i[\beta, x_i]}{\mathcal{P}_i(x_i) Z_i[-\beta, x_{i+1}]} - dI^T \beta. \tag{25}$$

This equation has been symmetrized by including the corresponding entropy decrease for i given that x_i was inferred from x_{i+1} using the forward step probability but with reversed forces, $-\beta$. Support for this form is given in the next section, where it is also shown that $\Delta S_{\text{tot}} \geq 0$ using the Gibbs inequality.

Nevertheless, Eq. 24 can already be applied to Langevin and Brownian motion with uniform diffusion constants. As another example, applying Eq. 24 to a process taking any point to the equilibrium distribution shows that the expected entropy change for this process is the usual system entropy difference plus $\beta_0 Q_{\text{ext}} = -\beta_0(\langle E|B\rangle - \langle E|A\rangle) - \Delta\tilde{I}^T\tilde{\beta} = (I(A) - I(B))^T\beta$. Because the end-point entropies are fixed, alternate processes for transforming $A \to B$ are restricted to varying $\Delta S_{\text{ext}} = \int \beta_0 dQ_{\text{ext}}$. We may give a direct microscopic interpretation for the Clausius form of the second law for such processes.

$$\Delta S_{\text{tot}} = \Delta\mathcal{H} + \int \beta_0 dQ_{\text{ext}} \geq 0 \tag{26}$$

Here $\Delta\mathcal{H}$ is a function of the end-points, and β_0 and dQ_{ext} are fully variable along the path. Choosing a path from a specified distribution, A, to a maximum entropy distribution, $I(A)$, and then to the ending maximum entropy distribution $I(B)$, we may employ a quasistatic, "reversible," path between the two maximum entropy distributions so that $\min \int \beta_0 dQ_{\text{ext}}(A \to I(B)) = \min \int \beta_0 dQ_{\text{ext}}(A \to I(A))$. The heat evolved in this best-case process has its origin in the re-classification of information that occurs during the coarsening of A, in accordance with conclusions on Maxwell's demon.[36]

The equilibrium theory is therefore contained in the present development in the form of a slow process that quickly relaxes to a canonical distribution at each moment. This perspective shows the intimate connection between coarse-graining that assumes infinitely fast relaxation of the reservoir and the traditional theory of quasi-static processes. The complete theory also permits an investigation of both relaxation processes and entropy production in time and history-dependent systems moved arbitrarily far from equilibrium by coupling to simple thermostatic reservoir systems.

5. Path Perturbation and Connection to Entropy Production Theorems

Free energy perturbation techniques are well-known methods for computationally estimating free energy differences to high accuracy.[37] This method carries directly over to nonequilibrium ensembles and allows us to progress beyond linear perturbation theories.

Any two processes on the same path space, $\{\Gamma\}$, can be compared using a likelihood,

$$e^{l_{A\to B}(\{x\}_0^S)} \equiv \mathcal{P}_B\left(\{x\}_0^S\right)/\mathcal{P}_A\left(\{x\}_0^S\right) \tag{27}$$
$$\Rightarrow \langle l\rangle_B = \mathcal{D}[B|A] \geq 0,$$

where $\mathcal{D}$ is the Kullback-Leibler divergence (and necessarily positive). The likelihood obeys $\left\langle e^{l[\Gamma]}\right\rangle_A = 1$, and (by the Gibbs inequality) $\langle l\rangle_A \leq 0 \leq \langle l\rangle_B$. The distribution of the weight satisfies the large deviation formula[38]

$$\begin{aligned} e^L \mathcal{P}_A\left(L = l[\Gamma]\right) &= \int e^{l[\Gamma]}\delta(L - l[\Gamma])\mathcal{P}_A\left(\Gamma\right)\, d\Gamma \\ &= \int \delta(L - l[\Gamma])\mathcal{P}_B\left(\Gamma\right)\, d\Gamma \\ &= \mathcal{P}_B\left(L = l[\Gamma]\right), \end{aligned}$$

where $d\Gamma$ represents integration over all possible paths.

It is possible to express the path average of a functional, $a[\Gamma]$, using this weighting factor as

$$\langle a[\Gamma]\rangle_B = \left\langle a[\Gamma]e^{l[\Gamma]}\right\rangle_A. \tag{28}$$

In the equilibrium theory, however, a constant related to the free energy difference is usually cancelled on the right-hand side of this expression. There is no such constant in the prior equation because we have not identified an appropriate extensive variable.

It has been claimed that entropy production can be gauged by the ratio of forward to reverse path probabilities.[14, 11] For the process derived in Sec. 3, we can define a reverse process by inverting the sign of the generalized forces, β_i, and normalizing the distribution separately for each x_{i+1}. This reversal corresponds to an attempt at guessing whether energy has been added or subtracted during a step $i \to i+1$, with the action deviation constraint, C (Eq. 14), unchanged. Thus, we can define a ratio

$$\begin{aligned} e^{dS_i} &\equiv \frac{\mathcal{P}\left(x_{i+1}|x_i, \beta\right)\mathcal{P}\left(x_i\right)}{\mathcal{P}\left(x_i|x_{i+1}, -\beta\right)\mathcal{P}\left(x_{i+1}\right)} \\ &= \frac{\mathcal{P}\left(x_i\right)p_0(x_{i+1}|x_i)Z[-\beta, x_{i+1}]}{\mathcal{P}\left(x_{i+1}\right)p_0(x_i|x_{i+1})Z[\beta, x_i]}e^{-dI^T\beta}. \end{aligned} \tag{29}$$

Since $\frac{p_0(x_{i+1}|x_i)}{p_0(x_i|x_{i+1})} = \frac{p_0(x_{i+1})}{p_0(x_i)}$ by Bayes' theorem, this result exactly matches Eq. 25 arrived at through thermodynamic reasoning.

Moreover, several steps can be concatenated to give

$$e^{\sum_{i=0}^{S-1} dS_i} = \frac{\mathcal{P}\left(\Gamma|x_0, \beta\right)\mathcal{P}\left(x_0\right)/p_0(x_0)}{\mathcal{P}\left(\Gamma|x_S, -\beta\right)\mathcal{P}\left(x_S\right)/p_0(x_S)}. \tag{30}$$

This equation can therefore be viewed as a statistical basis for the second law of thermodynamics. It is physically motivated by observing that the entropy increase is attributed to a combination of environmental entropy changes, dQ_{ext}, and information-like entropy changes,

$$-\ln\frac{\mathcal{P}\left(x_{i+1}\right)Z[\beta, x_i]/p_0(x_{i+1})}{\mathcal{P}\left(x_i\right)Z[-\beta, x_{i+1}]/p_0(x_i)}.$$

This term implies an extra contribution to the total entropy change beyond Eq. 24. To understand this contribution, consider a simple one-state system, to be transformed into a two-state system through the (unbiased) transition probability $\mathcal{P}\left(x_{i+1}|x_i\right) = p_0(x_{i+1}) = 1/2$. The presence of the normalization in Eq. 25 leads to an additional factor of $-\ln Z[x_i]/p_0(x_{i+1}) = -\ln 2$ in the entropy. This term is canceled by the entropy of the resulting state, $\mathcal{P}(x_{i+1}) = 1/2$, so that the total entropy change for this process is zero. If instead we perform a bit-set operation by going in the opposite direction, the entropy will increase if $\mathcal{P}(x_{i+1})$ is any non-uniform state – corresponding to our loss of information on discarding the bit. A term of this form is exactly what we should have expected when writing down Eq. 24. Applying this equation to a situation where a particle is added to the system, we find that the re-normalization will physically compensate for the expansion of phase space, which leads to a difference in entropy measures. Mathematically, this implies that the "default" measure $p_0(x_{i+1})$ is replaced by the re-normalized measure $p_0(x_{i+1})Z^*_{\text{tr}}(x_{i+1})/Z_{\text{tr}}(x_i)$ if we have information about the previous state when calculating the entropy at state $i+1$.

5.1. *Jarzynski's Equality*

Jarzynski equations can be derived in the special case where a stationary distribution (denoted $x \sim$ ss) is known during each time-propagation step. The stationary distribution is not required to have a physical temperature and may vary between time-steps. At each step, it can be any single-time distribution that remains a fixed-point of the transition operator. The derivation uses Eq. 27 to re-weight each state between transitions,

$$\langle a(x(T))|\text{ss}, T\rangle = \left\langle a(x(T)) \prod_{t=1}^{T} \frac{\mathcal{P}_{\text{ss}}(x(t), t)}{\mathcal{P}_{\text{ss}}(x(t), t-1)} |x_0 \sim \text{ss}, 0 \right\rangle. \tag{31}$$

The first average represents a single-time average over the steady-state distribution used to propagate the system during the transition $T \to T+1$. The second average is a path average over stochastic trajectories beginning in the steady-state at time zero. For the particular choice $a = 1$, the normalization constants cancel between successive $\mathcal{P}_{\text{ss}}(x(t), t)$, leaving only the ratio between the end-point partition functions to be expressed as a path average.

The proof[39] is by recursion from the property of the stationary distribution under time-propagation,

$$\mathcal{P}_{\text{ss}}(x(t+1), t) = \int \mathcal{P}_{\text{tr}}(x(t+1)|x(t), t)\, \mathcal{P}_{\text{ss}}(x(t), t)\; dx(t).$$

Because the transition probability, $\mathcal{P}_{\text{tr}}(\cdot)$, is not physically constrained, equality of the starting and ending temperatures is not required. Combinations of Eq. 27 can give a very large number of fluctuation theorems.

6. Conclusions

In this article, we have given a generalization of the theory for driven, irreversible processes. Our formulation of maximum transition entropy in Sec. 3 provides several distinct advantages for the derivation of nonequilibrium relations. This theory does not require assuming a Boltzmann-Gibbs form for the starting distribution, nor does it rely on any particular definition of a stationary distribution. Instead, the time-evolution for the system is determined directly from a set of transition probabilities. These are obtained from applying maximum entropy methods, paralleling the canonical form of equilibrium thermostatics (Table 1). Non-anticipating stochastic trajectories for classical particle and field motion can be cast in this form. The action functional interpretation gives a physical method for defining conserved quantities exchanged between the two during energy conversion processes (Sec. 4). The energy cost associated with transfers of these quantities from an external environment or experimental apparatus can easily be specialized by an appropriate choice of constraints and computed.

The Lagrangian and Langevin equations are seen seen to form two limits of the theory, with the rate of energy exchange with the bath as the control parameter. Deterministic dynamics are recovered from the Langevin equation when the deviation of the action functional is strongly constrained to zero (Sec. 3). In this limit, the external forces, which appeared as statistical in the stochastic approach, become the mechanical Lagrange-D'Alembertian forces. Because both limits appear here, and because the Lagrange multipliers can be set to produce constant force or constant flux transitions,[28] the fluctuation-dissipation theorems derived as Gibbs relations from taking second derivatives of Eq. 9 are applicable in the case of both thermal and mechanical driving forces. These equations are completely general because they apply not only arbitrarily far from equilibrium, but also during transient processes that do not possess a steady-state.

A particularly useful aspect of this approach is that it directly connects multiple length and time scales. The formulation of the equations has been in terms of particle motion, but coarse-grained relations are easy to define as appropriate ensemble averages over these motions. Examples of such averages include centers of mass for polymer units or average density and velocity fields. The coarse equations of motion will then lead to polymer coarse-graining models[40] or non-local hydrodynamic models.[41, 42, 43] For the time-evolution of average quantities, we expect the thermodynamic limit argument[44] to apply when the number of averaged degrees of freedom is large, so that the path realized by the system under a given set of constraints will fall arbitrarily close to the maximum entropy solution an overwhelming majority of the time. The present work is therefore a suitable foundation for the theory and analysis of nonequilibrium molecular dynamics.

Our method is free of projection operators and Fourier transforms that can often obscure the physical picture of driven dynamical systems. As a consequence, applications to simplified, standard examples such as circuit theory are easily accomplished.[28] In addition, we find the physical basis of entropy production has become much more apparent. The Joule heating of a resistor, for example, can be seen from Eq. 20 as fundamentally arising from the difference between the velocity added to each ion individually *vs.* the usable energy in the average ion velocity. Because the energy added to the system in driving the ions is not expressible in terms of the average velocity alone, spreading in the distribution of ion velocities becomes heat. The same remarks follow for driven convective transport, where a spreading in the distribution of forward fluid momentum leads to increases in the local temperature.

Connections of this theory to the formal structure of maximum entropy thermodynamics and Bayesian inference have been elaborated upon in Ref. 28. These connections allow the definition of thermodynamic cycles expressing differences between driving protocols using the same free energy techniques commonly employed in the equilibrium theory. Some examples have already appeared in the literature for path re-weighting[45, 46, 47]. It is expected that expression in terms of thermodynamic cycles will greatly simplify the derivation and interpretation of these studies.

We have identified a new generalization of the second law for irreversible processes by including the instantaneous information entropy. A traditional analysis shows that the total entropy increase (Eq. 25) is dependent on the details of system dynamics and exchange of conserved quantities with an external system. Connecting this with the fluctuation theorem (Eq. 29) gives a microscopic form for the second law of thermodynamics. The physical device of tracking work performed on individual particles as well as external reversible work sources allows us to track the flow of each type of work (and heat) through the system. Because these changes come directly from the forces on each degree of freedom, this analysis does not depend on arbitrary decompositions of energy functions or definitions of steady-states.

From an informational perspective, entropy increase comes about from discarding information and/or from the information loss associated with coupling to external reservoirs. This is distinguished from the entropy production functional of local equilibrium theory in that the entropy functionals developed here include long-range correlations and are thus not simply functions of locally defined quantities such as composition or temperature.[48] Instead, we have shown that temperature should not be thought of as a property of the system, but of the external bath. This frees us to define more complicated baths, for example placing constraints on $\partial\mathcal{A}/\partial x_a(t)$ only for particles, a, in a range of space can be used to model local thermal contacts for imposing thermal or pressure gradients.[49] It should be noted that the analysis in Sec. 4 showed that increasing the number of controllable variables decreases the number of degrees of freedom associated with heat production.

Many questions remain open for attack by this method. Feynman's path action formulation of quantum mechanics[50] uses the action functional as a rotation operator, whereas the present relations are of the diffusion-type. Is there an appropriate limiting process whereby the latter may be obtained and the multipliers C identified with quantum fluctuations?[51, 52, 53] Ref. [54] gives a Lagrangian related to the Clebsch parametrization for the field equations of Eulerian fluid mechanics. Could application of this derivation method yield novel turbulence models? Addition of electric fields to this picture should result in derivations for the dielectric response in nanoscale fluid mechanics[55, 56]. The present report has not been specialized to cases where near-equilibrium approximations may be applied, and these should also be investigated for the potential of generating new relations in that regime.

Acknowledgment
This work was supported by Sandia's LDRD program. Sandia National Laboratories is a multi-program laboratory operated by Sandia Corporation, a wholly owned subsidiary of Lockheed Martin Corporation, for the U.S. Department of Energy's National Nuclear Security Administration under contract DE-AC04-94AL85000.

References

[1] Kjelstrup S and Bedeaux D 2008 *Non-Equilibrium Thermodynamics Of Heterogeneous Systems (Series on Advances in Statistical Mechanics)* (World Scientific) ISBN 9812779132

[2] Zwanzig R 1961 *Phys. Rev.* **124** 983–992

[3] Zwanzig R, Nordholm K S J and Mitchell W C 1972 *Phys. Rev. A* **5** 2680–2682 correction to Zwanzig, 1961

[4] Mori H 1965 *Prog. Theor. Phys.* **33** 423–455

[5] Zwanzig R 1965 *Ann. Rev. Phys. Chem.* **16** 67–102

[6] Lebowitz J L 1993 *Phys. Today* **46** 32–38

[7] Jaynes E T 1980 *Ann. Rev. Phys. Chem.* **31** 579–601

[8] Jaynes E T 1985 *Complex Systems - Operational Approaches* ed Haken H p 254

[9] Jaynes E T 1979 *The Maximum Entropy Formalism* ed Levine R D and Tribus M (M.I.T Press, Cambridge) p 498 ISBN 0262120801,9780262120807

[10] Kubo R, Toda M and Hashitzume N 2003 *Statistical Physics II: Nonequilibrium Statistical Mechanics* (Springer) ISBN 354053833X 2nd ed. Translated by M. Toda, R. Kubo, N. Saito, and N. Hashitsume

[11] Maes C 1999 *J. Stat. Phys.* **95** 367–392

[12] Lebowitz J L and Spohn H 1999 *J. Stat. Phys.* **95**(1) 333–365 ISSN 0022-4715

[13] Dewar R 2003 *J. Phys. A: Math. Gen.* **36** 631–641

[14] Crooks G E 1999 *Phys. Rev. E* **60** 2721–2726

[15] Robin W A 1990 *J. Phys. A: Math. Gen.* **23** 2065–2085

[16] Robertson B 1993 *Physics and Probability: essays in honor of Edwin T. Jaynes* ed Grandy W T (Cambridge Univ. Press) pp 251–260

[17] Luzzi R, Áurea R Vasconcellos and Ramos J G 2002 *Predictive statistical mechanics: a nonequilibrium ensemble formalism* (Kluwer, Dordrecht) ISBN 1402004826

[18] Stock G, Ghosh K and Dill K A 2008 *J. Chem. Phys.* **128** 194102 ISSN 00219606

[19] Press S, Ghosh K and Dill K A 2011 *J. Phys. Chem. B* **115** 6202–6212

[20] Baiesi M, Maes C and Neton K 2009 *Journal of Statistical Physics* **135** 57–75 ISSN 0022-4715

[21] Filyukov A A and Karpov V Y 1967 *J. Eng. Phys. Thermophys.* **13**(6) 416–419 ISSN 1062-0125

[22] Jou D, Casas-Vázquez J and Lebon G 1988 *Rep. Prog. Phys.* **51** 1105–1179

[23] Jou D, Casas-Vázquez J and Lebon G 1999 *Rep. Prog. Phys.* **62** 1035–1142

[24] Attard P 2006 *J. Chem. Phys.* **124** 024109

[25] Attard P 2009 *J. Chem. Phys.* **130** 194113 (pages 10)

[26] Gill J 2002 *Bayesian Methods* (Chapman & Hall/CRC) ISBN 1584882883,978-1584882886

[27] Bobrowski A 2005 *Functional Analysis for Probability and Stochastic Processes: An Introduction* (Cambridge University Press) ISBN 0521831660

[28] Rogers D M, Beck T L and Rempe S B 2011 *J. Stat. Phys.* **145** 385–409

[29] Marsden J E and West M 2001 *Acta Numerica* **10** 357–514

[30] Noether E 1971 *Transp. Theory Stat. Phys.* **1** 186–207 English translation by M. A. Tavel

[31] Risken H 1996 *The Fokker-Planck Equation: methods of solution and applications* (Springer) ISBN 354061530X 2nd ed.

[32] Frank T D 2007 *Phys. Lett. A* **360** 552–562 ISSN 0375-9601

[33] Feller S E, Zhang Y, Pastor R W and Brooks B R 1995 *J. Chem. Phys.* **103** 4613–4621

[34] Callen H B 1985 *Thermodynamics and an Introduction to Thermostatistics* (Wiley, New York) ISBN 0471862568 2nd ed.

[35] Robertson B and Mitchell W C 1971 *J. Math. Phys.* **12** 563–568

[36] Leff H S and Rex A F 2003 *Maxwell's demon 2: entropy, classical and quantum information, computing* (IOP Publishing, London) ISBN 0750307595

[37] Chipot C and Pohorille A (eds) 2007 *Free Energy Calculations* (Springer) ISBN 3540384472

[38] Bennett C H 1976 *J. Comput. Phys.* **22** 245–268 ISSN 0021-9991

[39] Jarzynski C 1997 *Phys. Rev. Lett.* **78** 2690–2693

[40] Akkermans R L C and Briels W J 2000 *J. Chem. Phys.* **113** 6409–6422

[41] Grandy W T 1987 *Foundations of Statistical Mechanics* (Kluwer, Boston) ISBN 902772489X, 9789027724892

[42] Kubo R 1966 *Rep. Prog. Phys.* **29** 255

[43] Lehoucq R and Von Lilienfeld-Toal A 2010 *J. Elasticity* **100**(1) 5–24 ISSN 0374-3535

[44] Jaynes E T 1957 *Phys. Rev.* **108** 171–190

[45] Zuckerman D M and Woolf T B 1999 *J. Chem. Phys.* **111** 9475–9484

[46] Minh D D L and Chodera J D 2009 *The Journal of Chemical Physics* **131** 134110

[47] Chodera J D, Swope W C, Noe F, Prinz J, Shirts M R and Pande V S 2011 *J. Chem. Phys.* **134** 244107–14

[48] Jaynes E T 1992 *Maximum Entropy and Bayesian Methods* ed Smith C R, Erickson G J and Neudorfer P O (Kluwer, Dordrecht) pp 1–22 ISBN 978-0792320319

[49] Zhu F, Tajkhorshid E and Schulten K 2004 *Biophys. J.* **86** 50–57 ISSN 0006-3495

[50] Dyson F 2007 *Advanced Quantum Mechanics* (World Scientific) ISBN 9812706224 transcribed from 1951 Lecture Notes by David Derbes

[51] Christopeit N, Helmes K, Kohlmann M and Yasue K 1986 *Stochastic Differential Systems* (*Lecture Notes in Control and Information Sciences* vol 78) (Springer Berlin/Heidelberg) pp 352–361

[52] Davies I M 1989 *J. Phys. A: Math. Gen.* **22** 3199–3203 ISSN 0305-4470

[53] Dammak H, Chalopin Y, Laroche M, Hayoun M and Greffet J J 2009 *Phys. Rev. Lett.* **103** 190601

[54] Mobbs S D 1982 *Proc. Royal Soc. London. A. Math. Phys. Sci.* **381** 457–468

[55] Hubbard J and Onsager L 1977 *The Journal of Chemical Physics* **67** 4850–4857

[56] Ballenegger V, Blaak R and Hansen J P 2006 *Computer Simulations in Condensed Matter Systems: From Materials to Chemical Biology Volume 2* (*Lecture Notes in Physics* vol 704) ed Ferrario M, Ciccotti G and Binder K (Springer Berlin / Heidelberg) pp 45–63

Efficient non-equidistant FFT approach to the measurement of single- and two-particle quantities in continuous time Quantum Monte Carlo methods

Peter Staar[1], **Thomas A. Maier**[2], **and Thomas C. Schulthess**[1,2]

[1] Institute for Theoretical Physics, Wolfgang Pauli Strasse 27, ETH Zurich, 8093 Zurich, Switzerland

[2] Computer Science and Mathematics Division and Center for Nanophase Materials Sciencens, Oak Ridge National Laboratory, Oak Ridge, TN 37831

Abstract. Continuous time cluster solvers allow us to measure single- and two-particle Greens functions in the Matsubara frequency domain with unprecedented accuracy. Currently, the usage of the two-particle functions is limited due to a lack of an efficient measurement method that can deal with the random times of the vertices. In this paper, we show how the Non-equidistant Fast Fourier Transform (NFFT) algorithm can be modified in order to obtain a very efficient measurement algorithm. For the single particle case, we propose a delayed-NFFT (d-NFFT) scheme, which reduces the arithmetical operations from $\mathcal{O}(N\log(N))$ in NFFT to $\mathcal{O}(N)$, a huge improvement compared to the standard $\mathcal{O}(N^2)$ of the Non-equidistant Discrete Fourier Transform (NDFT), currently used in most continuous time cluster solvers. For the two-particle case, we discuss how the NFFT can be applied to measure the two-particle Greens functions and how to exploit its properties to further optimize the NFFT. We then apply these algorithms to the half-filled 2D Hubbard model at $U/t = 8$ in order to study the anti-ferromagnetic transition. In particular, we confirm the logarithmic decay of the Neel-temperatures versus cluster-sizes in accordance with the Mermin-Wagner theorem.

1. Introduction

The classification of different phases in Fermionic lattice models with strong repulsive interactions is one of the fundamental questions in solid state theory. Many methods have been developed to address this question, each with its own strengths and weaknesses. Exact diagonalization is adequate to investigate these models for small systems, but cannot be used to study the thermodynamic limit, and is thus not usable for the study of, e.g. critical temperature in phase transitions. The Dynamical Matrix Renormalization Group can analyze larger system sizes, but going beyond quasi 1 dimensions has proven difficult. To bridge the gap between dimensionality and thermodynamics, the Dynamical Cluster Approximation [1] was developed over the last ten years. Used in combination with a Quantum Monte Carlo (QMC) cluster-method, it has become the approach of choice to investigate phase transition with non-local order-parameters [2]. Although it is plagued by the fermionic sign problem at low temperatures, it has proven to predict phase transitions such as d-wave superconductivity in the 2D Hubbard model [3] and anti-ferromagnetism in the 3D Hubbard model [4].

Most of the results produced with the DCA have been obtained with a Hirsch-Fye QMC cluster method [5]. The Hirsch-Fye algorithm is based on a Trotter-decomposition of the

Published under licence by IOP Publishing Ltd

Hamiltonian and is thus only exact in the limit of an infinite number of time-slices. Therefore, each physical quantity obtained with a Hirsch-Fye cluster method needs to be carefully extrapolated towards zero time-step size [4]. Recently, a new generation of continuous time cluster-methods has been introduced [6], which overcome this problem. These methods are more accurate, and make extrapolations to zero time-step unnecessary.

Th.e continuous time methods do not suffer from the discretization problem of the Hirsch-Fye algorithm because the vertices are allowed to reside on randomly chosen times. This extra freedom leads to a better sampling of the partition function, but it also introduces a new set of problems when measuring physical quantities quantities. In continuous time Monte Carlo methods, quantities are measured in either imaginary time or in Matsubara frequencies. Measuring in imaginary time is very fast but it introduces errors via the binning, since the locations along the time axis for which observables are being measured are changing between measurements. Measuring in Matsubara frequencies on the other hand is exact, but slow, since each measurement requires a discrete Fourier transform. Efficient algorithms such as FFT can not be applied, since they require data on an equispaced time-grid. The measurement of the Greens functions and other related functions are therefore generally slow for continuous time methods, especially in the case of the two-particle Greens functions. This inefficiency limits drastically the usage of the two-particle Greens functions to detect phase-transitions, and hence the DCA as such.

In the implementation section of this paper, we will present methods to accelerate the measurements of single- and two-particle functions in the context of continuous time cluster methods. The focus in this paper lies on the CT-AUX cluster algorithm [7, 8], but the findings can be applied to other continuous time methods without complications. For the single particle Greens function, we optimize the existing algorithm Non-equispaced Fast Fourier Transform (NFFT) [9] for a Monte Carlo integration such that measurements scale linear in the expansion order $\langle k\rangle$, and not as $\sim \mathcal{O}(\langle k\rangle \log(\langle k\rangle))$ of the NFFT. For the two-particle Greens function, we discuss various methods that exploit the properties of the two particle Greens functions to accelerate the NFFT on these two-dimensional domains.

Due to the improvements on the measurements, we were able to conduct a systematic study over a wide range of cluster sizes of the half-filled 2D single-band Hubbard model. This model is defined by its Hamiltonian

$$H = -t \sum_{\langle i,j\rangle,\sigma} c^{\dagger}_{i,\sigma} c_{j,\sigma} + U \sum_{i} (n_{i,\uparrow} - \frac{1}{2})(n_{i,\downarrow} - \frac{1}{2}), \tag{1}$$

wherel, $c_i^{(\dagger)}$ destroys (creates) an electron with spin σ on site i, $n_{i,\sigma}$ corresponds to the number operator, t is the hopping amplitude between nearest neighbors i and j, and U the on-site Coulomb repulsion. At half filling, the QMC sign problem is absent, which allows us to investigate very large clusters with large interaction strength U. The large cluster runs show the effectiveness of our proposed methods. Here we are interested in the antiferromagnetic behavior of this model. The Mermin-Wagner theorem states that magnetic order cannot exist at finite temperatures in the 2D model. The DCA method, however, because of its mean-field character for finite cluster size, predicts magnetic order at finite temperatures. Here we show that the antiferromagnetic transition temperature T_N decreases logarithmically with increasing cluster size and give a heuristic argument to explain this behavior.

The paper is structured as follows: In Sec. 2, we will briefly review the use of the two-particle Greens functions in the context of the DCA to determine the phase-transitions. In Sec. 3, we will first briefly discuss the role of the new methods within the implementation of the DCA. Next, the d-NFFT-measuring method for single-particle quantities is explained in detail. Both accuracy and speed-up of the d-NFFT are compared to the exact NDFT method. For the two-particle measurements, we will first discuss how the 2D-NFFT can be used, and then discuss various optimization techniques to improve the speed-up. The optimization techniques will be motivated both from the physics and from the algorithm viewpoint. Sec. 5 contains the summary.

2. Theory

The ability to measure non-local two-particle Greens-functions is an important advantage of the DCA towards other methods which study correlated electron systems. With the two-particle Greens-functions, we can solve the Bethe-Salpeter (BS) equation, and accurately determine the critical temperature of a phase transitions. Moreover, the solutions of the BS equation will reveal the internal momentum and frequency structure of the process, which drives the phase-transition. Currently, the use of the two-particle Greens function in a DCA with continuous time cluster solvers is limited due to its time-consuming nature. Up till now, it has therefore mainly been used in fixed time algorithms [5], where the two-particle measurements can be efficiently implemented [10]. Here, we revisit briefly how the single and two-particle Greens function are defined in a continuous-time algorithm.

In the DCA, the infinite lattice-problem is reduced to a finite size quantum cluster impurity with periodic boundary conditions, embedded into a self-consistent mean-field. This reduction is achieved via a coarsegraining procedure of the Greens functions, in which the Brillouin zone is divided into N_c patches and the self-energy Σ is assumed to be constant on these patches. In this way, all correlations within the cluster can be dealt with exactly, while long-range correlations outside the cluster are described via a mean-field. In this picture, the single-particle Greens-function G^I and two-particle Greens-function G^{II} are translation invariant and therefore defined as

$$\begin{aligned}
G^I_\sigma[k] &= \frac{1}{N_c}\sum_{i,j} e^{i\,k\,(r_j - r_i)}\langle T_\tau\{c^\dagger_{r_j,\sigma} c_{r_i,\sigma}\}\rangle, \\
G^{II}_{\sigma_1,\sigma_2}[k_1,k_2,q] &= \frac{1}{N_c}\sum_{r_1,r_2,r_3,r_4} e^{i\,(k_1+q)\,r_1} e^{-i\,k_1\,r_2} \\
&\times\; e^{i\,k_2\,r_3}\, e^{-i\,(k_2+q)\,r_4}\langle T_\tau\{c^\dagger_{r_1,\sigma_1} c_{r_2,\sigma_1} c^\dagger_{r_3,\sigma_2} c_{r_4,\sigma_2}\}\rangle,
\end{aligned}$$

where we have used the conventional notation $k = \{\vec{k}, \varpi\}$ and $r = \{\vec{r}, \tau\}$ [11]. The susceptibility χ is now computed from the single and two-particle Greens function in a two step process. First, one obtains the cluster vertex function Γ_c via inversion of the Bethe-Salpeter equation on the cluster

$$\Gamma_c = \left[G_0^{II}\right]^{-1} - \left[G^{II}\right]^{-1}.$$

Here, G_0^{II} represents the noninteracting cluster-susceptibility, which is obtained from a pair of interacting Greens functions. The lattice susceptibility is now computed by inverting a second time the Bethe-Salpeter equation, which is defined on the infinite lattice and uses the Γ_c as an approximation for the vertex function. The noninteracting lattice-susceptibility χ_0 corresponds now to a coarsegrained pair of interacting Greens-functions, and the lattice susceptibility is computed as

$$\chi = \chi_0\left[1 - \Gamma_c\,\chi_0\right]^{-1}.$$

Since a phase transition occurs at the divergence of χ, it is sufficient to look at the leading eigenvalue λ_0 of the matrix $\Gamma_c\,\chi_0$, and observe that it crosses 1 as the temperature is lowered. In this paper, we will consider the anti-ferromagnetic transition, in which case we compute the two-particle Greens function that corresponds to the magnetic particle-hole channel. Its corresponding two-particle Greens function $G^{(m),II}$ and noninteracting lattice-susceptibility $\chi_0^{(m)}$ are defined respectively as,

$$G^{(m),II}[k_1,k_2,q] = \frac{1}{4}\sum_{\sigma_1,\sigma_2=\pm 1}(\sigma_1\,\sigma_2)\,G^{II}_{\sigma_1,\,\sigma_2}[k_1,k_2,q],$$
$$\chi_0^{(m)}[k_1,k_2,q] = \frac{\delta_{k_1,k_2}}{N_c}\int_{\mathbf{k}\in k_1} d\mathbf{k}\,G(k)G(k+q).$$

3. Implementation.

The DCA algorithm is schematically represented in Fig. (1). For a good implementation, it is crucial to have a fast QMC-cluster solver, since other sections only take a marginal amount of the run-time. All continuous time QMC-solvers consists of 2 parts that are executed consecutively order many times during the sampling of the Greens-function G^I (see Fig (1)). In the first part, one stochastically alters the expansion order $\langle k\rangle$ of the partition function in a step by step process, until the sample is uncorrelated from the original sample. In the CT-AUX solver, these steps are random additions and removals of Hubbard Stratonovitch spins in the Hubbard Stratonovitch configuration [7]. For a Hybridization solver [12, 13], the steps are implemented by random insertions or removals of pairs of operators. In the second part, we measure the Greens-functions which correspond to the sample of the partition function. Recently, a submatrix-update procedure [8] was presented, and it accelerated the sampling of the partition function by a factor of ~ 10 in the CT-AUX. This fast sampling lead to a quick thermalization, and allowed us to linearly scale large cluster-problems to tens-of-thousands of processors [8]. It also meant that the time spent in the measurements became disproportionally large. This motivated us to search for fast and efficient methods to measure the Greens-functions, both in the single- and two-particle case. We present here methods based on NFFT to accelerate the measurements of Greens functions in the context of continuous time cluster solvers. The focus in this paper lies on the CT-AUX cluster solver, but the findings can be applied to other continuous time solvers without complications, e.g. such as the Hybridization-solver.

3.1. measurement of single particle quantities

There exist two types of single particle quantities. The most common ones, such as the self-energy and the single-particle Greens function G^I are translation invariant in space and time. The non-invariant single particle Greens functions are only used during the computation of the two-particle Greens function. We focus first on the space-time translation invariant quantities. In the CT-AUX algorithm, the single-particle Greens function G^I is not directly obtained via the Monte Carlo integration, but derived from an other accumulated quantity $\langle M[\mathbf{k}, i\,\varpi]\rangle$. In the latter, $\langle ....\rangle$ represents the Monte Carlo integration. Following closely the notation in [7], we can write down the quantity M as a function of the N-matrix and its corresponding Hubbard-Stratonovitch spin configuration $\{v\}$,

$$G^I[k] = G^0[k] - G^0[k]\langle M[k]\rangle G^0[k], \tag{2}$$
$$M^{\{v\}}[\mathbf{r}, i\,\varpi] = \sum_{\lambda_1,\lambda_2\,\in\{v\}} e^{i\,\varpi\,(\tau_{\lambda_2}-\tau_{\lambda_1})} M^{\{v\}}[\lambda_1,\lambda_2]$$
$$M^{\{v\}}[\lambda_1,\lambda_2] = \delta_{\mathbf{r},\mathbf{r}_{\lambda_2}-\mathbf{r}_{\lambda_1}}(e^{V\{v_{\lambda_1}\}}-1)N[\lambda_1,\lambda_2].$$

In the case of the Hybridization algorithm, one directly accumulates the single particle Greens

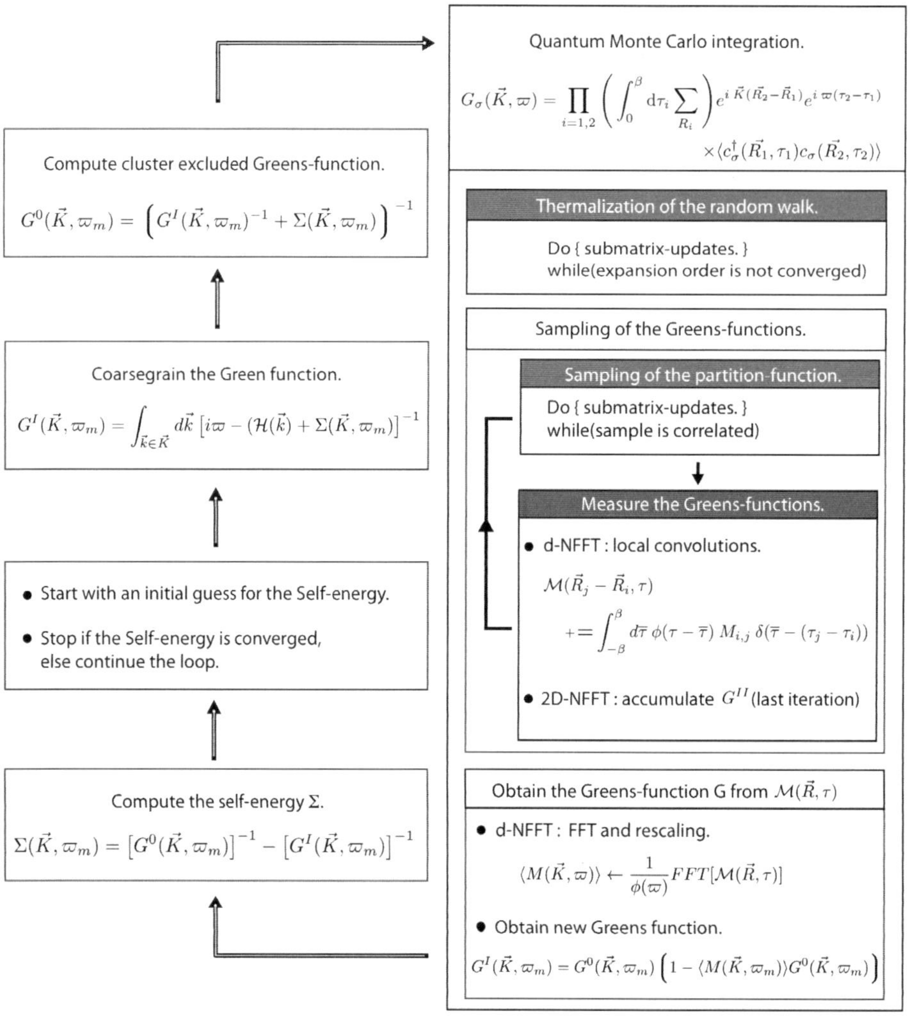

Figure 1: A schematic representation of the DCA self-consistency loop, in which a CT-AUX cluster solver is used. The sections affected by recent optimizations are high-lighted (color online). The thermalization of the random walk and the sampling of the partition function (green) have been accelerated by the submatrix update-procedure [8]. The measurement of the single and two-particle functions (red) are accelerated respectively by the d-NFFT and 2D NFFT procedures, which are discussed in this paper. Notice that the d-NFFT algorithm spans over two section of the Monte Carlo integration: In the sampling-phase of the Greens-function, it will convolute the samples of the function $M(\vec{r},\tau)$ with a localized window-function ϕ to obtain $\mathcal{M}$ and accumulate the latter. After the sampling phase, it will Fourier transform this convoluted function $\mathcal{M}$ and rescale it by the appropriate Fourier coefficient of the window-function, in order to obtain the averaged Fourier transform of $\langle M(\vec{r},\tau)\rangle$. The latter is then used to obtain the interacting Greens function G^I, from which a new Self-Energy Σ can be obtained.

function G^I according to Eq. (3) from the inverted Hybridization matrix $M = [F(\tau_j - \tau_j)]^{-1}$,

$$G^I = \sum_{i,j} M_{i,j} \exp^{i\,\varpi\,(\tau_j - \tau_i)}\,. \tag{3}$$

In both solvers, the measurement boils down to a Fourier transform of non-equispaced data since the vertices reside on randomly chosen imaginary times. This eliminates the use of the fast Fourier transform (FFT) because the latter requires equispaced data. In a straightforward implementation of the non-equispaced discrete Fourier transform (NDFT), each measurement would be a two step process. One would first compute the Fourier coefficient matrix $C_{m,l} = e^{i\,\varpi_m\,\tau_l}$, which involves the evaluation of many expensive trigonometric functions. Next, the actual transform would be carried out via a matrix-vector multiplication. Both operations scale as $\mathcal{O}(N_\varpi N_\tau)$, where N_ϖ and N_τ represent, respectively, the number of positive matsubara frequencies and the number of vertices.

We present here a measurement algorithm which scales linear in N_τ, and requires only one Fourier transform at the end of each Monte-Carlo integration. The dependency of our measurement algorithm on N_ϖ is therefore virtually eliminated and a speed-up of order N_ϖ can be expected. It is instructive to begin with a short review on the non-equispaced fast Fourier transform (NFFT) algorithm [9]. The NFFT algorithm is based on the convolution theorem, shown in Eq. (4). It states that the Fourier transform of a convolution is equal to the point-wise product of the Fourier transforms:

$$f_\varpi\,\phi_\varpi = \int_0^\beta d\tau e^{i\,\varpi\,\tau}\Big[\int_0^\beta d\lambda\,\phi(\tau-\lambda)f(\lambda)\Big]. \tag{4}$$

The NFFT algorithm makes use of this theorem in the following manner. First, the non-equispaced data-set $\{\tau_i, f_i\}$ is projected onto an equispaced grid, with mN_ϖ intervals. Here, m is called the oversampling factor and is typically in the range of 4 to 10. This projection is achieved via a convolution with a localized, translation invariant kernel ϕ, such as a Gaussian distribution function. The localization of the kernel is in the range of $[-m, m]$ lattice spacings, which insures that the time needed for the convolution scales linearly with the data size. Next, the FFT algorithm is applied on the projected, now equispaced data and one obtains its representation in Matsubara frequencies. At last, each data-point in Matsubara frequencies is renormalized by its corresponding Fourier-coefficient of the kernel ϕ, and hence we recover the Fourier component f_ϖ of the original non-equispaced data-set. The NFFT algorithm can be summarized as,

$$f_\varpi \leftarrow \frac{1}{\phi_\varpi}\mathrm{FFT}\Big[\{\overline{f}_l = \sum_\lambda \phi(\tau_\lambda - \frac{l\,\beta}{N})\,f_\lambda\,|\,l \in 1,...,m\,N_\varpi\}\Big] = \mathrm{NFFT}(\{\tau_\lambda, f_\lambda\}) \tag{5}$$

From Eq. (5), it is clear that the NFFT algorithm scales as $\mathcal{O}(m\,N_\varpi\,\log(m\,N_\varpi) + m\,N_\tau)$. An important property of the NFFT-algorithm is that for particular kernels the maximum relative error on f_ϖ can be proven to decrease exponentially as a function of the oversampling parameters m [9]. In the case of a gaussian window function ϕ_g, the maximum relative error E_2 is

$$E_2 = \frac{|f_\varpi^{NFFT} - f_\varpi^{NDFT}|}{||f_\varpi^{NDFT}||_1} = 4\,e^{-m\,\pi\,(1-1/(2\sigma-1))}.$$

In the case of a Monte-Carlo integration with a continuous time solver, the latter will produce many data-sets, all on a different set of randomly chosen times. Since the FFT is a linear operation, we can delay this operation until the end of each Monte-Carlo integration. In this

Algorithm 1 The delayed-NFFT algorithm, which computes the single-particle Greens-function G^I:

Require: N-samples $\{v\} = \big\{\{\tau_1, M(\tau_1)\}\big\}$ of function $M(\tau)$ at random times in the interval $[-\beta, \beta]$
Ensure: Return the average of the Fourier-transformed samples $\langle M(\varpi)\rangle$.

Define the function $\mathcal{M}(\tau_1)$ on the oversampled grid $\{\tau_1 = -\beta + l_1\,\Delta\tau \mid l_1 \in \{0, \ldots, m\,N_\varpi\}\}$ and with $\Delta\tau = \frac{2\beta}{m\,N_\varpi}$

for $i = 1 \to N$ **do**
  for $\{\tau_1, M(\tau_1)\} \in \{v\}_i$ **do**
    $I = \frac{\tau+\beta}{\Delta\tau}$ {find the lower-index of τ_i such that $\tau(I) \leq \tau_1 < \tau(I+1)$}
    for $\lambda = -m \to m$ **do**
      $\phi_{interp} \leftarrow$ interpolate ϕ at $\tau(I+\lambda) - \tau_1$
      $\mathcal{M}(I+\lambda) \leftarrow \mathcal{M}(I+\lambda) + \phi_{interp} * M(\tau_1)$
    end for
  end for
end for

Compute $\mathcal{M}(\varpi) \leftarrow FFT[\mathcal{M}(\tau)]$
Compute $\langle M\rangle \leftarrow \frac{\mathcal{M}(\varpi)}{N*\phi(\varpi)}$
Compute $G^I(\varpi) \leftarrow G^0(\varpi)\,[1 + \langle M(\varpi)\rangle G^0(\varpi)]$

way, one accumulates the convoluted function $M\big[\mathbf{r}, \tau_l\big]$ during the Monte-Carlo integration, instead of $M\big[\mathbf{r}, i\,\varpi\big]$.

$$M\big[\mathbf{r}, \tau_i\big] = \sum_{\lambda_1, \lambda_2 \in \{v\}} \phi((\tau_{\lambda_2} - \tau_{\lambda_1}) - \tau_i)\, M^{\{v\}}\big[\lambda_1, \lambda_2\big] \quad \text{with} \quad \tau_i = \frac{2\,i\,\beta}{m\,N_\varpi} \tag{6}$$

As a result, only a single Fourier transform and rescaling has to be performed on $M\big[\mathbf{r}, \tau_l\big]$ to obtain $\langle M\big[\mathbf{k}, \varpi\big]\rangle$ at the end of each Monte Carlo iteration. From the latter, G^I can be obtained in the usual manner according to Eq. (2). As shown in Fig. (1), this delayed-NFFT (d-NFFT) will now span over two sections in the Monte-Carlo integration. In the accumulating phase, it will convolute the function-samples on the oversampled NFFT-grid. As a consequence, each measurement now boils down to a convolution, and the desired scaling linear $\mathcal{O}(m\,N_\tau)$ is obtained. Only after the Monte-Carlo integration, we will Fourier transform the oversampled function $M\big[\mathbf{r}, \tau_l\big]$. Notice that the delay of the FFT will not affect the accuracy, and the relative error will still decrease exponentially as a function of the oversampling. A second optimization can be obtained by replacing the exact evaluation of the kernel ϕ with an interpolated value. Evaluating a Gaussian distribution function involves the evaluation of trigonometric functions, which require many cycles to compute. A drastic improvement in the speed-up can be obtained when ϕ is interpolated, but generally at a loss of accuracy. The accuracy can be improved via an adequate tuning of the oversampling factor m. This is illustrated in Fig. (2), which shows the relative error E_2 for different interpolation schemes. As expected from the theory, one can observe an exponential decline if the kernel is evaluated exactly. For linear and cubic interpolated kernels, we observe that the error saturates as a function of m, due to systematic errors. For linear interpolation, the errors saturates at $\approx 10^{-4}$, while in the cubic regime it saturates at machine precision. In Fig. (3) we show the speed-up of d-NFFT over the NDFT.

Notice that the speed-up is of the order $\mathcal{O}(N_{\varpi}) \sim 10^3$. Considering both the need for accuracy and speed-up, the most optimal choice for QMC cluster solvers seems to be a cubic interpolation scheme at $m = 5$. In this way, the error on the Fourier transform will always be two to three orders of magnitude smaller than a typical statistical error of a Monte-Carlo integration. The fully optimized d-NFFT algorithm is summarized in Alg. (1).

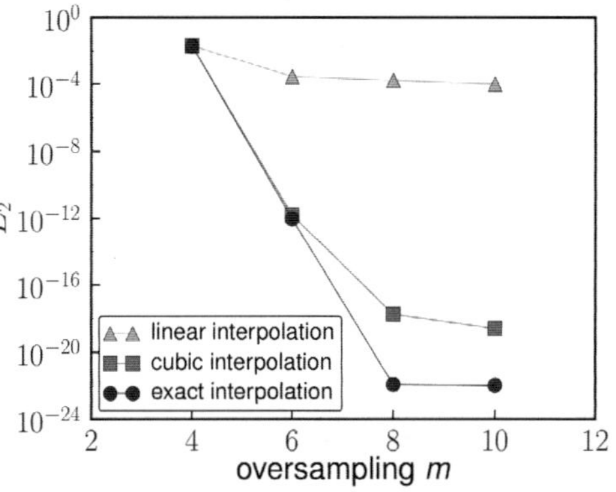

Figure 2: The relative error E_2 versus the oversampling factor m. The error decreases exponentially for increasing m, as expected from the theory. When the kernel ϕ is approximated by an interpolating value, the error becomes independent of the m. In the linear regime, the error saturates at $\approx 10^{-4}$, while in the cubic regime it saturates ate machine precision $\approx 10^{-16}$.

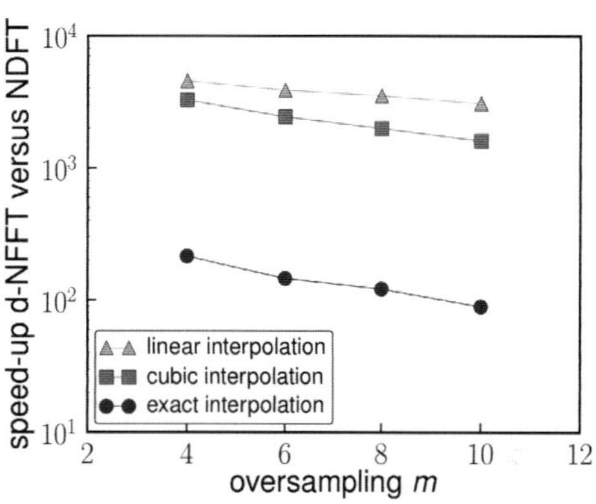

Figure 3: The speed-up of d-NFFT compared to DFT, versus the oversampling parameter m. Replacing the exact evaluation of the kernel ϕ with its interpolated value results clearly in a drastic speed-up. For this figure, we used typical parameters for a Monte Carlo simulation ($N_{\varpi} = 512, N_{\tau} = 1024$).

3.2. measurement of two-particle Greens-function G^{II}

The two-particle Greens-function G^{II} is obtained via a Quantum Monte Carlo Integration of a product of translation-noninvariant[1] single-particle TNSP Greens functions $G^I_\sigma[k_1, k_2]$,

$$G^{II}_{\sigma_1,\sigma_2}[k_1, k_2, q] = \Big\langle G^I_{\sigma_1}[k_1, k_1 + q] G^I_{\sigma_2}[k_2 + q, k_2] - \delta_{\sigma_1,\sigma_2} G^I_{\sigma_1}[k_1, k_2]\, G^I_{\sigma_2}[k_2 + q, k_1 + q]\Big\rangle. \tag{7}$$

Here, $\langle \ldots \rangle$ represents the Monte Carlo Integration. Just as in the single-particle case, the (TNSP) Greens functions $G^{\{v\}_\sigma}[k_1, k_2]$ can be computed straightforwardly from the N-matrix and its Hubbard-Stratonovitch configuration $\{v\}$.

$$\begin{aligned} G^{\{v\}_\sigma}[k_1, k_2] &= \delta_{k_1,k_2}\, G^0[k_1] - G^0[k_1] M^{\{v\}_\sigma}[k_1, k_2] G^0[k_2], \\ M^{\{v\}_\sigma}[k_1, k_2] &= \sum_{\lambda_1,\lambda_2 \in \{v\}_\sigma} (e^{V\{\phi\}_\sigma}_{\lambda_1,\lambda_2} - 1) N[\lambda_1, \lambda_2] \times e^{i\,(\omega_2\,\tau_{\lambda_2} - \omega_1\,\tau_{\lambda_1})} e^{i\,(\mathbf{k}_2\,\mathbf{r}_{\lambda_2} - \mathbf{k}_1\,\mathbf{r}_{\lambda_1})}. \end{aligned} \tag{8}$$

[1] in time and space

For this, the NFFT algorithm has to be generalized to two dimensions, which is straightforward. One constructs an fine 2D grid of the size mN_{ϖ} in both dimensions and convolutes the M-function with a 2D localized kernel. In our case, we simply used the product of two 1D Gaussian distributions. Just as in the 1D case, we apply the FFT on this fine equispaced grid , and renormalize each function-value in the Matsubara representation by the associated Fourier component of the localized kernel.

Algorithm 2 A 2D-NFFT algorithm to compute the TNSP Greens-function in the CT-AUX.

Require: A set of samples $\{v\} = \big\{\{\tau_i, \tau_j, M_{i,j}\}\big\}$ of the 2D function $M(\tau_1, \tau_2)$-function at random times in the interval $[-\beta, \beta] \times [-\beta, \beta]$.
Ensure: Return the corresponding TNSO Greens-function $G(\varpi_1, \varpi_2)$.

Define the function $\mathcal{M}(\tau_1, \tau_2)$ on the 2D-oversampled grid $\big\{\{\tau_1 = -\beta + l_1\,\Delta\tau, \tau_2 = -\beta + l_2\,\Delta\tau\} \mid l_1, l_2 \in \{0, \ldots, m\,N_{\varpi}\}\big\}$ and with $\Delta\tau = \frac{2\beta}{m\,N_{\varpi}}$.

for $\{\tau_i, \tau_j, M_{i,j}\} \in \{v\}$ **do**
  $I \leftarrow \frac{\tau_i+\beta}{\Delta\tau}$ {find the lower-index of τ_i such that $\tau(I) \leq \tau_i < \tau(I+1)$}
  $J \leftarrow \frac{\tau_j+\beta}{\Delta\tau}$ {find the lower-index of τ_i such that $\tau(J) \leq \tau_j < \tau(J+1)$}
  for $\lambda_i = -m \rightarrow m$ **do**
    $\phi_i \leftarrow \phi(\tau(I+\lambda_i) - \tau_i)$
    for $\lambda_j = -m \rightarrow m$ **do**
      $\phi_j \leftarrow \phi(\tau(J+\lambda_j) - \tau_j)$
      $\mathcal{M}(I+\lambda_i, J+\lambda_j) \leftarrow \mathcal{M}(I+\lambda_i, J+\lambda_j) + \phi_i * \phi_j * M_{i,j}$
    end for
  end for
end for

Compute $\mathcal{M}(\varpi_1, \varpi_2) \leftarrow FFT[\mathcal{M}(\tau_1, \tau_2)]$
Compute $M(\varpi_1, \varpi_2) \leftarrow \frac{\mathcal{M}(\varpi_1, \varpi_2)}{\phi_1(\varpi_1) * \phi_2(\varpi_2)}$

Compute $G^I(\varpi_1, \varpi_2) \leftarrow \delta_{\varpi_1, \varpi_2} G^0(\varpi_1) - G^0(\varpi_1) M(\varpi_1, \varpi_2) G^0(\varpi_2)$

In contrast to the 1D case, it is not possible to use a delayed FT scheme here. The TNSP Greens function needs to be computed every single time in order to evaluate Eq. (7). As a consequence, the measurement of a two-particle function is generally determined by the speed at which the FFT can be performed on the oversampled grid. We will therefore devote the rest of this section to tricks, which can speed up the FFT. First, we can take advantage of the fact that the function-values are real. This means that the function values in Matsubara representation will have a conjugate symmetry, and it is thus sufficient to compute only half of the Matsubara frequencies. This should speed up the FFT by a factor of two [14]. Another trick that should be used is pruning. Notice that we are actually only interested in a small subset of Fourier components of order $\mathcal{O}(N_{\varpi}^2)$ instead of all them $\mathcal{O}\big(m^2 N_{\varpi}^2\big)$. This essentially implies that we can skip the $(m-1)\,N_{\varpi}$ Fourier Transforms in the second dimension. The FFT in 2D is therefore not of the order $\mathcal{O}\big(m^2 N_{\varpi}^2 \log(m^2 N_{\varpi}^2)\big)$, but rather of $\mathcal{O}\big(m^2 N_{\varpi}^2 \log(m\,N_{\varpi}) + m\,N_{\varpi}^2 \log(m\,N_{\varpi})\big)$. The pruning of our FFT leads generally to an additional speed-up of roughly 2. In Fig. (4), we show you the combined speed-up of the NFFT method in 2D versus the NDFT. Due to its better scaling, the NFFT is clearly a good improvement in the case of a large number of vertices and a large number of Matsubara frequencies. The algorithm will therefore be hugely beneficial in the case of low temperatures and strong correlation.

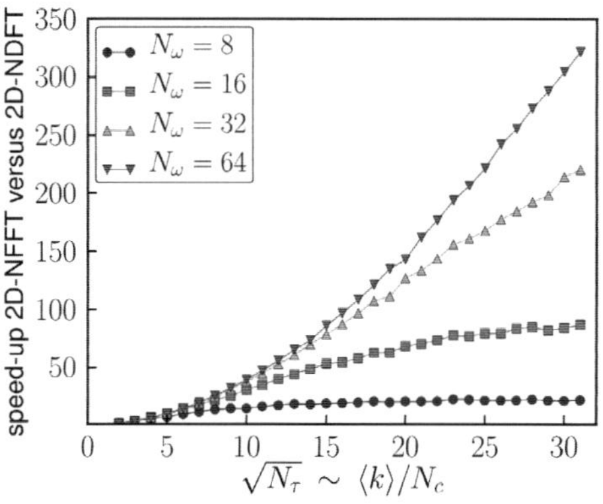

Figure 4: The speed-up of an optimized 2D-NFFT compared to NDFT in 2D for an oversampling factor $m = 6$, which corresponds to a relative error of $E_2 \sim 10^{-6}$. The method is clearly advantageous for large number of Matsubara frequencies ($N_{\varpi} \gg 1$) in the strongly correlated region $\langle k\rangle \gg 1$.

4. Study of the anti-Ferromagnetic transition in the 2D repulsive Hubbard model.

We will illustrate the effectiveness of our newly proposed measurement techniques by investigating the spin-spin correlation length l_c in the undoped 2D single band Hubbard model (SBHM) defined in Eq. (1). In particular we will look at the growth of the spin-spin correlation length as a function of the temperature. In this way, we can better understand the physics of the spin-fluctuations, which have been found to mediate the pairing interaction between electrons which is responsible for strong d-wave superconducting correlations in the 2D Hubbard model [15].

There exists no straightforward and direct way to obtain the correlation length l_c, as it does not have a strict definition. In previous work [3], an artificial antiferromagnetic transition temperature, the Neel-temperature T_N of the cluster was used to obtain a measure of the correlation length. The motivation for this strategy is the following: As the temperature is lowered, the correlation length grows and eventually reaches the linear cluster size $\sqrt{N_c}$. At that length scale, the DCA does not describe fluctuations and therefore predicts an antiferromagnetic transition in the cluster, i.e.

$$l_c(T_N) = \sqrt{N_c}\,. \tag{9}$$

This transition is artificial, since it contradicts the Mermin-Wagner theorem which demands $T_N = 0$.

The method of operation to determine the temperature-dependence of the correlation length $l_c(T)$ is now clear. We will compute the Neél-temperature T_N for a set of increasingly larger clusters and look at its decay as a function of the cluster size. By simply inverting the curve, we can obtain the growth of the spin-spin correlation length as the temperature is lowered.

This procedure demands a very effective implementation of the DCA, since many large clusters will have to be solved for increasingly lower Neel-temperatures. Previously, the award winning *DCA++* code [10] which implements a Hirsch-Fye solver was able to compute the Neel-temperature up to clusters of $N_c = 32$. Due to the efficient measurement techniques introduced in this paper and the submatrix updates, we were able to push the maximum cluster size to $N_c = 100$. We show the Neel-temperatures obtained with the CT-AUX in Fig. (5), and compare it to the values extracted from [3].

The decrease of the Neél-temperature T_N as a function of cluster size is clearly visible. Fig. (5) shows that the Hirsch-Fye captures the physics qualitatively well, but overestimates

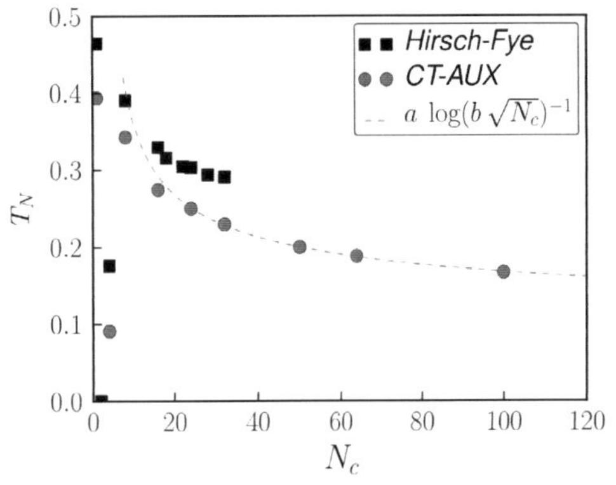

Figure 5: Neel-temperature T_N versus cluster size for different Monte Carlo solvers at $U/t = 8$. The squares are computed with a Hirsch-Fye solver, without extrapolation towards a zero Monte Carlo time step (data extracted from [3]). The dots are obtained with the CT-AUX. The Hirsch-Fye cluster solver clearly overestimates T_N.

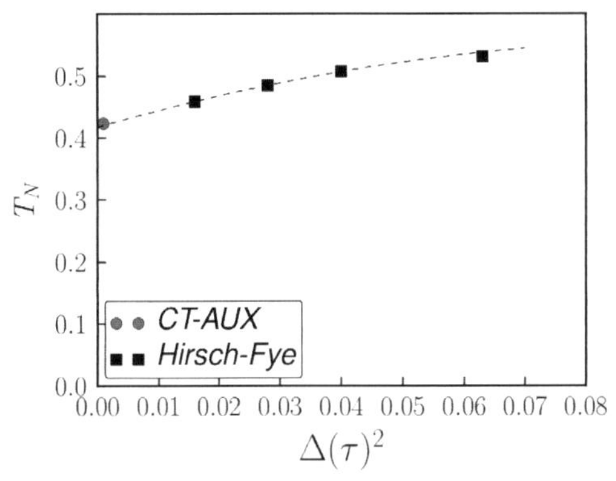

Figure 6: The Neel-temperature of an 18-site 3D cluster at $U/t = 8$ for different Monte Carlo time steps with the Hirsch-Fye algorithm (data extracted from [4]). Using a quadratic extrapolation, we obtain a T_N which lies very close to the anti-ferromagnetic transition temperature computed with the CT-AUX.

the transition temperatures in general. This discrepancy between the two solvers grows as clusters get larger and temperatures lower, due to the Trotter approximation. Via a controlled extrapolation towards infinite time slices, it is possible to avoid this discrepancy between the Hirsch-Fye and CT-AUX cluster solver. This is illustrated in Fig (6), where the extrapolated Neél-temperature from a Hirsch-Fye solver [4] is compared to the one obtained with CT-AUX for a 3D $N_c = 18$ site cluster. Due to limited computational resources, this extrapolation was not done for the Neel-temperatures presented in Fig (5), which explains the quantitative difference in the results.

Since spin-spin correlations develop exponentially in temperature in a 2D Heisenberg model, the strong coupling limit of the half-filled 2D Hubbard model, we expect the correlation length $l_c(T)$ to have an exponential temperature dependence $l_c(T) = b\,e^{1/T}$. Together with Eq. (9), this means that

$$T_N = \frac{a}{\log(b\sqrt{N_c})}. \tag{10}$$

In Fig. (5), we show a fit of the Neél-temperatures obtained with the CT-AUX using Eq. (10). The agreement is remarkably good for large clusters ($N_c \geq 32$). This confirms that the DCA does not violate the Mermin-Wagner theorem in its exact limit $N_c \to \infty$ and confirms that the correlation length l_c grows exponentially as a function of inverse temperature in the 2D Hubbard model.

With the solutions from the Bethe-Salpeter equation on the $N_c = 100$ cluster, we can get insight on the nature of the interaction leading to the antiferromagnetic correlations. The momentum and frequency structure of the eigenvector ϕ, associated with the leading eigenvalue, is shown for $\vec{k}$-points along the Fermi-surface in Fig (7). The leading eigenvector ϕ has very little momentum structure. The anti-ferromagnetic transition can thus be characterized as a condensation of local particle-hole pairs (local magnetic moments). In addition, the

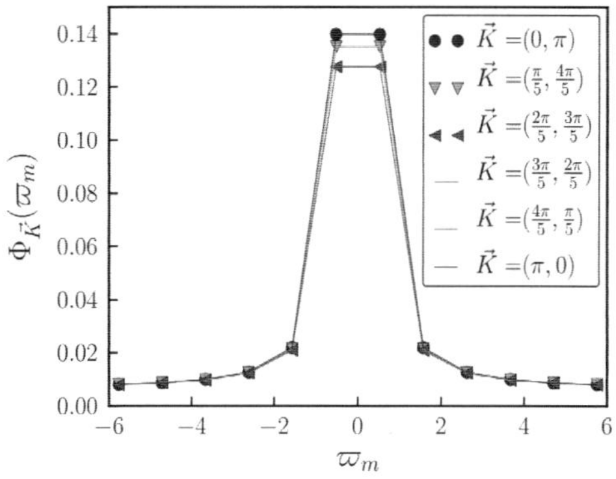

Figure 7: The eigenfunction $\phi_k(\varpi)$ for different k-points on the Fermi-surface ($N_c = 100$ and $\beta = 6$). The eigenfunction has very little momentum dependence, indicating that the antiferromagnetic order parameter is essentially a local particle-hole pair. Notice also that the eigenfunction has a finite asymptotic value for large frequencies, reflecting the role of the instantaneous Hubbard U in the moment formation.

eigenfunction has a clear finite asymptotic value. This reflects the instantaneous contribution of the Hubbard U to the effective interaction that leads to the moment formation.

5. Summary

In summary, we have presented methods based on NFFT which significantly speed up the measurements in the case of continuous time QMC cluster solvers. These methods were then applied to analyze the magnetic behavior of the half-filled 2D Hubbard model with interaction $U/t = 8$. The results are consistent with an exponential divergence of the antiferromagnetic spin-spin correlation length in inverse temperature, and therefore compliant with the Mermin-Wagner theorem.

Acknowledgement

Work at Oak Ridge National Laboratory (ORNL) was supported by the Laboratory Directed Research and Development fund and the Center for Nanophase Materials Sciences, which is sponsored by the Division for Scientific User Facilities of the Office of Basic Energy Sciences. Computational resources at the National Center for Computational Sciences (NCCS) of ORNL and the Swiss National Supercomputing Center (CSCS) in Lugano, Switzerland were used for the simulations.

References

[1] M. Jarrell, T. Maier, and C. Huscroft, "Quantum Monte Carlo algorithm for nonlocal corrections to the dynamical mean-field approximation," *Physical Review B*, 2001.

[2] T. Maier, M. Jarrell, T. Pruschke, and M. Hettler, "Quantum cluster theories," *Reviews of Modern Physics*, vol. 77, no. 3, pp. 1027–1080, 2005.

[3] T. Maier, M. Jarrell, T. Schulthess, P. Kent, and J. White, "Systematic Study of d-Wave Superconductivity in the 2D Repulsive Hubbard Model," *Physical Review Letters*, vol. 95, p. 237001, Nov. 2005.

[4] P. Kent, M. Jarrell, T. Maier, and T. Pruschke, "Efficient calculation of the antiferromagnetic phase diagram of the three-dimensional hubbard model," *Physical Review B*, vol. 72, p. 060411, Aug. 2005.

[5] J. E. Hirsch and R. M. Fye, "Monte carlo method for magnetic impurities in metals," *Phys. Rev. Lett.*, vol. 56, pp. 2521–2524, Jun 1986.

[6] E. Gull, A. J. Millis, A. I. Lichtenstein, A. N. Rubtsov, M. Troyer, and P. Werner, "Continuous-time monte carlo methods for quantum impurity models," *Rev. Mod. Phys.*, vol. 83, pp. 349–404, May 2011.

[7] E. Gull, P. Werner, O. Parcollet, and M. Troyer, "Continuous-time auxiliary-field monte carlo for quantum impurity models," *EPL (Europhysics Letters)*, vol. 82, no. 5, p. 57003 (6pp), 2008.

[8] E. Gull, P. Staar, S. Fuchs, P. Nukala, M. S. Summers, T. Pruschke, T. C. Schulthess, and T. Maier, "Submatrix updates for the continuous-time auxiliary-field algorithm," *Phys. Rev. B*, vol. 83, p. 075122, Feb 2011.

[9] J. Keiner, S. Kunis, and D. Potts, "Using NFFT3 - a software library for various nonequispaced fast Fourier transforms," *ACM Transactions on Mathematical Software*, vol. 36, pp. 19:1–19:30, Aug. 2009.

[10] G. Alvarez, M. S. Summers, D. E. Maxwell, M. Eisenbach, J. S. Meredith, J. M. Larkin, J. Levesque, T. A. Maier, P. R. C. Kent, E. F. D'Azevedo, and T. C. Schulthess, "New algorithm to enable 400+ TFlop/s sustained performance in simulations of disorder effects in high-Tc superconductors," in *SC '08: Proceedings of the 2008 ACM/IEEE conference on Supercomputing*, (Piscataway, NJ, USA), pp. 1–10, IEEE Press, 2008.

[11] A. A. Abrikosov, L. P. Gorkov, and I. Y. Dzyaloshinskii, *Quantum field theoretical methods in statistical physics, International series of monographs in natural philosophy.* Pergamon Press, Oxford, 1965.

[12] P. Werner, A. Comanac, L. de' Medici, *et al.*, "Continuous-time solver for quantum impurity models," *Phys. Rev. Lett.*, vol. 97, no. 7, p. 076405, 2006.

[13] P. Werner and A. J. Millis, "Hybridization expansion impurity solver: General formulation and application to kondo lattice and two-orbital models," *Phys. Rev. B*, vol. 74, no. 15, p. 155107, 2006.

[14] M. Frigo and S. G. Johnson, "The design and implementation of FFTW3," *Proceedings of the IEEE*, vol. 93, no. 2, pp. 216–231, 2005. Special issue on "Program Generation, Optimization, and Platform Adaptation".

[15] T. A. Maier, M. S. Jarrell, and D. J. Scalapino, "Structure of the pairing interaction in the two-dimensional hubbard model," *Phys. Rev. Lett.*, vol. 96, p. 047005, Feb 2006.

Electron Transfer on Impurity doped Graphene Nanoribbon

Hiroyoshi Tsuyuki, Shoichi Sakamoto and Mitsuyoshi Tomiya

Department of Materials and Life Science, Seikei University, Kichijyoji-Kitamachi 3-3-1, Musashino-shi, Tokyo, 180-8633, JAPAN

dd106102@cc.seikei.ac.jp

Abstract: Electronic transport properties in armchair shaped edges graphene nanoribbons (AGNRs) doped various impurities have been simulated by the non-equilibrium Green's function approach combined with the first principle calculation based on the density functional theory. We have observed that impurity levels appear in electronic structures, and that the quantization of transmission function is moderated for doped AGNRs. The I-V characteristic can be computed from the transmission function. Our simulation results show that AGNRs doped impurities have higher conductance than the non-doped one.

1. INTRODUCTION

Recently, a graphene has been extensively studied due to its unique physical properties originated from the unique two-dimensional (2D) honeycomb lattice structure [1]. Graphene nanoribbons (GNRs) are narrow strips of a single layer graphene with nanometer sized width and armchair or zigzag shaped edges.

A graphene sheet is half-metallic, whereas GNRs have ether metallic or semiconducting feature by the shape of edge [1-3], which is however different from that in the carbon nanotube (CNT) [4]. Electronic properties of the GNR are controlled by the width and the edge shape [2, 3]. A graphene sheet does not have a bandgap so that a conduction band and a valence band touch at K-point. On the other hand, the GNR with armchair shaped edges (AGNR) which has semiconducting feature has a bandgap by the quantum confinement effect to the width direction. The bandgap in the AGNR can be controlled by the ribbon width [3].

To apply for various devices, it is important to dope impurities on semiconductor. Boron (B) and nitrogen (N) atoms are typical dopants. It is expected to have various functions and interesting physical properties by the doping chemically for new device applications using carbon based materials. For carbon materials such as the GNR and the CNT, electronic properties, e.g., the electronic structure, the conductance etc., in doping impurities are now being studied intensively [5-8]. For example, B atom as a substitutional atom is often considered for p-type doping, N atom is also adopted as n-type doping. We are interested in effects of dopants to the electronic transfer on GNRs.

In this work, we focus on the AGNR. We have numerically investigated electronic transport properties on AGNRs doped some of atoms and molecules on the surfaces by the non-equilibrium Green's function (NEGF) formalism [9] combined with the standard first principle calculation. Electronic structures of doped AGNRs have first been computed by the first principle calculation. For electronic transport, I-V characteristics are finally obtained by calculating the transmission function considering to apply the bias voltage.

2. METHOD

Our calculation for the electronic structure is done by the standard first principle method, which is based on the density functional theory (DFT). We have used the DFT program package SIESTA. The pseudo potential for the Kleinman-Bylander (KB) potential, and the generalized gradient approximation (GGA) for the exchange correlation term is adopted. For our simulation, we take following conditions; the k-point sampling is 256 points, and the energy cutoff is 350 Rydbergs.

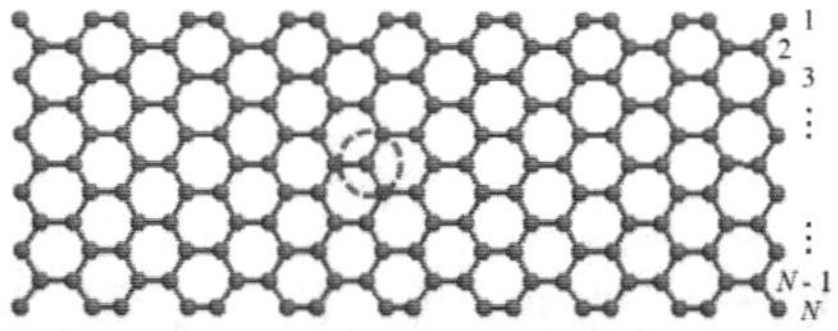

Figure 1: The carbon atomic structure of AGNR modeling as the super cell. N is the number of dimer lines, N = 11, across the ribbon width. At the center of the ribbon surface, as shown dashed line circle, some impurity has been attached on site. This structure is relaxed by the CG optimization.

We have made the AGNR structure in which the number of carbon atoms of the AGNR is 176 atoms, and the bond length between carbons is 1.48 Angstroms, as shown Figure 1. This bond length is obtained by the optimization of the structure with the conjugate gradient (CG) method. In the DFT calculation, the bond length is slightly longer.

We have also considered connections to electrodes which was used same type GNRs. Then the quantum transmission is computed by the NEGF theory combined with the first principle calculation using the TRANSIESTA package. For the NEGF formalism, the three parts; the left electrode, the contacting part, and the right electrode, are required.

In the NEGF approach, the Green's function at energy E is calculated as

$$G(E)=(E-H)^{-1}, \quad (1)$$

after the Hamiltonian H is obtained by the DFT calculation with SIESTA. Considering the non-equilibrium situation, the bias voltage is defined as the difference of chemical potentials μ_L and μ_R in the left and right electrodes, $eV=\mu_L-\mu_R$. The transmission is calculated as

$$T(E)=tr[\mathbf{t}^{\dagger}\mathbf{t}](E) = tr[\mathbf{\Gamma}_L(E)\mathbf{G}^{\dagger}(E)\mathbf{\Gamma}_R(E)\mathbf{G}(E)], \quad (2)$$

with the transmission amplitude matrix,

$$\mathbf{t}(E)=\mathbf{\Gamma}_R(E)^{1/2}\mathbf{G}(E)\mathbf{\Gamma}_L(E)^{1/2}, \quad (3)$$

where,

$$\mathbf{\Gamma}_{L,R}\equiv i\left[\mathbf{\Sigma}_{L,R}(E)-\mathbf{\Sigma}^{\dagger}_{L,R}(E)\right]/2, \quad (4)$$

and $\mathbf{\Sigma}_{L,R}$ is the self-energy in the left or right electrode. By the Landauer formulation, the current is calculated from the transmission function,

$$I=\frac{2e^2}{h}\int_{\mu_L}^{\mu_R}d\varepsilon[n_F(\varepsilon-\mu_L)-n_F(\varepsilon-\mu_R)]T(\varepsilon), \quad (5)$$

where n_F is the Fermi distribution function.

3. RESULTS

Four kinds of impurities; hydrogen (H) atom, hydroxide (OH), methyl (CH_3), and sodium (Na) atom are considered here. Just one atom or molecule is attached on (not substituted) the site of the AGNR surface (as shown Figure 1). The distance between the carbon atom of the AGNR surface and attached atoms or molecules is set 1.0 Angstroms before the structures are relaxed by the CG method. Then the total energy of the system becomes the local minimum.

First, the electronic structure of the non-doped AGNR has been simulated. The Fermi energy E_F is -6.105eV, and the bandgap, which is about 0.4eV, exists at K-point in this semiconducting AGNR (Figure 2(a)). When the H atom and CH_3 are attached on the AGNR surface, these energy levels of impurity states appear in the AGNR bandgap (Figure 2(b), (c)). These levels are equivalent to the Fermi-energy; E_F = -5.965eV for H doped and E_F = -5.989eV for CH_3 doped, which are slightly higher than that of the non-doped AGNR. Band structures of other impurities; OH and Na, doped AGNRs are also shows in Figure 2(c)-(e). In case of OH impurity, the impurity level also appears in the bandgap of the AGNR. Fermi energy, E_F = -6.215eV, is slightly lower than the non-doped AGNR. On the contrary, the Na impurity level appears in the conduction band, E_F = -5.252eV, thus the Na-doped AGNR is no longer the semiconductor.

In order to research properties of electronic transport for AGNRs doped impurities, transmission functions have been calculated. It is shown in Figure 3 for the zero-bias situation. For the non-doped AGNR, it is confirmed the transmission function is quantized against energies (Figure 3(a)). In contrast, the quantization of transmission functions are moderated for doped AGNRs (Figure 3(b)-(e)).

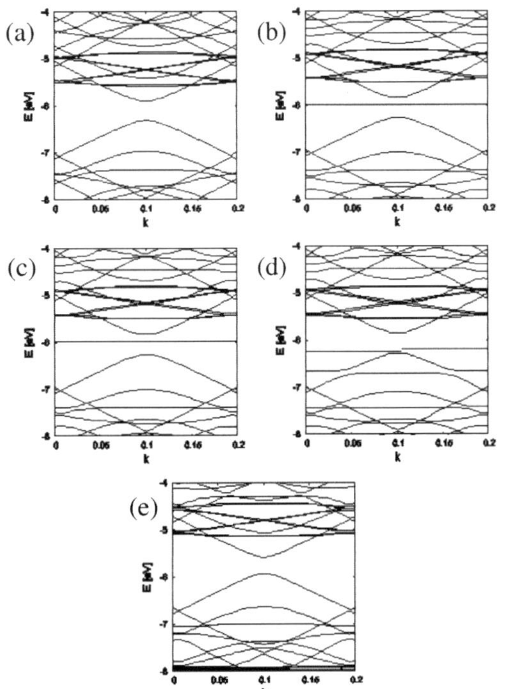

Figure 2: Band structures for (a) non-doped, (b) H-atom, (c) CH3, (d) OH, and (e) Na-atom, doped AGNRs. At K-point, the energy gap exist between the conduction and the valence band.

In cases of H atom, OH, and CH_3 impurities doping, the quantization of transmission functions is moderated. On the DOS, the contribution of the impurity level is clearly observed in the bandgap for OH doped AGNR. It is observed that dips appear on the transmission function, where impurity states exist (Figure 3(d)) [7]. However, contributions of H atom and CH_3 impurity levels are almost not observed (Figure 3(b)-(c)). For these impurities doping, no transmission exist at Fermi-levels. For Na impurity doping, the quantization of the transmission is broken (Figure 3(e)).

I-V characteristics of doped AGNRs are also simulated using transmission functions on finite biases (Figure 4). The bias voltage is applied from 0.0eV to 2.0eV. Since this AGNR is a semiconductor, the current starts to flow from about 0.4eV. It is observed that the conductance of AGNR semiconductor is increased by doping these impurities.

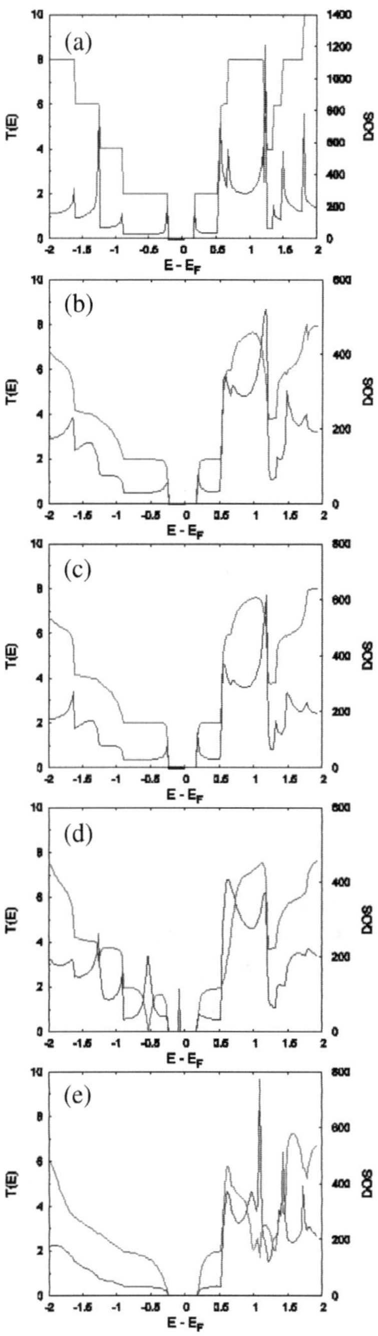

Figure 3: Transmission functions as shown the red line around the Fermi energy for (a) non-doped, (b) H-atom, (c) CH3, (d) OH, and (e) Na-atom, doped AGNRs. The blue line shows the density of states. At the Fermi energy, there exist little impurity states in doped one.

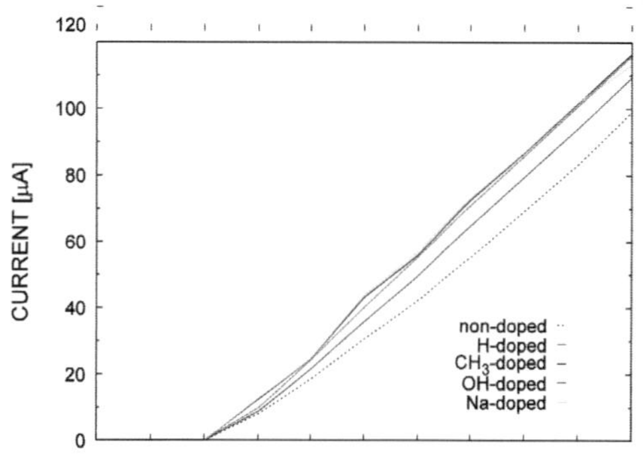

Figure 4: I-V characteristics of AGNRs. The conductance in doped AGNR is higher than that of non-doped AGNR denoted by the dashed line.

4. CONCLUSION

We have numerically investigated electronic properties of the AGNR doped impurity. When some atom or molecule is attached on the AGNR surface, impurity levels appear in bandgaps of AGNRs, whereas no impurity level of metal appears in the bandgap. It is observed that the transmission function is quantized against energies in non-doped AGNR, but the quantization is moderated in some doped AGNRs. I-V characteristics of doped AGNRs are evaluated. Especially, we have found that AGNRs doped impurities have higher conductance than the non-doped AGNR.

References

[1] A. H. C. Neto, F. Guinea, N. M. R. Peres, K. S. Novoselov, and A. K. Geim, Rev. Mod. Phys. **2009**; 81, 109-162.

[2] S. S. Yu, Q. B. Wen, W. T. Zheng, and Q. Jiang, Molecular Simulation, **2008**; 34, 1085-1090.

[3] Y. –W. Son, M. L. Cohen, and S. G. Louie, Phys. Rev. Lett. **2006**; 97, 216803.

[4] M. Ezawa, Phys. Rev. **2006**; B73, 045432.

[5] S. S. Yu, W. T. Zheng, and Q. Jiang, IEEE T. Nanotechnology, **2010**; 10, 926-930.

[6] T. C. Li, and S. –P. Lu, Phys. Rev. **2008**; B77, 085408.

[7] Y. Ouyang, S. Sanvito, and J. Guo, Surf. Sci. **2011**; 605, 1643-1648.

[8] B. Huang, Phys. Lett. **2011**; A375, 845-848.

[9] M. Brandbyge, J. -L. Mozos, P. Ordejo´n, J. Taylor, and K. Stokbro, Phys. Rev. **2002**; B65, 165401.

The Universal Edge Physics in Fractional Quantum Hall Liquids

Zi-Xiang Hu[1], R. N. Bhatt[1], Xin Wan[2], Kun Yang[3]
[1]Department of Electrical Engineering, Princeton University, Princeton, New Jersey 08544, USA
[2] Zhejiang Institute of Modern Physics, Zhejiang University, Hangzhou 310027, P.R. China
[3]National High Magnetic Field Laboratory and Department of Physics, Florida State University, Tallahassee, Florida 32306, USA

E-mail: zihu@princeton.edu

Abstract. The chiral Luttinger liquid theory for fractional quantum Hall edge transport predicts universal power-law behavior in the current-voltage (I-V) characteristics for electrons tunneling into the edge. However, it has not been unambiguously observed in experiments in two-dimensional electron gases based on GaAs/GaAlAs heterostructures or quantum wells. One plausible cause is the fractional quantum Hall edge reconstruction, which introduces non-chiral edge modes. The coupling between counterpropagating edge modes can modify the exponent of the I-V characteristics. By comparing the $\nu = 1/3$ fractional quantum Hall states in modulation-doped semiconductor devices and in graphene devices, we show that the graphene-based systems have an experimental accessible parameter region to avoid the edge reconstruction, which is suitable for the exploration of the universal edge tunneling exponent predicted by the chiral Luttinger liquid theory.

1. Introduction

The fractional quantum Hall (FQH) effect has been observed in GaAs-based semiconductor heterostructures and quantum wells [1] for almost 30 years. The fractional quantum Hall states have a lasting fascination due to their nontrivial topological properties, which lie beyond the paradigm of the Fermi-liquid theory, and the potential applications in topological quantum computation [2, 3, 4]. Probing excitations at the edge of a FQH droplet is the most accessible approach for the detection of the topological properties in the bulk thanks to the edge-bulk correspondence in the topological system. The edge excitations are gapless in the thermodynamic limit, contrast to gapped quasihole/quasiparticle excitations in the bulk. Owing to the existence of edge states, current exists between two contacts connected by an edge channel as electrons can be injected into or removed from the FQH edge without costing energy. The standard theory for the FQH edge physics is the chiral Luttinger liquid (CLL) theory [5]. The theory predicts that a FQH droplet exhibits power-law bahavior in the I-V characteristics ($I \propto V^{\alpha}$) when electrons are tunneling through a barrier into the edge, e.g., from a three-dimensional Fermi liquid. For the celebrated Laughlin state at a filling factor $\nu = 1/3$ in the lowest Landau level, the CLL theory predicts a tunneling exponent $\alpha = 1/\nu = 3$ [5]. However, the exponent α measured in experiments are sample dependent with a numerical value smaller than 3, which are nonuniversal [6, 7, 8, 9]. One of the possible causes of this discrepancy is

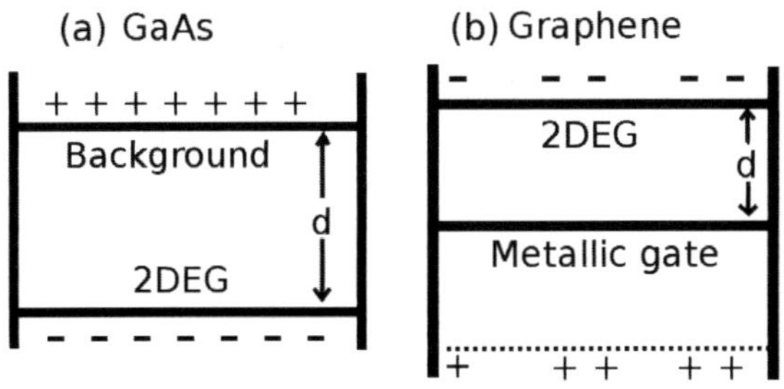

Figure 1. Simplified models to mimic the experimental setups in (a) GaAs-based and (b) graphene-based systems. The GaAs model consists of an electron layer and a homogeneous impurity charge layer, which confines electrons. The graphene model consists of layered Dirac electrons with an open edge and their image charge due to metallic gate; both confines electrons.

existence of counterpropagating edge modes, which result from edge reconstruction [9, 10, 11]. The edge reconstruction in the electron density profile manifests the competition between electron-electron interaction, which tends to spread out the electron density, and the electrostatic confinement at the edge, which holds the electrons in the interior of the sample. Numerical calculations [12, 13] indicate that the edge reconstruction in FQH states is unavoidable in the state-of-art GaAs-based samples based on modulation doping, which greatly enhances sample mobility but unintentionally introduces a large region of fringe field at the edge that favors edge electron redistribution. This effect hinders the observation of the universal behavior in edge transport.

In recent years new two-dimensional materials, such as graphene, emerged. The FQH effect has already been observed in graphene [14, 15, 16]. However, integer and fractional quantum Hall effects in graphene differ from those in normal semiconductor-based devices, because electrons in graphene obey the relativistic Dirac equation, instead of the nonrelativistic Schrödinger equation often encountered in solid state materials. Here we compare the FQH states in both GaAs-based samples and in graphene samples with the emphasis on the difference of the competition at the edge. We find that there is an experimental accessible parameter region in graphene-based samples, in which edge reconstruction can be avoided and, therefore, universal FQH edge physics can be observed. The universality is reflected in the calculation of the equal-time edge Green's function, which is related to the I-V characteristics in edge tunneling.

2. FQH states in GaAs-based systems

An experimental sample based on a modulation-doped GaAs/GaAlAs heterostructure [17] can be simplified to a two-dimensional electron layer, which is located at the interface between GaAs and GaAlAs, and a homogeneous positive background charge layer, which originates from the remote dopants. This is illustrated in Fig. 1 (a). The confinement potential for electrons due to background charge competes with the Coulomb interaction between electrons. This competition is the driving force for edge reconstruction. The density of the background charge σ is determined by the electron filling factor ν due to charge neutrality and, therefore, the background confinement potential is a single-body potential in the FQH Hamiltonian. The complete Hamiltonian can be written as

$$H = \frac{1}{2}\sum_{mnl} V^l_{mn} c^+_{m+l} c^+_n c_{n+l} c_m + \sum_m U_m c^+_m c_m, \tag{1}$$

where the c^+_m is the electron creation operator for the lowest LL single electron state with an angular momentum m.

$$V^l_{mn} = \int d^2 r_1 \int d^2 r_2 \phi^*_{m+l}(r_1)\phi^*_n(r_2)\frac{e^2}{r}\phi_{n+l}(r_2)\phi_m(r_1), \tag{2}$$

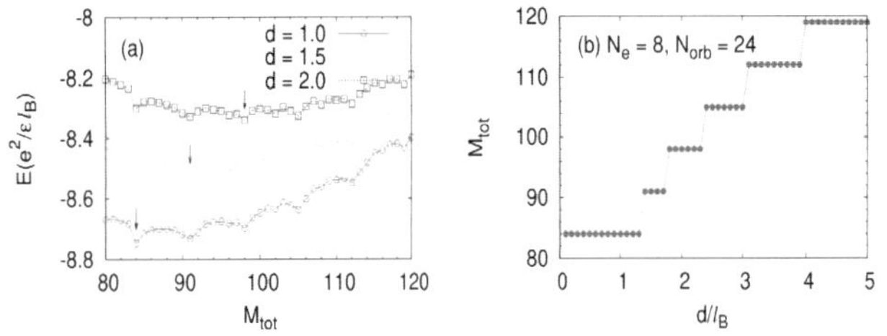

Figure 2. (a) The lowest energy in each total angular momentum subspace for 8 electrons in 24 orbitals with different ds in the GaAs model. The global ground state is labeled by an arrow. (b) The corresponding angular momentum of the global ground state as a function of d.

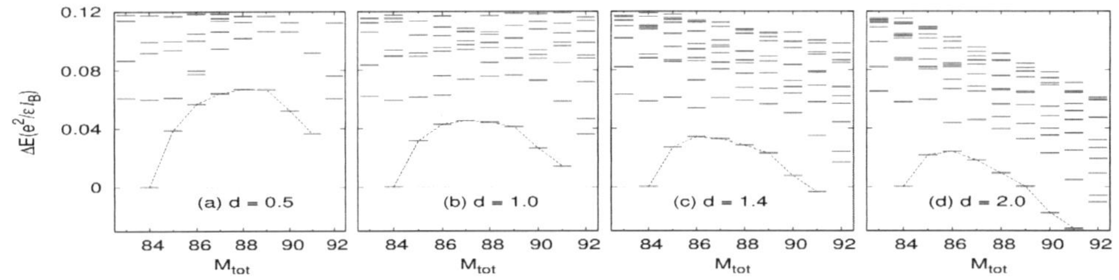

Figure 3. The low-energy spectrum for 8 electrons in 26 orbitals with (a) $d = 0.5\ l_B$, (b) $d = 1.0\ l_B$, (c) $d = 1.4\ l_B$, and (d) $d = 2.0\ l_B$ in the GaAs model. Edge reconstruction, signaled by the ground state angular momentum change, occurs around $d_c \sim 1.4\ l_B$.

represents the electron-electron interaction and

$$U_m = e\sigma \int \int_{r_2<R} d^2r_1 d^2r_2 \frac{1}{\sqrt{d^2 + |\vec{r_1} - \vec{r_2}|^2}} |\phi_m(r_1)|^2 \tag{3}$$

the confinement potential. The distance d between the electron layer and the background charge layer is the parameter that controls the relative strength of the confinement potential to the electron-electron interaction. When d is small, the confinement is strong and electrons tend to stay in the region defined by photo lithography process ($r < R$ for the disk sample we consider here). However, the confinement becomes weak when d is large and thus the electrons near the edge tend to spread out, leading to edge reconstruction.

Because of the rotational symmetry in the geometry and the gauge we choose, the total angular momentum M_{tot} is a good quantum number and thus we can diagonalize the Hamiltonian in each angular momentum subspace. In Fig. 2(a), we plot the lowest energy in each of the total angular momentum subspace with different ds for 8 electrons in 24 orbitals. As discussed in Ref. [10, 12], when d is smaller than the critical value $d_c = 1.5 \pm 0.1\ l_B$, the global ground state has the same total angular momentum $M_{tot} = M_0 = 3N(N-1)/2$ as that of the Laughlin state. When $d > d_c$, edge electrons spread out under the weak confinement and, therefore, the total angular momentum of the global ground state increases as shown in Fig. 2(b). To further reveal how the edge reconstruction happens with increasing d, we plot the low-lying energy spectrum in Fig. 3 for 8 electrons in 26 orbitals with different ds. The larger number (26 rather than 24 or 22) of orbitals is chosen to avoid unnecessary mixing of the bulk and edge modes in a finite system. The chiral bosonic edge mode (dashed line), which describes the edge electron density deformation, is linear for small $\Delta M = M_{tot} - M_0$ but develop downward curvature as ΔM increases [12]. As d increases, the bosonic mode bends down further and eventually the total ground state momentum jumps from $M_{tot} = 84$ to $M_{tot} = 91$ above the critical value d_c;

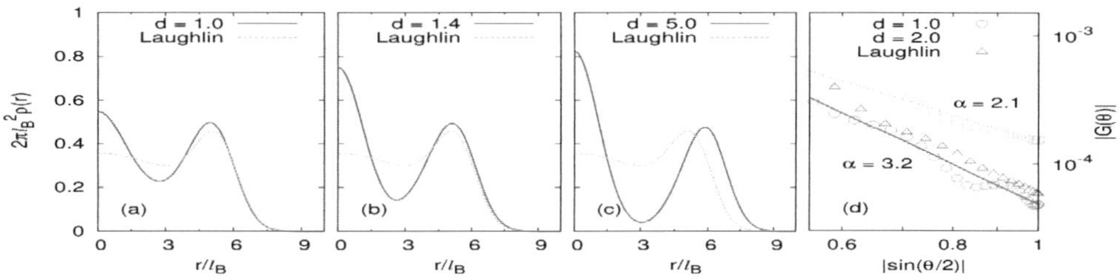

Figure 4. The electron density profile for the ground state at (a) $d = 1.0\ l_B$, (b) $d = 1.4\ l_B$ and (c) $d = 5.0\ l_B$ in the GaAs model. The dashed line is the density profile for the Laughlin state. (d) The edge Green's function before and after edge reconstruction; the slopes in the log-log plot yield $\alpha = 3.2$ and 2.1, respectively.

hence edge reconstruction occurs. Fig. 4(a)-(c) shows the evolution of the electron density profile for the ground state as d increases. For large enough d, the electron density profile develops an almost detached edge piece, as shown in Fig. 4(c), suggesting the reconstructed edge can accommodate three counterpropagating edge modes. The edge transport thus depends on the coupling of the edge modes, hence no universality can be expected except in the strong coupling limit.

To explore the universality of the FQH edge states, we calculate the equal-time edge Green's function,

$$G(\vec{r} - \vec{r'}) = \frac{\langle\psi|\Psi^+(\vec{r})\Psi(\vec{r'})|\psi\rangle}{\langle\psi|\psi\rangle}, \tag{4}$$

where $\Psi^+(\vec{r})$ [or $\Psi(\vec{r})$] creates (annihilates) one electron at $\vec{r}$ at the edge of the FQH droplet. In the disk geometry with $|r| = |r'| = R$, the distance $|\vec{r} - \vec{r'}| = 2R\sin(\theta/2)$ where θ is the angle between $\vec{r}$ and $\vec{r'}$. The CLL theory predicts that in the large distance limit, the edge Green's function has an asymptotic behavior

$$|G(\vec{r} - \vec{r'})| \sim |\vec{r} - \vec{r'}|^{-\alpha} \propto |\sin(\theta/2)|^{-\alpha}. \tag{5}$$

For the $\nu = 1/3$ Laughlin state, the CLL theory predicts an exponent $\alpha = 3$. Fig. 4(d) plots the edge Green's function for 8 electrons in 24 orbitals with Coulomb interaction for two ds on the opposite sides of the edge reconstruction transition. For comparision, the edge Green's function for the variational Laughlin wavefunction is also presented. We find an exponent $\alpha = 3.2 \pm 0.2$ [18] at $d = 1.0\ l_B < d_c$, which is consistent with the CLL theory prediction. The exponent for the variational Laughlin wavefunction is the same as that of the Coulomb ground state before reconstruction. However, when $d = 2.0\ l_B > d_c$, the exponent $\alpha = 2.1$ is noticeably smaller than 3. This suggests that the long-range Coulomb interaction does not directly lead to nonuniversal edge behavior, but the consequence of its competition with the confinement potential, i.e., edge reconstruction, does. In a typical GaAs-based sample, d is of order 1000 $\mathring{A}$, i.e., about 10 l_B [17]. While the numerical calculation for such a large d is not practical due to the system-size constraint, we speculate that multiple edge reconstructions can occur, leading to more than three counterpropagating edge modes. Therefore, such a device is always in the edge reconstructed regime, which does not guarantee the observation of a universal edge tunneling exponent.

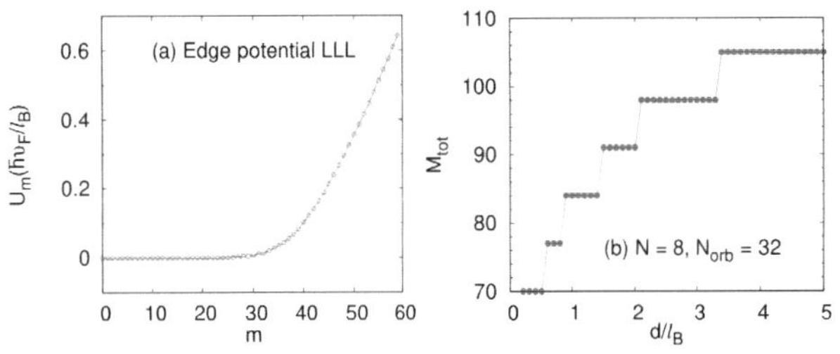

Figure 5. (a) The single-particle lowest Landau level energy in the graphene model with an open edge as a function of the angular momentum in a disk sample with 60 orbitals. (b) The angular momentum of the many-body ground state as a function of d for 8 electrons in 32 orbitals.

3. FQH states in graphene-base systems

The experimental setup for graphene samples is illustrated in Fig. 1(b) (and Ref. [21]). The graphene layer is situated at a distance d from a metallic gate (boron-nitride) [16]. Therefore, electrons have their image charge at a distance $2d$ from the graphene layer. This modifies the first term in the Hamiltonian in Eq. (1), which now contains both the electron-electron interaction and the electron-image charge interaction:

$$V^l_{mn} = \int d^2r_1 \int d^2r_2 \phi^*_{m+l}(r_1)\phi^*_n(r_2)(\frac{e^2}{r} - \frac{e^2}{\sqrt{r^2+(2d)^2}})\phi_{n+l}(r_2)\phi_m(r_1). \tag{6}$$

The combined interaction behaves like dipole-dipole interaction with $1/r^3$ dependence at short distances and thus the interaction pseudopotentials (V_ms) [19] are dominated by those with small ms. The single-body term in Eq. (1) comes from the edge confinement potential in the graphene due to the open boundary. The solution of the Dirac equation for a semi-infinite graphene with an open edge at $x = 0$ satisfies equation $D_\mu(-\sqrt{2}x_c) = 0$ [20] where $D_\mu(x)$ is the parabolic cylinder function and $x_c = kl_B$, $\mu = \epsilon^2/2$; k is the wave vector parallel to the edge and ϵ the energy in units of $\hbar v_F/l_B$. In the disk geometry with N_{orb} orbitals, the edge is located at $r_c = \sqrt{2N_{orb}}l_B$, thus the solution satisfies $D_\mu[-\sqrt{2}(x_c - r_c/l_B)] = 0$. Solutions are allowed for $\mu = 0, 1, 2, \cdots$ in the bulk, hence the energy for the nth LL is $\epsilon_n = \sqrt{2n}$. As plotted in Fig. 5(a), the eigenenergy of the lowest Landau level increases rapidly as approaching the open edge, which keeps electrons from spreading out. Since in the disk geometry the mth orbital is located at $\sqrt{2m}l_B$ along the radial direction, the Landau orbitals closer to the edge are denser than those closer to the center. The edge potential influences a large portion of the orbitals in the momentum space. Therefore, in a finite-size calculation we need to include significantly more orbitals than needed for the Laughlin state to prevent the edge potential from destroying the bulk properties.

Here we consider the $1/3$ filling factor in the lowest Landau level of graphene, or $\nu = 2 + 1/3$ with the spin-valley symmetry. In Fig. 5(b) we plot the ground-state angular momentum for 8 electrons in 32 orbitals as a function of d. Note that for small enough d the ground state can have an angular momentum smaller than $M_0 = 84$, in sharp contrast to the GaAs model in Fig. 2(b), in which the Laughlin phase remains stable down to $d = 0$. This is because that the interaction is too weak to compete with the edge potential when d is sufficiently small, thus electrons are repelled from the edge and the ground-state angular momentum decreases. Strictly speaking, a fair comparison requires us to use a sharp cutoff in real space in the GaAs model as well, such that the single-particle energy bends up and one needs to introduce Landau level mixing in a more complete treatment (see Ref. [12] for more detail), which is irrelevant to our main results so we skip here. Essential to our discussion, the Laughlin phase is nearly system size independent when N_{orb} is large as analyzed in Ref. [21]. As pointed out there, we believe that one has a good chance of realizing the Laughlin phase in graphene without reconstructed edge for

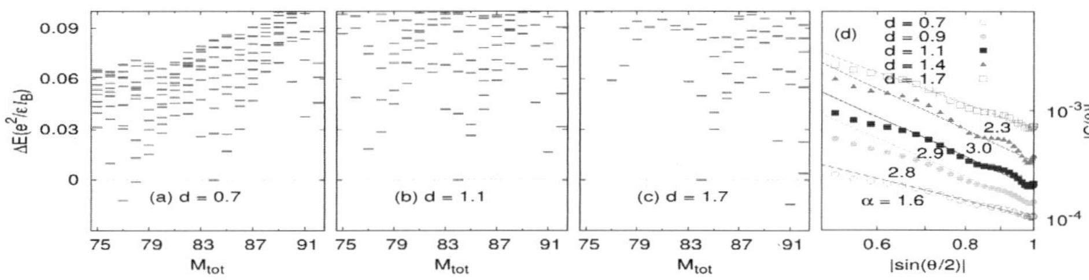

Figure 6. (a)-(c) The low-lying energy spectrum for 8 electrons in 32 orbitals in the graphene model with various ds. The corresponding Laughlin state has a $M_{tot} = 84$, the same as in (b). (d) The exponent α in the edge Greens function of the ground states for various ds.

$0.5 < d/l_B < 1.5$ [21]. The upper bound of the window in which the unreconstructed Laughlin phase can be realized in the graphene model agrees well with that in the GaAs model [10, 12, 13]. However, in the graphene model, e.g., on boron nitride substrate between the graphene layer and the gate, the parameter d can be chosen as small as a few nanometers or a small fraction of l_B [16], thus the universal exponent in tunneling experiments may be easily accessible in graphene.

In Fig. 6(a)-(c) we plot the low-lying energy spectra for $d = 0.7$, 1.1, and 1.7 l_B, whose ground-state angular momenta are 76, 84 (Laughlin), and 91, respectively. In these cases the edge mode is not well separated from other states in the spectrum (similar to Fig. 8 in Ref. [12]). To demonstrate the universal tunneling exponent, we compute the edge Green's function for 8 electrons in 32 orbitals, as shown in Fig. 6(d). We find that α for the three ground states are quite different. They are 1.6 and 2.3 for $d = 0.7\ l_B$ and $d = 1.7\ l_B$, respectively; neither is in the Laughlin phase. In these casses, the Laughlin state still suffers from the edge reconstruction instability, which results in charge accumulation similar to that studied earlier based on Hartree-Fock types of calculations [22, 23, 24, 25]. When $d = 0.9, 1.1$ and $1.4 l_B$ within the Laughlin phase $d/l_B \in [0.9, 1.4]$ as shown in Fig. 5(b), we find $\alpha \sim 2.9 \pm 0.1$. Therefore, the universal tunneling exponent can be expected if the Laughlin phase without reconstructed edges is confirmed in graphene.

4. Conclusions and discussions

We compare the FQH states at $\nu = 1/3$ in models for GaAs-based and graphene-based devices. Although the CLL theory predicts a tunneling current $I \propto V^\alpha$ with a universal $\alpha = 3$, the exponent can be affected by edge reconstruction, which exists in modulation-doped GaAs-based samples. However, in a graphene-based device, edge reconstruction can be avoided by using a very thin dielectric between the graphene layer and the metallic gate. We demonstrated the universal exponent in the edge Green's function calculation. The exponent is always about 3 once the ground state is in Laughlin region and smaller than 3 outside of the Laughlin region which consists with the CLL theory and the experiments in GaAs. We notice that there is a recent experiment show that the edge reconstruction does not happen at the edge of the integer quantum Hall states [26] and the implement of the fractional quantum Hall effect is in progress. In summary, we conclude that the non-universality in the FQH edge physics is due to the edge reconstruction, which induces more edges and breaks the chirality. In graphene system, the experiment can reach the parameter region which does not break the chirality of the edge; therefore, the universal edge physics is detectable. Our arguments also suitable for the other quantum Hall systems, such as integer or fractional FQH states in bilayer or trilayer graphene system.

5. Acknowledgments
This work is supported by DOE grant No. DE-SC0002140 and the 973 Program under Project No. 2009CB929100.

References
[1] D. C. Tsui, H. L. Stormer, and A. C. Gossard. Phys. Rev. Lett. **48**, 1559 (1982).
[2] A. Kitaev, Ann. Phys. **303**, 2 (2003).
[3] M. H. Freedman, A. Kitaev, and Z. Wang, Commun. Math. Phys. **227**, 587 (2002).
[4] M. H. Freedman, M. Larsen, and Z. Wang, Commun. Math. Phys. **227**, 605 (2002).
[5] X.-G. Wen, Int. J. Mod. Phys. B **6**, 1711 (1992).
[6] M. Grayson, D. C. Tsui, L. N. Pfeiffer, K. W. West, and A. M. Chang, Phys. Rev. Lett. **80**, 1062 (1998).
[7] A. M. Chang, L. N. Pfeiffer, and K. W. West, Phys. Rev. Lett. **77**, 2538 (1996).
[8] A. M. Chang, M. K. Wu, C. C. Chi, L. N. Pfeiffer, and K. W. West, Phys. Rev. Lett. **86**, 143 (2001).
[9] C. de C. Chamon and X.-G. Wen, Phys. Rev. B **49**, 8227 (1994).
[10] X. Wan, K. Yang, and E. H. Rezayi, Phys. Rev. Lett. **88**, 056802 (2002).
[11] K. Yang, Phys. Rev. Lett. **91**, 036802 (2003).
[12] X. Wan, E. H. Rezayi, and K. Yang, Phys. Rev. B **68**, 125307 (2003).
[13] S. Jolad and J. K. Jain, Phys. Rev. Lett. **102**, 116801 (2009). S. Jolad, D. Sen, and J. K. Jain, Phys. Rev. B **82**, 075315 (2010).
[14] K. I. Bolotin, F. Ghahari, M. D. Shulman, H. L. Stormer, and P. Kim, Nature, **462**, 196 (2009).
[15] X. Du, I. Skachko, F. Duerr, A. Luican, and E. Y. Andrei, Nature, **462**, 192 (2009).
[16] C. R. Dean,A. F. Young, P. Cadden-Zimansky, L. Wang, H. Ren, K. Watanabe, T. Taniguchi, P. Kim, J. Hone, and K. L. Shepard, Nature Physics, **7**, 693 (2011).
[17] L. Pfeiffer, K. W. West, H. L. Stormer, and K. W. Baldwin, Appl. Phys. Lett. **55**, 1888 (1989).
[18] X. Wan, F. Evers, and E. H. Rezayi, Phys. Rev. Lett. **94**, 166804 (2005).
[19] F. D. M. Haldane, Phys. Rev. Lett. **51**, 605 (1983).
[20] I. Romanovsky, C. Yannouleas, and U. Landman, Phys. Rev. B **83**, 045421 (2011).
[21] Z. X. Hu, R. N. Bhatt, X. Wan, and K. Yang, arXiv:1109.5994.
[22] J. Fernández-Rossier, J. J. Palacios, and L. Brey, Phys. Rev. B **75**, 205441 (2007);
[23] P. G. Silvestrov and K. B. Efetov, Phys. Rev. B **77**, 155436 (2008);
[24] A. A. Shylau and I. V. Zozoulenko, Phys. Rev. B **84**, 075407 (2011).
[25] A. H. Castro Neto, F. Guinea, and N. M. R. Peres, Phys. Rev. B **73**, 205408 (2006).
[26] G. H. Li, A. Luican, and Eva Y. Andrei, in preparation.

Numerical linked-cluster expansion for the distorted kagome lattice Heisenberg model

Ehsan Khatami and Marcos Rigol
Department of Physics, Georgetown University, Washington DC, 20057, USA
E-mail: ehsan@physics.georgetown.edu

Abstract. Motivated by experimental results for the thermodynamic properties of the $Rb_2Cu_3SnF_{12}$ material and the discovery of its valence-bond solid ground state, we utilize the numerical linked-cluster expansions (NLCEs) and devise an expansion tailored to solve the Heisenberg model on a pinwheel-distorted kagome lattice. Using the exchange interactions that are relevant to $Rb_2Cu_3SnF_{12}$, we calculate its uniform spin susceptibility and find a very good agreement with experiment. Next, we focus on the ground state of a simplified model of the distorted kagome lattice and take advantage of a zero-temperature Lanczos-based NLCE to study the approach to the ground state of the uniform kagome lattice Heisenberg model using a tuning parameter. Together with results from exact diagonalization of finite clusters, we find evidence that a phase transition occurs before the uniform limit is reached.

1. Introduction

Frustrated magnets are systems in which the combination of lattice geometry and the nature of interactions between spins prevent simultaneous minimization of the energy on all the bonds for any arrangement of spins (in a classical picture). Examples of these are the antiferromagnetic (AF) spin models on the kagome, triangular, checkerboard and pyrochlore lattices. They all share a common feature. That is, they consist of frustrating plaquettes (triangles or tetrahedra). In the (classical) Ising model, frustration results in an exponentially large ground state degeneracy and thus, a finite zero-temperature entropy [1, 2]. In the quantum mechanical models, such as the AF Heisenberg model, the large number of low lying states often leads to unusual low-temperature properties and exotic ground states, such as the resonating valence bond (RVB) spin-liquids [3, 4, 5, 6, 7, 8, 9].

Among different frustrated geometries, the kagome lattice Heisenberg model (KLHM) is of particular interest since, according to numerical calculations, it has energy gaps that are an order of magnitude smaller than for the other geometries and exponentially large number of singlet states below the lowest triplet state [10, 11, 4]. On the other hand, experiments on materials with almost perfectly decoupled kagome layers, such as Herbertsmithitse, show no magnetic order down to very low temperatures and no evidence of a spin gap [12, 13, 14, 15]. However, when studying these materials, one has to be careful about the role of magnetic impurities and Dzyaloshinskii-Moria anisotropy, which might ultimately determine the nature of the ground state [16, 17, 18].

Numerical studies have played an important role since the beginning of the search for the nature of the ground of the KLHM. Most recently, a large-scale density matrix renormalization

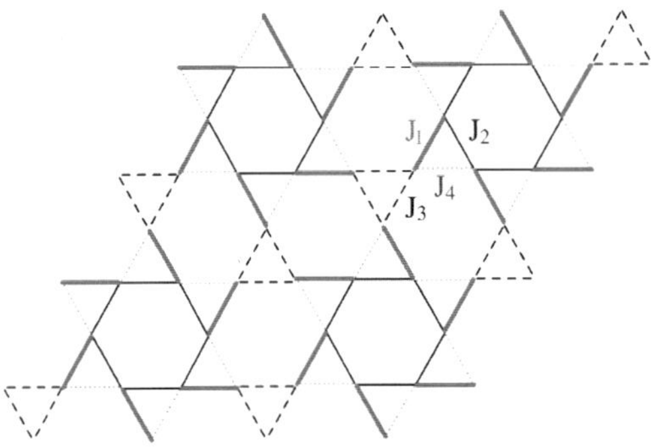

Figure 1. The anisotropic kagome lattice, representing the model for $Rb_2Cu_3SnF_{12}$ with four different exchange interactions, $J_1 \ldots J_4$.

group (DMRG) study by Yan *et al.* [9], suggested that the ground state of this model in the thermodynamic limit is a fully gapped singlet. This is despite the fact that previous series expansion studies had found indications of the 36-site unit cell valence-bond solid (VBS) being the ground state [6, 7]. By introducing a parameter, λ, which takes one from the distorted kagome to the uniform lattice, they predicted that such dimer order is picked up according to the perturbative expansion of the energies in powers of λ. However, in this scheme, other VBS orders can, in principle, be energetically favored in higher orders of λ.

In this work, we first focus on the finite-temperature properties of the material $Rb_2Cu_3SnF_{12}$, which is known to have distorted kagome layers and a gapped valence-bond solid (VBS) ground state with a 12-site unit cell [19, 20]. We report on the results from a numerical linked-cluster expansion (NLCE) [21, 22] that is designed to capture not only the high temperature properties of this material in the thermodynamic limit, but also the low-temperature physics associated with its VBS order. The advantage of the NLCE used here, over the regular series expansions, is that in NLCE the properties of linked-clusters in the series are calculated exactly, as opposed to perturbatively, using exact diagonalization techniques. We find that using the experimental values for the four different exchange interactions in the model for $Rb_2Cu_3SnF_{12}$ (shown in Figure 1), our NLCE yields a uniform spin susceptibility that is in very good agreement with the experiment at low temperatures, while results from exact diagonalization of small periodic clusters exhibit very large finite-size effects and lead to a different result. In order to approach the ground state of the uniform KLHM, we simplify the model Hamiltonian and use a tuning parameter, λ, that takes one from the distorted to the uniform KLHM and study the evolution of an order parameter for the VBS as λ is varied. Together with results from finite clusters, we find evidence for a phase transition before the uniform limit is reached and discuss its implications for the search for the properties of the KLHM ground state.

2. The Model

We write the KLHM Hamiltonian as:

$$\hat{H} = \sum_{\langle ij \rangle_\alpha} J_\alpha \, \hat{\mathbf{S}}_i \cdot \hat{\mathbf{S}}_j \tag{1}$$

where $\langle .. \rangle$ denotes nearest neighbors, $\hat{\mathbf{S}}_i$ is the spin-1/2 vector on site i, and α is the bond type. We consider up to four different bond types as depicted in Figure 1. The unit of energy is set to the largest exchange interaction (J_1).

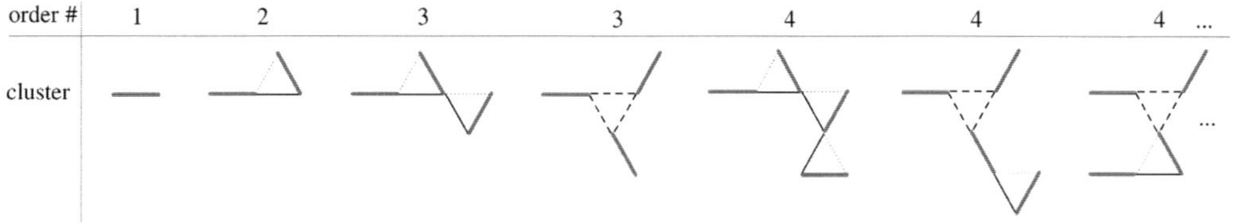

Figure 2. Topological clusters in the expansion with up to four pinwheel dimers (fourth order). Bond types are the same as in Fig. 1.

order number	1	2	3	4	5	6	7	8	9	10	11	12	13
number of clusters	1	1	2	3	7	14	25	50	102	216	442	929	1969

Table 1. Number of topological clusters in each order of the expansion.

3. Numerical Linked-Cluster Expansion

Linked-cluster expansions are methods in which an extensive property of the lattice model is expanded in terms of contributions from all the clusters, up to a certain size, that one can embed in the lattice [23]. This can be summarized in the following equation:

$$P = \sum_c L(c) W_{\mathcal{P}}(c), \tag{2}$$

where P is the property per site in the thermodynamic limit, $L(c)$ is the number of ways a cluster c can be embedded in the lattice, per site, and $W_{\mathcal{P}}(c)$ is the weight of cluster c for property $\mathcal{P}$. While $L(c)$ depends only on the geometry of the lattice, to calculate the weights, one has to compute the property of interest for different linked clusters, and use the inclusion-exclusion principle:

$$W_{\mathcal{P}}(c) = \mathcal{P}(c) - \sum_{s \subset c} W_{\mathcal{P}}(s), \tag{3}$$

where the sum is over all the subclusters of c. In high temperature expansions, $\mathcal{P}(c)$ are calculated using a perturbative expansion in powers of the inverse temperature. In that case, at low temperatures, the series can diverge even if the correlations in the model are short-ranged. A similar perturbative approach has also been used to calculate the ground state properties of a pinwheel distorted KLHM by using λ as the small parameter [6, 7]. Thus, the contributions of the clusters to each power of λ are properly added to the series.

In contrast, in NLCE, $\mathcal{P}(c)$ is calculated exactly using full diagonalization techniques and so, λ does not have to be small. The convergence of NLCE is controlled instead by the correlations in the system. Namely, NLCE converges if the correlation lengths, or the fluctuations around an ordered state, remain short-ranged. Moreover, the properties of each cluster are computed in all orders of λ, therefore, by increasing the size of the clusters in the series, one can systematically improve the results. There are many ways one can generate the topological clusters in the NLCE. Examples of that for the kagome lattice are, the site expansion, where the building block for generating clusters is a single site, bond expansion, triangular expansion, etc [21, 22].

For the pinwheel-distorted KLHM, one way to capture the low-temperature physics, associated with the VBS order, is to use the dimers of the pinwheel (strong bonds in Figure. 1, indicated by J_1) as the building blocks and generate the clusters in the expansion by attaching together more of those dimers through the other three types of bonds. In Figure. 2, we show all the topologically distinct clusters in our expansion that have four dimers or less. In Table 1, we enumerate the number of clusters in each order up to the 13th order. In the above expansion, we

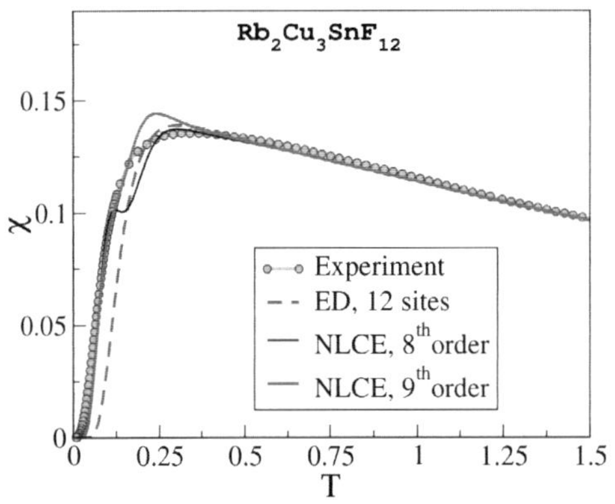

Figure 3. Comparison of the uniform spin susceptibility per site from our expansion to the experimental results (from Ref. [19]). The exchange interactions used are $J_2 = 0.95J_1$, $J_3 = 0.85J_1$ and $J_4 = 0.55J_1$, consistent with experimental finding for $Rb_2Cu_3SnF_{12}$. We have also included results from ED on a periodic 12-site cluster.

have restricted ourselves to those clusters that have the dimers connected through complete triangles. This is because at zero temperature, the contribution from any other cluster is zero [6, 7]. Even though this is not nessarily true at finite temperatures, this choice helps significantly reduce the number of topological clusters, which can be a limiting factor in large orders, while providing a consistent expansion scheme.

The limit on the number of orders in the expansion that can be treated by exact diagonalization is set either by memory or time requirements. In each order, the size of the clusters is twice the number of dimers, i.e., we need to solve the Heisenberg model on 18-site clusters in the 9th order. We use full diagonalization at finite temperatures and carry out the expansion to the 9th order. After block-diagonalizing the Hamiltonian matrix using the conservation of total spin in the z direction, the largest matrix belongs to the spin sector with equal number of spin-ups and spin-downs. More specifically, in the 9th order alone, we have to diagonalize matrices of size $\binom{18}{9} = 48,620$, or less, 102 times (the number of topological clusters in the 9th order) for each parameter set. For the zero-temperature calculations, we are interested only in the lowest eigenenergy and its eigenfunction. Therefore, we utilize the Lanczos algorithm to compute the ground state properties. Unlike for the finite-temperature case, here, we are not restricted by memory, but rather by the exponentially larger number of clusters in the higher orders and by the time it takes to diagonalize each of them. We carry out the calculations to the 13th order where there are 1969 clusters to diagonalize and the Hilbert space for their largest spin sector is $\binom{26}{13} = 10,400,600$.

4. Results

Employing NLCE with a pinwheel dimer expansion as explained above, we first explore the uniform susceptibility of $Rb_2Cu_3SnF_{12}$ that has been measured in experiments [19]. In Figure 3, we show the comparison between the experimental results and the results from NLCE with exchange interactions that are believed to be appropriate for this material ($J_2 = 0.95J_1$, $J_3 = 0.85J_1$ and $J_4 = 0.55J_1$) [19, 20]. We use $J_1 = 234$K, from experiments and match the two results at the highest temperature available experimentally with the g-factor, $g = 2.46$. We observe that NLCE converges at high temperatures and down to $T \sim 0.5$, as well as in the low-temperature window of $0 < T \lesssim 0.1$. Convergence in the latter region was expected since our NLCE is

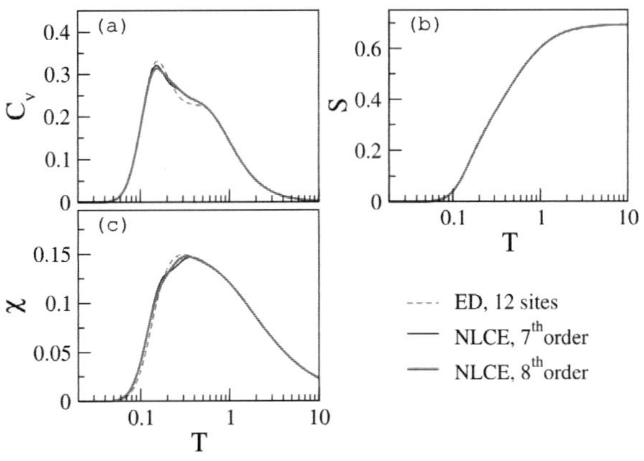

Figure 4. Specific heat, entropy, and uniform spin susceptibility per site for the pinwheel distorted Kagome lattice of Fig. 1 with $J_2 = J_3 = J_4 = \lambda J_1$ and $\lambda = 0.70$ using NLCE. Dashed lines are results from ED for the periodic 12-site cluster.

tailored to capture the VBS ground state of this material. However, to achieve convergence in the intermediate temperature region of $0.1 < T < 0.5$, where the thermal fluctuations around the VBS phase grow and eventually destroy it, one needs to consider higher orders (larger clusters) in NLCE, something that is computationally beyond our reach at this point. By comparing the ED results for a periodic 12-site cluster with the same exchange interactions used in NLCE, one can see that the discrepancy between ED and experiments at low temperatures is not due to the inaccurate exchange values, or their temperature dependence, but rather is caused by finite-size effects, which are absent in NLCEs.

To be able to define a tuning parameter that systematically transforms the pinwheel-distorted KLHM with VBS ground state to the uniform KLHM, we simplify the model and assume $J_2 = J_3 = J_4 = \lambda J_1$ [24]. When λ is sufficiently small, the ground state will still be the pinwheel VBS order with a 12-site unit cell. $\lambda = 1$ on the other hand, corresponds to the uniform KLHM. We are interested in the evolution of the thermodynamic properties of this model as λ is tuned from small values to close to unity. In Figure 4, we show the specific heat, entropy and the susceptibility for the case of $\lambda = 0.7$. The NLCE clearly converges at almost all temperature regions with this λ. We also point out the clear deviation of the ED results with the 12-site cluster from NLCE at $0.1 < T < 0.5$ [see e.g., Figure 4(a)]. The results for other values of λ [25] show that the convergence in NLCE at low temperatures is lost before λ reaches unity. Therefore, it is natural to ask, what is the value of λ beyond which NLCE no longer converges at low temperatures? and, what does that tell us about the robustness of the VBS order and the nature of the ground state of the KLHM?

To answer these questions, we focus on the ground state only and take advantage of the Lanczos algorithm to speed up the diagonalization process (since we are interested only in the lowest energy state), reduce the required memory, and therefore, be able to carry out the series to up to the 13th order in the dimer expansion. We calculate the bond energies, B_α, for the four different types of bonds that exist on the lattice. The difference between B_1 (corresponding to the pinwheel dimers) and any other bond can be used as the VBS order parameter [25]. Here, we choose this difference to be between B_1 and the average of the other three, $(B_2+B_3+B_4)/3-B_1$. In the small-λ region, $|B_1| > |(B_2 + B_3 + B_4)/3|$ and so, the order parameter is finite. On the other hand, at $\lambda = 1$, by symmetry, all B_α are the same and the order parameter for the uniform KLHM has to be zero.

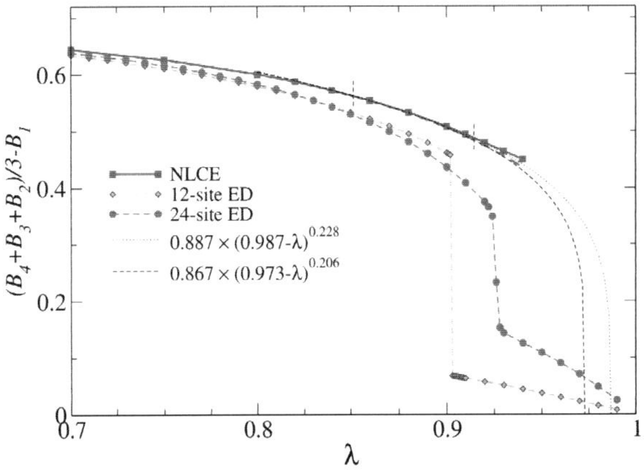

Figure 5. Order parameter at $T = 0$ vs the anisotropy parameter, λ, for $J_2 = J_3 = J_4 = \lambda J_1$. We use the Wynn sum with four cycles of improvement [21, 22] in NLCE. Results for the last two orders of the former are fit to $A(\lambda_c - \lambda)^B$ (thin dotted and dashed lines). Vertical solid lines show the region used for the fit.

The behavior of the order parameter by increasing λ at $T = 0$ is shown in Figure 5. As expected, the NLCE results converge well for small λ. However, beyond $\lambda = 0.9$, the series do not converge anymore. This signals that the VBS may not be the ground state phase in the thermodynamic limit when the pinwheel distortion is small. Moreover, the results from ED with the 12-site and 24-site periodic clusters show a clear first-order phase transition (level crossing) to another phase at or slightly above $\lambda = 0.9$. These findings provide strong evidence that in the thermodynamic limit, a phase transition likely occurs at a critical λ_c as the distortion is decreased and before the uniform KLHM limit is reached. We point out that there is also the possibility for a second-order, or weak first-order, phase transition at a λ_c. For this reason, we fit the high-λ region of the NLCE results to a power-law form to deduce λ_c in such a scenario and find that $\lambda_c < 1$ (see Figure 5).

Note that if in fact the transition is first order, λ_c will be smaller than that of the second-order transition found above. Moreover, if the transition happens at $\lambda = 1$, it is likely to a gapless spin liquid, since in that case, any dimerization perturbation of this type would immediately take the system from the spin liquid to the VBS phase. However, such scenario is not in agreement with the recent DMRG study [9] that argues the ground state of the KLHM is a *gapped* spin liquid.

5. Conclusions

In conclusion, we implemented a numerical linked-cluster expansion in terms of strong bonds on a pinwheel-distorted kagome lattice with a 12-site unit cell that describes the material $Rb_2Cu_3SnF_{12}$. Using the appropriate exchange parameters for this material, we calculate the uniform susceptibility that is in very good agreement with the experimental results. We define a tuning parameter, λ, that takes one from the distorted to the uniform kagome lattice in a simplified model and focus on the properties of the ground state as λ is varied. NLCE results, along with the results from exact diagonalization of small clusters with periodic boundary condition, point to a phase transition that likely occurs in the thermodynamic limit before the uniform limit is reached. This implies that the KLHM may not spontaneously dimerize in such pattern and that the ground state phase is stable to small dimerization perturbations of this type.

Acknowledgments
We are grateful to Rajiv R. P. Singh for fruitful discussions. This work was supported by NSF under Grant No. OCI-0904597.

References
[1] Lhuillier C 2005 *arXiv:cond-mat/0502464*
Misguich G and Lhuillier C 2004 in *Frustrated spin systems* edited by H. T. Diep (Hackensack NJ: World Scientific)
[2] Huse D A and Rutenberg A D 1992 *Phys. Rev. B* **45** 7536
[3] Zeng C and Elser V 1990 *Phys. Rev. B* **42** 8436
Marston J B and Zeng C 1991 *Journal of Applied Physics* **69** 5962
Sachdev S 1992 *Phys. Rev. B* **45** 12377
Leung P W and Elser V 1993 *Phys. Rev. B* **47** 5459
Hastings M B 2000 *Phys. Rev. B* **63** 014413
Syromyatnikov A V and Maleyev S V 2002 *Phys. Rev. B* **66** 132408
Nikolic P and Senthil T 2003 *Phys. Rev. B* **68** 214415
Budnik R and Auerbach A 2004 *Phys. Rev. Lett.* **93** 187205
Ran Y, Hermele M, Lee P A and Wen X G 2007 *Phys. Rev. Lett.* **98** 117205
Wang F and Vishwanath A 2006 *Phys. Rev. B* **74** 174423
Jiang H C, Weng Z Y and Sheng D N 2008 *Phys. Rev. Lett.* **101** 117203
Hao Z and Tchernyshyov O 2009 *Phys. Rev. Lett.* **103** 187203
Evenbly G and Vidal G 2010 *Phys. Rev. Lett.* **104** 187203
Poilblanc D, Mambrini M and Schwandt D 2010 *Phys. Rev. B* **81** 180402
[4] Mila F 1998 *Phys. Rev. Lett.* **81** 2356
[5] Singh R R P and Huse D A 1992 *Phys. Rev. Lett.* **68** 1766
[6] Singh R R P and Huse D A 2007 *Phys. Rev. B* **76** 180407
[7] Singh R R P and Huse D A 2008 *Phys. Rev. B* **77** 144415
[8] Yang B J and Kim Y B 2009 *Phys. Rev. B* **79** 224417
[9] Yan S, Huse D A and White S R 2011 *Science* **332** 1173
[10] Sindzingre P and Lhuillier C 2009 *EPL (Europhysics Letters)* **88** 27009
[11] Sindzingre P, Misguich G, Lhuillier C, Bernu B, Pierre L, Waldtmann C and Everts H U 2000 *Phys. Rev. Lett.* **84** 2953
[12] Fukuyama H 2008 *Journal of the Physical Society of Japan* **77** 111013
[13] Helton J S, Matan K, Shores M P, Nytko E A, Bartlett B M, Yoshida Y, Takano Y, Suslov A, Qiu Y, Chung J H, Nocera D G and Lee Y S 2007 *Phys. Rev. Lett.* **98** 107204
Olariu A, Mendels P, Bert F, Duc F, Trombe J C, de Vries M A and Harrison A *Phys. Rev. Lett.* **100** 087202
Imai T, Nytko E A, Bartlett B M, Shores M P and Nocera D G 2008 *Phys. Rev. Lett.* **100** 077203
de Vries M A, Kamenev K V, Kockelmann W A, Sanchez-Benitez J and Harrison A 2008 *Phys. Rev. Lett.* **100** 157205
Zorko A, Nellutla S, van Tol J, Brunel L C, Bert F, Duc F, Trombe J C, de Vries M A, Harrison A and Mendels P 2008 *Phys. Rev. Lett.* **101** 026405
[14] Chanier T, Opahle I, Sargolzaei M, Hayn R and Lannoo M 2008 *Phys. Rev. Lett.* **100** 026405
[15] Bert F, Bono D, Mendels P, Ladieu F, Duc F, Trombe J C and Millet P 2005 *Phys. Rev. Lett.* **95** 087203
[16] Rigol M and Singh R R P 2007 *Phys. Rev. Lett.* **98** 207204
[17] Rigol M and Singh R R P 2007 *Phys. Rev. B* **76** 184403
[18] Cépas O, Fong C M, Leung P W and Lhuillier C 2008 *Phys. Rev. B* **78** 140405
[19] Morita K, Yano M, Ono T, Tanaka H, Fujii K, Uekusa H, Narumi Y and Kindo K 2008 *Journal of the Physical Society of Japan* **77** 043707
[20] Matan K, Ono T, Fukumoto Y, Sato T J, Yamaura J, Yano M, Morita K and Tanaka H 2010 *Nature Physics* **6** 865
[21] Rigol M, Bryant T and Singh R R P 2006 *Phys. Rev. Lett.* **97** 187202
[22] Rigol M, Bryant T and Singh R R P 2007 *Phys. Rev. E* **75** 061118
[23] Oitmaa J H C and Zhang W 2006 *Series Expansion Methods for Strongly Interacting Lattice Models* (Cambridge University Press, Cambridge, England)
[24] Note that our model differs from Ref. [8] in which a $J_1 = J_2 = J_3 > J_4$ pattern is considered for exchanges.
[25] Khatami E, Singh R R P and Rigol M 2011 *Phys. Rev. B* **84** 224411

An Efficient Algorithm for Simulating the Real-Time Quantum Dynamics of a Single Spin-1/2 Coupled to Specific Spin-1/2 Baths

M A Novotny[1,2,5], **M Guerra**[1,2,3], **H De Raedt**[4], **K Michielsen**[5,6], **F Jin**[5]

[1] Department of Physics and Astronomy, Mississippi State University, Mississippi State, MS 39762-5167, USA
[2] HPC^2 Center for Computational Sciences, Mississippi State University, Mississippi State, MS 39762-5167, USA
[3] Department of Sciences & Mathematics, Mississippi University for Women, 1100 College Street, MUW-100, Columbus, MS 39701, USA
[4] Department of Applied Physics, Zernike Institute of Advanced Materials, University of Groningen, Nijenborgh 4, NL-9747 AG Groningen, The Netherlands
[5] Institute for Advanced Simulation, Jülich Supercomputing Centre, Forschungszentrum Jülich, D-52425 Jülich, Germany
[6] RWTH Aachen University, D-52056 Aachen, Germany

E-mail: man40@ra.msstate.edu

Abstract. An efficient algorithm for the computation of the real-time dependence of a single quantum spin-1/2 coupled to a specific set of quantum spin-1/2 baths is presented. The specific spin baths have couplings only with the spin operators S^x between bath spins and the central spin. We calculate spin expectation values, the quantum purity, the von Neumann entropy, and the off-diagonal components of the reduced density matrix for the central spin once the bath spins have been traced out. The algorithm does not require the storage of any vector larger than of size 2, even though the size of the Hilbert space is 2^{N+1}, where N is the number of bath spins. Results are presented for the central spin connected to different sizes and types of spin baths, and for different initial states for the central spin and for the bath spins. Results are also compared to those for more general baths.

1. Introduction

The real-time (as opposed to imaginary time) quantum dynamics of systems is becoming more and more of interest. This interest is driven by the increasingly precise experiments that are being performed on quantum systems, and by the long-range goal of building a quantum computer [1, 2]. The calculation that needs to be performed is to solve the time-dependent Schödinger equation (TDSE) [3]

$$\mathcal{H}\left|\Psi(t)\right\rangle = -\frac{\hbar}{i}\frac{\partial}{\partial t}\left|\Psi(t)\right\rangle , \tag{1}$$

where the Hamiltonian $\mathcal{H}$ is assumed to be independent of time and the initial state $|\Psi(0)\rangle$ is given. Furthermore, if the closed quantum system is thought of as being composed of a subsystem S and an environment or bath B, then the Hamiltonian can be written as

$$\mathcal{H} = \mathcal{H}_S + \mathcal{H}_B + \mathcal{H}_{SB}, \tag{2}$$

with the first term the Hamiltonian describing the subsystem S, the second term describes the Hamiltonian of the bath B, and the third term the Hamiltonian describing the coupling between the subsystem S and the bath B. Spin baths have been studied since at least the time of Fermi [4], and are also of recent interest [5, 6, 7]. In this paper we assume that the subsystem S consists of one spin-1/2 particle, also called the central spin, and that the bath B is composed of N spin-1/2 particles. The dimension of the Hilbert space for the bath is $D_B = 2^N$ and the dimension of the complete Hilbert space is $2D_B = 2^{N+1}$. The dimension of the vector $|\Psi(t)\rangle$ in Eq. (1) is 2^{N+1}. The state of the subsystem S is described by the reduced density matrix

$$\rho(t) = \mathrm{Tr}_B \left[\rho_{S+B}(t)\right], \tag{3}$$

where ρ_{S+B} is the density matrix of the complete $N+1$ spin system, and the trace is over the N spins in the bath. Thus $\rho(t)$ is a 2×2 matrix. However, to calculate $\rho(t)$ one first needs to propagate the state

$$|\Psi(t)\rangle = \exp\left(-\frac{i\mathcal{H}t}{\hbar}\right)|\Psi(0)\rangle, \tag{4}$$

for all $N+1$ spins according to the TDSE, and then trace out the bath spins.

The dimension of $|\Psi(t)\rangle$ presents a large difficulty in computer calculations of spin systems. We want to have N large, but memory of computers is limited. For example, the most powerful computer in the world on the November 2011 Top 500 list [8], the K computer at the RIKEN Advanced Institute for Computational Science in Japan, has 705,024 cores and operates at 10.5 Petaflop/s on the Linpack benchmark. Each compute node of the K computer has 16 Gbyte of memory, so the memory of the K computer is about 1.2×10^{16} bytes, which means that the number of spins for the largest vector $|\Psi(t)\rangle$ that can be stored in the K computer is about $N \approx 50$. Even the stated goal for exascale computing that the memory should be on the order of a few exabytes, limits the size of the vector $|\Psi(t)\rangle$ to about $N \approx 60$.

Clearly, algorithmic advances that could increase N without requiring storage of vectors of size 2^{N+1} would be extremely advantageous. In this article we present such an algorithm, where the only storage required is of vectors of size 2. This algorithm works only for specific types of baths. Although the type of bath can be generalized, in this paper we concentrate on one type of bath, which we call an x-bath, detailed explicitly in the next section. This type of bath can also be used for more than one spin in the subsystem of interest, but in order to describe the algorithm and explore the efficiency of this type of bath we limit ourselves to a single central spin. We assume that an efficient algorithm to calculate the TDSE for any Hamiltonian is available, such as the efficient method based on Chebyshev polynomials [9, 10, 11, 12]. We also present results for more general baths using the Chebyshev algorithm [12], comparing with our x-bath results. A recent paper reported the use of up to 36 spin-1/2 particles to study decoherence and thermalization [13] storing the full vectors of size 2^{36} required for the computations. The number of spins that can be included in computer calculations is small compared to the size of spin baths in laboratory experiments. For example, recent experiments studied a single electron coupled to a bath of about 10^9 spins [14, 15].

2. Methodology

We consider a single spin-1/2 particle coupled to a quantum bath of N spin-1/2 particles. The dimension of the Hilbert space is 2^{N+1}. Let the central spin be numbered spin 0, and the x-bath spins 1 through N. The Pauli spin matrices σ^x, σ^y, and σ^z are given by

$$s^x = \frac{\hbar}{2}\sigma^x = \frac{\hbar}{2}\begin{pmatrix} 0 & 1 \\ 1 & 0 \end{pmatrix}, \quad s^y = \frac{\hbar}{2}\sigma^y = \frac{\hbar}{2}\begin{pmatrix} 0 & -i \\ i & 0 \end{pmatrix}, \quad s^z = \frac{\hbar}{2}\sigma^z = \frac{\hbar}{2}\begin{pmatrix} 1 & 0 \\ 0 & -1 \end{pmatrix}. \tag{5}$$

We define

$$S_m^\ell = I_2 \otimes I_2 \otimes \cdots \otimes I_2 \otimes s^\ell \otimes I_2 \otimes \cdots \otimes I_2, \tag{6}$$

with I_2 the 2×2 identity matrix and s^ℓ the 2×2 Pauli spin matrix ($\ell \in \{x, y, z\}$) at the mth position in the Kronecker (direct) product of $N+1$ matrices of size 2×2. Here $m \in \{0, \cdots, N\}$ with $m = 0$ referring to the central spin and $m \in \{1, \cdots, N\}$ referring to the bath spins.

From now on we set $\hbar = 1$. The system Hamiltonian that we consider contains the following terms

$$\mathcal{H}_S = \omega_0 S_0^z, \tag{7}$$

where ω_0 denotes the strength of the external magnetic field that is applied to the central spin in the z-direction,

$$\mathcal{H}_B = \sum_{i_1=0}^{1} \sum_{i_2=0}^{1} \cdots \sum_{i_N=0}^{1} \widetilde{J}_{i_1,i_2,\cdots,i_N} \left(S_1^x\right)^{i_1} \left(S_2^x\right)^{i_2} \cdots \left(S_N^x\right)^{i_N}, \tag{8}$$

where $\widetilde{J}_{i_1,i_2,\cdots,i_N}$ denotes the strength of the coupling between the N spins in the x-bath. Furthermore, we assume that

$$\mathcal{H}_{SB} = S_0^x \left[\sum_{i_1=0}^{1} \sum_{i_2=0}^{1} \cdots \sum_{i_N=0}^{1} J_{i_1,i_2,\cdots,i_N} \left(S_1^x\right)^{i_1} \left(S_2^x\right)^{i_2} \cdots \left(S_N^x\right)^{i_N} \right] \tag{9}$$

where $J_{i_1,i_2,\cdots,i_N}$ denotes the strength of the coupling between the central spin and the N spins in the x-bath. Note that the type of spin interaction (2-spin, 3-spin, $\cdots$) is specified by the number of ones in the subscript of J or $\widetilde{J}$ since it follows from Eq. (6) that $(S_i^x)^0 = I_{\text{F}}$ where I_F denotes the $2^{N+1} \times 2^{N+1}$ identity matrix. For later, we define J_2 (J_3) as the strength of the two(three)-body interactions between the central spin and a (couple of) bath spin(s) in the case that they are all equal. Similarly, J_M denotes the strength of the N-body interactions between the central spin and the $N-1$ tuples of bath spins (only one bath spin is not in the sum) times 2^N, again for the case that the respective interactions are all equal. In addition, J_A denotes 2^{N+1} times the strength of the $(N+1)$-body interactions between the central spin and all bath spins. (The additional factors of 2 in the definitions of J_M and J_A are present for convenience. They are inserted because of the factor of 1/2 in $s^\ell = \sigma^\ell/2$. No factors of 2 are present in our definitions of J_2 or J_3.)

We first introduce the matrix

$$p_x = \frac{1}{\sqrt{2}} \begin{pmatrix} 1 & -1 \\ 1 & 1 \end{pmatrix}, \tag{10}$$

having the properties $p_x p_x^{-1} = I_2$ with $p_x^{-1} = p_x^{\text{T}} = p_x^\dagger$. The matrix p_x also exhibits the property $p_x^\dagger s^x p_x = s^z$. In addition

$$p_x^\dagger \begin{pmatrix} 0 \\ 1 \end{pmatrix} = \frac{1}{\sqrt{2}} \begin{pmatrix} 1 \\ 1 \end{pmatrix}. \tag{11}$$

We next introduce the matrix

$$P_p = I_2 \otimes p_x \otimes p_x \otimes \cdots p_x \,, \tag{12}$$

with the position 0 in the Kronecker product the 2×2 identity matrix I_2 and the positions 1 through N in the Kronecker product occupied by p_x. It follows that $P_p^\dagger P_p = I_F$, where I_F

denotes the $2^{N+1} \times 2^{N+1}$ identity matrix. By performing a unitary transformation with P_p, we obtain the transformed Hamiltonian

$$\begin{aligned}\widetilde{\mathcal{H}} &= P_p^\dagger \mathcal{H} P_p \\ &= \omega_0 S_0^z + \sum_{i_1=0}^{1}\sum_{i_2=0}^{1}\cdots\sum_{i_N=0}^{1} \widetilde{J}_{i_1,i_2,\cdots,i_N} (S_1^z)^{i_1} (S_2^z)^{i_2} \cdots (S_N^z)^{i_N} \\ &+ S_0^x \left[\sum_{i_1=0}^{1}\sum_{i_2=0}^{1}\cdots\sum_{i_N=0}^{1} J_{i_1,i_2,\cdots,i_N} (S_1^z)^{i_1} (S_2^z)^{i_2} \cdots (S_N^z)^{i_N} \right] .\end{aligned} \tag{13}$$

Thus the Hamiltonian $\widetilde{\mathcal{H}}$ is block diagonal with 2^N blocks of 2×2 matrices labeled by the index j and all having the form

$$\widetilde{\mathcal{H}}_j = \begin{pmatrix} \Omega_{zj} + \omega_0 & \Omega_{xj} \\ \Omega_{xj} & \Omega_{zj} - \omega_0 \end{pmatrix} = \Omega_{zj} I_2 + 2\vec{\Omega}_j \cdot \vec{s}, \tag{14}$$

with the vector $\vec{\Omega}_j$ having the components $(\Omega_{xj},\ 0,\ \omega_0)$ and $\left|\vec{\Omega}_j\right| = \sqrt{\Omega_{xj}^2 + \omega_0^2}$. The eigenvalues of $\widetilde{\mathcal{H}}_j$ are$\Omega\ _{zj} \pm \left|\vec{\Omega}_j\right|$. Note that Ω_{zj} depends on the $\widetilde{J}$ couplings only, and that$\Omega\ _{xj}$ depends only on the J couplings. The Hamiltonian $\widetilde{H}$ can thus also be written as

$$\widetilde{\mathcal{H}} = \widetilde{\mathcal{H}}_1 \oplus \widetilde{\mathcal{H}}_2 \oplus \cdots \oplus \widetilde{\mathcal{H}}_{2^N} . \tag{15}$$

As in [13] we assume that the central spin is decoupled from the bath for times $t < 0$, and that the central spin and the bath spins are brought into contact at time $t = 0$. We assume that at time $t = 0$ the central spin is in the state

$$|\psi(0)\rangle = \begin{pmatrix} \sin\alpha_0 \\ \cos\alpha_0 \end{pmatrix} \tag{16}$$

so that if $\alpha_0 = 0$ the central spin starts in the down state, and it starts in the up state for $\alpha_0 = \frac{\pi}{2}$. Note that the unitary transformation in Eq. (12) does not change the initial state of the central spin at time $t = 0$. After the unitary transformation by P_p at time $t = 0$ the state of the bath can be written as

$$\left|\widetilde{\Phi}(0)\right\rangle = \sum_{j=1}^{2^N} c_j \left|\varphi_j\right\rangle \qquad \text{with} \qquad \sum_{j=1}^{2^N} |c_j|^2 = 1, \tag{17}$$

with some complex coefficients c_j and $\{|\varphi_j\rangle\}$ the complete orthonormal set of states taken to be the usual direct products of the spin up and down states of the N bath spins. In practice one can regard the c_j as the coefficients obtained after starting with the $t = 0$ bath state $|\Phi(0)\rangle$ in the orthonormal basis of the eigenvectors of σ^x for the x-bath spins, which with the unitary transformation by $P_p^\dagger$ gives the environment spin state of Eq. (17).

The expectation values $\langle S_0^\ell(t)\rangle$ are independent of the values of $\widetilde{J}_{i_1,i_2,\cdots,i_N}$. With the initial state $|\Psi(0)\rangle = |\psi(0)\rangle \otimes |\Phi(0)\rangle$ this independence is because these terms only enter in$\Omega\ _{zj}$, and furthermore,

$$\begin{aligned}\left\langle S_0^\ell(t)\right\rangle &= \langle\Psi(0)|\exp\left(i\mathcal{H}t\right)\left(s^\ell \otimes I_2 \otimes I_2 \otimes \cdots \otimes I_2\right)\exp\left(-i\mathcal{H}t\right)|\Psi(0)\rangle \\ &= \left\langle\widetilde{\Psi}(0)\right|\exp\left(i\widetilde{\mathcal{H}}t\right)\left(s^\ell \otimes I_2 \otimes I_2 \otimes \cdots \otimes I_2\right)\exp\left(-i\widetilde{\mathcal{H}}t\right)\left|\widetilde{\Psi}(0)\right\rangle,\end{aligned} \tag{18}$$

with $\left|\widetilde{\Psi}(0)\right\rangle = P_p^\dagger \left|\Psi(0)\right\rangle$. The reason is that

$$\mathcal{H} = P_p \widetilde{\mathcal{H}} P_p^\dagger, \quad \text{so} \quad \exp\left[i\mathcal{H}t\right] = P_p \exp\left[i\widetilde{\mathcal{H}}t\right] P_p^\dagger , \tag{19}$$

and the constant termsΩ_{zj} in each block cancel because they commute with everything. One Ω_{zj} term comes from each of the blocks $\widetilde{H}$ in Eq. (19), and $\exp\left[i\Omega_{zj}t\right]\exp\left[-i\Omega_{zj}t\right] = 1$.

Next we want to trace out the bath spins to be left with the density matrix $\rho(t)$ for the central spin. At $t = 0$ the density matrix of the central spin is

$$\rho(0) = |\psi(0)\rangle\langle\psi(0)| . \tag{20}$$

Explicitly

$$\begin{aligned}
\rho(t) &= \mathrm{Tr}_B\left[e^{-i\mathcal{H}t}\left(\rho(0)\otimes\rho_B(0)\right)e^{i\mathcal{H}t}\right] && (21)\\
&= \mathrm{Tr}_B\left[P_p e^{-i\widetilde{\mathcal{H}}t}P_p^\dagger\left(\rho(0)\otimes\rho_B(0)\right)P_p e^{i\widetilde{\mathcal{H}}t}P_p^\dagger\right] && (22)\\
&= \mathrm{Tr}_B\left[e^{-i\widetilde{\mathcal{H}}t}\left(\rho(0)\otimes\widetilde{\rho}_B(0)\right)e^{i\widetilde{\mathcal{H}}t}\right] . && (23)
\end{aligned}$$

Here $\widetilde{\rho}_B(0) = |\widetilde{\Phi}(0)\rangle\langle\widetilde{\Phi}(0)|$. Writing Eq. (23) in terms of the c_j of Eq. (17) and the block diagonal matrices $\mathcal{H}_j$ of Eq. (14) gives the final equation for the reduced density matrix for the central spin,

$$\begin{aligned}
\rho(t) &= \left(|c_1|^2 e^{-i\widetilde{\mathcal{H}}_1 t}\rho(0)e^{i\widetilde{\mathcal{H}}_1 t}\right) + \left(|c_2|^2 e^{-i\widetilde{\mathcal{H}}_2 t}\rho(0)e^{i\widetilde{\mathcal{H}}_2 t}\right) + \cdots + \left(|c_{2^N}|^2 e^{-i\widetilde{\mathcal{H}}_{2^N} t}\rho(0)e^{i\widetilde{\mathcal{H}}_{2^N} t}\right)\\
&= |c_1|^2\rho_1(t) + |c_2|^2\rho_2(t) + \cdots + |c_{2^N}|^2\rho_{2^N}(t) . && (24)
\end{aligned}$$

Equation (24) shows that the reduced quantum density matrix $\rho(t)$ for the central spin is the sum of 2^N different 2×2 density matrices $\rho_j(t)$. Notice that the initial state of the bath spins enters only in the terms $|c_j|^2$. The 2×2 matrix exponentials can easily be calculated using the relation

$$\exp\left[\pm i\left(\vec{a}\cdot\vec{\sigma}\right)t\right] = I_2\cos\left(at\right) \pm i\frac{\vec{a}\cdot\vec{\sigma}}{a}\sin\left(at\right) , \tag{25}$$

with $a = |\vec{a}|$.

For any expectation value of the central spin, Eq. (24) gives

$$\left\langle S_0^\ell(t)\right\rangle = \sum_{j=1}^{2^N} |c_j|^2\,\mathrm{Tr}\left[s^\ell\rho_j(t)\right] . \tag{26}$$

Thus the final result is that for the type of Hamiltonian defined in Eq. (13) (one connected to an x-bath) the time dependence of the expectation value for any spin operator reduces to the sum over the expectation values of 2^N evolution equations with different Hamiltonians $\widetilde{\mathcal{H}}_j$. Equations (24) and (26) are the central results that will be exploited for our efficient algorithm. Note in particular that only 2×2 matrices must be worked with, and hence stored in memory. However, computationally we do not get something for nothing, in that we have to solve the TDSE for 2^N different Hamiltonians $\widetilde{\mathcal{H}}_j$ and then sum to get the final expectation value.

Journal of Physics: Conference Series **402** (2012) 012019 doi:10.1088/1742-6596/402/1/012019

3. Measured Quantities

We are particularly interested in measuring four types of quantities. The first are the time-dependent expectation values of the central spin, given by Eq. (26).

The second quantity of interest is the quantum purity for the central spin,

$$\mathcal{P}(t) = \mathrm{Tr}\left[\rho(t)^2\right] . \tag{27}$$

Because $\rho(t)$ is a 2×2 matrix we can write it as

$$\rho(t) = a_\rho(t) I_2 + \vec{b}_\rho(t) \cdot \vec{\sigma} \tag{28}$$

with the vector $\vec{b}_\rho(t)$ having components $(b_{\rho x}(t), b_{\rho y}(t), b_{\rho z}(t))$ and $b_\rho(t) = \sqrt{b_{\rho x}^2(t) + b_{\rho y}^2(t) + b_{\rho z}^2(t)}$. Using $\left[\vec{b}_\rho(t) \cdot \vec{\sigma}\right]^2 = b_\rho^2(t) I_2$ gives

$$\mathcal{P}(t) = 2a_\rho^2(t) + 2b_\rho^2(t) . \tag{29}$$

The third quantity of interest is the von Neumann entropy

$$\mathcal{S}(t) = -k_B \mathrm{Tr}\left[\rho(t) \log_2 \rho(t)\right] , \tag{30}$$

where from now on we also set Boltzmann's constant $k_B = 1$. The eigenvalues of $\rho(t)$ are

$$\lambda_\pm(t) = a_\rho(t) \pm b_\rho(t), \tag{31}$$

so that

$$\mathcal{S}(t) = -\,\lambda_+(t) \log_2\left[\lambda_+(t)\right] \;-\; \lambda_-(t) \log_2\left[\lambda_-(t)\right] . \tag{32}$$

The fourth quantity of interest, as defined in Ref. [13], is the sum of the off-diagonal elements of $\rho(t)$

$$\sigma(t) = \sqrt{\sum_{i=1}^{2^{N_s}-1} \sum_{j=i+1}^{2^{N_s}} \left|\rho_{ij}(t)\right|^2} , \tag{33}$$

if there were N_s spins in the subsystem. Note $\sigma(t)$ should not be confused with the Pauli spin matrices σ^ℓ. For general N_s, in order for the subsystem to thermalize toward a canonical distribution a necessary condition is that $\sigma(t)$ become very small [13]. In our case $N_s = 1$ so that

$$\sigma(t) = \left|\rho_{12}(t)\right| . \tag{34}$$

Since none of these measured quantities depend on the $\widetilde{J}$ couplings, without loss of generality we set all $\widetilde{J} = 0$.

4. Symmetry in interactions with the x-bath spins

Equations (24) and (26) are the central results, and are very general. For the general case where all J values are different one has to sum over 2^N different density matrices $\rho_j(t)$ to obtain $\rho(t)$. If there is any symmetry in the couplings between the central spin and the x-bath spins, then this symmetry can be exploited to reduce the number of terms that must be summed. The reduction comes because the symmetry in the Hamiltonian can make a number of the $\rho_j(t)$ identical. As long as some of the $\rho_j(t)$ are identical, the initial state of the bath need not have any symmetry

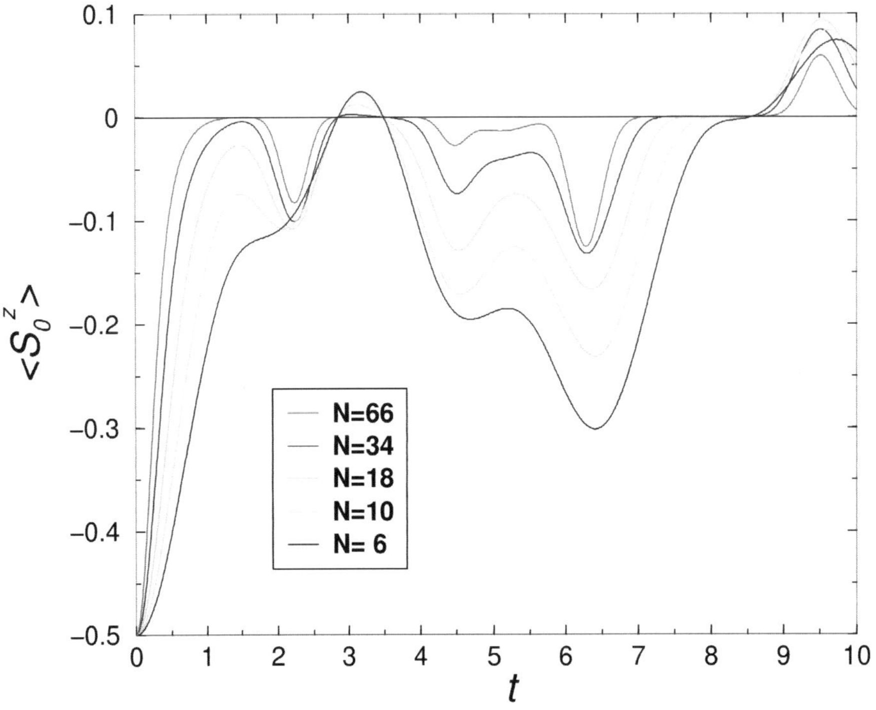

Figure 1. The central spin expectation value $\langle S_0^z(t)\rangle$ as a function of t is shown for different numbers of bath spins N. The N bath spins are divided into two equal baths with $N_1 = N_2 = N/2$. One bath has $J_2 = 0.9$ and $J_M = 0.1$, while the other bath has $J_2 = 1.1$ and $J_M = -0.1$. All spins start initially with spin down. Reading from top to bottom near $t = 0$ the values of N are 66, 34, 18, 10, and 6.

in order to combine terms in the sum of Eqs. (24) or (26). For example, if all $\rho_j(t)$ are identical for $j \in \{i, \cdots, i+k\}$ then in Eq. (26) one can group together the terms that enter the sum to be

$$\left\langle S_0^\ell(t)\right\rangle = \cdots + \left\{ \left(\sum_{j=i}^{i+k} |c_j|^2\right) \operatorname{Tr}\left[s^\ell \rho_{\rm sym}(t)\right]\right\} + \cdots \qquad (35)$$

with the definition $\rho_j(t) = \rho_{\rm sym}(t)$ for $j \in \{i, \cdots, i+k\}$. Therefore in this example one only has to solve the TDSE for one block Hamiltonian, rather than for k different block Hamiltonians.

Consider the special case of only two-body interactions between the central spin and the x-bath spins, all with the same coupling J_2. The Hamiltonian of the system (with $\tilde{J} = 0$), after the unitary transformation with P_p, reads

$$\widetilde{\mathcal{H}}_j = \omega_0 S_0^z + J_2 S_0^x \left[\sum_{i=1}^{N} S_i^z\right]_j = \vec{S}_0 \cdot \vec{\Omega}_j\,, \qquad (36)$$

where we have defined the vector

$$\vec{\Omega}_j = \left(J_2 \left[\sum_{i=1}^{N} S_i^z \right]_j , \, 0, \, \omega_0 \right) . \tag{37}$$

We now make one additional assumption in order to compare directly with the results presented in a recent paper by Rao *et al.* [16]. We assume that the initial state of all the spins (the central spin and all bath spins) is the down state, $|\Psi(0)\rangle = |\psi(0)\rangle \otimes |\Phi(0)\rangle = |\downarrow\downarrow \cdots \downarrow\rangle$. After the transformation with P_p this makes all the $|c_j|^2$ equally probable, so $|c_j|^2 = 2^{-N}$. Then we can organize the sum over the 2^N x-bath spins into ones that are identical, namely classified by the projection m onto the z-axis of the N spins (classified by the magnetization m). There are

$$\lambda_{N,m} = \left(\begin{array}{c} N \\ N/2 - m \end{array} \right) = \frac{N!}{(N/2-m)!\,(N/2+m)!}, \tag{38}$$

identical vectors $\vec{\Omega}_m = (mJ_2, \, 0, \, \omega_0)$ where $-N/2 \le m \le N/2$ with m having integer steps. The Hamiltonian of each 2×2 block has the form as given in Eq. (14), which here is explicitly

$$\mathcal{H}_m = \left(\begin{array}{cc} \omega_0 & mJ_2 \\ mJ_2 & -\omega_0 \end{array} \right) . \tag{39}$$

The result for $\langle S_0^z(t) \rangle$ is thus the sum not over 2^N terms but of only $N+1$ terms that have the form, with$\Omega_{\;xm} = mJ_2$,

$$\left(\begin{array}{cc} 0 & 1 \end{array} \right) \exp\left[i\widetilde{\mathcal{H}}_m t \right] \frac{1}{2} \left(\begin{array}{cc} 1 & 0 \\ 0 & -1 \end{array} \right) \exp\left[-i\widetilde{\mathcal{H}}_m t \right] \left(\begin{array}{c} 0 \\ 1 \end{array} \right) = \frac{1}{2} \frac{\omega_0^2 + \Omega_{xm}^2 \cos\left[\sqrt{\Omega_{xm}^2 + \omega_0^2} t \right]}{\Omega_{xm}^2 + \omega_0^2} . \tag{40}$$

Note that if $m = 0$ then $\Omega_{xm} = mJ_2 = 0$, and if the applied field $\omega_0 = 0$ the term in Eq. (40) equals $-1/2$. Since $\alpha_0 = 0$ the initial central spin density matrix is

$$\rho(0) = |\downarrow\rangle\langle\downarrow| = \left(\begin{array}{c} 0 \\ 1 \end{array} \right) \left(\begin{array}{cc} 0 & 1 \end{array} \right) = \left(\begin{array}{cc} 0 & 0 \\ 0 & 1 \end{array} \right) . \tag{41}$$

Since we have defined $\vec{\Omega}_m = (mJ_2, 0, \, \omega_0)$,$\Omega_{\;m} = \sqrt{m^2 J_2^2 + \omega_0^2}$. Then the result for the expectation value of the central spin is

$$\langle S_0^z(t) \rangle = \frac{1}{2^{N+1}} \sum_{m=-N/2}^{N/2} \lambda_{N,m} \frac{\left[m^2 J_2^2 \cos\left(\Omega_m t \right) \right] + \omega_0^2}{\Omega_m^2} . \tag{42}$$

As in reference [16], the quantum purity $\mathcal{P}(t)$ for our case is given by

$$\mathcal{P}(t) = \frac{1}{2} \exp\left(-\frac{N J_2^2 t^2}{4} \right) + \frac{1}{2} . \tag{43}$$

Thus, in the absence of disorder in the values of J_2, the quantity $2\mathcal{P}(t) - 1$ decays to its asymptotic value exponentially in $N J_2^2 t^2$ for one x-bath. We will test whether this is true also when there is disorder in the couplings J_2 of the 2-body interactions with the central spin.

The results can be generalized to the case where there are two x-baths that have only 2-body interactions with the central spin. Explicitly, we consider N_1 x-bath spins with coupling J_2 and N_2 x-bath spins with coupling J_2' with $N = N_1 + N_2$. Then rather than summing over

2^N different terms to find $\rho(t)$, symmetry allows us to sum only over $(N_1+1)(N_2+1)$ terms. Explicitly, one has

$$\langle S_0^z(t)\rangle = \frac{1}{2^{N+1}} \sum_{m_1=-N_1/2}^{N_1/2} \sum_{m_2=-N_2/2}^{N_2/2} \lambda_{N_1,m_1}\lambda_{N_2,m_2} \frac{\left[(m_1 J_2 + m_2 J_2')^2 \cos\left(\Omega_{m_1 m_2} t\right)\right] + \omega_0^2}{\Omega_{m_1 m_2}^2} \tag{44}$$

with the definition $\vec{\Omega}_{m_1 m_2} = (m_1 J_2 + m_2 J_2',\ 0,\ \omega_0)$. Of course this relation can be generalized so with different J_2 couplings between the central spin and n different sets of bath spins there would be a sum over $(N_1+1)(N_2+1)\cdots(N_n+1)$ different terms rather than $2^N = 2^{N_1+N_2+\cdots+N_n}$ terms.

The results of Eqs. (42) and (44) also can be generalized for the cases when there are more than 2-body couplings between the central spin and the spins in a bath. As long as the interactions are symmetric under the exchange of the label of any two spins in a given bath, again this reduces to a sum over $(N_1+1)(N_2+1)\cdots(N_n+1)$ different terms rather than $2^N = 2^{N_1+N_2+\cdots+N_n}$ terms. For example, Eq. (42) holds when all N different couplings between the central spin and the bath spins have the same strength J_2. If all N different N-body couplings had the same strength J_M the symmetry would again hold. The symmetry also holds if there is a $N+1$-body coupling J_A (the central spin is coupled to all N bath spins). If all possible $N(N-1)/2$ 3-body interactions had strength J_3 then the same symmetry holds. Therefore, for any values of J, if the symmetry of relabeling any two spins in every bath holds, then there is only a need to sum over $(N_1+1)(N_2+1)\cdots(N_n+1)$ terms.

Use of this symmetry allows calculations for $\langle S_0^\ell\rangle$. In Fig. 1 results are shown for $\langle S_0^z\rangle$, with $N_1 = N_2 = N/2$. One bath has $J_2 = 0.9$ and $J_M = 0.1$ while the other has $J_2' = 1.1$ and $J_M' = -0.1$. The values of N were chosen such that the N_1- and N_2-body interactions have an even number of spins so that $\langle S_0^x\rangle$ is always zero since we start with all spins down. The results in Fig. 1 for $N \le 34$ could be obtained on today's computers using methods such as those described in references [9, 10, 11, 12, 13], because the size of the 2^{N+1} vector that needs to be stored in computer memory is only $2^{35} \approx 3\times 10^{10}$. However, the $N = 66$ result in Fig. 1 would require storage of a vector of $2^{67} \approx 1.5\times 10^{20}$ words, or about 150 exawords. The stated goal for exascale computing is that the memory should be on the order of a few exabytes, so even the next generation of supercomputers could not perform the $N = 66$ calculations of Fig. 1 using general methods. Furthermore, without use of the symmetry, for the $N = 66$ data of Fig. 1 one would have to sum over $2^{66} \approx 7\times 10^{19}$ different 2×2 matrices $\rho_j(t)$ to obtain $\rho(t)$. With the symmetry there are only $67^2 = 4489$ different $\rho_j(t)$ values over which the sum is performed. With the algorithm described in this paper obviously calculations for large N values can be performed. For large values of N the limitation on N may be due to the precision of the computer calculations, rather than due to limitations in computer run time or in computer memory. For example, the next bath size in the sequence of Fig. 1 would be $N = 130$, but using double precision the quantity $2\mathcal{P}(t) - 1$ becomes negative due to rounding errors after a time of about $t = 1.1$, while this quantity cannot ever be negative.

The reduction in the number of terms to sum to get $\rho(t)$ only requires symmetry in the Hamiltonian $\mathcal{H}$, not in the initial bath vector $\left|\tilde{\Phi}(0)\right\rangle$ [see Eqs. (17) and (35)]. If $\mathcal{H}$ has the symmetry, then each term of the sum in Eqs. (42) or (44) is the sum over all the $|c_j|^2$ that have magnetization m. In other words, the replacement (assuming that the initial state is a direct product of the vectors for the N_n baths) just requires the replacement

$$\frac{1}{2^{N_j}}\lambda_{N_j,m_j} \longleftrightarrow \sum_{i=1}^{2^N}{}' |c_i|^2 \ , \tag{45}$$

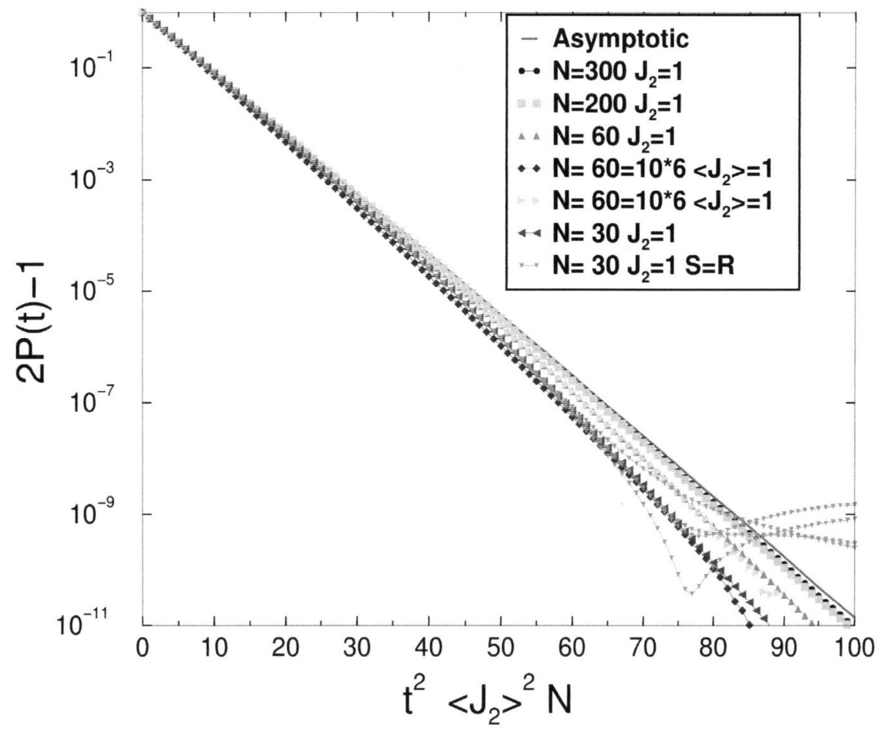

Figure 2. The quantity $2\mathcal{P}(t) - 1$ for the quantum purity $\mathcal{P}(t)$ is shown as a function of $t^2\langle J_2\rangle^2 N$; with $J_3 = 0$, $J_M = 0$ and $J_A = 0$. The asymptotic result is the solid red line from Eq. (43). Values for all $J_2 = 1$ are shown for $N = 30$, 60, 200, and 300 all starting with all spins down. For $N = 60$, as well as for 60 spins with $J_2 = 1$ [brown, $\triangle$], two different results are shown for $\langle J_2\rangle = 1$, but with ten baths each of six spins with the values [blue, $\diamond$, $J_2 = (1.5, 1.1, 1.03, 1.01, 1.001, 0.999, 0.99, 0.97, 0.9, 0.5)$] and [cyan, $\triangleright$, $J_2 = (1.01, 1.001, 1.0, 1.0, 1.0, 1.0, 1.0, 1.0, 0.999, 0.99)$]. For $N = 30$ five different random starting states for $|\Phi(0)\rangle$ are shown, all with $\alpha_0 = 0$ (orange), to compare with the other points which have all bath spins and the central spin intially down.

where the prime on the sum restricts the sum only to the terms that have the z-component of the magnetization of the bath spins equal to m_j. In the calculations performed in this paper, whenever a random state for the x-bath(s) is used, the c_j are chosen to be real and distributed uniformly between 0 and 1. After all the c_j are chosen randomly, the initial vector $\left|\widetilde{\Phi}(0)\right\rangle$ is normalized to unity. However, for large N the sum over so many random numbers is both time consuming and gives results that are subject to propagating round-off errors. Therefore, invoking the central limit theorem, the computer program was written so that when the number of terms in the primed sum of Eq. (45) was larger than 50 the sum was replaced by a random number drawn from a Gaussian distribution with the exact first and second moments expected from the sum over the squares of the required number of random terms.

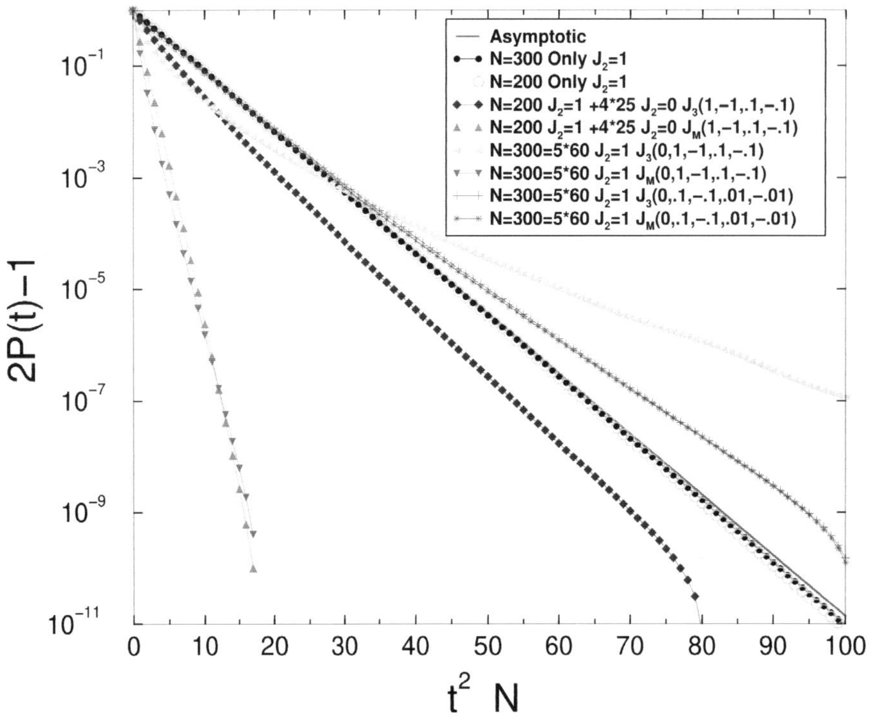

Figure 3. The quantity $2\mathcal{P}(t) - 1$ for the quantum purity $\mathcal{P}(t)$ is shown as a function of t^2N. The asymptotic result is the solid red line from Eq. (43). The starting state of all bath spins and the central spin is down. For $N = 300$ all baths have $J_2 = 1$. The $N = 300$ data (black, $\bullet$) is for one bath, and the others have five baths of 60 spins each with interactions [cyan, $\triangleleft$, $J_3 = (0, 1.0, 0.1, -0.1, -1.0)$], [magenta, $\bigtriangledown$, $J_M = (0, 1.0, 0.1, -0.1, -1.0)$], [violet, $+$, $J_3 = (0, 0.1, 0.01, -0.01, -0.1)$], and [indigo, $*$, $J_M = (0, 0.1, 0.01, -0.01, -0.1)$]. Also shown are results for five baths, one with only $J_2 = 1$ and 200 bath spins, and four baths of 25 spins each (so $N = 300$) with interactions [blue, $\diamond$, $J_3 = (1.0, 0.1, -0.1, -1.0)$] and [orange, $\triangle$, $J_M = (1.0, 0.1, -0.1, -1.0)$]. The case for $N = 200$ with only $J_2 = 1$ is also shown (green, $\bigcirc$).

5. Results and Analysis

We present test results to show that the algorithm does indeed work. One goal is to compare with the exact results of reference [16], and to extend these results for random J interactions, as well as for different starting vector configurations.

In Fig. 2 we show results for how the quantum purity $\mathcal{P}(t)$ approaches its asymptotic limit for the case where there is symmetry in the Hamiltonian under interchange of any two bath spins. The largest Hilbert space calculated in Fig. 2 is $2^{301} \approx 4 \times 10^{90}$. Clearly without the use of the algorithm in this paper and the symmetry of the Hamiltonian the calculations would be impossible. The largest number of 2×2 terms summed over in Fig. 2 is $7^{10} \approx 2.8 \times 10^8$ for the case of ten different x-baths. Fig. 2 shows that the quantum purity decays to its asymptotic $N \to \infty$ result of Eq. (43) even when the initial starting state is random or when there are a number of different x-baths with different couplings. In other words, in all cases except for finite

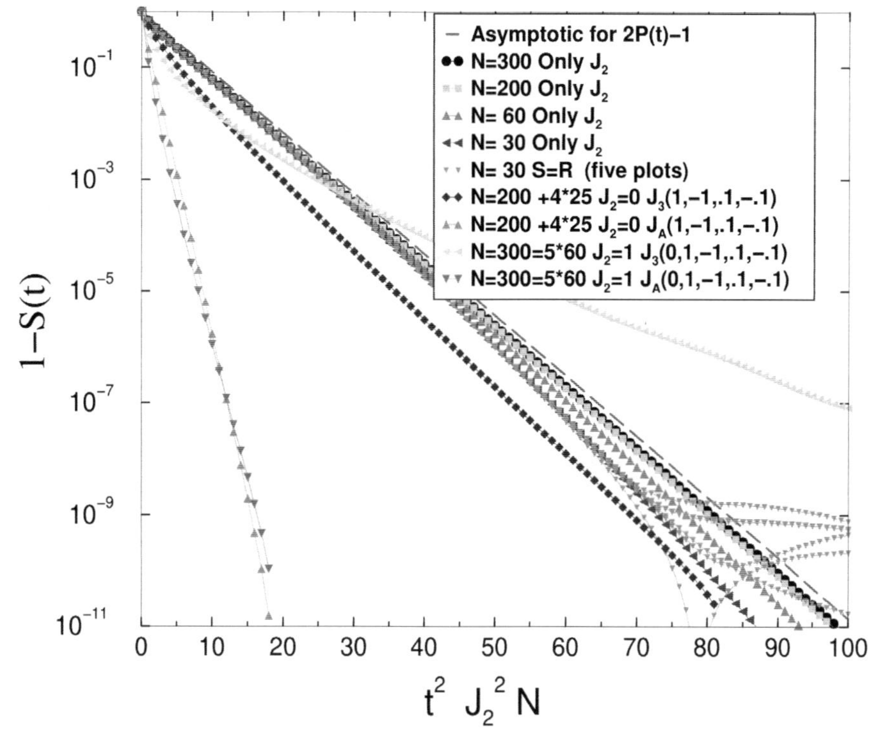

Figure 4. The von Neumann entropy $\mathcal{S}(t)$ subtracted from unity is shown as a function of $t^2\langle J_2\rangle^2N$; with $J_3 = 0$, $J_M = 0$ and $J_A = 0$. The asymptotic result for $2\mathcal{P}(t) - 1$ is the dashed red line from Eq. (43), only to guide the eye. Values for all $J_2 = 1$ are shown for $N = 30$, 60, 200, and 300 all starting with all spins down. For $N = 30$ five different random starting states for $|\Phi(0)\rangle$ are shown, all with $\alpha_0 = 0$ (orange) [different random starting states than in Fig. 2]. The $N = 300$ data (black, •) is for one bath, and the others have five baths of 60 spins with interactions [cyan, $\triangleleft$, $J_3 = (0, 1.0, 0.1, -0.1, -1.0)$] and [magenta, $\bigtriangledown$, $J_M = (0, 1.0, 0.1, -0.1, -1.0)$]. Also shown are results for five baths, one with only $J_2 = 1$ and 200 bath spins, and four baths of 25 spins each (so $N = 300$) with interactions [blue, $\diamond$, $J_3 = (1.0, 0.1, -0.1, -1.0)$] and [orange, $\triangle$, $J_M = (1.0, 0.1, -0.1, -1.0)$]. Except for the $N = 30$ points labeled S=R (orange), all spins start down.

N effects and small effects due to the random starting state of the bath(s), the quantum purity decays exponentially in t^2 and N when the couplings are only 2-body between the bath(s) and the central spin.

In Fig. 3 the approach to the asymptotic value for the quantum purity $\mathcal{P}(t)$ is shown when the baths also may include the 3-body couplings with strength J_3 and N_i-body couplings J_A ($i = 1, \cdots, 5$). Again, in all cases the decay of $\mathcal{P}(t)$ is exponential in t^2. The largest number of 2×2 matrices summed over in Fig. 3 is for five baths of 60 spins each, which requires $61^5 \approx 8\times10^8$ terms in the sum. Again, in all cases except for finite N effects, for large N the quantum purity decays exponentially in t^2 to its asymptotic value.

In Fig. 4 we show how the von Neumann quantum entropy $\mathcal{S}(t)$ approaches its asymptotic

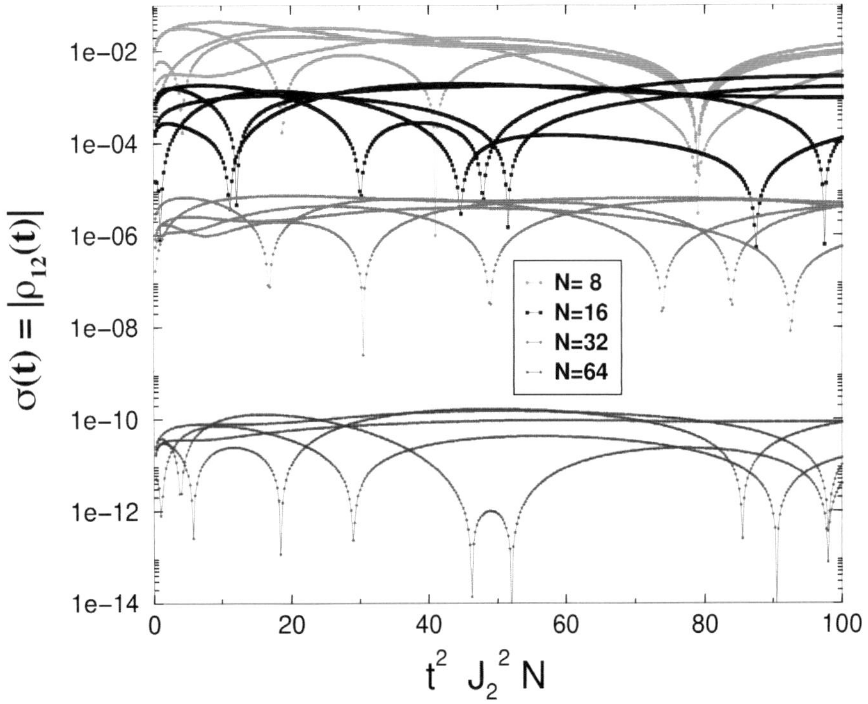

Figure 5. The quantity $\sigma(t) = |\rho_{12}(t)|$ is shown as a function of $NJ_2^2t^2$ for different values of N. The results are for random initial bath configurations $\left|\widetilde{\Phi}(0)\right\rangle$, and five different random states are shown for each value of N. The results are for one x-bath with only 2-body interactions of strength $J_2 = 1$.

value of unity. In all cases, except for finite N effects and small effects due to the random starting state of the bath(s), the quantum entropy decays exponentially in t^2. Comparing results for different baths of 300 spins in Fig. 3 and Fig. 4 shows that the interactions between the bath spins lead to quantitative but not qualitative differences in the behavior of $\mathcal{P}(t)$ and $\mathcal{S}(t)$.

In Fig. 5 we show the quantity $\sigma(t) = |\rho_{12}(t)|$ as a function of $NJ_2^2t^2$ for different values of N. A recent paper examined $\sigma(t)$ for a 4-spin cluster coupled to spin baths with up to $N = 32$ spins to study decoherence and thermalization [13]. The results shown in Fig. 5 are for the only non-zero couplings the 2-body coupling $J_2 = 1$. For an initial state of the bath with all spins down, the quantity $|\rho_{12}(t)|$ does not decay at all with t. Only for random initial states of the bath does $|\rho_{12}(t)|$ decay, and then it decays in an extremely short time and fluctuates about some average value. The average value of $|\rho_{12}(t)|$ decreases rapidly as N increases. The fact that $|\rho_{12}(t)|$ does not decay at all if all spins are started in the down (or the up) configuration illustrates the importance of a random initial state for the bath.

In Fig. 6 we examine one bath with only $J_2 = 1$ and an initial state for the bath with all spins up. The central spin is started at an angle α_0, as in Eq. (16). We see that the quantum purity no longer decays to zero if α_0 is not zero (or 90°). Thus our x-bath does not act effectively for decoherence of all initial configurations of the central spin. This is a limitation of the x-baths.

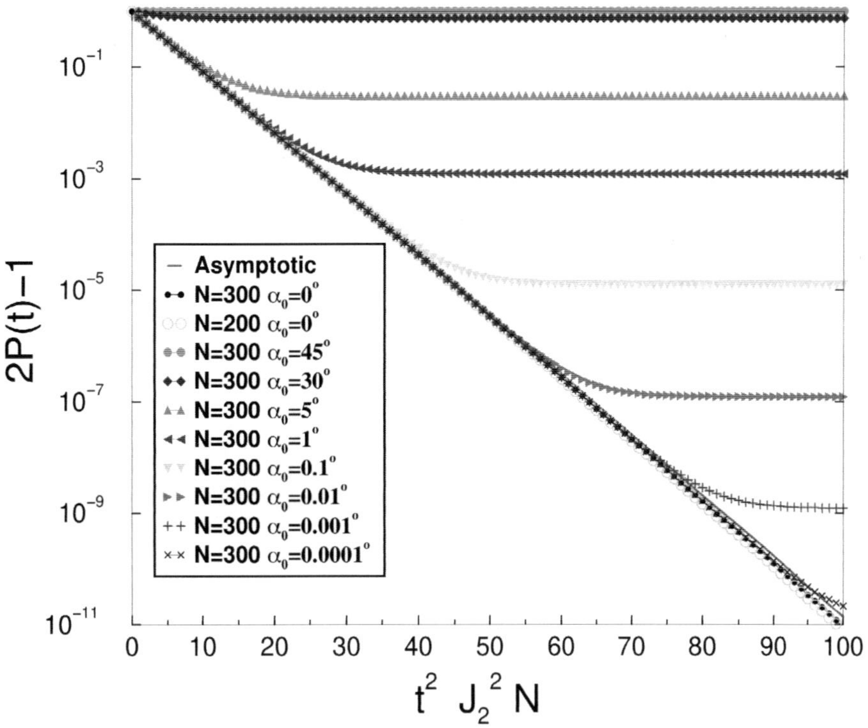

Figure 6. The quantity $2\mathcal{P}(t)-1$ for the quantum purity $\mathcal{P}(t)$ is shown as a function of $t^2 J_2^2 N$. The asymptotic result for $\alpha_0 = 0^o$ is the solid red line from Eq. (43). The starting state of all bath spins is down, and all $J_2 = 1$. For $N = 300$ the central spin has a starting state with different values of $\alpha_0 = 45^o$, 30°, 5°, 1°, 0.1^o, 0.01^o, 0.001^o, and 0.0001^o. Also shown is $N = 200$ for $\alpha_0 = 0^o$.

In Fig. 7 we show the quantum purity (normalized and with its asymptotic value subtracted) $2\mathcal{P}(t) - 1$ as a function of the number of bath spins N times t^2. The coupling between the central spin and bath spins only has a coupling of the form $J_{2,i}S_0^xS_i^x$ with $J_{2,i}$ uniformly distributed in the range $[1-r, 1+r]$, and hence would qualify as an x-bath. However, here we do not limit ourselves to only x-baths, in that we allow intra-bath couplings of the form $\mathcal{H}_B = \sum_{i,j}^N \left(J_{ij}^x S_i^x S_j^x + J_{ij}^y S_i^y S_j^y + J_{ij}^z S_i^z S_j^z\right)$ with $J_{ij}^\ell \in [-r, +r]$ with $i = 1, \cdots, N$. Hence for $r = 0$ the results reduce to those of an x-bath, and hence are compared with Eq. (43) for asymptotic N and to a sum over only $N+1$ different terms in Eq. (24). Fig. 7 shows results for $N = 10$, 20, and 30. For non-zero r, we have used the Chebyshev algorithm of [12, 13] storing vectors of size 2^{N+1}. The asymptotic result is the dashed line from Eq. (43). In all cases even for this more general bath we find that the quantum purity decays exponentially in Nt^2, except for small finite-N and random coupling effects. The numerical results for zero disorder ($r = 0$) agree with our algorithmic results and the exact results of ref. [16]. Note in this case the $\mathcal{H}_B$ contains operators that do not commute with S^x, and therefore the bath has some internal dynamics. Even for the generalized bath and for the random interactions an exponential decoherence in t^2 is observed. Also note that for small values of r the results are closer to the asymptotic result

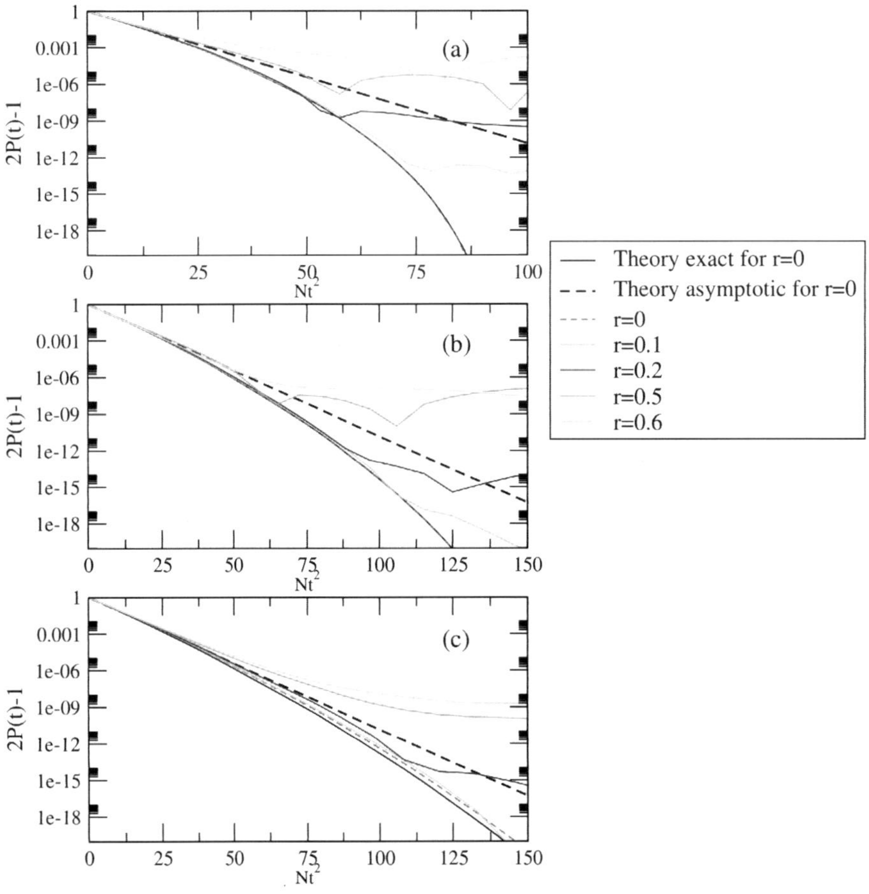

Figure 7. The quantity $2\mathcal{P}(t) - 1$ for the quantum purity $\mathcal{P}(t)$ is shown as a function of Nt^2 is shown for $J_i \in [1-r, 1+r]$ with $i = 1, \cdots, N$ and (a) $N = 10$, (b) $N = 20$, and (c) $N = 30$. The asymptotic result is the dashed line from Eq. (43). Note, the bath is only an x-bath for $r = 0$. This figure is related to that of reference [17].

than for the case with $r = 0$.

6. Conclusions and Discussion

We have shown that for a single spin coupled to specific types of spin baths, which we denote as x-baths, an efficient algorithm can be devised that eliminates the computer storage of large vectors. The x-baths are those that have in the Hamiltonian only the operators S_k^x $(k = 1, \cdots, N)$. Since we are not interested in the dynamics of the bath spins, we trace out the bath spins to obtain the reduced density matrix $\rho(t)$ of the central spin(s). To obtain the real-time propagation using the time dependent Schrödinger equation (TDSE) of a central spin coupled to N bath spins, one would normally be required to store one or more vectors of size the dimension of the Hilbert space, 2^{N+1}. We find that the dynamics of x-baths is such that the intra-bath Hamiltonian $\mathcal{H}_B$ does not affect the dynamics of the central system. For the x-baths with a single central spin, no vectors of dimension larger than 2 must be stored. Furthermore, from Eq. (24) the density matrix $\rho(t)$ of the central system is a sum over 2^N density matrices $\rho_j(t)$ that each have their own Hamiltonian. Once $\rho(t)$ has been obtained, in the normal fashion using Eq. (26) expectation values of operators can be calculated. An example for $\langle S_0^z(t)\rangle$ is shown in Fig. 1.

The algorithm requires that one solves the TDSE for 2^N different Hamiltonians of the size of the subsystem studied. In this paper the subsystem studied is a single spin, so the TDSE propagation of 2^N vectors of length 2 is required. Symmetries in the Hamiltonian can be used to further reduce the number of different TDSE Hamiltonians that one needs to sum to obtain $\rho(t)$. The reduction in the number of different terms in the sum for $\rho(t)$ only requires symmetry in the spin-bath Hamiltonian $\mathcal{H}_{SB}$, not in the initial vector of the bath (provided that the central spin is brought into contact with the bath at $t = 0$). The largest Hilbert space calculated in this paper is $2^{301} \approx 4 \times 10^{90}$, for one central spin and a bath of $N = 300$. For this type of bath, which has maximum symmetry in the bath, the number of different Hamiltonians one needs to propagate with the TDSE is only $N + 1$. Clearly without the use of the algorithm in this paper, and the symmetry of the Hamiltonian, the calculations performed in this paper would be impossible. If there is less symmetry in the bath, a smaller reduction in the number of different terms needed in the sum for $\rho(t)$ are obtained. For example, in Fig. 2 with $N = 60$ there are data points with ten symmetric baths with six spins each. For these points the Hilbert space is $2^{61} \approx 2 \times 10^{18}$ but results for $\rho(t)$ is from the sum of $7^{10} \approx 2.8 \times 10^8$ different 2×2 terms $\rho_j(t)$.

We used this algorithm to study a single spin coupled to different types of x-baths. If the central spin is either up or down initially, the quantum purity $\mathcal{P}(t) = \mathrm{Tr}\left[\rho^2(t)\right]$ decays exponentially in t^2 times N to its asymptotic value, as seen in Fig. 2 and Fig. 3. Furthermore, in this case the von Neumann entropy $\mathcal{S}(t)$ decays to its asymptotic value exponentially in t^2 and N as seen in Fig. 4. In addition, if the initial vector(s) of the spins in the bath(s) is (are) random, the off-diagonal component of the density matrix for a single spin in the central system $|\rho_{12}(t)|$ decreases with N as seen in Fig. 5. These exponential approaches to the asymptotic $N \to \infty$ limit are very good for finite N values up to a particular time t, where the size of the bath and the starting state of the bath becomes important (see Fig. 2, Fig. 3, and Fig. 4).

Unfortunately, the x-baths are not sufficient to always provide adequate decoherence and thermalization of the central system. This is seen in one instance in that $|\rho_{12}(t)|$ is time independent if the initial state of the bath is the state with all spins down or all spins up. The lack of ability of the x-baths to cause full decoherence of the central system is also illustrated in Fig. 6. Fig. 6 illustrates that if the central spin is not initially in either the up or down state the quantum purity $\mathcal{P}(t)$ saturates to a finite value. We have also used a Chebyshev algorithm [12, 13] for the more general case where the spin-bath Hamiltonian only has couplings of the form of an x-bath, namely $\mathcal{H}_{SB} = \sum_{i=1}^{N} J_i^x S_0^x S_i^x$, but the bath Hamiltonian $\mathcal{H}_B$ has general random two-body couplings with the operators $S_i^x S_j^x$, $S_i^y S_j^y$, and $S_i^z S_j^z$. As seen in Fig. 7 the approach of $\mathcal{P}(t)$ to its asymptotic values is still exponential in t^2 and N. The bath in Fig. 7 has its own internal dynamics even when it is not coupled to the central spin(s), and therefore should allow for decoherence and thermalization of the central spin(s) for all initial spin configurations.

Our x-bath algorithm can be expanded in several different directions. One is to also allow simultaneously x-baths, y-baths, and z-baths. The x-baths have only terms in the Hamiltonian using the operators S_k^x. Similarly the y-baths (z-baths) would only have terms in the Hamiltonian using the operators S_k^y (S_k^z). It has recently been shown that for a single central spin coupled to both an x-bath and a y-bath, the approach of the quantum purity $\mathcal{P}(t)$ to its asymptotic value becomes a power law in t and N [16]. This contrasts with the exponential dependence in t^2 and N for only an x-bath. A physically interesting question that can be studied with the expanded algorithm is under which circumstances the approach to the asymptotic value for the quantum purity and von Neumann entropy is power law or is exponential for different combinations of x-, y-, and z-baths [18]. For y-baths one could use

$$p_y = \frac{1}{\sqrt{2}} \begin{pmatrix} -i & i \\ 1 & 1 \end{pmatrix}, \tag{46}$$

which has the same properties as p_x of Eq. (10) for x-baths. Another direction of improving the algorithm would be to apply it to a larger number of spins in the central system. If there were M spin-1/2 particles in the central system, the calculation of the density matrix would reduce solving the TDSE for 2^N Hamiltonians that have a Hilbert space of size 2^M. Such calculations could be accomplished even for the case where the central subsystem Hamiltonian $\mathcal{H}_S$ is time dependent as long as the $2^M \times 2^M$ matrix can be propagated efficiently via the TDSE.

Our efficient algorithm for solving the TDSE for central spins coupled to x-bath spins should pave the way for other efficient algorithms for the real-time development of spin systems. Such algorithms would allow one to study quantum decoherence, thermalization of quantum systems coupled to spin baths, and may be applicable to applications such as quantum encryption and quantum computing.

Acknowledgements

This material is based upon work supported in part for MAN by the US National Science Foundation under Grant No. DMR-1206233. MAN also acknowledges support from the Jülich Supercomputing Centre (JSC).

References

[1] Mermin D.N., *Quantum Computer Science: An Introduction* (Cambridge University Press, Cambridge, UK, 2007).
[2] Kaye P., Laflamme R., Mosca M., *An Introduction to Quantum Computing* (Oxford University Press, Oxford, UK, 2007).
[3] Shankar R., *Principles of Quantum Mechanics*, Second Edition (Plenum Press, London, 1994).
[4] Fermi E., 1932 *Rev. Mod. Phys.* **4**, 87.
[5] Prokof'ev N.V. and Stamp P.C.E., 2000 *Rep. Prog. Phys.* **63**, 669.
[6] Yang W. and Liu R.-B., 2009 *Phys. Rev. B* **79**, 115320.
[7] Xu J., Jing J., and Yu T., 2011 *J. Phys. A: Math. and Theor.* **44**, 185304.
[8] http://www.top500.org.
[9] Tal-Ezer H., and Kosloff R., 1984 *J. Chem. Phys.* **81**, 3967.
[10] Leforestier C., Bisseling R., Cerjan C., Feit M., Friesner R., Guldberg A., Hammerich A., Jolicard G., Karrlein W., Meyer H.-D., Lipkin N., Roncero O., and KosloffR ., 1991 *J. Comput. Phys.* **94**, 59.
[11] Iitaka T., Nomura S., Hirayama H., Zhao X., Aoyagi Y., and Sugano T., 1997 *Phys. Rev. E* **56**, 1222.
[12] Dobrovitski V.V., and De Raedt H., 2003 *Phys. Rev. E* **67**, 056702.
[13] Jin F., De Raedt H., Yuan S., Katsnelson M.I., Miyashita S., and Michielsen K., 2010 *J. Phys. Soc. Japan* **79**, 123005.
[14] Danon J., Vink I.T., Koppens F.H.L., Nowack K.C., Vandersypen L.M.K., and Nazarov Y.V., 2009 *Phys. Rev. Lett.* **103**, 046601.
[15] Vink I.T., Nowack K.C., Koppens F.H.L, Danon J., Nazarov Y.V., and Vandersypen L.M.K., 2009 *Nature Physics* **5**, 764.
[16] Rao D.D.B., Kohler H., and Sols F, 2008 *New J. Physics* **10**, 115017.
[17] Guerra M.L., De Raedt H., and Novotny M.A., 2011 *Physics Procedia* **15**, 33.
[18] Novotny, M.A., Guerra, M.L., De Raedt H., Michielsen, K., and Jin, F., 2012 *Physics Procedia* **34**, 90.

Numerical studies of the fractional quantum Hall effect in systems with tunable interactions

Z. Papić[1], D. A. Abanin[2], Y. Barlas[3], and R. N. Bhatt[1,2]

[1] Department of Electrical Engineering, Princeton University, Princeton, NJ 08544, USA
[2] Princeton Center for Theoretical Science, Princeton University, Princeton, NJ 08544, USA
[3] National High Magnetic Field Laboratory and Department of Physics, Florida State University, FL 32306, USA

Abstract. The discovery of the fractional quantum Hall effect in GaAs-based semiconductor devices has lead to new advances in condensed matter physics, in particular the possibility for exotic, topological phases of matter that possess fractional, and even non-Abelian, statistics of quasiparticles. One of the main limitations of the experimental systems based on GaAs has been the lack of tunability of the effective interactions between two-dimensional electrons, which made it difficult to stabilize some of the more fragile states, or induce phase transitions in a controlled manner. Here we review the recent studies that have explored the effects of tunability of the interactions offered by alternative two-dimensional systems, characterized by non-trivial Berry phases and including graphene, bilayer graphene and topological insulators. The tunability in these systems is achieved via external fields that change the mass gap, or by screening via dielectric plate in the vicinity of the device. Our study points to a number of different ways to manipulate the effective interactions, and engineer phase transitions between quantum Hall liquids and compressible states in a controlled manner.

1. Introduction

Strongly correlated phases of electrons confined to move in the plane, subject to a perpendicular magnetic field, have attracted significant attention since the discovery of fractionally quantized Hall conductivity [1]. The profound role of topology in this extreme quantum limit leads to the presence of quasiparticles that carry a fraction of electron charge [2], and fractional (Abelian, or possibly non-Abelian) statistics [2, 3]. The prospect of excitations possessing non-Abelian statistics has motivated different schemes for topological quantum computation [4] based on these systems.

These remarkable phenomena occur in the fractional quantum Hall (FQH) regime, when the number of electrons, N_e, is comparable to the number of magnetic flux quanta N_Φ through the two-dimensional electron system (2DES). Correlated FQH liquid states appear at certain partial filling $\nu = N_e/N_\Phi$ of the "active" Landau level (LL). In traditional semiconductor heterostructures, the physics of a partially-filled $n = 1$ LL differs significantly from that of $n = 0$ LL, due to the node in the single-particle wavefunction [5]. As a consequence, the hierarchy/composite fermion states [6, 7], ubiquitous in the lowest Landau level (LLL), are significantly weakened in $n = 1$ LL, and some of the more exotic states, such as the Read-Rezayi (RR) parafermion states [8], are likely to be favored. A number of studies have focused on the simplest, $k = 2$ non-Abelian member of the RR sequence – the Moore-Read (MR) "Pfaffian" state [3], believed to describe the FQH plateau at $\nu = 5/2$ [9]. Quasiparticles of the MR and higher RR states obey the non-Abelian statistics [3] which is of interest for topological quantum computation [4].

One of the main disadvantages of the GaAs-based devices is that their 2DES is buried inside a larger, three-dimensional structure. This unfortunate fact fixes the effective interactions at values that are often not optimal for some of the most interesting FQH states, including the RR series. For example, the MR state is found to lie very close to the boundary with a compressible phase [10, 11]. Another problem stems from strong dielectric screening and finite well-width [12] in GaAs, which weaken the electron-electron interactions and make FQH states fragile. This has been a major obstacle in the studies of the non-Abelian states, which could only be observed in ultra-high-mobility samples [9]. Thus, it is desirable to find an alternative high-mobility 2DES with strong effective Coulomb interactions that are adjustable in a broad range.

Recently, a new class of such high-mobility 2DES which host chiral excitations with non-trivial Berry phases, has been discovered. These "chiral" materials include graphene and bilayer graphene [13], and, more recently, topological insulators [14], as well as certain quantum wells [15]. The chiral nature of the quasiparticles gives rise to new electronic properties, including an unusual LL sequence, the anomalous Hall effect, and suppression of weak localization [13]. When these chiral materials are subject to a perpendicular magnetic field, the kinetic energy quenches into discrete Landau levels, similar to the usual semiconductors with non-chiral carriers. However, due to the chiral band structure [16] and the fact that the surface of these materials is exposed [17], they offer new possibilities to tune the effective interactions and explore strongly correlated phases.

In this paper we review two practical ways [17, 16] of tuning the effective interactions in chiral 2DES, and the effect this has on the FQH states. One of the attractive features of the chiral 2DESs is that they allow for a more robust realization of certain FQH states in *several* LLs, as opposed to a single LL in GaAs. Additional insights can be obtained by driving transitions between the incompressible (FQH) states and the compressible, stripe and bubble, phases. Such transitions can be implemented in chiral and *massive* 2DESs by varying the external field. Overall, the tunability of the chiral 2DES allows one to explore a larger region of the effective interactions than has been achieved in GaAs.

2. Model

We begin this Section by providing a brief overview of the one-body problem in a magnetic field, as well as the summary of the standard recipe of exact diagonalization applied to quantum Hall systems.

Consider a single electron moving in a plane, subject to a perpendicular magnetic field $B\hat{z} = \nabla \times \mathbf{A}(\mathbf{r})$. The corresponding Hamiltonian can be written in a "covariant" notation as

$$K = \frac{1}{2m} g^{ab} \pi_a \pi_b, \tag{1}$$

where $\pi_a = p_a - \frac{e}{c} A_a(\mathbf{r})$ $(a = x, y)$ represents the dynamical momentum, and $\hat{g}$ is the band mass tensor. In the usual isotropic case $(g = I)$, we can obtain the single-particle energies (Landau levels) by choosing, for example, the symmetric gauge $A_x = By/2, A_y = -Bx/2$. In this case, the dynamical momenta become $\pi_x = -i\hbar\frac{\partial}{\partial x} + \frac{\hbar}{2\ell_B^2} y$ and $\pi_y = -i\hbar\frac{\partial}{\partial y} - \frac{\hbar}{2\ell_B^2} x$, in terms of the magnetic length $\ell_B = \sqrt{\hbar/eB}$. Hamiltonian can be transformed into a diagonal form $K = \frac{\hbar\omega_c}{2}\left(a^\dagger a + \frac{1}{2}\right)$ with the help of "ladder" operators

$$a \propto \pi_x + i\pi_y, \tag{2}$$

$$a^\dagger \propto \pi_x - i\pi_y. \tag{3}$$

However, for each value of $a^\dagger a$, there remains a residual degeneracy equal to the number of magnetic flux quanta, N_Φ. This degeneracy is resolved by another pair of operators $b, b^\dagger$ that commute with $a, a^\dagger$ and depend on the *guiding center* coordinates of the electron, $R^a = r^a - \frac{\epsilon^{ab}}{\hbar}\pi_b \ell_B^2$. Operators $b^\dagger$ create the (unnormalized) single particle eigenstates of the lowest, $n = 0$ Landau level (LLL),

$$\phi_l(z) \propto z^l e^{-z^* z/4\ell_B^2}, \tag{4}$$

with $z = x + iy$ being the complex coordinate of an electron in the plane (and z^* denoting its complex-conjugate). The quantum number l is an eigenvalue of the angular momentum, and the single particle states are localized along concentric rings around the origin. Single-particle states in higher LLs are constructed by the repeated action of $a^\dagger$, thus completing the solution of the one-body problem in a rotationally-invariant case.

2.1. Many-body problem: Exact diagonalization

An interacting problem of a finite number of electrons can be studied numerically, using exact diagonalization which has been remarkably successful in unravelling many of the essential physical properties of FQH systems, for systems as small as 10 electrons [5, 6]. A crucial approximation that makes exact diagonalization practical is to neglect the excitations between LLs and restrict the Hilbert space to a single, "active" LL, which has a degeneracy of N_Φ. This approximation is physically justified in high magnetic fields where such excitations indeed become very costly. One can then construct a many-body Hilbert space, consisting of Slater determinants $|l_1, l_2, \ldots\rangle$ built from the states (4), using a suitable choice of boundary condition. In the study of the FQH effect, two kinds of surfaces are available that preserve the translational invariance of an infinite 2DES: sphere [6] and torus [18]. The two choices of boundary conditions illustrate the specific features of a many-body FQH state under investigation: on a sphere, the FQH state couples to the curvature of the manifold, which is characterized by the topological number called shift [19]. The shift produces a small offset between $N_\Phi = N/\nu$ and the magnetic monopole, whose strength is denoted by $2S$, placed in the center of a sphere. As a consequence, different FQH states "live" in different Hilbert spaces and in principle can be directly compared only after extrapolation to the thermodynamic limit. On the other hand, the flat surface of a torus leads to a unique definition of N_Φ for given N and ν, such that different candidate states describing the same filling ν are all realized in the same Hilbert space. The caveats of this geometry are the additional geometric parameters, the angle and aspect ratio of the torus; the Hamiltonian of a finite-size system depends on these parameters and their specific values favor one FQH phase over the others. Thus, an analysis is slightly more involved but the gain is that the ground-state degeneracy can be used to identify topologically-ordered states. In contrast, on the sphere FQH states always appear as non-degenerate, zero angular momentum ground states.

Once the Hilbert space is defined and properly adapted to the symmetry group/boundary condition, the remaining task is to represent the Hamiltonian and diagonalize the corresponding matrix using a variant of Lanczos algorithm [20] (or Jacobi/Householder routine [21] for very small systems). The advantage of symmetry, in particular rotational invariance, is that Wigner-Eckart theorem applies, and leads to the definition of the so-called Haldane pseudopotentials [5]. Any two-body interaction, such as Coulomb potential projected to a single LL, can be written in the usual second-quantized form $H = \sum_{\{m_i\}} V_{m_1 m_2 m_3 m_4} c^\dagger_{m_1} c^\dagger_{m_2} c_{m_4} c_{m_3}$. Furthermore, assuming rotational invariance, the matrix element factorizes into

$$V_{m_1 m_2 m_3 m_4} = \sum_{L=0}^{2S} V_L \sum_{M=-L}^{L} C^{SSL}_{m_1 m_2 M} C^{SSL}_{m_3 m_4 M}, \tag{5}$$

i.e. into a product of a geometrical factor involving the Clebsch-Gordan coefficients $C^{J_1 J_2 J}_{m_1 m_2 m}$, and the Haldane pseudopotential V_L. The finite set of Haldane pseudopotentials contains all the information about the interaction projected to a single LL and thus determines all the physical properties of a FQH system. Haldane pseudopotentials can be conveniently evaluated from the Fourier transform of the interaction, $V(q)$.

2.2. Modification of the interaction due to the spinor nature of the wavefunction in massive Dirac materials

Once the Hilbert space is defined and symmetry-adapted, the Coulomb Hamiltonian can be represented in the matrix form. In the LLL, the interaction is simply given by $V(q) = 1/q$, which is the Fourier transform of the Coulomb potential in two dimensions. However, if we consider the case when the chemical potential is such that LLs from 1 through $n-1$ are completely filled and "inert", while the nth LL is partially filled and "active", the effective interaction becomes slightly modified and assumes the form $1/q \times |F_n(q)|^2$, where the *form factor* [5] $F_n(q) = L_n(q^2\ell_B^2/2)$ can be evaluated using the algebra of operators $a, a^\dagger$, Eq. (2) – see, e.g., Ref. [7] (L_n is the nth Laguerre polynomial). Similarly, if the 2DES has a finite extent in the z-direction, the interaction becomes modified via an appropriate form-factor that results from integrating out the z-component of the wavefunction.

The modification of the effective interaction may also result from the spinor nature of the wavefunction [22]. Such wavefunctions arise naturally in a family of 2D materials that have non-trivial Berry phases. One such material is monolayer graphene, a high-mobility atomically thick 2DES [13], where recently several FQH states of the type $\nu = m/3$ have been discovered [23]. A closely related material, bilayer graphene [13] has similarly high mobility, and exhibits interaction-induced quantum Hall states at integer filling factors at low magnetic fields [24]. Graphene and its bilayer are characterized by Berry phase π (graphene-like) and 2π (bilayer-graphene-like with an energy gap), respectively. More explicitly, we refer to the Dirac materials as those described by a family of 2×2 Hamiltonians with Berry phase π and 2π [16]. The case of π-carriers is realized in graphene, topological insulators, special quantum wells [15]; the case of 2π-carriers occurs in bilayer graphene, where the energy gap can be controlled in a wide range by a perpendicular electric field [13]. Details on how the effective 2×2 Hamiltonian can be derived from a more complete tight-binding model can be found in the review [25].

For π carriers, the single-particle wavefunctions are given by spinors [16] $\psi_n = (\cos\theta_n \phi_{|n|-1}, \sin\theta_n \phi_{|n|})$, where ϕ_n is the wave function of the nth non-relativistic LL ($n = \pm 1, \pm 2, \ldots$), and parameter θ depends on the ratio $\Delta/(\hbar v_0/\ell_B)$, where Δ is the mass gap, and v_0 the Fermi velocity. We use the notation $(\sigma\pi, n)$ to denote the nth LL for $\sigma\pi$ carriers ($\sigma = 1, 2$). As a consequence of the spinor wavefunction, the effective form factor [5] $F_n^\pi(q)$ that describes the interaction projected to a (π, n) LL is given by

$$F_n^\pi(q) = \cos^2\theta L_{|n|-1}\left(\frac{q^2\ell_B^2}{2}\right) + \sin^2\theta L_{|n|}\left(\frac{q^2\ell_B^2}{2}\right), \tag{6}$$

where for simplicity we omitted the index of θ. The form-factor is a mixture of the $(|n|-1)$th and $|n|$th LL form-factors in a non-relativistic 2DES with parabolic dispersion. At $\theta = \pi/4$, the above equation reduces to the form-factor of graphene [22], however, by varying $\Delta/(\hbar v_0/\ell_B)$ in the experimentally accessible range [16], one can realize any value of $\theta \in (0; \pi/2)$.

Similarly, for carriers with Berry phase 2π, the single-particle wavefunctions are $\psi_n = (\cos\theta_n \phi_{|n|-1}, \sin\theta_n \phi_{|n|+1})$, and the form-factor is a mixture of standard $(|n|-1)$th and $(|n|+1)$th form-factors,

$$F_n^{2\pi}(q) = \cos^2\theta L_{|n|-1}\left(\frac{q^2\ell_B^2}{2}\right) + \sin^2\theta L_{|n|+1}\left(\frac{q^2\ell_B^2}{2}\right). \tag{7}$$

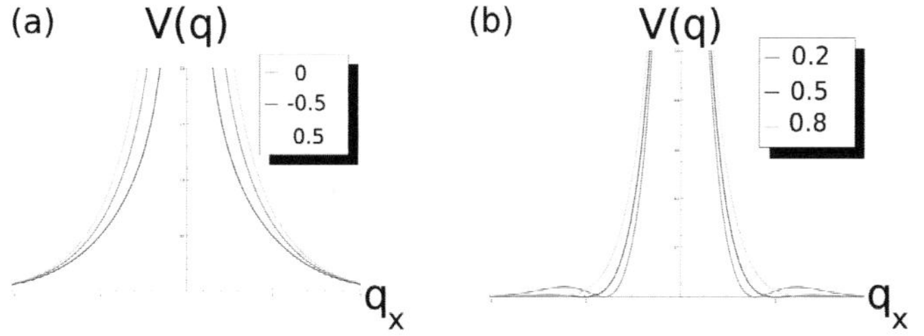

Figure 1. (Color online) Two ways of modifying the effective interaction $V(q)$ in the FQH regime: (a) via dielectric screening, for several values of α given in the inset (using Eq.(8)) for $d = \ell_B$, (b) via chiral band structure, for several values of $\cos^2\theta$ given in the inset (in $(\pi, 1)$ LL). Because the interaction $V(q)$ is isotropic, it is plotted as a function of q_x only. The comparison of the corresponding Haldane pseudopotentials for cases (a) and (b) can be found in Ref. [26].

The tunable form of the effective interactions Eqs.(6,7) provides a way to engineer transitions between strongly correlated phases *in situ* by changing the field, as illustrated in Fig. 1(b).

2.3. Modification of the effective interaction due to dielectric screening

Up to this point, we have discussed the change in the effective interaction $V(q)|F(q)|^2$ that resulted from a modification of the single-particle wavefunctions, and therefore $F(q)$. The method, proposed in Ref. [17], allows one to directly change $V(q)$. Chiral materials, such as graphene, are often exposed to the environment, which allows for dielectric material to be deposited on top of them. We consider a setup where graphene sample is situated in a dielectric medium with permittivity ϵ_1, and a semi-infinite dielectric plate with permittivity $\epsilon_2 \neq \epsilon_1$ is placed at a distance $d/2$ away from the graphene sheet. The effective interactions between electrons in graphene change due to the surface charges induced at the boundary between dielectrics [17]:

$$V(r) = \frac{e^2}{\epsilon_1 r} + \alpha \frac{e^2}{\epsilon_1\sqrt{r^2+d^2}}, \text{ where } \alpha = \frac{\epsilon_1 - \epsilon_2}{\epsilon_1 + \epsilon_2}. \quad (8)$$

The ratio d/ℓ_B controls the effective interactions within a partially filled LL. However, the overall energy scale is also modified and this has an impact on the magnitude of the excitation gap. The gap should be multiplied by a factor $\epsilon_{\mathrm{GaAs}}/\epsilon_1$ if comparison is to be made with GaAs 2DES. Again, an important advantage of this setup is that the interactions can be tuned *in situ* by varying the magnetic field B, which modifies the ratio d/ℓ_B. The consequences for the many-body system are illustrated in Fig. 1(a).

3. Results

We now present some exact-diagonalization results for the interaction models introduced in Sec.2. The focus is on the physical features resulting from the variation of the effective interaction; as stated in the introduction, we follow the traditional assumption of neglecting the excitations between different LLs, which is expected to be quite reasonable in GaAs, but perhaps less so in chiral materials. Much of previous theoretical work on chiral materials has been restricted to graphene, exploring the consequences of the four-fold LL degeneracy (valley and spin) that leads to new SU(2) and SU(4)-symmetric fractional states [27]. Instead, we focus on the high-field limit, neglecting the multicomponent degrees of freedom, and examine the effects originating from the interplay of the Coulomb interaction and band structure. We consider a family of

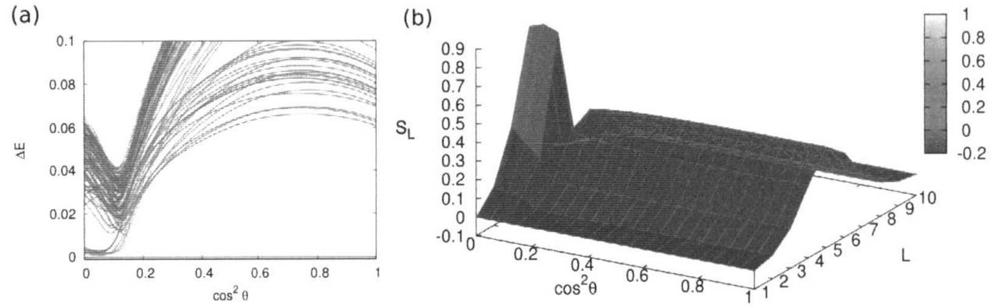

Figure 2. (Color online) Phase transition between the Laughlin and bubble phase in $(2\pi, 1)$ LL. (a) Low-lying energy spectrum on the torus as a function of θ. (b) Projected structure factor S_L as a function of angular momentum L on the sphere and parameter θ. Large ground-state degeneracy for small $\cos^2\theta$, as well as the sharp peak in S_L, correspond to the bubble phase.

band structures introduced in Ref. [16], which describe a number of high-mobility materials, including graphene with *massive* carriers (mass is generated either spontaneously, or as a result of sublattice symmetry breaking [13]), topological insulators [14], bilayer-graphene [13], trilayer graphene [28], and similar materials. Pristine graphene, which hosts massless Dirac-like fermions, is contained in this model as a particular case. Note that we explore fairly large variations of the effective interactions among fully polarized electrons; more subtle effects due to LL mixing [29] and multicomponent degrees of freedom are left for future study.

We show the results for both spherical and torus geometry. A number of useful insights can be inferred from the study of the energy spectrum; in addition, we use overlap calculations to compare an exact, many-body ground state Ψ_{exact}, with a numerical representation of a trial wavefunction, Ψ_{trial}. The overlap is defined as a scalar product between two normalized vectors, $\mathcal{O} = |\langle\Psi_{\text{trial}}|\Psi_{\text{exact}}\rangle|$. If $\mathcal{O}$ is consistently close to unity for a number of system sizes considered, we consider the trial wavefunction to be a faithful description of a FQH phase. From the knowledge of a ground-state wavefunction, we also evaluate the (projected) static structure factor [31]. Sharp peaks in the structure factor indicate the onset of compressible phases [30, 32].

The simplest example of the phase transition can be realized between the Laughlin state and a bubble phase in $(2\pi, 1)$ LL of the chiral materials. In Fig. 2(a) we show the low-lying part of the energy spectrum obtained by exact diagonalization in the torus geometry, as a function of parameter θ introduced in Sec. 2. For values $\cos^2\theta \geq 0.2$, the ground state of the system is non-degenerate and has a large overlap with the Laughlin wavefunction. When $\cos^2\theta$ is less than 0.2, the effective interaction resembles that of $n = 2$ non-relativistic LL, which favors the bubble phase at filling $\nu = 1/3$. This is manifested by a large ground-state degeneracy developing in the spectrum [32]. Similarly, the evolution of the projected structure factor S_L on the sphere as a function of θ develops a discontinuity around $\cos^2\theta \approx 0.2$ and a sharp peak below that value, signaling the compressibility of the ground state.

Previously [16], it was emphasized that the tunability of the interactions in chiral materials can lead to the more stable non-Abelian states. One such example of considerable experimental interest is the Moore-Read Pfaffian state [3]. In Fig. 3 we show the energy spectrum for $N = 12$ electrons on a torus in $(\pi, 1)$ LL. At $\cos^2\theta = 0$, the interaction reduces to $n = 1$ non-relativistic Coulomb interaction, which leads to a fragile Pfaffian state. As shown in the inset of Fig. 3, the overlap of the exact ground state with the Moore-Read wavefunction slightly increases as $\cos^2\theta$ is increased from zero. However, due to the particle-hole symmetry which is present on the torus at filling $\nu = 1/2$, a better indicator is to compare the exact ground state with a *particle-hole symmetrized* version of the Moore-Read wavefunction [11]. Indeed, in this case the overlap rises to nearly unity right before the transition at $\cos^2\theta \approx 0.14$. At this point, the

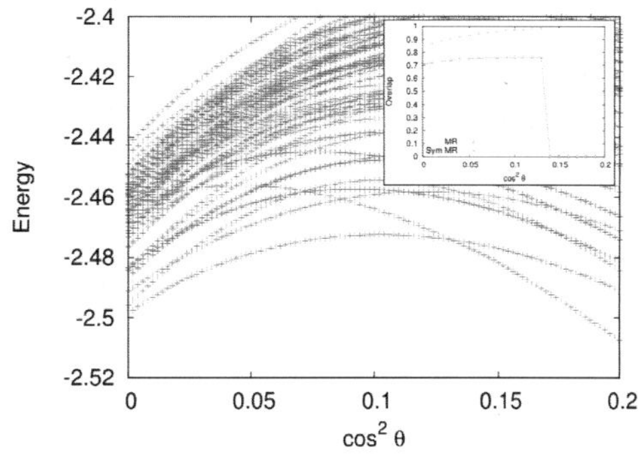

Figure 3. (Color online) Enhanced Pfaffian state and the transition to the CFL. Energy spectrum of $N = 12$ electrons on the torus in $(\pi, 1)$ LL is plotted as a function of θ. Inset shows the ground-state overlap with the Moore-Read wavefunction and its particle-hole symmetrized version.

level crossing occurs, and a composite Fermi liquid (CFL) state becomes the ground state of the system. We emphasize that although the level crossing would suggest a first-order transition, we cannot reliably rule out a possibility for the second-order transition in the thermodynamic limit. Note that, in addition to the considerable increase in the overlap with the Moore-Read state, also the excitation gap increases towards the transition to the CFL state, adding to the overall stability of the Pfaffian. As shown in Ref. [16], Pfaffian correlations can also be realized in $(\pi, 2)$ LL, and other states such as $k = 3$ RR state can be realized in several LLs [26].

Phase transitions can also be studied without reference to a particular wavefunction by evaluating the excitation gap of the system, Fig. 4. To be more general, we consider a model that involves a superposition of $n = 0$, $n = 1$ and $n = 2$ LL non-relativistic form factors [16]:

$$F_n(q) = \cos^2\theta L_{|n|}\left(\frac{q^2\ell_B^2}{2}\right) + \sin^2\theta\cos^2\phi L_{|n|+1}\left(\frac{q^2\ell_B^2}{2}\right) + \sin^2\theta\sin^2\phi L_{|n|+2}\left(\frac{q^2\ell_B^2}{2}\right). \quad (9)$$

This model generalizes the case of $(\pi, 1)$, $(\pi, 2)$ and $(2\pi, 1)$ LLs studied earlier, and naturally arises in systems such as trilayer graphene [28]. In Fig. 4 we plot the charge gap at $\nu = 1/3$ and $\nu = 1/2$ as a function of θ and ϕ. Because the calculation is performed in the spherical geometry, we fix the shift to be -3 (corresponding to the Laughin state at $\nu = 1/3$ and the Moore-Read state at $\nu = 1/2$). Along certain lines, indicated by arrows, the phase diagram reduces to one of the cases mentioned above. Note that proper finite-size scaling needs to be performed in order to get the correct values for the gap; however, it was previously found [17, 16] that this rigorous analysis produces values that are roughly in agreement with the ones shown in Fig. 4. At $\nu = 1/3$ the Laughlin state is realized for a wide range of θ and ϕ. The maximum of the gap occurs for the effective interaction that is a mixture of $n = 0$ and $n = 1$ non-relativistic form-factors. For $\cos^2\theta \leq 0.2$ the excitation gap is significantly reduced and a bubble phase is realized. The point of transition to the bubble phase is insensitive to the value of ϕ, unless $\cos^2\phi \approx 1$. On the other hand, at $\nu = 1/2$ the behavior of the gap suggests the existence of three phases. On the right side (large $\cos^2\theta$), the gaps are close to zero – this is the region corresponding to the CFL phase. In this parameter regime, the ground state of the system also has a different shift from the one chosen in Fig. 4. On the left side, one can discern two distinct branches – the middle one corresponds to the Moore-Read state because the overlap with the Pfaffian wavefunction is high in this region [16]. Note that this region is fairly large and encompasses the effective interactions which are quite different from the $n = 1$ non-relativistic

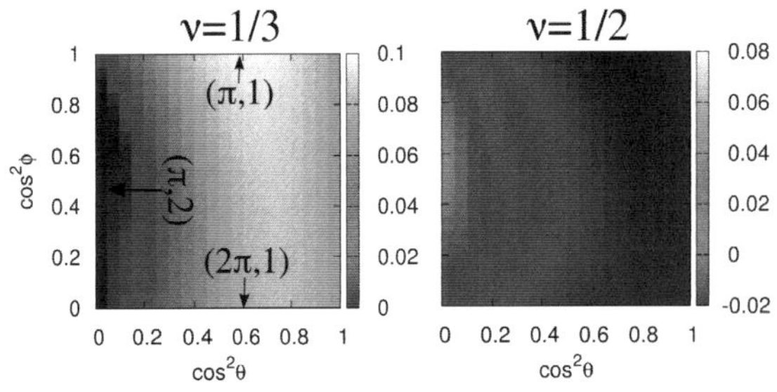

Figure 4. (Color online) Charge gap (in units of $e^2/\epsilon\ell_B$) at $\nu = 1/3$ and $\nu = 1/2$ as a function of the band-structure parameters θ and ϕ. The calculation is performed in the spherical geometry, for $N = 10$ particles at $\nu = 1/3$ and $N = 14$ particles at $\nu = 1/2$, with a fixed shift of -3.

Coulomb interaction, known to give rise to the Pfaffian state. On the far left there is a stripe phase; the excitation gap is seemingly large in this region, however this is only an artefact of the spherical geometry that cannot adequately accommodate the broken-symmetry state such as the stripe. We expect this gap to be significantly reduced as one approaches the thermodynamic limit. Note that the gaps of the incompressible states can be further increased using a dielectric setup proposed in Ref. [17].

4. Conclusion

In this paper we have discussed two practical ways of tuning the effective interactions in fractional quantum Hall systems. We emphasized an important role of chiral 2DESs where the interactions can be varied in a broader range than in GaAs, and more robust non-Abelian states are likely to be found. Moreover, these non-Abelian states are expected to exist in the new regimes of the effective interactions that are significantly different from the ones in GaAs. Finally, their excitation gaps can be tuned, and quantum phase transitions can be engineered *in situ* by varying the magnetic field.

5. Acknowledgements

This work was supported by DOE grant DE-SC0002140. Y.B. was supported by the State of Florida.

References

[1] D. C. Tsui, H. L. Stormer, and A. C. Gossard, Phys. Rev. Lett. **48**, 1559 (1982).
[2] R. B. Laughlin, Phys. Rev. Lett. **50**, 1395 (1983).
[3] G. Moore and N. Read, Nucl. Phys. B **360**, 362 (1991).
[4] C. Nayak *et al.*, Rev. Mod. Phys. **80**, 1083 (2008).
[5] *The Quantum Hall Effect*, 2nd ed., edited by R. E. Prange and S. M. Girvin, Springer-Verlag, New York, 1990.
[6] F. D. M. Haldane, Phys. Rev. Lett. **51**, 605 (1983).
[7] J. K. Jain, *Composite fermions*, (Cambridge University Press, 2007).
[8] N. Read and E. Rezayi, Phys. Rev. B **59**, 8084 (1999).
[9] R. Willett *et al.*, Phys. Rev. Lett. **59**, 1776 (1987).
[10] R. H. Morf, Phys. Rev. Lett. **80**, 1505 (1998).
[11] E. H. Rezayi and F. D. M. Haldane, Phys. Rev. Lett. **84**, 4685 (2000).
[12] F. C. Zhang and S. Das Sarma, Phys. Rev. B **33**, 2903 (1986).
[13] A. H. Castro Neto *et al.*, Rev. Mod. Phys. **81**, 109 (2009).
[14] M. Z. Hasan and C. L. Kane, Rev. Mod. Phys. **82**, 3045 (2010).

[15] B. A. Volkov and O. A. Pankratov, Pisma Zh. Eksp. Teor. Fiz. 42, 145; JETP Lett. **42**, 178 (1985); B. Buttner *et al.*, Nature Phys. **7**, 418 (2011).
[16] Z. Papić, D. A. Abanin, Y. Barlas, and R. N. Bhatt, arXiv:1108.1339.
[17] Z. Papić, R. Thomale, and D. A. Abanin, Phys. Rev. Lett. **107**, 176602 (2011).
[18] D. Yoshioka, B. I. Halperin, and P. A. Lee, Phys. Rev. Lett. **50**, 1219 (1983); F. D. M. Haldane in Ref. [5].
[19] X. G. Wen and A. Zee, Phys. Rev. Lett. **69**, 953 (1992).
[20] J. Cullum and R. A. Willoughby, J. Comp. Phys. **44**, 329 (1981).
[21] W. H. Press, S. A. Teukolsky, W. T. Vetterling, and B. P. Flannery, *Numerical recipes*, 2nd ed. (Cambridge University Press, Cambridge, 1992).
[22] K. Nomura and A. H. MacDonald, Phys. Rev. Lett. **96**, 256602 (2006); M. O. Goerbig, R. Moessner, and B. Douçot, Phys. Rev. B **74**, 161407 (2006).
[23] X. Du *et al.*, Nature **462**, 192 (2009); K. Bolotin *et al.*, Nature **462**, 196 (2009); C. R. Dean *et al.*, arXiv:1010.1179.
[24] B. E. Feldman, J. Martin, A. Yacoby, Nature Phys. **5**, 889 (2009).
[25] Y. Barlas, K. Yang, and A. H. MacDonald, arXiv:1110.1069
[26] D. A. Abanin, Z. Papić, Y. Barlas, and R. N. Bhatt (submitted to New Journal of Physics)
[27] V. M. Apalkov and T. Chakraborty, Phys. Rev. Lett. **97**, 126801 (2006); C. Tőke and J. Jain, Phys. Rev. B **75**, 245440 (2007); Z. Papić, M. O. Goerbig, and N. Regnault, Phys. Rev. Lett. **105**, 176802 (2010).
[28] W. Bao *et al.*, arXiv:1103.6088 (2011); T. Taychatanapat *et al.*, arXiv:1104.0438 (2011); A. Kumar *et al.*, arXiv:1104.1020 (2011); C. H. Lui *et al.*, arXiv:1105.4658 (2011).
[29] W. Bishara and C. Nayak, Phys. Rev. B **80**, 121302 (2009); A. Wojs, C. Tőke, and J. K. Jain, Phys. Rev. Lett. **105**, 096802 (2010); E. H. Rezayi and S. H. Simon, Phys. Rev. Lett. **106**, 116801 (2011).
[30] A. A. Koulakov, M. M. Fogler, and B. I. Shklovskii, Phys. Rev. Lett. **76**, 499 (1996); R. Moessner and J. T. Chalker, Phys. Rev. B **54**, 5006 (1996).
[31] S. He, S. H. Simon and B. I. Halperin, Phys. Rev. B **50**, 1823 (1994).
[32] E. H. Rezayi, F. D. M. Haldane, and K. Yang, Phys. Rev. Lett. **83**, 1219 (1999); F. D. M. Haldane, E. H. Rezayi, and K. Yang, Phys. Rev. Lett. **85**, 5396 (2000).

The fractional quantum Hall effect at filling factor 5/2: numerically searching for non-abelian anyons

Michael R. Peterson

Department of Physics & Astronomy, California State University Long Beach, 1250 Bellflower Blvd., Long Beach, California 90840, USA

E-mail: michael.peterson@csulb.edu

Abstract. We review some theoretical progress towards understanding the origin of the fractional quantum Hall effect at filling factor 5/2 using numerical techniques. The leading candidate thought to describe the physics is the Moore-Read Pfaffian state which supports non-abelian anyonic excitations. Interestingly, these excitations have possible applications in topologically protected quantum information processing. Specifically, we discuss how the finite-thickness of realistic quasi-two-dimensional systems stabilizes the Moore-Read Pfaffian in the fractional quantum Hall effect at 5/2 and the use of bilayer systems in the second Landau level to investigate the possible quantum phase transition between abelian and non-abelian fractional quantum Hall states.

1. Introduction

The prediction and discovery of new particles is usually reserved for particle physics, and yet, condensed matter physics has a very real possibility of observing never-before-observed non-abelian anyonic particles in the fractional quantum Hall effect [1] (FQHE) at filling factor $\nu = 5/2$. Beyond being of fundamental interest, non-abelian anyons, should they exist, could provide a platform for fault-tolerant topological quantum information processing [2, 3].

The search for non-abelian anyons was initiated with the Moore-Read Pfaffian [4] wavefunction in 1991 as a possible many-body description for the FQHE at $\nu = 5/2$, first observed in 1987 [5]. The Moore-Read (MR) Pfaffian wavefunction is written, for an even number of electrons N_e, as

$$\Psi_{\mathrm{MR}} = \mathrm{Pf}\left\{\frac{1}{z_i - z_j}\right\}\prod_{i<j}^{N_e}(z_i - z_j)^2 e^{-\sum_{i=1}^{N_e} z_i^2/4l^2} \tag{1}$$

where $z = x - iy$ is the electron position in the complex plane (l is defined below).

$$\mathrm{Pf}\left\{\frac{1}{z_i - z_j}\right\} = \mathcal{A}\left\{\frac{1}{z_1 - z_2}\frac{1}{z_3 - z_4}\cdots\frac{1}{z_{N_e-1} - z_{N_e}}\right\} \tag{2}$$

defines the Pfaffian operator where $\mathcal{A}$ is the antisymmetrization operator. A remarkable property of Ψ_{MR} is that it is a representative of a universality class with low-energy excitations that obey non-abelian braiding statistics [4, 6, 7]. The exotic braiding statistics of Ψ_{MR} and the FQHE at $\nu = 5/2$ were tied to the field of topologically protected quantum computation [2, 3] by Das Sarma, Freedman, and Nayak in 2005 by applying theoretical ideas of Kitaev [8] and Freedman [9] to a real experimental system, i.e., the FQHE, and the search for non-abelian anyons took on renewed urgency. Specifically, reference [2] proposed that one could braid the world lines of the non-abelian anyonic-particle excitations of the 5/2 FQHE and process quantum information in a topologically protected way.

The FQHE occurs when electrons are confined to a quasi-two-dimenensional plane, such as at the interface between the two semiconductors (e.g., GaAs/AlGaAs heterostructures), in the presence of a strong perpendicular magnetic field of strength B. The Hamiltonian describing the physics is

$$\mathrm{H}_{\text{Realistic}} = \sum_{i=1}^{N_e} \frac{(\mathbf{p} + \frac{e}{c}\mathbf{A})^2}{2m} + \sum_{i<j}^{N_e} \frac{e^2}{\varepsilon l |\mathbf{r}_i - \mathbf{r}_j|} + g\mu\mathbf{B}\cdot\mathbf{S} + (\text{other terms}) \tag{3}$$

where ε is the dielectric constant of the host semiconductor (in GaAs/AlGaAs $\varepsilon \approx 12$), m is the effective electron mass (in GaAs/AlGaAs, $m = 0.07m_0$, where m_0 is the bare electron mass), $l = \sqrt{c\hbar/eB}$ is the magnetic length (hence $\mathbf{r}_i$ above is dimensionless), $\mathbf{A}$ is the vector potential such that $\nabla \times \mathbf{A} = B\hat{z}$, and $\mathbf{S}$ is the total spin of the N_e system with $g = -0.44$ (again for GaAs/AlGaAs). The kinetic energy is quantized into Landau levels (LLs) with energy $\hbar\omega(n + 1/2)$, where $\omega = eB/mc$ is the cyclotron frequency and $n = 0, 1, 2, \ldots$, each with a macroscopically large degeneracy of $1/(2\pi l^2) = eB/hc = B/\phi_0$, where $\phi_0 = hc/e$ is the magnetic flux quantum. Thus, the degeneracy is equal to the number of flux quanta N_Φ per area piercing the quasi-two-dimensional surface. The filling factor $\nu = \rho 2\pi l^2 = N_e/N_\Phi$ (ρ is the electron density) defines how many LLs are filled (completely or fractionally/partially). The lowest Landau level (LLL) has $n = 0$ and, due to the electron spin, corresponds to $0 < \nu < 2$. The second LL (SLL) obtains for $2 < \nu < 4$ since the LLL for spin-up and spin-down electrons is completely filled, hence, $\nu = 5/2 = 2 + 1/2$ concerns the half-filled second LL.

When the filling factor in the highest unoccupied Landau is less than unity all the electrons have the same kinetic energy and the first term in equation 3 is an irrelevant constant. The last term in equation 3 labeled "other terms" denote single-particle interactions between the electrons and the ions of the semiconductor as well as disorder. All FQHE experiments are done at very high magnetic fields ($B = 3 - 40$ T), at very low temperatures ($T < 500$ mK), and in very clean samples with mobilities mostly greater than 10^7 cm^2/(Vs). Hence, it is usually a good approximation to take the large field limit ($\hbar\omega = \hbar eB/mc \to \infty$) and neglect the "other terms". Furthermore, the large magnetic field likely polarizes the electron spin making $g\mu\mathbf{B}\cdot\mathbf{S}$ another irrelevant constant [1]. This approximation leaves only the electron-electron Coulomb interaction

$$\mathrm{H}_0 = \sum_{i<j}^{N_e} \frac{e^2}{\varepsilon l |\mathbf{r}_i - \mathbf{r}_j|} \tag{4}$$

as the Hamiltonian describing N_e electrons fractionally filling the highest unoccupied LL. The FQHE occurs when the strongly interacting electrons with filling factor p/q (in the highest unoccupied LL), with p and q integers, condense into a new type of emergent incompressible quantum fluid with fractionally charged anyonic excitaitons [12, 13, 14]. This quantum fluid, being incompressible, is a uniform-density unique ground state with a finite energy gap to excitations. Sample disorder converts this energy gap into a mobility gap and the phenomena of

[1] Full spin polarization at $\nu = 5/2$ is not experimentally established definitively, however, theoretically the ground state of H_0 has been shown to be spin polarized [10, 11]. Thus we take this approximation throughout this work.

the FQHE can be understood [15, 16]. While there have been approximately 70 different odd-denominator FQHE states (q odd) observed in the lowest LL (i.e., $n = 0$ and $\nu < 1$) the FQHE in the second LL is relatively rare, and more fragile (small energy gaps), with approximately 8 FQHE states so far observed, including the *even*-denominator state at $\nu = 5/2$ which is the *only* even-denominator FQHE experimentally observed in single-layer systems.

In terms of strongly or weakly correlated problems, H_0 defines an essentially infinitely strongly interacting problem; there is only the interaction term in the Hamiltonian, no small parameter, and no normal state from which to construct a perturbation theory. Thus, only non-perturbative methods are applicable. One method is to propose variational wavefuctions–the approach of Laughlin [12], the Composite Fermion theory of Jain [14, 16], and the MR Pfaffian theory.

Another method to solve equation 4 is exact diagonalization for finite-sized systems. This can be done for systems up to approximately 20 electrons. Exact (Lanczos) diagonalization can be readily achieved for Hamiltonian matrices of dimension up to $10^6 \times 10^6$. For a finite sized system the filling factor is $\nu = \lim_{N_\Phi \to \infty} N_e/N_\Phi$ where N_Φ is the number of magnetic flux quanta piercing the surface of the sample, so the Hamiltonian matrix dimension depends on filling factor. In this review, we will report on results where both the wavefunction and exact diagonlization methods are used in combination to illucidate the physics and, needless to say, both methods rely heavily on numerics.

Although suggested earlier [17], the MR Pfaffian description of the FQHE at $\nu = 5/2$ gained serious traction after the work of Morf [10] in 1998. Reference [10] found that the Coulomb Hamiltonian H_0, for electrons in the second LL (SLL), is very likely spin-polarized. This is a necessary condition for the MR Pfaffian to be a viable ansatz. (However, Read and Green [18] find that full spin-polarization might not be totally necessary for the FQHE at $\nu = 5/2$ to be in the same non-abelian universality class of the MR Pfaffian, see their figure 1.) Further numerical work by Feiguin *et al.* [11] have more definitively shown that H_0 for the SLL is spin-polarized and, theoretically at least, the full spin-polarization of the ground state for an ideal, strictly two-dimensional, Coulomb Hamiltonian in the SLL is very well established. The second thing coming out of Morf's work [10] was that the wavefunction overlap (which is, of course, a quantitative measure of how similar one wavefunction is to another, with a value of zero being completely dissimilar, or orthogonal, and a value of unity being completely similar) between the exact ground state of the Coulomb Hamiltonian in the SLL and the MR Pfaffian was impressively large at $\sim 0.8 - 0.9$ and could be made equal to essentially unity by perturbing the Hamiltonian very slightly–namely, increasing the first Haldane pseudopotential V_1 by approximately 10%. While Morf's result considered the spherical geometry, this result was soon corroborated using the torus geometry [19]. Furthermore, recently it has been demonstrated that the spin-polarized ground state of the Coulomb Hamiltonian in the SLL is adiabatically connected to the MR Pfaffian state [20] and that the ground state displays even-odd oscillations as a function of particle number [21] indicative of quasiparticle pairing.

Nagging questions still remained. While an overlap of $\sim 0.8 - 0.9$ is certainly impressive since a randomly chosen wavefunction would have an overlap of the order of 1/(number of basis states)~ 0, FQHE theoreticians had gotten used to overlaps with variational ansatz being ~ 0.99, for example, the Laughlin wavefunction for ν =1/(odd) and Jain composite fermion wavefunctions describing $\nu = n/(2pn \pm 1)$ (n and p integers). Furthermore, while hardening the short-range interaction, increasing the Haldane pseudopoential V_1, makes the overlap between the ground state and the MR Pfaffian ~ 0.99 there is no physical mechanism for which one can increase V_1, i.e., experimentalists cannot simply increase V_1 by some Haldane pseudopotential "knob". But, certain realistic effects of quasi-two-dimensional systems that are ignored in equation 4 such as finite-thickness (which we discuss at length here), Landau level mixing, etc., can change the values of the V_m's with respect to their values for the strictly two-dimensional system. However, these effects change *all* the V_m's and it is impossible to change the relative strength of *only* V_1 with respect to the other V_m's. Hence, while the MR Pfaffian is seemingly a good description of a slightly perturbed Coulomb Hamiltonian in the SLL, perhaps H_0 is not the correct Hamiltonian to describe the physics in the SLL.

The plan of this review is as follows: in section 2 we report on recent results [22, 23, 24] finding that the finite-thickness of realistic quasi-two-dimensional systems stabilizes the MR Pfaffian description of the FQHE at $\nu = 5/2$. That is, the MR Pfaffian turns out to be an exceptionally good description of the physics for a more *realistic* Hamiltonian. We also briefly touch on the role of particle-hole symmetry breaking [25]. In section 3, we report on recent investigations [26, 27, 28] concerning the competition between abelian and non-abelian FQH states in bilayer systems as an alternative scenario to probe the possible non-abelian nature of the 5/2 FQHE.

2. Finite-thickness effects in single-layer systems

A realistic effect of quasi-two-dimensional systems that we review in this work is the finite-thickness effect. The quasi-two-dimensional plane in which the electrons reside (densities are $\sim 1 \times 10^{11}$ cm^{-2}) in GaAs/AlGaAs quantum wells have a typical thickness of $\sim$ 30 nm. The electrons in the direction perpendicular to the two-dimensional surface have only some probability of being within some length w of the interface, thus, finite-thickness softens the Coulomb interaction between electrons at short distances, leading to an effective electron-electron interaction $V(|\mathbf{r}_i - \mathbf{r}_j|)$. Starting with equation 4 we replace the pure Coulomb potential $(e^2/(\varepsilon l r))$ with a general potential $V(r)$. This general Hamiltonian H can be paramterized by Haldane pseudopoentials (alluded to above)

$$\mathrm{H} = \sum_{i<j}^{N_e} V(|\mathbf{r}_i - \mathbf{r}_j|) = \sum_{k=0}^{\infty} V_{2k+1}^{(n)} \sum_{i<j}^{N_e} \hat{P}_{2k+1}(m_{ij}) \tag{5}$$

where $\hat{P}_m(m_{ij})$ projects onto states of relative angular momentum $m_{ij} = m$.

$$V_m^{(n)} = \int_0^\infty dqq \left[L_n\left(\frac{q^2}{2}\right)\right]^2 L_m(q^2) e^{-q^2} \left\{ \frac{1}{2\pi} \int d\mathbf{r} e^{i\mathbf{q}\cdot\mathbf{r}} V(r) \right\} \tag{6}$$

are the Haldane pseudopotentials defined as the energy of a pair of electrons with relative angular momentum m confined to the n-th LL interacting via a potential $V(r)$.

In order to model the effect of finite-thickness a variety of models are possible. For this review, we focus on the model where the electrons are confined to the quasi-two-dimensional plane in an infinite-square-well potential [29]. Encouragingly, we find that our results do not qualitatively depend on the finite-thickness model (in reference [23] we considered the Fang-Howard heterostructure model [30, 31] and the Zhang-Das Sarma model [32]). Furthermore, if a more direct experimental comparison is desired, such as comparing the calculated energy gaps to the experimentally measured activation gaps, one can use the local density approximation to self-consistently find the single-particle wavefunction in a quantum-well of finite height as has been done previously [33]. However, as we are only interested in qualitative features and trends, the infinite-square-well serves our purposes.

To consider finite-thickness effects one needs the effective interaction depending on the finite-thickness model to calculate the Haldane pseudopotentials using equation 6. For the infinite-square-well, the Fourier transform of the effective interaction is

$$\tilde{V}(q) = \frac{e^2 l}{\varepsilon} \frac{1}{q} \frac{\left[3qw + \frac{8\pi^2}{qw} - \frac{32\pi^4[1-\exp(-qw)]}{(qw)^2[(qw)^2+4\pi^2]}\right]}{(qw)^2 + 4\pi^2} \tag{7}$$

which is found by using the single-particle states of the infinite square-well $\eta(z) = \sqrt{2/w}\cos(\pi z/w)$ and calculating

$$\tilde{V}(q) = \frac{e^2 l}{\varepsilon q}\int_0^d dz \int_0^d dz' |\eta(z)|^2 |\eta(z')|^2 e^{-q|z-z'|} . \quad (8)$$

We briefly mention the spherical and torus geometries in which we work. In the spherical geometry, the electrons are confined to the surface of a sphere with radius $R = \sqrt{N_\Phi/2}$ where a magnetic monopole of strength $N_\Phi/2$ (where N_Φ is an integer according to Dirac) is placed at the center of the sphere producing a radial magnetic field everywhere perpendicular to the surface with total magnetic flux of $N_\Phi \phi_0$. Total angular momentum L is a good quantum number and totally rotationally symmetric uniform states have $L = 0$. In the spherical geometry, the filling fraction is defined as $\nu = \lim_{N_e \to \infty} N_e/N_\Phi$ and $N_\Phi = \nu^{-1} N_e + S$ where S is known as the "shift"–a topological quantum number that arises due to the curved surface of the sphere and allows one to distinguish potential FQH states of different topological order [34]. In the torus geometry, the electrons are confined to a period rectangular plane with unequal sides a and b (we define the aspect ration as a/b). The energy is a function of pseudomomenta ($\mathbf{K}_x = 2\pi\hbar n_x/a, , \mathbf{K}_y = 2\pi\hbar n_y/b$) in a two-dimensional Brillouin zone with N_0^2 points ($n_x,\ n_y = 0, \ldots, N_0$) with N_0 being the greatest common factor of N_e and N_Φ. The filling factor on the torus is simply $\nu = N_e/N_\Phi$ and there is no "shift". Instead, states have a ground state degeneracy that depends on their topological origin [35]. We utilize both geometries and use the spherical "shift" and ground state degeneracy as complimentary diagnostics of possible FQH states.

2.1. Overlaps

In this section we discuss how the finite-thickness of realistic quasi-two-dimensional systems, in which the FQHE at $\nu = 5/2$ is observed, stablizes the non-abelian MR Pfaffian state.

Figure 1a shows the overlap between the exact ground state of H (equation 5) and the MR Pfaffian wavefunction as a function of finite-thickness w for $N_e = 8$, 10, and 12 electrons, respectively. As mentioned above the overlap for the strictly two-dimensional system ($w = 0$) is encouragingly large. However, inclusion of finite-thickness produces a ground state with a larger overlap with the MR Pfaffian than zero thickness with a maximum near $w = 4l$ that is largely insensitive to the confinement model and particle number [22, 23]. Hence, this effect is very likely a real effect that will survive the thermodynamic limit. It should be pointed out that other "standard" FQH states, i.e., the Laughlin and Composite Fermion states, are uniformly supressed with finite-thickness. In contrast, softening the Coulomb interaction via the finite-thickness effect, an effect that is realized in nature, produces just the right interaction to stabilize and favor the MR Pfaffian description of the FQHE at $\nu = 5/2$.

In figure 1a, on the overlap curve, we also indicate examples of the quantum-well widths of experimental samples where the FQHE at $\nu = 5/2$ has been observed. Interestingly, the experimental sample width is uniformly *sub-optimal*. Perhaps it is possible to experimentally "engineer" samples of width $w \sim 4l$ for which the FQHE at $\nu = 5/2$ would be more likely to be described by the MR Pfaffian wavefunction.

Before discussing topological degeneracy of the ground state in the torus geometry we point out that the first two pseudopotentials, $V_1^{(n)}$ and $V_3^{(n)}$, do not by themselves determine the physics. Needless to say, finite-thickness effects change *all* the pseudopotentials and the first two $V_m^{(n)}$'s do not determine the physics and manipulating the ratio of $V_1^{(n)}/V_3^{(n)}$ is not akin to changing the thickness of the two-dimensional layer. To illustrate this, in figure 1b we reproduce an example shown in reference [23] where we arbitrarily varied $V_3^{(1)}$ and $V_5^{(1)}$ away from their strictly two-dimensional Coulomb values. It was found that there is a region of decreasing $V_3^{(1)}$

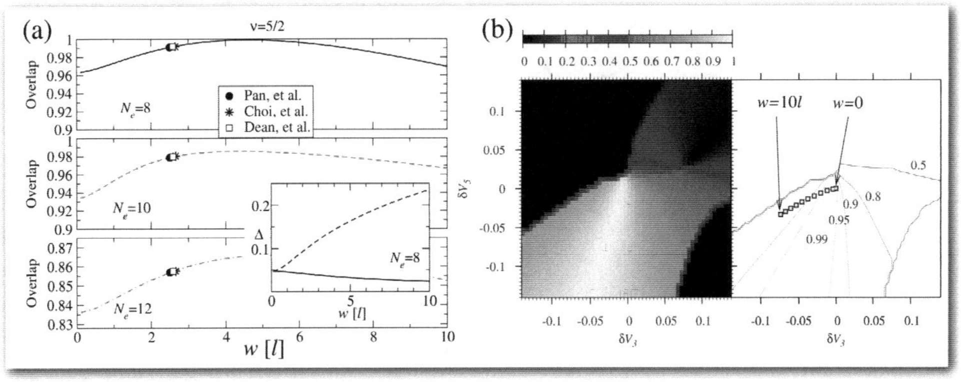

Figure 1. (a) Overlap between the exact ground state at $\nu = 5/2$, for a quasi-two-dimensional system modeled by an infinite square-well potential, and the MR Pfaffian as a function of finite width w for $N_e = 8$, 10, and 12 electrons. The experimental widths of references [36] (solid circle), [37] (star), and [38] (open square) are also indicated. The inset shows the excitation gap Δ for $N_e = 8$ in units of $e^2/\varepsilon l$ (solid) and $e^2/(\varepsilon\sqrt{l^2+d^2})$ (dashed) (From reference [22]). (b) The overlap between the exact ground state of H_0 projected into the second LL (i.e., zero-thickness) as a color contour plot (left panel) and contour lines (right panel), respectively, where we have arbitrarily varied the Haldane pseudopoentials away from their Coulombic values, i.e., $V_3^{(1)} \to V_3^{(1)} + \delta V_3$ and $V_5^{(1)} \to V_5^{(1)} + \delta V_5$. Also shown on the right panel is the path traced out in the δV_3-δV_5 plane using the infinite square-well potential from $w = 0$ to $10l$. The small squares are color-coded according to the value of the overlap (From reference [23]).

and $V_5^{(1)}$ where the overlap is nearly unity, hence, one can leave $V_1^{(1)}$ completely alone and simply vary the $m = 3$ and $m = 5$ pseudopotentials and yet still drive the system into the MR Pfaffian phase. On the same figure, we show the overlap for the square-well potential at different values of thickness from $w = 0$ to $10l$. Since finite-thickness changes all the $V_m^{(1)}$'s we only show the path traced out in the $V_3^{(1)}$-$V_5^{(1)}$ plane. Thus, it is clear that one cannot unambiguously parameterize the finite-thickness effect by changing only the first one or two pseudopotentials.

2.2. *Topological Degeneracy*

Another measure of stability of the MR Pfaffian state is found when one considers the torus geometry. On the torus, quantum states with topological order display ground state degeneracies. For the MR Pfaffian wavefunction the degeneracy is three-fold [18]. Hence, an approximately three-fold degeneracy of the ground state of our Hamiltonian (equation 5) allows us to distinguish it from other candidate states with different topological order. For example, the Composite Fermion fermi liquid is expected to show shell effects that are very sensitive to the aspect ratio and no particular ground state degeneracy. This is what is theoretically found [39]. Another candidate wavefunction is Halperin's 331 state [40] (which is only applicable for two-component systems, either bilayer systems or spin) which has a completely different ground state degeneracy signature compared to the MR Pfaffian. The three-fold ground state degeneracy of the MR Pfaffian occur at pseudompmenta $\mathbf{K} = (N_0/2, 0)$, $(0, N_0/2)$, $(N_0/2, N_0/2)$.

For $w = 0$, we find that the spectra is very sensitive to aspect ratio for all system sizes studied, indicative of a Composite Fermion fermi liquid ground state. Remarkably, upon increasing the thickness w to a value where we see the largest overlap, i.e., $w \sim 4l$, we discover the appearance

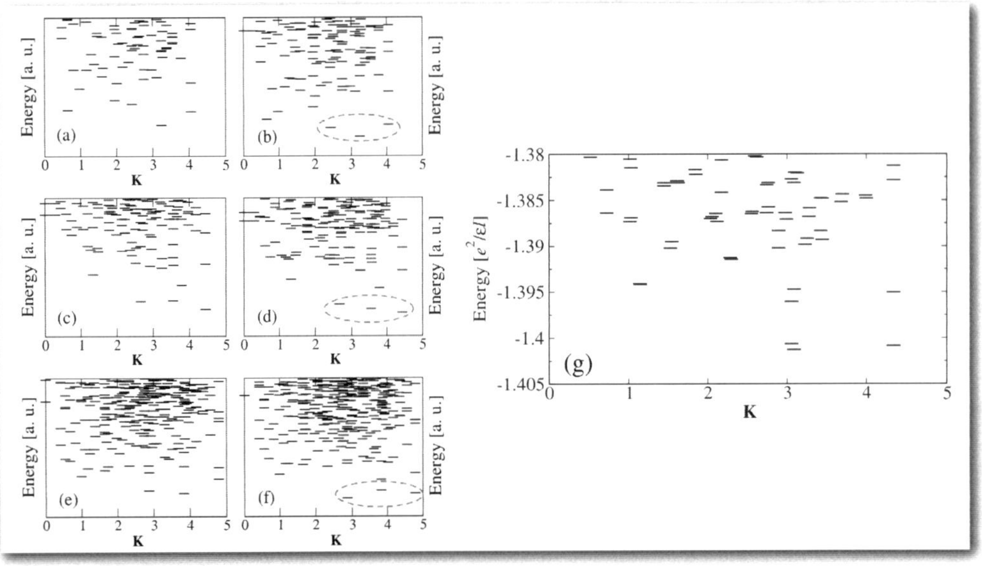

Figure 2. Energy (in arbitrary units) as a function of (pseudo)momentum **K** for zero thickness ($d = 0$) and $N_e = 10$, 12 and 14 electrons (panels (a), (c), and (e)) and $w = 4l$ for $N_e = 10$, 12, and 14 electrons (panels (b), (d), and (f)). The aspect ratio is chosen to be 0.75. For finite-thickness we see the appearance of an approximate three-fold degeneracy (circled in red) at the (psuedo)momenta $\mathbf{K} = (N_0/2, 0)$, $(0, N_0/2)$, and $(N_0/2, N_0/2)$. (g) shows energy vs. **K** for $w = 4l$, $N_e = 12$ electrons, and a nearly square unit cell (aspect ratio of 0.99). The approximate two-fold degeneracy is clearly evident for the Moore-Read Pfaffian/anti-Pfaffian (pseudo)momenta (From reference [22]).

of a three-fold degeneracy at the quantum numbers corresponding to the Moore-Read Pfaffian. This is shown in figure 2a-f. To the best of our knowledge, this is the first time this effect has been seen in a topological system generated from a Hamiltonian consisting of only two-body interactions.

Another interesting aspect of the MR Pfaffian wavefunction is that it is not particle-hole symmetric. In the spherical geometry, the MR Pfaffian exists when the particle number versus magnetic flux relationship is $N_\Phi = 2N_e - 3$ and -3. Since the shift is an order one effect it vanishes in the thermodynamic limit. A half-filled LL is particle-hole symmetric but under particle-hole symmetry the system at $N_\Phi = 2N_e - 3$ transforms to $N_\Phi = 2N_e + 1$ since $N_e + N_h = N_\Phi + 1$ (where N_h is the number of holes). The particle-hole conjugate has a shift of $+1$, different from -3, hence, the state with $N_\Phi = 2N_e + 1$ has been termed the anti-Pfaffian [41, 42] and is topologically distinct from the MR Pfaffian.

On the torus, the shift is absent and the existence of the anti-Pfaffian would lead to a two-fold degeneracy in the energy spectrum. However, in a finite size system this exact degeneracy is slightly lifted. For a nearly square unit cell, i.e., $a/b \approx 0.99$, if the MR Pfaffian is a good description of the physics, one would expect to see an approximately two-fold degeneracy for the ground state energy at the **K** values corresponding to the MR Pfaffian/anti-Pfaffian. For $w \sim 4l$ we see this approximate two-fold degeneracy (see figure 2g) while for $w = 0$ it is absent.

The observation of the finite-thickness effect providing stabilizing effects for the MR Pfaffian state for the FQHE at $\nu = 5/2$ in both the torus and spherical geometries gives us great confidence that this effect is robust and real and not a finite-sized artifact. We also encourage experimentalists to take seriously our suggestion of "engineering" systems with $w \sim 4l$ to produce systems more likely to have a MR Pfaffian ground state.

2.3. Particle-hole symmetry breaking

The explicit particle-hole symmetry breaking of the MR Pfaffian wavefunction was briefly mentioned above. This quality lay relatively dormant [43, 6, 44] until recently when it was realized that the consequences of this rather simple fact, due to the MR Pfaffian being the exact ground state of a three-body potential, are interestingly far reaching. The Hamiltonian H_3

$$\mathrm{H}_3 = \sum_{i<j<k}^{N_e} \hat{P}_{ijk}\left(\frac{3N_\Phi}{2} - 3\right) \tag{9}$$

produces the Moore-Read Pfaffian as an exact zero-energy ground state [17] (here H_3 is written in the spherical geometry). $\hat{P}_{ijk}(L)$ projects onto electron triplets with total angular momentum L. The particle-hole transform of H_3, denoted $\overline{\mathrm{H}}_3$, produces the anti-Pfaffian as it's zero energy ground state and consists of a term exactly the minus of H_3 as well as a two-body term, a chemical potential, and a constant. In contrast, the particle-hole conjugate of the two-body Coulomb Hamiltonian $\overline{\mathrm{H}} = \mathrm{H}$ and, hence, it respects particle-hole symmetry. Thus, if the MR Pfaffian is to be a good description of the ground state of H then perhaps H spontaneously breaks particle-hole symmetry.

If we consider the total ground state energy to be similar to a *Landau free energy* and the electron number N_e as the *order parameter* (strictly speaking we should take the electron number away from half-filling) then if the total ground state energy exhibits a "mexican hat"-like structure as a function of N_e this would indicate spontaneously broken particle-hole symmetry. Specifically, we would expect that whenever N_e is even and corresponds to the MR Pfaffian ($N_\Phi = 2N_e - 3$) or anti-Pfaffian ($N_\Phi = 2N_e + 1$) the total ground state energy will be lowest. We can now attempt to answer whether the two-body Coulomb interaction, with or without finite-thickness effects, spontaneously breaks particle-hole symmetry.

In the left panels of figure 3 we show the total ground state energy as a function of N_e for the Coulomb Hamiltonian H in the SLL for total values of magnetic flux $N_\Phi = 13$ to 17. The MR Pfaffian (anti-Pfaffian) has $N_\Phi = 2N_e - 3$ ($N_\Phi = 2N_e + 1$) and N_e must be an even integer. Hence, only $N_\Phi = 13$, 15, and 17 contain even integer N_e solutions to $N_\Phi = 2N_e - 3$ and $N_\Phi = 2N_e + 1$, namely, $N_e = 8$ and 10 for the MR Pfaffian and $N_e = 6$ and 8 for the anti-Pfaffian. It is found that the overall structure of the total ground state energy versus N_e is nearly a parabola around half-filling reflecting the charging energy of a finite-sized system and no sign of spontaneously broken particle-hole symmetry is observed. This result obtains even when we include finite-thickness effects. Thus, the two-body electronic Coulomb Hamiltonian does *not* spontaneously break particle-hole symmetry *with or without* finite-thickness. Experimentally, this could mean that the true ground state responsible for the FQHE at $\nu = 5/2$ is not strictly in the same universality class of the Pfaffian (Moore-Read or anti-Pfaffian). Or, the true ground state is a superposition of the MR Pfaffian and anti-Pfaffian forming a new particle-hole symmetric state with necessarily different topological order, and most likely abelian excitations, and edge physics. Or, something external explicitly breaks particle-hole symmetry such as disorder or Landau level mixing. Note that Landau level mixing is known to produce three-body terms perturbatively that explicitly break particle-hole symmetry [45]. Recently, theoretical investigations of LL mixing in the FQHE at 5/2 have been done with no consistent conclusion so far [46, 47].

Interestingly, however, adding H_3 and $\overline{\mathrm{H}}_3$ cancels the particle-hole symmetry breaking three-body terms and produces a particle-hole symmetric Hamiltonian which we label H_2. It turns

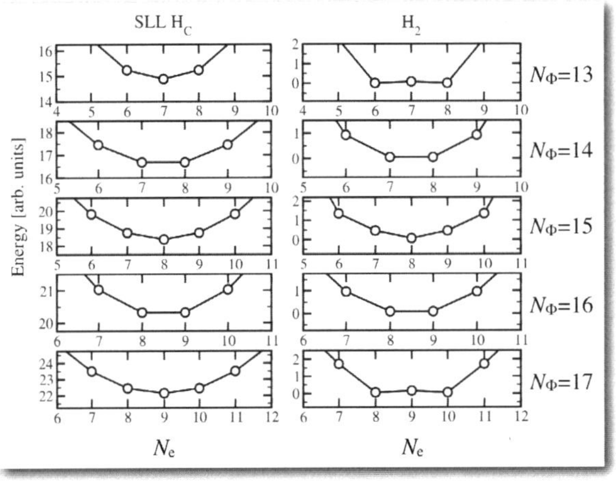

Figure 3. Total ground state energy of the Coulomb Hamiltonian in the SLL (with zero-thickness) and H_2 as functions of N_e near half-filling for different values of magnetic flux N_Φ. It is found that the ground state of the two-body Coulomb Hamiltonian for $\nu = 5/2$ does *not* spontaneously break particle-hole symmetry *with or without* finite-thickness while the ground state of H_2 does spontaneously break particle-hole symmetry. (From reference [25]).

out that H_2 spontaneously breaks particle-hole symmetry, as can be seen in the right panels of figure 3. Here a clear "mexican-hat"-like structure is observed in the total ground state energy as a function of N_e. Perhaps the ground state of H_2 might provide an interesting ansatz for the FQHE at $\nu = 5/2$ since it spontaneously breaks particle-hole symmetry and yet arises as a ground state solution to a particle-hole symmetric two-body Hamiltonian. The topological order of the ground state of H_2 has yet to be determined, so the nature of quasiparticle excitations are not known.

We briefly mention that Biddle *et al.* [24] have coraborated the finite-thickness stabilization of the MR Pfaffian state by investigating entanglement measures (both entanglement entropy and spectra) recently applied to the FQHE [48, 49].

3. Competition between abelian and non-abelian phases in bilayer FQH systems

The current experimental status of the FQHE at $\nu = 5/2$ is as follows. Besides being observed many times by different groups over the years, recent experimental work has measured the fractionally charged excitations of $e/4$ expected from the MR Pfaffian theory through both direct shot noise [2] measurements [51, 52, 50] and local measurements with comparison to other FQH states [53]. Observations of interference oscillations of quasiparticles [54] is also consistent with fractional charge $e/4$ and counter propagating neutral modes expected from theories in the same universality class as the MR Pfaffian have been recently observed [55]. The FQHE at $\nu = 5/2$ needs to be nearly spin-polarized in order for the MR Pfaffian theory to be a viable explanation and, yet, recent experimental work [56, 57] points to the possibility of an un-polarized ground state. Completely definitive experimental investigations of spin-polarization are currently lacking.

[2] We point out, however, that the interpretation of the shot noise measurements is evidently more complicated than originally thought [50].

The experimental situation, along with recent experimental developments [58, 59, 60] of the observation of the FQHE at $\nu = 1/2$ and $1/4$ in very wide quantum wells at high magnetic fields, have led us to consider alternative scenarios where one could probe the non-abelian nature of the FQHE at $\nu = 5/2$. In particular, we have recently investigated bilayer systems which may provide an avenue for the proposed non-abelian physics to show itself in new and unexplored ways. We investigate two questions: (1) whether the analog of the non-abelian state at $\nu = 5/2$ might also exist in the LLL at $\nu = 1/2$ for single-layer systems, and (2) what does the bilayer FQHE at total filling factor $\nu = 5/2$ teach us about abelian to non-abelian quantum phase transitions.

Bilayer, or two-component, systems come in two experimental varieties. One is an actual bilayer consisting of two parallel quasi-two-dimensional systems separated from one another by a separation d with an experimentally adjustable tunneling barrier [61]. The other variety is a single wide-quantum-well [59, 60, 62, 63, 64]. In a wide-quantum-well, with a typical width of ~ 70 nm, the electrostatic electron interaction creates an effective bilayer by depleting the electron density in the center of the wide-quantum-well. See reference [27] for an in-depth discussion.

Our simple model for bilayer FQHE most closely resembles that of a *true* bilayer [61]:

$$\mathrm{H_{Bilayer}} = \sum_{i<j}^{N_e} [V_{\mathrm{intra}}(|\mathbf{r}_i - \mathbf{r}_j|) + V_{\mathrm{intra}}(|\tilde{\mathbf{r}}_i - \tilde{\mathbf{r}}_j|) + V_{\mathrm{inter}}(|\mathbf{r}_i - \tilde{\mathbf{r}}_j|)] + t\, \mathrm{H_{Tunneling}} \,. \tag{10}$$

$V_{\mathrm{inter}}(r)$ is the electron-electron interaction between electrons in a single-layer with position $\mathbf{r}_i$ and $\tilde{\mathbf{r}}_j$ for, say, the left and right layer, respectively. We model the finite-thickness of each individual layer, taken to be of the same thickness w, using the Zhang-Das Sarma potential [32]. $V_{\mathrm{inter}}(r) = e^2/(\varepsilon\sqrt{r^2 + d^2})$ is the interaction between electrons in individual layers separated by a distance d, $d > w$ by definition. The last term $\mathrm{H_{Tunneling}}$ controls the tunneling of electrons between layers, i.e., it is a pseudo-spin operator, with large t denoting strong tunneling.

In order to discuss the physics of the bilayer problem let us concentrate first on the well known $\nu_{\mathrm{total}} = \nu_{\mathrm{left}} + \nu_{\mathrm{right}} = 1/4 + 1/4 = 1/2$ bilayer problem. For very small d and large t we have a single-layer problem with filling factor $\nu_{\mathrm{total}} = 1/2$ which is known to be a Composite Fermion fermi liquid state [65, 66] that, while exotic in it's own right, does not display the FQHE. For very large separation d we have two independent layers each with $\nu_{\mathrm{right/left}} = 1/4$ and each layer is again known to form Composite Fermion fermi liquid states that do not display the FQHE. However, at intermediate values of layer separation $d \sim 2l$ and very weak tunneling the FQHE is observed [67, 68, 69, 62, 61]. The 331 state

$$\Psi_{331} = \prod_{i<j}^{N_e/2} (z_i - z_j)^3 \prod_{i<j}^{N_e/2} (\tilde{z}_i - \tilde{z}_j)^3 \prod_{i}^{N_e/2} \prod_{j}^{N_e/2} (z_i - \tilde{z}_j) e^{-\sum_i^{N_e/2} (z_i^2/4l^2 + \tilde{z}_i^2/4l^2)} \tag{11}$$

due to Halperin [40] is known to be a good description of the FQHE for bilayer systems with total filling $\nu_{\mathrm{total}} = 1/2$ at intermediate d and weak tunneling. For intermediate d and strong tunneling the system is basically a wide-quantum-well of width $w + d$, albeit single-component, with total filling factor $1/2$ and again a non-FQHE Composite Fermion fermi liquid obtains. We also point out that the excitations of the Halperin 331 state are abelian and would not support topologically protected quantum information processing.

Instead of a bilayer at $\nu_{\mathrm{total}} = 1/2$, we start with a system with vanishing separation d and strong tunneling d with total filling factor $\nu_{\mathrm{total}} = 5/2$. This is a single-component system that is known to display the FQHE thought to be described by the non-abelian MR Pfaffian state that can be stabilized by finite-thickness w. If we now increase the separation and decrease the tunneling strength we eventually are left with two independent layers with

$\nu_{\rm right/left} = 5/4 = 1 + 1/4$ which would presumably form a Composite Fermion fermi liquid in each layer and not display the FQHE (see section VD in reference [26] for a discussion on the bilayer FQHE in higher Landau levels and the additional complications that are present compared to bilayer FQHE in the LLL).

In order to study this problem we exactly diagonalize $\rm H_{Bilayer}$, in the spherical geometry, for different values of layer thickness w, layer separation d, and tunneling strength t. Due to the complexity of the problem and large size of the Hilbert space we only consider one system size where the total number of electrons is $N_e = 8$ and the total magnetic flux is $N_\Phi = 13$. This, as noted above, corresponds to a shift of "-3" and since both the MR Pfaffian and the 331 state have a shift of "-3" we can compare both ansatz to the exact ground state of $\rm H_{Bilayer}$ on an equal footing. After obtaining the exact ground state we ask ourselves two questions: (1) what is the physics?, and (2) will the system display the FQHE? To answer the first we numerically calculate the wave function overlap between the exact ground state and the MR Pfaffian or 331 state, then we operationally define the system to be in the MR Pfaffian or 331 "phase" depending upon which overlap is larger. To answer the second question, we numerically obtain the excitation gap (provided the ground state is a uniform state with total angular momentum $L = 0$) and recall that in order for a system to display the FQHE it must have a non-zero energy gap.

Figure 4 shows the FQHE energy gap as a function of layer separation d and tunneling strength t for single-layer widths of $w = 0$ for the lowest and second LLs [3]. Superimposed on these plots is a dashed line that represents the "phase" boundary between the 331 or MR Pfaffian phases (the MR Pfaffian phase is labeled Pf on these figures) where this boundary is defined above. Regions of parameter space with a sizable FQHE energy gap are expected to display the FQHE whereas regions with small or vanishing gaps will not.

What is immediately clear from these results is that the FQHE energy gap as a function of d and t is qualitatively different in the SLL compared to the LLL. Specifically, in the LLL there is a ridge of large energy gap that is greatest for weak tunneling and intermediate layer separation, i.e., when the system is largely two-component. This region of large FQHE energy gap is described well by Halperin's abelian 331 state. Furthermore the ridge of positive gap monotonically increases curving upward in a parabolic fashion and, notably, is always on the 331 side of the "phase" boundary line. The symbols superimposed on the LLL FQHE energy gap figures correspond to different bilayer experiments, see caption of figure 4. The groups of symbols with circles around them correspond to experiments where the FQHE at $\nu_{\rm total} = 1/2$ was observed whereas in the experiments corresponding to the non-circled symbols no FQHE was observed. While there is some ambiguity in assigning an experiment to a single point in the d-t phase space it is qualitatively clear that the FQHE observed in bilayer systems at $\nu_{\rm total} = 1/2$ are described by the abelian 331 state as opposed to the non-abelian MR Pfaffian state.

In the SLL the result is qualitatively different from the LLL. Here we see *two* ridges of sizable FQHE energy gap as a function of d and t. One ridge is qualitatively similar to the ridge observed in the LLL (where the system is largely two-component) and on the 331 side of the "phase" boundary, while the other ridge is in the small layer separation and increasing tunneling region, i.e., where the system is largely one-component, and on the MR Pfaffian side of the "phase" boundary. Hence, a bilayer system with $\nu_{\rm total} = 5/2$ has a much richer quantum phase diagram than the LLL bilayer with $\nu_{\rm total} = 1/2$. In particular, it appears possible to experimentally prepare a system with sizable tunneling t and small layer separation d that displays the FQHE at $\nu_{\rm total} = 5/2$ described by the non-abelian MR Pfaffian state and tune experimental parameters in such a way to drive a quantum phase transition into an abelian FQHE described by Halperin's 331 state. While it is difficult to definitively determine, via finite

[3] Three other single-layer widths w are shown in reference [26] but, qualitatively, they are similar so we do not show them here.

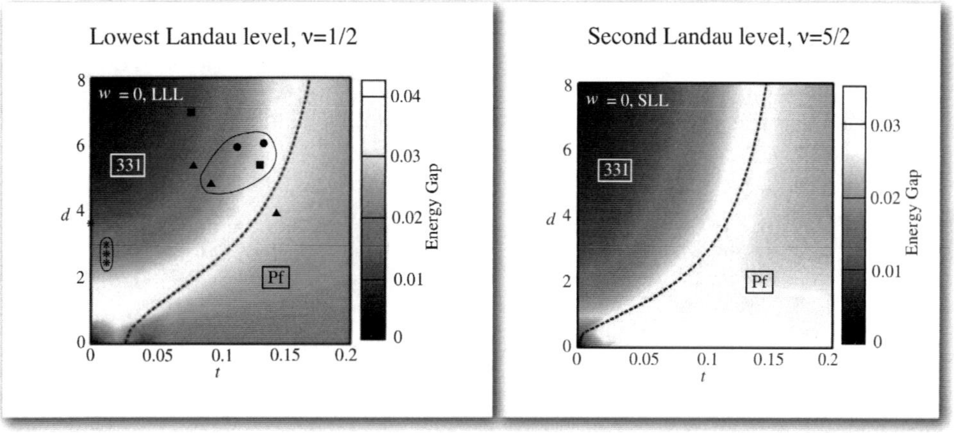

Figure 4. The quantum "phase" diagram and FQHE energy gap for bilayer systems with $\nu_{\text{total}} = 1/2$ and $5/2$, left and right panels, respectively. The 331 and MR Pfaffian phases are separated by a black-dashed line and defined by whichever overlap with the exact ground state is larger. The color contour plot is the FQHE energy gap with the light/dark color being a large/small gap. The square, circle, triangle, and asterisk symbols superimposed on the LLL FQHE energy gap figures correspond to different experiments, namely, square [58], circle [59], triangle [62], and asterisk [61]

size exact diagonalization, whether there would be a quantum phase transition or crossover going from the non-abelian phase to the abelian phase, our results are consistent with those of Read and Green [18] who obtained a (topologically) similar quantum phase diagram. Theoretically, the FQHE energy gap necessarily collapses across the quantum phase boundary since the two states, 331 and MR Pfaffian, are topologically distinct. Thus, we suspect a true quantum phase transition to occur.

In reference [27], we also investigated the possible quantum phase transition between the abelian 331 and non-abelian MR Pfaffian state in the torus geometry. By looking for the topological ground state degeneracy signature of the two states we corroborated the results presented above for the spherical geometry.

Before concluding, we mention that in reference [28] Scarola *et al.* looked at the bilayer problem under an electrostatic quantum-well tilt [60]. In these tilted samples, there is substantial subband mixing and it was found [28] that a generalization of the Halperin 331 state due to Scarola and Jain [70] provide a number of partial subband polarized abelian FQHE states for even-denominator filling factors $\nu = 1/2$ and $1/4$.

Hence, bilayer FQHE systems provide a wide variety of possibilities for observing new FQHE states as well as shedding light on the nature of the FQHE at $\nu = 5/2$.

4. Conclusions

We have provided a brief review of some theoretical searches for non-abelian anyons in the FQHE using numerical techniques. Specifically, we reported on results that showed that the finite-thickness of realistic quasi-two-dimensional systems stabilizes the MR Pfaffian non-abelian description of the FQHE at $\nu = 5/2$ and suggested the possibility of engineering wavefunctions that would be more likely to experimentally display a non-abelian FQHE. This result was bolstered by the fact that it was achieved using the complimentary spherical and torus geometries and was largely independent of the finite-thickness model implemented. Lastly, we reported on the use of bilayer systems in the second Landau level as alternative experimental platforms to investigate the possible quantum phase transition between abelian and non-abelian FQHE states. Needless to say, definitive experimental observation of non-abelian anyons in the FQHE would be an extremely exciting and important discovery.

Acknowledgments

I acknowledge my collaborators in this work: S. Das Sarma, Th. Jolicoeur, K. Park, J. Biddle, Z. Papić, V. Scarola, C. May, and M. Troyer. I also thank DARPA QuEST and Microsoft Station Q for funding as well as California State University Long Beach Start-up Funds.

References

[1] Tsui D C, Stormer H L and Gossard A C 1982 *Phys. Rev. Lett.* **48** 1559
[2] Das Sarma S, Freedman M and Nayak C 2005 *Phys. Rev. Lett.* **94** 166802
[3] Nayak C, Simon S H, Stern A, Freedman M and Das Sarma S 2008 *Rev. Mod. Phys.* **80** 1083
[4] Moore G and Read N 1991 *Nucl. Phys. B* **360** 362
[5] Willett R, Eisenstein J P, Stormer H L, Tsui D C, Gossard A C and English J H 1987 *Phys. Rev. Lett.* **59** 1776
[6] Nayak C and Wilczek F 1996 *Nucl. Phys. B* **479** 529
[7] Bonderson P, Gurarie V and Nayak C 2011 *Phys. Rev. B* **83** 075303
[8] Kitaev A Y 2003 *Ann. Phys. (N.Y.)* **303** 2
[9] Freedman M H 1998 *Proc. Natl. Acad. Sci. U.S.A.* **95** 98
[10] Morf R H 1998 *Phys. Rev. Lett.* **80** 1505
[11] Feiguin A E, Rezayi E, Yang K, Nayak C and Das Sarma S 2009 *Phys. Rev. B* **79** 115322
[12] Laughlin R B 1983 *Phys. Rev. Lett.* **50** 1395
[13] Halperin B I 1984 *Phys. Rev. Lett.* **52** 1583
[14] Jain J K 1989 *Phys. Rev. Lett.* **63** 199
[15] Das Sarma S and Pinczuk A (eds) 1997 *Perspectives in quantum Hall effects : novel quantum liquids in low-dimensional semiconductor structures* (New York: Wiley)
[16] Jain J 2007 *Composite fermions* (Cambridge University Press)
[17] Greiter M, Wen X G and Wilczek F 1991 *Phys. Rev. Lett.* **66** 3205
[18] Read N and Green D 2000 *Phys. Rev. B* **61** 10267
[19] Rezayi E H and Haldane F D M 2000 *Phys. Rev. Lett.* **84** 4685
[20] Storni M, Morf R H and Das Sarma S 2010 *Physical Review Letters* **104** 076803
[21] Lu H, Das Sarma S and Park K 2010 *Phys. Rev. B* **82**(20) 201303
[22] Peterson M R, Jolicoeur T and Das Sarma S 2008 *Phys. Rev. Lett.* **101** 016807
[23] Peterson M R, Jolicoeur T and Das Sarma S 2008 *Phys. Rev. B* **78** 155308
[24] Biddle J, Peterson M R and Das Sarma S 2011 *Phys. Rev. B* **84**(12) 125141
[25] Peterson M R, Park K and Das Sarma S 2008 *Phys. Rev. Lett.* **101** 156803
[26] Peterson M R and Das Sarma S 2010 *Phys. Rev. B* **81** 165304
[27] Peterson M R, Papić Z and Das Sarma S 2010 *Phys. Rev. B* **82** 235312
[28] Scarola V W, May C, Peterson M R and Troyer M 2010 *Phys. Rev. B.* **82** 121304
[29] Das Sarma S and Mason B A 1985 *Ann. Phys. (N.Y.)* **163** 78
[30] Ando T, Fowler A B and F S 1982 *Rev. Mod. Phys.* **54** 437
[31] Stern F and Das Sarma S 1984 *Phys. Rev. B* **30** 840
[32] Zhang F and Das Sarma S 1986 *Phys.Rev. B* **33** 2903
[33] Ortalano M W, He S and Das Sarma S 1997 *Phys. Rev. B* **55** 7702
[34] Wen X G and Zee A 1992 *Phys. Rev. B* **46** 2290
[35] Wen X G and Niu Q 1990 *Phys. Rev. B* **41** 9377
[36] Pan W, Xia J S, Stormer H L, Tsui D C, Vicente C, Adams E D, Sullivan N S, Pfeiffer L N, Baldwin K W and West K W 2008 *Phys. Rev. B* **77**(7) 075307
[37] Choi H C, Kang W, Das Sarma S, Pfeiffer L N and West K W 2008 *Phys. Rev. B* **77** 081301
[38] Dean C R, Piot B A, Hayden P, Das Sarma S, Gervais G, Pfeiffer L N and West K W 2008 *Phys. Rev. Lett.* **100**(14) 146803
[39] Haldane F D M, Rezayi E H and Yang K 2000 *Phys. Rev. Lett.* **85** 5396
[40] Halperin B I 1983 *Helv. Phys. Acta* **56** 75

[41] Lee S S, Ryu S, Nayak C and Fisher M P A 2007 *Phys. Rev. Lett.* **99** 236807
[42] Levin M, Halperin B I and Rosenow B 2007 *Phys. Rev. Lett.* **99** 236806
[43] Ho T L 1995 *Phys. Rev. Lett.* **75** 1186
[44] Read N and Rezayi E 1996 *Phys. Rev. B* **54** 16864
[45] Bishara W and Nayak C 2009 *Phys. Rev. B* **80** 121302
[46] Wójs A, Tőke C and Jain J K 2010 *Phys. Rev. Lett.* **105** 096802
[47] Rezayi E H and Simon S H 2011 *Phys. Rev. Lett.* **106** 116801
[48] Zozulya O S, Haque M, Schoutens K and Rezayi E H 2007 *Phys. Rev. B* **76**(12) 125310
[49] Li H and Haldane F D M 2008 *Phys. Rev. Lett.* **101**(1) 010504
[50] Dolev M, Gross Y, Chung Y C, Heiblum M, Umansky V and Mahalu D 2010 *Phys. Rev. B* **81** 161303
[51] Dolev M, Heiblum M and Mahalu D 2008 *Nature* **452** 829
[52] Radu I P, Miller J B, Marcus C M, Kastner M A, Pfeiffer L N and West K W 2008 *Science* **320** 899
[53] Venkatachalam V, Yacoby A, Pfeiffer L and West K 2011 *Nature* **469** 185
[54] Willett R L, Pfeiffer L N and West K W 2009 *Proc. Natl. Acad. Sci. U.S.A.* **106** 8853
[55] Bid A, Ofek N, Inoue H, Heiblum M, Kane C L, Umansky V and Mahalu D 2010 *Nature* **466** 585
[56] Rhone T D, Yan J, Gallais Y, Pinczuk A, Pfeiffer L and West K 2011 *Phys. Rev. Lett.* **106** 196805
[57] Stern M, Plochocka P, Umansky V, Maude D K, Potemski M and Bar-Joseph I 2010 *Phys. Rev. Lett.* **105** 096801
[58] Luhman D R, Pan W, Tsui D C, Pfeiffer L N, Baldwin K W and West K W 2008 *Phys. Rev. Lett.* **101** 266804
[59] Shabani J, Gokmen T and Shayegan M 2009 *Phys. Rev. Lett.* **103** 046805
[60] Shabani J, Gokmen T, Chiu Y T and Shayegan M 2009 *Phys. Rev. Lett.* **103** 256802
[61] Eisenstein J P, Boebinger G S, Pfeiffer L N, West K W and He S 1992 *Phys. Rev. Lett.* **68** 1383
[62] Suen Y W, Engel L W, Santos M B, Shayegan M and Tsui D C 1992 *Phys. Rev. Lett.* **68** 1379
[63] Suen Y W, Santos M B and Shayegan M 1992 *Phys. Rev. Lett.* **69** 3551
[64] Suen Y W, Manoharan H C, Ying X, B S M and Shayegan M 1994 *Phys. Rev. Lett.* **72** 3405
[65] Halperin B I, Lee P A and Read N 1993 *Phys. Rev. B* **47** 7312
[66] Rezayi E and Read N 1994 *Phys. Rev. Lett.* **72** 900
[67] He S, Xie X C, Das Sarma S and Zhang F C 1991 *Phys. Rev. B* **43** 9339
[68] He S, Das Sarma S and Xie X C 1993 *Phys.. Rev. B* **47** 4394
[69] Nomura K and Yoshioka D 2004 *J. Phys. Soc. Jpn.* **73** 2612
[70] Scarola V W and Jain J K 2001 *Phys. Rev. B* **64**(8) 085313

Spin waves in the classical Heisenberg antiferromagnet on the kagome lattice

Stefan Schnabel and David P. Landau

Department of Physics and Astronomy, University of Georgia, Athens, GA 30602, USA

E-mail: stefanschnabel@physast.uga.edu

Abstract. We investigate spin waves in the classical Heisenberg antiferromagnet on the kagome lattice by means of Monte Carlo and spin dynamics simulations. Using sophisticated sampling algorithms we were able to establish thermal equilibrium at temperatures as low as $T = 10^{-6}|J|/k_{\mathrm{B}}$, and from the dynamic structure factor we could precisely determine frequencies of spin waves. Results agree well with theoretical predictions for acoustic modes. Measurements in different coordinate systems as well as theoretical considerations showed that optical modes, which were apparently found in a recent study [1], are in fact artifacts and do not indicate the existence of additional spin wave modes.

1. Introduction

Although geometrically frustrated magnets are not easily investigated, they are subject of strong interest since their behavior is often both surprising and counter-intuitive. In these systems the spatial arrangement of spins prevents the simultaneous minimization of all interactions which usually leads to a degenerate ground state and/or glass-like behavior at low temperature. One example is the Heisenberg kagome antiferromagnet which has been investigated in a number of studies [2–5]. The classical Hamiltonian for Heisenberg model is given by

$$\mathcal{H} = -J\sum_{\langle ij \rangle} \mathbf{s}_i \cdot \mathbf{s}_j, \tag{1}$$

where the sum goes over nearest neighbor pairs $\langle ij \rangle$. The spins $\mathbf{s}_i$ are three-dimensional vectors of length unity and the interaction constant is negative ($J < 0$), i.e., two interacting spins have minimal energy if they are pointing in opposite directions. The lattice (Fig. 1) includes hexagons and triangles, which causes frustration, since antiparallel alignment of all spin pairs of a triangle is impossible. The ground state is highly degenerate without long-range order. However, it has been found that at low, but finite, temperatures ($T < 5 \times 10^{-3}|J|/k_{\mathrm{B}}$) in a process called 'order by disorder' [6], coplanar configurations are selected due to their higher entropy. It has been shown analytically that this state gives rise to a degenerate acoustic branch and a dispersionless branch of soft modes. A recent numerical study [1] also seemed to suggest that a branch of nearly dispersionless optical modes exists as well.

The paper is organized as follows: In the next section we will give a short overview of the low temperature state of the system, which is followed in section 2 by a description of the methods we used. In section 3 and 4 we present results of Monte Carlo and spin dynamics simulations

Published under licence by IOP Publishing Ltd

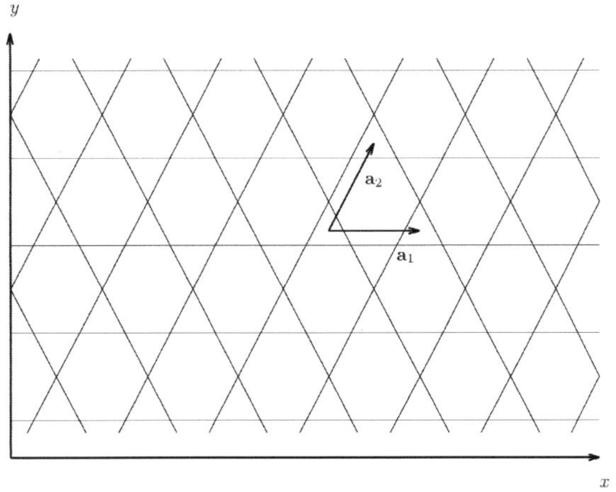

Figure 1. The kagome lattice with lattice vectors $\mathbf{a}_1$ and $\mathbf{a}_2$.

and in section 4 we will discuss how the apparent conflict between theory and simulation is resolved.

2. Coplanar state

It is easy to see that the energy of one spin triangle is minimal if the sum of the three spins is zero, i.e. if the three spins lie within one plane and pairwise include angles of $2\pi/3$. Applying this condition to every triangle in the lattice still leaves the system under-determined, hence, the ground state is highly degenerate. However, it turns out that the harmonic constraints for certain degrees of freedom vanish if all spins lie within the same plane. These degrees of freedom then provide additional 'soft' modes which strongly increase the system entropy, and such spin configurations are, therefore, favored at low finite temperatures. We will call these configurations 'coplanar' and the macrostate comprising them the 'coplanar state'. Since the Hamiltonian is symmetric under arbitrary rotations we are free to orient the system as necessary, and in the following we assume that the spin plane coincides with the xy-plane. In the coplanar state each spin fluctuates around one of only three possible basic directions [Fig.2(a)] and labels $\{\sigma\}$ can be assigned to each spin such that $\sigma_i = 1, 2, 3$ if $\mathbf{s}_i$ points in the first, second or third direction. For neighboring spins $\mathbf{s}_i, \mathbf{s}_j$ which always point in different directions $\sigma_i \neq \sigma_j$. This means that $\{\sigma\}$ is also a ground state configuration of the three-state Potts model, although, in contrast to that model, due to the entropic effects the probability distribution over all $\{\sigma\}$ that fulfill this condition is non-uniform [2].

In order to understand these effects it is helpful to consider the so-called Weathervane loops. These are created by connecting adjacent spins which belong to two of the three different types [Fig.2(b)], i.e. neighboring spins $\mathbf{s}_i, \mathbf{s}_j$ are connected if $\sigma_i \neq \nu$ and $\sigma_j \neq \nu$, with $\nu = 1, 2$ or 3. Each loop is surrounded only by spins pointing in the third direction ν; and its own spins can, therefore, rotate collectively around this direction with low energy cost. In consequence, fluctuations in these rotational degrees of freedom are large increasing the entropy of the coplanar state. It also explains the deviation from the ground state of the three state Potts model: Since additional loops provide additional entropy there is a bias towards configurations $\{\sigma\}$ with many loops. On the other hand, there are few configurations with high loop numbers. For example the so-called $\sqrt{3}\times\sqrt{3}$ state (Fig. 3) providing the maximum number of loops - one on every hexagon - is only six-fold degenerate. This means that there is a counter effect reducing entropy with

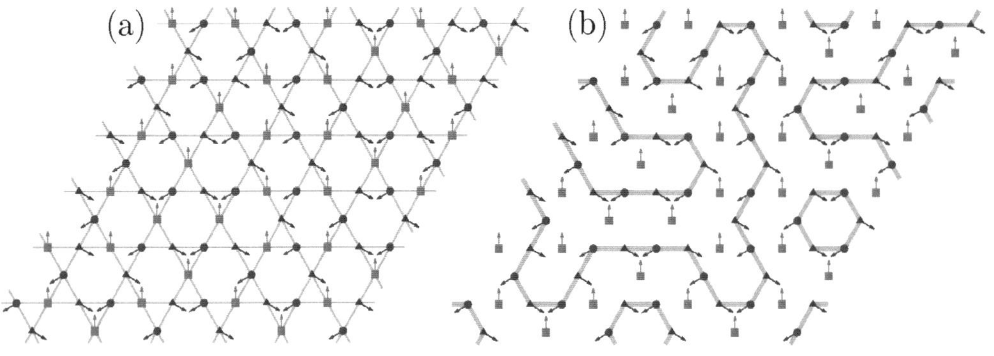

Figure 2. (Color online) (a) A typical configuration of the coplanar state. The three different colors and symbols have no physical meaning, but indicate spins which fluctuate around the same basic direction. (b) Weathervane loops connect spins which point in two of the three basic directions and are separated by spins pointing in the third direction.

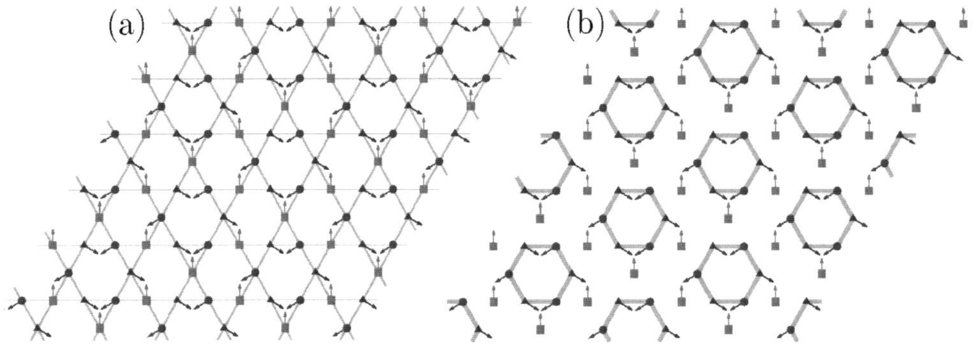

Figure 3. (Color online) (a) The $\sqrt{3} \times \sqrt{3}$ state possesses a periodic structure. (b) It provides a maximum number of Weathervane loops. Note that only one third of all loops is displayed.

increasing loop number and in consequence, the system neither shows the uniform probability distribution of the Potts ground-state nor does it solely occupy the $\sqrt{3} \times \sqrt{3}$ configurations.

3. Method

We investigated systems of different sizes, although most of the data presented in the following were obtained from simulations of a system containing $N = 3888$ spins with periodic boundary conditions.

In order to perform spin dynamic simulations at low temperatures we first have to perform a Monte Carlo simulation that produces equilibrium configurations at that temperature. Since ergodicity is surely broken for Metropolis [7] simulations at these temperatures, the application of a generalized ensemble method is obligatory. For the classical Heisenberg antiferromagnet on the kagome lattice there is no phase transition involving latent heat [8] and it is not necessary to overcome barriers in energy. This gives us the freedom to rely on methods like simulated tempering [9] or parallel tempering [10] which are based on canonical ensembles. Canonical

Figure 4. (Color online) At low temperature larger changes to the configuration can still be achieved when using the loop-flip update which flips all spins in one Weathervane loop.

ensembles on the other hand allow for the application of the Heatbath algorithm [11, 12] for which single spin updates are always accepted. As a consequence, we do not have to worry about cone opening angles or acceptance rates at low T. During the simulated or parallel tempering simulation the system is required to change its temperature frequently. These steps in temperature space can only be performed if there is a large enough overlap between the energy distributions of neighboring temperatures. From the previous knowledge of the specific heat [8] we can estimate the width of these distributions and we find that we need a few thousand canonical ensembles in order to cover many orders of magnitude in temperature. To keep the simulation simple and stable we chose the simulated tempering technique which requires a preliminary iteration for weight determination. The algorithm can, however, be performed entirely on a single processor core. We used 10000 temperatures such that $-6 \leq \log_{10}(T_i k_B/|J|) \leq 3$ with constant intervals in the logarithmic temperature scale.

Although spin-freezing occurs only at extremely low temperatures (we still observe sporadic spin-flips at $T < 10^{-5}|J|/k_B$) the dynamics within the coplanar state is slow, especially changes in $\{\sigma\}$ are rare. In order to reduce autocorrelation times we exploited the specific properties of the coplanar state to design a novel Monte Carlo move. In this update all spins of a particular Weathervane loop are flipped such that their directions σ are exchanged (Fig. 4). It turned out that simple π rotations around the direction of the neighboring spins result in low acceptance rates. Spins pointing above the spin plane become spins pointing below the plane and vice versa, which distorts the second loop to which each spin belongs. Instead the spins of the loop have to be reflected in a mirror plane that contains the neighboring spins and is simultaneously perpendicular to the spin plane. We find that such an update is accepted with a probability of roughly 10% independently of temperature. Thereby, most of the flipped loops are small and involve less than 30 spins; but sometimes loops of more than 100 spins are flipped as well.

In order to capture the dynamics of the system we integrated the coupled equations of motion for multiple configuration using a 4th order Adams-Bashforth predictor- Adams-Moulton corrector calculations [13] which were initialized with the Runge-Kutta method. The derived trajectories $\{\mathbf{s}(t)\}$ were used to obtain the space-time correlation function:

$$C(\mathbf{r}, t) = \langle \mathbf{s}(\mathbf{r}', 0) \cdot \mathbf{s}(\mathbf{r}' + \mathbf{r}, t) \rangle_{\mathbf{r}'}, \tag{2}$$

where $\mathbf{r}'$ goes over the three points in one unit cell, i.e. one triangle of the lattice. A Fourier transform provides the dynamic structure factor

$$S(\mathbf{k}, \omega) = \sum_{\mathbf{r}} \int dt\, C(\mathbf{r}, t) e^{-i\mathbf{r}\cdot\mathbf{k}} e^{-i\omega t}, \tag{3}$$

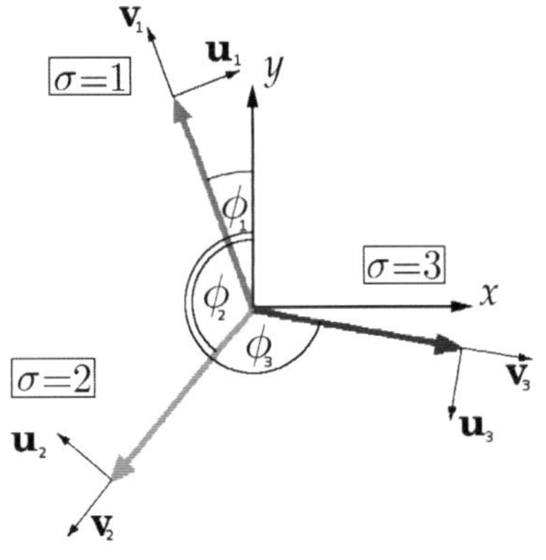

Figure 5. (Color online) Spins fluctuate around three basic directions (arrows). Angles ϕ_σ are used to define alternative in-plane coordinate systems u_σ, v_σ (4), which allow to measure transversal and longitudinal fluctuations separately.

with the wave vector $\mathbf{k}$ and frequency ω. In the context of neutron scattering experiments $S(\mathbf{k}, \omega)$ corresponds to the scattering function, k to the momentum transfer, and ω to energy.

In the transition to the coplanar state rotational symmetry is broken, hence, we cannot expect the in-plane components $\{s^x(t)\}$, $\{s^y(t)\}$ and the out-of plane components $\{s^z(t)\}$ to show the same collective excitations. We, therefore, also calculate correlations C^x, C^y, C^z and dynamic structure factors S^x, S^y, S^z for single components only.

Considering the unique geometry of the coplanar state it is apparent that the system is poorly described within a single Cartesian coordinate system. While the out-of-plane excitation can well be measured in the z-direction, the in-plane behavior is inadequately reflected in the x and y-components (Fig. 5). If, for instance, an average is taken over spins which rotate collectively around the z-axis, the different contributions will add up to zero if an equal number of spins belongs to each of the three basic spin directions. It is, therefore, helpful to introduce in-plane coordinates that are parallel (s^u) and perpendicular (s^v) to the respective basic directions similarly to the coordinates used in [5]:

$$\begin{pmatrix} s_i^u \\ s_i^v \\ s_i^z \end{pmatrix} = \begin{pmatrix} \cos\phi_{\sigma_i} & \sin\phi_{\sigma_i} & 0 \\ -\sin\phi_{\sigma_i} & \cos\phi_{\sigma_i} & 0 \\ 0 & 0 & 1 \end{pmatrix} \begin{pmatrix} s_i^x \\ s_i^y \\ s_i^z \end{pmatrix} - \begin{pmatrix} 0 \\ 1 \\ 0 \end{pmatrix}, \qquad (4)$$

where ϕ_{σ_i} denotes the angle between the y-axis and the basic direction of the ith spin, i.e. its equilibrium direction. These coordinates have two advantages. Once the spin-plane is aligned with the xy-plane (i.e. uv-planes) further orientation is not necessary, which means that ϕ_{σ_i} don't have to be adjusted, and the system is at minimal energy $E_{\min} = -N|J|$ if all $(s^u, s^v, s^z)^T = \mathbf{0}$. We also calculated correlations and dynamic structure factors S^u, S^v for these coordinates, although there is no direct link to experimentally accessible quantities.

4. Monte Carlo results

During the simulated tempering simulations we recorded the energy distribution and derived the density of states. Since the simulation covers nine orders of magnitude in temperature and since the investigated systems contain thousands of degrees of freedom, the density of states extends over ten-thousands of orders of magnitude. In return we use the density of states to calculate the specific heat (Fig. 6) and obtain a relatively precise result considering the size of the system or the large temperature interval. In the high temperature region the system is driven by entropy, changes in temperature have little influence on the internal energy and the specific heat is, therefore, small. In the intermediate region spins on each triangle are correlated and energy is minimized, but the overall symmetry remains unbroken. At very low temperatures we observe a good agreement with the theoretical prediction: $\lim_{T\to 0} C(T)/N = 11/12$ [4] which indicates the existence of soft modes and proves that the system has settled in the coplanar state. It should be mentioned here that methods like simulated tempering allow a more direct access to thermodynamic quantities at the temperatures used in the simulation. The approach based on the density of states, however, provides data for all intermediate temperatures as well, while at the same time using all existing data for each data point. It is, therefore, more flexible and more precise.

To observe the formation of the coplanar state we also determined averages of absolute in-plane $\langle |s^u| \rangle$ and out-of-plane $\langle |s^z| \rangle$ components as functions of temperature (Fig. 6). It should be noted here that both are defined only in the oriented system, i.e. within a coordinate system with the xy-plane identical to the spin plane. Of course the system can only be oriented when in the coplanar state, since there is no spin plane otherwise, i.e., although there are finite values of $\langle |s^u| \rangle$ and $\langle |s^z| \rangle$ for $T > 10^{-2}|J|/k_B$ these have no physical meaning, because the coplanar state does not exist at these temperatures. We find that in the coplanar state $\langle |s^u| \rangle$ and $\langle |s^z| \rangle$ show a power law behavior and, as expected, the in-plane excitations are much smaller and decrease faster than the out-of-plane excitations. This is not surprising, since the latter include the rotations of the Weathervane loops which are comparatively weakly constrained.

5. Spin dynamics results

In this section we will present data for structure factors $S(\mathbf{k}_l, \omega)$, with $\mathbf{k}_l = l\mathbf{k}_{11}$, where the vector $\mathbf{k}_{11}$ is perpendicular to bonds of the lattice, i.e. $\mathbf{k}_{11} \parallel (\mathbf{a}_1 + \mathbf{a}_2)$ or $\mathbf{k}_{11} \parallel (2\mathbf{a}_1 - \mathbf{a}_2)$ or $\mathbf{k}_{11} \parallel (-\mathbf{a}_1 + 2\mathbf{a}_2)$. The length of $\mathbf{k}_{11}$ is chosen such that $l = 1$ at the edge of the Brillouin zone. From the analytical work [5] the dispersion relation for the acoustic branch is known, and we expect maxima in $S(\mathbf{k}_l, \omega)$ for

$$\omega = 2\sin(\pi l). \tag{5}$$

All measurements were taken at $T = 10^{-6}|J|/k_B$.

The results agree well with the aforementioned numerical study [1]. We, too, observe the apparent, novel optical mode (Fig. 7) and the acoustic branch. But, additionally we find that the acoustic branch only appears in the out-of-plane component S^z, while the 'optical branch' manifests only in the in-plane component of the dynamic structure factor $S^{xy} = S^x + S^y$ together with a number of dispersionless signals which form a washboard-like pattern. Furthermore, if we use the alternative coordinates u, v (Fig. 8), we observe the acoustic branch in S^u while no maxima are obtained in S^v. This is consistent with the picture of spins precessing collectively around their equilibrium positions - the respective basic direction - while maintaining a constant angle with respect to it. We conclude that the features of S^{xy} are either artifacts that arise in the transition from s^u, s^v to s^x, s^y or, less likely, different unpredicted modes that leave no sign in S^{uv} while at the same time the acoustic modes are invisible in S^{xy}.

6. Optical illusion

In this section we briefly present results of considerations that explain the features of S^{xy}. Detailed calculations will soon be published elsewhere.

To understand which signals the acoustic modes cause in the in-plane dynamic structure factor S^{xy} we calculated how these modes shape the correlation function. We could show that for the $\sqrt{3} \times \sqrt{3}$ state the correlations in $\{\sigma\}$ cause a shift in the Brillouin zone such that

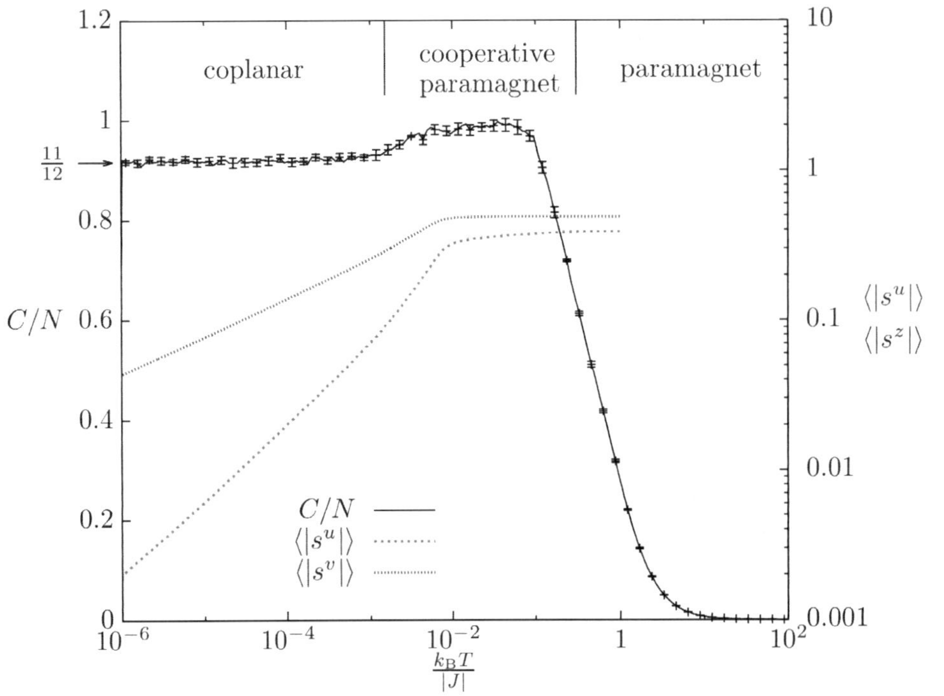

Figure 6. (Color online) Specific heat and averaged absolute deviations from ground state.

signals found in $S^{xy}(\mathbf{k}, \omega)$ belong to modes with wave vectors $\mathbf{k} \pm \pi \mathbf{a}_1 / 3|\mathbf{a}_1|^2$. Since we know the dispersion relation for the acoustic branch for the entire zone from analytical studies [5] we could calculate the resulting positions of maxima in $S^{xy}(\mathbf{k}_l, \omega)$:

$$\omega = \sqrt{\frac{7}{2} + \cos(2\pi l)}, \tag{6}$$

which is in good agreement with the 'optical' branch in Fig. 7.

The coplanar state, however, differs from the $\sqrt{3} \times \sqrt{3}$ state. The correlations are basically the $\sqrt{3} \times \sqrt{3}$ - correlations enveloped in a power law, which causes additional signals. The Fourier transform of the central peak leads to contributions of each mode in $S^u(\mathbf{k}, \omega)$ to $S^{xy}(\mathbf{k}', \omega)$ at each point $\mathbf{k}'$. In consequence, regardless of wave vectors the frequency of every mode of the system can in principle be measured everywhere leading to the dispersionless wash-board-like pattern in $S^{xy}(\mathbf{k}, \omega)$. In the thermodynamic limit the short-range $\sqrt{3} \times \sqrt{3}$ correlations will have no important influence and no maxima related to acoustic modes will be observed in S^{xy}. Next-nearest neighbor interactions on the other hand can stabilize the $\sqrt{3} \times \sqrt{3}$ structure [5] and in that case we expect the described effects to be detectable in experiments.

7. Conclusions

Combining various advanced simulation techniques we could sample the kagome Heisenberg antiferromagnet at temperatures much lower than previously reached. Thermodynamic and structural quantities were in good agreement with previous analytical and numerical studies.

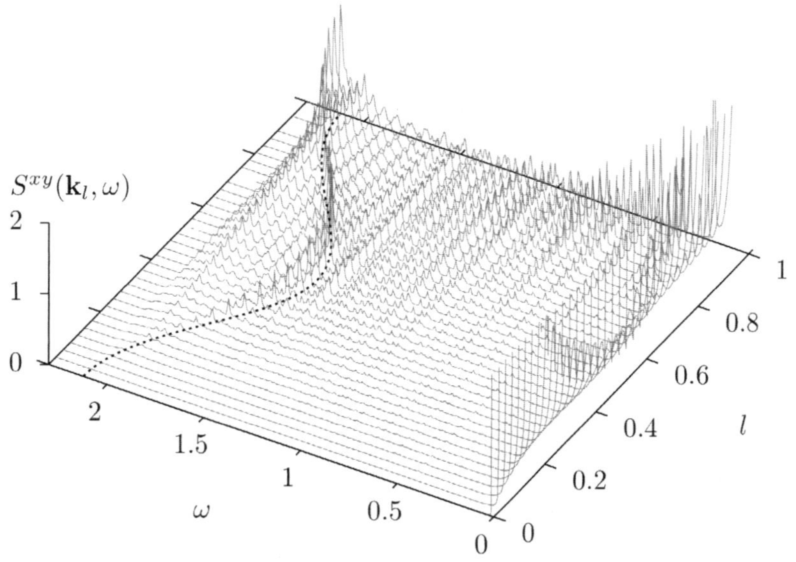

Figure 7. (Color online) Rescaled in-plane component of the dynamic structure factor. The wave vector $\mathbf{k}_l = l\mathbf{k}_{11}$ with $l = 1$ at the end of the Brillouin zone. The dispersion relation, Eqn.(6), is drawn with a dotted line.

Spin dynamic simulations were used to obtain dynamic structure factors with a very good signal-noise ratio due to the low temperature. We found that appropriate coordinates must be chosen to directly observe the predicted acoustic modes within the spin plane. Analytical calculations showed that signals found when using global coordinates do not belong to a novel optical mode, but origin from a different part of the Brillouin zone and are displaced due to the special structure of the coplanar state.

8. Acknowledgements

We thank Shan-Ho Tsai for helpful discussions. This work was funded by the NSF under grant no. DMR-08/0223.

References

[1] Robert J, Canals B, Simonet V and Ballou R 2008 *Phys. Rev. Lett.* **101** 117207
[2] Huse D A 1992 *Phys. Rev. B* **45** 7536
[3] Zeng C and Elser V 1990 *Phys. Rev. B* **42** 8436
[4] Chalker J T, Holdsworth P C W and Shender E F 1992 *Phys. Rev. Lett.* **68** 855
[5] Harris A B, Kallin C and Berlinsky A J 1992 *Phys. Rev. B* **45** 2899
[6] Reimers J N and Berlinsky A J 1993 *Phys. Rev. B* **48** 9539
[7] Metropolis N, Rosenbluth A W, Rosenbluth M N, Teller A H and Teller E 1953 *J. Chem. Phys.* **21** 1087
[8] Zhitomirsky M E 2008 *Phys. Rev. B* **78** 094423
[9] Marinari E and Parisi G 1992 *Europhys. Lett.* **19** 451
[10] Swendsen R H and Wang J S 1986 *Phys. Rev. Lett.* **57** 2607

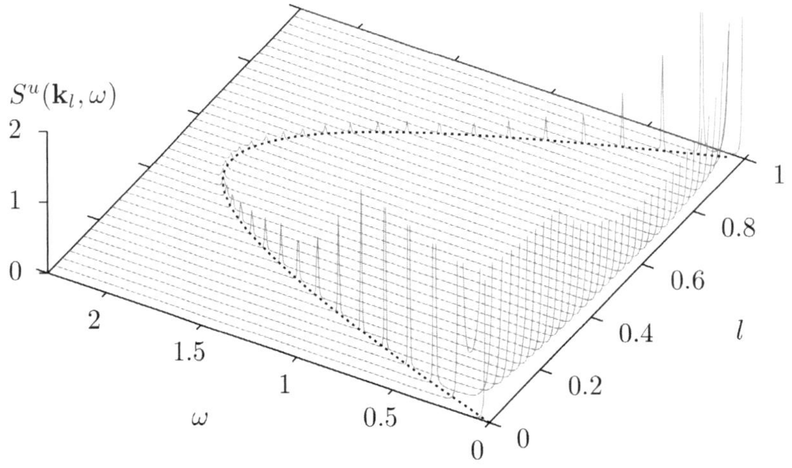

Figure 8. (Color online) Rescaled in-plane component of the local coordinate structure factor. Signals are one hundred times larger compared to Fig. 7. The dispersion relation is given by Eqn.(5).

[11] Landau D P and Binder K 2005 *A Guide to Monte Carlo Simulation in Statistical Physics* (Cambridge: Cambridge University Press)
[12] Miyatake Y, Yamamoto M, Kim J J and Nagai O 1986 *J. Phys. C* **19** 2539
[13] Burden R L, Faires J D and Reynolds A C 1981 *Numerical Analysis* (Boston: Prindle, Weber & Schmidt) pp 205,219

On Simulating Type Ia Supernovae

A. C. Calder[1,2], B. K. Krueger[1], A. P. Jackson[1,3], D. M. Townsley[4], E. F. Brown[5,6], and F. X. Timmes[7,6]

[1] Department of Physics and Astronomy, Stony Brook University, Stony Brook, NY 11794-3800
[2] New York Center for Computational Sciences, Stony Brook University, Stony Brook, NY, 11794-3800
[3] present address: Laboratory for Computational Physics & Fluid Dynamics, Naval Research Laboratory, Washington, DC, 20375
[4] Department of Physics and Astronomy The University of Alabama, Tuscaloosa, AL, 35487-0324
[5] Department of Physics and Astronomy, Michigan State University, East Lansing, MI 48824-2320
[6] The Joint Institute for Nuclear Astrophysics, Notre Dame, IN 46556-5670
[7] School of Earth and Space Exploration, Arizona State University, Tempe, AZ, 85287-1404

E-mail: acalder@mail.astro.sunysb.edu

Abstract. Type Ia supernovae are bright stellar explosions distinguished by standardizable light curves that allow for their use as distance indicators for cosmological studies. Despite their highly successful use in this capacity, the progenitors of these events are incompletely understood. We describe simulating type Ia supernovae in the paradigm of a thermonuclear runaway occurring in a massive white dwarf star. We describe the multi-scale physical processes that realistic models must incorporate and the numerical models for these that we employ. In particular, we describe a flame-capturing scheme that addresses the problem of turbulent thermonuclear combustion on unresolved scales. We present the results of our study of the systematics of type Ia supernovae including trends in brightness following from properties of the host galaxy that agree with observations. We also present performance results from simulations on leadership-class architectures.

1. Introduction

1.1. Type Ia Supernovae

Type I supernovae are bright stellar explosions characterized by a lack of hydrogen in the observed spectrum. The type Ia sub-classification depends on the observation of a specific silicon line [1, 2]. These events are thought to be the result of a thermonuclear explosion consuming roughly one and a half solar masses of degenerate stellar material composed principally of C and O. The peak brightness of a type Ia supernova is set not by the explosion energy, but by the synthesis in the explosion of radioactive ^{56}Ni, which decays via the chain ^{56}Ni to ^{56}Co to ^{56}Fe releasing the observed energy. The vast majority of type Ia supernovae obey a correlation in which the peak brightness is positively correlated with the timescale over which the lightcurve decays from its maximum. This "brighter is broader" relation is known as the Phillips relation [3] and it allows the peak brightnesses to be calibrated so that these events may be treated as "standard candles" for determining distances. The correlation is understood

Published under licence by IOP Publishing Ltd

physically as stemming from having both the luminosity and opacity being set by the mass of ^{56}Ni synthesized in the explosion [4, 5, 6]. This relation has been exploited to make type Ia supernovae the premier distance indicators for cosmological studies.

While it is accepted that type Ia supernovae result from a thermonuclear runaway as described above, the progenitor systems of these events remain the subject of debate. Motivated largely by interest in cosmology, observational campaigns are gathering information about these events at an unprecedented rate. In particular, observations are finding a surprising range in the variation of the intrinsic brightness of these events, including very bright events that suggest more material burned than could originate from a single white dwarf [7, 8, 9, 10]. Contemporary research explores the efficacy of several progenitor systems for producing events with a range of brightnesses. The Phillips relation explains the first-order variations in peak brightness, but current research also aims to understand higher-order effects on the variation of the light curves and the physics behind the Phillips relation.

The progenitor system we assume is that of a single white dwarf composed principally of carbon and oxygen that has gained mass from a companion star [2]. The process of the white dwarf gaining mass from the companion is known as accretion. As the white dwarf gains mass, it is compressed and the temperature increases, which ignites thermonuclear reactions. Initially the reactions proceed relatively slowly, which drives convection in the core [11, 12]. As the temperature rises, the thermonuclear reactions proceed faster and faster, and eventually a subsonic flame is born that will eventually disrupt the star. Models that incorporate a transition from the subsonic deflagration to a supersonic detonation, allowing some expansion of the star prior to the detonation, best match some spectral features and the abundance stratification observed in the ejecta [13, 14, 15]. Our models assume this deflagration-to-detonation transition (DDT) paradigm.

1.2. Trends in Observations of Type Ia Supernovae

Many contemporary observations address correlations between the event and properties of the host galaxy such as its composition, age, and mass. Of particular interest are correlations between the brightness of an event and the isotopic composition and age, measured by the intensity of star formation, of the host galaxy. The proportion of material that has previously been processed in stars (i.e. elements other than hydrogen and helium, which are collectively referred to as "metals") is a measurable property of the galaxy. The relative abundance of these elements is referred to as the galaxy's "metallicity." The presence of these elements in a progenitor white dwarf influences the outcome of the explosion by changing the path of nuclear burning, which influences the amount of ^{56}Ni synthesized in an event [16]. Because the decay of ^{56}Ni powers the light curve, metallicity can therefore directly influence the brightness of an event. Metallicity can also influence the outcome of an explosion in other ways, including changes in the structure of the white dwarf that result from metallicities influence on stellar evolution, sedimentation within the white dwarf, the nuclear flame speed, and additional sources of opacity [17].

Observational results to date are consistent with a shallow dependence of brightness on metallicity, with dimmer events in metal-rich galaxies, but are unable demonstrate a conclusive trend [18, 19, 20, 21]. Determining the metallicity dependence is challenging because the effect appears to be small, it is difficult to measure, and there are systematic effects associated with the mass-metallicity relationship within galaxies [19]. This effect is also difficult to decouple from the apparently stronger effect of the age of the parent stellar population on the mean brightness of type Ia supernovae [19, 21, 22]. Howell, et al. [21] note that the scatter in brightness of this observed relation is unlikely to be explained by the effect of metallicity.

The composition of a galaxy changes as stars form, consume hydrogen, and produce metals. Newly-formed galaxies are rich in hydrogen gas and undergo a period of intense star formation.

Observations target correlations between the brightness of an event and the age of a galaxy measured by the delay time (the elapsed time from the period of intense star formation). Some results indicate that the dependence of the Ia rate on delay time is best fit by a bimodal distribution with a prompt component less than 1 Gyr after star formation and a tardy component several Gyr later [23, 24]. Other studies only indicate a correlation between the delay time and the brightness of type Ia supernovae, with dimmer events occurring at longer delay times [19, 21, 20, 25].

2. Methodology

Our investigation into type Ia supernovae proceeds by isolating facets of the problem and subjecting these to study via statistically well-controlled ensembles of multidimensional simulations. We are trying to understand the physical mechanism of the explosion and isolating one effect at a time allows us to understand these processes. The parts of the problem we consider are properties of the progenitor white dwarf (such as structure and composition) that follow from properties the host galaxy (such as age and metallicity). The goal is to identify and quantify systematic trends in the brightness of events that follow from properties of the host galaxy. There are many possible systematic effects (outlined in [26]), and for each study we attempted to isolate all but one effect. Eventually, we hope to consider the interdependence of all of these effects in the construction of the full theoretical picture. Another feature of the type Ia supernova problem is that the explosion depends sensitively on the initial conditions, which are largely unknown.

We construct parametrized, hydrostatic massive white dwarf progenitor models with a variety of thermal and compositional structures thought to follow from the properties of the host galaxy and the accretion history [27]. The simulations are performed with a customized version of the FLASH simulation code [28, 29, 30] that includes a model flame and energetics scheme. We perform suites of simulations in which we vary properties of the progenitor white dwarf. The suites of simulations are controlled by a theoretical framework we developed that allows for statically well-controlled studies to determine any systematic effects due to properties of the progenitor that may follow from properties of the host galaxy.

Simulations with FLASH provide the bulk energetics of the explosion and an estimate of the yield of an event. We calculate detailed nucleosynthetic yields from the simulations by post-processing the thermodynamic histories of Lagrangian tracer particles embedded in the flow with a detailed nuclear network [31, 32].

In the subsections below, we highlight the flame model, sub-grid-scale turbulence and turbulence-flame interaction models, and the statistical framework we employ. Complete details of the methodology can be found in previously published results [31, 33, 34, 26, 32].

2.1. Model Flame and Energetics Scheme

The tremendous range in length between the white dwarf radius ($\sim 10^9$ cm) and the laminar nuclear flame width ($\approx [\kappa T/(\rho\varepsilon)]^{1/2} < 1$ cm for typical values of the thermal conductivity κ, density ρ, temperature T, and mass-specific heating rate ε), prohibits direct numerical simulation of type Ia supernovae. The problem is so severe that even an adaptive mesh simulation cannot resolve the actual diffusive flame front. For these reasons, simulations must rely on some sort of "model" flame. The structure of a one-dimensional laminar burning front is reasonably well understood [35], but the interaction of turbulence with this very thin flame front introduces small-scale structure in a way that is not well understood, even for flames in the laboratory. Capturing the net effect of this unresolved flame structure is necessary for a realistic model.

Thermonuclear flames in FLASH are tracked with an advection-diffusion-reaction scheme [36, 37, 34] that propagates an artificially broadened model flame. This model is based on the evolution of a reaction progress variable ϕ, where $\phi = 0$ indicates unburned fuel and $\phi = 1$

indicates burned ash, satisfying the advection-diffusion-reaction equation

$$\partial_t \phi + \boldsymbol{u} \cdot \boldsymbol{\nabla} \phi = \kappa \nabla^2 \phi + \frac{1}{\tau} R(\phi)\,. \tag{1}$$

Here $R(\phi)$ is a single chosen function, used everywhere, and κ and τ are parameters that are tuned locally so that the reaction front propagates at the physical speed of the real flame in that region [35] and is just wide enough to be resolved in our simulation. We use a modified version of the KPP reaction rate discussed by [37], in which $R(\phi) \propto (\phi - \epsilon)(1 - \phi + \epsilon)$, where $\epsilon \simeq 10^{-3}$, which is acoustically quiet and gives a unique flame speed [34].

In addition to the model flame, a realistic supernova model must accurately describe the energy release at the front and in the burned material, which is in a dynamic equilibrium. We performed a detailed study of the nuclear processes occurring in a flame in the interior of a white dwarf and developed an efficient and accurate scheme for numerical simulation [33, 34]. Tracking many nuclear species is computationally prohibitive, so simulations reproduce the energy release with an abstracted model with a tabulated state of the burned material produced from earlier directed simulations. This method accurately captures the thermal history of the material as it burns and evolves, which enables embedded particles to obtain accurate Lagrangian density and temperature histories. Detailed abundances can then be recovered by post-processing these time histories with a nuclear network including hundreds of nuclides. The nuclear processing can be well approximated as a three-stage process: Initially carbon is consumed, followed by oxygen, which creates a mixture of Si group and light elements that is in quasi-statistical equilibrium [38, 39] (also known as nuclear statistical quasi-equilibrium [40, 41, 42]); finally the Si group nuclei are converted to Fe group, reaching full NSE. In both of these equilibrium states, the capture and creation of light elements (via photodisintegration) is balanced, so that energy release can continue by changing the relative abundance of light (low nuclear binding energy) and heavy (high nuclear binding energy) nuclides, an action that releases energy as burned material rises and expands. Neutronziation in the flame and dynamic ash, mainly via electron captures, is explicitly accounted for with tabulated rates from NSE calculations with 443 nuclides that include modern weak reaction rates [43]. Complete details of the NSE calculations may be found in [44] and the details of the implementation in our simulations may be found in [33, 34].

2.2. Sub-grid-scale Models

The model flame with its advection-diffusion-reaction scheme requires an input flame speed. For the one-dimensional laminar case, flame speeds are readily available from direct numerical simulations [45, 35] In the multidimensional case, turbulence on unresolved scales boosts the flame speed. Realistic simulations must therefore include a method for determining the effect of this turbulence-flame interaction (TFI).

Initially we employed a method proposed by Khokhlov [36], that assumes perturbations to the flame front are dominated by the Rayleigh-Taylor instability. The method sets the flame front speed to $s = \max(s_\ell, 0.5\sqrt{\mathrm{At}\, gm\Delta})$ where s_ℓ is the laminar flame speed, $\mathrm{At} = (\rho_{\rm fuel} - \rho_{\rm ash})/(\rho_{\rm fuel} + \rho_{\rm ash})$ is the Atwood number, g is the local gravity, Δ is the grid resolution, and m is a calibrated constant. This prescription effectively keeps the flame speed above some lower limit that depends on the resolution of the simulation. [34]. Another approach that accounts for both unresolved Rayleigh-Taylor instability and background turbulence, originally proposed by Niemeyer and Hillebrandt [46] and developed in detail by Schmidt et al. [47, 48], relies on a dynamic measure of the local turbulent energy on sub-grid scales to enhance the local flame front propagation. The model sets the flame speed to $s = \sqrt{s_\ell^2 + C_t q^2}$, where q is a velocity that characterizes the sub-grid turbulence energy content and C_t is a constant taken to be 4/3 [49].

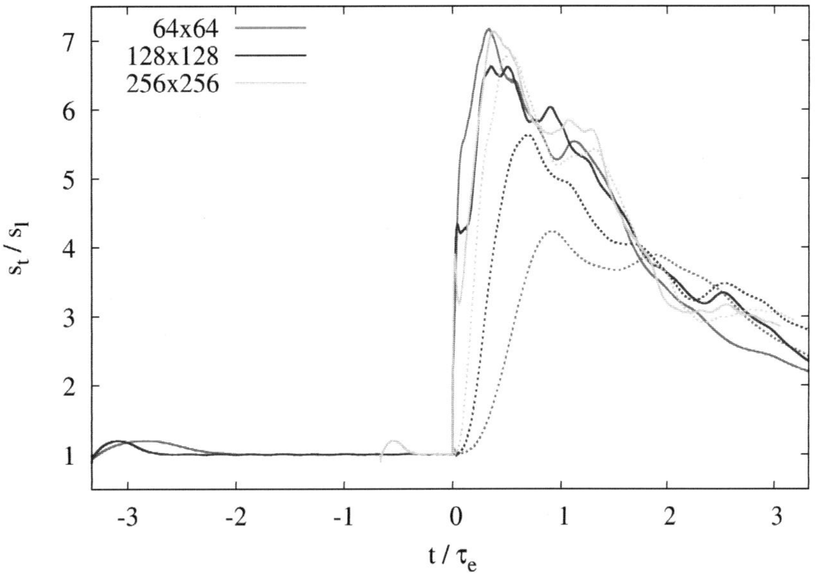

Figure 1. Convergence properties with resolution, with and without a TFI model. Shown is the ratio of the turbulent flame speed to the laminar flame speed as a function of the simulation time scaled by the turbulence eddy turnover time at $t = 0$, τ_e, in a channel with decaying turbulence. Three different resolutions are shown using a TFI model based on Charlette et al. [51] (solid lines) and without a TFI model (dotted lines). Convergence with resolution is achieved with the TFI model.

We have recently implemented a TFI model based on Colin et al. [50] and Charlette et al. [51] that reasonably predicts the behavior of turbulent flames in terrestrial experiments [52]. The method utilizes a local, instantaneous measure of the turbulence, in contrast to the dynamic turbulence model employed by Schmidt et al. [47]. In addition, the relation between the turbulent flame speed and the turbulent intensity is more carefully considered. Particular care has been taken to ensure that the turbulent flame speed predicted by the TFI model shows good convergence properties with resolution for the case of Kolmogorov turbulence. Figure 1 shows the ratio of the turbulent flame speed to the laminar flame speed for a flame propagating in a channel with decaying turbulence having an initial eddy turnover timescale τ_e. The ratio of the turbulent flame speed to the laminar flame speed is plotted as a function of simulation time scaled by τ_e. The solid lines show results from the TFI model following [51], while the dotted lines show results without the inclusion of a TFI model. Three tests were performed with cubic cells that cover the channel cross section with 64x64 cells (red), 128x128 cells (blue), and 256x256 cells (green) to demonstrate the necessity of a TFI model to obtain convergence with resolution.

2.3. Statistical Framework

The statistical framework we developed is designed to explore systematic trends in the brightness of type Ia supernovae. The ideas, however, are general and the same approach may be taken for other multi-physics applications. Our approach allows the evaluation of the average dependence of properties of the model on underlying parameters by constructing a theoretical sample based on probabilistic initial conditions.

For our type Ia supernova research, we utilize two- and three-dimensional simulations in the DDT paradigm within the framework to evaluate the average dependence of the brightness of an event on parameters such as composition of the progenitor. The theoretical sample is based on probabilistic initial ignition conditions for the deflagration. Such sample-averaged dependencies are particularly important to the type Ia supernova problem because we seek to understand how models may explain features of the observed sample, particularly samples generated by large dark energy surveys utilizing type Ia supernovae as distance indicators.

The theoretical sample is constructed to represent statistical properties of the observed sample of type Ia supernovae such as the mean inferred ^{56}Ni yield and its variance. Within the DDT paradigm, the variance in ^{56}Ni yields can be explained by the development of fluid instabilities during the deflagration phase of the explosion. Choice of the initial configuration of the flame influences the growth of fluid instabilities, which result in varying amounts of ^{56}Ni synthesized during the explosion. While there are many sources of uncertainty, particularly in the progenitor white dwarf's composition and structure, variations of the initial configuration of the flame can be used to introduce fluid instabilities that produce the variance of ^{56}Ni yields seen in observations.

The physical initial conditions at ignition are not well known and most likely involve scales our simulations cannot resolve. Our initial conditions consist of an initially burned region that is resolved on the simulation grid. We found that perturbing a spherical flame surface (with radius $r_0 = 150$ km) with spherical harmonic modes (Y_l^m) between $12 \leq l \leq 16$ with random amplitudes (A) normally distributed between $0 - 15$ km and, for three-dimensions, random phases (δ) uniformly distributed between $-\pi$ and π best characterized the mean inferred ^{56}Ni yield and sample variance from observations [26]. We write the perturbation as

$$r(\theta) = r_0 + \sum_{l=l_{\min}}^{l_{\max}} A_l Y_l(\theta) \tag{2}$$

$$r(\theta, \phi) = r_0 + \sum_{l=l_{\min}}^{l_{\max}} \sum_{m=-l}^{l} \frac{A_l^m e^{i\delta_l^m}}{\sqrt{2l+1}} Y_l^m(\theta, \phi). \tag{3}$$

With a suitable random-number generator, a sample population of progenitor WDs is constructed by defining the initial flame surface for a particular progenitor.

Our theoretical framework takes advantage of the fact that the outcome of an explosion is very sensitive to the initial conditions to produce the sample population. We note that other aspects of the problem such as the DDT density can significantly influence the ^{56}Ni yield as well. The mechanism by which a DDT occurs is poorly understood, but it might also contribute to the variation seen in the observations.

3. Performance Results

FLASH is a parallel, adaptive-mesh simulation code for multidimensional compressible reactive flows in astrophysical environments. At its heart is an explicit hydrodynamics method, and FLASH also includes solvers for the Poisson equation of self-gravity and the advection-diffusion-reaction scheme for propagating a model flame as described above. FLASH also provides an equation of state for a degenerate ionized plasma to describe stellar material. FLASH uses a customized version of the PARAMESH library [53, 54] to manage a block-structured adaptive grid, adding resolution elements in areas of complex flow. FLASH and PARAMESH use MPI for parallel communication.

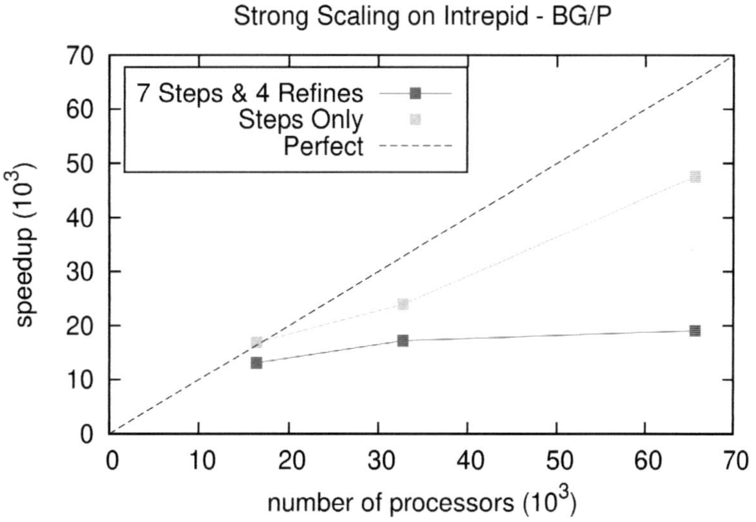

Figure 2. Strong scaling results for the ANL BG/P machine illustrating parallel speedup. Plotted is the speedup vs. number of processors for 16384, 32768, and 65536 processors. Also shown is the corresponding ideal scaling case.

FLASH has always demonstrated almost perfect weak scaling, winning the SC2000 Gordon Bell Prize (special category) [29], for scaling to 6420 processors of the Intel ASCI Red machine at LANL in 2000. As an example of strong scaling for production simulations, we present a performance test run on the BG/P machine at ANL. The results shown in Figure 2 are for a three-dimensional simulation under conditions typical for a production research run.

Simulations of type Ia supernovae utilize adaptive mesh refinement to track critical features of the simulation, particularly the location of the flame, at the highest resolutions. As the star explodes, the size of the simulation increases with time as the burned volume of the star increases. Accordingly, the proportion of simulation domain at the highest resolution grows, drastically increasing the size of the simulation. For this reason, the timing for the scaling study considered only a small part of the actual evolution. The test measured the amount of wall clock time taken to evolve a supernova simulation for a fixed period of simulation time (corresponding to 7 time steps, with mesh adaptations and load balancing every two time steps for a total of 4 adaptations). The study was run on 16384, 32768, and 65536 processors, and the simulation was for a late-stage of the deflagration, and is thus representative of a fully developed simulation. Figure 2 illustrates the parallel speedup,

$$\text{speedup} = \frac{16384 t_{16384}}{t_N}$$

where t_N is the simulation time run on $N = 16384$–65536 processors. In this case, perfect scaling corresponds to the number of processors. Work is needed to improve the cost of mesh refinement.

4. Astrophysical Results

In Townsley et al. [26] we investigated the effect of the metallicity of the host galaxy by assuming it determined the initial neutron excess of the progenitor white dwarf. Weak interactions during nuclear burning, particularly electron capture, lead to the production of neutron-rich elements. The neutron excess of these elements is thought to drive the explosion yield toward stable iron-group elements. As a result, there is relatively less radioactive ^{56}Ni in the yield of iron-group elements, which results in a dimmer event [16]. Thus the introduction of neutron-rich metals into the progenitor white dwarf is thought to influence the brightness of an event by influencing the ^{56}Ni yield.

We investigated the role of metallicity by introducing ^{22}Ne into the progenitor white dwarf as a proxy for neutron-rich metals. The presence of ^{22}Ne influences the progenitor structure, the energy release of the burn, and the flame speed. The study was designed to measure how the ^{22}Ne content influences the competition between rising plumes and the expansion of the star, which determines the yield. We performed a suite of 20 two-dimensional DDT simulations varying only the initial ^{22}Ne in a progenitor model, and found a negligible effect on the pre-detonation expansion of the star and thus the yield of iron-group elements. The neutron excess sets the amount of material in NSE that favors stable iron-group elements over radioactive ^{56}Ni. Our results were consistent with earlier work calculating the direct modification of ^{56}Ni mass from initial neutron excess [16].

We expanded the study of Townsley et al. [26] to include the indirect effect of metallicity in the form of the ^{22}Ne mass fraction through its influence on the density at which the DDT takes place in Jackson et al. [27]. The study consisted of 30 "realizations" or sets of initial conditions (the randomized sample), and for each performed two-dimensional simulations with 5 transition densities between $1 - 3 \times 10^7$ g cm^{-3} for a total of 150 simulations. We found a quadratic dependence of the iron-group yield on the log of the transition density, which is determined by the competition between rising unstable plumes and stellar expansion. By then considering the effect of metallicity on the transition density, we found that the iron-group yield decreases slightly with metallicity, but that the ratio of the ^{56}Ni yield to the overall iron-group yield does not change significantly. Observations testing the dependence of the yield on metallicity remain somewhat ambiguous, but the dependence we found is comparable to that inferred from [55]. We also found that the scatter in the results increases with decreasing transition density, and we attribute this increase in scatter to the nonlinear behavior of the unstable rising plumes.

In Krueger et al. [22] we investigated the effect of central density on the explosion yield. We performed a suite of simulations from 30 realizations, and for each performed two-dimensional simulations with 5 central densities between $1 - 5 \times 10^9$ g cm^{-3}. We found that the overall production of iron-group material did not change, but there was a definite trend of decreasing ^{56}Ni production with increasing progenitor central density (consistent with earlier studies [56, 57]). We attribute this result to higher rates of weak interactions (electron captures) that produce a higher proportion of neutronized material when the burning occurred at higher density. Similarly to the influence of metals, more neutronization means less symmetric nuclei like ^{56}Ni, and, accordingly, a dimmer event.

This result may explain the observed decrease in the brightness of events with increasing age measured as delay time from star formation. For the accreting white dwarf progenitor, only a narrow window in the range of possible accretion rates will produce a massive progenitor in which carbon can be centrally ignited, avoiding far off-center ignitions and subsequent gravitational collapse due to high electron-capture rates. Spherically symmetric models generally find central densities of $1.8 - 13.0 \times 10^9$ g cm^{-3}, depending on the mass of the accreting WD [58, 59, 60]. This narrow window of rates implies a similar accretion duration for all progenitor systems.

Given that all progenitor systems seem to have the same duration of accretion, differences in progenitor age must then follow from differences in the time scale of evolution prior to the onset of accretion. If there is a long period of cooling before the onset of mass transfer, the central density of the progenitor will be higher when the core reaches the carbon ignition temperature [61], thereby producing less ^{56}Ni and thus a dimmer event. Using the results of Lesaffre [61], we were able to relate our results to the progenitor age and compare our results to observations, shown in Figure 3. In addition, in this study we found considerable variation in the trends from some

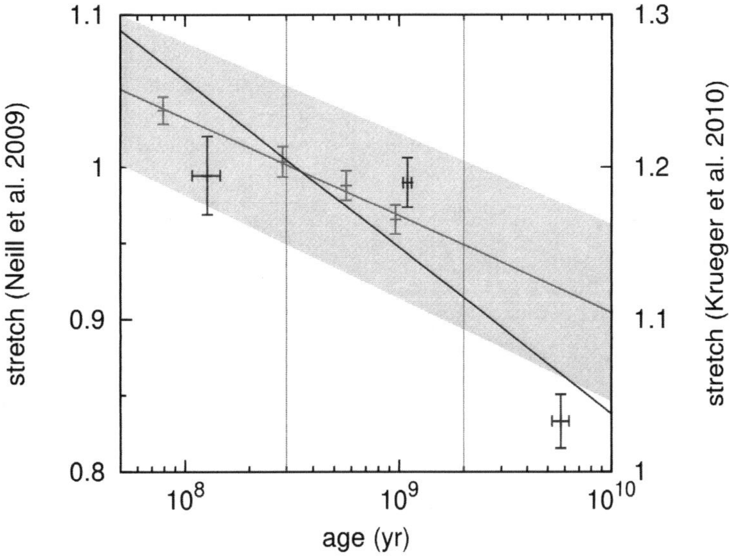

Figure 3. Plots of stretch (s) [21], a measure related to brightness [62], vs. age, comparing our trend in brightness with progenitor age to observational results. In red are results from our study, based on variations in ρ_c, along with the standard deviation (shaded region). In blue are the binned and averaged points from Figure 5 of [20], along with a best-fit line. Note the different scales: the overall offset to larger stretch in the simulations is due to the choice of DDT density. Adapted from [22].

realizations, stressing the importance of statistical studies [63].

5. Summary and Conclusions

We presented a snapshot of our research into the systematics of type Ia supernovae at the time of the Conference on Computational Physics held in Gatlinburg, TN, in the fall of 2011. We outlined our methodology and presented some highlights from our research into the systematics of type Ia supernovae. Our goal for this research it to develop sophisticated models that can reliably address issues like the intrinsic scatter of type Ia supernovae, an issue critical for the use of these events as cosmological distance indicators. Our contemporary research focuses on identifying trends in the brightness properties of the progenitor white dwarf that follow from the properties such as age and composition of the host galaxy. We developed a theoretical framework allowing study of the systematics of type Ia supernovae via statistically well-controlled suites of simulations [34], and we applied this method to studies of the composition and structure of the progenitor white dwarf.

Our results suggest that the direct effect of metallicity on the outcome of our multidimensional explosion models is very much in keeping with previous one-dimensional studies, e.g. [16]. A suite of simulations showed a negligible effect on the expansion of the white dwarf prior to the detonation, and hence the yield [26]. We find that considering the indirect effect of metallicity on flame speeds and the deflagration to detonation transition produces a stronger trend. We found a quadratic dependence of the yield on the log of the transition density, which is determined by the competition between plume rise and stellar expansion during the deflagration phase. By considering the effect of metallicity on the transition density, we obtained a relationship between brightness and metallicity [27].

Our results from variations of the central density of the progenitor provide a theoretical explanation of the observed trend that type Ia supernovae from older host galaxies are systematically dimmer. Previous work [64, 65, 66, 67, 68] addressed the question of the impact of central density on brightness, with different studies reaching different conclusions. We found a strong trend of decreasing brightness with increasing central density due to increased rates of weak interactions that drive the burning toward more neutron rich products. By relating these results to the accretion history of the progenitor, we were able to obtain a relationship between brightness and age of the system [22].

A result drawn from all of the studies is dependence of the problem on the morphology of the deflagration. The deflagration phase of type Ia supernovae is strongly influenced by fluid instabilities and has a very nonlinear evolution. These results stress the necessity of a statistical study in order to capture the true trend.

Our conclusion is that we are beginning to understand the systematics of these events, a necessary step prior to addressing the issue of the intrinsic scatter. Future work will continue in this direction, with a study of the effect of the C/O ratio in the progenitor in progress. We also continue to refine and develop our models, particularly the sub-grid-scale turbulence and turbulence-flame interaction models. We have explored much of the parameter space of white dwarf progenitors and will perform targeted three-dimensional simulations to support our present results.

Acknowledgments

This work was supported by the Department of Energy through grants DE-FG02-07ER41516, DE-FG02-08ER41570, and DE-FG02-08ER41565, and by NASA through grant NNX09AD19G. ACC also acknowledges support from the Department of Energy under grant DE-FG02-87ER40317. DMT received support from the Bart J. Bok fellowship at the University of Arizona for part of this work. The authors acknowledge the hospitality of the Kavli Institute PHY05-51164, during the programs "Accretion and Explosion: the Astrophysics of Degenerate Stars" and "Stellar Death and Supernovae." The software used in this work was in part developed by the DOE-supported ASC/Alliances Center for Astrophysical Thermonuclear Flashes at the University of Chicago. This work was also supported in part by the U.S. Department of Energy, Office of Nuclear Physics, under contract DE-AC02-06CH11357 and utilized resources at the New York Center for Computational Sciences at Stony Brook University/Brookhaven National Laboratory which is supported by the U.S. Department of Energy under Contract No. DE-AC02-98CH10886 and by the State of New York. Finally, the authors acknowledge and thank the organizers and committees of the Conference on Computational Physics held October 30 – November 3, 2011 in Gatlinburg, TN for putting together a delightful and informative conference.

References

[1] A. V. Filippenko. Optical spectra of supernovae. *ARA&A*, 35:309–355, 1997.
[2] W. Hillebrandt and J. C. Niemeyer. Type Ia supernova explosion models. *ARA&A*, 38:191, 2000.
[3] M. M. Phillips. The absolute magnitudes of type Ia supernovae. *ApJ*, 413:L105, 1993.
[4] W. D. Arnett. Type I supernovae. I - analytic solutions for the early part of the light curve. *ApJ*, 253:785, 1982.

[5] P. A. Pinto and R. G. Eastman. The physics of type Ia supernova light curves. I. analytic results and time dependence. *ApJ*, 530:744–756, 2000.

[6] D. Kasen and S. E. Woosley. On the Origin of the Type Ia Supernova Width-Luminosity Relation. *ApJ*, 656:661–665, February 2007.

[7] D. A. Howell, M. Sullivan, P. E. Nugent, R. S. Ellis, A. J. Conley, D. Le Borgne, R. G. Carlberg, J. Guy, D. Balam, S. Basa, D. Fouchez, I. M. Hook, E. Y. Hsiao, J. D. Neill, R. Pain, K. M. Perrett, and C. J. Pritchet. The type Ia supernova SNLS-03D3bb from a super-Chandrasekhar-mass white dwarf star. *Nature*, 443:308–311, September 2006.

[8] R. A. Scalzo, G. Aldering, P. Antilogus, C. Aragon, S. Bailey, C. Baltay, S. Bongard, C. Buton, M. Childress, N. Chotard, Y. Copin, H. K. Fakhouri, A. Gal-Yam, E. Gangler, S. Hoyer, M. Kasliwal, S. Loken, P. Nugent, R. Pain, E. Pécontal, R. Pereira, S. Perlmutter, D. Rabinowitz, A. Rau, G. Rigaudier, K. Runge, G. Smadja, C. Tao, R. C. Thomas, B. Weaver, and C. Wu. Nearby Supernova Factory Observations of SN 2007if: First Total Mass Measurement of a Super-Chandrasekhar-Mass Progenitor. *ApJ*, 713:1073–1094, April 2010.

[9] F. Yuan, R. M. Quimby, J. C. Wheeler, J. Vinkó, E. Chatzopoulos, C. W. Akerlof, S. Kulkarni, J. M. Miller, T. A. McKay, and F. Aharonian. The Exceptionally Luminous Type Ia Supernova 2007if. *ApJ*, 715:1338–1343, June 2010.

[10] M. Tanaka, K. S. Kawabata, M. Yamanaka, K. Maeda, T. Hattori, K. Aoki, K. Nomoto, M. Iye, T. Sasaki, P. A. Mazzali, and E. Pian. Spectropolarimetry of Extremely Luminous Type Ia Supernova 2009dc: Nearly Spherical Explosion of Super-Chandrasekhar Mass White Dwarf. *ApJ*, 714:1209–1216, May 2010.

[11] S. E. Woosley, S. Wunsch, and M. Kuhlen. Carbon Ignition in Type Ia Supernovae: An Analytic Model. *ApJ*, 607:921–930, June 2004.

[12] M. Zingale, A. S. Almgren, J. B. Bell, A. Nonaka, and S. E. Woosley. Low Mach Number Modeling of Type IA Supernovae. IV. White Dwarf Convection. *ApJ*, 704:196–210, October 2009.

[13] A. M. Khokhlov. Delayed detonation model for type IA supernovae. *A&A*, 245:114–128, May 1991.

[14] P. Höflich and A. Khokhlov. Explosion Models for Type IA Supernovae: A Comparison with Observed Light Curves, Distances, H 0, and Q 0. *ApJ*, 457:500, February 1996.

[15] D. Kasen and T. Plewa. Spectral Signatures of Gravitationally Confined Thermonuclear Supernova Explosions. *ApJ*, 622:L41–L44, March 2005.

[16] F. X. Timmes, E. F. Brown, and J. W. Truran. On variations in the peak luminosity of type Ia supernovae. *ApJ*, 590:L83, 2003.

[17] I. Domínguez, P. Höflich, and O. Straniero. Constraints on the progenitors of type Ia supernovae and implications for the cosmological equation of state. *ApJ*, 557:279, 2001.

[18] J. S. Gallagher, P. M. Garnavich, P. Berlind, P. Challis, S. Jha, and R. P. Kirshner. Chemistry and Star Formation in the Host Galaxies of Type Ia Supernovae. *ApJ*, 634:210–226, November 2005.

[19] J. S. Gallagher, P. M. Garnavich, N. Caldwell, R. P. Kirshner, S. W. Jha, W. Li, M. Ganeshalingam, and A. V. Filippenko. Supernovae in Early-Type Galaxies: Directly Connecting Age and Metallicity with Type Ia Luminosity. *ApJ*, 685:752–766, October 2008.

[20] J. D. Neill, M. Sullivan, D. A. Howell, A. Conley, M. Seibert, D. C. Martin, T. A. Barlow, K. Foster, P. G. Friedman, P. Morrissey, S. G. Neff, D. Schiminovich, T. K. Wyder, L. Bianchi, J. Donas, T. M. Heckman, Y.-W. Lee, B. F. Madore, B. Milliard, R. M. Rich, and A. S. Szalay. The Local Hosts of Type Ia Supernovae. *ApJ*, 707:1449–1465, December 2009.

[21] D. A. Howell, M. Sullivan, E. F. Brown, A. Conley, D. LeBorgne, E. Y. Hsiao, P. Astier, D. Balam, C. Balland, S. Basa, R. G. Carlberg, D. Fouchez, J. Guy, D. Hardin, I. M. Hook, R. Pain, K. Perrett, C. J. Pritchet, N. Regnault, S. Baumont, J. LeDu, C. Lidman, S. Perlmutter, N. Suzuki, E. S. Walker, and J. C. Wheeler. The Effect of Progenitor Age and Metallicity on Luminosity and ^{56}Ni Yield in Type Ia Supernovae. *ApJ*, 691:661–671, January 2009.

[22] B. K. Krueger, A. P. Jackson, D. M. Townsley, A. C. Calder, E. F. Brown, and F. X. Timmes. On Variations of the Brightness of Type Ia Supernovae with the Age of the Host Stellar Population. *ApJ*, 719:L5–L9, August 2010.

[23] F. Mannucci, M. Della Valle, and N. Panagia. Two populations of progenitors for Type Ia supernovae? *MNRAS*, 370:773–783, August 2006.

[24] C. Raskin, E. Scannapieco, J. Rhoads, and M. Della Valle. Prompt Ia Supernovae are Significantly Delayed. *ApJ*, 707:74–78, December 2009.

[25] T. D. Brandt, R. Tojeiro, É. Aubourg, A. Heavens, R. Jimenez, and M. A. Strauss. The Ages of Type Ia Supernova Progenitors. *AJ*, 140:804–816, September 2010.

[26] D. M. Townsley, A. P. Jackson, A. C. Calder, D. A. Chamulak, E. F. Brown, and F. X. Timmes. Evaluating Systematic Dependencies of Type Ia Supernovae: The Influence of Progenitor ^{22}Ne Content on Dynamics. *ApJ*, 701:1582–1604, August 2009.

[27] A. P. Jackson, A. C. Calder, D. M. Townsley, D. A. Chamulak, E. F. Brown, and F. X. Timmes. Evaluating Systematic Dependencies of Type Ia Supernovae: The Influence of Deflagration to Detonation Density. *ApJ*, 720:99–113, September 2010.
[28] B. Fryxell, K. Olson, P. Ricker, F. X. Timmes, M. Zingale, D. Q. Lamb, P. MacNeice, R. Rosner, J. W. Truran, and H. Tufo. FLASH: An adaptive mesh hydrodynamics code for modeling astrophysical thermonuclear flashes. *ApJS*, 131:273–334, 2000.
[29] A. C. Calder, B. C. Curtis, L. J. Dursi, B. Fryxell, G. Henry, P. MacNeice, K. Olson, P. Ricker, R. Rosner, F. X. Timmes, H. M. Tufo, J. W. Truran, and M. Zingale. High-performance reactive fluid flow simulations using adaptive mesh refinement on thousands of processors. In *Proceedings of Supercomputing 2000*, page http://sc2000.org, 2000.
[30] A. C. Calder, B. Fryxell, T. Plewa, R. Rosner, L. J. Dursi, V. G. Weirs, T. Dupont, H. F. Robey, J. O. Kane, B. A. Remington, R. P. Drake, G. Dimonte, M. Zingale, F. X. Timmes, K. Olson, P. Ricker, P. MacNeice, and H. M. Tufo. On validating an astrophysical simulation code. *ApJS*, 143:201–229, 2002.
[31] E. F. Brown, A. C. Calder, T. Plewa, P. M. Ricker, K. Robinson, and J. B. Gallagher. Type Ia supernovae: Simulations and nucleosynthesis. *Nucl. Phys. A*, 758:451, 2005.
[32] D. M. Townsley, F. X. Timmes, A. P. Jackson, A. C. Calder, and E. F. Brown. Nucleosynthesis During a Deflagration Detonation Transition Supernova Type Ia. *ApJ*, 2012. to be submitted.
[33] A. C. Calder et al. Capturing the Fire: Flame Energetics and Neutronization for Type Ia Supernova Simulations. *ApJ*, 656:313–332, February 2007.
[34] D. M. Townsley, A. C. Calder, S. M. Asida, I. R. Seitenzahl, F. Peng, N. Vladimirova, D. Q. Lamb, and J. W. Truran. Flame Evolution During Type Ia Supernovae and the Deflagration Phase in the Gravitationally Confined Detonation Scenario. *ApJ*, 668:1118–1131, October 2007.
[35] D. A. Chamulak, E. F. Brown, and F. X. Timmes. The Laminar Flame Speedup by ^{22}Ne Enrichment in White Dwarf Supernovae. *ApJ*, 655:L93, February 2007.
[36] A. M. Khokhlov. Propagation of turbulent flames in supernovae. *ApJ*, 449:695, 1995.
[37] N. Vladimirova, G. Weirs, and L. Ryzhik. Flame capturing with an advection-reaction-diffusion model. *Combust. Theory Modelling*, 10(5):727–747, 2006.
[38] D. Bodansky, D. D. Clayton, and W. A. Fowler. Nuclear Quasi-Equilibrium during Silicon Burning. *ApJS*, 16:299, November 1968.
[39] S. E. Woosley, W. D. Arnett, and D. D. Clayton. The Explosive Burning of Oxygen and Silicon. *ApJS*, 26:231, November 1973.
[40] V. S. Imshennik, S. S. Filippov, and A. M. Khokhlov. The Conditions for Establishment of Nuclear Statistical Equilibrium in Stellar Interiors. *Soviet Astronomy Letters*, 7:121, February 1981.
[41] A. M. Khokhlov. Nuclear-reaction kinetics in a quasiequilibrium approximation. *Soviet Astronomy Letters*, 7:410–413, December 1981.
[42] A. M. Khokhlov. Deflagration-Front Structure in a Degenerate Carbon Core. *Soviet Astronomy Letters*, 9:160, April 1983.
[43] K. Langanke and G. Martínez-Pinedo. Rate tables for the weak processes of pf-shell nuclei in stellar environments. *At. Data Nucl. Data Tables*, 79:1, 2001.
[44] I. R. Seitenzahl, D. M. Townsley, F. Peng, and J. W. Truran. Nuclear statistical equilibrium for Type Ia supernova simulations. *Atomic Data and Nuclear Data Tables*, 95:96–114, January 2009.
[45] F. X. Timmes and S. E. Woosley. The conductive propagation of nuclear flames. I - Degenerate C + O and O + NE + MG white dwarfs. *ApJ*, 396:649–667, September 1992.
[46] J. C. Niemeyer and W. Hillebrandt. Turbulent Nuclear Flames in Type IA Supernovae. *ApJ*, 452:769, October 1995.
[47] W. Schmidt, J. C. Niemeyer, and W. Hillebrandt. A localised subgrid scale model for fluid dynamical simulations in astrophysics. I. Theory and numerical tests. *A&A*, 450:265–281, April 2006.
[48] W. Schmidt, J. C. Niemeyer, W. Hillebrandt, and F. K. Röpke. A localised subgrid scale model for fluid dynamical simulations in astrophysics. II. Application to type Ia supernovae. *A&A*, 450:283–294, April 2006.
[49] A. Pocheau. Scale invariance in turbulent front propagation. *Phys. Rev. E*, 49:1109–1122, February 1994.
[50] O. Colin, F. Ducros, D. Veynante, and T. Poinsot. A thickened flame model for large eddy simulations of turbulent premixed combustion. *Physics of Fluids*, 12:1843–1863, July 2000.
[51] F. Charlette, C. Meneveau, and D. Veynante. A power-law flame wrinkling model for les of premixed turbulent combustion part i: non-dynamic formulation and initial tests. *Combustion and Flame*, 131(1-2):159–180, 2002.
[52] Valeria Di Sarli, Almerinda Di Benedetto, and Gennaro Russo. Sub-grid scale combustion models for large eddy simulation of unsteady premixed flame propagation around obstacles. *Journal of Hazardous Materials*, 180(1-3):71 – 78, 2010.

[53] P. MacNeice, C. Olson, K. M. Mobarry, R. de Fainchtein, and C. Packer. Paramesh: A parallel adaptive mesh refinement community toolkit. *NASA Tech. Rep. CR-1999-209483*, 1999.

[54] P. MacNeice, C. Olson, K. M. Mobarry, R. de Fainchtein, and C. Packer. Paramesh: A parallel adaptive mesh refinement community toolkit. *Comput. Phys. Commun.*, 126:330–354, 2000.

[55] E. Bravo, I. Domínguez, C. Badenes, L. Piersanti, and O. Straniero. Metallicity as a Source of Dispersion in the SNIa Bolometric Light Curve Luminosity-Width Relationship. *ApJ*, 711:L66–L70, March 2010.

[56] K. Iwamoto, F. Brachwitz, K. Nomoto, N. Kishimoto, H. Umeda, W. R. Hix, and F. Thielemann. Nucleosynthesis in Chandrasekhar mass models for type Ia supernovae and constraints on progenitor systems and burning-front propagation. *ApJS*, 125:439–462, 1999.

[57] F. Brachwitz, D. J. Dean, W. R. Hix, K. Iwamoto, K. Langanke, G. Martínez-Pinedo, K. Nomoto, M. R. Strayer, F. Thielemann, and H. Umeda. The role of electron captures in Chandrasekhar-mass models for type Ia supernovae. *ApJ*, 536:934–947, 2000.

[58] K. Nomoto and I. Iben, Jr. Carbon ignition in a rapidly accreting degenerate dwarf - A clue to the nature of the merging process in close binaries. *ApJ*, 297:531–537, October 1985.

[59] M. Hernanz, J. Isern, R. Canal, J. Labay, and R. Mochkovitch. The final stages of evolution of cold, mass-accreting white dwarfs. *ApJ*, 324:331–344, January 1988.

[60] E. Bravo, A. Tornambe, I. Dominguez, and J. Isern. Clues to Type IA SN progenitors from degenerate carbon ignition models. *A&A*, 306:811–+, February 1996.

[61] P. Lesaffre, Z. Han, C. A. Tout, P. Podsiadlowski, and R. G. Martin. The C flash and the ignition conditions of Type Ia supernovae. *MNRAS*, 368:187–195, May 2006.

[62] G. Goldhaber, D. E. Groom, A. Kim, G. Aldering, P. Astier, A. Conley, S. E. Deustua, R. Ellis, S. Fabbro, A. S. Fruchter, A. Goobar, I. Hook, M. Irwin, M. Kim, R. A. Knop, C. Lidman, R. McMahon, P. E. Nugent, R. Pain, N. Panagia, C. R. Pennypacker, S. Perlmutter, P. Ruiz-Lapuente, B. Schaefer, N. A. Walton, and T. York. Timescale Stretch Parameterization of Type Ia Supernova B-Band Light Curves. *ApJ*, 558:359–368, September 2001.

[63] A. C. Calder, B. K. Krueger, A. P. Jackson, D. M. Townsley, F. X. Timmes, E. F. Brown, and D. A. Chamulak. Evaluating systematic dependencies of type ia supernovae. In *Proceedings of SciDAC 2010*, page http://computing.ornl.gov/workshops/scidac2010/2010_SciDAC_Proceedings.pdf, Oak Ridge, 2011.

[64] E. Bravo, J. Isern, R. Canal, and J. Labay. Nucleosynthesis in SNIa. *Ap&SS*, 169:19–24, July 1990.

[65] F. K. Röpke, M. Gieseler, M. Reinecke, C. Travaglio, and W. Hillebrandt. Type Ia Supernova Diversity in Three-Dimensional Models. *A&A*, 453:203, 2006.

[66] R. Fisher, D. Falta, G. Jordan, and D. Lamb. The Role of Variations of Central Density of White Dwarf Progenitors upon Type IA Supernovae. In J. Luo, Z.-B. Zhou, H.-C. Yeh, & J.-P. Hsu, editor, *Gravitation and Astrophysics*, pages 335–344, April 2010.

[67] P. Höflich, K. Krisciunas, A. M. Khokhlov, E. Baron, G. Folatelli, M. Hamuy, M. M. Phillips, N. Suntzeff, L. Wang, and NSF07-SNIa Collaboration. Secondary Parameters of Type Ia Supernova Light Curves. *ApJ*, 710:444–455, February 2010.

[68] I. R. Seitenzahl, F. Ciaraldi-Schoolmann, and F. K. Röpke. Type Ia supernova diversity: white dwarf central density as a secondary parameter in three-dimensional delayed detonation models. *MNRAS*, 414:2709–2715, July 2011.

Conservative Initial Mapping For Multidimensional Simulations of Stellar Explosions

Ke-Jung Chen[1], Alexander Heger[1], and Ann Almgren[2]
[1] Minnesota Institute for Astrophysics, University of Minnesota, Minneapolis, MN 55455
[2] Center for Computational Sciences and Engineering, Lawrence Berkeley National Lab, Berkeley, CA 94720

E-mail: kchen@physics.umn.edu

Abstract. Mapping one-dimensional stellar profiles onto multidimensional grids as initial conditions for hydrodynamics calculations can lead to numerical artifacts, one of the most severe of which is the violation of conservation laws for physical quantities such as energy and mass. Here we introduce a numerical scheme for mapping one-dimensional spherically-symmetric data onto multidimensional meshes so that these physical quantities are conserved. We validate our scheme by porting a realistic 1D Lagrangian stellar profile to the new multidimensional Eulerian hydro code CASTRO. Our results show that all important features in the profiles are reproduced on the new grid and that conservation laws are enforced at all resolutions after mapping.

1. Introduction

Multidimensional simulations shed light on how fluid instabilities arising in supernova explosions mix ejecta [1, 2, 3, 4]. Unfortunately, computing the full self-consistent three-dimensional (3D) stellar evolution initial models for the explosion setup is still beyond the realm of contemporary computational power. One alternative is to first evolve the main sequence star in a 1D stellar evolution code in which the equations of momentum, energy and mass are solved on a spherically symmetric Lagrangian grid, such as KEPLER [5] or MESA [6]. Once the star reaches the pre-supernova phase, its 1D profiles can then be mapped into multidimensional hydro codes such as CASTRO [7, 8] or FLASH [9] and continue to be evolved until the star explodes.

Differences between codes in dimensionality and coordinate mesh can lead to numerical issues such as violation of conservation of mass and energy when profiles are mapped from one code to another. A first, simple approach could be to initialize multidimensional grids by linear interpolation from corresponding mesh points on the 1D profiles. However, linear interpolation becomes invalid when the new grid fails to resolve critical features in the original profile such as the inner core of a star. This is especially true when porting profiles from 1D Lagrangian codes, which can easily resolve very small spatial features in mass coordinate, to a fixed or adaptive Eulerian grid. Besides conservation laws, some physical processes such as nuclear burning are very sensitive to temperature, so slight errors in mapping can lead to very different outcomes for the simulation. Only a few studies have examined mapping 1D profiles to 2D or 3D meshes [10], and none address the conservation of physical quantities by such procedures.

We investigate these issues and introduce a new scheme for mapping 1D data sets to multidimensional grids. We first describe our mapping algorithm in § 2 and then present results of porting a massive star model from KEPLER to CASTRO in § 3. Finally, we conclude in § 4.

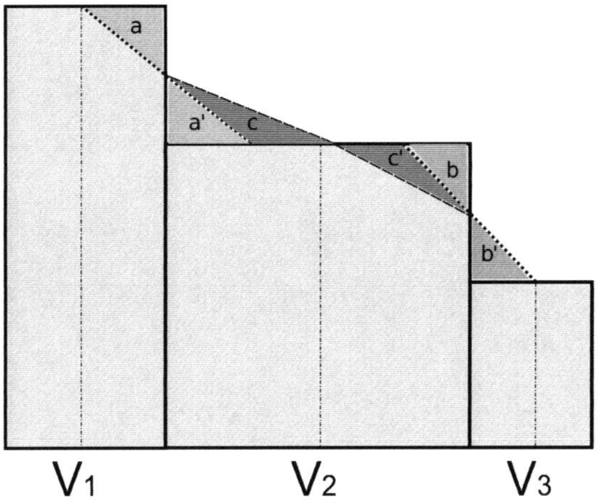

Figure 1. Constructing a conservative profile. The rectangular bins illustrate the original 1D profile. The areas of different colors represent conserved quantities such as mass and internal energy. Note that uniform zones in mass lead to nonuniform bins in volume coordinate, as shown above.

2. Method

Since the star is very nearly in hydrostatic equilibrium but we also map explosions where we want to conserve the total energy, care must be taken in mapping its profile from the non-uniform Lagrangian grid in mass coordinate to the new Eulerian spatial grid. Our method preserves conservation of quantities such as mass and energy that are analytically conserved in the evolution equations on the new mesh. Our mapping method does not guarantee that a 1D hydrostatic star will be fully hydrostatic after mapping onto 2D or 3D meshes because the hydrostatic setup usually needs to break down the conservation laws [10]. However, our method is physically motivated and is sufficient for stellar explosion problems which are less sensitive to hydrostatic equilibrium. The algorithm we describe below is specific to our stellar models but can be easily generalized to other mappings of 1D data to higher dimensions.

First, we construct a continuous (C^0) function that conserves the physical quantity when it is mapped onto the new grid. An ideal choice for interpolation is the volume coordinate V, the volume enclosed by a given radius from the center of the star. Then, integrating a density ρ_X (which can represent mass or internal energy density) with respect to the volume coordinate yields a conserved quantity X

$$X = \int_{V_1}^{V_2} \rho_X \, \mathrm{d}V, \tag{1}$$

such as the total mass or internal total energy lying in the shell between V_1 and V_2.

Next, we define a piecewise linear function in volume V that represents the conserved quantity ρ_X, preserves its monotonicity (no new artificial extrema), and is bounded by the extrema of the original data. The segments are constructed in two stages. First, we extend a line across the interface between adjacent zones that either ends or begins at the center of the smaller of the two zones, as shown in Figure 1 (note that uniform zones in mass coordinate do not result in uniform zones in V). The slope of the segment is chosen so that the area trimmed from one zone by the segment (a and b) is equal to the area added under the segment in the neighboring bin ($a' = a$ and $b' = b$).

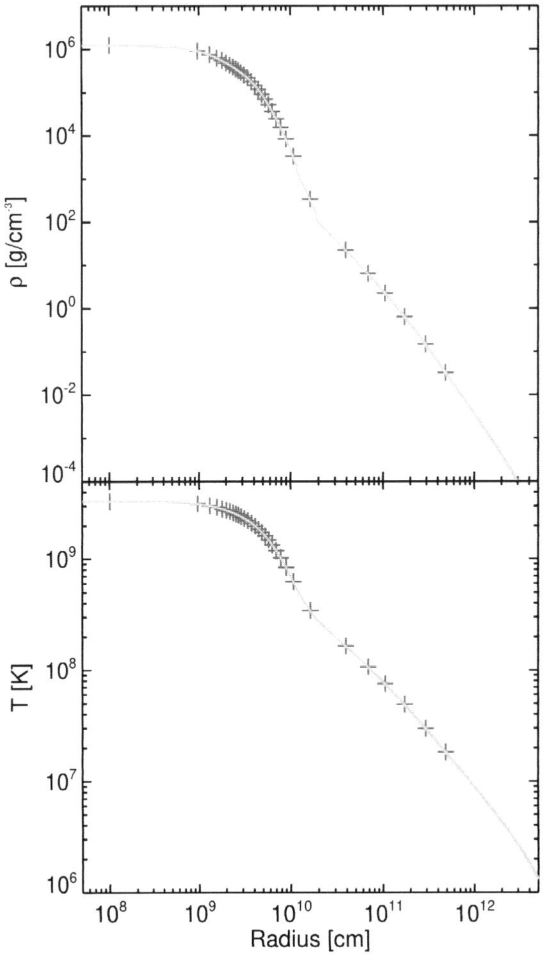

Figure 2. KEPLER densities and temperatures (red crosses) and our piecewise linear fits (green lines). Since we map internal energy (a conserved quantity) rather than temperature, we calculate T from the equation of state using density, element abundance, and internal energy.

If the two segments bounding a and a' and b and b' are joined together by a third in the center zone in Figure 1, two "kinks", or changes in slope, can arise in the interpolated quantity there; plus, the slope of the flat central segment is usually a poor approximation to the average gradient in that interval. We therefore construct two new segments that span the entire central zone and connect with the two original segments where they cross its interfaces, as shown in Figure 1. The new segments join each other at the position in the central bin where the areas c and c' enclosed by the two segments are equal (note that they in general have different slopes). After repeating this procedure everywhere on the grid, each bin will be spanned by two linear segments that represent the interpolated quantity ρ_X at any V within the bin and have no more than one kink in ρ_X across the zone. Our scheme introduces some smearing (or smoothing) of the data, but it is limited to less than or equal to the width of one zone on the original grid.

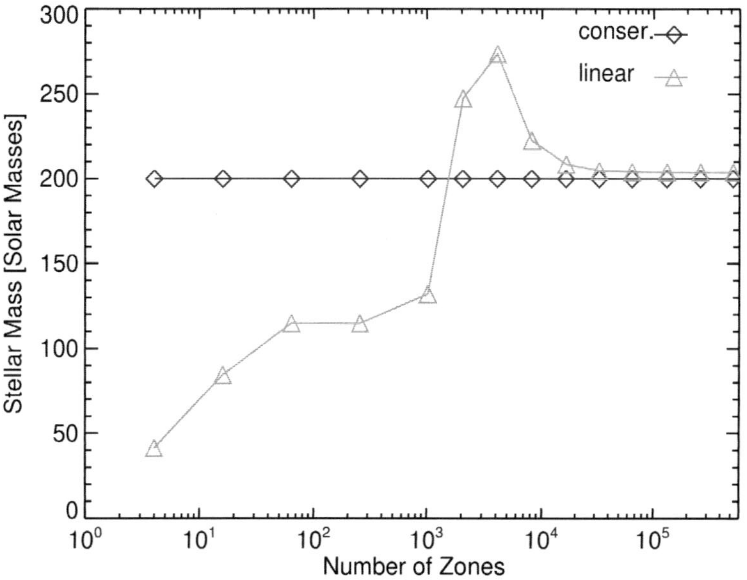

Figure 3. Total mass of the star on the new 1D CASTRO grid vs. number of zones. Conservative mapping (blue) preserves the mass of the star at all resolutions, while linear interpolation (orange) converges to 200 $M_\odot$ at a resolution of $\sim$ a few $\times 10^4$, when the grid begins to resolve the core of the star ($\sim 10^9$ cm). Even using a very high resolution, the linear interpolation still fails to resolve the density gradient between the inner core and the outer envelope (see regions; $10^9 \sim 10^{11}$ cm in Figure 2). So the curve of linear interpolation is saturated at zone number $\sim 10^5$, its results are still off by a few %.

The result of our interpolation scheme is a piecewise linear reconstruction in V of the original profile in mass coordinate, for which the quantity ρ_X can be determined at any V, not just the radii associated with the Lagrangian grid. We show this profile as a function of the radius associated with the volume coordinate V for a zero-metallicity 200 $M_\odot$ star with $r \sim 2 \times 10^{13}$ cm from KEPLER [11, 12] in Figure 2.

We populate the new multidimensional grid with conserved quantities from the reconstructed stellar profiles as follows. First, the distance of the selected mesh point from the center of the new coordinate grid is calculated. We then use this radius to obtain its V to reference the corresponding density in the piecewise linear profile of the star. The density assigned to the zone is then determined from adaptive iterative subsampling. This is done by first computing the total mass of the zone by multiplying its volume by the interpolated density. We then divide the zone into equal subvolumes whose sides are half the length of the original zone. New V are computed for the radii to the center of each of these subvolumes and their densities are again read in from the reconstructed profile. The mass of each subvolume is then calculated by multiplying its interpolated density by its volume element. These masses are then summed and compared to the mass previously calculated for the entire cell. If the relative error between the two masses is larger than some predetermined tolerance, each subvolume is again divided as before, masses are computed for all the constituents comprising the original zone, and they are then summed and compared to the zone mass from the previous iteration. This process continues recursively until the relative error in mass between the two most recent consecutive

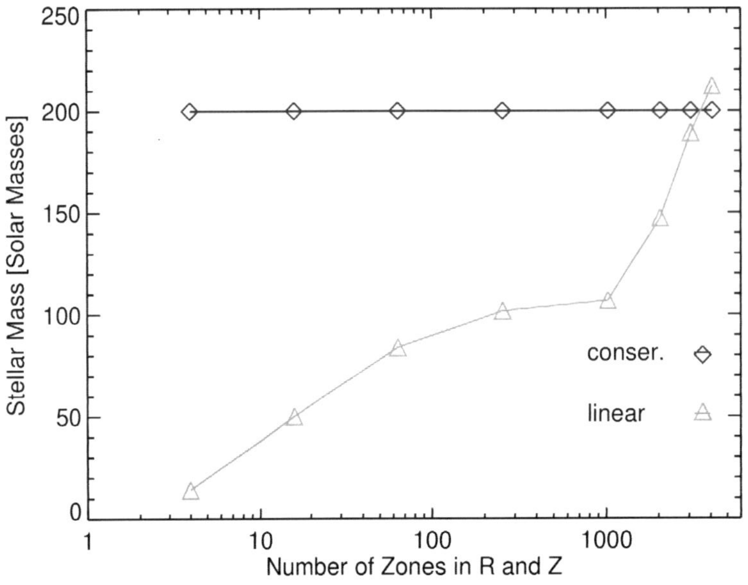

Figure 4. Total mass of the star on the new 2D CASTRO grid vs. number of zones in both r and z. Conservative mapping (blue) recovers the mass of the star at all resolutions and linear interpolation (orange) approaches 200 $M_{\odot}$ at a resolution of $\sim 2048^2$.

iterations falls within an acceptable value, typically 10^{-4}. The density we assign to the zone is just this converged mass divided by the volume of the entire cell. This method is used to map internal energy density and the partial densities of the chemical species to every zone on the new grid. The total density is then obtained from the sum of the partial densities; pressure and temperature in turn are determined from the equation of state. This method is easily applied to hierarchy geometry of the target grid.

3. Results

We port a 1D stellar model from KEPLER into CASTRO to verify that our mapping is conservative. As an example here, we use the zero-metalicity pre-supernova 200 $M_{\odot}$ star whose profiles are shown in Figure 2. KEPLER is a Lagrangian code that evolves stars in mass coordinate so its mesh is nonuniform in space, and CASTRO has an Eulerian grid with uniform spatial zones.

We compare our piecewise linear fits to the KEPLER data in Figure 2, which shows that they reproduce the original stellar profile. Because our fits smoothly interpolate the block histogram structure of the KEPLER bins (especially at larger radii), they reduce the number of unphysical sound waves that would have been introduced in CASTRO by the discontinuous interfaces between these bins in the original data. The density profile is key to the hydrodynamic and gravitational evolution of the explosion, and the temperature profile is crucial to the nuclear burning that powers the explosion.

We first map the profile onto a 1D grid in CASTRO and plot the mass of the star as a function of grid resolution in Figure 3. The mass is independent of resolution for our conservative mapping because we subsample the quantity in each cell prior to initializing it, as described above. In contrast, the total mass from linear interpolation is very sensitive to the number of grid points but does eventually converge when the number of zones is sufficient to resolve the core of the

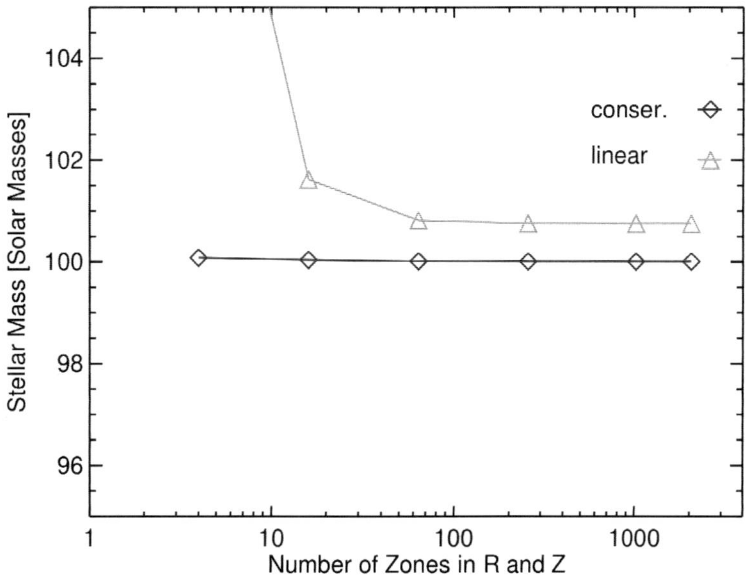

Figure 5. Total mass of the He core on the 2D CASTRO grid vs. number of zones in both r and z. Conservative mapping (blue) preserves the original mass of the core at all resolutions while linear interpolation (orange) begins to converge to 100 $M_\odot$ at a resolution of 64^2 but is still off by $\sim 1\%$ even as the resolution approaches $\sim 2048^2$.

star, in which most of its mass resides.

We next map the KEPLER profile onto a 2D cylindrical grid (r, z) and a 3D cartesian grid (x, y, z) in CASTRO. The only difference between mapping to 1D, 2D, and 3D is the form of the volume elements used to subsample each cell, which are $4\pi r^2 dr$, $2\pi r dr dz$, $dx dy dz$, respectively. We show the mass of the star as a function of resolution in Figure 4. Conservative mapping again preserves its mass at all grid resolutions. In 2D, more zones are required for linear interpolation to converge to the mass of the star. To further validate our conservative scheme, we map just the helium core of star ($\sim$ 100 $M_\odot$ with $r \sim 10^{10}$ cm) onto the 2D grid. The helium core is crucial to modeling thermonuclear supernovae because it is where explosive burning begins. We show its mass as a function of resolution in Figure 5. We again recover all the mass of the core at all resolutions while linear interpolation overestimates the mass by at least ~ 1 %, even with large numbers of zones.

Conservative mapping is effective in 3D but requires much more computational time to subsample each cell to convergence. Furthermore, an impractical number of zones is needed for linear interpolation to reproduce the original mass of the star, so we defer its comparison to our scheme in 3D to a later study. We note that our method also works with adaptive mesh refinement (AMR) grids because both V and the interpolated quantities can be determined and subsampling can be performed on every grid in the hierarchy.

4. Conclusion

Multidimensional stellar evolution and supernova simulations are numerically challenging because multiple physical processes (hydrodynamics, gravity, burning) occur on many scales in space and time. For computational efficiency, 1D stellar models are often used as initial conditions in 2D and 3D calculations. Mapping 1D profiles onto multidimensional grids can introduce serious numerical artifacts, one of the most severe of which is the violation of conservation of physical quantities. We have developed a new mapping algorithm that guarantees that conserved quantities are preserved at any resolution and reproduces the most important features in the original profiles. Our method is practical for 1D and 2D calculations, and we are now developing integral methods (an explicit integral approach instead of using volume subsampling) that are numerically tractable in 3D.

Acknowledgments

The authors thank Daniel Whalen for reviewing the earlier manuscript and providing many insightful comments, the members of the CCSE at LBNL for help with `CASTRO`, and Hank Childs for assistance with *VisIt*. We also thank Laurens Keek, Volker Bromm, Dan Kasen, Adam Burrows, and Stan Woosley for many useful discussions. All numerical simulations were performed with allocations from the University of Minnesota Supercomputing Institute and the National Energy Research Scientific Computing Center. This work has been supported by the DOE SciDAC program under grants DOE-FC02-01ER41176, DOE-FC02-06ER41438, and DE-FC02-09ER41618, and by the US Department of Energy under grant DE-FG02-87ER40328.

References

[1] Herant M and Woosley S E 1994 *ApJ* **425** 814–828
[2] Joggerst C C, Woosley S E and Heger A 2009 *ApJ* **693** 1780–1802 (*Preprint* 0810.5142)
[3] Joggerst C C, Almgren A, Bell J, Heger A, Whalen D and Woosley S E 2010 *ApJ* **709** 11–26 (*Preprint* 0907.3885)
[4] Joggerst C C and Whalen D J 2011 *ApJL* **728** 129 (*Preprint* 1010.4360)
[5] Weaver T A, Zimmerman G B and Woosley S E 1978 *ApJ* **225** 1021–1029
[6] Paxton B, Bildsten L, Dotter A, Herwig F, Lesaffre P and Timmes F 2011 *ApJS* **192** 3 (*Preprint* 1009.1622)
[7] Zhang W, Howell L, Almgren A, Burrows A and Bell J 2011 *ApJS* **196** 20 (*Preprint* 1105.2466)
[8] Almgren A S, Beckner V E, Bell J B, Day M S, Howell L H, Joggerst C C, Lijewski M J, Nonaka A, Singer M and Zingale M 2010 *ApJ* **715** 1221–1238 (*Preprint* 1005.0114)
[9] Fryxell B, Olson K, Ricker P, Timmes F X, Zingale M, Lamb D Q, MacNeice P, Rosner R, Truran J W and Tufo H 2000 *ApJS* **131** 273–334
[10] Zingale M, Dursi L J, ZuHone J, Calder A C, Fryxell B, Plewa T, Truran J W, Caceres A, Olson K, Ricker P M, Riley K, Rosner R, Siegel A, Timmes F X and Vladimirova N 2002 *ApJS* **143** 539–565 (*Preprint* arXiv:astro-ph/0208031)
[11] Heger A and Woosley S E 2002 *ApJ* **567** 532–543 (*Preprint* arXiv:astro-ph/0107037)
[12] Heger A and Woosley S E 2010 *ApJ* **724** 341–373 (*Preprint* 0803.3161)

Advancements in modeling self-consistent core collapse supernovae with CHIMERA

M.A. Chertkow[1], O.E.B. Messer [2,1], W.R. Hix[1,2], C.T. Lee[2], K. Yakunin[3], P. Marronetti[3], S.W. Bruenn[3], E.J. Lentz[1,2], J. Blondin[4], A. Mezzacappa[2]

[1] Department of Physics & Astronomy, University of Tennessee, Knoxville, TN 37996-1200
[2] Physics Division, Oak Ridge National Laboratory, Oak Ridge, TN 378316354
[3] Physics Department, Florida Atlantic University, Boca Raton, FL 33431-0991
[4] Department of Physics, North Carolina State University, Raleigh, NC 27695-8202

E-mail: chertkow@utk.edu

Abstract.
We discuss advancements in modeling core-collapse supernovae with our code CHIMERA. We describe the status and details of our tracer particle method and its uses for post-processing nucleosynthesis and as a tool for broad core-collapse supernova (CCSN) model analyses. We also introduce our progress towards expanding a nuclear reaction network beyond the alpha-network for the purpose of accurate in-situ nucleosynthesis, not only for core-collapse supernovae (CCSNe) in general, but for a special sub-class of supernovae called electron capture supernovae (ECSNe), which stem from progenitors stars between 8 and 10 $M_{\odot}$. By using an advanced nuclear reaction network, our 2D and 3D code will allow for unparalleled studies of CCSNe and ECSNe ejecta.

1. Introduction

The final stage of stellar evolution for stars between roughly 8 and 25 solar masses ($M_{\odot}$) marks a singularly unique class of phenomena in the observable universe. Within this mass range, a core-collapse supernova (CCSN) occurs when a star's degenerate core is unable to support itself against gravitational collapse. Once the core reaches super-nuclear densities, the collapse halts and a rebound shock forms, which eventually stalls as its energy is sapped via neutrino losses and nuclear dissociation. Self-consistent computer simulations attempt to describe how neutrino energy deposition, aided by the convection of neutrino-heated material between the stalled shock and gain radius, eventually re-energizes the stalled shock towards explosive energies. This process accelerates as instabilities increase the size of the heating region until enough matter is energized to turn the collapse around and become gravitationally unbound — thus creating a supernova. Without question, we know CCSNe are complex multi-dimensional events; and years of theoretical and computational development have shown that the codes we develop to model them must follow suit.

The ORNL/UTK group, in strong collaboration with Florida Atlantic University and NC State University, have built a tightly coupled radiation hydrodynamic code called CHIMERA. The code's primary modules are designed to evolve the stellar gas hydrodynamics (VH1), compute the "ray-by-ray-plus" neutrino transport (MGFLD-TRANS), and solve the thermonuclear kinetics (XNet) [1]. These three heads are augmented by a sophisticated equation of state for nuclear matter (e.g. Lattimer & Swesty [2]) and a self-gravity solver capable of an approximation to general-relativistic gravity. A more general description of CHIMERA and general features of our recent two-dimensional simulations have been reported in Messer et al. (2007) [3] and Bruenn et al. (2010) [4].

This report features our use of tracer particle to expand the CHIMERA models and highlights two CHIMERA advancements not addressed in Bruenn et al. (2010). These advancements are (1) the development of a larger nuclear reaction network beyond the popular 14-species alpha-chain within CHIMERA, which allows for (2) the recent progress toward modeling electron capture supernovae (ECSNe) of $8 - 10\ M_{\odot}$ progenitors with ONeMg cores. ECSNe are unique in that they require more detailed burning treatments to address their initial composition of oxygen, neon, and magnesium, which differ from larger iron core progenitors. We want to model ECSNe because theoretical and observational studies are still subject to many uncertainties [5, 6, 7, 8].

Regardless of the initial mass, the prime motivation for modeling realistic CCSNe abundances is to reconstruct observables for the purpose of understanding the origin of the elements. This means we cannot simply "follow" chemical abundances throughout the evolution, but we must properly treat the feedback of thermodynamic energy generation upon the overlaying hydrodynamic model. Since this energy generation can have a local effect upon the hydrodynamics, potentially affecting whether individual mass elements are ejected, it should be directly included among the physical processes associated with the explosion mechanism [9, 10].

2. The Tracer Particle Method (TPM)

CHIMERA's Eulerian grid makes it difficult to obtain the necessary data to calculate realistic ejecta abundances via post-processing, as what is needed are time histories of thermodynamic conditions for mass elements. A Lagrangian grid would provide us with this useful information, but these schemes quickly become tangled in the multi-dimensional hydrodynamic flows inherent in any realistic supernova calculation. As a solution to this problem, we have recovered a Lagrangian perspective by implementing massless tracer particles to keep track of thermodynamic histories which we use for post-processing nucleosynthesis using thousands of nuclear species [11]. We are currently running state-of-the-art 2D models with CHIMERA towards explosions and we will have a plethora of tracer particle data to analyze. Although the tracer particles were primarily developed to perform post-processing nucleosynthesis calculations, we have also found use for them as a tool to search for possible code problems manifested in physical quantities during code development.

2.1. Tracer Data

The tracer particles record thermodynamic, hydrodynamic, gravitational energy, nuclear binding energy, and neutrino energy deposition rates and flux quantities for the parcels of matter which they represent throughout the evolution of the supernova. To gain a representative picture of our simulations, we initially distribute tracer particles uniformly within the inner core layers from iron to oxygen (in our iron core models). Depending on the size of the progenitor and the dimensionality of the simulation, we distribute between several thousand tracer particles in 2D and hundreds of thousands in 3D. As an example, a run containing $\sim 8,000$ tracer particles produces approximately 400 GB of processed data, of which 47 GB will be used for post-processing, while the rest is used for general analysis. Each tracer particle is represented by one file since the tracer's physical movement causes it to move from processor to processor through CHIMERA's domain decomposition. This can mean we must handle upwards of 100,000 individual files. For these reasons, we must carefully consider proper data management, efficient processing, and long-term storage.

2.2. The Tracer Particle Reader

To handle the tracer particle data management and processing, we have developed a tracer particle "reader," (TPR) to (1) read-in the binary tracer files; (2) digest and pre-process the data; (3) perform automated analysis of CHIMERA not necessarily associated with nucleosynthesis; and (4) organize the resultant information into suitable formats for post-processing nucleosynthesis by our general purpose nuclear kinetic solver XNet. The TPR also helps us perform many service tasks such as finding bad tracers (tracers whose data may have become corrupt due to I/O or file system issues), and hunting down tracers of particular interest. Without this effective data management, sorting through $8000 - 100,000$ files for the location of, for example, the non-spherical shock and the location of the boundary between NSE & non-NSE matter to determine which particles (an thus what parcel of matter) becomes unbound, would be a daunting task.

There are many ways of utilizing tracer particles as a tool to understand and interpret our models, but we highlight three examples here. Figure 1 shows tracers superimposed over the entropy of the matter in a simulation of a 15 $M_{\odot}$ star approximately 300 ms after core bounce. The tracers are shown with their color code indicating the amount of neutrino heating or cooling. The pattern of the tracers illustrates how infalling matter is concentrated into the low entropy downflows as it flows towards the proto-neutron star. The neutrino heating occurs predominately in the inner 100 km, but the heated matter, with its entrained tracers, rises to form low density plumes which push the stalled shock outward. Figure 2 shows the location of tracers at $\sim$ 1000 ms post-bounce for our 25 $M_{\odot}$ simulation. The color coding of the tracers indicates the layer in the progenitor from which that tracer originated. This illustrates the difficulty in translating the "mass-cut" of spherically symmetric models to multi-dimensions, pointing out that the term is really a misnomer. This figure also indicates the complexity of mixing issues relative to the initial composition distributions. CHIMERA overestimates microscopic mixing due to zone interpolation, while the tracers underestimate fluid mixing because of their Lagrangian nature. Figure 3 provides a different perspective on this issue, which highlights the trajectories of six tracer particles from the same initial radius but different angular latitudes; two of these tracers are captured by the proto-neutron star (PNS) while the other four escape to possibly become ejecta — depicting clearly that one-dimensional models can not capture the complex behavior of the inner most supernova ejecta.

2.3. Alternate Use of Tracers: Gravitational Waves

The TPM method can also be used to explore multi-dimensional effects, in general, that stem from anisotropies, instabilities, and fluid mixing. We have used the unique Lagrangian perspective of the tracer particles to assist in the analysis of gravitational wave (GW) signatures from our non-parameterized, 2D, $12 - 25$ $M_{\odot}$ core-collapse supernovae models [12]. In these models, each particle is assigned a constant mass ($1/4000^{th}$ – $1/8000^{th}$ of the mass of the progenitor in the region of interest where the tracers were deployed), and the GW signal their movement produces is calculated. Comparing the GW signal corresponding to a given group of tracers with the signal produced by the bulk-matter motion (captured on the Eulerian grid) has allowed us to identify what part of the fluid generates a specific GW feature. We have calculated the contribution to the signals produced by both baryonic matter motion and anisotropic neutron emissions up to 530 ms after bounce for all four progenitors. This analysis matches the predictions based on the parameterized models by Murphy et al. [13]. However, given the development of non-parameterized explosions in our models, we were able to compute the waveforms through explosion and to determine more precisely the per-explosion amplitudes and timescales. In doing so, we have decomposed the GW signal and determined which mass motions contribute to the various components of the waveforms. As a result, we have been able to explore the prompt signal, which is generated by two independent phenomena. Prompt

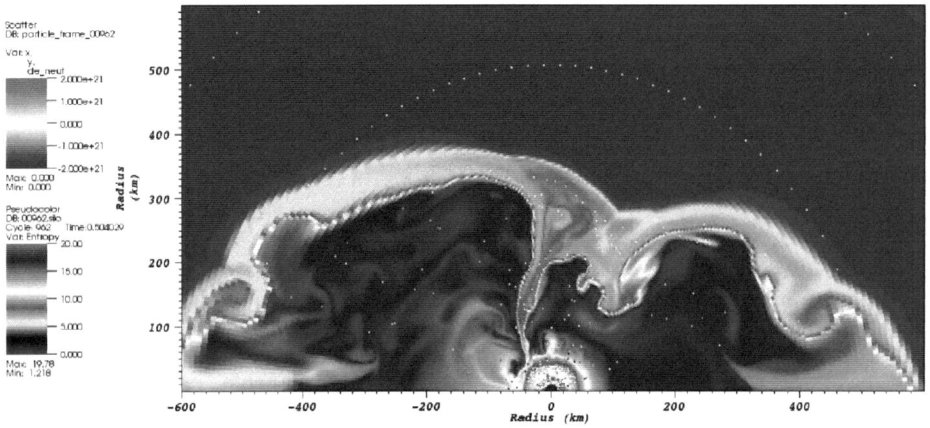

Figure 1. Tracer particles plotted over an entropy colormap. Here, the tracers are colored to represent the superposition of neutrino-heating.

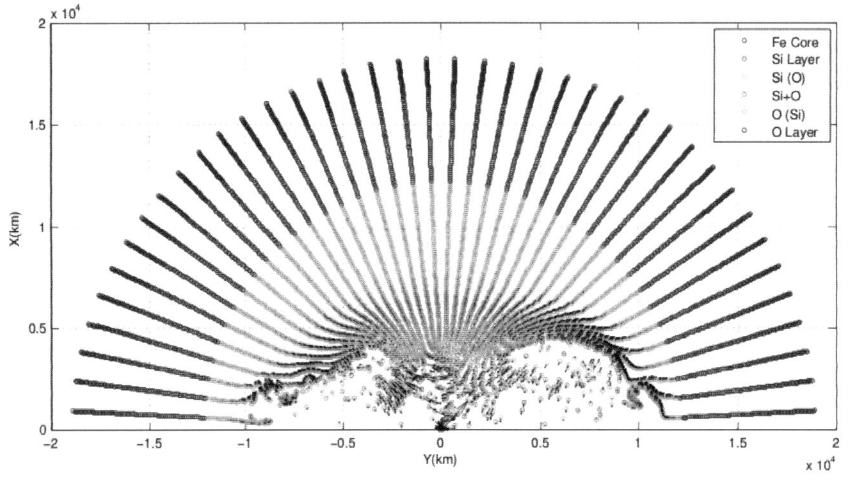

Figure 2. Tracer particles from a 25 $M_{\odot}$ model at $\sim$ 1000 ms post-bounce have diverged from initial spherical symmetry illustrating that the definition of a "mass-cut" is not well defined in multi-dimensions. This figure also indicates the complexity of mixing issues relative to the initial composition distributions.

convection inside the PNS generates a high-frequency signal that is superimposed on a lower frequency component. There is a hint of this in the inset of Figure 4, where the signal for our 15 $M_{\odot}$ run has been split into the contributions from two different regions, but it is in the tracer analysis of Figure 5 (right) where this becomes most evident. The matter-generated GW (solid red) is closely tracked by the GW generated by the infalling tracer particles deflected by the shock (dashed blue), some of which are shown in the left panel. Although the hydrodynamic resolution (256 radial and 256 angular zones) for this suite of runs was significantly finer than the

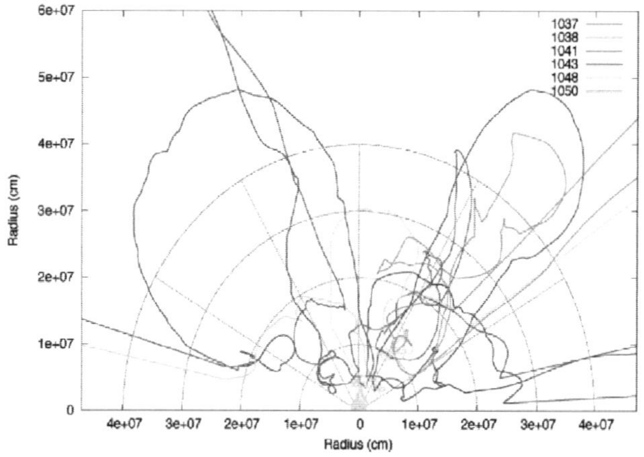

Figure 3. The time history of 6 example tracers from an initial spherically symmetric distribution within a 15 $M_\odot$ star. Two tracers are captured onto the PNS while the other four are possibly ejected through convection

tracer-matter resolution (50×100 tracers), we identified a low-frequency signal from 20 to 60 ms after bounce originating at the shock radius, which is at $\sim$ 100 km at this time and well outside the PNS. The results presented here are preliminary: a new set of 2D simulations performed with an enhanced version of our CHIMERA code is currently ongoing [4], and will explore the use of tracer particles to investigate the GW signal at lower frequencies with post-bounce dynamics.

3. Nuclear Networks

As detailed in Timmes et al. (2000) [18], since the implicit solution of a nuclear reaction network is, at its heart, the solution of a matrix equation, the computational cost grows non-linearly with the number of nuclei included in the calculation. For this reason, an alpha-network, or even a simpler "flashing" scheme, is typically used to evolve the composition where conditions are not in nuclear statistical equilibrium (NSE). Because the alpha-network neglects reaction flows involving neutron-rich (or proton-rich) nuclei, it cannot fully represent accurate compositions, and it provides incomplete estimates of the energy generation from nuclear burning encountered in CCSNe. Detailed nucleosynthesis calculations can be performed by post-processing methods, but only if the temporal evolution of matter is present (see section 2). To avoid the need for post-processing, accurate nucleosynthesis in-situ can be obtained by evolving 150, or more, nuclear isotopes throughout the matter that is ultimately ejected. A large network would also explore the importance of composition mixing, and facilitate the testing of post-processed results.

As additional motivation for improving the nuclear kinetics within CHIMERA, Pruet et al. [19, 20] and Frölich et al. [17, 21] have studied the inner ejecta layers showing that neutrino-powered explosions (using spectral neutrino transport) result in decreased neutronization, and that the electron fraction (Y_e) has a dominate effect on nucleosynthesis. This finding is at odds with the prediction of previous models, all of which neglect this important piece of physics. The decreased neutronization removes the overproduction of neutron-rich iron and nickel isotopes commonly seen in parameterized bomb and piston models [22, 23]. The simulations of Pruet et al. and Frölich et al. also show enhanced production of Sc, Cu, and Zn, elements which observations of metal-poor stars (see, [24, 25]) suggest are $3 - 10$ times more abundant than previous models predicted. Finally, a significant neutrino driven flow to proton-rich nuclei above $A > 64$ is seen, suggesting that the innermost ejecta of core-collapse supernovae may be the production site of the light p-process nuclei [17]. This agreement with galactic and solar νp-process abundances at later times generates a small abundance of neutrons, which helps explain things like strontium in otherwise metal-poor stars [17].

Within our ongoing runs, the equation of state (EOS) of Lattimer and Swesty [2] is employed in CHIMERA for matter at high densities. For smaller densities, an electron-positron EOS, with arbitrary degeneracy and degree of relativity, spans the entire density-temperature regime of interest. For regions *not* in nuclear statistical equilibrium (NSE), we use an EOS with a nuclear component consisting of a 14-species alpha-network (^{4}He to ^{60}Zn) which carries the protons, neutrons, and an iron-like nucleus. Here, the composition is evolved by XNet, the

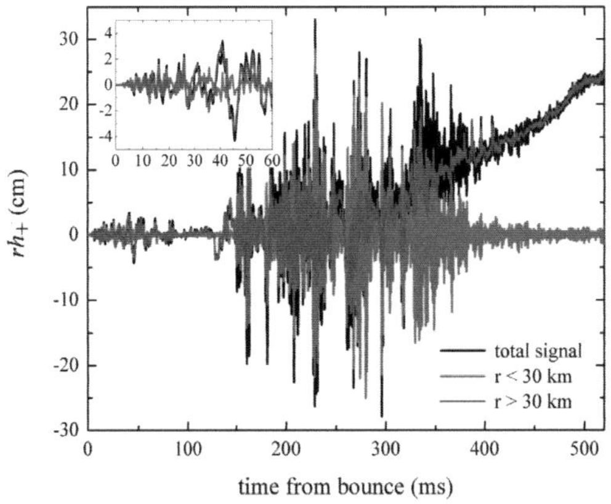

Figure 4. Contributions to the matter-generated GW signal from two different regions for the 15 $M_\odot$ model: the PNS ($r < 30$ km) and the region above the PNS ($r > 30$ km). The latter includes the region of neutrino-driven convection, the SASI, and the shock. [12]

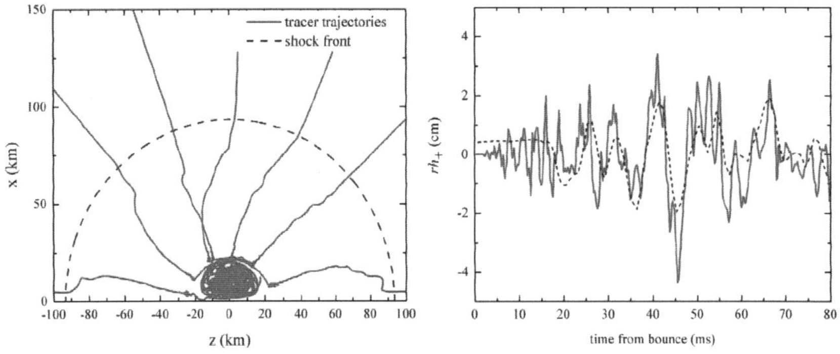

Figure 5. Left: trajectories of the tracer particles. The clear deflection of infalling particles through the shock that collectively produce a low-frequency high-amplitude component of the GW signal shown on the right panel. Right: comparison between the matter signal (solid red) and signal calculated using the tracers (dashed blue). Both panels correspond to our 15 $M_\odot$ simulations. [12]

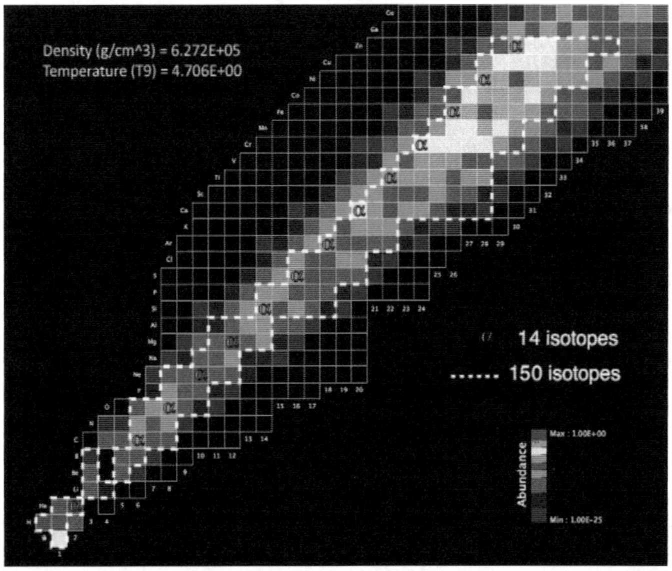

Figure 6. Another way of looking the alpha-network's inability to represent abundance values at a constant temperature and density profile. As indicated, the thick dashed line representing the 150-isotope network chosen to accommodate neutron-rich nuclei of significant abundances.

thermonuclear reaction network [14]. Data for these reactions is drawn from the REACLIB compilations [15].

To improve the nucleosynthesis results while reducing the energy-generation uncertainty, we have improved CHIMERA's nuclear network by allowing for the substitution of the alpha-network with a larger, 150-isotope network. Figure 6 shows the alpha-network's inability to represent abundance values for the given temperature and density profile. It is important to note that the three most abundant species — protons, ^{54}Fe, and ^{58}Ni — are not members of an alpha-network. It is also clear that the alpha-network does not span the entire regime of interest whereas a 150-isotope network is much more inclusive. The downside is that a simple replacement of the 14-species alpha-network with 150-isotopes in the fully-implicit backward-Euler integration scheme represents a several hundred-fold increase in the computational cost. Given the current $\sim$ 5% of execution time required by the alpha network in the current simulations, this cost is prohibitive if implemented naïvely. Figure 7 illustrates two 15 $M_\odot$ radial composition profiles of the significant nuclear species using an alpha-network (solid lines) and a large-network (dashed lines) within CHIMERA. While the alpha-network reproduces the general abundance patterns, there are many examples where it simply does not represent the composition.

To increase the physical fidelity, yet have calculations that complete in a reasonable time, we have improved the speed of these calculations by using the shared-memory parallel-programming API OpenMP. We plan to further decrease our computational time by adaptively managing the nucleosynthesis by taking advantage of quasi-statistical equilibrium (QSE), a precursor to NSE, to reduce the size of the system of equations where conditions apply. This adaptive QSE-network has been developed in-house and is expected to be become a CHIMERA module within the coming year [9]. With our improvements to CHIMERA's nuclear kinetic module, we hope to expand upon the recent work on by Pruet et al. (2005, 2006) [19, 20] and Frölich et al.(2006a,

2006b) [17, 21]

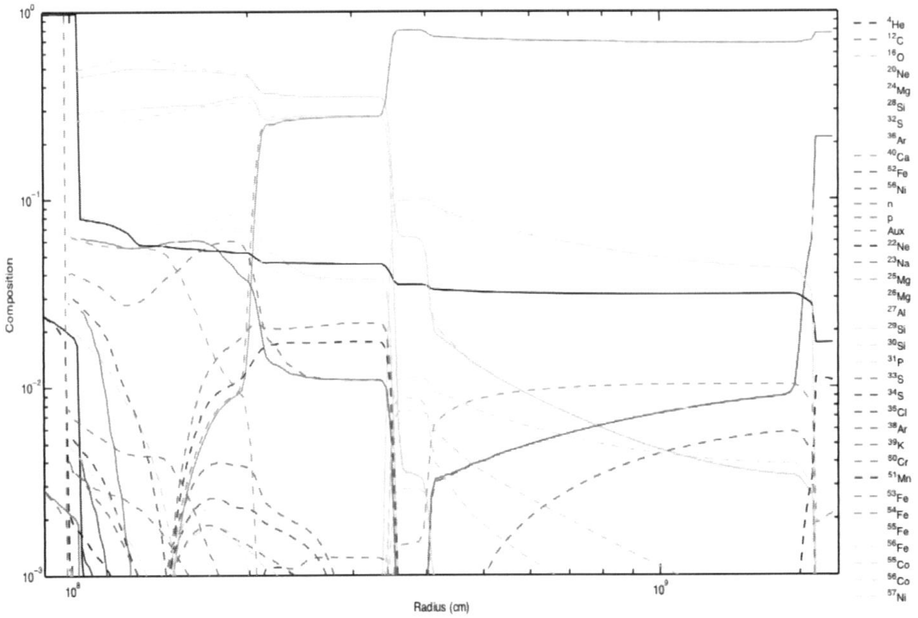

Figure 7. An example of a radial composition profile of significant nuclear species highlighting the difference between an alpha (solid) and 150-isotopes (dashed) network early during the collapse our 15 $M_{\odot}$ model.

4. ECSN and ONeMg Progenitors

With our self-consistent, multi-dimensional, tracer-laden code, and a network extended beyond 14-species (to 150 species enabling us to follow the neutronization) we can expand the scope of our studies with progenitors of various masses and compositions. One such investigation is exploring the lower mass range ($\sim 8-10\ M_{\odot}$) of CCSN progenitors whose cores are composed of oxygen, neon, and magnesium. These supernova are called electron capture supernovae (ECSNe) since they can undergo collapse via extreme deleptonization within the degenerate core, where stellar oxygen burning is not favored. There is a large interest in resolving the specific explosion mechanism of ECSNe as they have been suggested as the source of the r-process [26], though more recent models suggest only a light r-process as the origin of elements heavier than Zn, potentially up to Cd [8].

The large network will play a unique role in these models because the core is not in NSE and are composed of oxygen, neon, sodium, and magnesium, as previously stated. Lacking significant mass for gravitational contraction, the core cannot progress beyond the production of magnesium and it becomes physically inert. The H- and He-burning shells continue to deposit mass until the core reaches the Chandrasekhar mass ($M_{CH} \sim 1.37\ M_{\odot}$) [7]. Here, extreme deleptonization begins through sequential electron captures on ^{24}Mg, ^{23}Na, ^{20}Ne, and the core begins to collapse [27]. The coincident oxygen deflagration is unable to stop the collapse and the core reaches nuclear densities. The subsequent explosion mechanism is moderately under debate, though most recent studies point towards a requirement for neutrino energy deposition behind a stalled shock [6, 8]. Observationally, the explosion associated with these less massive stars is relatively weak ($\sim 10^{50}$ ergs), they have a larger Ni/Fe ratio ($\sim 1-2$) than their iron core counterparts, and they eject a much smaller amount of ^{56}Ni [7, 27].

Kitaura et al. (2008) [6] have performed spherically symmetric ECSN models up to one second after bounce finding a delayed neutrino-driven explosion instead of the prompt-shock mechanism as described by Hillebrandt et al. (1984) [28]. Although Kitaura et al. carefully considered a description of weak interaction, they were still limited by a seven species alpha-chain network. As with all alpha-networks, important abundances that may be present in the core, such as ^{23}Na [29], are simply omitted. This of course ignores important reaction channels for nucleosynthesis. Kitaura et al. project ECSNe to produce neutron-rich eject with Y_e as low as ~ 0.2 and entropies down to $\sim 10k_B$ [6].

The 2D model of Janka et al. (2007) [5] was somewhat more energetic due to convective overturn in the gain layer while using the same network of Kitaura et al. [6]. Although they determine that neutrino energy deposition within the gain region is the primary source for reviving a temporarily stalled shock, their model fails to create conditions to support the r-process scenario by having too modest entropies and/or not sufficient temperatures. Janka et al. [5] point out that their thermonuclear energy production is approximate and a full reaction network is desired in order to take electron capture rates into consideration.

Also in contrast to Kitaura et al. [6], Wanajo et al. (2010) [8] found that a multi-dimensional model produces neutron-rich ejecta due to convective overturn that could otherwise not be seen in 1D cases. They performed the first nucleosynthesis study on the basis of self-consistent 2D models (while using tracer particles for post-processing), and have found neutron-rich lumps with electron fractions down to 0.40, which can lead to nuclei up to $A \sim 80$. Ultimately, they hypothesize that a 3D model may eject a tiny amount of matter with a Ye as low as $\sim 0.35-0.30$ leading to the creation of Cd. Wanajo et al. were also unfortunately limited by resolution and dimensionality constraints, as well as the small network, and could not exclude weak r-process elements beyond $N = 50$ up to Pd, Ag, and Cd [8].

We are presently modeling the collapse and explosion of the ONeMg core of Nomoto et al. (1987) [27] — the same pre-collapse core used in the Prometheus-Vertex models of Hillebrandt et al. [28], Kitaura et al. [6], Janka et al. [5], and Wanajo et al. [8]. Previous 2D, $12-25\ M_\odot$ models in CHIMERA have resolved up to 512×256 zones and we can extend this resolution to ECSNe. We are working on the evolution of a 1D model out to ~ 1000 ms post-bounce to ensure CHIMERA's compatibility with a progenitor it was not explicitly designed to evolve. This study will use an alpha-network as a baseline for comparison to the 150-isotope network. The alpha-network model alone will produce physically meaningful results, but the 150-isotope network will go far beyond any previous study.

As previously stated, a large network in 2D is too taxing given the current scheme in CHIMERA, so our in-situ alpha-network will rely on the post-processing of tracers to model the nucleosynthesis. After the adaptive QSE-network is installed we can perform a nucleosynthesis study between an alpha-network and a large-network in the context of a multi-dimensional simulation. In parallel to these two network runs, we would like to perform a detailed study on how the thermonuclear energy from burning and nucleon recombination feeds back into the hydrodynamics. This will help us understand the effect thermonuclear energy release has on the outward propagating shock and supernova dynamics.

Ultimately, we would like to run ECSNe with an adaptive QSE-network and a yin-yang grid — a pole-free 3D grid unlike the typical spherical-polar grid. As with any realistic hydrodynamic model, a self-consistent 3D model would provide a verification of 2D results. But most importantly, as stated by Wanajo et al. [8] and Pumo et al. [7], a 3D run is required to not only study the creation of nuclei within ECSN, but to understand the composition of the ejecta due to a more realistic determination of convective processes afforded by the move to 3D.

5. Conclusion

We seek to extend existing core-collapse supernova (CCSN) models to examine nucleosynthesis in a self-consistent fashion. Self-consistency requires, among other things, a code that evolves explosions without parameterized methods to obtain delayed neutrino heated shock revival and late time ejecta that match CCSNe morphologies and galactic-abundance contributions from CCSNe. The tracer particles have been (and continue to be) a vital tool in our efforts to improve the physical fidelity of our models. Since observations can only be confronted by computational investigations, over the next few of years, the CHIMERA team has an ambitious agenda of running many 2D and 3D models to examine the effects of progenitor mass, rotation rate, and micro-physics input on CCSN evolution and the explosion mechanism. It must be stressed that these analyses will be incomplete without the improvements and efforts highlighted in this paper to our nucleosynthesis studies.

6. Acknowledgement

The simulations described herein were conducted at the Oak Ridge Leadership Computing Facility at ORNL and at the National Institute for Computational Sciences. This work was supported in part by NSF grants NSF-AST-0653376, NSF-PHY-0855315, NSF-OCI-0749242, NSF-OCI-0749204 & NASA grant 07-ATFP07-0011. WRH and AM acknowledge support from the Office of Nuclear Physics, US Department of Energy, and OEBM and AM acknowledge support from the Office of Advanced Scientific Computing Research, US Department of Energy.

References

[1] Messer O E B, Bruenn S W, Blondin J M, Chertkow M A, Hix W R, Lee C, Lentz E J, Marronetti P and Yakunin K N 2010 *Proceedings of Nuclei in the Cosmos XI*
[2] Lattimer J and Swesty F D 1991 *Nucl. Phys. A* **535** 331–376
[3] Messer O E B, Bruenn S W, Blondin J M, Hix W R, Mezzacappa A and Dirk C J 2007 **78** 1–5
[4] Bruenn S W, Mezzacappa A, Hix W R, Blondin J M, Marronetti P, Messer O E B, Dirk C J and Yoshida S 2010 *February* (*Preprint* `arXiv:1002.4914`)
[5] Janka H T, Mueller B, Kitaura F S and Buras R 2007 *Astronomy & Astrophysics* 10
[6] Kitaura F, Janka H T, and Hillebrandt W, 2008 *Astronomy & Astrophysics* 6
[7] Pumo M L, Turatto M, Botticella M T, Pastorello A, Valenti S, Zampieri L, Benetti S and Patat F 2009 *October*
[8] Wanajo S, Janka H T, and Müller B, 2010 *ApJ* **726** L15
[9] Hix W R, Parete-Koon S T, Freiburghaus C and Thielemann F K 2007 *ApJ* **667** 476–488
[10] Fröhlich C, Liebendörfer M, Thielemann F K, Martínez-Pinedo G, Langanke K, Zinner N, Hix W and Bravo E 2006 *Proceedings of Nuclei in the Cosmos -IX* ed A Mengoni e a (SISSA Proceedings of Science) p 33
[11] Lee T C 2008 *The Applications of the Tracer Particle Method to Multi-Dimensional Supernovae Siulatons* Ph.D. thesis University of Tennessee Knoxville
[12] Yakunin K N, Marronetti P, Mezzacappa A, Bruenn S W, Lee C T, Chertkow M A, Hix W R, Blondin J M, Lentz E J, Messer O E B and Yoshida S 2010 *Classical and Quantum Gravity* **27** 194005
[13] Murphy J W, Ott C D and Burrows A 2009 *The Astrophysical Journal* **707** 1173–1190
[14] Hix W R and Thielemann F 1995 *BAAS* **27** 1423
[15] Rauscher T, and Thielemann F K, 2000 *Atomic Data and Nuclear Data Tables* **75** 1–351
[16] Pruet J, and Woosley SE, and Buras R., and Janka H.T., and Hoffman R.D, *The Astrophysical Journal* **400** 325–336
[17] Fröhlich C, Hix W R, Martínez-Pinedo G, Liebendörfer M, Thielemann F K, Bravo E, Langanke K and Zinner N T 2006 *New Astronomy Review* **50** 496–499
[18] Timmes F X and Woosley S E 2000 *Astronomy* 377–398
[19] Pruet J, Woosley S E, Buras R, Janka H T and Hoffman R D 2005 *ApJ* **623** 325–336
[20] Pruet J, Hoffman R D, Woosley S E, Janka H T and Buras R 2006 *ApJ* **644** 1028–1039
[21] Fröhlich C, Martínez-Pinedo G, Liebendörfer M, Thielemann F K, Bravo E, Hix W R, Langanke K and Zinner N T 2006 *Phys. Rev. Lett.* **96** 142502
[22] Woosley S E and Weaver T A 1995 *ApJS* **101** 181
[23] Thielemann F K, Nomoto K and Hashimoto M 1996 *ApJ* **460** 408

[24] Cayrel R, Depagne E, Spite M, Hill V, Spite F, François P, Plez B, Beers T, Primas F, Andersen J, Barbuy B, Bonifacio P, Molaro P and Nordström B 2004 *A&A* **416** 1117–1138
[25] Gratton R G and Sneden C 1991 *A&A* **241** 501–525
[26] Ning H, Qian Y Z and Meyer B S 2007 *The Astrophysical Journal* **667** L159–L162
[27] Nomoto K 1987 *ApJ* **322** 206
[28] Hillebrandt W, Nomoto K and Wolff R 1984 *Astronomy and Astrophysics* **133** 175–184
[29] Ritossa C, Garcia-Berro E and Iben Icko J 1996 *ApJ* **460** 489

Phase diagram of carbon-oxygen plasma mixtures in white dwarf stars

AS Schneider[1], CJ Horowitz[1], J Hughto[1] and DK Berry[2]

[1]Department of Physics and Nuclear Theory Center, Indiana University, Bloomington, IN 47405, USA

[2]University Information Technology Services, Indiana University, Bloomington, IN 47408, USA

E-mail: andschn@indiana.edu

Abstract.
The liquid-solid phase-diagram of dense carbon-oxygen plasma mixtures found in white dwarf stars interiors is determined from molecular dynamics (MD) simulations. Our MD simulations consist of boxes with 55296 ions with different carbon to oxygen ratios. Finite size effects are estimated comparing the new MD simulations results to previous smaller simulations. We use bond angle metric to identify whether an ion is in the solid, liquid or interface and study non-equilibrium effects by obtaining the diffusion coefficients in the different phases. Our phase diagram agrees with predictions from Medin and Cumming obtained by an independent method.

1. Introduction

White Dwarf (WD) stars are planet sized remains of conventional stars that no longer burn nuclear fuel. The vast majority of WD stars have carbon and oxygen in their cores as they are the evolutionary end of stars not massive enough to synthesize heavier elements. Observations of the cooling curves of these stars provide important information on the evolution and age of a stellar population [1]. The core of a WD is a plasma of charged ions immersed in a degenerate electron gas. As the star cools by radiating away its internal heat this plasma crystallizes. This can delay the WD cooling [2]. Recently, Winget et al. observed effects from the latent heat of crystallization on the luminosity function of WDs in the globular cluster NGC 6397 [3]. These observations allow one to determine the melting temperature of the carbon and oxygen mixtures inside WD stars, which impose constrains on the ratio of carbon to oxygen in the core.

In this paper we determined from molecular dynamics (MD) simulations the melting temperature and other properties of the liquid-solid phase diagram for carbon-oxygen plasma mixtures. Phase diagrams for these mixtures have been calculated before assuming a local density model for the free energy of the solid [4] or based on Monte Carlo and MD simulations of free energies for both the liquid and the solid phases[5–8]. Recently Potekhin et al. have made accurate calculations of the free energy of liquid mixtures [9, 10] and Medin and Cumming calculated the phase diagram for both binary mixtures such as C/O and much more complicated multicomponent mixtures [11].

All of these works determine liquid-solid phase equilibria by equating liquid and solid free energies that have been calculated separately. This procedure allows the use of smaller Monte Carlo or MD simulations where only a single phase is present at a time. However, it may be very sensitive to any small errors in the free energy difference between liquid and solid phases. Indeed for the C/O system, Segretain et al. [4] predict higher melting temperatures and a spindle type phase diagram while both Ogata et al. [5, 6] and Medin and Cumming [11] predict lower melting temperatures and an azeotrope type phase diagram. In a spindle-type phase diagram the melting temperature of the mixture is always greater than the melting temperature of pure carbon, while in an azeotrope type phase diagram the melting temperature of the mixture can be lower than that of pure carbon. This difference in phase diagrams could be due to small errors in Segretain's solid free energies.

Furthermore, equating the free energies of liquid and solid phases provides no information on the dynamics of the phase transition nor are interface properties addressed. For multicomponent systems there is, in general, a gradient in composition across the liquid-solid interface. To our knowledge the spatial extent of this gradient has not been determined. Recently, we performed direct two phase molecular dynamics simulations of liquid-solid phase equilibria both for carbon/oxygen mixtures in WD [12] and for a complex 17 component mixture modeling the crust of an accreting neutron star [13]. These simulations have both liquid and solid phases present simultaneously which allows for a direct determination of the melting temperature, and the composition of the liquid and solid phases from a single simulation. One of the advantages of this method is that phase equilibria for very complicated systems can be simulated. However, direct simulations need to address potential systematic errors from finite size effects and from the lack of thermodynamic equilibration. Finite size effects are potentially important because one must fit not only liquid and solid phases but also two liquid-solid interfaces within the simulation volume. This, in general, requires a larger simulation volume than for simulations of only a single phase. Nevertheless, recent computer advances reduced dramatically the computational limitations on these larger simulations. It is now easy to simulate much larger systems than have typically been run in the past.

One must run these two-phase simulations long enough to ensure that the phases have come into thermodynamic equilibrium. This may require impurities to diffuse throughout the solid phase. However, diffusion in the solid phase is relatively fast, for these Coulomb systems, because the ions do not have hard core interactions. There is only a relatively soft $1/r$ interaction between ions. As a result, ions can move past one another. We have studied diffusion in Coulomb crystals in a recent paper [14]. If finite size and non-equilibirum effects are addressed, direct two phase simulations should yield accurate results. The systematic errors from two-phase simulations are likely very different from previous free energy calculations. Therefore, comparing the two methods provides an important check on both approaches.

In this paper we perform larger MD simulations of C/O mixtures with 55296 ions and compare them to the 27648 ions we used previously [12]. This allows us to study finite size effects. We discuss our MD formalism in section 2, including a method based on local order parameter [15] to determine whether ions are in the solid liquid or interface, present results in section 3 and conclude in section 4.

2. Formalism

In section 2.1 we describe the formalism used in our MD simulations. The formalism is essentially the same we used in a previous work on carbon oxygen mixtures in WD [12]. In section 2.2 we discuss the procedure used to determine whether a given ion is in a liquid or solid phase.

2.1. MD Formalism

In our simulations we address a system of ions, of two different species, and electrons. Since the system is under great pressure all ions are fully ionized and electrons form a degenerate Fermi gas. The electronic structure of this relativistic electron gas is very simple and, therefore, does not need to be directly computed in the simulation. Ions interact with each other via a long ranged Coulomb interactions, screened by the electron background. The potential between two ions, labeled i and j, is

$$v_{ij}(r) = \frac{Z_i Z_j e^2}{r} e^{-r/\lambda}. \tag{1}$$

Z_i and Z_j are the ion charges, r is their separation and λ is the Thomas-Fermi screening length. For cold relativistic electrons $\lambda^{-1} = 2\alpha^{1/2} k_F / \pi^{1/2}$ where α is the fine structure constant and k_F the electron Fermi momentum, $k_F = (3\pi n_e)^{1/3}$. The electron density n_e is equal to the ion charge density, $n_e = \langle Z \rangle n$ where n is the ion density and $\langle Z \rangle$ the average ion charge. Our simulations are classical and the electron mass has been neglected in order to be consistent with our previous work. However, the electron mass is important at the lower densities in WD and this may change our results slightly [16]. In addition, quantum effects could play some role at high densities [17]. The simulations can be characterized by an average Coulomb parameter

$$\Gamma = \frac{\langle Z^{5/3} \rangle e^2}{a_e T}. \tag{2}$$

Above $\langle Z^{5/3} \rangle$ is an average over the ion charges, T is the temperature and $a_e = (3/4\pi n_e)^{1/3}$ the electron sphere radius. Time can be measured in units of one over the plasma frequency ω_p, which depends on the ion charge Z and mass M. Long wavelength fluctuations in the charge density can undergo oscillations at the plasma frequency. A hydrodynamical plasma frequency $\bar{\omega}_p$ from the simple averages of Z and M was defined for mixtures,

$$\bar{\omega}_p = \left[\frac{4\pi e^2 \langle Z \rangle n_e}{\langle M \rangle} \right]^{1/2}. \tag{3}$$

Choices for the average over composition in Eq. 3 are not unique. However, $\bar{\omega}_p$ is not expected to change significantly for different choices. All of our simulations are run for the same electron density of $n_e = 5.026 \times 10^{-4}\text{fm}^{-3}$. Since the pressure is dominated by the electronic contribution, constant electron density corresponds, approximately, to constant pressure. The density can be scaled to any desired value by also changing the temperature T so that the value of Γ, see Eq. 2, remains the same.

2.2. Interface Finding

All of our MD simulations consist of two phases, a solid phase and a liquid phase. Though determining whether an ion is solid-like or liquid-like may be simple by a visual inspection, this determination is not straightforward numerically. Also, as the system evolves ions can diffuse from one phase to the other or simply melt/solidify. Since we want to keep track of this and obtain quantities such as the composition and volume of each phase we need a numerical approach. The phase of an entire system can be determined by computing the global order parameter Q_l [18]. The phase of individual ions was determined using a prescription laid out in reference [15].

For each ion i, an ion j is defined as its neighbors if it is within a given radius $r_{min} = 4a$, where $a = (3/4\pi n)^{1/3}$ is the ion sphere radius. The vectors $\hat{\mathbf{r}}_{ij}$ joining neighbors are called bonds and their direction can be described by the angles θ_{ij} and ϕ_{ij} in the frame of ion i. The local structure around ion i can be characterized using spherical harmonics $Y_{lm}(\theta_{ij}, \phi_{ij})$ by

$$\bar{q}_{lm}(i) = \frac{\sum_{j=1}^{N_b(i)} \alpha(r_{ij}) Y_{lm}(\theta_{ij}, \phi_{ij})}{\sum_{i=1}^{N_b(i)} \alpha(r_{ij})} \tag{4}$$

where $N_b(i)$ is the number of ions bonded with ion i and we chose

$$\alpha(r_{ij}) = \left(\frac{r_{ij} - 4a}{2a}\right)^2$$

if $r < 4a$ and zero otherwise.

These local order parameters can be large in both the solid and the liquid. The order parameter with $l = 6$ was chosen since is the most sensitive to the bcc structure of the solid on our simulations. The global order parameter

$$Q_6 = \left[\frac{4\pi}{13}\sum_{m=-6}^{6} q_{6m}^{*} q_{6m}\right]^{1/2}, \quad \text{where} \quad q_{6m} = \frac{1}{N}\sum_{n=1}^{N} \bar{q}_{6m}(i), \tag{5}$$

is large in the solid due to the fact that the $\bar{q}_{6m}(i)$ add up coherently. In the liquid, $\bar{q}_{6m}(i)$ add incoherently, so Q_6 is small. This coherence is exploited to determine local order.

For each $\bar{q}_{6m}(i)$ a normalization is applied,

$$\tilde{q}_{6m}(i) \equiv \frac{\bar{q}_{6m}(i)}{\left[\sum_{m=-6}^{6} |\bar{q}_{6m}(i)|^2\right]^{1/2}} \tag{6}$$

and a dot product of the vectors $\mathbf{q}_6$ is defined for neighboring particles i and j,

$$\mathbf{q}_6(i) \cdot \mathbf{q}_6(j) \equiv \sum_{m=-6}^{6} \tilde{q}_{6m}(i)\tilde{q}_{6m}^{*}(j). \tag{7}$$

Note that, by construction, $\mathbf{q}_6(i) \cdot \mathbf{q}_6(i) = 1$.

As in reference [15] we define two particles to be connected whenever $\mathbf{q}_6(i) \cdot \mathbf{q}_6(j) > 0.5$. Then, we define a threshold on the number of connections to determine if an ion is solid-like or liquid-like. On average an ion in our simulations has 62 neighbors. The threshold value used was 20 connections. Thus, an ion connected to more (less) than 20 neighboring ions is labeled as solid-like (liquid-like). This criterion is met for most of the ions within a given phase. However, due to diffusion and thermal fluctuations, ions in the solid (liquid) may temporarily be connected to less (more) than 20 neighbors and be incorrectly labeled. Thus, we add an extra step to our algorithm which also enables us to establish where the liquid-solid interface is. Deep in the solid, a vast majority of the ions are identified as solid-like. In the bulk of the liquid a similar majority is identified as liquid-like. Along the interface, there is a mixture of solid-like and liquid-like ions. For this reason, we tag an ion as being in the solid or liquid if a large majority (> 0.80) of ions within four lattice spacings are in the same phase. If this criterion is not met, then the ion is determined to be in the interface. Notice that ions determined to be in the interface are found where one would expect them (see figure 1), along the border of the solid and liquid phases.

3. Results

In this section we present our results for the evolution of three large MD simulations of liquid-solid systems. Each system had a different carbon to oxygen ratio and from their final compositions the phase diagram was obtained. Diffusion coefficients were estimated in order to monitor non-equilibrium effects.

The MD simulations consisted of boxes with 55296 ions which were twice as long in the z direction as in the x and y directions. As in reference [12], we used periodic boundary conditions and the potential between ions was obtained from the single interaction between the nearest periodic image. Note that the rectangular geometry increases the distance between the two liquid-solid interfaces compared to cubical simulation volume. We expect this to minimize finite size effects. To decrease simulation time we used a cut-off distance $r_{cut} = 8\lambda$ when calculating the potential between ions. Taking the cut-off into account the potential between ions is altered to

$$v'_{ij}(r) = v_{ij}(r) - v_{ij}(r_{cut}). \tag{8}$$

This cut-off is large enough that the phase diagram should not be altered by it and, still, significantly smaller than the sides of the box so that Ewald sums over periodic images do not have to be included. We also point out that we do not include gravity in our simulations as the sides of the boxes are of the order of 1 000 fm while the radius of a white dwarf star is 19 orders of magnitude larger, about 10 000 km.

The initial conditions consisted of two cubes with 27648 ions, one in the liquid phase and one in the solid phase, stacked together along the z direction. Throughout the paper, for convenience, we will refer to the top (bottom) of the box as the fraction of the box along the z direction where the liquid (solid) cube was initially placed. We note here that this naming scheme was chosen arbitrarily since, because of periodic boundary conditions, the simulation volumes have no physical discernible top or bottom. Initial composition of each phase and temperature of the system was based on our results for previous simulations with cubic boxes [12].

3.1. The 75% oxygen run

The first simulation to be discussed had 75% oxygen and 25% carbon by number, or $x_O = 0.75$ and $x_C = 0.25$. Liquid and solid phases were prepared independently and then combined to form the initial 55296 ions system. Based on previous work suggesting that the solid will be enriched in oxygen we prepared the solid phase to be composed of 80% oxygen, $x^s_O = 0.80$. Here s denotes the solid phase. Since each volume contains half of the 55296 particles, the liquid phase was prepared enriched in carbon; $x^l_O = 0.70$, where l denotes the liquid.

In order to obtain the 27648 crystal we started with a 3456 ion box. Ions where placed in the box with random positions and velocities and oxygen number fraction of $x_O = 0.80$. The system was cooled until it crystallized and, then, 8 copies of the crystal were combined to form a 27648 system. This system was evolved for $\bar{\omega}_p t = 8900$ at a $\Gamma = 213.1$.

The liquid box was obtained in a similar way. We started from a 3456 ion box with random positions and velocities, but a oxygen fraction of $x_O = 0.70$. This system was evolved at the same Γ for $\bar{\omega}_p t = 4400$. Because of the larger carbon fraction the system does not crystallize at the same Γ as the solid. Then, 8 copies of the system were combined to form the 27648 liquid. This system was evolved for a further $\bar{\omega}_p t = 8900$ and then placed on top of the solid box to form the initial 55296 ion configuration.

This initial condition is not equilibrated for a few reasons to be discussed. First, by just placing the liquid box on top of the solid one the interface may acquire a very high energy: ions in the liquid and the solid may end up very close to each other. Also, we expect the composition gradient across the solid-liquid interface to be smooth. That is not the case when the boxes are just placed on top of each other. The solid may not be equilibrated as diffusion in the solid phase is slow and the crystallized system was only evolved for a short time. Finally, the composition of each phase and the equilibrium temperature may be wrong, we need to wait for the phases to interact for some in order for it to reach equilibrium.

The full system was evolved for a total time of $2.2 \times 10^6 \bar{\omega}_p t$ using a velocity Verlet [19] algorithm with a time step of $0.177/\bar{\omega}_p$. During the run we rescaled the velocities every 100 time steps to keep the chosen temperature approximately constant. We also slowly adjusted the temperature to ensure the fractions of liquid and solid would stay approximately the same. Though this does not conserve energy during the run, our main goal in this paper was to minimize finite size effects by keeping the two liquid-solid interfaces as far from each other as possible. We employed a simple hybrid OpenMP/MPI computer code and used about 768 cores on the Cray XT5 system Kraken [20]. The total run time was about two weeks.

Using the VMD software [21] we show in figure 1 the projection along the yz plane of the final configuration. On the left figure particles are colored based on whether they represent carbon or oxygen ions. The bottom half of the simulation volume is seen to contain a crystalline region. On the right figure we used the procedure of Sec. 2.2 to determine which parts of the system are liquid, solid, or belong to the two liquid-solid interfaces and colored particles accordingly. Note that the interface regions are not rectangular and show some fluctuations. These fluctuations were observed to change with time, even though the volume of the interface did not (see figure 2).

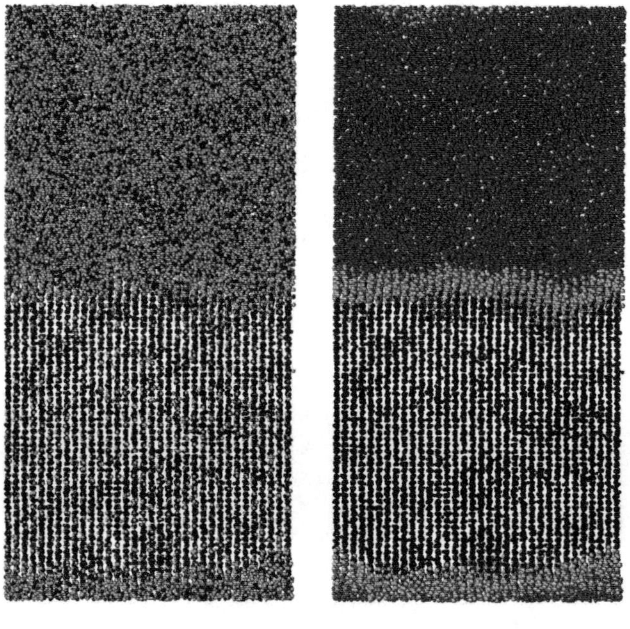

Figure 1. (Color on line) Projection along the yz plane for the final configuration of our 75% oxygen 25% carbon simulation.
Left: Ions were colored based on whether they were carbon (light pink) or oxygen (black).
Right: Ions were colored based on whether they belonged to the solid (black), liquid (light pink) or interface (dark purple).

In figure 2 we show the fluctuations in temperature

$$\delta\Gamma = \frac{\Gamma - \bar{\Gamma}}{\bar{\Gamma}} \tag{9}$$

where $\bar{\Gamma}$ is the average of Γ over the last third of the run, see table 1. Initially the temperature had to be kept low (high Γ) in order to keep the non-equilibrated solid from melting. After a short time the system equilibrates and the fluctuations in temperature are only of the order of half of a percent.

Also shown in the same figure are the volume fractions of the solid and liquid phases and the interface. Note that the interface volume fraction barely changes over the total run. Its volume is mainly dependent on how we defined the interface, see algorithm described in section 2.2. Meanwhile, the total liquid and solid volumes fluctuate in about 10% of the total simulation

Table 1. Equilibrium compositions of our 55296 ion runs. Oxygen number fraction of the whole system is x_O, Coulomb parameter averaged over the last third of the run $\bar{\Gamma}$ (determined from the final temperature and x_O). The composition of the solid is x_s, the liquid is x_l, and the interface regions is x_i. Statistical errors are quoted in the last digit in parentheses.

x_O	$\bar{\Gamma}$	x_O^s	x_O^l	x_O^i
0.75	203.9(6)	0.806(1)	0.703(3)	0.744(6)
0.50	221.6(7)	0.552(1)	0.454(3)	0.494(8)
0.25	213.3(6)	0.250(2)	0.249(2)	0.255(6)

volume, even if their oxygen compositions (see figure 3) change only by 1%. In table 1 are also shown the average oxygen fractions of each phase for the last third of the run. We note that in this oxygen rich system the solid phase is enriched in oxygen, $x_O^s \approx 0.80$, the liquid phase becomes depleted in it, $x_O^l \approx 0.70$, and the interface has a value that is similar to the total oxygen fraction of the system, $x_O^i \approx 0.75$.

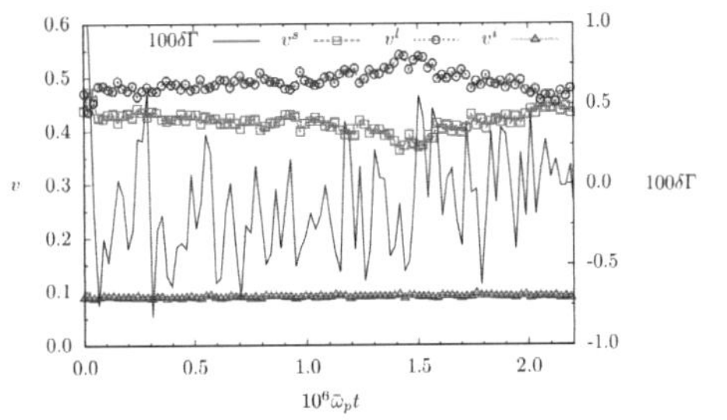

Figure 2. (Color on line) Fractional fluctuations in Coulomb parameter $\delta\Gamma$ (solid black line and right hand scale), see equation 9, vs simulation time in units of $10^6\bar{\omega}_p t$ for a 55296 ion system with $x_O = 0.75$. Also shown are the fraction of the system that is solid (red squares), liquid (blue circles) or interface (brown triangles).

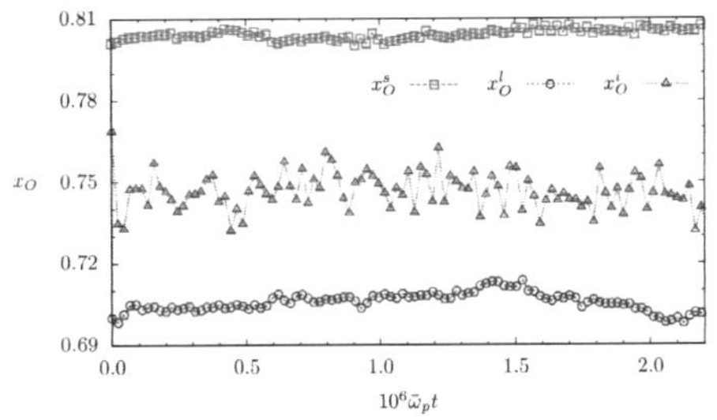

Figure 3. (Color on line) Number fraction of oxygen for solid (red squares), liquid (blue circles) or interface (brown triangles) vs simulation time in units of $10^6\bar{\omega}_p t$ for a 55296 ion system with $x_O = 0.75$.

In figure 4 we show different quantities as a function of the height z in the simulation volume averaged over the last one third of the run. We divided the simulation volume into fifty slices along the z coordinate, with slice 1 being at the bottom of figure 1 and slice 50 being at the top. We show how (from top to bottom) the diffusion coefficients, oxygen number fraction and phase fraction change as a function of the height in the box. It is easy to see that slices that are mainly solid, 6 through 21, have higher oxygen fraction than the slices that are mainly liquid, slices 30 to 48. Also important to note is the composition gradient across both interfaces and the fact that composition in the liquid is much more uniform than in the solid. The later is due to the fact that the liquid equilibrates faster than the solid. This happens because ions diffuse much faster in the liquid, as discussed next.

Diffusion coefficients were calculated using the method discussed in [14] in order to check on equilibration. The coefficients plotted in figure 4 have been normalized to a reference value

$$D_0 = \frac{3\bar{\omega}_p a^2}{\Gamma^{4/3}} \tag{10}$$

for both carbon and oxygen ions along the box, see also table 2. Diffusion for carbon in the liquid, D^l_C, is slightly larger than the for oxygen, D^l_O. This agrees with the result obtained in reference [22]. In the solid, as expected, diffusion is much slower than in the liquid, with D^s_C being more than 70 times smaller than D^l_C. This result is similar to the diffusion determined for one component solids near the melting temperature in [14]. Moreover, in the solid D^s_O is much smaller than D^s_C. Therefore, we expect the equilibration time of the system to depend more on D^s_C than on D^s_O. This is because the system will equilibrate by having carbon ions diffuse out of the the solid to increase x^s_O or into it to increase x^s_O.

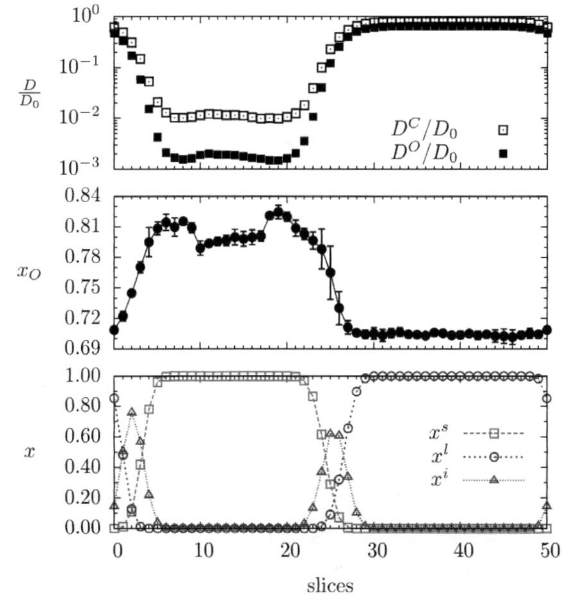

Figure 4. (Color on line) Diffusion, oxygen fraction and phase fractions for the 55296 ion simulation that is 75% oxygen and 25% carbon as a function of the height z of the simulation volume. Slices 0 and 50 contain the same data.
Top: Diffusion coefficients for carbon D_C (open squares) and oxygen D_O (closed squares) as a function of the height of the box.
Center: Oxygen number fraction as a function of the height in the box.
Bottom: Fraction of ions in a given slice that are in the liquid region (blue circles), solid (red squares) and interface (brown triangles).

Table 2. Diffusion coefficients of liquid and solid phases averaged over the last third of the runs. Results are expressed as D^p_X in units of D_0, see equation 10. The letter p denotes the phase, s for solid and l for liquid, while X stands for ion species, C for carbon and O for oxygen. Statistical errors are quoted in the last digit in parentheses.

x_O	D^l_C	D^l_O	D^s_C	D^s_O
0.75	0.80	0.65	0.011(1)	0.0018(1)
0.50	0.68	0.55	0.0074(4)	0.0013(1)
0.25	0.64	0.52	0.0032(3)	0.0007(1)

3.2. The 50% oxygen run

In this section we discuss the 55296 ion run with 50% oxygen and 50% carbon. This run was prepared much in the same way as the previous one, with the exception that now the solid was initially prepared to contain 55% oxygen and the liquid 45%. These compositions were also chosen based on our previous 27648 simulations [12].

We show in figure 5 the fluctuations in temperature $\delta\Gamma$ and the volume fractions of each phase as a function of simulation time. The evolution of the composition of each phase is shown in figure 6. The time dependence of the oxygen fraction in each phase is small. This suggests that the starting composition is very close to the desired equilibrium composition or that equilibrium is reached very slowly. Longer runs and runs starting from different initial conditions should be able to settle this.

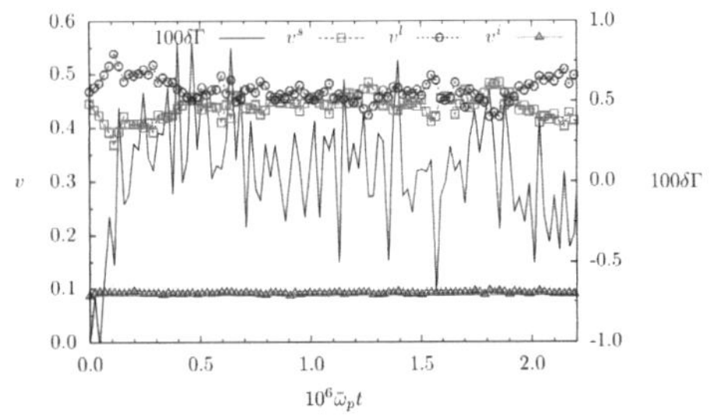

Figure 5. (Color on line) Fractional fluctuations in Coulomb parameter $\delta\Gamma$ (solid black line and right hand scale), see equation 9, vs simulation time in units of $10^6\bar{\omega}_p t$ for a 55296 ion system with $x_O = 0.50$. Also shown are the fraction of the system that is solid (red squares), liquid (blue circles) or interface (brown triangles).

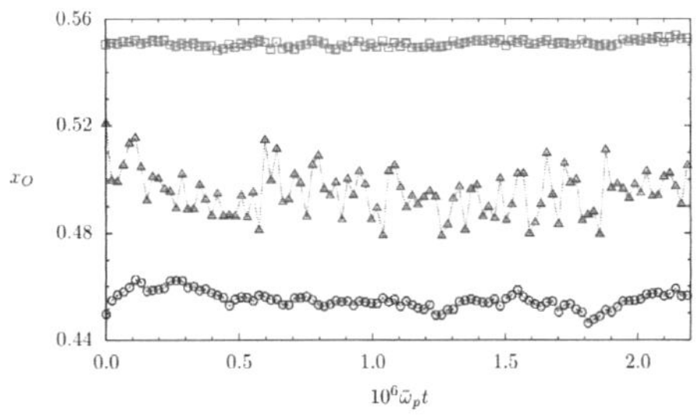

Figure 6. (Color on line) Number fraction of oxygen for solid (red squares), liquid (blue circles) or interface (brown triangles) vs simulation time in units of $10^6\bar{\omega}_p t$ for a 55296 ion system with $x_O = 0.50$.

In figure 7 we show the diffusion coefficients, oxygen number fraction and the fraction of each phase as a function of the height z in the box. The qualitative discussion above for the 75% oxygen run is also valid for the 50% oxygen run. Diffusion in the liquid is much faster than in the solid and there are large fluctuations in the solid composition. The average composition for each phase in this run is quoted in table 1 while diffusion coefficients were quoted in table 2.

3.3. The 25% oxygen run

In this section we discuss the 55296 ion run with 25% oxygen and 75% carbon. This run was prepared much in the same way as the previous ones, with the exception that now both solid and liquid were initially prepared to contain 25% oxygen. This is because on our previous 27648 simulations the chemical separation between these two phases was shown to be small [12].

Figure 8 shows the fluctuations in temperature $\delta\Gamma$ and the volume fractions of each phase as a function of simulation time. As in the previous simulations the interface volume is approximately constant and the temperature fluctuations are within half of a percent of the average temperature for the last third of the run. The evolution of the composition of each phase is shown in figure 9. Unlike the richer oxygen systems, the solid phase is not enriched in oxygen, see table 1. The oxygen composition of the liquid and solid phases fluctuate less than one percent, figure 9, even if the total fraction of these phases fluctuate in about 20% of the total simulation volume, figure

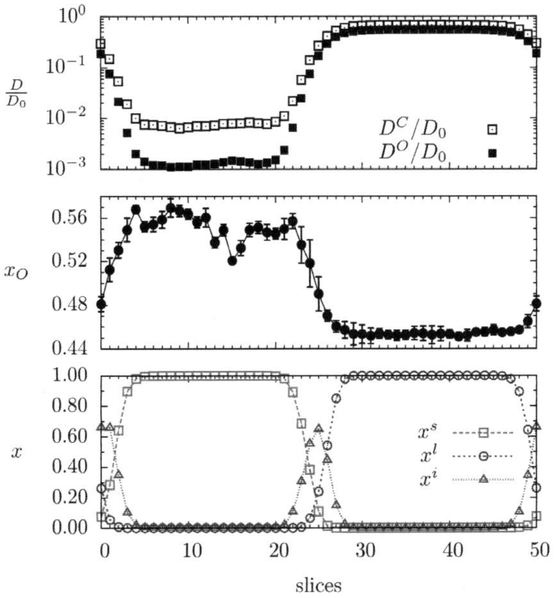

Figure 7. (Color on line) Diffusion, oxygen fraction and phase fractions for the 55296 ion simulation that is 50% oxygen and 50% carbon as a function of the height z of the simulation volume. Slices 0 and 50 contain the same data.
Top: Diffusion coefficients for carbon D_C (open squares) and oxygen D_O (closed squares) as a function of the height of the box.
Center: Oxygen number fraction as a function of the height in the box.
Bottom: Fraction of ions in a given slice that are in the liquid region (blue circles), solid (red squares) and interface (brown triangles).

8. The interface composition undergoes larger oscillations and is slightly oxygen enriched with respect to the solid and liquid phases. In the thermodynamic limit, which is the one relevant for the phase diagram, the interface composition is not relevant. However, the simulation might be showing a real effect where solid and liquid have very similar compositions and the interface is slightly oxygen rich.

In figure 10 we show the diffusion coefficients, oxygen number fraction and the fraction of each phase as a function of the height z in the box. Again diffusion in the liquid is much faster than in the solid and there are large fluctuations in the solid composition. The interface again is only about 10% of the simulation volume, however looking at the bottom plot in figure 10 it seems that it stretches for a larger volume. This may be the case if the fluctuations in the interface are larger for this run than the other two. The average composition for each phase in this run is quoted in table 1 while diffusion coefficients were quoted in table 2.

3.4. The carbon oxygen phase diagram

This section discussed the phase diagram obtained from the compositions of the runs discussed in sections 3.1, 3.2 and 3.3. Figure 11 shows the phase diagram as a function of the oxygen composition. For historical reasons, the vertical axis is shown as the temperature divided by the melting temperature of pure carbon, assumed to happen at $\Gamma = 178.4$ [12]. This temperature is slightly different than the one obtained for one component plasmas because we include the effect of electron screening.

The points plotted in figure 11 are the ones shown in table 1. In the same figure we plot the points obtained for 27648 ion simulations [12]. When comparing the points of the two simulations we notice a few differences.

First, for the 55296 ion 25% oxygen simulation the composition of the liquid and solid phases seem to be about the same, whether in our previous run the solid seem to be somewhat enriched in oxygen. In addition, the equilibrium temperature for the larger system is slightly larger

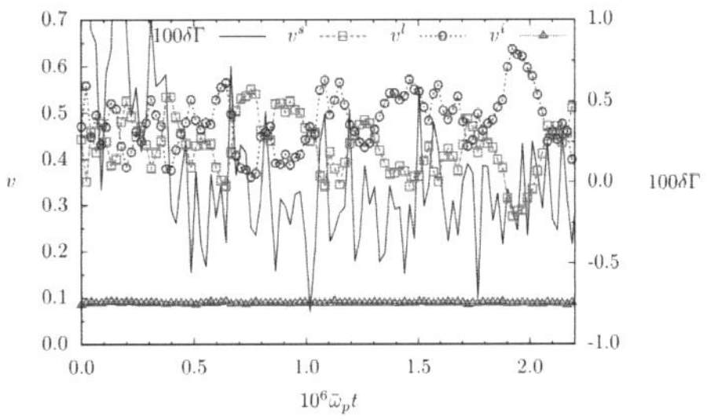

Figure 8. (Color on line) Fractional fluctuations in Coulomb parameter $\delta\Gamma$ (solid black line and right hand scale), see equation 9, vs simulation time in units of $10^6\bar{\omega}_p t$ for a 55296 ion system with $x_O = 0.25$. Also shown are the fraction of the system that is solid (red squares), liquid (blue circles) or interface (brown triangles).

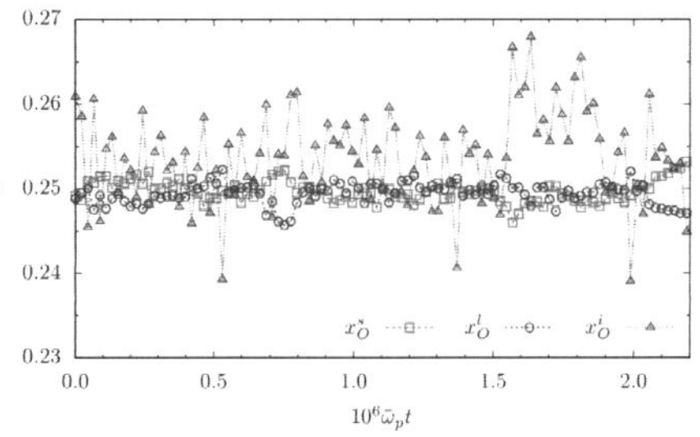

Figure 9. (Color on line) Number fraction of oxygen for solid (red squares), liquid (blue circles) or interface (brown triangles) vs simulation time in units of $10^6\bar{\omega}_p t$ for a 55296 ion system with $x_O = 0.25$.

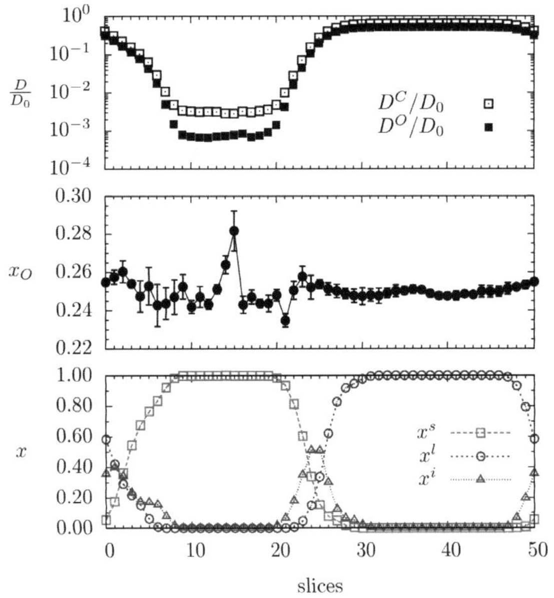

Figure 10. (Color on line) Diffusion, oxygen fraction and phase fractions for the 55296 ion simulation that is 25% oxygen and 75% carbon as a function of the height z of the simulation volume. Slices 0 and 50 contain the same data.
Top: Diffusion coefficients for carbon D_C (open squares) and oxygen D_O (closed squares) as a function of the height of the box.
Center: Oxygen number fraction as a function of the height in the box.
Bottom: Fraction of ions in a given slice that are in the liquid region (blue circles), solid (red squares) and interface (brown triangles).

for the 55296 ion simulation. The 50% oxygen simulation with 55296 ions seem to have a larger difference in solid and liquid composition than the 27648 ion simulation, though almost the same equilibrium temperature. Finally, the 75% oxygen simulation with 55296 ions has a slightly higher melting temperature than the 27648 ion one. The differences between the 55296 and 27648 ion simulations suggest that there are some finite size effects playing a role in these runs. However, the differences are small enough that we can conclude that finite size effects are

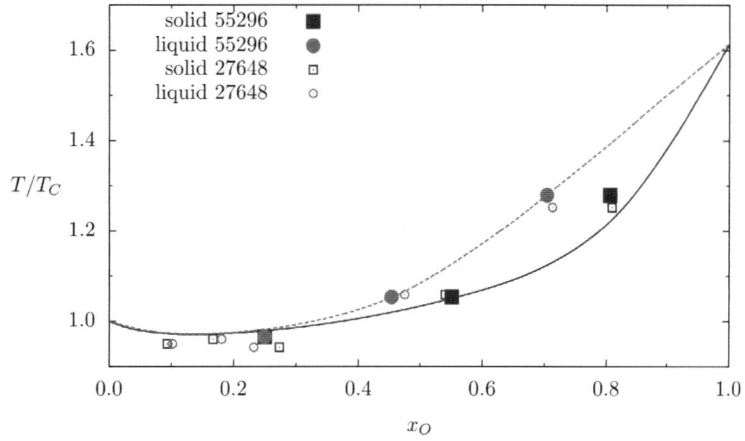

Figure 11. (Color on line) Phase diagram for carbon-oxygen plasma mixtures. The horizontal axis represents the oxygen fraction x_O while the vertical one represents the temperature scaled to a pure carbon plasma melting temperature, $\Gamma = 178.4$. The composition of the liquid phase (upper dashed red curve or red circles) that is in equilibrium with the solid phase (lower solid blue curve or squares). Our present results from 55296 ion simulations are filled symbols while the open symbols are our previous results with 27648 ions from reference [12]. The curves are the results of Medin and Cumming [11].

well under control.

Alongside our MD simulation results in figure 11 are the curves obtained by Medin and Cumming [11]. The agreement between their model and our simulations is excellent. All three liquid points and two of the three solid compositions from table 1 are very close to Medin and Cumming's results. There is a small difference in the solid composition of the 75% oxygen simulation. This difference could be due to small systematic errors in our simulations, our system still not being in equilibrium, or on small errors in the model of liquid and solid mixtures free energies used by Medin and Cumming. In any case, the small finite size corrections in going from a 27648 ion to a 55296 ion simulation improved the agreement between our results and Medin and Cumming's. This suggests that the remaining errors due to finite-size and non-equilibrium effects in our direct MD simulation are small. It also suggests that the model free energies employed by Medin and Cumming are good, at least for carbon-oxygen systems.

We stress here that finite size and non-equilibrium effects might still be playing a role in the final result. Although small, finite size effects seem to have different effects on all three runs. First, the 25% oxygen run has a lower melting temperature and smaller composition difference between the liquid and solid phases. These points are in a region of the phase diagram where melting temperature, according to the model of Medin and Cumming, is almost independent of composition. Because of that, small errors may drastically change the composition of each phase.

For the 50% oxygen run the main difference is the larger composition difference between liquid and solid phases. This may arise if the gradient in the composition across the interface stretches over a distance comparable to the total size of the box. Because of that for the 27648 ion simulation the two interfaces may get significantly close to each other and the liquid and solid may end up with similar compositions.

There could also be a statistical component to the errors coming from a variety of non-equilibrium effects. For example, fluctuations in the location of an interface could create a new solid region and the composition of this region might not have time to equilibrate. Alternatively, there could be composition changes from large fluctuations. One could test for a variety of errors of this type by simply repeating these simulations a number of times with different initial conditions. Unfortunately, we have not had time to do this for the present paper.

We found that direct two-phase MD simulations can accurately determine liquid-solid phase equilibria. This result is very useful because direct MD simulation can be applied to many other systems, including very complex ones. Furthermore, direct MD simulations for a few compositions may provide very helpful benchmarks for simpler models. Note that simulations with a somewhat large number of particles may be necessary and these may have to be run for extended simulation times. However, rapid advances in computer power should make such simulations even easier in the future.

4. Summary and conclusions

The liquid-solid phase diagram for carbon-oxygen plasma mixtures using two-phase MD simulations was determined. Liquid, solid, and interface regions were identified in our simulations using a bond angle metric described in section 2.2. To study finite size effects, we performed 55296 ion simulations and compared to earlier 27648 ion results. To help monitor non-equilibrium effects, we calculated diffusion coefficients. We find that oxygen ions in the solid diffuse much slower than carbon ions and that both diffusion constants are 80 or more times smaller than diffusion constants in the liquid phase. There is excellent agreement between our phase diagram and that predicted by Medin and Cumming [11]. This suggests that errors from finite size and non-equilibrium effects are small, and that the carbon-oxygen phase diagram is now accurately known. This improved phase diagram can be used to explore the the evolution of cool white dwarfs, as done in [23]. In future work we will present phase diagrams for the three component carbon-oxygen-neon system and for the two component oxygen-selenium system. The oxygen-selenium system, with a larger ratio of charges than carbon-oxygen, can serve as a simple two component model of the complex rapid proton capture ash composition on an accreting neutron star [13].

Acknowledgments

We thank Z. Medin for helpful discussions. This research was supported in part by DOE grant DE-FG02-87ER40365 and by the National Science Foundation through TeraGrid resources provided by the National Institute for Computational Sciences under grant TG-AST100014.

References

[1] Fontaine G, Brassard P and Bergeron P 2001 *PASP* **113** 409
[2] Salaris M, Dominguez I, Garíca-Berro E, Hernanz M, Isern J and Mochkovitch R 1997 *ApJ* **486** 413
[3] Winget D E, Kepler S O, Campos F, Montgomery M H, Girardi L, Bergeron P and Williams K 2009 *ApJ* **693**
[4] Segretain L and Chabrier G 1993 *A&A* **271**
[5] Ogata S and Ichimaru S 1993 *Phys. Rev. E* **48** 1344
[6] Ichimaru S, Iyetomi H and Ogata S 1988 **334**
[7] DeWitt H and Slattery W 2003 *Contrib. Plasma Phys* **43** 279
[8] DeWitt H, Slattery W and Chabrier G 1996 *Physica B* **228** 21
[9] Potekhin A Y, Chabrier G and Rogers F J 2009 *Phys. Rev. E* **79** 016411
[10] Potekhin A Y, Chabrier G, Chugunov A I and DeWitt H E 2009 *arXiv:0909.3990*
[11] Medin Z and Cumming A 2010 *Phys. Rev. E* **81** 036107
[12] Horowitz C J, Schneider A S and Berry D K 2010 *Phys. Rev. Let.* **104** 231101
[13] Horowitz C J, Berry D K and Brown E F 2007 *Phys. Rev. E* **75** 066101
[14] Hughto J, Schneider A S, Horowitz C J and Berry D K 2011 *Phys. Rev. E* **84** 016401

[15] ten Wolde P R, Ruiz-Montero M J and Frenkel D 1996 *J. Chem. Phys.* **104** 9932–9947
[16] Potekhin A Y and Chabrier G 2000 *Phys. Rev. E* **62** 8554
[17] Jones M D and Ceperley D M 1996 *Phys. Rev. Lett.* **76** 4572
[18] Steinhardt P J, Nelson D R and Ronchetti M 1983 *Phys. Rev. B* **28** 784–805
[19] Verlet L 1967 *Phys. Rev.* **159** 98
[20] National Institute for Computational Sciences U T O R N L
[21] Humphrey W, Dalke A and Schulten K 1996 *Journal of Molecular Graphics* **14** 33–38
[22] Horowitz C J, Hughto J, Schneider A S and Berry D K 2010 *Phys. Rev. E* **82** 066401
[23] Althaus L G, García-Berro E, Isern J, Córsico A H and Miller Bertolami M M 2012 *A&A* **537** A33

Journal of Physics: Conference Series **402** (2012) 012027 doi:10.1088/1742-6596/402/1/012027

Turbulent magnetic field amplification from spiral SASI modes in core-collapse supernovae

E Endeve[1], C Y Cardall[2,3], R D Budiardja[2,3,4] and A Mezzacappa[1,2,3]

[1]Computer Science and Mathematics Division, Oak Ridge National Laboratory (ORNL), Oak Ridge, TN 37831-6354, USA
[2]Physics Division, ORNL, Oak Ridge, TN 37831, USA
[3]Department of Physics and Astronomy, University of Tennessee, Knoxville, TN 37996, USA
[4]Joint Institute for Heavy Ion Research, ORNL, Oak Ridge, TN 37831, USA

E-mail: `endevee@ornl.gov`

Abstract. We describe the initial implementation of magnetohydrodynamics (MHD) in our astrophysical simulation code GenASiS. Then, we present MHD simulations exploring the capacity of the stationary accretion shock instability (SASI) to generate magnetic fields by adding a weak magnetic field to an initially spherically symmetric fluid configuration that models a stalled shock in the post-bounce supernova environment. Upon perturbation and nonlinear SASI development, shear flows associated with the spiral SASI mode contributes to a widespread and turbulent field amplification mechanism. While the SASI may contribute to neutron star magnetization, these simulations do not show qualitatively new features in the global evolution of the shock as a result of SASI-induced magnetic field amplification.

1. Introduction

Shortly after the discovery of pulsars [23], the potential role of magnetic fields in the core-collapse supernova (CCSN) explosion mechanism began to be investigated (e.g., [28, 3, 35, 45]). In principle, a differentially rotating proto-neutron star (PNS) could give rise to magnetically powered explosions. An early conclusion, however, was that (unrealistically) rapid rotation *and* strong magnetic fields would be needed at the pre-collapse stage for magnetic fields to play a principal role in the explosion dynamics [28, 45].

More recently, interest in strong magnetic fields has returned in connection with a number of observables related to CCSNe, including asymmetries in the explosion ejecta [51], natal neutron star kick velocities [27], and the high-energy electromagnetic activity connected to some neutron stars known as magnetars, or Anomalous X-ray Pulsars (AXPs) and Soft Gamma Repeaters (SGRs) (e.g., [15, 48, 24, 52]). The theoretical discovery of the magneto-rotational instability (MRI) [2] and its application to CCSNe [1] has also contributed to renewed interest in the possible role of magnetic fields in the explosion of some supernovae (i.e., those from rapidly rotating progenitor cores; e.g., [51, 40, 39, 10, 46]). Still, magneto-rotationally driven CCSNe likely would be peculiar events, since magnetic progenitor cores tend to rotate slowly [22].

Leaving aside the explosion mechanism, the relationship between the formation of neutron star magnetic fields and CCSNe is still an open and interesting question, particularly in the case of AXPs and SGRs [33] (although magneto-rotational processes are likely involved [47, 6]).

The lack of sufficient rotational energy in magnetized pre-collapse progenitor cores, as predicted by stellar evolution models, has sparked some recent interest in MHD processes in non-rotating CCSN environments [16, 19, 41]. In particular, Endeve et al. [16] studied magnetic field amplification by the stationary accretion shock instability (SASI [4]). The SASI may play an important role in neutrino-powered explosions [8, 9, 36, 34]. Thus, magnetic fields may be an important part of a supernova model if the SASI is found to be sensitive to their presence.

This paper complements the investigations initiated in [16]. We begin with a description of the numerical methods used (the initial implementation of MHD in our astrophysical simulation code GenASiS), including results from some standard tests. Then, with a new set of high-resolution simulations, we investigate the growth and impact of magnetic fields during operation of the SASI. We also attempt to quantify the levels of neutron star magnetization that may result from SASI dynamics. We find that the magnetic energy may grow exponentially in turbulent flows driven by the spiral SASI mode. (The magnetic energy growth time, based on the turnover time of SASI-driven turbulence, is estimated to be a few milliseconds.) The presence of amplified magnetic fields results in less kinetic energy on small spatial scales, but we find no impact of magnetic fields on global shock dynamics. However, magnetic field evolution remains sensitive to numerical resolution. We argue that MHD processes associated with the SASI may contribute significantly to strong, small-scale PNS magnetic fields, and provide a connection between the magnetic fields of neutron stars at birth and supernova dynamics. The saturation energies may be sufficient to power flaring activity of AXPs, and possibly SGRs. Moreover, their formation does not require progenitor rotation.

2. Equations and numerical methods

To study magnetic field amplification from SASI-driven flows we solve the equations of ideal MHD. The fluid mass, momentum and energy densities obey conservation laws with sources

$$\frac{\partial \mathbf{U}}{\partial t} + \nabla \cdot \mathbf{F} = \mathbf{S}, \tag{1}$$

and the evolution of the magnetic field is governed by Faraday's law (the induction equation)

$$\frac{\partial \mathbf{B}}{\partial t} = -\nabla \times \mathbf{E}, \tag{2}$$

where the vector of conserved variables is $\mathbf{U} = [\rho, \rho\mathbf{u}, E]^T$, the vector of fluxes is $\mathbf{F} = [\rho\mathbf{u}, \rho\mathbf{u}\mathbf{u}+\mathbf{I}P^\star - \mathbf{B}\mathbf{B}, (E+P^\star)\mathbf{u}-\mathbf{B}(\mathbf{B}\cdot\mathbf{u})]^T$, and the vector of sources is $\mathbf{S} = [0, -\rho\nabla\Phi, \rho\mathbf{u}\cdot\nabla\Phi]^T$. The electric field is $\mathbf{E} = -\mathbf{u}\times\mathbf{B}$ (assuming a perfectly conducting fluid). We denote the combined vector of evolved variables by $\mathbf{W} = [\mathbf{U}, \mathbf{B}]^T$. (Thus, $\mathbf{F} = \mathbf{F}(\mathbf{W})$ and $\mathbf{E} = \mathbf{E}(\mathbf{W})$.) Variables ρ, $\mathbf{u}$, $E = e + \rho u^2/2 + B^2/2$, $\mathbf{B}$ denote mass density, fluid velocity, fluid energy density, and magnetic field, respectively. (We adopt units where the vacuum permeability is $\mu_0 = 1$.) The total pressure (thermal plus magnetic) is $P^\star = P + B^2/2$. In this study, the internal energy density is related to the pressure via the ideal gas law $e = P/(\gamma - 1)$, where γ is the adiabatic exponent (ratio of specific heats).

We report on simulations of flows around a compact object (the PNS), and we take the gravitational potential to be given by the point-mass formula $\Phi = -GM/r$, where G is Newton's constant, M the mass of the central object, and r the radial distance from the center.

2.1. Magnetohydrodynamics scheme in GenASiS

We adopt an explicit finite volume method (e.g., [29]) for the simultaneous time-integration of Eqs. (1) and (2). Cartesian coordinates are used to represent points in a three-dimensional computational domain with sides L_x, L_y, and L_z, and volume V. The computational domain is subdivided into $N_x \times N_y \times N_z$ computational cells with sides Δx, Δy, and Δz, and volume ΔV. The x-coordinate of the interface separating cells indexed i and $i+1$ is denoted $x_{i+1/2} = x_{1/2} + i\,\Delta x$, with $i = 0, \ldots, N_x$, and the corresponding cell center coordinate is $x_i = (x_{i-1/2} + x_{i+1/2})/2$. Similar definitions apply for the other coordinate dimensions. The time step Δt increments time from t^n to $t^{n+1} = t^n + \Delta t$.

Integrating Eq. (1) over the spacetime volume element $\Delta V \times \Delta t$, applying Gauss' theorem, and representing time integrals of the fluxes and sources with the rectangle rule results in the time-explicit finite volume update formula

$$\begin{aligned} \langle \mathbf{U} \rangle_{ijk}^{n+1} &= \langle \mathbf{U} \rangle_{ijk}^{n} - \frac{\Delta t}{\Delta x}\left[\langle \mathbf{F}_x \rangle_{i+\frac{1}{2}jk}^{n} - \langle \mathbf{F}_x \rangle_{i-\frac{1}{2}jk}^{n}\right] - \frac{\Delta t}{\Delta y}\left[\langle \mathbf{F}_y \rangle_{ij+\frac{1}{2}k}^{n} - \langle \mathbf{F}_y \rangle_{ij-\frac{1}{2}k}^{n}\right] \\ & -\frac{\Delta t}{\Delta z}\left[\langle \mathbf{F}_z \rangle_{ijk+\frac{1}{2}}^{n} - \langle \mathbf{F}_z \rangle_{ijk-\frac{1}{2}}^{n}\right] + \Delta t \,\langle \mathbf{S} \rangle_{ijk}^{n}. \end{aligned} \tag{3}$$

Angle brackets on $\mathbf{U}$ and $\mathbf{S}$ (centered on the geometric centers of computational cells) denote volume averages, while angle brackets on the fluxes (centered on the faces of computational cells) denote area averages. We use the HLL Riemann solver [21] to compute the interface fluxes, given left and right states $\mathbf{U}^{\mathrm{L}}$ and $\mathbf{U}^{\mathrm{R}}$, respectively. The structure between the fastest left and right propagating waves (fast magnetosonic waves with propagation speeds denoted $-\alpha^-$ and α^+, respectively) in the so-called Riemann fan is then represented by a single average state $\mathbf{U}^\star$. The HLL flux (for example in the x-direction $\mathbf{F}_x^\star$) is obtained from the Rankine-Hugoniot jump conditions cross the waves separating the left and right states: $-\alpha_x^-(\mathbf{U}^{\mathrm{L}} - \mathbf{U}^\star) = (\mathbf{F}_x^{\mathrm{L}} - \mathbf{F}_x^\star)$ and $\alpha_x^+(\mathbf{U}^\star - \mathbf{U}^{\mathrm{R}}) = (\mathbf{F}_x^\star - \mathbf{F}_x^{\mathrm{R}})$. The resulting flux is used in Eq. (3); i.e.,

$$\langle \mathbf{F}_x \rangle^n = \frac{\alpha_x^+ \mathbf{F}_x^{n,\mathrm{L}} + \alpha_x^- \mathbf{F}_x^{n,\mathrm{R}}}{\alpha_x^+ + \alpha_x^-} - \frac{\alpha_x^+ \alpha_x^-}{\alpha_x^+ + \alpha_x^-}\left[\mathbf{U}^{n,\mathrm{R}} - \mathbf{U}^{n,\mathrm{L}}\right], \tag{4}$$

where the fluxes $\mathbf{F}_x^{n,\mathrm{L(R)}}$ are evaluated from $\mathbf{W}^{n,\mathrm{L(R)}}$. Similar expressions are obtained for the other flux components. The wave speeds are computed from the left and right states of the interface [11] $\alpha_x^\pm = \max[0, \pm\lambda_x^\pm(\mathbf{W}^{\mathrm{L}}), \pm\lambda_x^\pm(\mathbf{W}^{\mathrm{R}})]$, where $\lambda_x^\pm = u_x \pm c_{f,x}$ and $c_{f,x}$ is the fast magnetosonic speed. The HLL Riemann solver is among the simplest approximate Riemann solvers available for MHD. It relies on minimal information about the characteristic structure of the underlying hyperbolic system (only two eigenvalues of the flux Jacobian matrix). As such, the HLL flux has been frequently used to construct simple, efficient, and robust solvers for hyperbolic systems, including classical MHD [32, 53], and special *and* general relativistic MHD [12, 18, 14, 13]. The averaging over the Riemann fan results in diffusive evolution of intermediate waves (e.g., contact and Alfvén). However, the intermediate waves can be restored in the HLL framework [49, 31, 20, 30, 38, 37].

For second-order spatial accuracy the left and right states needed by the Riemann solver are reconstructed from cell-centered volume averages with monotonic linear interpolation: $\mathbf{U}_{i+1/2}^{n,\mathrm{L}} = \langle \mathbf{U} \rangle_i^n + \mathcal{D}_x[\langle \mathbf{U} \rangle_i^n](x_{i+1/2} - x_i)$ and $\mathbf{U}_{i+1/2}^{n,\mathrm{R}} = \mathbf{U}_{i+1}^n - \mathcal{D}_x[\mathbf{U}_{i+1}^n](x_{i+1} - x_{i+1/2})$. Monotonic interpolation of a variable f_i^n is achieved with the the generalized minmod slope (e.g., [25])

$$\mathcal{D}_x\left[f_i^n\right] = \Delta x^{-1} \,\mathrm{minmod}\left[\vartheta\left(f_i^n - f_{i-1}^n\right), 0.5\left(f_{i+1}^n - f_{i-1}^n\right), \vartheta\left(f_{i+1}^n - f_i^n\right)\right], \tag{5}$$

where $\vartheta \in [1, 2]$. We use $\vartheta = 1.4$ in the simulations presented in this paper.

The magnetic field is evolved with the constrained transport (CT) method of Evans & Hawley [17]. Magnetic field components are collocated on the faces of computational cells. Integrating the induction equation (Eq. 2) over cell faces and Δt, applying Stoke's theorem, and replacing time integrals of electric field components with the rectangle rule results in time-explicit update formulae for area averaged magnetic field components

$$\begin{aligned}
\langle B_x\rangle^{n+1}_{i+\frac{1}{2}jk} &= \langle B_x\rangle^{n}_{i+\frac{1}{2}jk} \\
&+\frac{\Delta t}{\Delta z}\left[\langle E_y\rangle^n_{i+\frac{1}{2}jk+\frac{1}{2}} - \langle E_y\rangle^n_{i+\frac{1}{2}jk-\frac{1}{2}}\right] - \frac{\Delta t}{\Delta y}\left[\langle E_z\rangle^n_{i+\frac{1}{2}j+\frac{1}{2}k} - \langle E_z\rangle^n_{i+\frac{1}{2}j-\frac{1}{2}k}\right] \quad (6)\\
\langle B_y\rangle^{n+1}_{ij+\frac{1}{2}k} &= \langle B_y\rangle^{n}_{ij+\frac{1}{2}k} \\
&+\frac{\Delta t}{\Delta x}\left[\langle E_z\rangle^n_{i+\frac{1}{2}j+\frac{1}{2}k} - \langle E_z\rangle^n_{i-\frac{1}{2}j+\frac{1}{2}k}\right] - \frac{\Delta t}{\Delta z}\left[\langle E_x\rangle^n_{ij+\frac{1}{2}k+\frac{1}{2}} - \langle E_x\rangle^n_{ij+\frac{1}{2}k-\frac{1}{2}}\right] \quad (7)\\
\langle B_z\rangle^{n+1}_{ijk+\frac{1}{2}} &= \langle B_z\rangle^{n}_{ijk+\frac{1}{2}} \\
&+\frac{\Delta t}{\Delta y}\left[\langle E_x\rangle^n_{ij+\frac{1}{2}k+\frac{1}{2}} - \langle E_x\rangle^n_{ij-\frac{1}{2}k+\frac{1}{2}}\right] - \frac{\Delta t}{\Delta x}\left[\langle E_y\rangle^n_{i+\frac{1}{2}jk+\frac{1}{2}} - \langle E_y\rangle^n_{i-\frac{1}{2}jk+\frac{1}{2}}\right]. \quad (8)
\end{aligned}$$

Angle brackets on electric field components (centered on the edges of computational cells) denote line averages. Note that with the divergence of the magnetic field defined by

$$\langle\nabla\cdot\mathbf{B}\rangle_{ijk} = \frac{\langle B_x\rangle_{i+\frac{1}{2}jk} - \langle B_x\rangle_{i-\frac{1}{2}jk}}{\Delta x} + \frac{\langle B_y\rangle_{ij+\frac{1}{2}k} - \langle B_y\rangle_{ij-\frac{1}{2}k}}{\Delta y} + \frac{\langle B_z\rangle_{ijk+\frac{1}{2}} - \langle B_z\rangle_{ijk-\frac{1}{2}}}{\Delta z}, \quad (9)$$

the magnetic field update given by Eqs. (6)-(8) results in $\langle\nabla\cdot\mathbf{B}\rangle^{n+1}_{ijk} = \langle\nabla\cdot\mathbf{B}\rangle^{n}_{ijk}$—independent of how the electric fields are computed. Thus, the magnetic field remains divergence-free, provided it is so initially.

We use HLL-type expressions [32] (see also [26]) to compute edge centered electric fields in Eqs. (6)-(8). The z-component centered on, say, $(x_{i+1/2}, y_{j+1/2}, z_k)$ is

$$\begin{aligned}
\langle E_z\rangle^n &= \frac{\alpha_x^+\alpha_y^+E_z^{n,\mathrm{SW}} + \alpha_x^+\alpha_y^-E_z^{n,\mathrm{NW}} + \alpha_x^-\alpha_y^+E_z^{n,\mathrm{SE}} + \alpha_x^-\alpha_y^-E_z^{n,\mathrm{NE}}}{(\alpha_x^+ + \alpha_x^-)(\alpha_y^+ + \alpha_y^-)} \\
&+\frac{\alpha_x^+\alpha_x^-}{(\alpha_x^+ + \alpha_x^-)}\left[B_y^{n,\mathrm{E}} - B_y^{n,\mathrm{W}}\right] - \frac{\alpha_y^+\alpha_y^-}{(\alpha_y^+ + \alpha_y^-)}\left[B_x^{n,\mathrm{N}} - B_x^{n,\mathrm{S}}\right], \quad (10)
\end{aligned}$$

where $\alpha_x^\pm$ and $\alpha_y^\pm$ are computed by taking the maximum among the four states surrounding the edge. The HLL electric field in Eq. (10) considers the four cells sharing the edge where the electric field is to be evaluated. It is derived from the principles used in the derivation of the HLL flux in Eq. (4), but applied in two dimensions. For the electric field centered on $(x_{i+1/2}, y_{j+1/2}, z_k)$, superscripts SW, NW, SE, and NE indicate that a quantity is obtained from values in cells with centers (x_i, y_j, z_k), (x_i, y_{j+1}, z_k), (x_{i+1}, y_j, z_k), and (x_{i+1}, y_{j+1}, z_k), respectively. Magnetic field components are centered on cell faces, and the cells with centers (x_i, y_j, z_k) and (x_i, y_{j+1}, z_k) share $\langle B_y\rangle_{ij+1/2k}$, which is labeled with superscript W. Similarly, the cells with centers (x_i, y_j, z_k) and (x_{i+1}, y_j, z_k) share $\langle B_x\rangle_{i+1/2jk}$, which is labeled with superscript S. Velocity components are cell centered and must be interpolated in two dimensions, while face centered magnetic field components are interpolated in one dimension only. The last two terms in Eq. (10) result from averaging over Riemann fans and introduce explicit dissipation of the magnetic field. They act to stabilize the time evolution in under-resolved regions, and can become dominant if the bulk of the magnetic fields evolve on spatial scales comparable to the grid spacing (see Section 3).

The update formulae given by Eq. (3) and Eqs. (6)-(8) are equivalent to the forward Euler method and result in first-order temporal accuracy. Second-order accuracy in time is achieved with a total variation diminishing (TVD) Runge-Kutta method (e.g., [43])

$$\begin{aligned}
\langle\mathbf{W}\rangle^\star &= \langle\mathbf{W}\rangle^n + \mathcal{L}\left[\langle\mathbf{W}\rangle^n\right] \\
\langle\mathbf{W}\rangle^{n+1} &= \langle\mathbf{W}\rangle^n + \left(\mathcal{L}\left[\langle\mathbf{W}\rangle^n\right] + \mathcal{L}\left[\langle\mathbf{W}\rangle^\star\right]\right)/2, \quad (11)
\end{aligned}$$

Table 1. L_1-error and convergence rate for the circularly polarized Alfvén wave test.

	TW ($t_f = 1$)		SW ($t_f = 1$)		TW ($t_f = 5$)		SW ($t_f = 5$)	
N_x	$L_1(B_z)$	Rate	$L_1(B_z)$	Rate	$L_1(B_z)$	Rate	$L_1(B_z)$	Rate
32	1.373E-01	–	2.279E-01	–	4.974E-01	–	7.361E-01	–
64	5.196E-02	−1.40	8.629E-02	−1.40	1.395E-01	−1.83	2.144E-01	−1.78
128	1.433E-02	−1.86	2.535E-02	−1.77	5.053E-02	−1.46	8.700E-02	−1.30
256	3.625E-03	−1.98	6.842E-03	−1.89	1.483E-02	−1.77	2.754E-02	−1.66
512	9.085E-04	−2.00	1.760E-03	−1.96	3.945E-03	−1.91	7.782E-03	−1.82
1024	2.242E-04	−2.02	4.491E-04	−1.97	1.015E-03	−1.96	2.059E-03	−1.92

which is a convex combination of two forward Euler steps. The spatial operators and sources are here represented by $\mathcal{L}[\langle \mathbf{W} \rangle]$. The time step is determined from the Courant-Friedrichs-Lewy (CFL) condition: $\Delta t = \mathrm{C} \times \min\left[\Delta t_x, \Delta t_y, \Delta t_z\right]$, where, for example, $\Delta t_x = \Delta x / \max[\alpha_x^+, \alpha_x^-]$, $\mathrm{C} \leq 1$ is the Courant number, and the minimum is taken among all computational cells in V.

We point out that our MHD scheme is similar to the MC-HLL-UCT scheme described in [32] and the MHD scheme in the NIRVANA code [53]. (See also the semidiscrete central-upwind schemes developed for hyperbolic conservation laws and Hamilton-Jacobi equations in [26].)

2.1.1. Numerical tests To demonstrate the correctness of our implementation of the MHD solver we present results from two well-known test problems, the circularly polarized Alfvén wave and the Orszag-Tang vortex, computed in two spatial dimensions. Both tests use periodic boundary conditions everywhere. The Courant number is set to $\mathrm{C} = 0.4$, the slope limiting parameter is set to $\vartheta = 1.4$, and adiabatic exponent $\gamma = 5/3$.

Circularly polarized (CP) Alfvén wave This test problem has been used by many authors (e.g., [50, 32, 44]). In particular, Tóth [50] used this test in an extensive comparison study of multidimensional MHD schemes. The CP Alfvén wave is an analytic nonlinear solution to the MHD equations. As such it provides a means to verify the formal order of the scheme through convergence testing. The wave propagates with an angle $\alpha = \tan^{-1}(2) \approx 63.4°$ relative to the x-axis. The problem is set up with constant background density $\rho_0 = 1$ and pressure $P_0 = 0.1$. The velocity (magnetic field) components are $u_x(B_x) = u_\parallel(B_\parallel)\cos\alpha - u_\perp(B_\perp)\sin\alpha$, $u_y(B_y) = u_\parallel(B_\parallel)\sin\alpha + u_\perp(B_\perp)\cos\alpha$, and $u_z(B_z) = A\sin(2\pi x_\parallel)$, with $u_\perp = B_\perp = A\sin(2\pi x_\parallel)$ and $x_\parallel = x\cos\alpha + y\sin\alpha$. The background magnetic field parallel to the direction of wave propagation is $B_\parallel = 1$ (i.e., the Alfvén speed v_A is unity). The parallel velocity $u_\parallel$ is set to zero for the traveling wave (TW) test and -1 for the standing wave (SW) test. The wave amplitude is $A = 0.01$. Periodic boundary conditions can be used if the domain is $[x, y] \in [0, 1/\cos\alpha] \times [0, 1/\sin\alpha]$ ($L_x/L_y = 2$, and we use $N_x = 2N_y$ for square cells).

Convergence results from the CP Alfvén wave test (TW and SW tests), run for one ($t = 1$) and five ($t = 5$) grid crossings, are listed in Table 1. The L_1-error norm is $L_1(B_z) = \sum |B_z(t = 0) - B_z(t = t_f)| / \sum |B_z(t = 0)|$, where the sums extend over all cells. The tests demonstrate second-order convergence with increasing resolution. The SW errors are larger (almost a factor 2) than the TW errors. (The convergence rate is given by $\log(L_1(B_z, N_x)/L_1(B_z, N_x/2))/\log(2)$.)

Orszag-Tang Vortex This is another test which has be used extensively to benchmark multidimensional MHD codes (e.g., [32, 44]). Following the initialization in [44], the computational domain is the unit square $[x, y] \in [0, 1] \times [0, 1]$, where the density and pressure are initially uniform $\rho = \frac{25}{36\pi}$ and $P = \frac{5}{12\pi}$ (resulting in sound speed $c_\mathrm{S} = 1$). The nonzero components of the velocity field are $u_x = -\sin(2\pi y)$ and $u_y = \sin(2\pi x)$, and the magnetic field

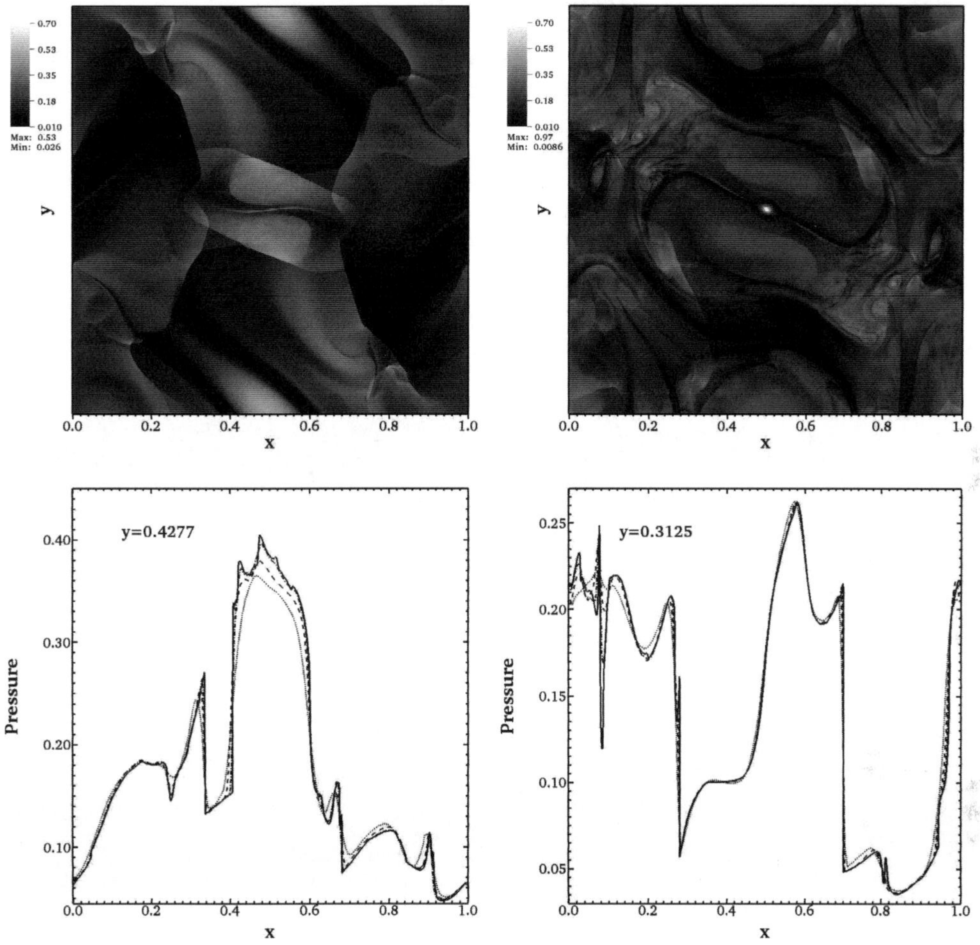

Figure 1. Results from running the Orszag-Tang vortex test with GenASiS. The upper panels shows the fluid pressure at times $t = 0.5$ (left) and $t = 1.0$ (right) for a model computed with 4096^2 cells. In the two lower panels we plot the thermal pressure versus x at $t = 0.5$ for $y = 0.4277$ (left) and $y = 0.3125$ (right) (cf. [44]). Profiles are plotted for runs with varying spatial resolutions: $N_x \times N_y = 128^2$ (······), 256^2 (- - - -), 512^2 (— · —), and 4096^2 (——).

is given by $B_x = -B_0 \sin(2\pi y)$ and $B_y = B_0 \sin(4\pi x)$ with $B_0 = 1/\sqrt{4\pi}$.

Shocks develop quickly in the flow for $t > 0$. The flow becomes very complex and develops MHD turbulence after multiple shock-shock interactions. Figure 1 displays results from the Orszag-Tang test computed with GenASiS. With the profiles displayed in the lower panels, we have computed the L_1-error norm of the lower-resolution runs with respect to the high-resolution run (4096^2). The error norm decreases with increasing resolution at the first order rate (as expected for flows containing discontinuities).

2.2. Gravitational source terms

The gravitational force and power must be specified for problems involving gravity. In particular, constructing source terms which result in good energy conservation properties is of interest. Multiplication of the discrete mass conservation equation in Eq. (3) with the potential $\langle\Phi\rangle_{ijk}$ results in (after some algebra and assuming a static gravitational potential)

$$\begin{aligned}\langle E_g\rangle_{ijk}^{n+1} = \; & \langle E_g\rangle_{ijk}^{n} - \frac{\Delta t}{\Delta x}\left[\langle F_x^{E_g}\rangle_{i+\frac{1}{2}jk}^{n} - \langle F_x^{E_g}\rangle_{i-\frac{1}{2}jk}^{n}\right] - \frac{\Delta t}{\Delta y}\left[\langle F_y^{E_g}\rangle_{ij+\frac{1}{2}k}^{n} - \langle F_y^{E_g}\rangle_{ij-\frac{1}{2}k}^{n}\right] \\ & - \frac{\Delta t}{\Delta z}\left[\langle F_z^{E_g}\rangle_{ijk+\frac{1}{2}}^{n} - \langle F_z^{E_g}\rangle_{ijk-\frac{1}{2}}^{n}\right] + \frac{\Delta t}{2}\left[\langle F_x^{\rho} g_x\rangle_{i-\frac{1}{2}jk}^{n} + \langle F_x^{\rho} g_x\rangle_{i+\frac{1}{2}jk}^{n}\right] \\ & + \frac{\Delta t}{2}\left[\langle F_y^{\rho} g_y\rangle_{ij-\frac{1}{2}k}^{n} + \langle F_y^{\rho} g_y\rangle_{ij+\frac{1}{2}k}^{n}\right] + \frac{\Delta t}{2}\left[\langle F_z^{\rho} g_z\rangle_{ijk-\frac{1}{2}}^{n} + \langle F_z^{\rho} g_z\rangle_{ijk+\frac{1}{2}}^{n}\right], \qquad (12)\end{aligned}$$

where the gravitational energy density is $\langle E_g\rangle_{ijk}^{n} = \langle\rho\rangle_{ijk}^{n}\langle\Phi\rangle_{ijk}$ and the x-component of the gravitational energy flux is $\langle F_x^{E_g}\rangle_{i+1/2jk}^{n} = \langle F_x^{\rho}\rangle_{i+1/2jk}^{n}(\langle\Phi\rangle_{ijk} + \langle\Phi\rangle_{i+1jk})/2$. We have also defined $\langle F_x^{\rho} g_x\rangle_{i+1/2jk}^{n} = \langle F_x^{\rho}\rangle_{i+1/2jk}^{n}(\langle\Phi\rangle_{i+1jk} - \langle\Phi\rangle_{ijk})/\Delta x$. (The x-component of the mass flux in Eq. (3) is denoted $\langle F_x^{\rho}\rangle_{i+1/2jk}^{n}$.) The last three terms in Eq. (12) suggest a finite volume representation of the gravitational power $\langle S^E\rangle_{ijk}^{n} = -\langle\rho\mathbf{u}\cdot\nabla\Phi\rangle_{ijk}^{n}$. This choice results in energy conservation to machine precision; i.e.,

$$\mathcal{E}^{n+1} = \sum_{i=1}^{N_x}\sum_{j=1}^{N_y}\sum_{k=1}^{N_z}\left(\langle E\rangle_{ijk}^{n+1} + \langle E_g\rangle_{ijk}^{n+1}\right)\Delta V = \mathcal{E}^{n} + \text{boundary energy fluxes}. \qquad (13)$$

This result relies on the assumption of a static gravitational potential, and is sufficient for the simulations presented in this paper. We will present a suitable generalized discretization of the gravitational power—valid for time-dependent potentials—in a forthcoming paper.

A finite volume representation of the x-component of the gravitational force is given by

$$\langle S^{\rho u_x}\rangle_{ijk}^{n} = -\frac{1}{2}\left[\langle\rho g_x\rangle_{i-\frac{1}{2}jk}^{n} + \langle\rho g_x\rangle_{i+\frac{1}{2}jk}^{n}\right], \qquad (14)$$

where $\langle\rho g_x\rangle_{i+1/2jk}^{n} = \langle\rho\rangle_{i+1/2jk}^{n}(\langle\Phi\rangle_{i+1jk} - \langle\Phi\rangle_{ijk})/\Delta x$. (Similar expressions are used for the y- and z-components.) For the density centered on $x = x_{i+1/2}$ we use the so-called HLL average

$$\langle\rho\rangle_{i+\frac{1}{2}} = \frac{\alpha_{x,i+\frac{1}{2}}^{+}\rho_{i+\frac{1}{2}}^{\mathrm{R}} + \alpha_{x,i+\frac{1}{2}}^{-}\rho_{i+\frac{1}{2}}^{\mathrm{L}} - [F_{x,i+\frac{1}{2}}^{\rho,\mathrm{R}} - F_{x,i+\frac{1}{2}}^{\rho,\mathrm{L}}]}{\alpha_{x,i+\frac{1}{2}}^{+} + \alpha_{x,i+\frac{1}{2}}^{-}}. \qquad (15)$$

Note that the expression for the gravitational power does *not* depend on the particular Riemann solver, while the force (with Eq. (15)) does. For a different Riemann solver, Eq. (15) should be replaced with the face density consistent with that solver.

3. Simulations of SASI-induced magnetic field amplification

In this section we describe high-resolution MHD simulations with GenASiS, exploring the capacity of nonlinear SASI-driven flows to amplify magnetic fields. With an idealized model we seek to investigate the character of SASI-induced magnetic field amplification. Interesting questions to explore include: does the presence of magnetic fields impact the evolution of the SASI?, what are the implications for observables associated with CCSNe, in particular PNS magnetization? However, with idealized simulations, we can only begin to scratch the surface of these complicated topics.

3.1. Model setup and parallel execution

Our initial conditions follow closely the adiabatic setup described in [4] and [5] (see also [16]): a spherical stationary accretion shock is placed at a radius $R_{\rm Sh}$ = 200 km. A highly supersonic flow is nearly free-falling towards the shock for $r > R_{\rm Sh}$. Between the shock and the PNS the flow settles subsonically—obeying the Bernoulli equation $u_r^2/2 + \gamma P/[\rho(\gamma-1)] - GM/r = 0$—in nearly hydrostatic equilibrium. Matter is allowed to flow through an inner boundary placed at $R_{\rm PNS} = 40$ km. The ratio of specific heats is set to $\gamma = 4/3$, the mass of the central object is held fixed at $M = 1.2\ M_\odot$, and the accretion rate ahead of the shock $\dot{M} = 0.36\ M_\odot\ {\rm s}^{-1}$ is also held fixed during the simulations. A radial magnetic field is superimposed on the initial condition: $B_r = {\rm sign}(\cos\theta) \times B_0(R_{\rm PNS}/r)^2$, where B_0 is the magnetic field strength at the surface of the PNS. We initiate the SASI with random pressure perturbations in the post-shock flow.

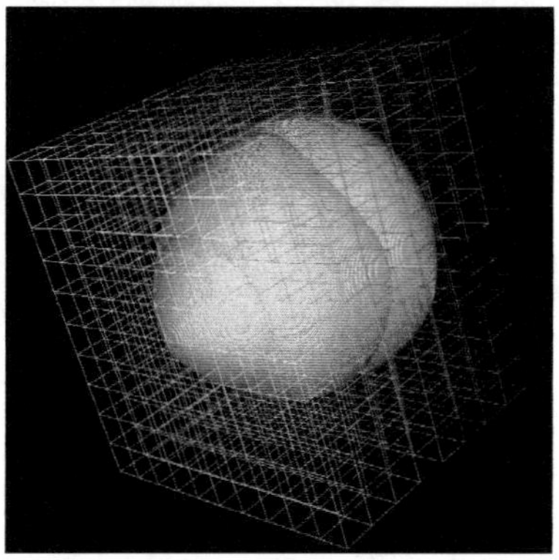

Figure 2. Domain composition from a 3D MHD-SASI simulation.

The MHD solver is parallelized using the Message Passing Interface (MPI), and the computational domain is subdivided into blocks containing an equal number of zones, which are distributed among MPI processes (see Figure 2). During the simulations we keep the number of zones per block (MPI process) fixed to 32^3. To conserve computational resources the simulations are initiated in a relatively small computational domain with sides $L_{\rm min}$ = 600 km covered by 512 zones ($\Delta x \approx$1.17 km). Once the SASI evolves into the nonlinear regime the volume encompassed by the shock grows, and the shock eventually interacts with the domain boundaries. When this happens, we expand the computational domain by adding a layer of 32^3-zones blocks (i.e., we add 64 zones in each coordinate direction) and restart the simulation from the last checkpoint written before the shock interacted with the boundary. We repeat this process, and run our simulations until the shock interacts with the boundary of the largest computational box $L_{\rm max}$ = 1500 km, or the simulation time reaches t = 1100 ms, whichever occurs first. Since we keep Δx fixed during the simulations, the largest computational domain is covered by 1280 zones in each spatial dimension. The weak scaling plot in Figure 3 shows the wall-time per time step, which remains reasonably constant as the computational domain expands to accommodate the growing shock volume. During a run, we write simulation output for analysis and visualization every 2 ms of physical time, resulting in tens of Terabytes of data from each model.

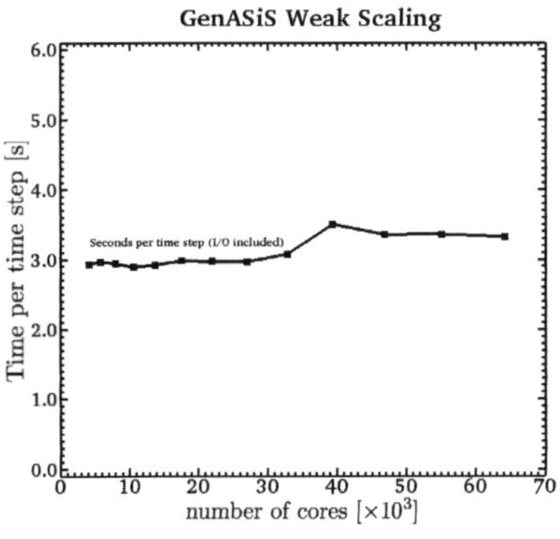

Figure 3. Weak scaling from a high-resolution 3D MHD-SASI simulation.

3.2. Simulation results

We have carried out simulations with varying initial magnetic field: $B_0 = 1 \times 10^{10}$ G (model B10 or "weak-field model"), $B_0 = 1 \times 10^{12}$ G (model B12), and $B_0 = 1 \times 10^{13}$ G (model B13 or "strong-field model"). The initial field can be considered weak in all the models in the sense

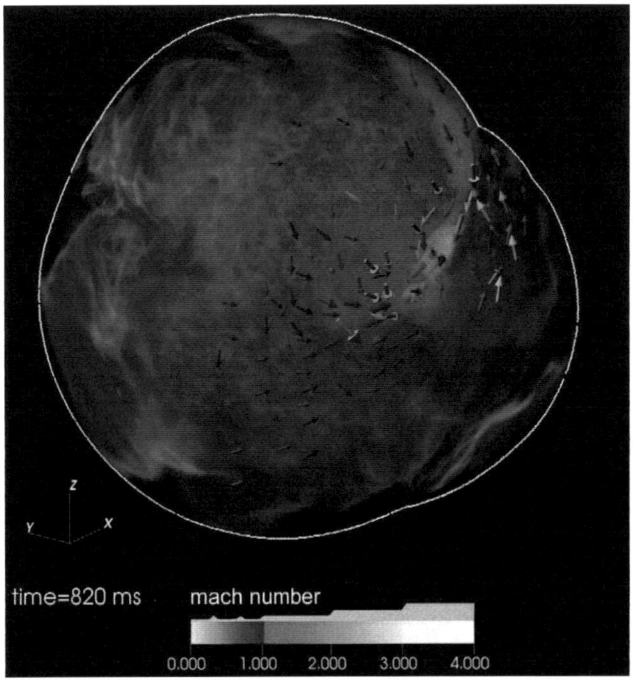

Figure 4. Volume rendering of the Mach number $\text{Ma} = |\mathbf{u}|/c_s$ with velocity vectors ($|\mathbf{u}| \geq 10^4$ km s^{-1}) superimposed. Angular momentum redistribution has occurred as a result of the SASI, and the presence of counterrotating flows is apparent. The shock triple-point [5] (a line segment extending across the shock surface), positioned to the upper right, is visible as the kink in the shock surface (white contour). It connects the pre-shock accretion flow and the two counterrotating post-shock flows, and moves on the shock surface in the counterclockwise direction. A layer of sheared flows extends from the triple-point and results in post-shock vorticity generation. Image courtesy of Ross Toedte, ORNL.

that the magnetic energy density is small compared to the kinetic and internal energy densities.

Upon perturbation and nonlinear development, all models eventually exhibit typical spiral mode flows. This is consistent with [5], who found the spiral mode to dominate the late-time evolution independent of the initial perturbation. Figure 4 illustrates the post-shock flows associated with the nonlinear spiral SASI mode. (The rendering is created from a snapshot of the strong-field model at $t = 820$ ms, but the hydrodynamic developments exhibited by this model are typical of all the computed models.)

The pre-shock accretion flow impinges on the shock at an oblique angle due to the aspherical shock and its off-center position. The sizable tangential velocity component (relative to the shock surface), which is preserved across the shock, leads to supersonic post-shock flows ahead of (and directed towards) the triple-point. Supersonic shear flows are directed down towards the PNS and result in vorticity generation. Vorticity is distributed in a large fraction of the post-shock volume during the operation of the spiral SASI mode. Strongly forced accretion-driven turbulence develops as a result of the SASI, and the post-shock flow becomes roughly divided into a supersonic (driving) component and a subsonic (volume-filling) turbulent component. These hydrodynamic developments result in turbulent magnetic field amplification [16].

The magnetic field amplifies in response to the hydrodynamics developments, and the magnetic energy becomes concentrated in thin intense magnetic flux tubes (or ropes). The volume rendering in Figure 5 shows streamlines tracing out the magnetic field below the shock and reveals a complicated magnetic field geometry. The magnetic field is "frozen-in" to the fluid in the ideal MHD limit. The large scale flows associated with the SASI result in relatively straight field lines, while the turbulent flows are responsible for a more "chaotic" structure.

Figure 6 contains further details from the numerical simulations. In Panel (A) we plot the relative change in total magnetic energy below the shock versus time for all models: B10 (——), B12 (- - - -), and B13 (······). (The change is scaled to the initial magnetic energy for easy

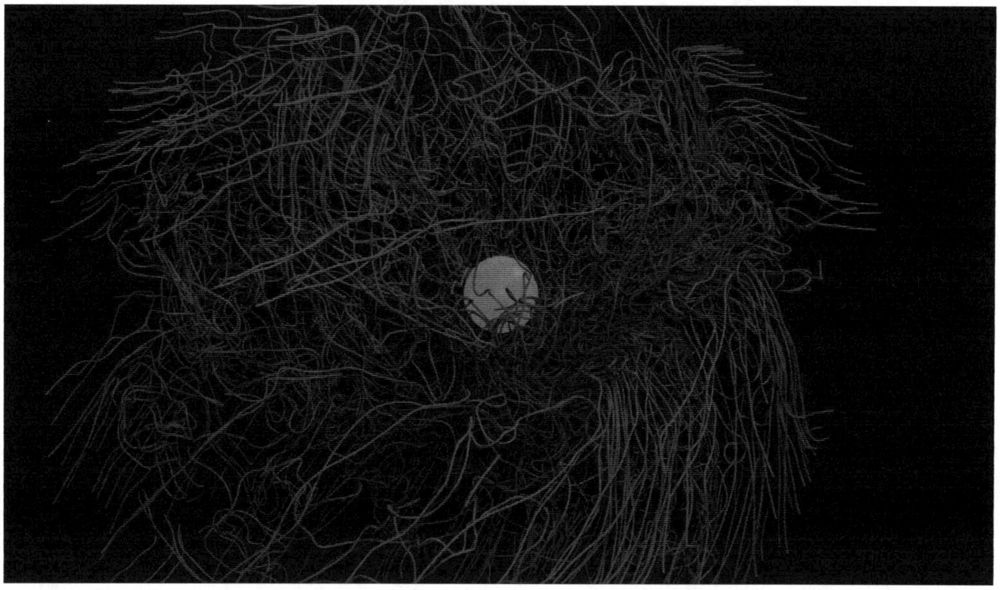

Figure 5. Streamlines tracing magnetic field lines in an MHD-SASI simulation during the nonlinear stage. Streamlines are randomly seeded on the shock surface. Image courtesy of Dave Pugmire, ORNL. (An animation from which this image is taken can be viewed at `http://events.cels.anl.gov/scidac11/visualization-night/visualization-night-winners/`.)

comparison across the models.) After an initial spurt, all the models experience an early period of exponential magnetic energy growth with essentially the same growth rate (cf. the temporal window from 650 ms to 780 ms). Exponential growth is typical during the kinematic regime of a turbulent dynamo, in which the magnetic field's back-reaction on the fluid is negligible (e.g., [7]), and the growth time is roughly the turnover time of the turbulent eddies τ_{eddy}. The magnetic energy in the weak-field model grows exponentially at a nearly constant rate, with growth time $\tau \approx 66$ ms (dash-dotted reference lines), until the end of the simulation ($t = 1100$ ms), and receives a total boost of about five orders of magnitude ($E_{\text{m}} \approx 1.8 \times 10^{-7}$ B). In the model with $B_0 = 1 \times 10^{12}$ G (B12), E_{m} also grows steadily until the end of the run. The magnetic energy in this model receives a boost of four orders of magnitude ($E_{\text{m}} \approx 2.3 \times 10^{-4}$ B). It grows initially at the same rate as the weak-field model, but the growth rate tapers off at later times ($t > 900$ ms). The strong-field model (B13, $B_0 = 1 \times 10^{13}$ G) exhibits exponential magnetic energy growth ($\tau \approx 66$ ms) early on. Then, around $t \approx 780$ ms, the growth rate drops almost discontinuously, and E_{mag} grows by only about 50% for the remainder of the simulation. Model B13 receives a total boost in magnetic energy of about a factor of 300 ($E_{\text{m}} \approx 8.9 \times 10^{-4}$ B). The abrupt drop in the magnetic energy growth rate observed in the strong-field model occurs when magnetic fields become dynamically important: concentrations of high $\beta_{\text{kin}}^{-1} (\equiv v_{\text{A}}^2/|\mathbf{u}|^2)$, which exceeds unity in localized regions, are scattered throughout the shock volume. About 50% of the total magnetic energy in model B13 resides in regions where $\beta_{\text{kin}}^{-1} \geq 10^{-1}$, and these magnetic fields occupy less than 10% of the total shock volume. Regions with $v_{\text{A}} \gtrsim |\mathbf{u}|$ appear and disappear in a highly intermittent manner during this stage.

The magnetic energy grows at the expense of turbulent kinetic energy below the shock. The turbulent kinetic energy $E_{\text{k}}^{\text{T}} = \int_{k_{\text{T}}}^{\infty} \hat{e}_{\text{k}}(k)\, dk$ is plotted in Panel (B). It is defined as flows

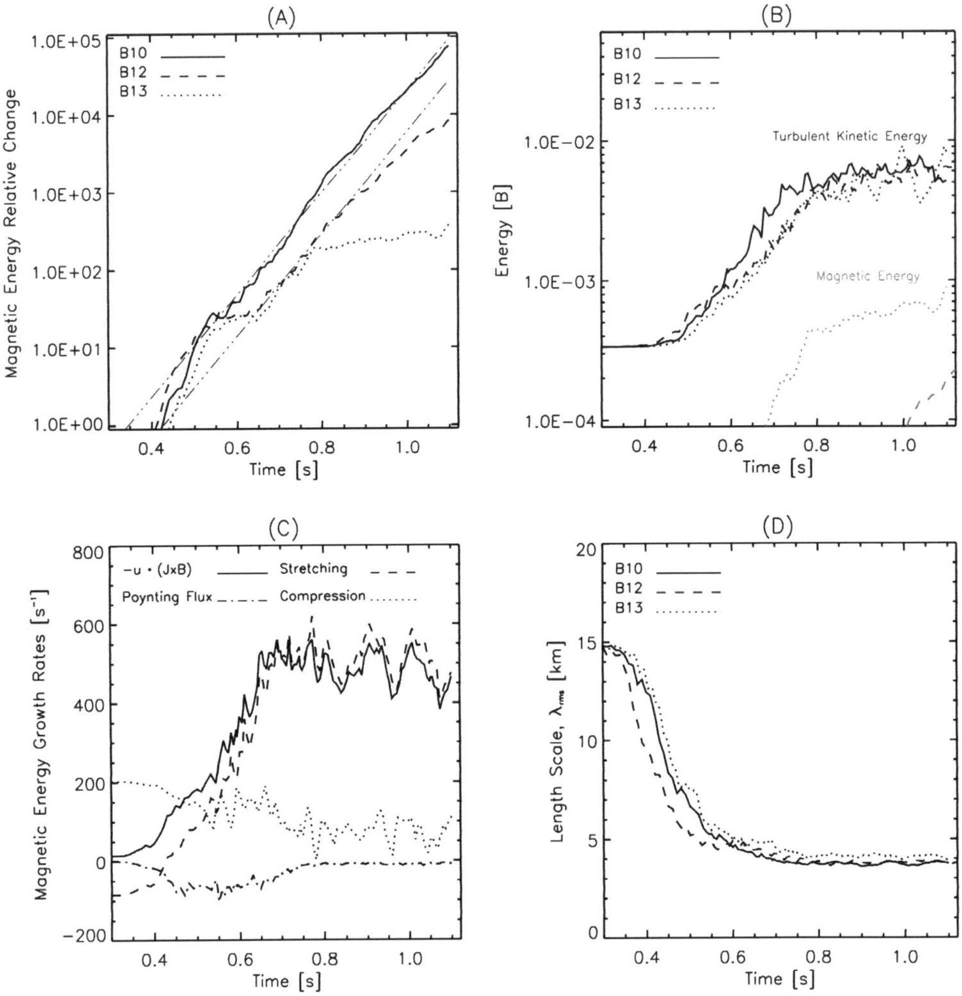

Figure 6. Summary of simulation results. Panel (A): magnetic energy versus time; panel (B): turbulent kinetic energy versus time; panel (C): growth rates due to various terms in Eq. (16) versus time for model B10; and panel (D): magnetic rms scale (average flux rope thickness) versus time. See text for further details and discussion. (1 B $\equiv 10^{51}$ erg.)

varying on scales with wavenumber $k \geq k_{\rm T} = 2\pi/\lambda_{\rm T}$, where $\lambda_{\rm T} \approx 30$ km. (The spectral kinetic energy density $\hat{e}_{\rm k}(k)$ is obtained from Fourier transforms of components of the kinetic energy.) The particular choice for $k_{\rm T}$ is motivated by several factors, including (1) magnetic field amplification occurs mostly on spatial scales with $k > k_{\rm T}$ (from magnetic energy spectra we obtain a characteristic magnetic field scale $\bar{\lambda}_{\rm m} \approx 20$ km), and (2) the Taylor microscale $(\langle u^2\rangle/\langle|\nabla \times \mathbf{u}|^2\rangle)^{1/2}$, which measures the average size of turbulent eddies, is comparable to $\lambda_{\rm T}$.

The turbulent kinetic energy grows rapidly during the ramp-up of the SASI and reaches a saturation level. After saturation, the time-averaged turbulent kinetic energy (averaged over the time interval from 800 ms to 1100 ms) in the respective models are found to be $\langle E_{\mathrm{k}}^{\mathrm{T}} \rangle_{0.8\,\mathrm{s}}^{1.1\,\mathrm{s}} = 5.90 \times 10^{-3}$ B (B10), 5.48×10^{-3} B (B12), and 5.11×10^{-3} B (B13). (The total kinetic energy below the shock E_{k} is about an order of magnitude larger during this stage.) We note that $E_{\mathrm{k}}^{\mathrm{T}}$ in model B12 and model B13 is reduced (relative to the weak-field model) by $\sim 7\%$ (4.2×10^{-4} B) and $\sim 13\%$ (7.9×10^{-4} B), respectively. The reduction in turbulent energy is comparable to the gain in magnetic energy in the respective models. Although there is a measurable reduction in the turbulent kinetic energy in the models attaining stronger B-fields, we find no impact of magnetic fields on large scale flows and SASI evolution. (The turbulent rms velocity u_{rms} inferred from these models exceeds 4000 km s^{-1}, and the associated eddy turnover time is $\tau_{\mathrm{eddy}} = \lambda_{\mathrm{m}}/u_{\mathrm{rms}} \approx 5$ ms.)

The saturation level for the turbulent kinetic energy may serve as an upper limit on the magnetic energy attainable in these simulations. We note that the magnetic energy in model B13 (dotted red line in Panel (B)) seems to saturate at about 10% of the turbulent kinetic energy. The magnetic fields in this model become dynamically significant, but are also influenced by numerical dissipation during the evolution (see below). The fluid in a developing CCSN is an excellent electrical conductor and magnetic energy dissipation is probably not important on the explosion time scale. Thus, the magnetic energy may possibly grow beyond the levels seen in our simulations, but probably not above the turbulent kinetic energy.

Assuming a non-ideal electric field $-(\mathbf{u} \times \mathbf{B}) + \eta \mathbf{J}$ with scalar resistivity η, the evolution equation for the magnetic energy density is easily obtained by dotting Eq. (2) with $\mathbf{B}$:

$$\begin{aligned} \frac{\partial e_{\mathrm{m}}}{\partial t} &= \mathbf{B} \cdot [(\mathbf{B} \cdot \nabla)\,\mathbf{u} - (\mathbf{u} \cdot \nabla)\,\mathbf{B} - \mathbf{B}\,(\nabla \cdot \mathbf{u}) - \nabla \times (\eta \mathbf{J})] \\ &= -\nabla \cdot \mathbf{P} - \mathbf{u} \cdot (\mathbf{J} \times \mathbf{B}) - \mathbf{B} \cdot \nabla \times (\eta \mathbf{J}), \end{aligned} \quad (16)$$

where $\mathbf{P} = [\mathbf{u}(\mathbf{B} \cdot \mathbf{B}) - \mathbf{B}(\mathbf{B} \cdot \mathbf{u})]$ and $\mathbf{J} = \nabla \times \mathbf{B}$. (We use scalar resistivity here for sake of simplicity of the discussion, but it is also appropriate due to the similarity between the last two terms in Eq. (10) and ηJ_z.) The first and third term on the right-hand-side of Eq. (16) (first line) describe magnetic field amplification due to stretching and compression, respectively. Magnetic energy growth rates due to compression (······) and stretching (- - - -) are plotted versus time for model B10 in Panel (C) of Figure 6. Stretching is the dominant magnetic field amplification mechanism in the nonlinear regime ($t \gtrsim 650$). In a turbulent flow, the separation between two (initially adjacent) fluid elements grows exponentially with time. If the fluid elements are connected by a weak magnetic field the frozen-in condition of ideal MHD results in stretching, and thereby strengthening, of the magnetic field and an increase in the magnetic energy [42].

The terms on the right-hand-side of Eq. (16) (second line) represent Poynting fluxes, work done against the Lorentz force (W_{L}) and magnetic energy decay due to resistive dissipation ($-Q_{\mathrm{J}}$), respectively. Kinetic energy is converted into magnetic energy if $W_{\mathrm{L}} > 0$. It is apparent from Eq. (16) that the magnetic energy growth rate is due to the sum of individual rates from accretion of magnetic energy (Poynting flux) through the surface enclosing the PNS ($\partial V_{\mathrm{PNS}}$), work done against the Lorentz force, and resistive dissipation. The Poynting flux through $\partial V_{\mathrm{PNS}}$ and resistive dissipation generally result in decay of the magnetic energy in the computational domain. The decay must be overcome by the Lorentz work term in order for the magnetic energy to increase. (The magnetic Reynolds number is $R_{\mathrm{m}} = |W_{\mathrm{L}}|/|Q_{\mathrm{J}}|$.) Growth rates due to work done against the Lorentz force (——) and losses due to the Poynting flux through $\partial V_{\mathrm{PNS}}$ (— · —) are plotted in Figure 6 (Panel (C)). Evidently, the growth due to W_{L} greatly exceeds that due to Poynting losses through $\partial V_{\mathrm{PNS}}$. Moreover, the growth rate is ≈ 480 s^{-1}, which implies a magnetic energy growth time of about 2 ms (comparable to τ_{eddy}!), as opposed to the 66 ms seen in Panel (A). This discrepancy in growth rates is due to numerical dissipation: the field evolves on small spatial scales in the turbulent flows and the dissipative terms in Eq. (10)

become nontrivial and affect the overall growth rate. This suggests that the magnetic energy may grow on a *millisecond time scale* in SASI-driven turbulence with high R_m.

The effect of numerical dissipation is illustrated in the evolution of magnetic flux tubes. In Panel (D) of Figure 6 we plot the average flux tube thickness $\lambda_{rms} = (\langle B^2 \rangle / \langle |\nabla B|^2 \rangle)^{1/2}$ versus time for all the models. During the ramp-up of the SASI, λ_{rms} decreases rapidly until it reaches about 4 km ($\sim 3\Delta x$) and stays relatively constant thereafter. The decrease in flux rope thickness is halted by (numerical) resistive dissipation when λ_{rms} reaches the resistive scale (a few$\times\Delta x$). In an environment with near perfect conductivity (i.e., the CCSN environment) we expect the field amplification to be halted by dynamical interactions with the fluid before the resistive scale is reached [47]. In our large-scale simulations, finite spatial resolution and associated numerical dissipation plays a decisive role for the magnetic field evolution: the attainable field strength *and* the magnetic energy growth rate are underestimated. By investigating individual terms in the induction equation during the late-time phase of the MHD-SASI simulations, we have verified that the dissipative terms in Eq. (10) result in magnetic energy decay with a decay time comparable (but smaller in magnitude) to the growth time due to W_L.

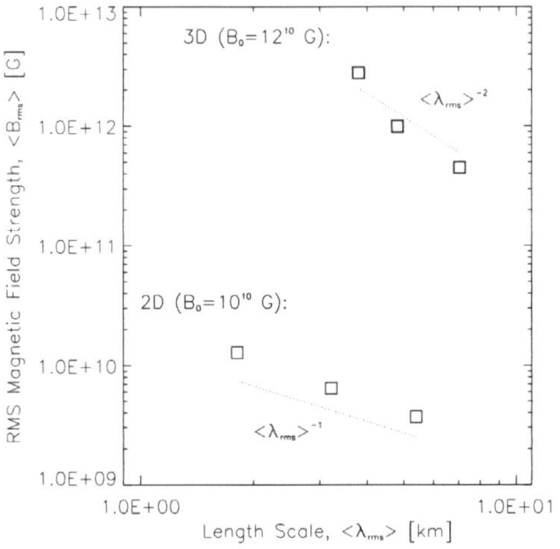

Figure 7. Plot of rms magnetic field B_{rms} versus λ_{rms}. Both quantities are averaged over an extended period in the nonlinear evolution.

The geometric concept of magnetic flux ropes is further illustrated in Figure 7, where we plot the time-averaged rms magnetic field $\langle B_{rms} \rangle$ versus time-averaged magnetic rms scale $\langle \lambda_{rms} \rangle$ from models where the spatial resolution has been varied ($\Delta x \approx 2.34, 1.56$, and 1.17 km). Higher spatial resolution accommodates the development of thinner flux ropes. The decreased cross-sectional area of a flux rope implies an increased field strength. In Figure 7, the increase in field strength is in direct proportion to the decrease in the cross-sectional area afforded by higher spatial resolution. Also, the slope observed in the 3D simulations is steeper than that observed in 2D (axisymmetric) simulations [16]. (We also observe that the exponential growth rate increases with increasing spatial resolution.)

The increase in magnetic energy in the volume occupied by the PNS due to Poynting flux through ∂V_{PNS} is 1.1×10^{44} erg (B10), 3.2×10^{47} erg (B12), and 4.5×10^{48} erg (B13). The results form models B12 and B13 imply PNS rms magnetic fields exceeding 10^{14} G. We expect the degree of PNS magnetization to be insensitive to the initial field in realistic settings with high R_m where dissipative processes can be ignored.

4. Summary and discussion

We have presented the initial implementation of a second-order finite volume MHD scheme in GenASiS. Using the MHD capabilities, we have performed 3D simulations of the evolution and amplification of magnetic fields in SASI-driven flows. The simulations are initiated from a configuration which resembles the early stalled shock phase in a CCSN, albeit with simplified physics that excludes critical components of a supernova model (e.g., neutrino transport, self-gravity, and the PNS itself). On the other hand, our simulations are computed with a spatial resolution that is currently inaccessible to state-of-the-art supernova models in three spatial dimensions, and they may therefore provide valuable insight into MHD developments in CCSNe.

Flows associated with the spiral SASI mode result in vigorous turbulence below the shock ($u_{\rm rms} \approx 4000$ km s^{-1}). The turbulence amplifies magnetic fields by stretching, and the magnetic energy grows exponentially with time as long as the kinematic regime obtains. The resulting magnetic fields display a highly intermittent flux rope structure. In the strong-field model, the magnetic energy saturates when the associated energy density becomes comparable to the kinetic energy density (i.e., $v_{\rm A} \sim |\mathbf{u}|$) in localized regions of the flow. The subsequent magnetic field evolution remains highly dynamic: strong fields emerge, are advected with the flow, are temporarily weakened, and then reemerge in a seemingly stochastic manner.

The presence of amplified magnetic fields does not result in noticeable effects on the global shock dynamics in our simulations, and this may be understood from considerations of the energetics: the turbulent kinetic energy—which powers SASI-driven field amplification—only amounts to about 10% of the total kinetic energy below the shock ($E_{\rm k} \sim 5 \times 10^{-2}$ B), and is not significant to the explosion (~ 1 B). These observations alone suggest a rather passive role of magnetic fields in the overall dynamics of non-rotating CCSNe.

Our simulations further suggest that the SASI may play a role in determining the strength of the magnetic field in proto-neutron stars and young pulsars. We estimate that the magnetic energy accumulated on the PNS may account for magnetic field strengths exceeding 10^{14} G.

The evolution of magnetic fields in our simulations remains sensitive to the spatial resolution. The magnetic energy attained and the rate at which the magnetic energy grows increase with increasing grid resolution. Data extracted from our simulations suggest that the magnetic energy may grow exponentially on a millisecond timescale ($\sim \tau_{\rm eddy}$) under more realistic physical conditions (i.e., huge magnetic Reynolds numbers), as opposed to the $\sim$ 50-60 ms timescale measured directly in our runs. Model B13 attains dynamically relevant magnetic fields, which are also subject to significant numerical dissipation. The MHD evolution displayed by this model during the saturated state is probably spurious and the dynamical impact of the B-fields may be underestimated. We therefore caution that the uncertainties associated with the sensitivity to numerical resolution prevents us from completely dismissing magnetic fields as unimportant to the explosion dynamics of weakly rotating progenitors.

In summary, we conclude from our simulations that magnetic fields in CCSNe may be amplified exponentially by turbulence driven by the spiral SASI mode. Details on the impact of SASI-amplified magnetic fields on explosion dynamics remain unclear, but on energetic grounds alone the role of these magnetic fields is likely sub-dominant. The simulations further suggest that small-scale PNS magnetic fields in the $10^{14} - 10^{15}$ G range may be formed, which may be sufficient to power some of the energetic activity that define AXPs and SGRs.

Acknowledgments

This research was supported by the Office of Advanced Scientific Computing Research and the Office of Nuclear Physics, U.S. Department of Energy. This research used resources of the Oak Ridge Leadership Computing Facility at the Oak Ridge National Laboratory provided through the INCITE program.

References

[1] Akiyama, S., Wheeler, C.J., Meier, D.L., and Lichtenstadt, I. 2003, *ApJ*, **584**, 954
[2] Balbus, S.A. and Hawley, J.F. 1991, *ApJ*, **376**, 214
[3] Bisnovatyi-Kogan, G.S., Popov, Y.P., and Samochin, A.A. 1976, *Ap&SS*, **41**, 287
[4] Blondin, J.M., Mezzacappa, A., and DeMarino, C. 2003, *ApJ*, **584**, 971
[5] Blondin, J.M. & Mezzacappa, A. 2007, *Nature*, **445**, 58
[6] Bonanno, A., Rezzolla, L., and Urpin, V. 2003, *A&A*, **410**, L33
[7] Brandenburg, A. & Subramanian, K. 2005, *Physics Reports*, **417**, 1

[8] Bruenn, S.W., J., Dirk, C.J., Mezzacappa A., Hayes, J.C., Blondin, J.M., Hix, W.R., and Messer, O.E.B. 2006, *J. Phys.: Conf. Series*, **46**, 393
[9] Buras, R., Janka, H.-Th., Rampp, M., and Kifonidis, K. 2006, *A&A*, **457**, 281
[10] Burrows, A., Dessart, L., Livne, E., Ott, C.D., and Murphy, J. 2007, *ApJ*, **664**, 416
[11] Davis, S.F. 1988, *SIAM, J. Sci. Comput.*, **9**, 445
[12] Del Zanna, L., Bucciantini, N., and Londrillo, P. 2003, *A&A*, **400**, 397
[13] Del Zanna, L., Zanotti, O., Bucciantini, N., and Londrillo, P. 2007, *A&A*, **473**, 11
[14] Duez, M.D., Liu, Y.T., Shapiro, S.L., Stephens, B.C. 2005, *Phys. Rev. D*, **72**, 024028
[15] Duncan, R.C. and Thompson, C. 1992, *ApJ*, **392**, L9
[16] Endeve, E., Cardall, C.Y., Budiardja, R.D., and Mezzacappa, A. 2010, *ApJ*, **713**, 1219
[17] Evans, C.R. & Hawley, J.F. 1988, *ApJ*, **332**, 659
[18] Gammie, C.F., McKinney, J., and Tóth, G. 2003, *ApJ*, **589**, 444
[19] Guilet, J., Foglizzo, T., and Fromang, S. 2011, *ApJ*, **729**, 71
[20] Gurski, K.F. 2004, *SIAM, J. Sci. Comput.*, **25**, 2165
[21] Harten, A., Lax, P.D., and Van Leer, B. 1983, *SIAM Review*, **25**, 35
[22] Heger, A., Woosley, S.E., and Spruit, H.C. 2005, *ApJ*, **626**, 350
[23] Hewish, A., Bell, S.J., Pilkington, J.D.H., Scott, P.F., and Collins, R.A. 1968, *Nature*, **217**, 709
[24] Hurley, K. et al. 2005, *Nature*, **434**, 1098
[25] Kurganov, A. & Tadmor, E. 2000, *JCP*, **160**, 241
[26] Kurganov, A., Noelle, S., and Petrova, G. 2001, *SIAM, J. Sci. Comput.*, **23**, 707
[27] Lai, D. & Qian, Y-Z. 1998, *ApJ*, **505**, 844
[28] LeBlanc, J.M. and Wilson, J.R. 1970, *ApJ*, **161**, 541
[29] LeVeque, R.J. 2002, *Finite Volume Methods for Hyperbolic Problems*, Cambridge University Press
[30] Li, S. 2005, *JCP*, **203**, 344
[31] Linde, T. 2002, *Int. J. Numer. Meth. Fluids*, **40**, 391
[32] Londrillo, P. & Del Zanna, L. 2004, *JCP*, **195**, 17
[33] Lorimer, D. & Kramer, M. 2005, *Handbook of Pulsar Astronomy*, Cambridge University Press
[34] Marek, A. & Janka, H.-Th. 2009, *ApJ*, **694**, 664
[35] Meier, D.L., Epstein, R.I., Arnett, D.W., and Schramm, D.N. 1976, it ApJ, **204**, 869
[36] Mezzacappa, A., Bruenn, S.W., Blondin, J.M., Hix, W.R., and Messer, O.E.B. 2007, in AIP Conf. Proc. 924, *The Multicolored Landscape of Compact Objects and Their Explosive Origins*, ed. L.A. Antonelli et al. (Melville, NY: AIP), 234
[37] Mignone, A., Ugliano, M., and Bodo, G. 2009, *MNRAS*, **393**, 1141
[38] Miyoshi, T. & Kusano, K. 2005, *JCP*, **208**, 315
[39] Moiseenko, S.G., Bisnovatyi-Kogan, G.S., and Ardeljan, N.V. 2006, *MNRAS*, **370**, 501
[40] Obergaulinger, M., Aloy, M.A., and Müller, E. 2005, *A&A*, **450**, 1107
[41] Obergaulinger, M & Janka, H.-Th. 2011, Magnetic field amplification in collapsing, non-rotating stellar cores, *Preprint* astro-ph/1101.1198
[42] Ott, E. 1998, *Phys. Plasmas*, **5**, 1636
[43] Shu, C-.W. 1997, *JCP*, **97**, 325
[44] Stone, J.M., Gardiner, T.A., Teuben, P., Hawley, J.F., and Simon, J.B. 2008, *ApJS*, **178**, 137
[45] Symbalisty, E.M.D. 1984 *ApJ*, **285**, 729
[46] Takiwaki, T., Kotake, K., and Sato, K. 2009, *ApJ*, *691*, 1360
[47] Thompson, C. and Duncan, R.C. 1993, *ApJ*, **408**, 194
[48] Thompson, C. and Duncan, R.C. 2001, *ApJ*, **561**, 980
[49] Toro, E.F., Spruce, M., and Speares W. 1994, *Shock Waves*, **4**, 25
[50] Tóth, G. 2000, *JCP*, **161**, 605
[51] Wheeler, J.C., Meier, D.L., and Wilson, J.R. 2002, *ApJ*, **568**, 807
[52] Woods, P.P. & Thompson, C. 2006, in *Compact stellar X-ray sources*, ed. W.H.G. Lewin & M. van der Klis (Cambridge: Cambridge Univ. Press), 547
[53] Ziegler, U. 2004, *JCP*, **196**, 393

IUPAP C20 Conference on Computational Physics (CCP 2011) IOP Publishing
Journal of Physics: Conference Series **402** (2012) 012028 doi:10.1088/1742-6596/402/1/012028

Ash-detonation interactions in multi-dimensional simulation of Type Ia supernovae

S T Parete-Koon[1], C R Smith[2], M W Guidry[12], W R Hix[12], O E B Messer[12]

[1] Oak Ridge National Laboratory, P.O. Box 2008, Oak Ridge, TN 37831
[2] University of Tennessee, Department of Physics and Astronomy, 1408 Circle Drive, Knoxville, TN 37996

E-mail: parete-koonst@ornl.gov

Abstract. Thermonuclear supernovae (Type Ia SNe) are believed to result from the complete disruption of a near-Chandrasekhar-mass white dwarf (WD) by explosive thermonuclear runaway. Their use as cosmological distance indicators as well as comprehension of their role in galactic chemical evolution ultimately depends on a reliable and robust understanding of the explosion mechanism. We have undertaken a study of turbulent thermonuclear combustion in these events, focusing on the effects of increased physical fidelity in simulations, through improved resolution and increased dimensionality. We begin with a series of detonation simulations at low WD densities that we will use for baseline metrics against which subsequent simulations will be measured. We have performed these simulations using a version of the FLASH code[1]. We show progress on a set of simulations in one, two, and three spatial dimensions designed to explore the impact of dimensionality on the evolution of the detonations, and thereby clearly delineate the role of the nuclear kinetics. By performing a suite of simulations incorporating an alpha network at several levels of maximum mesh refinement, we will be able to quantify the effects of resolution and network size on future results.

1. Introduction

Driven by the need to explain the physical mechanism that generates the one of the Universe's brightest standard candles, Type Ia supernovae, considerable effort has been expended to understand the physics of burning degenerate carbon and oxygen in exploding white dwarf stars. For Type Ia supernovae it is believed that the detonation, which drives the explosion, is preceded by a sub-sonic burning phase that pre-expands the white dwarf and, due to turbulence driven by that burning, leaves the carbon and oxygen in the interior of the white dwarf riddled with more evolved pockets of ash. The ash presents a likely damping obstacle to a subsequent detonation.

Maier & Niemeyer (2006), here after M&N, carried out a small number of 1-D and 2-D simulations in order to test the ability of detonations to pass through ash restrictions and thin funnels of fuel separating neighboring regions of ash [2] . Limited computational resources restricted the resolution of their 2-D study to 0.50 cm, which is not sufficient to fully resolve the detonation at the appropriate densities. We build on their preliminary studies, reproducing their results and running simulations at higher resolutions and in 3-D.

We begin with initial conditions similar to those of M&N [2]. Using the FLASH adaptive mesh hydrodynamics code [1], we set up a system of length 1024 cm and width of 128 cm with a reflecting boundary at 0 cm and an outflow boundary condition at 1024 cm. The long sides of the box have periodic boundary conditions. Our 3-D setup has a height of 128 cm. This setup allows the resolution to be equal in all dimensions. The FLASH code refines the mesh based on temperature and density differences between grid cells. In regions where there are small gradients between cells the code de-refines the mesh.

As initial conditions we choose a composition of 50% ^{12}C and 50% ^{16}O by mass at a temperature of 10^7 K and a material velocity of 0 cm s^{-1}. To ignite the detonation we set the initial conditions in a 25.6 cm "match head" to a temperature of 10^{10}K and a material velocity of 10^9 cm s^{-1}. Halfway down the tube, fuel funnels are formed by two opposing semi-circles of ash with a composition near NSE. The narrowest constriction formed by the two semi-circles will be referred to as the width of the fuel funnel. For the 3-D case we used "soup cans" of ash where two semi-cylinders formed the constriction for the fuel funnel so the constriction had the same width along the third dimention. For all simulations described here, we use a fuel density of $5.0 \cdot 10^7$g cm^{-3}. In all cases nuclear burning is turned off in the shock itself.

2. Resolution in 1-D

We ran 1-D simulations without ash to study the formation of the detonation. Figure 1 is in agreement with M&N and shows that the behavior of the maximum nuclear energy generation associated with the detonation is erratic for resolutions courser than 0.125 cm [2]. Though it is less erratic using 0.125 cm resolution, true convergence to a steady energy generation rate does not occur until the resolution is set to 0.031 cm. Thus the burning associated with the detonation, excluding the shock, is fully resolved at 0.031 cm resolution. The burning must be sufficiently resolved to conclusively determine the fate of the detonation.

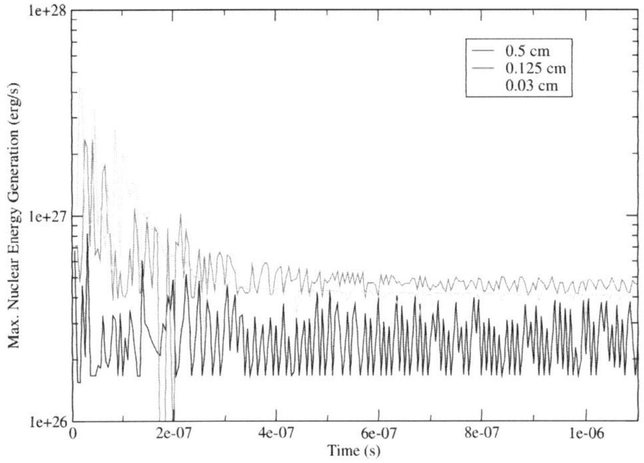

Figure 1. Maximum energy generation rate plotted over time for a detonation in 1-D at three resolutions. The behavior of energy generation rate associated with the detonation is erratic until the resolution is fine enough to resolve the burning.

Journal of Physics: Conference Series **402** (2012) 012028 doi:10.1088/1742-6596/402/1/012028

3. Resolution Studies in 2-D and 3-D

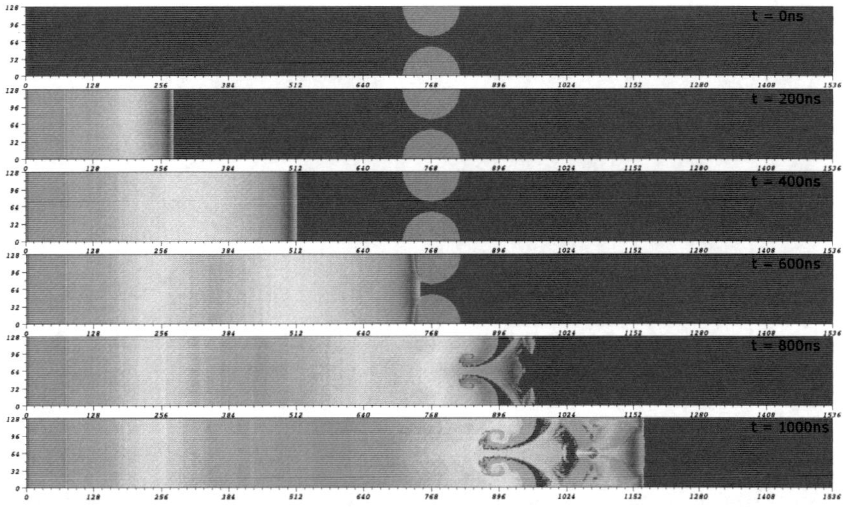

Figure 2. The survival of an under-resolved detonation shown in carbon abundance. Resolution is 0.5 cm.

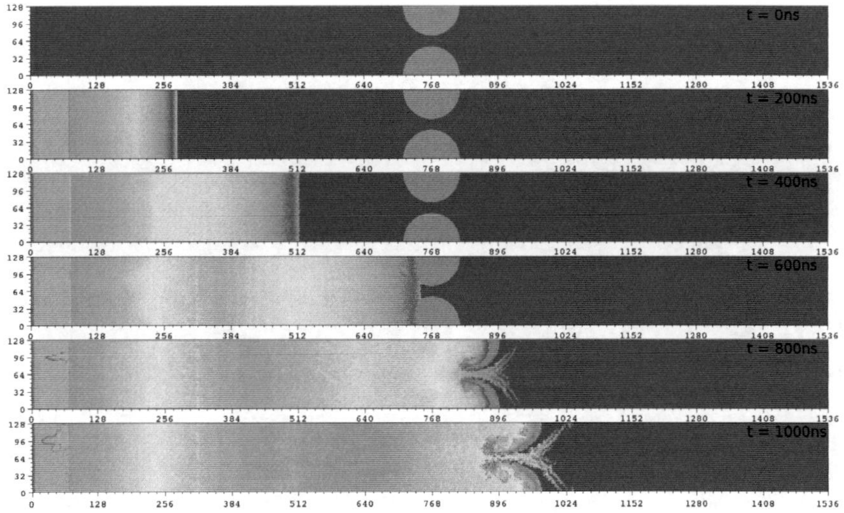

Figure 3. The death of a well resolved detonation shown in carbon abundance. Resolution is 0.031 cm.

We explored cases with ash constrictions in 2-D and in 3-D, with a number of different resolutions and configurations as shown in the table. Our results confirm the 2-D results of

Table 1. Resolution Study

Case Number	Resolution cm	Dimensions for Flame	Ash position cm	Fuel funnel width cm	Cellular Structure	Detonation Survived?
1	0.5	2	512	16	No	Yes
2	0.5	2	786	16	No	No
3	0.5	2	786	20	No	Yes
4	0.5	3	786	20	No	Yes
5	0.25	2	786	20	No	No
6	0.125	2	786	20	Yes	No
7	0.125	3	786	20		
8	0.031	2	786	20	Yes	No

M&N, at the insufficient resolutions they report [2]. For example, Figure 2, case 1 in the table, shows conditions similar to those reported by M&N with 0.5 cm resolution, where the detonation survives passing through the ash constriction. However, if the ash was located further from the match head as in case 2, the detonation did not survive unless the width of fuel funnel was extended to 20 cm, as in case 3. Based on their 1-D results, M&N expected that no conclusions could be drawn from under-resolved numerical experiments. Our preliminary study indicates that the detonation will not survive passing through the ash constriction if the resolution is sufficient to resolve the detonation. Figure 3 shows a time series of a case with 0.031 cm resolution. In this case the burning front becomes detached from the pressure front soon after the flame emerges from the ash and the detonation dies.

Figure 4 shows the setup and end state of a 3-D trial, case 4 in the table, run at 0.5 cm resolution. The gross behavior in 3-D is similar to that of its 2-D analog, in that the detonation survives. Figure 5, which shows a comparison of a slice of the 3-D case with its 2-D analog, demonstrates that the smaller scale structure in the two simulations differs. The 3-D flame front is slightly slower than that of the 2-D after it emerges from the ash. Both of these observations are consistent with more efficient turbulent cascade in the 3-D case.

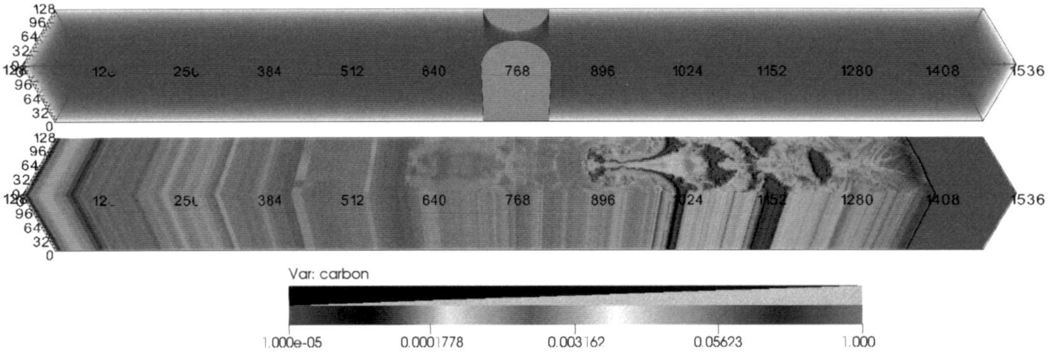

Figure 4. Initial setup and end state of a 3-D case at 0.05 resolution. The detonation survives the ash, but this resolution is insufficient to resolve the detonation.

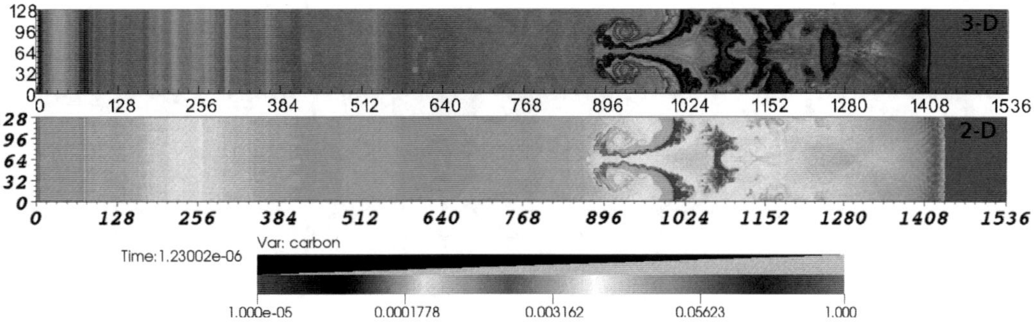

Figure 5. Slices of the 3-D case share the gross features of their 2-D analogs.

Future work will include a systematic study of the effects of resolution on this problem in 2 and 3 dimensions. We are also planning to investigate more complicated ash geometries for the 3-D cases.

4. Cell Structure

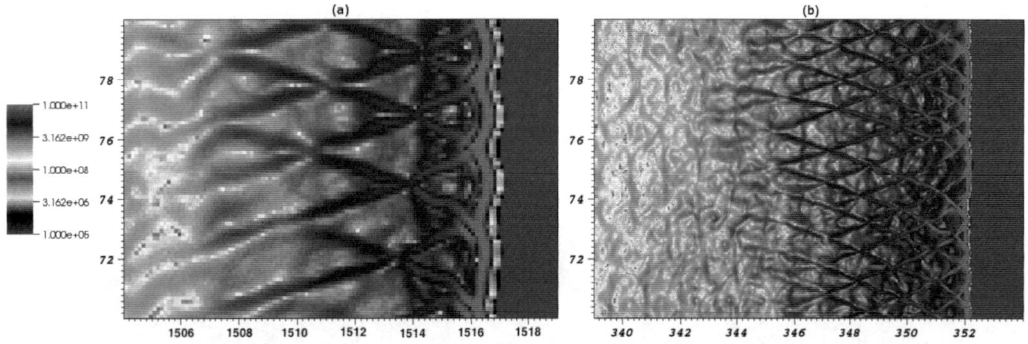

Figure 6. The carbon burning cells at 0.125 cm resolution, a, and 0.031 cm resolution, b. All distances given in cm.

Multi-D simulations allow the formation of cellular detonation structures that were first observed in terrestrial gaseous explosions and have since been studied theoretically and numerically for nuclear burning [3][4][5][6].

The cellular structure results from transverse instabilities in the detonation. This instability could impact Type Ia supernova models, by increasing the effective reaction length of the detonation compared to that predicted by 1-D models, thereby altering the detonation speed and final composition of the resulting ash. The transverse waves and the resulting unreacted pockets of fuel could also change the conditions for detonation failure due to the incomplete energy release [4][6].

Cell formation occurs in all of our trials with resolutions finer than 0.125 cm, consistent with the results of [5] and [4]. Figure 6 uses the magnitude of the pressure gradient to display the cellular structure that results from carbon burning for resolutions of 0.125 cm and 0.03 cm.

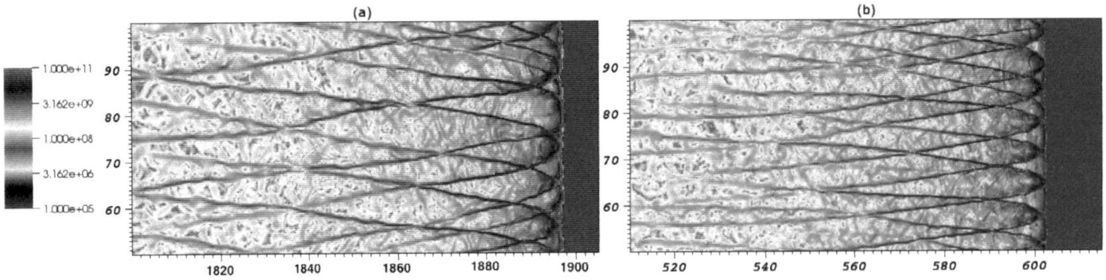

Figure 7. The oxygen burning cells at 0.125 cm resolution, a , and 0.031 cm resolution, b. All distance given in cm.

Figure 7 shows the oxygen cells burning cells which form on a longer timescale and are an order of magnitude larger than the carbon cells. Gamezo et al. (1999) found that the size of the cells as well as the time needed for them to form is greatly impacted by resolution [4]. However, Timmes et al. (2000) found that the size was independent of the spatial resolution once the cells were fully resolved [5]. Consistent with Timmes et al., we do not notice a large difference in the size of the oxygen burning cells between the different resolutions. However, the smaller carbon burning cells appear to be markedly larger in the 0.125 cm resolution case than in the 0.031 cm resolution case, likely indicating that we are not fully resolving carbon burning at 0.125 cm resolution.

5. Conclusions

Though this work is still ongoing, we have learned several important lessons. Our higher resolution study supports the conjecture of M&N [2], that a detonation front can not survive passage through ash or fuel constrictions caused by ash if the width of the constriction is close to the detonation thickness. This is an important lesson because the burning must be sufficiently resolved to conclusively determine the fate of the detonation. This study also implies that the 2-D models for this problem are reasonable predictors of the behavior of the analogous 3-D models, though the influence of demensionality is apparent in the size of the ash structures. This work supports the conjecture of Timmes et al. (2000) that the size of cellular structures that results from the transverse instability in the flame is independent of the resolution once the structures are sufficiently resolved [5] . In the absence of any additional data, for models that attempt to study the development of the thermonuclear runaway in Type Ia supernovae, this study suggests that detonation fronts are likely to be quenched when they encounter regions of ash.

Acknowledgments

This research used resources of the Oak Ridge Leadership Computing Facility, located in the National Center for Computational Sciences at Oak Ridge National Laboratory, which is supported by the Office of Science of the Department of Energy under Contract DE-AC05-00OR22725. S T Parete-Koon and O E B Messer aknowledge support from the Office of Advanced Scientific Computing Reseach of the U.S. Department of Energy, W R Hix and M W Guidry acknowledge support from the Office of Nuclear Physics U.S. Department of Energy.

References

[1] Fryxell B, Olson K, Ricker P, et al. 2000, ApJS, 131, 273
[2] Maier A, & Niemeyer, J C, 2006, A&A, 451, 207

[3] Denisov, Y N, & Troshin, Y K 1959, Dokl. Akad. Nauk SSSR, 125, 110
[4] Gamezo, V N, Wheeler, J C Khokhlov, A M, & Oran, E S, 1999, ApJ, 512, 827
[5] Timmes, F X et al. , 2000, ApJ, 543,938
[6] Boisseau, J R, Wheeler, J C, Oran, E S, & Khokhlov, A M, 1996, ApJ, 471, L99

Journal of Physics: Conference Series **402** (2012) 012029 doi:10.1088/1742-6596/402/1/012029

Parallel computation of Feynman loop integrals

E de Doncker[1] **and F Yuasa**[2]

[1] Department of Computer Science, Western Michigan University, Kalamazoo MI 49008, U. S.
[2] High Energy Accelerator Research Organization (KEK), Oho 1-1, Tsukuba, Ibaraki, 305-0801, Japan

E-mail: elise.dedoncker@@wmich.edu

E-mail: fukuko.yuasa@@kek.jp

Abstract. The need for large numbers of compute-intensive integrals, arising in quantum field theory perturbation calculations, justifies the parallelization of loop integrals. In earlier work, we devised effective multivariate methods by iterated (repeated) adaptive numerical integration and extrapolation, applicable for some problem classes where standard multivariate integration techniques fail through integrand singularities. In repeated integration, the function evaluations for the outer integral are independent integral computations for the next lower level and can be distributed to threads in a multi-core parallelization. The distribution level determines the number of dimensions in the inner integral below it and thus the granularity of the computation. We discuss a multi-threaded implementation over the OpenMP Application Program Interface.

1. Introduction

We present a multi-threaded approach for the computation of iterated integrals. An important application of (numerical) iterated integration has been the computation of Feynman loop integrals, which are needed in higher order perturbation calculations of the cross section for particle interactions. Thousands of integrals may be needed for one interaction.

Furthermore, Feynman loop integrals are generally affected by non-integrable singularities, through vanishing denominators in the interior and/or at the boundaries of the integration domain. In order to handle singularities inside the domain, a value for the integral is calculated by introducing a parameter $i\delta$ in the integrand denominator, effectively moving the singularity into the complex plane, and by taking the limit of the integral value as $\delta \to 0$. We make use of numerical iterated integration for computing an integral sequence as δ decreases. In view of the computational expense we propose a technique utilizing multi-threading for a parallel computation of the individual integrals.

The goal of an *automatic* integration procedure is to obtain an approximation $Q(f)$ to an integral

$$If = \int_{\mathcal{D}} f(\vec{x})\, d\vec{x} \tag{1}$$

and an error estimate $\mathcal{E}f$, in order to satisfy a specified accuracy requirement for the error $Ef = |Qf - If|$, so that

$$|\, Qf - If \,| \;\le\; \mathcal{E}f \;\le\; \max\{\, t_a \,,\, t_r \,|\, If \,|\, \} \tag{2}$$

for a given integrand function f, region $\mathcal{D}$ and (absolute/relative) error tolerances t_a and t_r, respectively. This can be interpreted as a *black box* method for the computation of the integral (1), where the inputs specify the integral dimension, the region $\mathcal{D}$, the function $f(\vec{x})$ and the requested accuracy. The output contains the result Qf and error estimate $\mathcal{E}f$. No specification of the integrand behavior is needed by the general automatic integration procedure.

Section 2 below gives an outline of global adaptive and iterated integration techniques in the framework of automatic integration. Our parallelization for a multi-threaded environment is explained in Section 3, and some background on Feynman loop integrals is provided in Section 4. Timing results using the shared memory programming OpenMP [1] Application Program Interface (API) are given in Section 5.

2. Automatic numerical integration

2.1. Global adaptive procedure

A versatile type of algorithm to implement the black box approach is by adaptive partitioning of the integration region. At each step, a region is subdivided, integral and error estimates are computed over the subregions, and the overall result and error estimate are updated. Various strategies are possible for the selection of the region to be subdivided at each step. A *global adaptive* strategy maintains a priority queue on the subregion collection (e.g., a linked list or a heap), keyed with the local error estimates of the subregions. Examples of global adaptive integrators are the 1D adaptive programs of QUADPACK [2], and the multivariate DCUHRE [3] for integration over an (N-dimensional) cube, which select the region with the highest (absolute) error estimate for subdivision at each step. As a result, sample points are concentrated in the vicinity of irregular integrand behavior such as singularities, discontinuities, peaks or ridges and troughs. The CUBPACK [4] package targets adaptive integration over a collection of cubes and simplices.

The local integral (over a subregion or an interval) is approximated by a (cubature or quadrature) rule which is a linear combination of function values of the form $\sum_{k=1}^{K} w_k f(\vec{x_k})$. By evaluating more than one rule over the subregion, a local error estimate can be obtained as a function of the difference between local integral approximations. For example, the (1D) program DQAG of the QUADPACK package uses a pair of approximations, given by an r-point Gauss rule and the interlacing $(2r+1)$-point Kronrod rule. The user has the option of choosing one of the pairs, with $r = 7, 10, 15, 20, 25$ or 30. The Kronrod rule supplies the local integral approximation, and the Gauss rule (together with the Kronrod rule) serves to obtain the local error estimate. The r-point Gauss rule is of polynomial degree of accuracy $2r-1$ (i.e., it integrates all polynomials of degrees $0, \cdots, 2r-1$ exactly, and there are polynomials of degree $2r$ which are not integrated exactly). The Kronrod rule is of polynomial degree $3r+1$ if r is even, and of degree $3r+2$ when r is odd (because of symmetry). The multivariate program DCUHRE applies an embedded sequence of cubature rules [5] in an adaptive setting.

2.2. Iterated integration

For integration over a product region $\mathcal{D} = \mathcal{D}_1 \times \ldots \times \mathcal{D}_\ell$, we consider the *iterated (repeated) integral*

$$If = \int_{\mathcal{D}_1} d\vec{x}^{(1)} \ldots \int_{\mathcal{D}_\ell} d\vec{x}^{(\ell)}\; f(\vec{x}^{(1)}, \ldots, \vec{x}^{(\ell)}).$$

The integration over the regions $\mathcal{D}_j$ can be performed with a multivariate or a 1D integration code, and different methods can be applied on different levels $j = 1, \ldots, \ell$. For example, assume that the computation of a 2D integral over the unit square requires the outer integration over a subinterval $[a, b] \subseteq [0, 1]$. The integral over $[a, b]$ is evaluated in the form

$$\int_a^b dx \int_0^1 dy\ f(x, y) \ \approx\ \sum_{k=1}^{K} w_k F(x_k), \tag{3}$$

where the w_k and x_k are the weights and abscissae of the local rule scaled to the interval $[a, b]$, applied in the x-direction. The function evaluation

$$F(x_k) = \int_0^1 f(x_k, y)\ dy, \qquad 1 \leq k \leq K, \tag{4}$$

is itself an integral in the y-direction, and is computed by the method for the inner integration. The latter may be adaptive, and subject to an error control condition of the form (2). Note that the error incurred in the inner integration will contribute to the overall integration error. Work on the integration error interface is reported in [6, 7, 8].

Subsequently, in Section 5, we give results obtained with 1D repeated integration by the program DQAG from QUADPACK, using the (10, 21)-points Gauss-Kronrod pair. Thus in that case, the weights and abscissae on the right of the approximation (3) represent those of the Kronrod rule scaled to $[a, b]$.

Figure 1 from [9] shows adaptive subdivision patterns which may be obtained when the integrand has a ridge or singular behavior along the (anti-)diagonal of the (square) integration domain. Figure 1(a) is that of a standard 2D subdivision where each selected subregion is partitioned into four subsquares of equal size. Figure 1(b) depicts an iterated integration. The integrand is evaluated at three points in the x-direction, and an adaptive subdivision is shown vertically at each of the abscissae. As noted in [9] and in low dimensions, the adaptive iterated integration is found to be more effective than the general multivariate adaptive partitioning, for handling singularities that do not line up with subdivision directions.

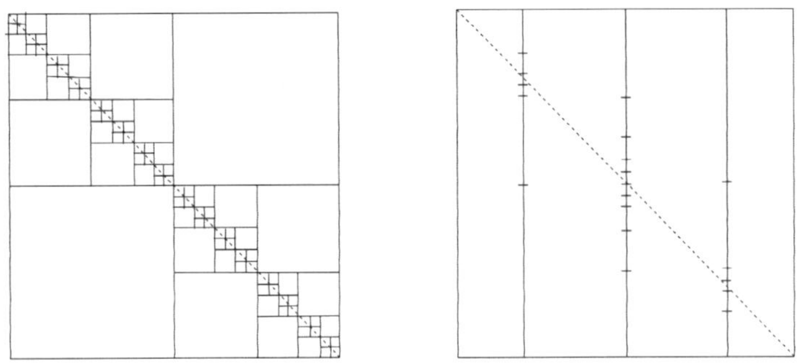

(a) (b)

Figure 1. (a) Standard subdivision; (b) Iterated adaptive strategy for irregular behavior on diagonal [9]

3. Parallel automatic integration approach

Non-adaptive methods such as Monte Carlo, Quasi Monte Carlo and lattice rules can be parallelized in a straightforward way, by assigning function evaluations to the available processes or threads. In the approximation of the integral (1) by

$$If = \int_{\mathcal{D}} f \approx \sum_k w_k\ f(\vec{x_k}), \tag{5}$$

the function evaluations $f(\vec{x_k})$ in the weighted sum can be performed in an *embarrassingly parallel* way. While these methods are available for high-dimensional integration, they have a slow rate of convergence.

Adaptive methods are suited for integrals of low to moderate dimensions (say, ≤ 10). These are more complex than the non-adaptive techniques, and usually parallelized on the subregion level. Typically a parallel task pool scheme is implemented in a master-slave approach, for maintaining the priority queue data structure over the subregion collection.

In shared memory applications the task pool is shared, so that some type of mutual exclusion is required for task pool updates and maintenance. This causes various parallel losses (analyzed, e.g., in our "archives" [10, 11, 12]). On distributed memory systems, the task pool is distributed and updated using message passing, as done in programs layered over the Message Passing Interface (MPI [13] or OPEN MPI [14]). Load balancing is then needed to control the size of the local priority queues for problems where the integration has hot spots due to intensive region partitioning, in the vicinity of singularities and other irregular behavior (see, e.g., the PARINT strategies [15, 16, 17] and [18, 19]). Apart from the communication times, breaking loss (incurred at the end of the computations when slaves are working during termination detection) is an important characteristic of these methods. This also leads to the fact that parallel executions may be very different from the sequential computation, in an irreproducible manner.

For iterated integration, it would be possible to use a parallelization of the outer integration on the subregion level. However, in this paper we propose a parallelization on the function evaluation level(s) of the outer integration. For example, with regard to (3) and (4), this means that the values of the inner integrals $F(x_k)$ are computed in parallel. More generally, if $\mathcal{D} = \mathcal{D}_1 \times \mathcal{D}_2$ and S is a selected subregion $\subseteq \mathcal{D}_1$, we approximate

$$\int_S d\vec{x}^{(1)}\ F(\vec{x}^{(1)}) \approx \sum_{k=1}^{K} w_k\, F(\vec{x_k}^{(1)}), \tag{6}$$

$$\text{with} \quad F(\vec{x_k}^{(1)}) = \int_{\mathcal{D}_2} d\vec{x}^{(2)}\ f(\vec{x_k}^{(1)}, \vec{x_k}^{(2)})$$

and the integrals $F(\vec{x_k}^{(1)})$ are computed in parallel.

Important properties of this method include that

(i) the granularity of the parallel integration is large, especially when the inner integrals $F(\vec{x_k}^{(1)})$ are of dimension two or greater;

(ii) the points $\vec{x_k}^{(1)}$ where the function F is evaluated in parallel are the same as those of the sequential evaluation; i.e., apart from possibly the order of the summation in (6), the parallel calculation is the same as the sequential one.

4. Feynman loop integrals

In addition to the lowest order or tree level, higher order corrections are required for an accurate theoretical prediction of an interaction cross section, and for checking its agreement with the data observed at colliders. Feynman loop diagrams need to be taken into account, necessitating the calculation of *loop integrals*.

A diagram with L loops and N propagators is given in Feynman parameter space as

$$\mathcal{I} = \frac{\Gamma\left(N - \frac{nL}{2}\right)}{(4\pi)^{nL/2}}(-1)^N \int_0^1 \prod_{j=1}^{N} dx_j\, \delta(1 - \sum x_j) \frac{C^{N-n(L+1)/2}}{(D - i\,\delta\, C)^{N-nL/2}}, \tag{7}$$

where C and D are polynomials determined by the topology of the corresponding diagram and the physical parameters at hand. Loop integrals are generally divergent when denominators vanish in the integration region. We assume $\mathcal{I}$ does not suffer from other divergences, such as infrared (IR) divergence, and exists in the limit as $\delta \to 0$. To calculate the limit numerically [20], we generate a sequence of $\mathcal{I} = \mathcal{I}(\delta)$ for (geometrically) decreasing values of δ, and apply convergence acceleration or extrapolation to the limit with the ε-algorithm [21, 22].

Automatic packages, based on symbolic computations, are available for one-loop integrals [23, 24, 25, 26, 27, 28, 29, 30, 31]. However, many diagrams are required for an interaction, and the symbolic reduction leads to large sets of integrals. Furthermore, analytic integration is not possible in general, for higher orders and for general mass configurations. Therefore we apply the numerical *Direct Computation Method (DCM)* of [20, 32, 33, 34] as a building block for the computation, using (automatic) iterated integration and convergence acceleration to the limit as $\delta \to 0$. The motivational intensive nature of the computations motivates a parallelization of the individual integrations, in particular for 3D and higher-dimensional integrals.

The *DCM* building block can also be used for integrals resulting from a reduction that is not necessarily carried through to the primitive level. For example, in [35] we implemented a reduction by sector decompositions (and dimensional regularization) proposed in [36] for one-loop integrals through the hexagon, and applied *DCM* to a resulting set of 2D (vertex) and 3D (box) integrals. Especially the box integrals may still be computational intensive.

5. Results

We implemented the method of Section 2, as a multi-threaded application layered over OpenMP [1], and obtained timing results on a multi-core Intel Xeon X5680@3.33GHz CPU with 6 cores/12 logical cores, for the integrals of $If_1(\delta)$, $If_2(\delta)$ and If_3 given below in (8-10), and for a non-scalar (box) loop integral, $M_4(f,g;\delta)$ from [16]. The function f_1 is given in [37] as a test function (Dice3), for the Monte Carlo integration program Dice, and f_2 is a 3D version of the function Dice1. We treated the non-scalar loop integral M_4 in [16], using a calculation that required the integral for values of $\delta = 1.2^{30-\ell}$, $\ell = 0, \cdots, 16$, which are large compared to the value $\delta = 0.01$ used below. We also gathered preliminary results for If_3, on a Power7@3.83GHz CPU of the HITACHI SR16000 system at KEK, Japan, using up to 16 threads (maximum) assigned to our OpenMP application on this system.

$$\begin{aligned} If_1(\delta) &= \int_{-1}^{1} dx \int_{-1}^{1} dy \int_{-1}^{1} dz \frac{\delta\sqrt{x^2+y^2+z^2}\;\theta(1-x^2-y^2-z^2)}{(x^2+y^2+z^2-a^2)^2+\delta^2}, \qquad (8) \\ & \delta = 0.01, \quad a = 0.8 \\ \theta(u) &= 1 \text{ if } u > 0 \\ & \frac{1}{2}, \text{ if } u = 0 \\ & 0, \text{ if } u < 0 \\ If_2(\delta) &= \int_{0}^{1} dx \int_{0}^{1} dy \int_{0}^{1} dz \frac{2\,\delta\, y}{(x+y+z-1)^2+\delta^2}, \qquad (9) \\ & \delta = 10^{-6} \\ I(f_3) &= \int_{0}^{1} dx \int_{0}^{1} dy \int_{0}^{1} dz \frac{1}{(x+y+z)^2} \qquad (10) \end{aligned}$$

and

$$\begin{aligned} M_4(f,g;\delta) &= \int_0^1 dx \int_0^{1-x} dy \int_0^{1-x-y} dz \left[\frac{f(x,y,z)}{(D_4 - i\delta)^2} - 2\frac{g(x,y,z)}{D_4 - i\delta}\right], \qquad (11) \\ & \delta = 0.1 \end{aligned}$$

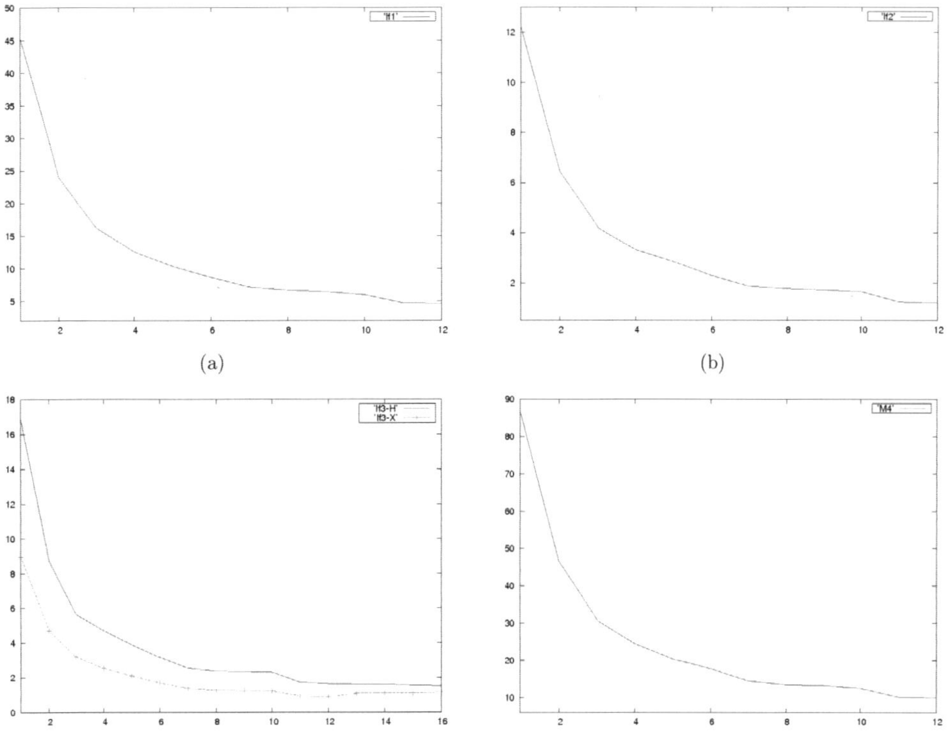

(a) (b) (c) (d)

Figure 2. Times (in s): (a) Time for If_1, $\delta = 0.01$, $a = 0.8$, $t_r = 10^{-8}$; (b) Time for If_2, $\delta = 10^{-6}$, $t_r = 10^{-14}$; (c) Time for If_3, $t_r = 10^{-12}$ on Xeon X5680 (If3-X) and on SR16000 (Hitachi) node (If3-H); (d) Time for $M_4(f, g, \delta)$, $\delta = 0.1$ $t_r = 10^{-8}$

where $D_4 = \mathbf{x}^T A\mathbf{x} + 2\mathbf{v} \cdot \mathbf{x} + C$, and $A_{j,k} = q_j \cdot q_k$, $q_1 = -p_{e^-}$, $q_2 = p_{e^+}$, $q_3 = p_{e^+} - p_{W^+}$, $C = M_0^2 = M_Z^2$, $v_k = \frac{1}{2}(-q_k^2 + M_k^2 - M_0^2)$ with $M_1 = m_e, M_2 = M_W, M_3 = m_e$.

The integrals $If_1(\delta)$ and $If_2(\delta)$ exhibit a behavior not unlike that of loop integrals, with a ridge that becomes steeper and narrower as $\delta \to 0$. The integrand f_1 furthermore has a discontinuity on the spherical surface $x^2 + y^2 + z^2 = 1$, The function f_2 has a peaked behavior at $z = 1 - x - y$, and f_3 has a singularity at the origin. The integrand of M_4 in (11) has a singularity on a hyperbolic surface where $D_4 = 0$ within the (unit simplex) integration region. Note that we treated the non-scalar loop integral $M_4(f, g, \delta)$ in [16], using a calculation that required the integral for values of $\delta = 1.2^{30-\ell}$, $\ell = 0, \cdots, 16$, which are large compared to the value $\delta = 0.1$ used below.

On the Intel machine, the Intel *ifort* Fortran XE compiler was used with flag *-openmp*. The f90 compiler with flag *-omp* was used on the SR16000 node. In all cases the main loop in the rule routine DQK21 of QUADPACK was parallelized (for the outer integration), by means of a *omp parallel do* pragma for the Gauss and Kronrod rule evaluations. On comparing loop schedules, we found that the dynamic loop schedule outperforms the default static schedule for the selected test problems. In the static schedule, the participating threads are assigned their loop iterations at the beginning of the loop, in a round-robin fashion; thus the assigned iterations are determined only by the thread ID and the total number of iterations. The dynamic schedule adjusts the assignments at runtime. The threads can request more work at the completion of their previously assigned iterations, which is beneficial in cases where the integrand behavior over the subregion is non-uniform.

The timings show a good speedup pattern for a number of threads within the available number of logical cores. The runs were performed while several other users were on the system. We were able to get preliminary results on a node of the Hitachi SR16000 supercomputer at KEK (just before the system was taken down – and will be down for a move during the coming month). The SR16000 is suited for multi-core computations on each node, and for message passing applications between nodes. We plan on performing hybrid runs for loop integral calculations in the future. It appears that our multi-threaded OpenMP application runs faster on the Xeon system for the same problem parameters.

6. Concluding remarks

For compute-intensive integral calculations we have developed a multi-core parallelization on the (outer) function evaluation level in the iterated integration procedure. In this paper we give timing results of the method implemented in OpenMP, which show good speedups for tests on a Xeon X5680 based system and on a (Hitachi) SR16000 system node.

So far, only the rule application was parallelized in the outer integration, with a Gauss-Kronrod scheme for the local integrations. Future extensions of the method include hybrid (distributed shared memory) parallelizations, where individual problems are distributed to multi-core nodes, and computed using multi-threading on the nodes. Furthermore, when more cores are available per node, the rules can easily be evaluated in parallel over subregions resulting from the local subdivision of a region. Nesting of parallel constructs is also possible, so that more than one level can be parallelized. GPUs can be used on the lower evaluation level(s), particularly for cubature rules (of higher dimensions) with many evaluation points. Our future work will naturally involve implementations and testing for more computational intensive loop integrals.

Acknowledgments

This work is partially supported by Grant-in-Aid for Scientific Research (20340063) of JSPS and by the CPIS program of Sokendai (Japan). We also acknowledge the support from Western Michigan University and the College of Engineering and Applied Sciences (CEAS).

References

[1] http://www.openmp.org, OpenMP web site
[2] Piessens R, de Doncker E, Überhuber C W and Kahaner D K 1983 *QUADPACK, A Subroutine Package for Automatic Integration* Springer Series in Computational Mathematics (Springer-Verlag)
[3] Berntsen J, Espelid T O and Genz A 1991 *ACM Trans. Math. Softw.* **17** 452–456
[4] Cools R and Haegemans A 2003 *ACM Transactions on Mathematical Software* **29** 287–296
[5] Berntsen J, Espelid T O and Genz A 1991 *ACM Trans. Math. Softw.* **17** 437–451
[6] Fritsch F N, Kahaner D K and Lyness J N 1981 *ACM TOMS* **7** 46–75
[7] Kahaner D, Moler C and Nash S 1988 *Numerical Methods and Software* (Prentice Hall)
[8] de Doncker E and Kaugars K 2010 *Procedia Computer Science* **1**
[9] Li S December 2005 *Online Support for Multivariate Integration* PhD dissertation Western Michigan University
[10] de Doncker E and Kapenga J 1987 *Numerical Integration; Recent Developments, Software and Applications* NATO ASI Series ed Keast P and Fairweather G (Reidel) pp 207–218
[11] de Doncker E and Kapenga J A 1987 *Proceedings of the Third SIAM Conference on Parallel Processing for Scientific Computing, Los Angeles* pp 109–113
[12] Kapenga J and de Doncker E 1988 *Parallel Computing* **7** 211–225
[13] http://www-unix.mcs.anl.gov/mpi/index.html, MPI web site
[14] http://www.open-mpi.org, Open MPI web site
[15] de Doncker E, Genz A, Gupta A and Zanny R 1998 *Supercomputing '98*

[16] de Doncker E, Kaugars K, Cucos L and Zanny R 2001 *Proc. of Computational Particle Physics Symposium (CPP 2001)* pp 110–119
[17] de Doncker E, Zanny R, Kaugars K and Cucos L 2001 *Lecture Notes in Computer Science* vol 2074 (Springer-Verlag) pp 118–127
[18] de Doncker E and Gupta A 1998 *Parallel Computing* **24** 1223–1244
[19] Achalla C, de Doncker E, Kaugars K and VanṼoorst J 2004 *Parallel and Distributed Computing and Systems*
[20] de Doncker E, Shimizu Y, Fujimoto J and Yuasa F 2004 *Computer Physics Communications* **159** 145–156
[21] Shanks D 1955 *J. Math. and Phys.* **34** 1–42
[22] Wynn P 1956 *Mathematical Tables and Aids to Computing* **10** 91–96
[23] van Oldenborgh G J and Vermaseren J A M 2000 *Nucl. Phys. Proc. Suppl.* **89** 231
[24] Bauer C 2002 *KEK Proceedings 2002-11* vol [MZ-TH/02-04] pp 179–185 do Hoang Son, Ph.D. thesis at the Physics Department, Johannes Gutenberg Univ., 2003
[25] Bélanger G, Boudjema F, Fujimoto J, Ishikawa T, Kaneko T, Kato K and Shimizu Y 2006 *Physics Reports* **430** 117–209
[26] Giele W T and Zanderighi G 2008 *JHEP* **0806** 038
[27] van Hameren A, Papadopoulos C G and Pittau R 2009 *JHEP* **0909** 106
[28] Hirschi V, Frederix R, Frixione S, Garzelli M V, Maltoni F and Pittau R Automation of one-loop qcd corrections arXiv:1103.0621 [hep-ph]
[29] Berger C F, Bern Z, Dixon L J, Febres Cordero F, Forde D, Gleisberg T, Ita H, Kosower D A and Maitre D 2009 *Phys. Rev. D* **80** 074036 arXiv:0907.1984 [hep-ph]
[30] Binoth T, Guillet J P, Heinrich G, Pilon E and Reiter T 2009 *Comput. Phys. Commun.* **180** 2317
[31] Heinrich G, Ossola G, Reiter T and Tramontano F 2010 *JHEP* **1010** 105 arXiv:1008.2441 [hep-ph]
[32] Yuasa F, Ishikawa T, Fujimoto J, Hamaguchi N, de Doncker E and Shimizu Y 2008 *XII Adv. Comp. and Anal. Tech. in Phys. Res.* PoS (ACAT08) 122; arXiv:0904.2823
[33] de Doncker E, Fujimoto J, Hamaguchi N, Ishikawa T, Kurihara Y, Shimizu Y and Yuasa F 2010 *Springer Lecture Notes in Computer Science (LNCS)* **6017** 139–154
[34] de Doncker E, Fujimoto J, Hamaguchi N, Ishikawa T, Kurihara Y, Shimizu Y and Yuasa F 2011 *Journal of Computational Science (JoCS)* doi:10.1016/j.jocs.2011.06.003
[35] de Doncker E, Fujimoto J, Kurihara Y, Hamaguchi N, Ishikawa T, Shimizu Y and Yuasa F 2010 *XIV Adv. Comp. and Anal. Tech. in Phys. Res.* PoS (ACAT10) 073
[36] Binoth T, Heinrich G and Kauer N 2003 *Nuclear Physics B* **654** 277 hep-ph/0210023
[37] Tobimatsu K and Kawabata S 1998 Multi-dimensional integration routine DICE Tech. Rep. 85 Kogakuin University

Microscopic nuclear mass table with high-performance computing

J. Erler[1,2], N. Birge[1], M. Kortelainen[1,2], W. Nazarewicz[1–3], E. Olsen[1], A. Perhac[1], M. Stoitsov[1,2]

[1]Department of Physics and Astronomy, University of Tennessee, Knoxville, Tennessee 37996, USA

[2]Physics Division, Oak Ridge National Laboratory, Oak Ridge, Tennessee 37831, USA

[3]Institute of Theoretical Physics, Warsaw University, ul. Hoża 69, PL-00681, Warsaw, Poland

E-mail: jerler@utk.edu

Abstract. We overview the methodology behind the large-scale mass table calculations based on the nuclear density functional theory (DFT) with Skyrme energy density functionals (EDFs). The calculations employing massively parallel computers are done in a large configuration space of an axially deformed harmonic oscillator. Nuclear mass tables tabulating global nuclear properties such as binding energies, radii, shape deformations, and pairing gaps are obtained for several Skyrme EDFs augmented by a mixed-type pairing functional, using an approximate particle number projection before variation. Specialty visualization tools have been developed to analyze the results. As an illustrative example of our current capabilities, we show some results for a wide range of even-even nuclei with $Z \leq 120$ and $N \leq 300$.

1. Introduction

Experiments employing beams of short-lived nuclei have dramatically expanded our knowledge of the nuclear landscape, including the remote regions of nuclei far off the valley of stability [1]. Modern radioactive ion beam facilities enable explorations of very exotic nuclei as one moves towards, and sometimes approaches, the particle drip lines where the nuclear binding ends. In this endeavor, of most interest are neutron-rich nuclei which are of critical importance for our understanding of a plethora of nuclear and astrophysical phenomena, as well as superheavy systems at the limits of nuclear mass and charge. From a theoretical point of view, the description of these species is a demanding task, as it requires the understanding and control of three crucial aspects of the nuclear many-body problem: interactions, many-body correlations, and coupling to the continuum of scattering and decay channels [2].

There is a well-delineated path towards a comprehensive description of the nucleus at the nucleonic level across the nuclear chart that merges three approaches: ab initio, configuration-interaction, and nuclear DFT. For complex nuclei with many valence particles, and mass table calculations throughout the nuclear chart, ab initio and configuration interaction methods become intractable due to exploding dimensions of the configuration space. Here the DFT becomes the tool of choice.

The modern nuclear DFT is based on the self-consistent mean-field approach rooted in the self-consistent Hartree-Fock-Bogoliubov (HFB), or Bogoliubov-de Gennes, problem [3]. The three most prominent nuclear DFT schemes are: the Skyrme-HFB method based on local densities and currents [3, 4, 5], the Gogny-HFB approach using the finite-range Gogny force [6, 7], and the relativistic mean-field model [8, 9]. All those schemes have been used to produce global nuclear mass tables, see Refs. [10, 11, 12, 13] and references quoted therein.

To take advantage of the steady increase in computing power, partnerships have been created in which physicists collaborate with mathematicians and computer scientists on a specific science challenge. In low-energy nuclear theory research, new paradigms have been opened up by the SciDAC Universal Nuclear Energy Density Functional (UNEDF) project [14, 15]. Some of the unique computational nuclear DFT infrastructure developed under UNEDF has recently been used to optimize the nuclear EDF [13, 16] and carry out large-scale nuclear mass table calculations [11, 12]. In this work, we briefly discuss such calculations and their content.

In our systematic calculations, we employ Skyrme EDFs. Their general form is more or less prescribed by a low momentum expansion [5]. Since the coupling constants of the nuclear EDF cannot yet be computed by means of ab-initio methods, it is customary to adjust these to experimental data using least-square optimization techniques [17, 13, 16]. As the parameters of currently used EDFs are primarily constrained by global properties of nuclei near the valley of stability and some nuclear matter parameters, dramatic extrapolations are involved when extending mass tables into unexplored regions, e.g., towards the neutron drip line. To assess corresponding statistical and systematic errors, it is inevitable to compare a variety of different functionals to get a robust prediction for observables and information on their theoretical uncertainties. In combination with that it is helpful to use linear regression techniques to obtain estimates on extrapolations uncertainties [17, 18]. Because of the number of nuclear configurations involved and complexities of the HFB problem, to optimize EDFs and calculate mass tables is a serious undertaking that requires a significant computational effort and resources. To this end, highly optimized codes and access to massively parallel supercomputers are needed to perform calculations in a manageable time.

This paper is organized as follows. Section 2 briefly reviews the Skyrme-HFB model. The computational framework employed, i.e., the HFB solver and adaptation to the supercomputer infrastructure, is discussed in Sec. 3. Results of systematic calculation of mass tables for even-even nuclei are reported in Sec. 4. Finally, Sec. 5 contains conclusions and prospects for future work.

2. The Skyrme-Hartree-Fock-Bogoliubov Approach

This section briefly outlines the self-consistent Skyrme-Hartree-Fock-Bogoliubov framework employed in this study. (For an in-depth presentation of the HFB theory, we refer the reader to Refs. [3, 19, 5] and references therein.) In the nuclear DFT, the total binding energy of the system can be written as a spatial integral:

$$E[\rho, \tilde{\rho}] = \int d^3\boldsymbol{r}\; \mathcal{H}(\boldsymbol{r}), \tag{1}$$

where ρ and $\tilde{\rho}$ are, respectively, the one-body density and pairing matrices, and $\mathcal{H}(\boldsymbol{r})$ is the quasi-local energy density. In practical applications, one considers local proton and neutron densities and currents built from single-particle wave functions, namely: particle density, kinetic density, spin-orbit density, current and spin density, as well as pairing densities describing nucleonic superconductivity.

The energy functional (1) can be decomposed into several terms. The key piece in the Skyrme-HFB approach is the Skyrme energy functional, which represents the effective nuclear interaction between nucleons and can be partly derived by considering a low momentum expansion of the density matrix [20, 21, 22]. Another crucial ingredient in a mean-field description of nuclear structure is pairing [23, 24]. A widely used mean-field approach to pairing is that provided by the HFB method, where pairing correlations are described by introducing the concept of

Bogoliubov quasiparticles [19]. Nuclear superconductivity is explicitly included by adding a pairing functional based on a zero-range two-body force combined with a density dependence [25]. The EDF also contains the kinetic term, the Coulomb term (consisting of the direct Hartree energy treated exactly and the exchange term in Slater approximation), and the correlation term that accounts for dynamical correlations due to, e.g., symmetries that are spontaneously broken on a mean-field level. How to account for such correlation energy is still an area of intense research. Meanwhile, it is often assumed that all correlations are already built into the EDF through its density dependence. Such a strategy has been recently pursued during the optimization of the functional UNEDF1 [16].

A fundamental consequence of the pairing ansatz is that the HFB states are not eigenstates of the particle-number operator. To restore the broken particle number symmetry and avoid artificial pairing collapse in the regions of the low single-particle density, we use the approximate Lipkin-Nogami scheme [26, 27], which – in most cases – yields results that are close to those obtained by means of the exact projection before variation [28].

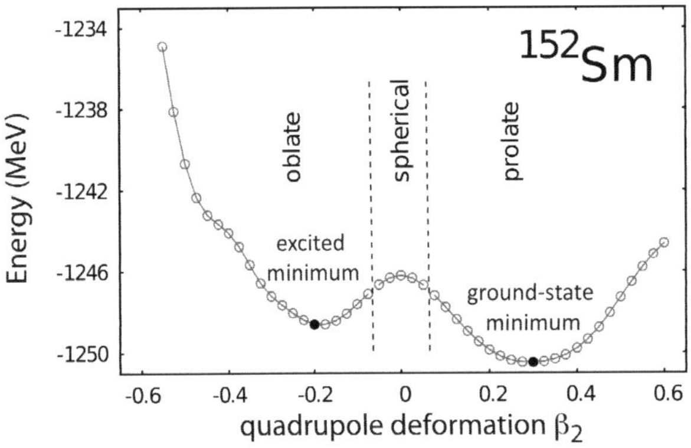

Figure 1. The potential energy surface of the deformed nucleus ^{152}Sm as a function of the quadrupole deformation β_2. Open circles mark three regions of nuclear shapes (spherical, oblate and prolate) where deformation-constrained HFB calculations are performed. The local minima (dots) are obtained by unconstrained HFB calculations initiated from the neighboring constrained solutions.

3. Computational Framework

The centerpiece of the present calculation is the HFB solver HFBTHO [29]. The code solves the non-linear HFB equations in configurational space by expanding self-consistent eigenstates in a large basis of the deformed harmonic oscillator. The axial symmetry of nuclear mean fields is imposed to reduce the dimension and complexity of the problem.

To cure the incorrect asymptotic behavior of harmonic oscillator wave functions when dealing with weakly bound neutron-rich nuclei near particle drip lines, an alternative approach based on the local scaling transformation of basis vectors was developed in Ref. [30]. Under UNEDF, the original HFBTHO solver has been replaced by that based on the modified Broyden's method [31]. The Broyden's scheme guarantees stable and fast convergence of self-consistent iterations; this is crucial for large-scale nuclear structure calculations.

The application of such advanced numerical techniques and further optimization of the HFBTHO solver carried out in Refs. [13, 16] have resulted in significant performance improvements. For instance, a ground state calculation of ^{120}Sn on a single processor, using

Journal of Physics: Conference Series **402** (2012) 012030 doi:10.1088/1742-6596/402/1/012030

a large transformed harmonic oscillator basis of 20 oscillator shells (1,771 basis states) can be done in about 1.5 minutes. To reduce the speed of a single HFB run is essential, as the number of nucleonic configurations that need to be computed during the EDF optimization process and mass table calculations can easily reach several millions.

As the general majority of nuclei are known to be axially and reflection symmetric in their ground states, in our study we consider only axial and parity-conserving intrinsic shapes. In spite of this simplification, one still has to explore configurations associated with different shapes (spherical, prolate, and oblate), since the ground-state shape is not known a priori [32]. This makes it necessary to scan the nuclear energy surface of each isotope in a wide range of shape deformations. The procedure is shown schematically in Fig. 1 for the deformed nucleus ^{152}Sm. To find the ground-state minimum, we divide the potential energy surface into three regions: spherical, prolate, and oblate. In each region, we calculate the total energy by constraining the total quadrupole moment Q_2 of the nucleus. If a local minimum is found, we carry out unconstrained minimization to pin down its energy, and the ground-state energy is eventually obtained by taking the lowest-energy solution.

As one HFBTHO run representing the single nuclear configuration can be performed on a single processor, mass table calculations are embarrassingly parallel. For the present large-scale mass table calculations, the HFBTHO solver has been extended with a minimal MPI communication in order to run in a parallel regime across the nodes. The scaling of the mass table calculation with the number of processors implies that a simple master-slave parallel architecture is sufficient.

The calculations were done in a wide range of nuclei in order to include all nuclei between the neutron and proton drip line for a given EDF parameterization. We used JAGUAR or KRAKEN Cray XT5 supercomputers housed at Oak Ridge National Laboratory's Leadership Computing Facility and the University of Tennessee's National Institute for Computational Sciences. Figure 2 shows all 4,508 even-even nuclei initially included in the mass table calculations. Using 9,060 processors of JAGUAR, computation of the entire even-even mass table took about 2 hours. The fast turnaround time is crucial, as the selected nuclear observables stored in the mass table are further used as input data for the EDF optimization [13, 16]. If odd-A nuclei are added to the table, several one-quasiparticle states around the Fermi level need to be considered for each nucleus to find the lowest-energy state [33]. Considering 6 blocking candidates (3 particle-like states and 3 hole-like states) and 50 constrained calculations for each configuration, the full nuclear mass table, corresponding to 5,000 even-even, 5,000 odd-Z, 5,000 odd-N, and 5,000 odd-odd nuclei, requires

$$N_{\mathrm{HFB}} \approx 5,000 \times (1+6+6+36) \times 50 \approx 12 \cdot 10^6 \tag{2}$$

HFB runs distributed over hundreds of thousands of nodes.

4. Large-Scale Nuclear Mass Tables

Nuclear properties obtained in large-scale DFT simulations can be directly confronted with specific experimental data. However, theoretical mass tables can also be used to search for systematic trends, especially in the regions where experimental data are not available. To facilitate analysis, specialty visualization tools have been developed to analyze various nuclear observables and compare predictions across different models. In this way, one can estimate systematic model errors and assess robustness of extrapolations.

As an illustrative example, Fig. 3 summarizes the results of systematic calculations of ground-state quadrupole deformation β_2 for all particle-bound even-even nuclei up to $Z = 120$ using the Skyrme EDFs SkM* [34], SkP [4], SLy4 [35], SV-min [17], UNEDF0 [13] and UNEDF1 [16]. To complete the picture, Fig. 3 also displays results from other published models; namely the Skyrme-HFB-21 model [36], Gogny D1S model [37], and finite-range droplet model FRDM [38]. While all models show similar global systematic trends (e.g., large quadrupole deformations in

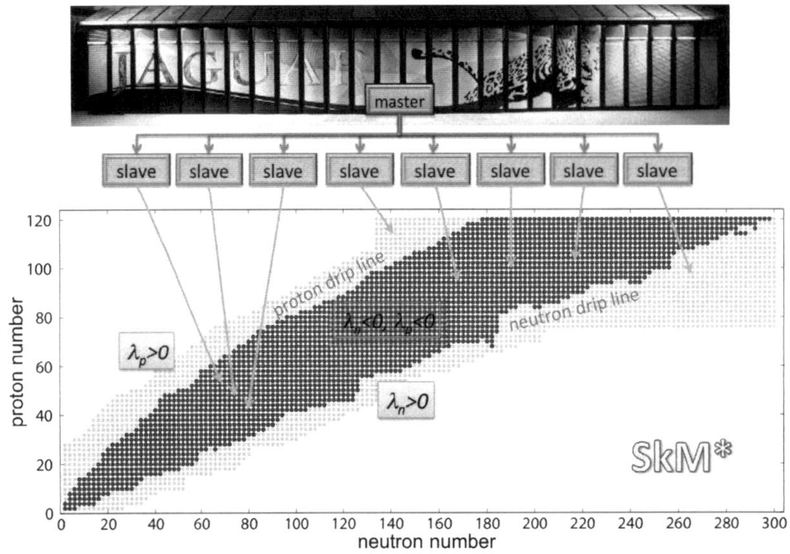

Figure 2. Computing nuclear mass table on JAGUAR XT5 supercomputer. The problem is embarrassingly parallel, as a single HFB run can be carried out on one slave node without communicating with other tasks. Simulations were performed using the SkM* EDF [34] for 4,508 even-even nuclei (gray circles). 2,342 of those nuclei (marked by dark dots) are predicted to be particle stable, i.e., they have negative proton and neutron chemical potentials λ_p and λ_n, respectively. The neutron-rich nuclei beyond the neutron drip line ($\lambda_n > 0$) are unstable to the ground-state neutron decay. The proton-rich nuclei beyond the proton drip line ($\lambda_p > 0$) are ground-state proton emitters. Image of JAGUAR courtesy of the National Center for Computational Sciences, Oak Ridge National Laboratory.

mid-shell nuclei), there are important local differences both in predictions for the location of the neutron drip line and deformations of neutron-rich and superheavy nuclei. Those differences are going to impact simulation involving the neutron-rich matter, especially in the context of astrophysical r-process and modeling nuclear reactions taking place in advanced fission reactors.

5. Conclusions

High-performance computing transforms the nuclear many-body problem in a profound way: it provides answers to questions that neither experiment nor analytic theory can address. Nuclear theorists seek to replace current phenomenological models of nuclear structure and reactions with a well-founded microscopic theory that delivers maximum predictive power with well-quantified uncertainties. To this end, advanced algorithms and extensive computational resources are needed.

This paper shows how global nuclear properties across the nuclear landscape can be rapidly computed using the nuclear density functional theory in its Skyrme-HFB version. Such calculations are crucial for interpreting experimental data, assessing importance and feasibility of planned measurements, predicting nuclear properties in the regions that are impossible to access experimentally (thus providing crucial nuclear physics input for astrophysics and applications), and defining future research directions at major experimental radioactive ion beam facilities worldwide.

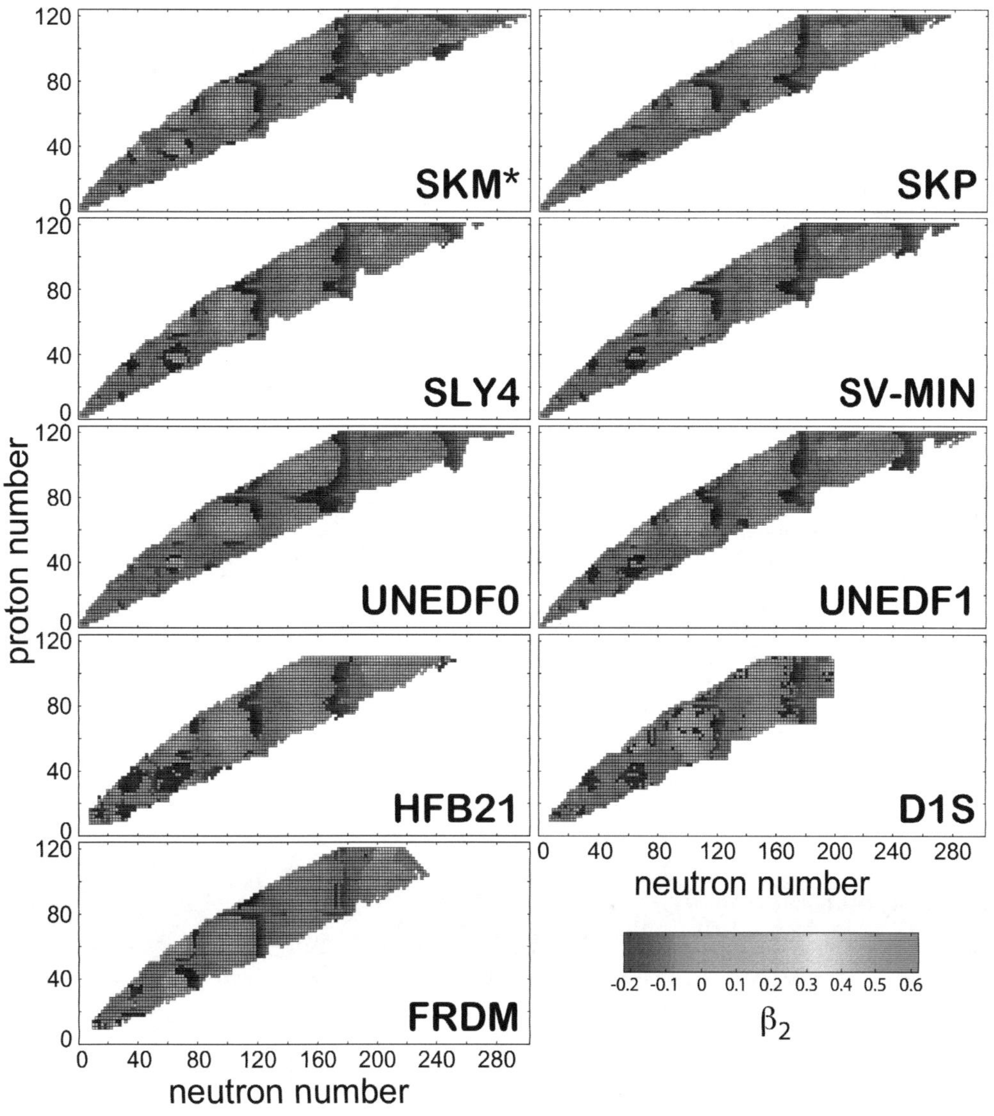

Figure 3. Ground-state quadrupole deformations β_2 of all even-even nuclei, from proton to neutron drip line up to proton number $Z = 120$, using the Skyrme EDFs SkM* [34], SkP [4], SLy4 [35], SV-min [17], UNEDF0 [13] and UNEDF1 [16]. Skyrme-HFB calculations were performed using the mixed pairing [25] and the Lipkin-Nogami procedure [30]. Results from other mass models are shown for comparison: Skyrme model HFB-21 [36], Gogny D1S model [37], and FRDM microscopic-macroscopic model [38].

6. Acknowledgments

This work was supported by the U.S. Department of Energy under Contract Nos. DE-FG02-96ER40963 (University of Tennessee) and DE-FC02-09ER41583 (UNEDF SciDAC

Collaboration). Computational resources were provided through an INCITE award "Computational Nuclear Structure" by the National Center for Computational Sciences (NCCS) and National Institute for Computational Sciences (NICS) at Oak Ridge National Laboratory.

7. References

[1] *Scientific Opportunities with a Rare-Isotope Facility in the United States* (The National Academies Press, Washington, DC, 2007)
[2] Dobaczewski J, Michel N, Nazarewicz W, Płoszajczak M and Rotureau J 2007 *Prog. Part. Nucl. Phys.* **59** 432
[3] Bender M, Heenen P H and Reinhard P G 2003 *Rev. Mod. Phys.* **75** 121
[4] Dobaczewski J, Flocard H and Treiner J 1984 *Nucl. Phys. A* **422** 103
[5] Perlińska E, Rohoziński S G, Dobaczewski J and Nazarewicz W 2004 *Phys. Rev. C* **69** 014316
[6] Dechargé J and Gogny D 1980 *Phys. Rev. C* **21** 1568
[7] Goriely S, Hilaire S, Girod M and Péru S 2009 *Phys. Rev. Lett.* **102** 242501
[8] Reinhard P G 1989 *Rep. Prog. Phys.* **52** 439
[9] Ring P 1996 *Prog. Part. Nucl. Phys.* **37** 193
[10] Lunney D, Pearson J M and Thibault C 2003 *Rev. Mod. Phys.* **75** 1021
[11] Stoitsov V, Dobaczewski J, Nazarewicz W and Borycki P 2006 *Int. J. Mass Spectrometry* **251** 243
[12] Stoitsov M, Nazarewicz W and Schunck N 2009 *Int. J. Mod. Phys. E* **18** 816
[13] Kortelainen M, Lesinski T, Moré J, Nazarewicz W, Sarich J, Schunck N, Stoitsov M V and Wild S 2010 *Phys. Rev. C* **82** 024313
[14] Bertsch G, Dean D and Nazarewicz W 2007 *SciDAC Review* **102** 42
[15] Furnstahl R 2011 *Nucl. Phys. News* **21**(2) 18
[16] Kortelainen M, McDonnell J, Nazarewicz W, Reinhard P G, Sarich J, Schunck N, Stoitsov M V and Wild S 2011 *Phys. Rev. C* Submitted; arXiv:1111.4344
[17] Klüpfel P, Reinhard P G, Bürvenich T J and Maruhn J A 2009 *Phys.Rev. C* **79** 034310
[18] Reinhard P G and Nazarewicz W 2010 *Phys. Rev. C* **81** 051303(R)
[19] Ring P and Schuck P 1980 *The Nuclear Many-Body Problem* (New York, Heidelberg, Berlin: Springer–Verl.)
[20] Negele J W and Vautherin D 1972 *Phys. Rev. C* **5** 1472
[21] Bogner S K, Furnstahl R and Schwenk A 2010 *Prog. Part. Nucl. Phys.* **65** 94
[22] Carlsson B G and Dobaczewski J 2010 *Phys. Rev. Lett.* **105** 122501
[23] Bohr A, Mottelson B R and Pines D 1958 *Phys. Rev.C* **110** 936
[24] Brink D M and Broglia R A 2005 *Nuclear Superfluidity:pairing in finite systems* (Cambridge University Press)
[25] Dobaczewski J, Nazarewicz W and Stoitsov M 2002 *Eur. Phys. J.* **15** 21
[26] Lipkin H J 1960 *Ann. Phys. (NY)* **9** 272
[27] Nogami Y 1964 *Phys. Rev.* **B134** 313
[28] Stoitsov M V, Dobaczewski J, Kirchner R, Nazarewicz W and Terasaki J 2007 *Phys. Rev. C* **76** 014308
[29] Stoitsov M, Dobaczewski J, Nazarewicz W and Ring P 2005 *Comput. Phys. Commun.* **167** 43
[30] Stoitsov M V, Dobaczewski J, Nazarewicz W, Pittel S and Dean D J 2003 *Phys. Rev. C* **68** 054312
[31] Baran A, Bulgac A, Forbes M M, Hagen G, Nazarewicz W, Schunck N and Stoitsov M V 2008 *Phys. Rev. C* **78** 014318
[32] Stoitsov M V, Dobaczewski J, Ring P and Pittel S 2000 *Phys. Rev. C* **61** 034311
[33] Schunck N, Dobaczewski J, McDonnell J, Moré J, Nazarewicz W, Sarich J and Stoitsov M V 2010 *Phys. Rev. C* **81** 024316
[34] Bartel J, Quentin P, Brack M, Guet C and Håkansson H B 1982 *Nucl. Phys. A* **386** 79
[35] Chabanat E, Bonche P, Haensel P, Meyer J and Schaeffer R 1998 *Nucl. Phys. A* **635** 231 ,*Nucl. Phys.* **A643**, 441(E)
[36] Goriely S, Chamel N and Pearson J M 2010 *Phys. Rev. C* **82** 035804
[37] Delaroche J P, Girod M, Libert J, Goutte H, Hilaire S, Péru S, Pillet N and Bertsch G F 2010 *Phys. Rev. C* **81** 014303
[38] Möller P, Nix J R, Myers W D and Swiatecki W J 1995 *At. Data Nucl. Data Tables* **59** 185

IUPAP C20 Conference on Computational Physics (CCP 2011) IOP Publishing
Journal of Physics: Conference Series **402** (2012) 012031 doi:10.1088/1742-6596/402/1/012031

Ab Initio Nuclear Structure Calculations of Light Nuclei

Pieter Maris

Department of Physics and Astronomy, Iowa State University, Ames, IA 50011, USA

E-mail: pmaris@iastate.edu

Abstract. We perform no-core configuration interaction calculations for nuclei in the p-shell. We show that for typical light nuclei, a truncation on the total number of quanta in the many-body system converges much more rapidly than a full configuration interaction (FCI) truncation, which is a truncation on the single-particle basis space. We present new results for the ground state energies of the Be isotopes with the nonlocal two-body potential JISP16, and discuss emerging phenomena such as clustering and rotational band structures in ^{9}Be. We also show that the anomalously suppressed beta decay of ^{14}C to the ground state of ^{14}N can be reproduced using two- and three-nucleon forces from chiral effective field theory. In particular the structure of the ground state of ^{14}N is sensitive to the three-nucleon force.

1. Ab initio nuclear physics

Solving for nuclear properties with the best available nucleon-nucleon (NN) potentials [1, 2], supplemented by three-nucleon forces (3NF) as needed [3, 4, 5], using a quantum many-particle framework that respects all the known symmetries of the potentials is referred to as an "ab initio" problem and is recognized to be computationally hard. Through the UNEDF SciDAC collaboration [6, 7, 8] of nuclear theorists, applied mathematicians, and computer scientists there has been a rapid development of ab initio methods for solving finite nuclei, which has opened a range of nuclear phenomena that can now be evaluated to high precision using realistic inter-nucleon interactions.

A commonly used approach in nuclear physics is the Configuration Interaction (CI) method for solving the many-body nuclear Hamiltonian in a (sufficiently large) single-particle basis space. In this approach, the many-body Schrodinger equation becomes a large sparse matrix problem. The eigenvalues of this matrix are the binding energies, and the corresponding eigenvectors the nuclear wave functions, which can be employed to evaluate experimental quantities. In order to reach numerical convergence for fundamental problems of interest, the matrix dimension often exceeds two billion; obtaining the lowest eigenvalues and eigenvectors of such large sparse matrices poses significant challenges even on today's leadership class facilities.

In this work, we discuss convergence rates of CI calculations, contrasting Full Configuration Interaction (FCI), which is a truncation on the single-particle basis space only, with the $N_{\max}$ truncation, which is a truncation on the many-body basis space. The latter is commonly used in nuclear structure calculations, not only because of its superior convergence rate for light nuclei, but also because it leads to an exact seperation of the center-of-mass (cm) motion and the translationally-invariant wavefunction.

We use a recently introduced technique for unfolding the cm motion from the one-body density matrices [9], which allows us to obtain translationally-invariant nucleon densities. With this technique we can reveal interesting phenomena such as clustering emerging from our ab initio caculations. In particular we show that the ground state of ^{9}Be consists of two clusters of protons and neutrons separated by a few fm, consistent with α-cluster models, with the 'extra' neutron forming a ring or donut shape. We also discuss emerging rotational band structure in ^{9}Be, and present new results for the ground state energies of the beryllium isotopes.

Finally, we discuss recent results for the anomalously long lifetime of ^{14}C, which can be reproduced by the inclusion of three-nucleon forces. We analyse in detail what the influence of the 3NF is, not only on the transition between the ground states of ^{14}C and ^{14}N, but also on the transition between the first excited 2^+ state of ^{14}C and the ground state of ^{14}N. We identify that it is the structure of the ground state of ^{14}N, rather than that of ^{14}C, that responds sensitively to the 3NF, and therefore leads to the anomalously small transition matrix element.

2. No-Core Configuration Interaction calculations of light nuclei

In a no-core CI approach, the wavefunction Ψ of a nucleus consisting of A nucleons (protons and neutrons) is expanded in an A-body basis of Slater Determinants of single-particle states; that is

$$\Psi(\vec{r_1},\ldots,\vec{r_A}) \quad = \quad \sum c_k \Phi_k(\vec{r_1},\ldots,\vec{r_A}) \,, \tag{1}$$

where Φ_k are A-body Slater Determinants. Conventionally, one uses a harmonic oscillator (HO) basis for the single-particle states, but it is straightforward to extend this approach to a more general single-particle basis [10].

We use the so-called M-scheme for our basis. The single-particle states are labelled by the quantum numbers n, l, j, and m_j, where n and l are the radial and orbital HO quantum numbers, with $N = 2n + l$ the number of HO quanta; j is the total single-particle spin, and m_j its projection along the z-axis. The many-body basis states have well-defined total spin-projection, which is simply the sum of m_j of the single-particle states, $M_j = \sum m_j$, hence the name M-scheme. However, in general the many-body basis states do not have a well-defined total J. Some of the advantages of this scheme is that it is very simple to implement, and that in two runs (for positive and negative parity), we get the complete low-lying spectrum, including the ground state, even if we do not know what the spin of the ground state is.

The interaction V, typically a two-body (NN), sometimes supplemented by a three-body (3NF) potential, is also expressed in this many-body basis, as is the kinetic energy T. Diagonalizing the resulting matrix $H = T + V$ gives us the eigenstates and eigenenergies. In general, we are only interested in the low-lying spectrum, so we do not need a complete diagonalization of this matrix, which can be rather large, exceeding 12 billion basis states in a recent application.

2.1. Truncation and Convergence

In principle, by expanding the nuclear wavefunction in a complete basis, we would obtain exact results for a given input interaction V. However, such a basis is infinite-dimensional, and practical calculations can only be done in a finite-dimensional truncation of a complete basis. In order to recover exact results, independent of the basis space truncation, we typically perform a series of calculations in increasingly large basis spaces. Based on the results of such a series of calculations, we either find that we have achieved numerical convergence (within a certain numerical uncertainty), or extrapolate to the complete basis (with a certain extrapolation uncertainty). Of course, for some observables it is much harder to achieve convergence than for others, as we shall see. In addition, one could employ renormalization techniques to accelerate convergence in relatively small basis spaces.

Full Configuration Interaction or FCI employs a widely-know truncation in which all many-body basis states are retained that can be constructed from a finite set of single-particle states. This method is commonly used in atomic and molecular physics.

In ab initio nuclear physics however, we generally use a truncation on the many-body basis space, rather than on the single-particle basis. To be specific, we construct a finite-dimensional basis by a truncation on the total number of quanta of the system. That is, the many-body basis for a nucleus consisting of A nucleons is limited to states that satisfy the condition

$$\sum_{k=1}^{A} N_k \quad \leq \quad N_0 + N_{\rm max} \,, \tag{2}$$

where N_k is the number of quanta of each of the single-particle states in the many-body basis state; N_0 is the minimal number of quanta for that nucleus; and $N_{\rm max}$ is the truncation parameter. For HO single-particle states, or more general, for any basis in which the single-particle states have radial and orbital quantum numbers n and l, we have $N_k = 2n_k + l_k$.

There are several advantages to this $N_{\rm max}$ truncation. One is that, in combination with a HO basis for the single-particle basis, it leads to an exact factorization of the cm motion and the relative motion. In nuclear physics, we deal with a selfbound system, and we have to either use relative coordinates for our calculations, or (as we do here) single-particle coordinates. Note that the wavefunctions in Eq. (1) are functions of A (independent) single-particle coordinates, not $(A-1)$ relative coordinates. This makes the anti-symmetrizations and bookkeeping in the numerical codes almost trivial, but it does mean that we have to somehow seperate out the cm wavefunction from the translationally-invariant wavefunction. A HO single-particle basis, in combination with $N_{\rm max}$ truncation, allows for an exact factorization of the cm wavefunction

$$\Psi(\vec{r_i}) \quad = \quad \Psi^{\omega}_{\rm cm}(\vec{R}) \otimes \Psi_{\rm ti} \,, \tag{3}$$

where $\vec{R} = (\frac{1}{A})\sum_{i=1}^{A} \vec{r_i}$ and $\Psi_{\rm ti}$ does not depend on the cm motion. In order to remove the states with cm excitations from the low-lying spectrum, we use the Lawson method [11] whereby we add a Lagrange multiplier term, $\lambda(H^{\omega}_{\rm cm} - \frac{3}{2}\hbar\omega)$, to the many-body Hamiltonian. The actual Hamiltonian that we are using thus takes the form

$$H^{\lambda,\omega} \quad = \quad T_{\rm rel} + \lambda_{\rm cm}\left(H^{\rm HO}_{\rm cm} - \frac{3}{2}\hbar\omega\right) + \sum_{i<j} V_{ij} + \sum_{i<j<k} V_{ijk} + \ldots \tag{4}$$

Note that although this Hamiltonian depends explicitly on the HO parameter $\hbar\omega$, as well as on λ, the low-lying spectrum (i.e. states with $E_{\rm excitation} \lesssim \lambda\hbar\omega$) only depends implicitly on $\hbar\omega$ through the basis space truncation, and becomes independent of $\hbar\omega$ in the infinite-basis limit.

In the infinite-basis limit both FCI and the $N_{\rm max}$ truncations should converge to the same exact result. It turns out that the numerical convergence is significantly faster with an $N_{\rm max}$ truncation than with an FCI truncation [12, 13] in our nuclear physics applications. In Fig. 1 we compare the convergence rate for two light nuclei, ^{4}He and ^{6}Li, with an FCI truncation and with an $N_{\rm max}$ truncation. These calculations were performed with a realistic two-body interaction derived from inverse scattering theory, JISP16 [14], which has been demonstrated to have good convergence rates for the ground state energies of nuclei with $A \leq 16$ [15]; and it gives a good description of light nuclei without explicit three-body forces.

The left panel of Fig. 1 shows that for ^{4}He both FCI and the $N_{\rm max}$ truncation have a similar convergence rate, at least in terms of the HO parameter $\hbar\omega$ and the truncation parameter $N_{\rm shell}$ and $N_{\rm max}$ respectively. As the truncation parameter increases, the dependence on $\hbar\omega$ decreases, and the succesive $N_{\rm shell}$ curves (dotted) and $N_{\rm max}$ curves (solid) converge to the same ground state energy of -28.299 MeV.

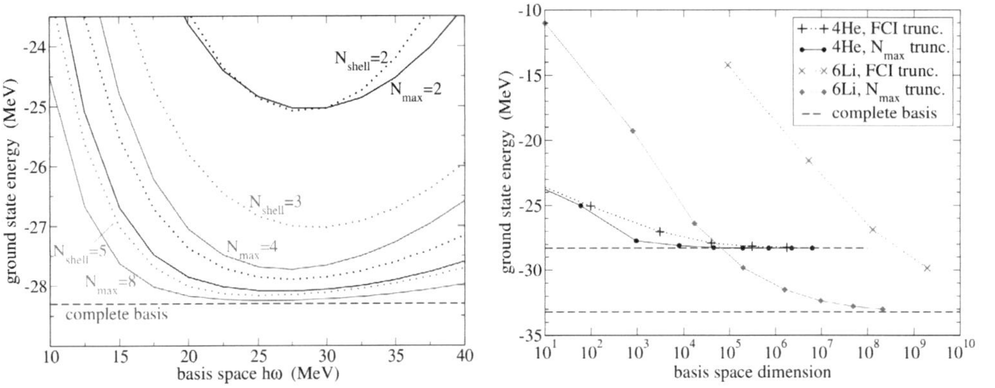

Figure 1. Comparison of the convergence rates with JISP16 using FCI (dotted) and $N_{\rm max}$ truncations (solid). Left: the ground state energy of ^{4}He as function of $\hbar\omega$; Right: the ground state energies of ^{4}He and ^{6}Li as function of the basis space dimension. The dashed line represents the exact result in a complete basis.

However, as function of the basis space dimension, the $N_{\rm max}$ truncation converges much more rapidly than the FCI calculations, as is illustrated in the right panel of Fig. 1. E.g., an $N_{\rm max} = 8$ calculation with a basis dimension of less than 50,000 for ^{4}He leads to a ground state energy that is well within 100 keV of the exact result, whereas for an FCI calculation $N_{\rm shell} = 6$ with a basis dimension of almost 2 million is needed in order to achieve a similar level of convergence. For ^{6}Li, an FCI calculation with a basis dimension of over a billion is still more than 3 MeV from the exact ground state energy, whereas an $N_{\rm max} = 10$ calculation with a basis dimension of about 10 million is within 1 MeV of the converged ground state energy. Also for other observables [13], such as radii and quadrupole moments, the convergence seems to be slower with the FCI truncation than with the $N_{\rm max}$ truncation, when viewed as function of the basis space dimension.

Finally, with the $N_{\rm max}$ truncation one may easily separate the positive parity spectrum and the negative parity spectrum. Two succesive shells have opposite parity, depending on whether the orbital angular momentum is even or odd. By increasing the total number of quanta of the many-body basis states by two quanta at a time, we automatically create a basis with a specific parity. The 'natural' parity of a nucleus is given by its valence space, for which our calculations start with $N_{\rm max} = 0$; for the 'unnatural' parity spectrum we would perform a series of calculations starting with $N_{\rm max} = 1$. In both cases, $N_{\rm max}$ is increased by two quanta as we increase the basis space. (Of course, it is straightforward to include a parity-constraint in FCI calculations, and in fact, the dimensions given here for FCI calculations do incorporate such a constraint.)

2.2. Extrapolation to the complete basis

For a very light nucleus like ^{4}He or ^{6}Li one can achieve convergence of the ground state energy and other observables by simply going to a sufficiently large basis, but for larger nuclei that is not practical. However, based on a series of calculations in finite basis spaces, we can extrapolate those results to a complete, infinite-dimensional, basis. Empirically [15, 16] the nuclear binding energies seem to converge exponentially with $N_{\rm max}$

$$E^N_{\rm binding} \;=\; E^\infty_{\rm binding} + a_1 \exp(-a_2 N_{\rm max})\,. \tag{5}$$

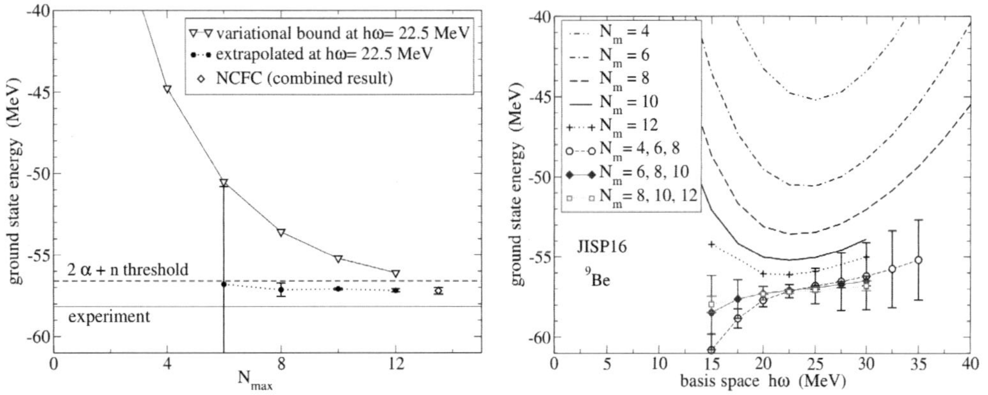

Figure 2. Ground state energy of ^{9}Be with JISP16 as function of $N_{\max}$ (left) and as function of $\hbar\omega$ for several $N_{\max}$ values (right), as well as extrapolated results with error estimates.

There are several different ways in which one can exploit this (empirical) convergence behavior. Here we use three consecutive $N_{\max}$ values at a fixed basis space parameter $\hbar\omega$ in order to estimate a converged binding energy at that $\hbar\omega$. Under the assumption that the convergence is indeed exponential, such an extrapolation should get more accurate as $N_{\max}$ increases; the difference between the extrapolated results from two consecutive sets of three $N_{\max}$ values is then used as an estimate of the numerical uncertainty associated with the extrapolation. We do this for a range of basis space $\hbar\omega$ values, in order to check the consistency of the extrapolated results and their error estimates. (We only use three consecutive $N_{\max}$ values for each extrapolation, since our results at finite basis spaces are accurate to at least five significant figures.)

This procedure is illustrated for ^{9}Be in Fig. 2, again with JISP16. In the left panel we show the ground state energy for a series of calculations at a fixed HO parameter $\hbar\omega$, as well as the extrapolated ground state energy. For the error estimate on the first extrapolated value, obtained from calculations at $N_{\max} = 2$, 4, and 6, we use the difference between the calculated result at $N_{\max} = 6$ and the extrapolated value; for higher $N_{\max}$ values we use the difference with the value obtained from the next-smaller basis spaces.

The extrapolated result appears to be very well converged (i.e. consistent within diminishing errors) starting from $N_{\max} = 8$, not only as function of $N_{\max}$, but also as function of the basis space parameter $\hbar\omega$, as can be seen from the right panel of Fig. 2. For our final No-Core Full Configuration (NCFC) result we combine the extrapolation results for a range of HO parameters at and slightly above the variational minimum to arrive at our best estimate for the results in the complete basis, independent of all basis parameters.

2.3. Need for HPC and Code performance

In order to reliably extrapolate to the complete basis, we need results in finite bases up to at least $N_{\max} = 8$ (9 for unnatural parity states), and preferably $N_{\max} = 10$ (11 for unnatural parity states) or higher. This leads to a computational challenge, because the basis space dimension increases rather rapidly with $N_{\max}$ and with the number of particles. Furthermore, the sparsity of the matrix depends on the 'rank' of the potential: a two-body potential leads to a much sparser many-body matrix than a three-body potential.

In Fig. 3 we show the basis space dimensions for a number of $N = Z$ nuclei in the p-shell (solid curves), as well as a few nuclei in the sd-shell. The basis space dimension for even the first couple of nuclei in the sd-shell (more than 8 protons and 8 neutrons) is of the order of 10 billion or more

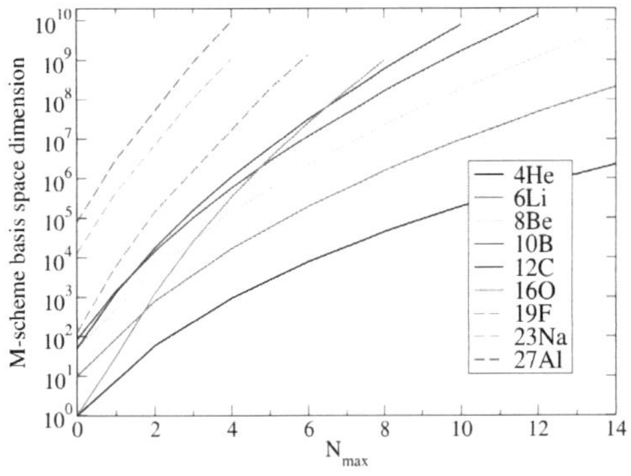

Figure 3. Basis space dimension as function of $N_{\max}$ for $N = Z$ nuclei in the p-shell (solid), as well as a few nuclei in the sd-shell (dashed).

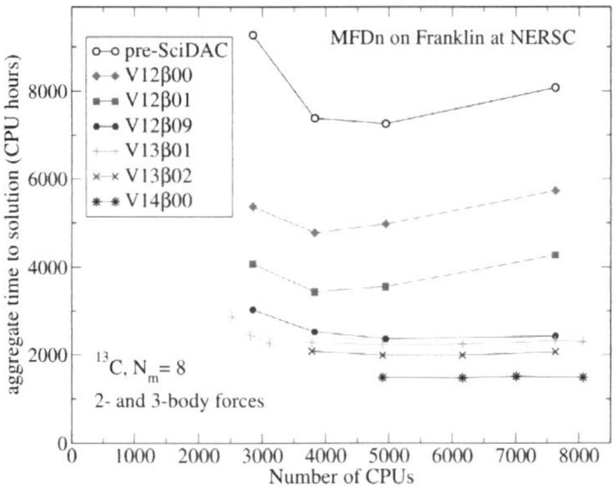

Figure 4. Performance improvements on Franklin (at NERSC) of the code MFDn for ^{13}C with 2- and 3-body forces.

at $N_{\max} = 8$. As the matrix size increases, there is a clear need for high-performance computing: obtaining the lowest 10 to 20 eigenvalues and eigenvectors of matrices with a dimension of the order of several billion is a nontrivial task, even if the matrix is extremely sparse.

For our calculations we use the code MFDn, which is a hybrid MPI/OpenMP parallel Fortran 90 code for nuclear structure calculations, that has been in development for over two decades. Significant improvements in its performance have been made over the last four years [17, 18, 19] under the UNEDF SciDAC program, as is illustrated in Fig. 4. On currently available machines, the largest many-body basis dimension that MFDn can handle is about 10 to 15 billion with two-body forces only, and about one billion with two- and three-body forces.

MFDn constructs the many-body basis states in M-scheme and the corresponding Hamiltonian matrix, and solves for the lowest eigenstates using an iterative Lanczos algorithm. At the end of a run, it writes the nuclear wavefunctions to file, and evaluates selected physical observables which can be compared to experimental data (the wavefunctions themselves are not experimentally observable). It also writes out the one-body density matrix elements (in the underlying single-particle basis), which can be used for further analysis.

The matrix is distributed in a 2-dimensional fashion over the processors, and we work with the lower triangle only, because the matrix is symmetric (and real). Because of the 2-dimensional distribution of the matrix, it runs on $n(n+1)/2$ processors, where n is the number of "diagonal" processors. Both the memory usage and the (local) CPU usage are distributed evenly over all processors. This load-balancing is achieved by a round-robin distribution of the M-scheme basis states over the "diagonal" processors. The disadvantage of this scheme is that it destroys any natural sparsity structure in the matrix. We are currently working on a new load-balancing scheme that retains some of the natural sparsity structure (Version 14 in Fig. 4). The Lanczos vectors are also stored in memory, distributed over all processors; this allows for fast and efficient re-orthoganalization after every matrix-vector multiplication. Note that the re-orthogonalization is done in double precision, even though the matrix and vectors themselves are stored in single precision. This turned out to be essential in order to maintain numerical accuracy over hundreds of Lanczos iterations with matrix dimensions in the billions.

The code has good scaling characteristics, as illustrated by the flatness of the curves for the more recent versions of the code in Fig. 4. For large runs on modern multi-core architectures such as the Cray XT4/5 and XE6, it is significantly more efficient to run the code in a hybrid mode, using OpenMP threading within a node, and MPI between nodes [19]. We are currently working on improving the performance on NUMA architectures by focussing on the data locality [20].

The increase in aggregate CPU time as the number of CPUs goes below 4,000 in Fig. 4 is due to the memory limitations. One of the recent code developments was focussed on reducing the memory footprint by partially re-generating the matrix 'on the fly'. This enabled us to perform the largest runs with two- and three-body forces for ^{14}C and ^{14}N on Jaquar at ORNL [21].

3. Emergence of clustering and rotational phenomena in Beryllium isotopes

Among the Be-isotopes there are several interesting and challenging problems for ab initio nuclear structure calculations: α-clustering is likely to play an important role, in particular for ^{8}Be and ^{9}Be; there are several rotational bands; and they are among the lightest nuclei for which both positive and negative parity states are known experimentally. With the NCFC approach described above, we should be able to perform reasonably accurate calculations for the Be-isotopes from ^{6}Be up to ^{12}Be, provided we use a suitably soft interaction such as JISP16.

3.1. Clustering

We have already presented our results for the ground state energy of ^{9}Be in Fig. 2. With JISP16 we find a ground energy of $E_{\rm gs} = -57.2(2)$ MeV, compared to an experimental value of $E_{\rm gs} = -58.167$ MeV, that is JISP16 underbinds ^{9}Be by about 1 MeV. Despite this underbinding, our result is below the threshold for two α particles (plus a neutron), which is at -56.6 MeV. Thus we are dealing with a true bound state, not a resonance state.

The next question that naturally arises is: what is the structure of this state? In particular, does it have a structure of two α particles plus a neutron, as one might expect? In an attempt to answer such questions, we have looked at the proton and neutron densities in coordinate-space.

The local density for a nucleus with wavefunction Ψ is given by

$$\rho(\vec{r}) \;=\; \int \Psi^\star(\vec{r},\vec{r}_2,\ldots,\vec{r}_A)\,\Psi(\vec{r},\vec{r}_2,\ldots,\vec{r}_A)\,d^3r_2\ldots d^3r_A\,. \qquad (6)$$

This corresponds to the probability of finding a nucleon at position $\vec{r}$. Since we use a basis in single-particle coordinates, rather than relative coordinates, our wavefunction $\Psi(\vec{r}_i)$ includes cm motion. The local density extracted from such wavefunctions therefore includes also contributions from the cm motion. However, because of the exact factorization of the cm wavefunction, see Eq. (3), this density is a convolution of the cm density and a translationally-invariant density $\rho_{\rm ti}$ relative to the cm of the entire nucleus

$$\rho^\omega(\vec{r}) \;=\; \int \rho_{\rm ti}(\vec{r}-\vec{R})\,\rho^\omega_{\rm cm}(\vec{R})\,d^3\vec{R}\,. \qquad (7)$$

The cm density $\rho^\omega_{\rm cm}$ is a simple Gaussian that smears out $\rho_{\rm ti}$ and obfuscates details of the local density[1]. The translationally-invariant density $\rho_{\rm ti}$ can be obtained by a deconvolution of the cm density using standard Fourier methods [9]

$$\rho_{\rm ti}(\vec{r}) \;=\; F^{-1}\left[\frac{F[\rho^\omega(\vec{r})]}{F[\rho^\omega_{\rm cm}(\vec{r})]}\right], \qquad (8)$$

where $F[f(\vec{r})]$ is the 3D Fourier transform of $f(\vec{r})$. Note that the densities on the RHS of this equation depend on the basis space parameter $\hbar\omega$, even in the infinite-basis limit, but the LHS is independent of the basis in this limit.

[1] Note that the cm motion also introduces a spurious dependence on the basis parameter $\hbar\omega$ into ρ^ω that masks the convergence. Even in the limit of a converged calculation in a complete basis, ρ^ω depends on $\hbar\omega$, whereas the translationally-invariant density $\rho_{\rm ti}$ becomes independent of the basis as one approaches convergence.

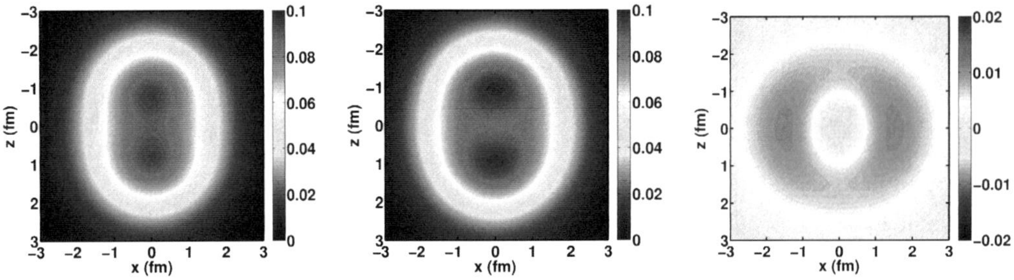

Figure 5. Proton (left), neutron (center), and the difference between the neutron and proton density (right) for the $J^\pi = \frac{3}{2}^-$ ground state of ^{9}Be, with $M_j = \frac{3}{2}$ using JISP16.

In Fig. 5 we show the proton and neutron densities for the $J^\pi = \frac{3}{2}^-$ ground state of ^{9}Be, with $M_j = \frac{3}{2}$, obtained with JISP16 in a $N_{\max} = 10$, $\hbar\omega = 12.5$ MeV basis space, where the RMS radii seem to be most converged. The spurious cm contribution has been removed as described above producing the translationally-invariant neutron and proton densities shown in Fig. 5. The figure shows the density in the xz-plane, and is symmetric along the z-axis (vertical axis). It clearly shows a clustering of both the protons and the neutrons into two compact dense regions, seperated by about 2 fm along the z-axis. Although this does not prove that we have α-clustering in our ab initio calculation, it is at least consistent with α-clustering. Furthermore, the difference of the neutron and proton density (right panel of Fig. 5) shows that the 'extra' neutron forms a ring or donut shape around the z-axis, rather than being centered in between the dominant proton and neutron clusters.

3.2. Rotation

Rotational bands arise quite naturally in α-cluster models for the Be isotopes [22], so now that we do see some evidence for clustering in ^{9}Be, we can also expect the corresponding rotational states in the spectrum. The rotational energy for states in a rotational band with an axial symmetry is given by [23]

$$E_{\text{rotational}} = \frac{\hbar^2}{2\mathcal{I}} \left(J(J+1) - K^2 \right) , \tag{9}$$

where $\mathcal{I}$ is the moment of inertia, K is the projection of J along the intrinsic symmetry axis, and J is the total angular momentum. Thus rotational bands show up as straight lines if we plot the excitation energies as function of $J(J+1)$.

For the ground state rotational band in ^{9}Be we have $K = J = \frac{3}{2}$. Indeed, the spectrum shown in the left panel of Fig. 6 suggests that the lowest $J = \frac{5}{2}$, $J = \frac{7}{2}$, and $J = \frac{9}{2}$ states form a rotational band with the ground state. Furthermore, the excitation energy of the members of this rotational band is very well converged, in particular for the $J = \frac{5}{2}$ and $J = \frac{7}{2}$ states. The calculated excitation energy also agrees quite well with the experimental spectrum, though we do predict a $J = \frac{9}{2}$ member of this rotational band, in addition to the confirmed experimental $J = \frac{5}{2}$ and $J = \frac{7}{2}$ states.

The spectrum by itself is not sufficient to claim that these states are indeed rotational states. The rotational model also predicts the quadrupole moments of a rotational band to all be related to an 'intrinsic' quadrupole moment Q_0 [23]

$$Q(J) = \frac{3K^2 - J(J+1)}{(J+1)(2J+3)} Q_0 . \tag{10}$$

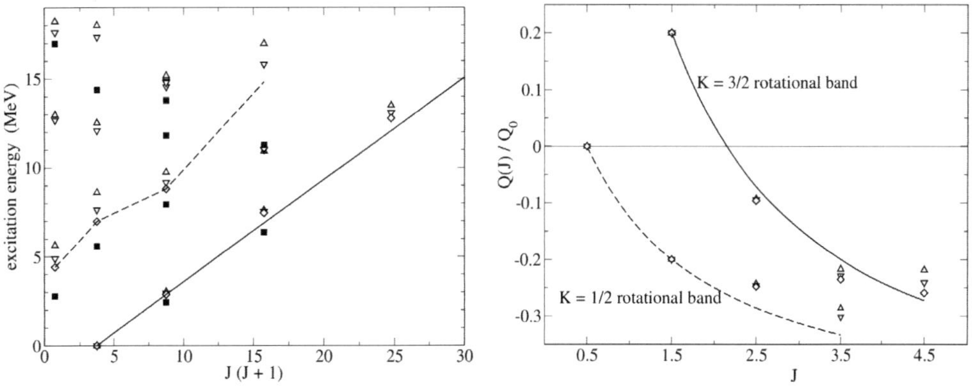

Figure 6. Evidence for the ground state $K = 3/2$ (——) and excited $K = 1/2$ (- - - -) rotational bands in ^{9}Be from ab initio calculations with JISP16 ($\triangle$: $N_{\rm max} = 8$, ∇: $N_{\rm max} = 10$, $\Diamond$: $N_{\rm max} = 12$) Left: Negative parity spectrum as function of $J(J+1)$ (■: experimental data from Ref. [25].); Right: Quadrupole moments of the rotational bands normalized by Q_0.

The strength of the B(E2) transitions within a rotational band also follow from this intrinsic quadrupole moment Q_0. Even though the quadrupole moments are not very well converged in our calculations[2], the *ratio* of the quadrupole moments does seem to be reasonably well converged. In the right panel of Fig. 6 we show the quadrupole moments, normalized by the intrinsic quadrupole moment, Q_0, extracted from the ground state. They agree quite well with the prediction of Eq. (10). Also the calculated B(E2) transitions are in reasonable agreement with the rotational model [24].

The other states in the low-lying negative-parity spectrum of ^{9}Be are not as well converged as the ground state rotational band, as can be seen from the left panel of Fig. 6. However, our calculations suggests that there is also a $K = \frac{1}{2}$ rotational band, starting from the first $J = \frac{1}{2}$ state. Note that it involves the third $J = \frac{7}{2}$ state, rather than the second. Not only are the excitation energies in agreement with predictions from a rotational model (including the staggering due to the Coriolis effect), the quadrupole moments also agree with Eq. (10). In addition, the convergence pattern of these four states is quite similar. The lowest three states of this band seem to converge towards the experimental excitation energies. More work is needed to determine whether these states indeed form a rotational band [24].

3.3. Ground state energies and parity splittings

With JISP16 we have not only looked at ^{9}Be, but also at the other Be-isotopes, from ^{6}Be up to ^{14}Be. Because of the variational principle, any result for the ground state energy in a finite basis truncation forms a strict upper bound for the exact ground state energy. Indeed, our calculated energies in finite basis spaces tend to be well above the experimental ground state energies, see Fig. 7. With an exponential extrapolation on the total binding energies we obtain results for the ground state energies (with numerical error bars due to the extrapolation uncertainties) that are in general much closer to the experimental ground state energies than these variational upper bounds. It turns out that with JISP16 all Be-isotopes are underbound (left panel of Fig. 7): the light isotopes, ^{6}Be and ^{7}Be, by a fraction of an MeV, and the neutron-rich isotopes ^{13}Be and

[2] The HO basis functions fall off like $\exp(-r^2)$ whereas the asymptotic behavior of the nuclear wavefunction is known to be an expontial, $\exp(-r)$, rather than a Gaussian. The quadrupole moments are sensitive to this long-range part of the nuclear wavefunction that is not very well represented in a finite HO basis.

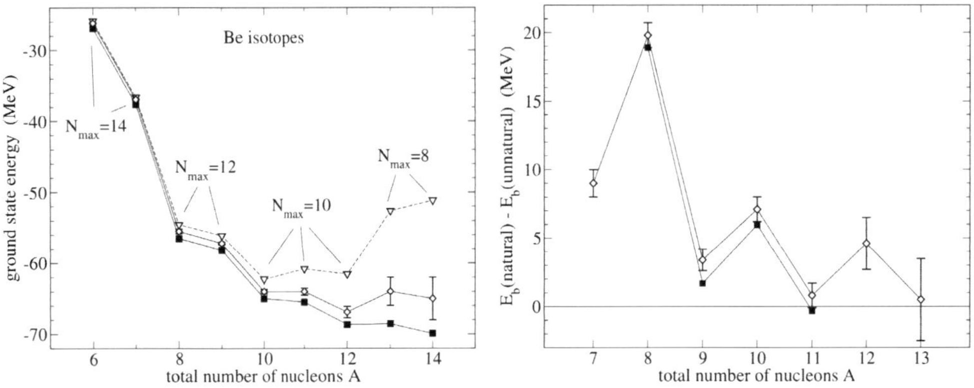

Figure 7. Ground state energies (left) and parity splittings (right) of Be-isotopes with JISP16 (variational upperbound: ▽, NCFC: ◇), together with experimental data (■) from Refs. [25, 26].

^{14}Be, by about 5 MeV. Nevertheless, the overall pattern of the ground state energies is in quite good agreement with the data. Also the spectra and several other observable quantities do seem to be in quite good agreement with the data, despite this overall underbinding.

Starting with ^{8}Be, states of both partities appear in the low-lying spectrum; and for ^{11}Be the lowest positive parity state is the ground state, contrary to the expectations based on the shell model, which predicts negative parity ground states for all odd p-shell nuclei. For ^{7}Be through ^{13}Be we performed our calculations for both natural parity states and for unnatural parity states. We then perform independent extrapolations for the lowest positive parity states and the for the lowest negative parity states.

In the right panel of Fig. 7 we show the difference between the binding energy of the lowest natural parity state and the lowest unnatural parity state. For most isotopes we expect this difference to be positive, but for isotopes with parity inversion it becomes negative. Although our results do not quite reproduce the observed parity inversion for ^{11}Be, parity inversion is within our numerical error estimates for this isotope. Furthermore, over the range of isotopes from ^{8}Be to ^{11}Be our results are in very good qualitative agreement with the data: with JISP16 we seem to underbind all unnatural parity states by a similar amount of about 1 MeV. Based on these results, we also predict parity inversion for ^{13}Be; experimentally, the parity of the ground state is not confirmed, though likely to be negative [26], which indeed implies parity inversion (^{13}Be has one neutron in the sd-shell, so the natural parity is positive).

It is not yet clear whether the relative underbinding of the unnatural parity states compared to the natural parity states in beryllium is a deficiency of the NN interaction, JISP16, or due to underestimating the numerical error in the extrapolation to the infinite basis, in particular for the unnatural parity states [27]. Also the general underbinding of the ground states hints at a deficiency of JISP16. Indeed, it is unlikely that this interaction is 'perfect': it was fitted to the NN scattering data, as well as to select observables in light nuclei, in particular to ^{6}Li. Nevertheless, the overall agreement between data and our calculation is quite encouraging, and we expect to be able to get even closer to the data with improved NN and 3NF potentials.

For a true ab initio calculation one would obtain the nuclear forces (NN, 3NF, etc.) from the underlying quantum field theory, QCD. Unfortunately, that is not yet possible without introducing free parameters. Based on QCD, one can derive nonrelativistic effective NN, 3NF, and even 4NF nuclear potentials using chiral perturbation theory (ChPT) [2, 4]. Such a derivation determines the structure of the nuclear interaction, but not the numerical constants that govern the strength of the different terms in the interaction – these have to be fitted to low-energy nuclear physics data. (In the future they may be calculable through lattice QCD.)

4. Understanding the Gamow–Teller transitions between ^{14}C and ^{14}N

The anomalously long lifetime of ^{14}C, 5730 ± 30 years, compared to lifetimes of other light nuclei undergoing the same decay process, allowed Gamow–Teller (GT) beta-decay, poses a major challenge to ab initio nuclear structure calculations. Not only is the number of particles (14) pushing the limit of what ab initio calculations can reliably calculate, the experimental lifetime can only be reproduced by an anomalously small matrix element for this transition. This suggests that delicate cancellations could play a crucial role, which always poses a challenge to numerical calculations. On the other hand, since the transition operator, in leading approximation, depends only on the nucleon spin and isospin but not on the spatial coordinate, this decay provides a precision tool to inspect selected features of the initial and final states.

Traditional realistic NN forces alone appear insufficient to produce the observed lifetime [28, 29], indicating that three-body forces may be needed. Indeed, in Ref. [21] we showed that the 3NF of ChPT plays a major role in producing a transition rate that is near zero, needed for the anomalous long lifetime. We used the chiral two- and three-body potentials of ChPT [2, 4] and the No-Core Shell Model (NCSM) using Lee–Suzuki (LS) renormalization [30] to improve the convergence. The non-perturbative coupling constants of the 3-body force, not fixed by $\pi - N$ or NN data, are c_D for the $N - \pi - NN$ contact term and c_E for the $3NF$ contact term. In Ref. [21] we presented results for $(c_D, c_E) = (-0.2, -0.205)$ and for $(-2.0, -0.501)$. Both sets fit the A=3 binding energies and are allowed by the 'naturalness criterium' for these parameters. The former also produces a precise fit to the triton half life [31]; the latter produces the triton half life within 20% of experiment but is preferred by the ^{14}C half life.

In the left panel of Fig. 8 we show the obtained binding energy of ^{14}N both with the chiral two- and three-body forces and with the phenomenological two-body interaction JISP16. For JISP16 we can use our extrapolation technique to extract the binding energy in the complete basis space: we find 116(5) MeV, that is, ^{14}N is overbound by about 10 MeV with JISP16. For the chiral two- and three-body forces we used a LS renormalization procedure on the truncated basis space, which means that the results are no longer variational upperbounds, and the convergence

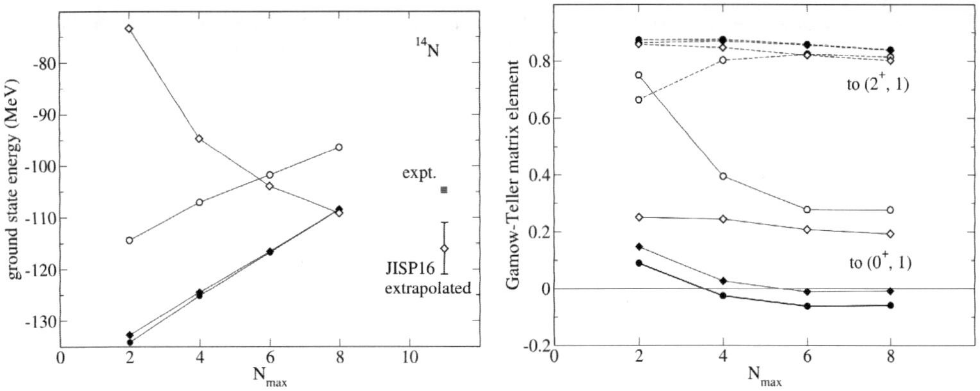

Figure 8. Ground state energies of ^{14}N (left) and GT matrix element (right) as function of N_{max} with JISP16 at $\hbar\omega = 27.5$ MeV (◇), with chiral two-body forces at $\hbar\omega = 14.0$ MeV (○), with chiral two- and three-body forces at $\hbar\omega = 14.0$ MeV with $(c_D, c_E) = (-0.2, -0.205)$ (●) and with $(c_D, c_E) = (-2.0, -0.501)$ (◆), and experimental data (■).

to the infinite basis space is not monotonic. Our results for the binding energy do not imply convergence with $N_{\max}$; however, the obtained spectrum at $N_{\max} = 8$ appears to be reasonably well converged [21] and in qualitative agreement with the experimental spectrum, both for ^{14}C and for ^{14}N. Furthermore, in general the LS ground state energies increase with $N_{\max}$ at small and moderate $N_{\max}$ values, but this trend tends to turn around at (very) large $N_{\max}$ values, and in the limit $N_{\max} \to \infty$ the ground state energy tends to converge from above, as is the case without LS renormalization.

The GT matrix element that governs the β-decay of ^{14}C is

$$M_{\rm GT} \;=\; \sum_{\alpha,\beta} \langle \beta | \sigma\tau_+ | \alpha \rangle \; \rho_{\alpha\beta} \,, \tag{11}$$

where $\langle \alpha | \sigma\tau_+ | \beta \rangle$ is the one-body matrix element between HO single-particle states α and β, which is non-vanishing only when both single-particle states are in the same shell, and the one-body density matrix

$$\rho_{\alpha\beta} \;\equiv\; \langle \Psi_f | a^\dagger_\beta a_\alpha | \Psi_i \rangle \,, \tag{12}$$

with $\langle \Psi_f |$ and $| \Psi_i \rangle$ the ^{14}N and ^{14}C ground states respectively. For comparison, we also calculate the GT transition between the first 2^+ state in ^{14}C and the ground state in ^{14}N. In order to reproduce the measured half life of $T_{1/2} \simeq 5730$ years, the GT matrix element must be anomalously small, $|M^{M}_{\rm GT}| \simeq 2 \times 10^{-3}$, in contrast with a conventional strong GT transition between states in light nuclei with $|M_{\rm GT}| \simeq 1$.

In the right panel of Fig. 8 we show GT matrix elements[3] both with the chiral two- and three-body forces and with JISP16. The results for the GT matrix element between the ground state of ^{14}N and the lowest $(J^\pi, T) = (2^+, 1)$ excited state of ^{14}C is reasonably well converged for the different interactions. Furthermore, it is insensitive to the interaction: JISP16 gives approximately the same result as the chiral interaction, with or without 3-body forces. Our calculations are in reasonable agreement with experiment: the measured transition rate from the ground state in ^{14}N to the low-lying 2^+ states in ^{14}C is $\sum B(GT) = 0.92(33)$ [29], compared to our calculated value for the transition rate of about 1.6.

The situation is very different for the GT matrix element between the ground state of ^{14}N and the $(0^+, 1)$ ground state of ^{14}C. The results with JISP16 do not depend much on the basis space truncation, and the matrix element is suppressed compared to a 'natural' value of the order of one, but not sufficiently suppressed in order to explain the long lifetime of ^{14}C. The results with the chiral NN forces alone depend much stronger on $N_{\max}$, but they do seem to be converged at $N_{\max} = 8$, also at a value that is significantly smaller than the 'natural' value of one, but not sufficiently suppressed in order to explain the long lifetime of ^{14}C. However, once we add the chiral 3NF, we do get a further suppression of the GT matrix element to almost zero, depending on the exact values of (c_D, c_E).

In order to reveal what drives this suppression, we show in Fig. 9 the decomposition of $M_{\rm GT}$ at $N_{\max} = 8$ into the contributions arising from each HO shell for two cases with the 3NF ($c_D = -0.2, -2.0$) and one without. On the left we show our results for the transition between the ground states of ^{14}N and ^{14}C (corresponding to β-decay of ^{14}C); on the right the transition between the ground state of ^{14}N and the first excited 2^+ state of ^{14}C. The largest effect occurs in the p-shell for the transition between the ground states of ^{14}N and ^{14}C, where the 3NF reduces the contributions by an order-of-magnitude from the result with the NN interactions only. This happens for both values of c_D; also note that the transition between the ground state of ^{14}N and

[3] For simplicity, these results are obtained by calculations by assuming isospin symmetry in ^{14}N, and considering transitions between the $(1^+, 0)$ ground state and the lowest $(0^+, 1)$ and $(2^+, 1)$ excited states in ^{14}N.

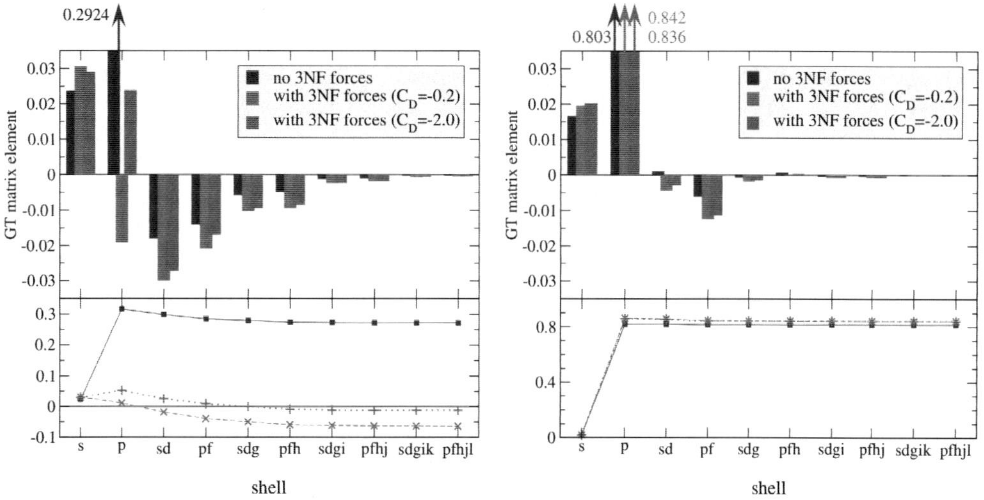

Figure 9. GT matrix element for the transitions ^{14}N to the 0^+ (left) and 2^+ (right) states of ^{14}C, decomposed per HO shell, obtained with two-body chiral interaction with (red and green) and without (blue) three-body forces (3NF) at $N_{\max} = 8$. Top panels display the contributions summed within the shell to yield a total for that shell. Bottom panels display the running sum of the GT contributions over the shells included in the sum. Note that the top panels have the same scale left and right, but the bottom panels do not.

the first excited 2^+ state of ^{14}C receives its largest contribution from the p-shell, and is barely affected by the 3NF. In addition, the three-body forces significantly enhance the contributions of each of the higher shells, for both transitions. In combination with the strong cancellations within the p-shell for the GT matrix element between the ground states of ^{14}N and ^{14}C, the contributions of the higher shells overwhelm that of the p-shell, whereas these higher shells contribute almost nothing to the value of the GT matrix element between the ground state of ^{14}N and the first excited 2^+ state of ^{14}C.

Finally, we calculated the GT matrix element using one wavefunction obtained with three-body forces, but the other wavefunction obtained without three-body forces. Fig. 10 clearly shows that it is the $(1^+, 0)$ ground state of ^{14}N, rather than the $(0^+, 1)$ ground state of ^{14}C, that causes the cancellations within the p-shell once the three-body forces are included. This seems to be a recurring theme of three-body forces in the p-shell nuclei – they manifest themselves in particular in the odd-odd nuclei [5].

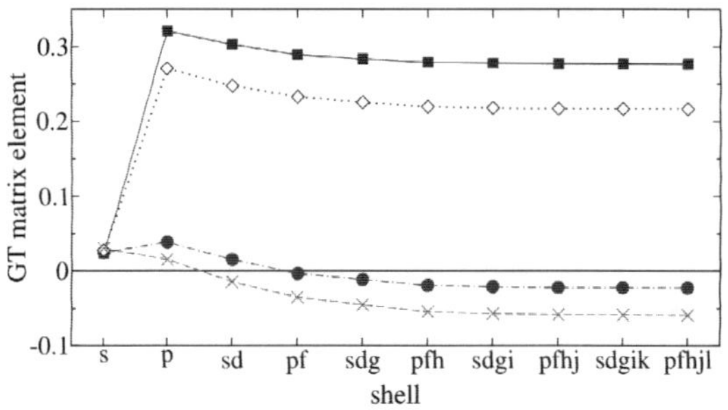

Figure 10. M_{GT} between the ground states of ^{14}N and ^{14}C, using the ^{14}N wavefunction obtained with 3NF, but the ^{14}C wavefunction obtained without 3NF (purple ●), and vice versa (maroon ◇). For comparison, we also include the results with (red cross) and without (blue ■) 3NF for both wavefunctions.

Of course, the precise value of the strongly suppressed GT matrix element between the ground states of ^{14}N and ^{14}C depends on details of the interaction, of the operator, and of the calculations, because of the cancellations between the dominant contributions. Although meson-exchange currents do not contribute significantly to GT transition with a 'natural' value of the order of one [32], they may be important for the actual value of anomalously long lifetime of ^{14}C. For a purely phenomenological interaction like JISP16 it is unclear what the appropriate corrections to the canonical operator would be; but with the chiral interactions one should use consistent meson-exchange currents, and apply the same renormalizaton procedure to the operators as to the interaction. These effects are currently under investigation.

Acknowledgments

We would like to thank Takashi Abe, Mark Caprio, Chase Cockrell, and James Vary for useful discussions. This work was supported in part by U.S. Department of Energy Grant DE-FC02-09ER41582 (SciDAC/UNEDF) and DE-FG02-87ER40371, and by the U.S. NSF grant 0904782. Computational resources were provided by the National Energy Research Supercomputer Center (NERSC), which is supported by the DOE Office of Science, and by an INCITE award, "Nuclear Structure and Nuclear Reactions", from the DOE Office of Advanced Scientific Computing. This research used resources of the Oak Ridge Leadership Computing Facility at ORNL, which is supported by the DOE Office of Science under Contract DE-AC05-00OR22725.

References

[1] Wiringa R B, Stoks V G J and Schiavilla R 1995 *Phys. Rev.* **C51** 38
[2] Entem D R and Machleidt R 2003 *Phys. Rev.* **C68** 041001
[3] Pieper S C, Pandharipande V R, Wiringa R B and Carlson J 2001 *Phys. Rev.* **C64** 014001
[4] Epelbaum E 2006 *Prog. Part. Nucl. Phys.* **57** 654–741
[5] Navratil P, Gueorguiev V G, Vary J P, Ormand W E and Nogga A 2007 *Phys. Rev. Lett.* **99** 042501
[6] Bertsch G F, Dean D J and Nazarewicz W 2007 *SciDAC Review* **6** 42
[7] Furnstahl R J (for the UNEDF Council) 2011 *Nuclear Physics News* **21** 18
[8] Nam H *et al.* 2012 *these proceedings* (*Preprint* 1205.227)
[9] Cockrell R, Maris P and Vary J P 2012 *Phys. Rev.* **C** in press (*Preprint* 1201.0724)
[10] Caprio M A, Maris P and Vary J P 2012 *Phys. Rev.* **C** in press (*Preprint* 1208.4156)
[11] Gloeckner D and Lawson R 1974 *Phys. Lett.* **B53** 313
[12] Abe T, Maris P, Ostuka T, Shimizu N, Ustuno Y and Vary J 2011 *AIP Conf. Proc.* **1355** 173.
[13] Abe T, Maris P, Otsuka T, Shimizu N, Utsuno Y and Vary J 2012 *Phys. Rev.* **C** in press (*Preprint* 1204.1755)
[14] Shirokov A M, Vary J P, Mazur A I and Weber T A 2007 *Phys. Lett.* **B644** 33
[15] Maris P, Vary J P and Shirokov A M 2009 *Phys. Rev.* **C79** 014308
[16] Bogner S K, Furnstahl R J, Maris P, Perry R J, Schwenk A and Vary J P 2008 *Nucl. Phys.* **A801** 21
[17] Sternberg P, Ng E G, Yang C, Maris P, Vary J P, Sosonkina M and Le H V 2008 *Proc. of the 2008 ACM/IEEE conference on Supercomputing* (Piscataway, NJ, USA: IEEE Press) pp 15:1–15:12 ISBN 978-1-4244-2835-9
[18] Vary J P, Maris P, Ng E, Yang C and Sosonkina M 2009 *J. Phys. Conf. Ser.* **180** 012083
[19] Maris P, Sosonkina M, Vary J P, Ng E and Yang C 2010 *Procedia Computer Science* **1** 97
[20] Srinivasa S, Sosonkina M, Maris P, and Vary J P 2012 *Procedia Computer Science* **9** 256
[21] Maris P Vary J P, Navratil P, Ormand W E, Nam H, and Dean D J 2011 *Phys. Rev. Lett.* **106** 202502
[22] Bohlen H G *et al.* 2008 *J. Phys.: Conf. Ser.* **111** 012021
[23] Rowe D J 1970 *Nuclear Collective Motion* (London: Methuen, London)
[24] Caprio M A, Maris P and Vary J P 2012 *in preparation*
[25] Tilley D, Kelley J, Godwin J, Millener D, Purcell J, Sheu C and Weller H 2004 *Nuclear Physics A* **745** 155
[26] Audi G, Bersillon O, Blachot J and Wapstra A 1997 *Nuclear Physics A* **624** 1
[27] Forssen C, Navratil P, Ormand W E and Caurier E 2005 *Phys. Rev.* **C71** 044312
[28] Aroua S *et al.* 2003 *Nucl. Phys.* **A720** 71
[29] Negret A *et al.* 2006 *Phys. Rev. Lett.* **97** 062502
[30] Suzuki K and S. Y. Lee S Y 1980 *Prog. Theor. Phys.* **64** 2091; Suzuki K 1982 *Prog. Theor. Phys.* **68**, 246; **68** 1999; Suzuki K and Okamoto R (1994) *Prog. Theor. Phys.* **92** 1045
[31] Gazit D, Quaglioni S and Navratil P 2009 *Phys. Rev. Lett.* **103** 102502
[32] Vaintraub S, Barnea N and Gazit D 2009 *Phys. Rev.* **C79** 065501

IUPAP C20 Conference on Computational Physics (CCP 2011) IOP Publishing
Journal of Physics: Conference Series **402** (2012) 012032 doi:10.1088/1742-6596/402/1/012032

Real-time calculations of many-body dynamics in quantum systems

Takashi Nakatsukasa

RIKEN Nishina Center, Wako, 351-0198, Japan

E-mail: nakatsukasa@riken.jp

Abstract. Real-time computation of time-dependent quantum mechanical problems are presented for nuclear many-body problems. Quantum tunneling in nuclear fusion at low energy is described using a time-dependent wave packet. A real-time method of calculating strength functions using the time-dependent Schrödinger equation is utilized to properly treat the continuum boundary condition. To go beyond the few-body models, we resort to the density-functional theory. The nuclear mean-field models are briefly reviewed to illustrate its foundation and necessity of state dependence in effective interactions. This state dependence is successfully taken into account by the density dependence, leading to the energy density functional. Photoabsorption cross sections in ^{238}U are calculated with the real-time method for the time-dependent density-functional theory.

1. Time-dependent approaches to quantum problems in nuclear physics

In this paper, we will report several applications of the real-time calculations. Although we concentrate our discussion on the nuclear physics problems, the real-time approaches are useful in many fields of many-body quantum systems. Before discussing applications, let us start from giving some reasons why we adopt the real-time calculations, instead of a more prevalent time-independent method.

1.1. Stationary solutions

It is customary to solve a quantum mechanical problem in a form of stationary equations, such as the eigenvalue problem of the Schrödinger equation:

$$H|\Phi_n\rangle = E_n|\Phi_n\rangle. \tag{1}$$

If a direct solution of this problem is achieved, the solution provides a wave function of each energy eigenstate. In principle, we may calculate all the physical observables using these wave functions. Amount of the computational task is roughly proportional to number of eigenstates $|\Phi_n\rangle$ to be calculated. Thus, for instance, when we are interested in the excitation spectra and/or strength functions in a wide range of energy, it should be computationally very demanding.

1.2. *Time-dependent solutions*

In contrast, a real-time solution of the time-dependent Schrödinger equation is, in general, a linear combination of energy eigenstates $|\Phi_n\rangle$:

$$|\Psi(t)\rangle = \sum_n c_n e^{-iE_n t} |\Phi_n\rangle. \tag{2}$$

The number of eigenstates and their coefficients c_n involved in Eq. (2) are simply determined by the initial state. A trivial but important feature of the time-dependent solution, Eq. (2), is that the state consists of many eigenstates and distribution of their weights, c_n, can be chosen as we wish, by selecting the initial wave packet. Therefore, we may obtain physical quantities for states at a variety of energies from a single time evolution of the wave packet, using the Fourier decomposition into different energies. Since the computational task does not strongly depend on the weights c_n, the real-time calculation for the time-dependent equation could be computationally more efficient than solving the time-independent eigenvalue equation (1) to obtain many eigenstates.

1.3. *Boundary condition for continuum states*

Another practical advantage of the real-time calculation is the fact that the method does not require the boundary condition for the description of the unbound continuum states. The solution of Eq. (1) for an unbound state at positive energy E_n cannot be uniquely determined without the asymptotic boundary condition. In contrast, the solution of the time-dependent equation, Eq. (2), is unique for a given initial state. In this sense, the boundary condition is naturally given by the time-dependent solution.

To illustrate this fact, let us consider scattering of two particles A and B, which are initially apart from each other, outside of the range of interaction V. These particles may be composite. The Hamiltonian H_0 describes two free particles, A and B, at the initial channel. The total Hamiltonian of the system is given by $H = H_0 + V$. The scattering solution for the energy eigenstate can be formally written as [1]

$$|\Psi^{(+)}(E)\rangle = \left(1 + \frac{1}{E - H + i\eta} V\right) |\Phi(E)\rangle, \tag{3}$$

where the initial (incident) state is an eigenstate of H_0, $H_0|\Phi(E)\rangle = E|\Phi(E)\rangle$. The Green's function, $(E - H + i\eta)^{-1}$, contains the infinitesimal $i\eta$ that determines the scattering boundary conditions. The state of Eq. (3) is obtained by acting the Møller's wave operator on the initial state.

$$|\Psi^{(+)}(E)\rangle = U(0, -\infty)|\Phi(E)\rangle = \left(1 - \int_{-\infty}^{0} dt' U(0, t') e^{-\eta|t'|} V_I(t')\right) |\Phi(E)\rangle, \tag{4}$$

where $U(t, t') = e^{iH_0 t} e^{-iH(t-t')} e^{-iH_0 t'}$ and $V_I(t) \equiv e^{iH_0 t} V e^{-iH_0 t}$. Here, the initial state $|\Phi(E)\rangle$ is given by a plane wave for the relative motion between A and B. Then, we need the convergence (adiabatic switch-off) factor $e^{-\eta|t'|}$, in order to remove the interaction between A and B at the limit of $t \to -\infty$. This trick is necessary when we treat the energy eigenstate, because the particles A and B are overlapped in the plain-wave solution and interacting each other. In contrast, the infinitesimal factor η becomes unnecessary [1] if we use a wave packet, $|\Phi_{\rm pac}\rangle$, instead of the eigenstate $|\Phi(E)\rangle$ in Eq. (4). Since the wave packet is spatially localized, we can construct the initial state in which the wave packets of A and B are far apart, not interacting each other ($V|\Phi_{\rm pac}\rangle = 0$). The scattering wave packet, which is a superposition of many eigenstates, can be calculated by propagating the initial wave packet. We may stop the time evolution at a finite period of time, because the interaction takes place only in the finite time. Once the wave packet is scattered away outside of the interacting region, the state is governed by the Hamiltonian H_0. After the scattering, the proper scattering asymptotic behavior should be automatically imposed for the wave packet.

1.4. Intuitive picture of quantum dynamics

The time-dependent description has not only these advantages in practical computation, but also provides an intuitive picture of the quantum dynamics. It is very helpful for our insight into complex systems to visualize movement of the wave packet. This will be demonstrated in the following applications.

1.5. Nuclear many-body problems

The nucleus is a self-bound quantum system which presents a rich variety of phenomena. It is composed of fermions of spin 1/2 and isospin 1/2, called nucleons (protons and neutrons), interacting with each other through a complex interaction with a short-range repulsive core [2]. Thus, the nucleus is a strongly correlated system. Furthermore, it is highly "quantum", which could be quantified by the magnitude of the zero-point kinetic energy relative to the potential energy, $\langle T\rangle/\langle V\rangle \approx 1$. This is roughly unity in nuclear systems: $\langle T\rangle/\langle V\rangle \approx 1$, which indicates that the nucleus has a very strong quantum nature. Even for the liquid helium, this ratio is smaller than the nuclear case.

Remarkable experimental progress in production and study of exotic nuclei requires us to construct a theoretical model with higher accuracy and reliability. Extensive studies have been made in the past, to introduce models and effective interactions to describe a variety of nuclear phenomena and to understand basic nuclear dynamics behind them [2, 3]. Simultaneously, significant efforts have been made in the microscopic foundation of those models. For light nuclei, the "first-principles" large-scale computation, starting from the bare nucleon-nucleon (two-body & three-body) forces, is becoming a current trend in theoretical nuclear physics. It is computationally very challenging, because, as we mentioned above, the nucleus is a highly quantum, strongly correlated Fermionic many-body system, interacting via the complex and "singular" nuclear force. Although these ab-initio-type approaches are still limited to nuclei with very small mass number, they have recently shown a significant progress. These issues are addressed by other contributions [4, 5, 6].

In this paper, we show a few nuclear-physics problems: The low-energy nuclear fusion reactions, the nuclear response in the continuum, and the time-dependent density-functional theory in the linear regime. Here, we would like to emphasize again that, although a particular model/theory is adopted for each calculation, the concept of the real-time method is quite general and applicable to other models and to other subfields of physics as well.

2. Real-time calculation of sub-barrier fusion reaction

In this section, we present the time-dependent wave-packet method and discuss tunneling dynamics in sub-barrier fusion process [7, 8, 9, 10, 11]. To illustrate the essential idea, we start from a simple two-body problem.

2.1. Two-body time-dependent wave-packet model

.

A simple potential model of the fusion reaction is constructed as follows: The Hamiltonian for the relative motion between two nuclei is given by ($\hbar = 1$)

$$H = -\frac{1}{2\mu}\nabla_R^2 + V(R) + iW(R) = -\frac{1}{2\mu}\left(\frac{1}{R}\frac{\partial}{\partial R}R\right)^2 + \frac{\vec{L}^2}{2\mu R^2} + V(R) + iW(R). \quad (5)$$

This leads to the radial Schrödinger equation in each partial wave:

$$i\frac{\partial}{\partial t}u_L(R,t) = \left\{ -\frac{1}{2\mu}\frac{d^2}{dR^2} + \frac{L(L+1)}{2\mu R^2} + V(R) + iW(R) \right\} u_L(R,t), \tag{6}$$

where $\psi(R,t) = \sum_{LM}(u_L(R,t)/r)Y_{LM}(\hat{R})$. The real potential, $V(R)$, consists of the repulsive Coulomb and attractive nuclear potentials. When the approaching two nuclei pass through the Coulomb barrier, this is regarded as the fusion. To calculate its probability, it is convenient to use the imaginary potential, $iW(R)$, that is non-zero only inside the Coulomb barrier. Then, we define the fusion by the flux loss caused by $iW(R)$. We assume here the collision of ^{10}Be ($Z = 4$) on ^{208}Pb ($Z = 82$). $W(R)$ and the nuclear part of $V(R)$ are assumed to be in the Woods-Saxon form.

The most common way to solve this problem is to find a stationary solution with a fixed energy E by integrating the time-independent radial Schrödinger equation. Then, in the asymptotic region, one can compare the flux of incoming and outgoing Coulomb waves, $u_L(R) \sim C_L^{\text{in}}u_L^{(-)}(R) + C_L^{\text{out}}u_L^{(+)}(R)$, to obtain the fusion probability, $P_L = (|C_L^{\text{in}}|^2 - |C_L^{\text{out}}|^2)/|C_L^{\text{in}}|^2$. Results obtained in this way are shown by circles in the bottom panel of Fig 2 for $L = 0$.

We demonstrate an alternative approach: the time-dependent wave-packet method. The initial wave function is of the Gaussian form

$$u_L(R, t = 0) = e^{-iKR}e^{-\gamma^2(R-R_0)^2}, \tag{7}$$

where R_0 indicate the initial position of the Gaussian center. The parameters, K and γ, are arbitrary, but should be chosen so as to cover the interested energy region. In the top panel of Fig. 1, the solid line shows the initial wave packet for $L = 0$, the head-on collision with average relative momentum K. In the present calculation, we set $K^2/2\mu = 28$ MeV and calculate the time evolution up to $t_f = 10$ MeV$^{-1} \approx 2000$ fm/c. Time evolution of the wave packet is calculated by recursive operation of the small-time evolution operator, $U(\Delta t) = e^{-i\Delta tH}$, that is simply approximated with the fourth-order Taylor expansion, $U(\Delta t) \approx \sum_{n=0}^{4}(-i\Delta tH)^n/n!$. The wave packet collides with the Coulomb barrier in the second and third panels in Fig. 1. Then, the wave packet is partially reflected back. In the bottom panel, the final wave packet contains only outgoing waves. The missing part of the wave packet passes through the barrier, entering the inner region and disappearing because of the absorbing imaginary potential, $iW(R)$. One can see, in the second and third panels, that a part of the wave packet actually penetrates the Coulomb barrier. This incoming component vanishes in the fourth and fifth panels.

Now, let us show the energy profile of the initial and final wave packets. We calculate the energy distribution of the wave packet at time t.

$$f_L(E;t) = \langle u_L(t)|\delta(E - H)|u_L(t)\rangle. \tag{8}$$

This quantity can be also calculated using the time propagation technique [7]. The dashed line in the top panel of Fig. 2 indicates the calculated energy distribution of the initial wave packet, $f_{L=0}(E, t_i)$. This is in a Gaussian form whose centroid is about 39 MeV that corresponds to the given kinetic energy of 28 MeV. The difference of 11 MeV is the Coulomb potential energy at $R_0 \sim 40$ fm. The final distribution, $f_0(E, t_f)$ shown in the solid line in Fig. 2 (top panel), is calculated from the wave packet at the bottom of Fig. 1. We have $f_0(E, t_f) \approx 0$ for $E > 42$ MeV. These high energy parts end up the fusion. On the other hand, the flux remains almost invariant ($f_0(E, t_f) \approx f_0(E, t_i)$) for $E < 36$ MeV, which suggests that the fusion cross section is negligible in this low energy region. These arguments can be quantified by calculating the fusion probability,

$$P_L(E) = \frac{f_L(E, t_i) - f_L(E, t_f)}{f_L(E, t_i)}. \tag{9}$$

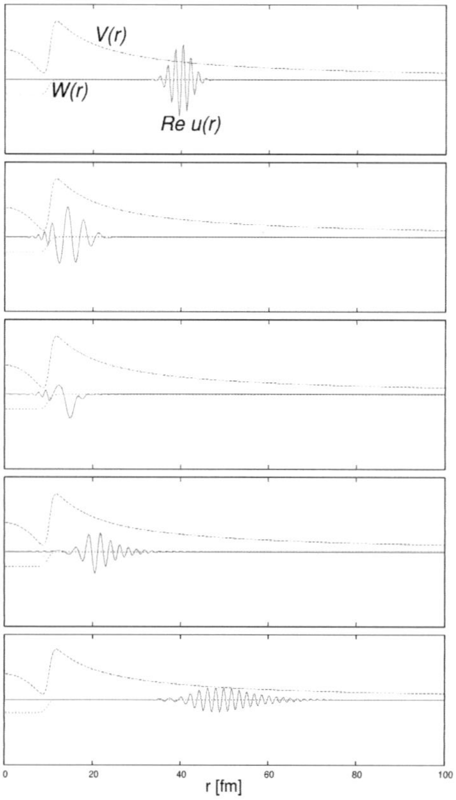

Figure 1. Snap shots of time-dependent wave functions, Re $u_{L=0}(r)$. The panels are ordered in time from the top to the bottom. The dashed (dotted) line indicates a shape of the real (imaginary) part of the potential. See text for details.

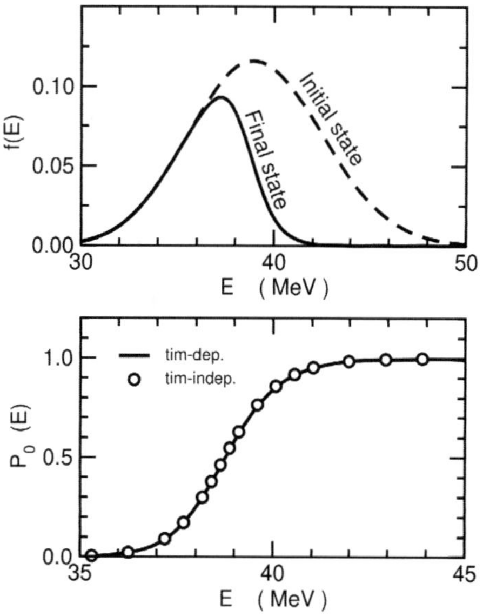

Figure 2. Top panel: Energy distribution of the initial (dashed) and final wave packets (solid line) corresponding to the top and bottom panels in Fig. 1, respectively. Bottom: Calculated fusion probability ($L = 0$) as a function of energy. The solid line is calculated with the time-dependent wave-packet method, while the circles indicate results obtained from time-independent (fixed energy) solutions of the Schrödinger equation.

This is shown in the bottom panel of Fig. 2 by solid line. The result calculated with the time-dependent wave-packet method perfectly agrees with that of the fixed-energy calculation (circles). This means that the time-dependent method provides an alternative method of accurate calculation for fusion cross section.

2.2. Three-body time-dependent wave-packet model

Let us upgrade the model to three bodies and discuss effects of a valence neutron on the fusion. There are many theoretical and experimental investigations on this, but the conclusion is somewhat elusive yet. Especially, the effect of weakly-bound valence neutrons with a spatially extended wave function, which is often called "neutron halo", is controversial. There were many arguments that the weakly-bound neutron enhances the fusion cross section, especially at sub-barrier energies [12, 13, 14, 15, 16]. However, as is shown in the followings, we have reached the opposite conclusion [7, 8, 9, 10, 11].

We assume that the projectile is a bound state of a neutron (n) and a core nucleus (C), and the target (T) is treated as a point particle. In this work, we neglect the intrinsic spin of neutron. The time-dependent Schrödinger equation of the three-body scattering is given as

$$i\hbar\frac{\partial}{\partial t}\Psi(\mathbf{R},\mathbf{r},t) = \left\{-\frac{1}{2\mu}\nabla^2_{\mathbf{R}} - \frac{1}{2m}\nabla^2_{\mathbf{r}} + V_{\mathrm{nC}}(r) + V_{\mathrm{CT}}(R_{\mathrm{CT}}) + V_{\mathrm{nT}}(r_{\mathrm{nT}})\right\}\Psi(\mathbf{R},\mathbf{r},t), \quad (10)$$

where we denote the relative n-C coordinate as $\mathbf{r}$ and the relative P-T coordinate as $\mathbf{R}$. The reduced masses of n-C and P-T motions are m and μ, respectively. The n-C potential, $V_{\mathrm{nC}}(r)$, is assumed to be real with the Woods-Saxon form. This potential must produce a bound state of the neutron around the core. The depth of the potential V_{nC} is varied according to the orbital energy simulating a projectile nucleus with a tightly-bound to weakly-bound neutron. The core-target potential, $V_{\mathrm{CT}}(R_{\mathrm{CT}})$, contains an imaginary part $iW(R_{\mathrm{CT}})$ that simulates the fusion. The n-T potential, $V_{\mathrm{nT}}(r_{\mathrm{nT}})$, is taken to be real. The imaginary potential is present only between the core and target nuclei. When the core and target become close enough, the wave function will vanish, to be counted as fusion. Therefore, since the final destination of the neutron is irrelevant, the present calculation does not distinguish complete and incomplete fusion. The total fusion probability (sum of complete and incomplete ones) is calculated below.

The wave-packet is constructed using the partial wave expansion in the body-fixed frame [17, 18]. This calculation is equivalent to the one in the space-fixed frame. However, the body-fixed frame has an advantage for calculations of states with non-zero total angular momentum $J \neq 0$ ($\vec{J} = \vec{L} + \vec{l}$). In the body-fixed frame, channels are characterized by the magnetic quantum number Ω (the projection of J on the z-axis) and the angular momentum l conjugate to the angle θ between $\mathbf{r}$ and $\mathbf{R}$. Coupling between different Ω channels is present only for $\Delta\Omega = \pm 1$, caused by the Coriolis term. In the present paper, we focus our discussion on the head-on collision ($J = 0$). See Ref. [10] for the complete calculation. The $J = 0$ state is characterized by $\Omega = 0$ and $L = l$, given by

$$\psi^{J=0}(R,r,\theta;t) = \sum_{l=0}^{l_{\max}} \frac{u_l(R,r,t)}{Rr}\frac{\sqrt{2l+1}}{4\pi}P_l(\cos\theta). \quad (11)$$

Here, we need to truncate the space by setting the maximum value of the partial waves l. To calculate the Coulomb breakup process, we may take $l_{\max} = 4$. However, when the n-T potential is included, nuclear breakup and the neutron transfer may take place, then, much larger $l_{\max}$ is necessary to obtain convergent results; $l_{\max} \sim 70$ [8, 10]. Since the present calculation is simply based on the discretization of the radial coordinate, it is straightforward to include these high partial waves. This is an advantage over the continuum-discretized coupled-channel calculations [16].

In this model, the projectile is assumed to have a simple structure that a neutron sits on the $2s$ orbital in the n-C potential V_{nC}: $\phi_0(\mathbf{r}) = \sqrt{1/4\pi}y_0(r)/r$. The initial wave packet at time $t = 0$ is prepared in the same manner as we have shown in Sect. 2.1. Namely, the P-T relative wave function is given by the boosted Gaussian wave packet whose centroid is at $R_0 = 25$ fm.

$$u_l(R,r,t=0) = e^{-iK_0R}e^{-\gamma^2(R-R_0)^2}y_0(r)\delta_{l0}. \quad (12)$$

2.2.1. Case (1) Well-bound projectile: First, we investigate fusion reaction of projectile with a well-bound neutron. For this case, we found that the fusion cross section is identical to the one without the valence neutron. Thus, the valence neutron essentially sticks to the core nucleus, and they move together as a single nucleus. This may be naturally expected.

However, there are exceptions. when the condition of energy matching is satisfied, which means that energies of the neutron orbital in the projectile and target are approximately

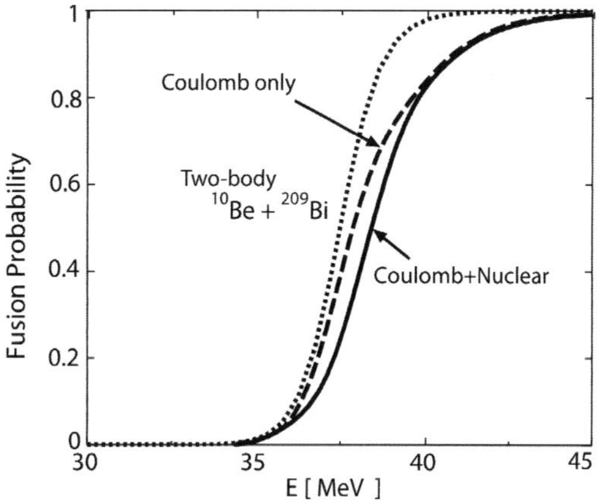

Figure 3. Fusion probability ($J = 0$) as a function of incident energy for halo ^{11}Be ($\epsilon_n = -0.6$ MeV) on ^{209}Bi. Dotted line indicates the probability for the projectile without the valence neutron. Dashed and solid lines indicate fusion probabilities calculated with and without V_{nT} potentials, respectively.

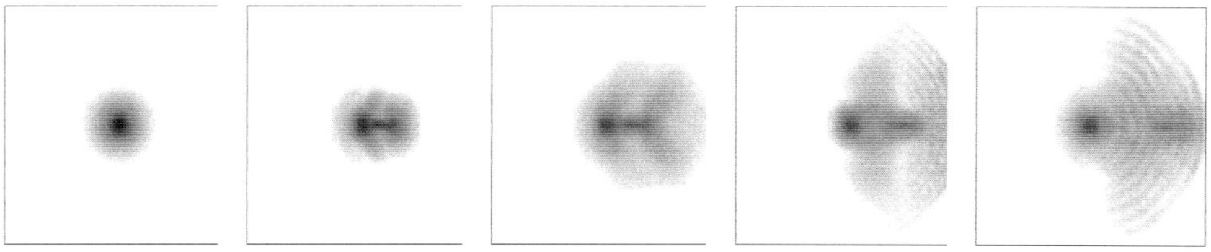

Figure 4. Time-dependent density distribution of the valence neutron, $\rho(r, \theta; t)$ in the collision process of ^{11}Be on ^{209}Bi. The panel at the left (right) end shows the initial (final) wave packet. The center of each panel, $r = 0$, indicates the position of the core, and the right is the direction for the target, $\theta = 0$.

degenerate, we observe a substantial enhancement of sub-barrier fusion cross section. The fusion probability is strongly correlated with the neutron transfer probability and sensitive to V_{nT}. This may be understood in terms of adiabatic dynamics of the valence neutron. See discussion in Ref. [7].

2.2.2. Case (2) Weakly-bound projectile (Neutron halo): Now, let us discuss fusion of weakly bound ^{11}Be on ^{209}Bi, namely effects of the neutron halo. As is mentioned above, there are theoretical arguments that the weakly-bound neutron enhances the sub-barrier fusion cross section [12, 13, 14, 15, 16]. We studied the same problem in a fully microscopic framework with the real-time method.

Calculated fusion probabilities at the head-on collision ($J = 0$) are shown in Fig. 3. The fusion probability is slightly suppressed by the presence of the halo neutron. It seems that the fusion dynamics for the neutron-halo projectile are mainly determined by the Coulomb breakup process. Time evolution of the neutron density distribution during the collision process is shown in Fig. 4. We plot the following quantity in the x-y plane with $x = r\cos\theta$ and $y = r\sin\theta$ for the head-on collision of $J = 0$;

$$\rho(r, \theta; t) = \int dR |\psi^{J=0}(R, r, \theta; t)|^2. \quad (13)$$

The direction of the target always corresponds to $\theta = 0$, which means that the target nucleus approaches from the right side on the x-axis, then returns back to the right. Figure 4 suggests the followings: When the core nucleus is decelerated by the target Coulomb field, the halo neutron behaves like a spectator, keeping its incident velocity and leaving the core nucleus. This yields the Coulomb breakup of the projectile, then reduces the collision energy between the core and the target. This spectator picture of the fusion dynamics for halo nuclei nicely explains why the total fusion cross section does not increase even though the radius of the neutron-halo nucleus is so large [19].

Our conclusion is that the presence of the halo neutron suppresses the fusion cross section, irrelevant to the collision energy. This contradicts other former predictions [12, 13, 14, 15, 16]. In fact, the model adopted in the coupled-channel calculation in Ref. [16] is very close to ours. We have found that the difference comes from the truncation of the model space [10]. Namely, the result in Ref. [16] did not reach the convergence with respect to the number of the partial wave $l_{\max}$ in Eq. (11). If we set $l_{\max} \leq 4$ as is done in Ref. [16], we also obtain an enhanced fusion cross section at sub-barrier energies. However, the cross section decreases as $l_{\max}$ increases, and finally it becomes even smaller than the two-body result without the valence neutron.

3. Real-time calculation of strength functions in the continuum

In this section, we discuss a method of calculating the strength distribution, in particular, the $E1$ strength distribution, from the time-dependent wave function.

The $E1$ strength distribution in the continuum is defined by

$$\frac{dB(E1; E)}{dE} \equiv \sum_{m,m'} \int dE' \delta(E - E') \left| \langle \Phi^{(+)}(E'; m') | M(E1; m) | \Phi_0 \rangle \right|^2. \quad (14)$$

Here, $M(E1; m)$ is the electric dipole operator with the magnetic quantum number m. The final states $|\Phi^{(+)}(E; m)\rangle$ should have the proper outgoing asymptotic form when unbound channels are open. The construction of the final state in the continuum is a difficult task in general, especially for many-body non-spherical systems. An alternative way of avoiding this difficulty is the time-dependent description. Let us show a simple example as an illustration of the method.

3.1. Illustrative example

The $E1$ strength function of Eq. (14) can be written in a time-dependent form:

$$\frac{dB(E1; E)}{dE} = -\frac{1}{\pi} \text{Im} \sum_m i \int_0^\infty dt e^{i(E+i\eta)t} \langle \Phi_0 | M^\dagger(E1; m) e^{-itH} M(E1; m) | \Phi_0 \rangle. \quad (15)$$

This is easily obtained by the fact that the time integration simply produces the Green's function $(E - H + i\eta)^{-1}$ whose imaginary part is nothing but the delta function $\delta(E - H)$. Therefore, we can calculate the $E1$ strength distribution as follows: (1) Construct the initial state by operating the $E1$ operator to the ground state, $|\Psi(t = 0)\rangle = M(E1; m)|\Phi_0\rangle$. (2) Calculate its time evolution $|\Psi(t)\rangle$ and its overlap with the initial state. (3) Its Fourier transform leads to $dB(E1; E)/dE$. In this time-dependent formulation, we do not need to construct the final states with the proper boundary condition. This is automatically taken into account in the time evolution of wave packets. When the $E1$ operator excites the ground state into unbound continuum states, the wave packet $|\Psi(t)\rangle$ will decays in time and the integrand of Eq. (15) will vanish at $t \to \infty$. Thus, we do not need the convergence factor $+i\eta$.

Let us discuss application of the time-dependent method to a simple single-particle model. We again use the Woods-Saxon potential of $V_{\rm nC}$ in Sect.2.2 to describe the ^{11}Be nucleus as a two-body system of a ^{10}Be core and a neutron. The neutron is assumed to sit on the $2s$ orbital. Multiplying it by the $E1$ operator with the recoil charge, we construct the initial state which is shown in the top panel of Fig. 5. We solve the time-dependent Schrödinger equation similar to Eq. (6) with $L = 1$. The imaginary potential $iW(R)$ is active at $R > 50$ fm to absorb outgoing waves. The time evolution is shown in the following panels in Fig. 5. The state with $L = 1$ propagates into $r \to \infty$ in time. After a certain period, the state no longer has an overlap matrix element with the initial state. Then, we stop the calculation. From this time evolution, we obtain $E1$ strength distribution (dotted line) shown in Fig. 6. There is a strong low-energy peak known as the threshold effect. This peak is drastically reduced if we change the energy of the $2s$ orbital. For instance, the solid line in Fig. 6 shows the result for the case that the energy of the $2s$ orbital is -2 MeV. Experimental data [20, 21, 22] indicate a similar shape but somewhat smaller strength.

This method can be extended into many-body problems in a straightforward manner. However, its computational task drastically increases as the number of particles increases. Thus, we resort to the density functional model of nuclei in the next section.

4. Density-functional approach to nuclei

The density functional theory is a leading theory for describing nuclear properties of heavy nuclei and perhaps the only theory capable of describing all nuclei and nuclear matter with a single universal energy density functional. In the nuclear physics, it is often called the self-consistent mean-field model, because of a historical development based on the Brueckner-Hartree-Fock theory and introduction of the effective interaction. However, it should be noted that there are both fundamental and practical differences between the naive mean-field theory and the mean-field model of the nucleus which is, in fact, conceptually analogous to the density-functional theory.

I briefly review basic properties of nuclei and show that those properties cannot be understood by a simple mean-field model. Here, the saturation property plays a key role.

4.1. Nuclear saturation in the mean-field model

Nuclei are known to be well characterized by the saturation property. Namely, they have an approximately constant density $\rho_0 \approx 0.17$ fm^{-3}, and a constant binding energy per particle $B/A \approx 16$ MeV.[1] In this section, I show that the nuclear saturation property has a great impact on nuclear models. Especially, it is inconsistent with the independent-particle model of nuclei with a "naive" average (mean-field) potential.

There are many evidences for the fact that the mean-free path of nucleons is larger than the size of nucleus. In fact, the mean free path depends on the nucleon's energy, and becomes larger for lower energy [2]. Therefore, it is natural to assume that the nucleus can be primarily approximated by the independent-particle model with an average one-body potential. The crudest approximation is the degenerate Fermi gas of the same number of protons and neutrons ($Z = N = A/2$). The observed saturation density of $\rho_0 \approx 0.17$ fm^{-3} gives the Fermi momentum, $k_F \approx 1.36$ fm^{-1}, that leads to the Fermi energy (the maximum kinetic energy), $T_F = k_F^2/2M \approx 40$ MeV.

[1] This is the extrapolated value for the infinite nuclear matter without the surface and the Coulomb energy. The observed values for finite nuclei are $B/A \approx 8$ MeV.

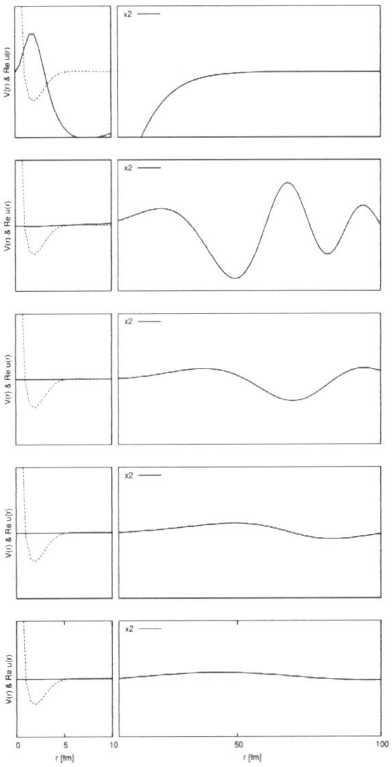

Figure 5. Time evolution of a state (real part of the wave function) initially excited by the dipole operator. The vertical scale for $r = 10 \sim 100$ fm is magnified by a factor of two compared to that for $r = 0 \sim 10$ fm. The dashed line indicate the shape of the p-wave potential.

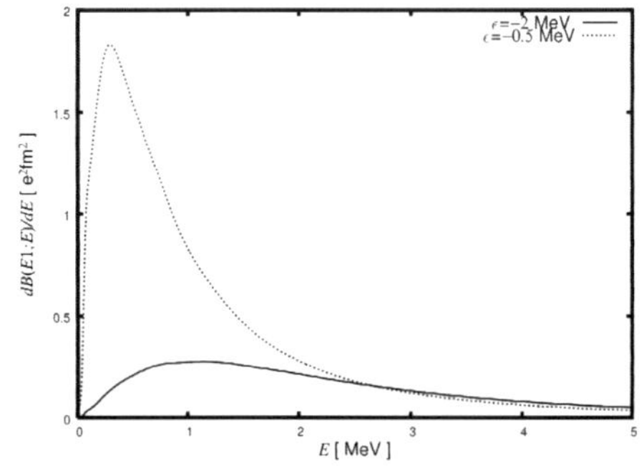

Figure 6. Calculated $E1$ strength distribution for ^{11}Be. See text for explanation.

First, I show that the independent-particle model with a constant attractive potential $V < 0$ cannot describe the nuclear saturation property. It follows from the simple arguments. The constancy of B/A means that it is approximately equal to the separation energy of nucleons, S. In the independent-particle model, it is estimated as

$$S \approx B/A \approx -(T_F + V). \tag{16}$$

Since the binding energy is $B/A \approx 16$ MeV, the potential V is about -55 MeV. It should be noted that the relatively small separation energy is the consequence of the significant cancellation between kinetic and potential energies. The total (binding) energy is given by

$$-B = \sum_{i=1}^{A} \left(T_i + \frac{V}{2} \right) = A \left(\frac{3}{5} T_F + \frac{V}{2} \right), \tag{17}$$

where we assume that the average potential results from a two-body interaction. The two kinds of expressions for B/A, Eqs. (16) and (17), lead to $T_F \approx -5V/4 \approx 70$ MeV, which is different from the previously estimated value ($\sim$ 40 MeV). Moreover, it contradicts the fact that the nucleus is bound ($T_F < |V|$).

To reconcile the independent-particle motion with the saturation property of the nucleus, the nuclear average potential should be state dependent. Allowing the potential V_i depend on the state i, the potential V should be replaced by that for the highest occupied orbital V_F in Eq. (16), and by its average value $\langle V \rangle$ in the right-hand side of Eq. (17). Then, we obtain the following relation:

$$V_F \approx \langle V \rangle + T_F/5 + B/A. \tag{18}$$

Therefore, the potential V_F is shallower than its average value.

Weisskopf suggested the momentum-dependent potential V, which can be expressed in terms of an effective mass m^* [23]:

$$V_i = U_0 + U_1 \frac{k_i^2}{k_F^2}. \tag{19}$$

Actually, if the mean-field potential is non-local, it can be expressed by the momentum dependence. Equation (19) leads to the effective mass, $m^*/m = (1 + U_1/T_F)^{-1}$. Using Eqs. (16), (18), and (19), we obtain the effective mass as

$$\frac{m^*}{m} = \left\{ \frac{3}{2} + \frac{5}{2}\frac{B}{A}\frac{1}{T_F} \right\}^{-1} \approx 0.4. \tag{20}$$

Quantitatively, this value disagrees with the experimental data. The empirical values of the effective mass vary according to the energy of nucleons, $0.7 \lesssim m^*/m \lesssim 1$, however, they are almost twice larger than the value in Eq. (20). As far as we use a normal two-body interaction, this discrepancy should be present in the mean-field calculation with any interaction, because Eq. (20) is valid in general for a saturated self-bound system. Therefore, the conventional models cannot simultaneously reproduce the most basic properties of nuclei; the binding energy and the single-particle property.

This problem can be solved in the density functional theory. In nuclear physics, basically the same solution was interpreted in terms of the density-dependent interaction. The variation of the total energy density functional $E[\rho]$ with respect to the density contains "re-arrangement potential", $\partial V_{\text{eff}}[\rho]/\partial\rho$, which appear due to the density dependence of the effective force $V_{\text{eff}}[\rho]$. These terms turn out to be crucial to obtain the saturation condition. Now, the expression for the total energy, Eq. (17), should be modified to include the re-arrangement effect. This resolves the previous issue, then provides a consistent independent-particle description for the nuclear saturation. This is important to understand basic nuclear properties and to describe nuclei in a universal framework.

Intensive studies in nuclear density functional models in recent years have produced numerous results and new insights into nuclear structure [24, 25]. However, it is impossible to review all of them in this paper. Thus, I will present results of our study with the (time-dependent) density functional approach, on the shape phase transition in the rare-earth nuclei and on the photoabsorption cross section in ^{238}U.

4.2. Density-functional theory for superfluid nuclei

A modern energy functional for nuclei is a functional of many kinds of density, such as kinetic $\tau(\vec{r})$ and spin-orbit density $\vec{J}(\vec{r})$. In addition, it is known that the heavy nuclei in the open-shell configurations show characteristic features of the superfluid systems [3]. To describe superfluid properties produced by pairing correlation among nucleons, we need to add the pair (abnormal) density $\kappa(\vec{r})$. These densities are collectively denoted as $\tilde{\rho}$ in the followings. Variation of the total energy, $E[\tilde{\rho}]$, leads to the following equation, known as the Hartree-Fock-Bogoliubov equation in nuclear physics [3]:

$$H[\Psi, \Psi^*]|\Psi_\mu\rangle = E_\mu|\Psi_\mu\rangle, \tag{21}$$

where E_μ and $|\Psi_\mu\rangle$ are quasi-particle energies and states, respectively [3]. $|\Psi_\mu\rangle$ is composed of two components; the upper $|U_\mu\rangle$ and the lower one $|V_\mu\rangle$. The solution of Eq. (21) defines the normal density $\rho(\vec{r}) = \sum_\mu V_\mu(\vec{r})V_\mu^*(\vec{r})$, the pair density $\kappa(\vec{r}) = \sum_\mu U_\mu(\vec{r})V_\mu(\vec{r})$, and other densities at the ground state. Since $h[\Psi, \Psi^*]$ depends on these densities, Eq. (21) must be solved in a self-consistent way. Minimization of the energy density functional may lead to a spontaneous breaking of symmetry. An example is given in Fig. 7 for Nd isotopes [26]. The intrinsic quadrupole moment calculated with the Skyrme functional of SkM* is compared with the experimental data. At $N = 82$, the nucleus at the ground state is spherical $Q_0 = 0$, while for $N = 86 \sim 92$, the deformation gradually develops. The observed ground-state deformations deduced from the transition probability $B(E2; 2^+ \to 0^+)$ are nicely reproduced. Note that there are no adjustable parameters in this calculation.

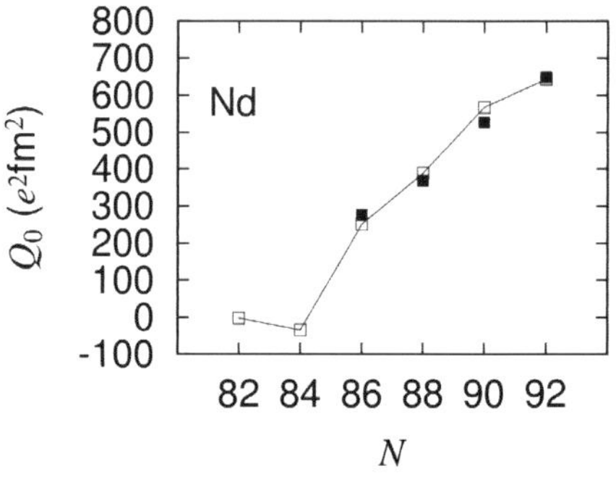

Figure 7. Calculated (open squares) and experimental (filled) intrinsic quadrupole moment for Nd isotopes [26].

4.3. Time-dependent density-functional theory for superfluid nuclei

For a description of the time-dependent phenomenon, we need to extend the energy functional to include the time-odd densities, such as the spin density $\vec{s}(\vec{r}, t)$ and the current density $\vec{j}(\vec{r}, t)$. Now, all the densities are time dependent. The time-dependent version of Eq. (21) is formally written as

$$i\frac{\partial}{\partial t}|\Psi_\mu(t)\rangle = H[\Psi(t), \Psi^*(t)]|\Psi_\mu(t)\rangle. \tag{22}$$

This is known as the time-dependent Hartree-Fock-Bogoliubov equation in nuclear physics.

The computation of Eq. (22) is very demanding because we need to calculate the time evolution of all the quasi-particle states $|\Psi_\mu\rangle$. The number of them is same as the dimension of the single-particle model space which is much larger than the particle number. Although we proposed a feasible approach to its linear regime, known as the finite amplitude method (FAM) [27, 28, 29, 30, 31], the FAM is based on a time-independent formulation. In this section, we present another approximate treatment of Eq. (22) to provide time-dependent equations.

The approximate equations alternative to Eq. (22) is called the canonical-basis TDHFB [32]. Assuming the diagonal property of the pair potential in the canonical pair of states k and $\bar{k}$, Eq. (22) can be approximated by the following set of equations.

$$i\frac{\partial}{\partial t}|\phi_k(t)\rangle = (h(t) - \eta_k(t))|\phi_k(t)\rangle, \qquad i\frac{\partial}{\partial t}|\phi_{\bar{k}}(t)\rangle = (h(t) - \eta_{\bar{k}}(t))|\phi_{\bar{k}}(t)\rangle, \tag{23a}$$

$$i\frac{d}{dt}\rho_k(t) = \kappa_k(t)\Delta_k^*(t) - \kappa_k^*(t)\Delta_k(t), \tag{23b}$$

$$i\frac{d}{dt}\kappa_k(t) = (\eta_k(t) + \eta_{\bar{k}}(t))\,\kappa_k(t) + \Delta_k(t)\,(2\rho_k(t) - 1)\,. \tag{23c}$$

These basic equations determine the time evolution of the canonical states, $|\phi_k(t)\rangle$ and $|\phi_{\bar{k}}(t)\rangle$, their occupation, $\rho_k(t)$, and pair probabilities, $\kappa_k(t)$. The real functions of time, $\eta_k(t)$ and $\eta_{\bar{k}}(t)$, are arbitrary and associated with the gauge degrees of freedom. The time-dependent pairing gaps, $\Delta_k(t)$, are defined in the same manner as the BCS pairing gap [3] except for the fact that the canonical pair of states are no longer related to each other by time reversal. It should be noted that the quantities in the Cb-TDHFB equations, (ρ, κ, Δ), are not matrixes but are only their diagonal elements with a single index for the canonical states k.

The Cb-TDHFB equations, (23), are invariant with respect to the gauge transformation with arbitrary real functions, $\theta_k(t)$ and $\theta_{\bar{k}}(t)$.

$$|\phi_k\rangle \to e^{i\theta_k(t)}|\phi_k\rangle \quad \text{and} \quad |\phi_{\bar{k}}\rangle \to e^{i\theta_{\bar{k}}(t)}|\phi_{\bar{k}}\rangle \tag{24}$$

$$\kappa_k \to e^{-i(\theta_k(t)+\theta_{\bar{k}}(t))}\kappa_k \quad \text{and} \quad \Delta_k \to e^{-i(\theta_k(t)+\theta_{\bar{k}}(t))}\Delta_k \tag{25}$$

simultaneously with

$$\eta_k(t) \to \eta_k(t) + \frac{d\theta_k}{dt} \quad \text{and} \quad \eta_{\bar{k}}(t) \to \eta_{\bar{k}}(t) + \frac{d\theta_{\bar{k}}}{dt}.$$

It is now clear that the arbitrary real functions, $\eta_k(t)$ and $\eta_{\bar{k}}(t)$, control time evolution of the phases of $|\phi_k(t)\rangle$, $|\phi_{\bar{k}}(t)\rangle$, $\kappa_k(t)$, and $\Delta_k(t)$.

In addition to the gauge invariance, the Cb-TDHFB equations possess the following properties.

(i) Conservation law
  (a) Conservation of orthonormal property of the canonical states
  (b) Conservation of average particle number
  (c) Conservation of average total energy
(ii) The stationary solution corresponds to the HF+BCS solution.
(iii) Small-amplitude limit
  (a) The Nambu-Goldstone modes are zero-energy normal-mode solutions.
  (b) If the ground state is in the normal phase, the equations are identical to the particle-hole, particle-particle, and hole-hole RPA with the BCS approximation.

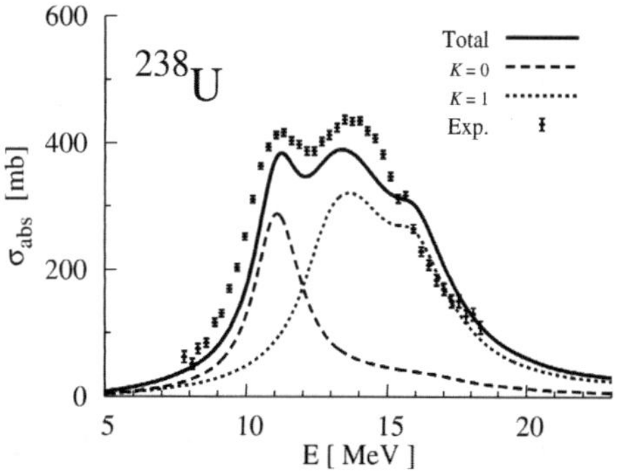

Figure 8. Calculated (solid line) and experimental (symbols) photoabsorption [33] cross sections for ^{238}U. The dashed and dotted lines indicate the contribution from the dipole oscillation parallel and perpendicular to the symmetry axis, respectively.

For numerical calculations, we extended the computational program of the TDHF in the three-dimensional (3D) coordinate-space representation [34] to include the BCS-type pairing correlations. The ground state is first constructed by the HF+BCS calculation. Then, we add a weak impulse electric dipole field $V_{E1}(t)$ to the ground state. $V_{E1}(t)$ is chosen as

$$V_{E1}(t) = -\eta F_i^{E1}\delta(t), \quad F_i^{E1} = \sum_{i:\text{protons}} (Ne/A)r_i - \sum_{i:\text{neutrons}} (Ze/A)r_i, \tag{26}$$

where $i = (x, y, z)$. We solve the Cb-TDHFB equations in real time and real space. To obtain the $E1$ strength function, we calculate the time evolution of the expectation value $\langle\Psi(t)|F^{E1}|\Psi(t)\rangle$

under the external field $V_{E1}(t)$ with small η. Then, the strength function $S(E1;E)$ can be obtained by the Fourier transform of $\langle\Psi(t)|F^{E1}|\Psi(t)\rangle$ with an artificial damping factor with $\Gamma = 1$ MeV: $\langle\Psi(t)|F^{E1}|\Psi(t)\rangle \rightarrow \langle\Psi(t)|F^{E1}|\Psi(t)\rangle e^{-\Gamma t/2}$.

The calculated $E1$ strength distribution is transformed into the photoabsorption cross section and shown in Fig. 8, for the ^{238}U nucleus. The ground state of ^{238}U is deformed in the prolate shape with the axial symmetry. Thus, the photoabsorption peak is split into two peaks: the lower peak represents an oscillation along the symmetry axis ($K = 0$) and the higher one corresponds to that perpendicular the the symmetry axis ($K = 1$). The total cross section is the sum of these, which well agree with the experimental data. This can be another indication of the deformation of the ^{238}U nucleus.

The computational task for solving Eq. (23) is significantly smaller than that for solving Eq. (22). This is due to the fact that the number of the calculated canonical states $k, \bar{k}$ is roughly the same order as the particle number ($N \sim 10^2$), which is much smaller than the dimension of the model space ($M \sim 10^4 - 10^5$).

5. Summary

The real-time calculation is a useful tool to calculate many-body dynamics, especially when we are interested in bulk properties in a wide range of energy, and when the complicated continuum boundary condition is necessary to construct the energy eigenstates. We have demonstrated its capability and usefulness in several examples, including the fusion reaction and the response in nuclei. The methodology is quite universal and applicable to a variety of sub-fields in many-body quantum physics.

Acknowledgments

Most of the results presented in this paper has been achieved by collaboration with S. Ebata, M. Ito, K. Yabana, and K. Yoshida. This work is supported by Grant-in-Aid for Scientific Research in Japan (Nos. 21340073 and 20105003).

References

[1] Messiah A 1961 *Quantum Mechanics* (Amsterdam: North Holland)
[2] Bohr A and Mottelson B R 1969 *Nuclear Structure, Vol. I* (New York: W. A. Benjamin)
[3] Ring P and Schuck P 1980 *The nuclear many-body problems* (New York: Springer-Verlag)
[4] Carlson J 2012 Contribution to this volume
[5] Quaglioni S 2012 Contribution to this volume
[6] Hagen G 2012 Contribution to this volume
[7] Yabana K 1997 *Prog. Theor. Phys.* **97** 437–450
[8] Yabana K, Ueda M and Nakatsukasa T 2003 *Nucl. Phys. A* **722** 261c–266c
[9] Nakatsukasa T, Yabana K, Ito M, Kobayashi M and Ueda M 2004 *Prog. Theor. Phys. Suppl.* **154** 85–91
[10] Ito M, Yabana K, Nakatsukasa T and Ueda M 2006 *Phys. Lett. B* **637** 53–57
[11] Ito M, Yabana K, Nakatsukasa T and Ueda U 2007 *Nucl. Phys.* **A787** 267c–274c
[12] Takigawa N and Sagawa H 1991 *Phys. Lett.* **B265** 23
[13] Hussein M S, Pato M P, Canto L F and Donangelo R 1992 *Phys. Rev. C* **46** 377–379
[14] Takigawa N, Kuratani M and Sagawa H 1993 *Phys. Rev. C* **47** R2470
[15] Hagino K, Vitturi A, Dasso C H and Lenzi S M 2000 *Phys. Rev. C* **61** 037602
[16] Diaz-Torres A and Thompson I 2002 *Phys. Rev. C* **65** 024606
[17] Pack T T 1974 *J. Chem. Phys.* **60** 633
[18] Imanishi B and von Oertzen W 1987 *Phys. Rep.* **155** 29
[19] Signorini C *et al.* 2004 *Nucl. Phys. A* **735** 329 – 344
[20] Nakamura T *et al.* 1994 *Phys. Lett. B* **331** 296–301
[21] Palit R *et al.* 2003 *Phys. Rev. C* **68** 034318
[22] Fukuda N *et al.* 2004 *Phys. Rev. C* **70**(5) 054606
[23] Weisskopf V F 1957 *Nucl. Phys.* **3** 423–432
[24] Bender M, Heenen P H and Reinhard P G 2003 *Rev. Mod. Phys.* **75** 121–180

[25] Lunney D, Pearson J M and Thibault C 2003 *Rev. Mod. Phys.* **75** 1021–1082
[26] Yoshida K and Nakatsukasa T 2011 *Phys. Rev. C* **83** 021304
[27] Nakatsukasa T, Inakura T and Yabana K 2007 *Phys. Rev. C* **76** 024318
[28] Inakura T, Nakatsukasa T and Yabana K 2009 *Phys. Rev. C* **80** 044301
[29] Inakura T, Nakatsukasa T and Yabana K 2011 *Phys. Rev. C* **84**(2) 021302
[30] Avogadro P and Nakatsukasa T 2011 *Phys. Rev. C* **84**(1) 014314
[31] Stoitsov M, Kortelainen M, Nakatsukasa T, Losa C and Nazarewicz W 2011 *Phys. Rev. C* **84**(4) 041305
[32] Ebata S, Nakatsukasa T, Inakura T, Yoshida K, Hashimoto Y and Yabana K 2010 *Phys. Rev. C* **82** 034306
[33] Berman B L and Fultz S C 1975 *Rev. Mod. Phys.* **47** 713–761
[34] Nakatsukasa T and Yabana K 2005 *Phys. Rev. C* **71** 024301 (pages 14)

UNEDF: *Advanced Scientific Computing Collaboration Transforms the Low-Energy Nuclear Many-Body Problem*

H Nam[1], M Stoitsov[1,2], W Nazarewicz[1−3], A Bulgac[4], G Hagen[1] M Kortelainen[1,2], P Maris[7], J C Pei[1,2], K J Roche[5], N Schunck[6], I Thompson[6], J P Vary[7], and S M Wild[8]

[1] Oak Ridge National Laboratory, P.O. Box 2008, Oak Ridge, TN 37831, USA
[2] Department of Physics and Astronomy, University of Tennessee Knoxville, TN 37996, USA
[3] Institute of Theoretical Physics, Warsaw University, ul. Hoża 69, PL-00681, Warsaw, Poland
[4] Department of Physics, University of Washington, Seattle, WA 98195-1560, USA
[5] Pacific Northwest National Laboratory, Richland, WA 99352, USA
[6] Lawrence Livermore National Laboratory, L-414, P.O. Box 808, Livermore, CA 94551, USA
[7] Department of Physics and Astronomy, Iowa State University, Ames, IA 50011-3160, USA
[8] Math & Computer Science Division, Argonne National Laboratory, Argonne, IL 60439, USA

E-mail: namha@ornl.gov

Abstract. The demands of cutting-edge science are driving the need for larger and faster computing resources. With the rapidly growing scale of computing systems and the prospect of technologically disruptive architectures to meet these needs, scientists face the challenge of effectively using complex computational resources to advance scientific discovery. Multi-disciplinary collaborating networks of researchers with diverse scientific backgrounds are needed to address these complex challenges. The UNEDF SciDAC collaboration of nuclear theorists, applied mathematicians, and computer scientists is developing a comprehensive description of nuclei and their reactions that delivers maximum predictive power with quantified uncertainties. This paper describes UNEDF and identifies attributes that classify it as a successful computational collaboration. We illustrate significant milestones accomplished by UNEDF through integrative solutions using the most reliable theoretical approaches, most advanced algorithms, and leadership-class computational resources.

1. Introduction

One of the discovery frontiers in physics is to explain the nature of atomic nuclei. Apart from a plethora of basic science interests, this is also an essential component of energy, medical, and biological research, and national security. Nuclear physicists are working toward a fundamental and unified description of nuclei based on the underlying theory of the strong interactions, quantum chromodynamics, and to transform descriptive and highly phenomenological models into predictive capability; in particular, allowing reliable extrapolations into regions that are not accessible by experiments. Ultimately, this would allow for accurate predictions of nuclear reactions, with significant impact on the development of advanced fission reactors and fusion energy sources, and in industrial and medical innovations through the use of stable isotopes and radioisotopes.

The UNEDF collaboration of nuclear theorists, applied mathematicians, and computer scientists (see Fig. 1) is making significant strides toward realizing this goal through a

Journal of Physics: Conference Series **402** (2012) 012033 doi:10.1088/1742-6596/402/1/012033

comprehensive study of all nuclei built on the latest advances in nuclear theory and scientific computing. UNEDF, which stands for "Universal Nuclear Energy Density Functional," is a five-year SciDAC ("Scientific Discovery through Advanced Computing") project [1, 2]. The SciDAC program (www.scidac.gov) has provided the opportunity for applied mathematicians and computer scientists to work collaboratively with physicists to develop and interconnect the most accurate knowledge of the strong nuclear interaction, high-precision theoretical approaches, scalable algorithms, and high-performance computing tools and libraries to enable scientific discoveries using leadership-class computing resources. Working toward a predictive theory, the UNEDF project emphasizes the verification of methods and codes, the estimation of uncertainties, and the assessment of results. An added and unexpected benefit of the UNEDF project has been the realization of new physics collaborations, identified through shared computational methods and needs. Here we present an overview of UNEDF, some significant milestones achieved through the UNEDF collaborative effort and the outlook for the future; more details and references can be found at the UNEDF website http://www.unedf.org.

Figure 1. UNEDF involves over 50 researchers from 9 universities and 7 national laboratories. Annually, it provides training to about 30 young researchers (postdocs and students).

2. UNEDF Overview

There are approximately 3,000 known nuclei, most of them produced in the laboratory, and an estimated 6,000 nuclei that could in principle still be created in nuclear laboratories and in the Cosmos. Understanding the properties of these nuclei is crucial for a complete nuclear theory, for element formation, for properties of stars, and for present and future energy and defense applications. Figure 2 shows the nuclear landscape as a function of neutron and proton number. Overlaying the nuclear landscape are the regions applicable for the major theoretical approaches and computational techniques utilized in the UNEDF collaboration: ab initio, configuration interaction, and density functional theory. Methods are applicable to a particular mass of nuclei and constrained by the computational resources available. Furthermore, by investigating the intersections and overlaps of these regions, UNEDF members gain valuable input for establishing a robust theory with high-quality predictive power.

2.1. Strategy Diagram

The UNEDF collaboration involves a synthesis of different perspectives to face the challenge of understanding the low-energy nuclear many-body problem within the landscape of high-performance computing. Accomplishing the scientific goals requires development of integrative solutions that extend beyond the capabilities of a single domain while maintaining a clear, unified strategy.

The UNEDF strategy diagram shown in Fig. 3 displays the cohesive view of the collaboration's strategy [2]. This diagram shows the major scientific focus areas and provides necessary granularity to the overlapping computational methods seen in Fig. 2. Through the strategy

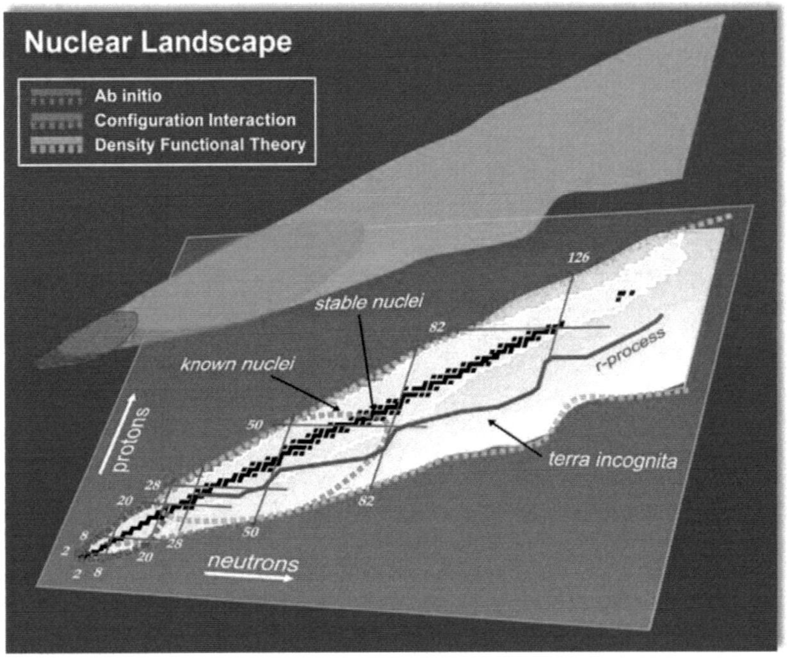

Figure 2. Theoretical methods and computational techniques to solve the nuclear many-body problem across the nuclear landscape. The thick dotted lines indicate domains of major theoretical approaches to the nuclear many-body problem. For the lightest nuclei, ab initio calculations based on the bare nucleon-nucleon and three-nucleon interactions, are possible (red). Medium-mass nuclei can be treated by configuration interaction (CI) techniques (interacting shell model (green)). For heavy nuclei, the density functional theory, based on self-consistent/mean-field theory (blue), is the tool of choice. The red vertical and horizontal lines show the magic numbers, reflecting regions where nuclei are expected to be more tightly bound and have longer half-lives. The anticipated path of the astrophysical r-process responsible for nucleosynthesis of heavy elements is also shown (purple line). Adapted from Ref. [1].

diagram, UNEDF conveys the interdependence of the various focus areas of the collaboration and identifies the challenges where multidomain expertise is necessary. It also provides a meaningful division of labor as well as a global perspective on the impact from individual efforts. Furthermore, the diagram in Fig. 3 helps to identify and foster unexpected cross-cutting physics research and shared computational challenges. Clearly outlining the efforts serves to maintain focus and heighten the collaborative spirit within the group.

2.2. Measures of Success

The success of the SciDAC UNEDF collaboration can be quantitatively measured by the scientific impact of the research performed. Figure 4 shows the number of publications resulting from UNEDF research over five years. Important to note are the high number of *Physical Review Letters* as the collaboration has matured; a *Science* highlight [5]

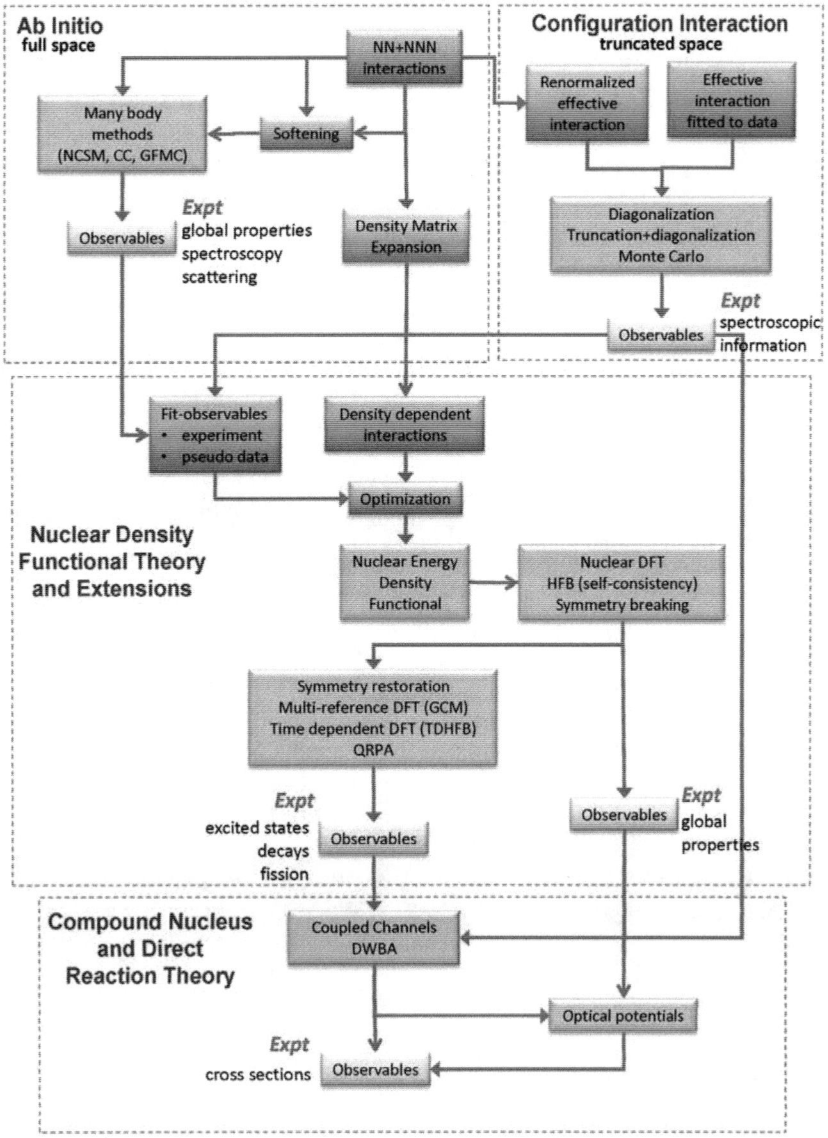

Figure 3. UNEDF strategy diagram identifying the interconnections between the four primary focus areas: ab initio, configuration interaction, nuclear density functional theory and extensions, and compound nuclear and direct reaction theory. Taken from http://unedf.org.

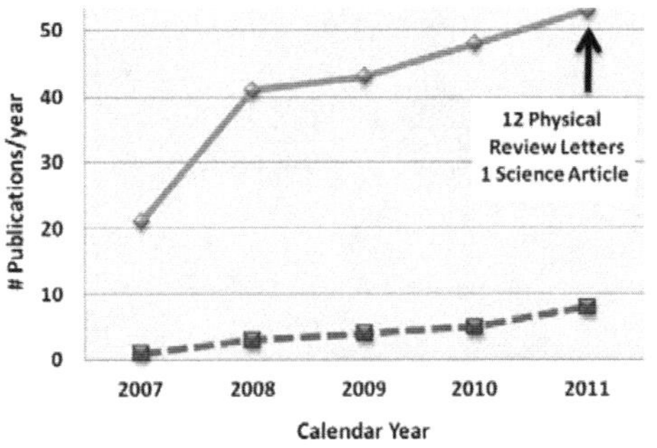

Figure 4. Publications produced yearly through the UNEDF collaboration. The solid line shows all publications and the dashed-line shows cross-domain authorship on publication. *Number of publications for 2011 is incomplete.*

in 2011; and cross-domain authorship in nuclear physics, physics, and computing publications.

The UNEDF effort has also placed great importance on recruiting and retaining the next generation leaders in low-energy nuclear physics, applied mathematics, computer science, and high-performance computing. Annually, it has provided training to approximately 30 young researchers, including postdocs and graduate students. Through experience in the UNEDF collaboration, researchers have received job placement in national laboratories and faculty positions at universities, as well as received various funding and awards. A list of publications, one-page highlights, and additional information on awards and appointments can be found at the UNEDF website.

Qualitatively, UNEDF success can be seen in the increased collaboration across domains. These connections help lay the foundation for future research and address the challenges posed by emerging architectures.

3. High-Performance Computing Resources

Access to leadership-class computing resources and large compute time allocations are critical to the scientific investigations of many UNEDF members. Through the competitive INCITE ("Innovative and Novel Computational Impact on Theory and Experiment") program, UNEDF members have been awarded large allocations on leadership-class computing resources at the Oak Ridge Leadership Computing Facility (OLCF) and the Argonne Leadership Computing Facility (ALCF). Computing resources include the following:

Intrepid (ALCF), an IBM Blue Gene/P system with 40 racks containing 1024 nodes per rack and 850 MHz quad-core processors and 2 GB RAM per node. Intrepid currently provides users with 163,840 cores, roughly 82 TB of memory, 7.6 PB of disk space, and 88 GB/s of disk bandwidth.

Jaguar (OLCF), a Cray XT with two partitions. The XT4 partition contains 7,832 compute nodes with quadcore AMD Opteron 1354 (Budapest) processors and 8 GB RAM per node, totaling 31,328 processing cores. The XT5 partition contains 18,688 compute nodes with dual hex-core AMD Opteron 2435 (Istanbul) processors and 16 GB RAM per node, totaling 224,256 processing cores. Jaguar currently provides users with a peak performance of 2.332 PF, 299 TB of system memory, 10 PB of disk space, and 240 GB/s of disk bandwidth. *Note: The XT4 partition was the primary resource in 2008 and was retired in 2011. The XT5 partition was available in 2009 with dual quad-core AMD Opteron 2356 (Barcelona) processors and was upgraded to hex-core in 2010.*

Through the UNEDF collaboration, members have been able to continuously scale codes to efficiently utilize these ever-increasing resources. Figures 5 and 6 show the INCITE allocations awarded to UNEDF collaborators and the CPU-hour utilization starting from 2008. These figures highlight the increasing demand for computing time in low-energy nuclear physics research. The combined utilization across Jaguar and Intrepid in 2008 was nearly 20 million CPU-hours and has increased more than threefold in 2011. For the 2012 calendar year, UNEDF members were granted the sixth largest allocation of the 60 INCITE projects awarded. These statistics show that low-energy nuclear physics research is dependent on high-performance computing and equally that low-energy nuclear physics is a scientific driver in HPC.

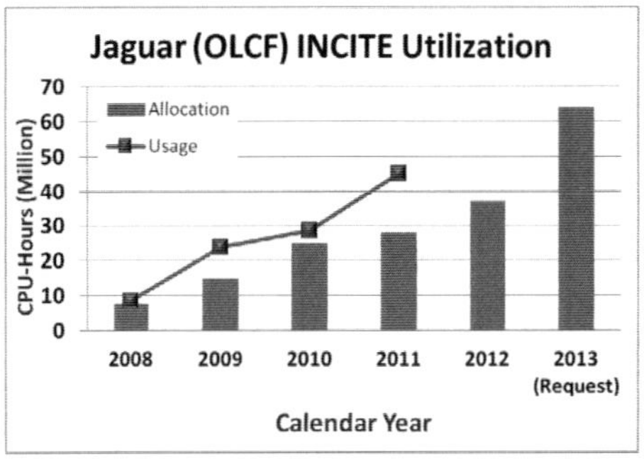

Figure 5. INCITE allocation and utilization of the Jaguar supercomputer at the OLCF (CY 2008-2013)

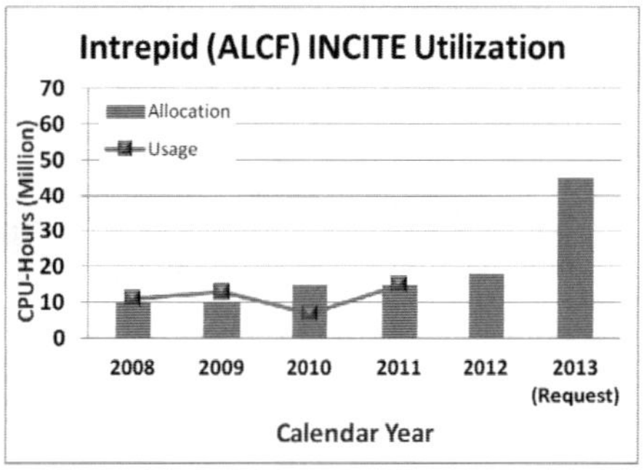

Figure 6. INCITE allocation and utilization of the Intrepid supercomputer at the ALCF (CY 2008-2013)

Leadership-class utilization is an important metric for assessing the need for and usability of capability computing resources. Capability computing is defined as "using the maximum computing power to solve a large problem in the shortest amount of time," thus solving "a problem of a size or complexity that no other computer can" [4]. This is in contrast with capacity computing, which uses "efficient cost-effective computing power to solve somewhat large problems or many small problems." At the OLCF, utilization of Jaguar is binned into three job-size categories: usage of less than 20%, between 20 and 60%, and greater than 60% of the computing resource for a single job. The typical scale of a code appropriate for a leadership-class system is utilization of greater than 20% of the resource for a single calculation. Job sizes less than 20% of the resource can typically fit onto smaller capacity systems.

Table 1. Utilization of Jaguar Supercomputer

Year	Allocation	Usage (CPU-Hours)	Leadership
2011	28,000,000	44,945,624	66%
2010	25,000,000	28,465,982	61%
2009	15,000,000	23,859,172	78%
2008	7,500,000	8,432,335	65%

Table 1 and Fig. 7 show utilization by UNEDF projects of the Jaguar supercomputer binned by job size. Table 1 shows that UNEDF projects consistently use over 60% of their allocation for leadership-size jobs. It is important to note that in 2008, when Jaguar was an XT4, leadership-class jobs used more than 6,266 cores, in 2009 for the Jaguar XT5 quad-core jobs used more than 29,901 cores, and since 2010 for the Jaguar XT5 hex-core jobs used more than 44,852 cores. Figure 7 provides additional granularity to show that UNEDF projects require 60% of the resource for a single computational run for nearly 25% of their usage. This shows the success of UNEDF collaborations to continually meet the changing architecture and growing size of computing systems.

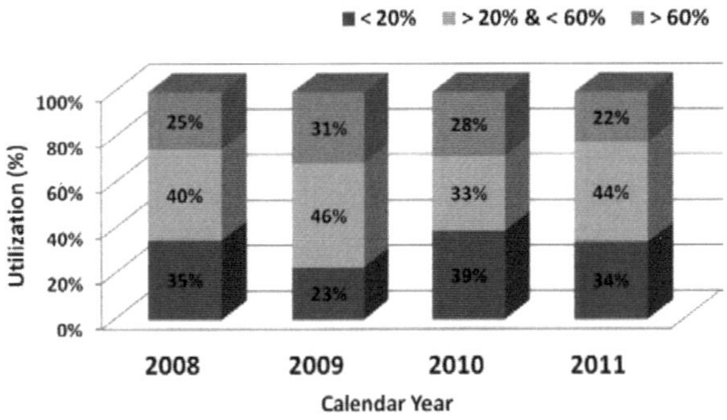

Figure 7. Utilization of Jaguar.

Additional computing time was provided in 2009 through the OLCF Early Science period prior to transitioning the general user population onto the Jaguar XT5 quad-core. At that time, the XT5 partition had 18,688 compute nodes with dual quadcore AMD Opteron 2356 (Barcelona) processors, totaling 149,504 processing cores. The XT5 partition became available to the larger user community in July 2009; for the first half of 2009, during its transition-to-operations period, it was open only to select Early Science users. UNEDF members were awarded 30 million CPU-hours for an Early Science project on the XT5. Figure 8 shows the utilization by job size of over 350 million CPU-hours over six months by 26 projects. The low-energy nuclear physics project labeled NPH009 shows that over 95% of its utilization was at leadership class [3].

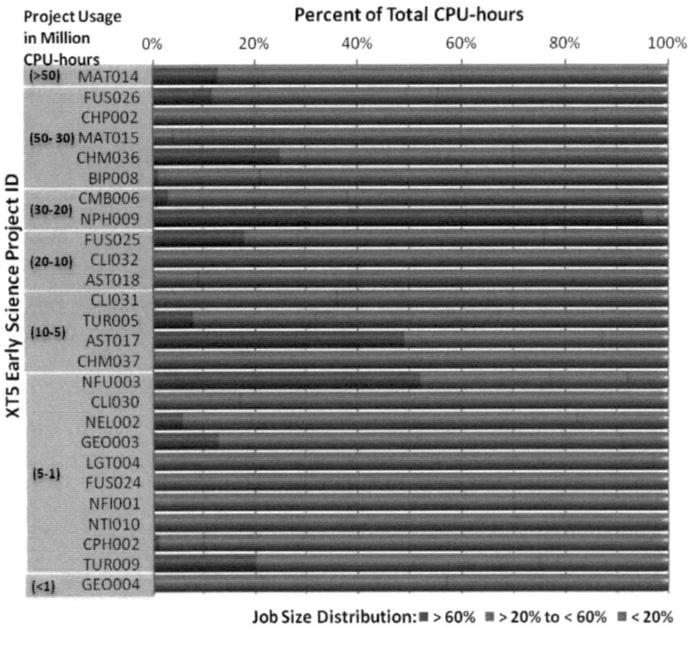

Figure 8. Utilization of Jaguar XT5 by job size for Early Science projects [3].

Close collaboration between nuclear physicists, applied mathematicians and computer scientists enable UNEDF research to effectively utilize high-performance computing resources, leading to the science highlights presented here.

Journal of Physics: Conference Series **402** (2012) 012033 doi:10.1088/1742-6596/402/1/012033

4. High-Performance Computing Enhances *Ab Initio* Nuclear Structure Calculations

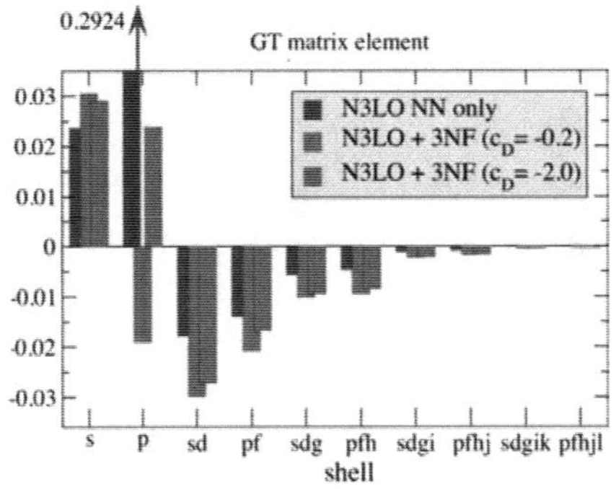

Figure 9. MFDn simulation shows three-body forces are necessary to explain the anomalously long half-life of isotope carbon-14 used in carbon dating [9].

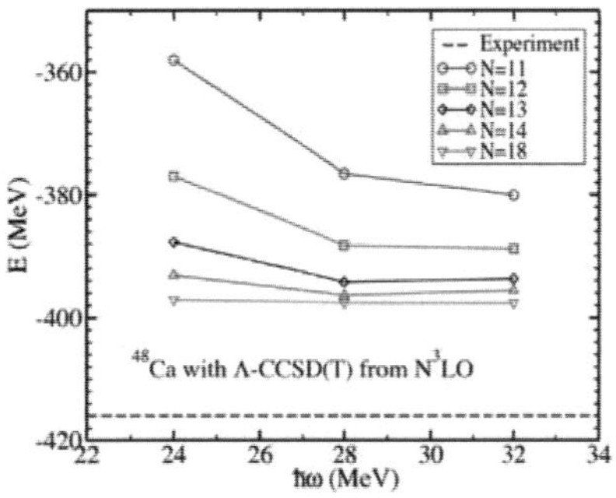

Figure 10. NUCCOR calculations for medium-mass nuclei calcium-48 shows three-nucleon forces account for the missing binding energy [8].

Ab initio, or from first principles, nuclear structure calculations have made major advances under UNEDF toward effectively utilizing high-performance computing resources and transforming to meet the challenges posed by emerging architectures. Ab initio techniques provide a fine-grained method for studying nuclei and the nuclear interaction, but they often come with a high computational cost. They are necessary to the UNEDF effort by providing "control data" to constrain more general functionals and test candidate energy density functionals even for systems not experimentally accessible. UNEDF collaborators continue to scale ab initio nuclear structure simulations and perform the largest and most accurate calculations currently possible on both Jaguar (ORNL) and Intrepid (ANL).

Examples of UNEDF-directed advances include development of the Asynchronous Dynamic Load Balancing (ADLB) software library by using *Green's function Monte Carlo* (GFMC) calculations as a testbed. The ADLB library has enabled GFMC to run efficiently on over 100,000 cores on Intrepid [6]. Utilizing Jaguar, UNEDF applications *Nuclear Coupled-Cluster - Oak Ridge* (NUCCOR) and *Many Fermion Dynamics-nuclear* (MFDn) have undergone considerable code and algorithm development. Improvements include implementation of a hybrid MPI and OpenMP approach for efficient memory management, memory-aware algorithms, and integration of libraries and tools to enable further scaling for higher-precision calculations.

Recent scientific breakthroughs include using NUCCOR to calculate medium-mass nuclei from the ground up starting from nucleon degrees of freedom, such as ^{48}Ca shown in Figure 10 [8]. Results show chiral nucleon-nucleon interactions perform remarkably well; the 400 keV per nucleon missing binding energy in ^{48}Ca can be attributed to chiral three-nucleon forces missing in calculations [8]. Another recent highlight explains the useful but anomalously long lifetime of ^{14}C by identifying the critical role of the three-nucleon force in its beta decay seen in Fig. 9 [9]. These calculations involve diagonalization of a Hamiltonian matrix of dimension 2 billion, using 214,668 cores on the Jaguar supercomputer at ORNL under the Early Science projects allocation of 30 million CPU-hours.

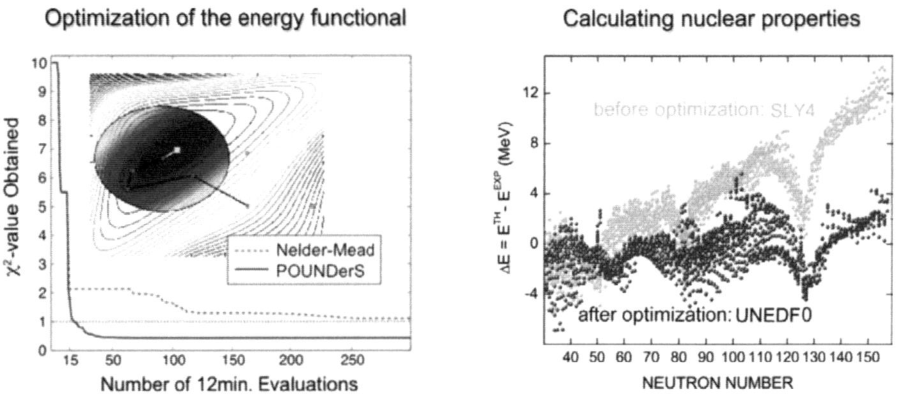

Figure 11. The optimization algorithm POUNDerS yields dramatic computational savings over alternative optimization methods (left). The resulting parameterization UNEDF0 obtained by using the POUNDerS algorithm provides a baseline of nuclear ground-state properties to compare with future functionals (right) [10].

5. Advanced Algorithms and Tools Define New Generation Energy Density Functionals

The UNEDF project has devoted considerable effort to develop and improve the algorithmic and computational infrastructure needed to optimize candidate energy density functionals (EDF). These developments have resulted in new optimization tools that are broadly available to other science domains. An example is the optimization algorithm POUNDerS, which not only provides a computational savings over other methods, as shown in Fig. 11, but greatly improves the time to solution to test candidate EDFs. With the derivative-free POUNDerS algorithm, the resulting parameterization of existing data yielded UNEDF0, which sets a solid baseline of nuclear ground-state properties to compare with future functionals shown in Figure 11 [10]. Continuing work

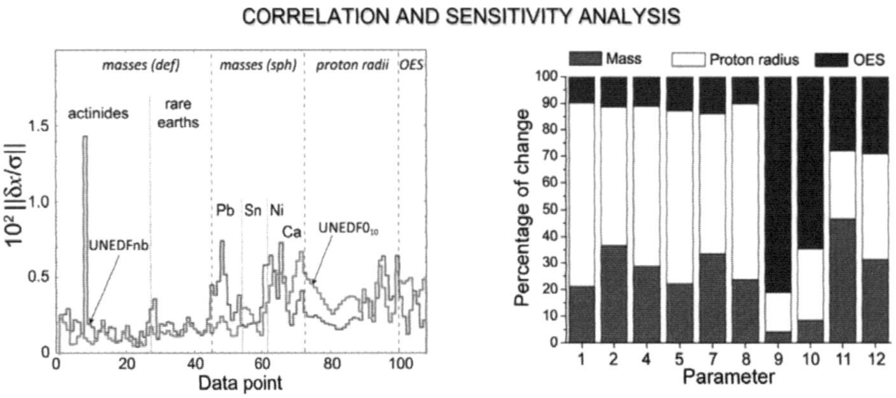

Figure 12. Statistical tools are used to deliver uncertainty quantification and error analysis for theoretical studies as well as to assess new experimental data [10].

involves utilizing this approach to study new hybrid functionals with microscopic input from chiral effective field theory [11].

These new tools enable for the first time, a consistent method for uncertainty quantification and correlation analysis to estimate errors and significance as a first step toward a formal process for future verification and validation. Included in the UNEDF project are the development and application of statistical tools, particularly important for directing future experiments by providing analysis of the significance of new experimental data. For example, the sensitivity of two optimized functionals to particular data is shown in Fig. 12 [10]. Such capabilities have not been previously available in the low-energy nuclear theory community but are increasingly important as new theories and computational tools are applied to new nuclear systems and to conditions inaccessible to experiment.

6. Massively Parallel Algorithms Open Cold Atoms as a Testing Ground

UNEDF theorists have made important contributions to the study of strongly coupled superfluid systems such as ultracold Fermi atoms, which show many similarities to the cold nuclear matter found in the crust of neutron stars. Cold atoms make excellent laboratories for testing and improving the computational methods to be used for nuclei. Cold-atom systems also allow for predictions of superfluid DFT that are testable against experiment.

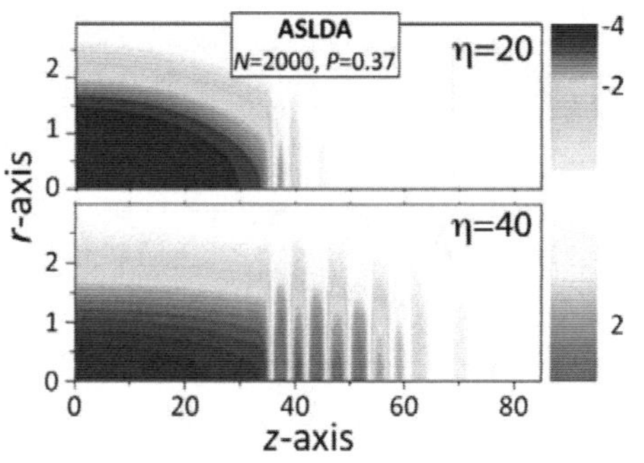

Figure 13. ASLDA simulation with strongly interacting spin-imbalanced atomic gases in extremely elongated traps [7].

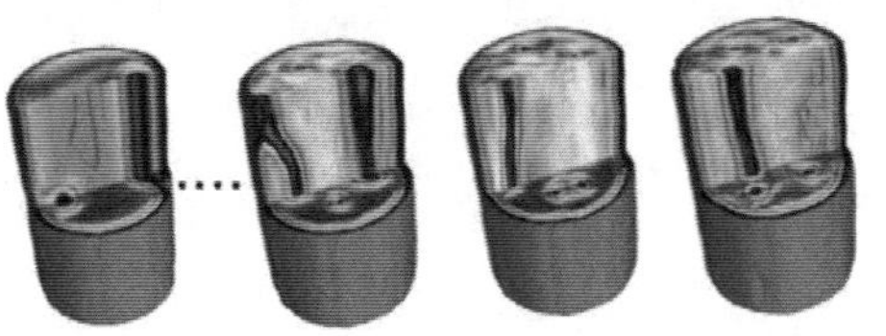

Figure 14. TDSLDA simulation shows the ball-and-rod excitation of a unitary fermi gas with vortex formation [5].

UNEDF developments using cold atoms as a testing ground include adding new algorithms to existing applications, such as adapting the antisymmetric superfluid local density approximation (ASLDA) to an existing massively parallel nuclear DFT code with strongly interacting spin-imbalanced atomic gases in extremely elongated traps, seen in Fig. 13 [7]. Another major UNEDF development is implementation of the time-dependent superfluid local density approximation (TDSLDA) on a 3D spatial lattice [5]. Unlike previous methods, the UNEDF implementation eliminates the need for matrix operations, allowing it to accommodate a basis set that is 2-3 orders of magnitude larger than other approaches. Calculations [5] were performed by using 97% of Jaguar to simulate the unitary gas (e.g., vortex formation) and a heavy nucleus under the action of various external fields, seen in Fig. 14. While still exploratory, these first-time simulations of this kind for fermion superfluids serve as proof of principle for an eventual treatment of neutron-induced fission.

7. HPC Empowers New Era for Nuclear Reaction Theory

One of the principal aims of the UNEDF project is to calculate nucleon-nucleus reactions crucial for

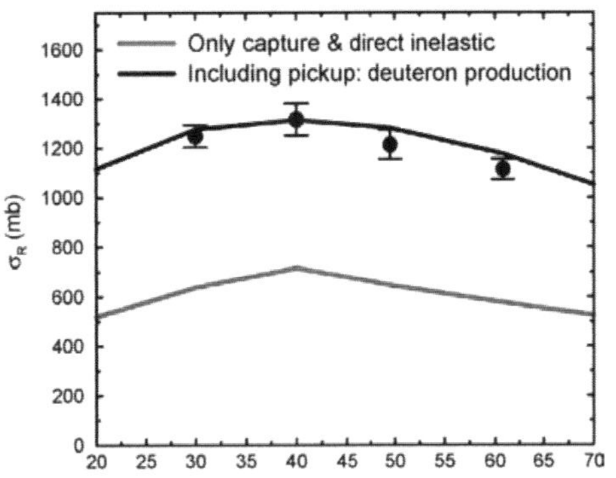

Figure 15. New methods for calculating reaction cross-sections (black) show good agreement with experimental data. Total reaction cross section as a function of the incident energy for the reaction p + 90Zr using the Gogny D1S force. The results are shown for couplings to the inelastic RPA states (red line), and to the inelastic and transfer channels with nonorthogonality corrections (black line) [12].

astrophysics, nuclear energy, radiobiology, and national security, for which extensions of standard phenomenology is insufficient. Under the UNEDF effort, neutron reactions on heavier nuclei are being modeled by using DFT results to predict not just bound states but also scattering states for nucleons. As shown in Fig. 15, the calculated reaction cross-sections agree well with experimental data. For the first time, a complete microscopic calculation using basic interactions between nucleons can be used to predict reaction observables with low-incident energy [12]. This technology provides the basis for future calculations of unstable species outside the range of experiment.

Another important capability for reactions is the calculation of level densities, which provides insight to the interactions inside the system. A new proton-neutron algorithm for the parallel JMoments code was recently designed and implemented, which scales to tens of thousands of cores and greatly increases the code's overall performance. This development opens the door to calculating accurate, nuclear level densities and reaction rates for a large class of nuclei [13].

8. Outlook

The UNEDF collaboration has provided fertile ground for new and continuing growth between applied mathematics, computer science, and nuclear physics. Over the past five years, the collaboration has established cross-disciplinary working relationships to facilitate future efforts and has matured to adequately address new challenges in verification and validation, workflow, visualization, and new programming models with changing architectures. Reaching next-generation science objectives requires computational resources several orders of magnitude beyond what is currently available. Adapting to these changes will take conscious planning and purposeful action. The members of the collaboration are well positioned to meet these disruptive changes through the close working relationship established through UNEDF. UNEDF, and similar future collaborations, will continue to develop key computational codes and algorithms for reaching the goal of solving the nuclear quantum many-body problem, thus paving the road to the comprehensive model of the atomic nucleus.

9. Acknowledgments

The UNEDF SciDAC collaboration is supported by the U.S. Department of Energy (DOE) under grant No. DOE-FC02-09ER41583. This work was also supported by DOE Contract Nos. DE-FG02-96ER40963 (University of Tennessee), DE-AC05-00OR22725 with UT-Battelle, LLC (Oak Ridge National Laboratory), DE-FG05-87ER40361 (Joint Institute for Heavy Ion Research), DE-AC02-06CH11357 (Argonne National Laboratory), DEFC02-09ER41582 (Iowa State University), and DE-FG02-87ER40371(Iowa State University). Computational resources were provided through an INCITE award "Computational Nuclear Structure" by the National

Center for Computational Sciences (NCCS) and National Institute for Computational Sciences (NICS) at Oak Ridge National Laboratory.

References

[1] Bertsch G F, Dean D J and Nazarewicz W 2007 *SciDAC Review* **6** 42
[2] Furnstahl R J (for the UNEDF Council) 2011 Nuclear Physics News **21** 18
[3] Joubert W, Kothe D and Nam H 2009 *Preparing for Exascale: ORNL Leadership Computing Facility Application Requirements and Strategy* (Oak Ridge: ORNL)
[4] Graham S L, Snir M, and Patterson C A 2005 *Getting Up to Speed: The Future of Supercomputing* (National Academies Press: Washington, D.C.)
[5] Bulgac A et al 2011 *Science* **332** 1288
[6] Lusk E et al 2010 *SciDAC Review* **17** 30
[7] Pei J C et al 2010 *Phys. Rev.* A **82** 021603(R)
[8] Hagen G et al 2010 *Phys. Rev.* C **82** 034330
[9] Maris P et al 2011 *Phys. Rev. Lett.* **106** 202502
[10] Kortelainen M et al 2010 *Phys. Rev.* C **82** 024313
[11] Stoitsov M et al 2010 *Phys. Rev.* C **82** 054307
[12] Nobre G et al 2011 *Phys. Rev.* C **84** 064609
[13] Senkov R and Horoi M 2010 *Phys. Rev.* C **82** 024304

Reflection-asymmetric nuclear deformations within the Density Functional Theory

E Olsen[1], J Erler[1,2], W Nazarewicz[1−3] and M Stoitsov[1,2]

[1] Department of Physics and Astronomy, University of Tennessee, Knoxville, Tennessee 37996, USA

[2] Physics Division, Oak Ridge National Laboratory, Post Office Box 2008, Oak Ridge, Tennessee 37831, USA

[3]Institute of Theoretical Physics, Warsaw University, ul. Hoża 69, PL-00681, Warsaw, Poland

E-mail: eolsen1@utk.edu

Abstract.
Within the nuclear density functional theory (DFT) we study the effect of reflection-asymmetric shapes on ground-state binding energies and binding energy differences. To this end, we developed the new DFT solver AXIALHFB that uses an approximate second-order gradient to solve the Hartree-Fock-Bogoliubov equations of superconducting DFT with the quasi-local Skyrme energy density functionals. Illustrative calculations are carried out for even-even isotopes of radium and thorium.

1. Introduction

The goal of low-energy nuclear theory is to provide a comprehensive theoretical framework to explain properties of atomic nuclei and their reactions at the nucleonic level. The roadmap towards this goal involves three theoretical strategies: ab initio methods, configuration-interaction techniques, and the nuclear DFT [1, 2]. Due to exploding dimensions of the configuration space, properties of complex heavy nuclei with many valence particles are best described by the nuclear DFT and its various extensions [3]. The main idea of DFT is to describe an interacting system of fermions via its densities rather than the many-body wave function. The energy of the many body system can be written as a density functional, and the ground state energy is obtained through the variational procedure. Figure 1 illustrates the strategy behind the nuclear DFT. A major thrust in this area is to enhance the predictive power of DFT by firmly connecting the nuclear energy functional (EDF) to the nuclear interactions and to optimize the low-energy coupling constants of the microscopically based EDF by comparing results of theoretical simulations with a selected set of observables [4, 5].

The nuclear DFT is based on the self-consistent mean-field approach rooted in the self-consistent Hartree-Fock (HF) or Hartree-Fock-Bogoliubov (HFB) problem. The self-consistent HFB equations allow us to compute the nuclear ground state and a set of elementary quasi-particle excitations that can be used to build excitations of the system. The HFB equations constitute a system of coupled integro-differential equations that can be written in a matrix form as an eigenvalue problem, where the dependence of the HFB matrix elements on the eigenvectors induces nonlinearities.

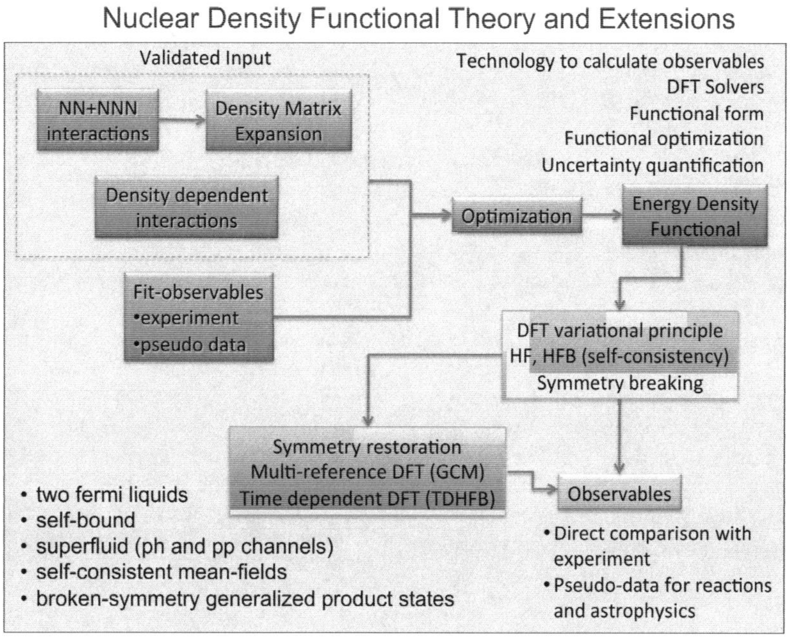

Figure 1. The nuclear DFT strategy diagram.

Since all the nucleons are treated on equal footing and contribute to single-particle and pairing mean fields, a few hundred nucleonic wave functions have to be considered which makes systematic calculations quite demanding. We have at our disposal a suite of optimized HFB solvers working in various geometries [6]. All our DFT solvers adopt an iterative approach to solving the HFB equations [7, 8, 9, 10, 11, 12]. Nuclear configurations (characterized by shape deformations, angular momentum, temperature, single-particle occupations, etc.) can be selected by means of external constraints [13]. By employing modern DFT infrastructure global nuclear properties can be rapidly computed across the nuclear landscape [14].

One of the fundamental properties of an atomic nucleus is its shape. The DFT description of nuclei is performed in the frame of reference of the nucleus, the intrinsic frame, in which the nucleus may acquire a deformed shape. The concept of shape deformation is ultimately related to the spontaneous symmetry breaking effect known in many areas of physics. Symmetry breaking solutions may appear variationally when, in a mean-field configuration respecting the original symmetries of the nuclear Hamiltonian, degenerate single-particle orbits are strongly coupled to collective vibrations [15, 16, 17, 18]. In heavy nuclei, the most important collective modes are the quadrupole and octupole vibrations, leading to intrinsic mean fields with non-zero quadrupole and octupole moments. If a nucleus has nonzero quadrupole moments, its non-spherical shape can be characterized by ellipsoidal deformations [19, 20]. The presence of octupole moments is indicative of reflection-asymmetric deformations [21]. While quadrupole deformations are common in ground states across the mass table, octupole deformations are more concentrated in particular regions of the chart of the nuclides [21, 22].

This paper introduces a new DFT solver AXIALHFB which solves the HFB problem by means of an approximate second-order gradient method [23]. Using AXIALHFB with Skyrme EDFs, we study the effect of reflection-asymmetric moments on nuclear binding energy in even-even Ra and Th nuclei. These nuclei were chosen for their well known strong octupole correlations. The binding energies obtained in the nuclear DFT are further used to compute mass filters such as two-nucleon separation energies and the double-difference indicator δV_{pn} [24, 25, 26].

This paper is structured as follows. In Sec. 2 we give a brief overview of HFB equations, introduce the new DFT solver AXIALHFB, and provide necessary details about our DFT calculations. The results for the mass filters are presented in Sec. 3. Finally, conclusions and prospects for future work are given in Sec. 4.

2. DFT Calculations

The following is a succinct description of the HFB approach. For a more in-depth discussion, the reader is referred to Refs. [3, 20]. The goal is to find the ground state energy of a nucleus (N, Z) subject to constraints. The energy functional is a three-dimensional spatial integral of local energy density $\mathcal{H}$ that is a real, scalar, time-even, and isoscalar function of local densities and their derivatives. The energy density consists of a kinetic term, the Skyrme energy functional representing the effective nuclear interaction between nucleons, and the Coulomb term. A Skyrme functional depends on a number of local densities: nucleonic densities, kinetic densities, spin densities, spin-kinetic densities, current densities, tensor-kinetic densities, and spin-current densities [27]. Since pairing correlations are considered, the EDF also includes pairing terms defined in terms of pairing densities. Variation of the binding energy leads to the HFB equations:

$$\begin{pmatrix} h-\lambda & \Delta \\ -\Delta^* & -h^*+\lambda \end{pmatrix}\begin{pmatrix} U_k \\ V_k \end{pmatrix} = E_k \begin{pmatrix} U_k \\ V_k \end{pmatrix}, \tag{1}$$

where h is the HF Hamiltonian containing the self-consistent field, Δ is the pairing field, λ is the Fermi level, E_k are one-quasiparticle energy eigenstates, and (U_k, V_k) are two-component HFB eigenvectors. To restore the particle symmetry spontaneously broken in the HFB approach we use the Lipkin-Nogami scheme [28, 29] implemented as in [30].

The present calculations were done using a parity-breaking axial DFT solver AXIALHFB recently developed by Stoitsov using the gradient method routine for solving the HFB equations (1) provided by L.M. Robledo [23]. Like in the parity-conserving axial HFB code HFBTHO [8], AXIALHFB expresses the HFB equations in a large configuration space of single-particle eigenstates of the deformed harmonic oscillator basis. Unlike HFBTHO, however, which employs the direct diagonalization technique, AXIALHFB uses an approximate second-order gradient method [23, 31]. This method is particularly well suited to deal with multiple constraints [20]; hence, it does not require special techniques such as the augmented Lagrangian method [13] implemented in HFBTHO.

As an individual constrained HFB problem can be solved on a single processor, the computation of a potential energy surface (PES) in a space of collective coordinates is embarrassingly parallel (see Fig. 2). This also makes it possible to produce deformed HFB mass tables – including reflection asymmetric shapes – on leadership-class supercomputers such as JAGUAR at Oak Ridge National Laboratory's Leadership Computing Facility.

The goal was to find the ground state (g.s.) energy, i.e. the global minimum on a 2D potential energy surface. To guarantee that a global, rather than local, minimum is found, one has to compute and scan the PES in two coordinates: the axial mass quadrupole moment Q_{20} (characterizing the elongation of the nucleus expressed in terms of deformation β_2) and the mass octupole moment Q_{30} (characterizing the mirror asymmetry, or pear-like deformation β_3). Figure 2 shows three energy surfaces for different isotopes of Radium calculated using the Skyrme EDF SLy4 [32] together with a density dependent mixed pairing functional [33]. All of the calculations here were done using a large oscillator basis of 21 oscillator shells (2024 basis states). We performed 525 constrained calculations for different combinations of Q_{20} and Q_{30} (indicated by white dots on the plots in Fig. 2). The range of Q_{30}-values was identical for

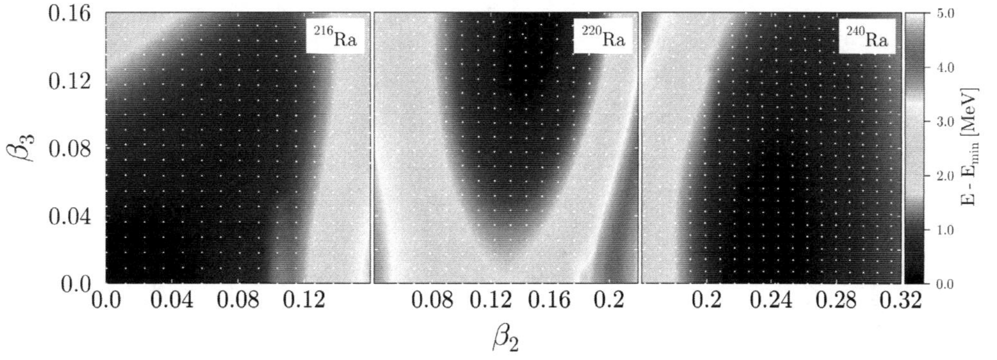

Figure 2. The contour maps of the 2D potential energy surfaces of ^{216}Ra (left), ^{220}Ra (middle), and ^{240}Ra (right) calculated by constraining both the quadrupole and octupole moments characterized by shape deformations β_2 and β_3, respectively. The mesh points at which calculations were performed are marked by white dots. The energy (in MeV) is shown relative to the ground-state minimum. The Skyrme functional SLy4 [32] and mixed pairing [33] were used.

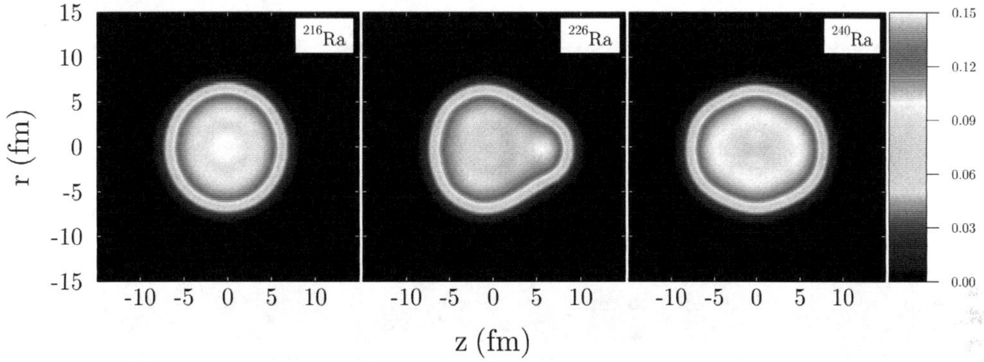

Figure 3. Nucleon density distributions (in cylindrical coordinates r and z) calculated for ^{216}Ra ($\beta_2 = 0.00$, $\beta_3 = 0.00$), ^{226}Ra ($\beta_2 = 0.20$, $\beta_3 = 0.198$), and ^{240}Ra ($\beta_2 = 0.24$, $\beta_3 = 0.0$) showing spherical, quadrupole and octupole deformed shapes.

each Q_{20}; they went from 0 to 6000 fm^3 in steps of 250 fm^3. Following the minimization on the resulting grid, unconstrained calculations have been carried out starting from the grid minimum: in this way the precise total g.s. binding energy was obtained and used to compute mass filters.

Figures 2 and 3 illustrate three situations typical to all nuclei considered [34, 35]. In general, for a given isotopic chain, a transition is expected with increasing neutron number from spherical shape to quadrupole well-deformed shapes, via reflection-asymmetric minima. The PES for ^{216}Ra is characteristic of a spherical system, which is fairly soft in the quadrupole-octupole direction. For ^{220}Ra, our calculations yield a well-developed minimum with stable quadrupole and octupole deformations. As seen in Fig. 3, in its ground state ^{220}Ra is predicted to have a pear-like shape. The nucleus ^{240}Ra is strongly deformed but reflection-symmetric. Its shape is indicative of appreciable quadrupole and hexadecapole g.s. moments.

3. Results

The objective of this work was to determine whether reflection-asymmetric g.s. moments may have an affect on binding energy differences (mass filters) of even-even Ra and Th isotopes.

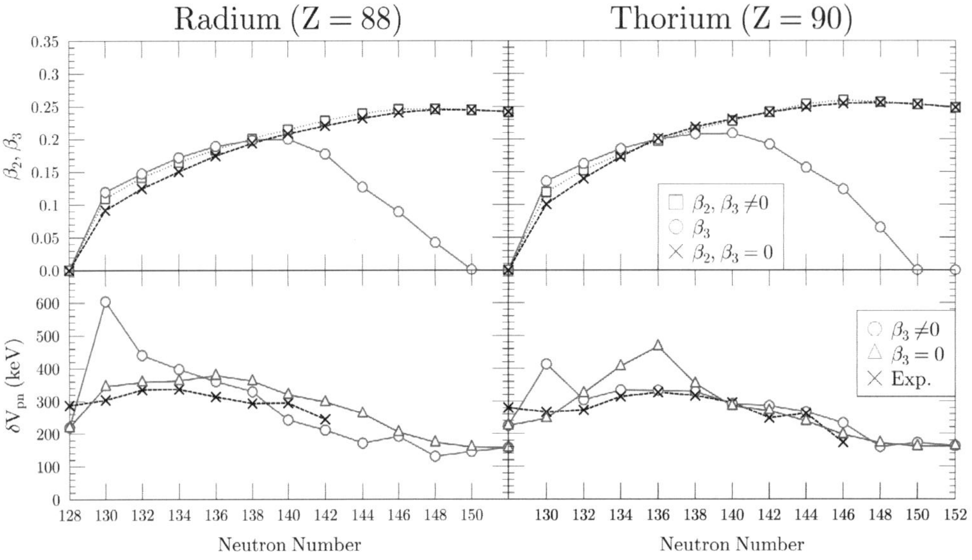

Figure 4. Results of HFB calculations using the EDF SLy4 [32] and mixed pairing [33] for the even-even isotopes of Ra and Th. Top: g.s. deformation parameters β_2 and β_3 obtained in the full minimization allowing reflection-asymmetric shapes ($\beta_3 \neq 0$) compared with those obtained by assuming mirror symmetry ($\beta_3 = 0$). Bottom: δV_{pn} for reflection-symmetric (triangles) and asymmetric (circles) shapes compared with experimental values ($\times$) of Ref. [36].

Figure 4 (top) shows the equilibrium deformations β_2 and β_3 predicted with the Skyrme functional SLy4. It is seen that the octupole deformation maximizes around N=138 and then gradually decreases with the neutron number. The effect of the octupole deformation on the quadrupole moment appears to be very small. This is evident by comparing the g.s. β_2-values obtained in the full minimization ($\beta_3 \neq 0$) with those obtained by assuming reflection symmetric shapes ($\beta_3 = 0$). The bottom part of Fig. 4 shows the filter δV_{pn}:

$$\delta V_{pn} = \frac{1}{4}\left[B(Z,N) - B(Z,N-2) - B(Z-2,N) + B(Z-2,N-2)\right], \tag{2}$$

which can be viewed as an approximation to the mixed partial derivative [26]:

$$\delta V_{pn}(Z,N) \approx \frac{\partial^2 B}{\partial Z \partial N}. \tag{3}$$

In general, the agreement with experiment for δV_{pn} is improved when octupole correlations are considered. The spike around $N = 130$ in the $\beta_3 \neq 0$ results is due to the rapid transition between spherical and deformed shapes; this effect – typical to mean-field calculations – is supposed to be washed out if beyond-mean-field effects are taken into account.

Finally, Fig. 5 shows predicted and experimental two-neutron and two-proton separation energies:

$$S_{2n} = B(Z,N) - B(Z,N-2), \quad S_{2p} = B(Z,N) - B(Z-2,N). \tag{4}$$

For these mass filters, the effect of octupole correlations is very small.

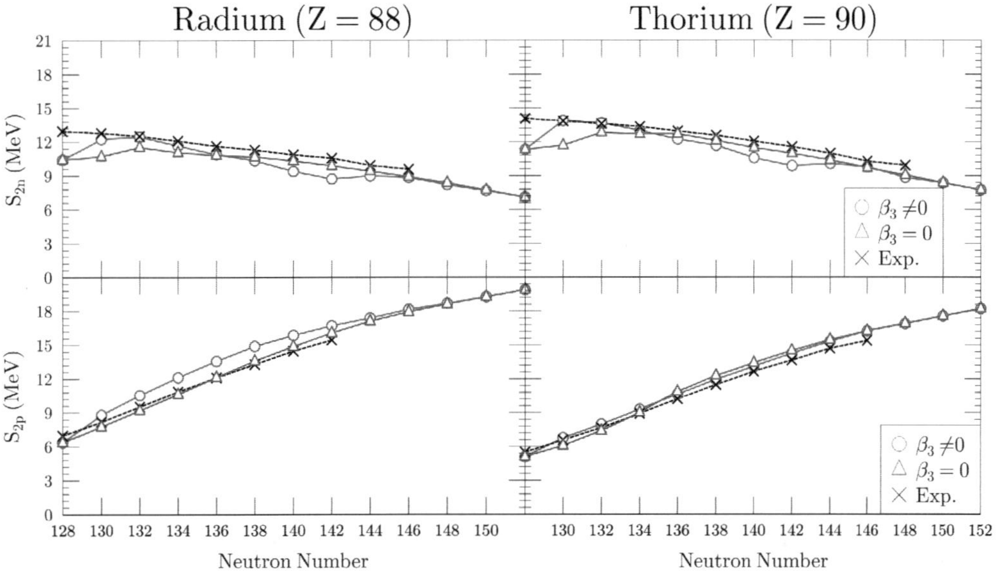

Figure 5. Similar to Fig. 4 except for two-neutron (top) and two-proton (bottom) separation energies.

4. Conclusions

The large-scale simulations utilizing leadership-class supercomputers have transformed the DFT-based description of nuclei. A close coupling of math and computer scientists with physicists, focused on a close coupling with present and future experiments and observations, has dramatically advanced the field [37].

This work presents selected results of large-scale DFT calculations including reflection-asymmetric shape degrees of freedom. We focus on the influence of parity-breaking intrinsic moments on the binding energy differences. Our calculations suggest that δV_{pn} is affected by octupole effects, bringing it closer to experiment, whereas S_{2n} and S_{2p} are not. Future work will involve performing similar calculations across the nuclear chart, also for odd-A nuclei. This will require extending AXIALHFB to the odd-particle case [38]. We also intend to study the model dependence of our predictions by considering different energy density functionals.

Acknowledgments

Discussions with Rick Casten and Burcu Cakirli are gratefully acknowledged. In particular we want to thank Luis Robledo for providing the conjugate gradient routine. This work was supported by the U.S. Department of Energy under Contract Nos. DE-FG02-96ER40963 (University of Tennessee) and DE-FC02-09ER41583 (UNEDF SciDAC Collaboration). Computational resources were provided through an INCITE award "Computational Nuclear Structure" by the National Center for Computational Sciences (NCCS) and National Institute for Computational Sciences (NICS) at Oak Ridge National Laboratory.

References

[1] Bertsch G, Dean D and Nazarewicz W 2007 *SciDAC Review* **102** 42
[2] Furnstahl R 2011 *Nucl. Phys. News* **21**(2) 18
[3] Bender M, Heenen P H and Reinhard P G 2003 *Rev. Mod. Phys.* **75** 123

[4] Kortelainen M, Lesinski T, Moré J, Nazarewicz W, Sarich J, Schunck N, Stoitsov M V and Wild S 2010 *Phys. Rev. C* **82** 024313
[5] Kortelainen M, McDonnell J, Nazarewicz W, Reinhard P G, Sarich J, Schunck N, Stoitsov M V and Wild S 2011 *Phys. Rev. C* submitted; arXiv:1111.4344
[6] Pei J C, Stoitsov M V, Fann G I, Nazarewicz W, Schunck N and Xu F R 2008 *Phys. Rev. C* **78** 064306
[7] Bennaceur K and Dobaczewski J 2005 *Comput. Phys. Commun.* **168** 96
[8] Stoitsov M, Dobaczewski J, Nazarewicz W and Ring P 2005 *Comput. Phys. Commun.* **167** 43
[9] Schunck N, Dobaczewski J, McDonnell J, Satuła W, Sheikh J, Staszczak A, Stoitsov M and Toivanen P 2012 *Comput. Phys. Commun.* **183** 166
[10] Fann G, Pei J, Harrison R, Jia J, Hill J, Ou M, Nazarewicz W, Shelton W and Schunck N 2009 *J. Phys. Conf. Ser.* **180** 012080
[11] Pei J C, Kruppa A T and Nazarewicz W 2011 *Phys. Rev. C* **84** 024311
[12] Pei J, Fann G, Harrison R, Nazarewicz W, Hill J, Galindo D and Jia J 2012 *J. Phys. Conf. Ser.*, this volume
[13] Staszczak A, Stoitsov M, Baran A and Nazarewicz W 2010 *Eur. Phys. J. A* **46** 85
[14] Erler J, Birge N, Kortelainen M, Nazarewicz W, Olsen E, Perhac A and Stoitsov M 2012 *J. Phys. Conf. Ser.*, this volume
[15] Reinhard P G and Otten E 1984 *Nucl. Phys. A* **420**
[16] Nazarewicz W 1992 *Prog. Part. Nucl. Phys.* **28** 307
[17] Nazarewicz W 1994 *Nucl. Phys. A* **574** 27c
[18] Frauendorf S 2001 *Rev. Mod. Phys.* **73** 463
[19] Bohr A and Mottelson B 1975 *Nuclear Structure, vol. II* (W A. Benjamin, Reading)
[20] Ring P and Schuck P 1980 *The Nuclear Many-Body Problem* (New York: Springer)
[21] Butler P A and Nazarewicz W 1996 *Rev. Mod. Phys.* **68** 351
[22] Ahmad I and Butler P 1993 *Ann. Rev. Nucl. Part. Sci.* **43** 71
[23] Robledo L and Bertsch G F 2011 *Phys. Rev. C* **84** 014312
[24] Zhang J Y, Casten R and Brenner D 1989 *Phys. Lett. B* **227** 1
[25] Cakirli R B, Brenner D S, Casten R F and Millman E A 2005 *Phys. Rev. Lett.* **94** 092501
[26] Stoitsov M, Cakirli R B, Casten R F, Nazarewicz W and Satula W 2007 *Phys. Rev. Lett.* **98** 132502
[27] Perlińska E, Rohoziński S G, Dobaczewski J and Nazarewicz W 2004 *Phys. Rev. C* **69** 014316
[28] Nogami Y 1964 *Phys. Rev. B* **134** 313
[29] Lipkin H J 1960 *Ann. Phys., NY* **9** 272
[30] Stoitsov M V, Dobaczewski J, Kirchner R, Nazarewicz W and Terasaki J 2007 *Phys. Rev. C* **76** 014308
[31] Egido J L, Lessing J, Martin V and Robledo L M 1995 *Nucl. Phys. A* **594** 70
[32] Chabanat E, Bonche P, Haensel P, Meyer J and Schaeffer R 1998 *Nucl. Phys. A* **635** 231, *Nucl. Phys. A* **643**, 441(E)
[33] Dobaczewski J, Nazarewicz W and Stoitsov M V 2002 *Eur. Phys. J. A* **15** 21
[34] Nazarewicz W, Olanders P, Ragnarsson I, Dudek J, Leander G, Möller P and Ruchowska E 1984 *Nucl. Phys. A* **429** 269
[35] Robledo L, Egido J, Berger J and Girod M 1987 *Phys. Lett. B* **187** 223
[36] Audi G, Wapstra A and Thibault C 2003 *Nucl. Phys. A* 337
[37] Nam H, Stoitsov M, Nazarewicz W, Bulgac A, Hagen G, Kortelainen M, Maris P, Pei J, Roche K, Schunck N, Thompson I, Vary J and Wild S 2012 *J. Phys. Conf. Ser.*, this volume
[38] Schunck N, Dobaczewski J, McDonnell J, Moré J, Nazarewicz W, Sarich J and Stoitsov M V 2010 *Phys. Rev. C* **81** 024316

Coordinate-Space Hartree-Fock-Bogoliubov Solvers for Superfluid Fermi Systems in Large Boxes

J C Pei[1,2], G I Fann[3], R J Harrison[3,4], W Nazarewicz[1,2,5], J Hill[3], D Galindo[3], J Jia[3]

[1]Department of Physics and Astronomy, University of Tennessee, Knoxville, TN 37996, USA
[2]Physics Division, Oak Ridge National Laboratory, P.O. Box 2008, Oak Ridge, TN 37831, USA
[3]Computer Science and Mathematics Division, Oak Ridge National Laboratory, Oak Ridge, TN 37830, USA
[4]Department of Chemistry, University of Tennessee, Knoxville, TN 37996 USA
[5]Institute of Theoretical Physics, Warsaw University, ul. Hoża 69, 00-681 Warsaw, Poland

E-mail: peij@ornl.gov

Abstract. The self-consistent Hartree-Fock-Bogoliubov problem in large boxes can be solved accurately in the coordinate space with the recently developed solvers HFB-AX (2D) and MADNESS-HFB (3D). This is essential for the description of superfluid Fermi systems with complicated topologies and significant spatial extend, such as fissioning nuclei, weakly-bound nuclei, nuclear matter in the neutron star rust, and ultracold Fermi atoms in elongated traps. The HFB-AX solver based on B-spline techniques uses a hybrid MPI and OpenMP programming model for parallel computation for distributed parallel computation, within a node multi-threaded LAPACK and BLAS libraries are used to further enable parallel calculations of large eigensystems. The MADNESS-HFB solver uses a novel multi-resolution analysis based adaptive pseudo-spectral techniques to enable fully parallel 3D calculations of very large systems. In this work we present benchmark results for HFB-AX and MADNESS-HFB on ultracold trapped fermions.

1. Introduction

The Hartree-Fock-Bogoliubov (HFB) equation of the Density Functional Theory (DFT) is suitable for describing superfluid Fermi systems by properly accounting for the self-consistent coupling between the particle-hole and particle-particle mean fields. Recently, we solved HFB equations in complicated geometries to describe nuclei and ultra-cold polarized Fermi gases [1]. The general HFB equation for a polarized system can be written as:

$$\begin{pmatrix} h_a(\mathbf{r}) - \lambda_a & \Delta(\mathbf{r}) \\ \Delta^*(\mathbf{r}) & -h_b(\mathbf{r}) + \lambda_b \end{pmatrix} \begin{pmatrix} u(\mathbf{r}) \\ v(\mathbf{r}) \end{pmatrix} = E \begin{pmatrix} u(\mathbf{r}) \\ v(\mathbf{r}) \end{pmatrix}, \tag{1}$$

where h_a and h_b are Hartree-Fock Hamiltonians for the two spin components, λ_a and λ_b are the corresponding chemical potentials, and Δ is the pairing potential.

In some cases, solving the HFB equation in large boxes in coordinate space is essential. Weakly-bound nuclei and the large-amplitude nuclear collective motion such as fission and fusion are examples of problems that require large-box calculations [2]. The ultra-cold atoms trapped in elongated optical traps also require a description involving large spatial dimensions [1], as well as HFB description of nucleonic pasta phases in the neutron star crusts [3]. All these problems are both interesting and important but the underlying calculations are challenging.

Many HFB solvers used in the nuclear physics context are based on the basis expansion method employing harmonic oscillator wave functions. The configuration-space method is efficient but offers a fairly poor accuracy for cases involving weakly-bound systems and large deformations, see discussion in Ref. [2]. On the other hand, solving HFB equations directly in coordinate-space can offer very precise results. Unfortunately, because of numerical challenges involved, HFB calculations in non-spherical geometries require high computation cost. In this context, multi-core processor architectures such as the Jaguar and Kraken Cray XT5 with 12 cores per node, or Hopper Cray XE6 supercomputer with 24 cores per node, promise to revolutionize the deformed HFB problem. To fully take advantage of unique computational capabilities, HFB solvers should be adapted and improved in terms of scalability.

To this end, we developed two parallel HFB codes: 2D HFB-AX for axially symmetric systems and MADNESS-HFB for fully 3D systems. Both solvers take advantage of modern multi-core architectures. The following sections, overview computational and numerical techniques employed by these two codes and present benchmark calculations for ultra-cold Fermi atoms. To illustrate the deformed HFB problem in coordinate space, we present an extreme application to novel pairing phases in polarized cold Fermi gases in extremely elongated traps [4].

2. Two-dimensional HFB solver HFB-AX

In HFB-AX [2], the wave functions are discretized on a 2-D grid (r_α, z_β) with the M-order B-splines:

$$\psi_{n\Omega^\pi}(r_\alpha, z_\beta) = \sum_{i,j} B_i^M(r_\alpha) B_j^M(z_\beta) C_{n\Omega^\pi}^{ij}, \tag{2}$$

where the coefficients C_n^{ij} can be obtained by diagonalizing the Hamiltonian matrix. To reduce

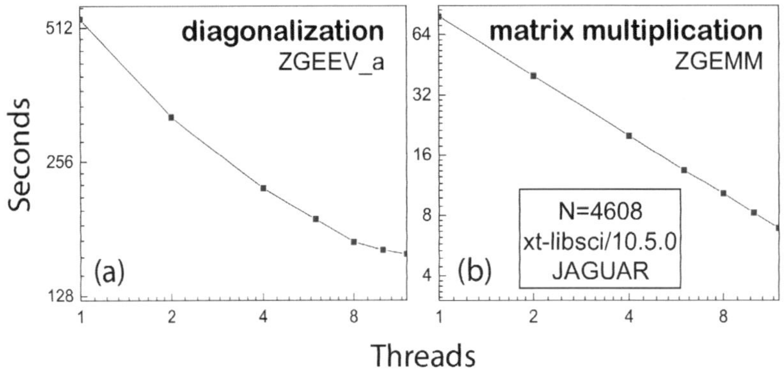

Figure 1. Left: performance of the complex eigensolver ZGEEV_a using different number of threads (ZGEEV_a is a modified version of LAPACK's ZGEEV). Right: performance of the complex matrix multiplication routine ZGEMM. The two routines are employed by HFB-AX and tested on Jaguar with the xt-libsci/10.5.0 library.

computational costs, we assume reflection symmetry; hence, the intrinsic parity π is a good quantum number that can be used to enumerate HFB eigenstates. Another quantum number preserved by HFB-AX is the angular momentum projection Ω on the symmetric axis. The boundary conditions are implemented through first- and second-order derivative operators. The precision of computations depends on the mesh size, order of B-splines, and the box size. Those parameters determine the numerical effort involved.

In HFB-AX, the diagonalization of the complex Hamiltonian matrix takes the bulk of the computing time. Diagonalization blocks of given Ω and π quantum numbers are assigned to computing cores through MPI communications. We modified the LAPACK diagonalization routine ZGEEV to calculate only the selected eigenvectors. For calculations of a heavy nucleus with the mesh size of 0.6 fm, 10-th order B-splines, and a box size of 24 fm, the rank of Hamilton matrix is N=4608. To take the advantage of multi-core architectures, we use the threaded LAPACK library, which can speed up the diagonalization by a factor of 3 with 6 threads, as shown in Fig. 1(a). For the matrix multiplication problem in Fig. 1(b), the routine ZGEMM scales perfectly. For other matrix operations, we use OpenMP for parallel calculations. The modified Broyden method is used to accelerate the iteration convergence [2]. The hybrid MPI and OpenMP programming have made HFB-AX an accurate and fast HFB solver for large box calculations. Besides the threaded library on Cray machines, one can also utilize the threaded AMD's ACML library or the threaded Intel MKL library. In the near future, the multi-core+GPU Cray XT6 machine is expected, and the corresponding MAGMA library [5] can further improve the capability of HFB-AX.

The precision of HFB-AX has been demonstrated in calculations of weakly-bound nuclei in which the coupling to scattering continuum is essential [2, 6]. In large-box calculations, very dense quasi-particle spectrum is obtained, and the non-resonant continuum contribution can be precisely taken into account by means of the direct integration [6]. Furthermore, HFB-AX is expected to provide precise solutions for the continuum QRPA description of excited states by means of the recently developed finite-amplitude method [7]. The recent success of HFB-AX is the prediction of rapidly oscillating pairing potential in highly elongated traps [4] by using the SLDA [8] and ASLDA [9] energy density functionals for polarized Fermi gases.

3. 3D MADNESS-HFB Solver

The software MADNESS-HFB [10, 11] is a fast $O(Nlog\epsilon)$ method based on the multiresolution analysis and low separation rank methods of representing and approximating functions and operators for solving Schrödinger and Lippman-Schwinger equations up to user-determined precision ϵ. MADNESS is an acronym for Multiresolution Adaptive Numerical Environment for Scientific Simulations [12]. The application of multiresolution analysis separates the behavior of functions and operators at different length scales in a functional way. Interesting mathematical and computational feature in our MADNESS library implementation is that each of the operators and wave functions has its own adaptive structure of refinement to achieve and guarantee the desired accuracy.

The representation of functions and operators in MADNESS is based on an adaptive pseudo-spectral representations of functions and operators using the discontinuous Alpert's multiwavelets [13, 14] and the low-separation rank representations and applications of Green's functions. Alpert's multiwavelets have compact support. For representing functions in 3D, we use tensor products of 1D multiwavelets. The two-scale relations provides for an adaptive approximation of the expansion in terms of multiwavelet basis as a function of accuracy in terms of the regularity of the solutions. The support of the bases functions in each level resembles that of structure adaptive refinement. A schematic figure of such a refinement process is shown in Fig. 2. Using two-scale relations, the multiwavelet basis provides a way of constructing an adaptive pseudo-spectral method. The multiwavelet basis is a local basis of degree k defined on the interval $(0,1)$ generated from scaling functions which are defined by shifted, rescaled and orthonormalized Legendre polynomials from degree 0 to $k-1$ from $(-1,1)$ with a value of 0 defined outside of $(0,1)$. The subspace, of $L_2((0,1))$, spanned by scaling functions of degree k is denoted by $V^k = \phi_k(x) = \sqrt{k+1/2}P_i(2x-1)$ *on* $(0,1)$ where $P_i(x)$ is the i-th Legendre

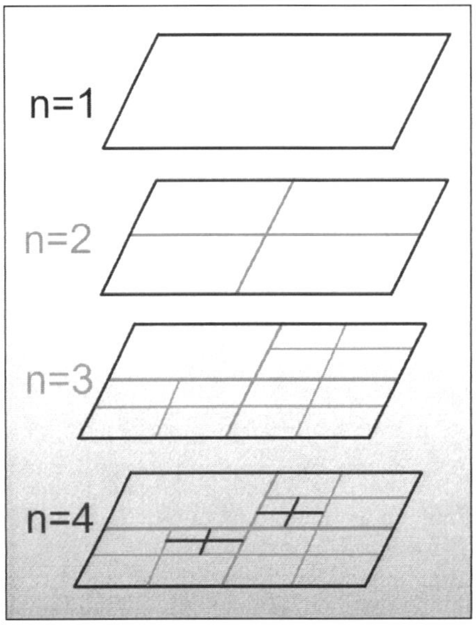

Figure 2. A schematic picture illustrating the two-scale generated refinement process of a multi-resolution representation of a function in 2D with 3-levels.

polynomial on $(-1, 1)$. For each level n, we further define an ascending sequence of the subspaces $V_n^k = \phi_{jl}^n(x) = 2^{n/2}\phi_j(2^n x - l), j = 0, ..., k-1,\ l = 0, ..., 2^n - 1$. Specifically, $V_k^0 \subset V_k^1$; the multiwavelets basis, denoted by W_k^0, is the orthogonal complement of V_k^0 in V_k^1. For each level n, we can further construct W_k^n in the same way of shifting and rescaling the basis functions of W_k^0 as in V_k^n to construct W_k^n. Thus,

$$\begin{aligned} &V_0^k \subset V_1^k \subset V_2^k \subset ... \subset V_n^k, \\ &V_n^k = V_0^k + W_0^k + W_1^k + ... + W_{n-1}^k. \end{aligned} \tag{3}$$

The notable features of the multiwavelet bases are that at each level the support of the basis functions are of width 2^{-n}, and that there is an exact algebraic relationship between the basis functions at level n and level $n+1$. In particular, the coefficients of a representation of a smooth function in the multiwavelet basis decays proportionally to the width of the interval at level n. Thus, we can estimate the accuracy of the representation and truncate coefficients below a specified threshold for a sparse representation.

In 3D, we use a tensor product basis generated from the 1D multiwavelet basis; the representation of operators uses a non-standard approximation or a low separation rank approximation based on Gaussians. The structure of the support of a function resembles that of an oct-tree. Each node of the oct-tree consists of a tensor of coefficients. The truncation of the sparse coefficients produce a pruned tree. The oct-tree is distributed across the nodes of a massively parallel computer referenced by a global hash-table for each node of the tree. The singular integral operators are represented using a low-separation rank approximation of the Green's functions [15] with details described in [16].

The MADNESS library uses a combined MPI and pthread parallel computing method in a task based computing model with a task graph scheduling and queue on each node. On each node of a parallel multicore computer, one core is devoted to processing internode communication via MPI, and one core is devoted to handling thread scheduling and task allocation and queue scheduling within the node. The code is polymorphic with the use of C++ templates. Each operation on functions and the application of operators are defined in terms of tasks to be performed on the nodes or tensors of the "oct-trees" in the representation the functions or operators. In addition, a data path and flow path dependency analysis is performed to determine when different steps of the code can be overlapped and scheduled in each node's task queue so that distributed multi-threading computation can be performed. Furthermore, since different functions and operations have different data and work-loads, a user directed dynamic load-balancing of data memory can be performed to permit more efficient data and work load distribution.

The solution methodology for the MADNESS-HFB consists of two main steps: (i) diagonalization of the Hamiltonian matrix using wave functions expanded in the multiwavelet bases in 3D, and (ii) computation of each of the wave functions using the associated Lippman-Schwinger equation. Recall that the Green's function for the operator $-\Delta - \lambda$ is $e^{-\lambda r}/r$, the Yukawa potential. The solving procedure is similar to that for solving Hartree-Fock problems [10]. In short, the solution algorithm is:

- Obtain a set of guess wave-function u_0 in the multiwavelet basis.
- Iterate until convergence:
  - Compute and diagonalize the Hamiltonian matrix $\mathcal{H}$ to obtain the latest orthogonal wave functions;
  - Compute densities, properties and gradients;
  - Compute the approximation of the Yukawa potential for each eigenvalue λ;
  - Solve the Lippman-Schwinger equation by convolving with the Green's function based on Yukawa potential, $u_{n+1} = G \star V u_n$.
- Compute observables.

We carried out benchmark calculations with HFB-AX and MADNESS-HFB for 100 ultra-cold fermionic atoms in elongated traps. The cold Fermions at the unitary limit can be described by the superfluid density functional SLDA [8]. The single-particle Hamiltonian of SLDA can be written as:

$$h(\mathbf{r}) = -\frac{\alpha\hbar^2\nabla^2}{2m} + \frac{\beta(3\pi^2\rho(\mathbf{r}))^{2/3}}{2} - \frac{|\Delta(\mathbf{r})|^2}{3\gamma\rho^{2/3}(\mathbf{r})} + V_{ext}(\mathbf{r}), \tag{4}$$

where the deformed external trap potential is:

$$V_{ext}(\mathbf{r}) = V_0\left[1 - \exp\left(-\frac{\omega^2(x^2+y^2+z^2/\eta^2)}{2V_0}\right)\right]. \tag{5}$$

The densities of spin-up ($\rho_\uparrow$) and spin-down ($\rho_\downarrow$) atoms, the pairing densities κ, and pairing gaps Δ can be expressed in terms of the HFB two-component eigenvectors (1):

$$\begin{aligned}
\rho_\uparrow(\mathbf{r}) &= \sum_i f_i|u_i(\mathbf{r})|^2, \quad \rho_\downarrow(\mathbf{r}) = \sum_i (1-f_i)|v_i(\mathbf{r})|^2, \\
\kappa(\mathbf{r}) &= \sum_i f_i u_i(\mathbf{r})v_i^*(\mathbf{r}), \quad \Delta(\mathbf{r}) = -g_{eff}(\mathbf{r})\kappa(\mathbf{r}),
\end{aligned} \tag{6}$$

where $\rho = \rho_\uparrow + \rho_\downarrow$ and $g_{eff}(\mathbf{r})$ is the regularized pairing strength. In Eq. (6), $f_i = [1+\exp(E_i/kT)]^{-1}$ and the temperature is kT=0.01. The parameters α, β and γ in Eq.(4) are taken as 1.14, -0.553 and $-1/0.0906$, respectively [8]. We work in trap units for which $\hbar = m = \omega = 1$. The trap aspect ratio η in (5) denotes the trap elongation. In experiments, the adopted optical trap is highly elongated with η up to 50 [17]. The highly elongated trap is interesting because it provides a connection to quasi one-dimensional systems. In MADNESS-HFB, we employed very large boxes: (x, y, z)[-100, 100] for $\eta = 5$ and [-160, 160] for $\eta = 16$. In HFB-AX, the 2D box sizes are 10.5×35 ($\eta = 5$) and 9.1×70 ($\eta = 16$). The potential depth V_0 is 12 and 10 for $\eta = 5$ and $\eta = 16$, respectively.

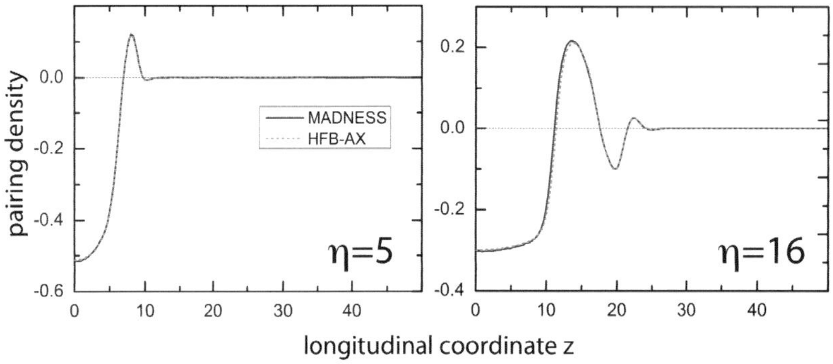

Figure 3. Pairing densities $\kappa(x = 0, y = 0, z)$ calculated by MADNESS-HFB and HFB-AX for 100 polarized atoms in an elongated trap with $\eta = 5$ (left) and $\eta = 16$ (right).

Figure 3 displays the calculated pairing densities $\kappa(x = 0, y = 0, z)$ for ultra-cold Fermi gas with polarization of 0.2. The energy cutoff is taken as 6.75 and 5.05 for $\eta = 5$ and 16, respectively. It is seen that the two solvers agree very well at η=5. At $\eta = 16$, there appears a very small difference between HFB solutions; this may be an indication that an even larger box is needed in HFB-AX. It is to be noted that at $\eta = 16$ there appear coexisting solutions which are close in energy [18, 4]. Figure 4 shows the 3D pairing density distribution computed by MADNESS-HFB for very elongated system with $\eta = 16$. The characteristic transversal oscillations of the pairing field are indicative of the Larkin-Ovchinnikov phase [4].

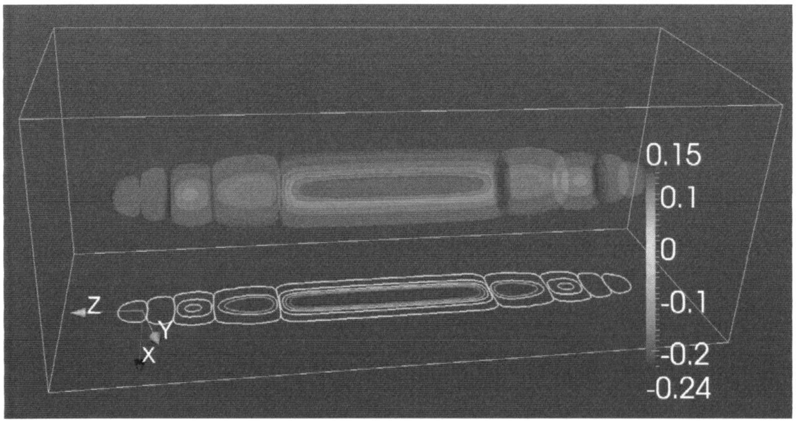

Figure 4. Pairing density $\kappa(x, y, z)$, calculated by MADNESS-HFB for the elongated trap with $\eta = 16$. The box scale in view is x[-12, 12], y[-12, 12] and z[-32, 32].

In general, the computational effort of 3D HFB calculations is significantly greater than in the 2D case. Thus, the adaptive representation using multiwavelet techniques and parallel computing are crucial for the success of MADNESS-HFB. To put things in perspective, MADNESS-HFB takes about 5 minutes per iteration on 5000 cores for $\eta = 5$ and $\eta = 16$, while HFB-AX takes about 10 minutes per iteration for $\eta = 5$ on 400 cores and 25 minutes for $\eta = 16$ on 800 cores. From this trend, we conclude that the parallel performance of MADNESS-HFB is superior to HFB-AX in large-box calculations.

4. Conclusions

The coordinate-space HFB solvers HFB-AX and MADNESS-HFB have been developed for large-box calculations of superfluid Fermi systems. The 2D solver HFB-AX has been adapted to the multi-core supercomputers by using a hybrid MPI and OpenMP programming model. The 3D solver MADNESS-HFB is based on multi-resolution multiwavelet techniques and more sophisticated hybrid MPI and pthreads based task parallelism programming methodologies. A scaling benchmark test for 100 ultracold Fermions in elongated traps has been carried out. Developments are underway to efficiently use the next generation of multi-core+GPU architectures for more demanding nuclear problems.

Acknowledgments

This work was supported in part by the U.S. Department of Energy under Contract Nos. DE-FG02-96ER40963 (University of Tennessee) and DE-FC02-09ER41583 (UNEDF SciDAC Collaboration). It is also partially sponsored by the Office of Advanced Scientific Computing Research; U.S. Department of Energy. The work was partially performed at the Oak Ridge National Laboratory, which is managed by UT-Battelle, LLC under Contract No. De-AC05-00OR22725.

Computational resources were provided through an INCITE award "Computational Nuclear Structure" by the National Center for Computational Sciences (NCCS) and National Institute for Computational Sciences (NICS) at Oak Ridge National Laboratory, and the National Energy Research Scientific Computing Center (NERSC).

References

[1] Pei J C, Nazarewicz W, and Stoitsov M 2009 *Eur. Phys. J.* A **42** 595.
[2] Pei J C, Stoitsov M V, Fann G I, Nazarewicz W, Schunck N and Xu F R 2008 *Phys. Rev.* C **78** 064306.
[3] Newton W G and Stone J R 2009 *Phys. Rev.* C **79** 055801.
[4] Pei J C, Dukelsky J, Nazarewicz W 2010 *Phys. Rev.* A **82** 021603(R).
[5] http://icl.cs.utk.edu/magma/
[6] Pei J C, Kruppa A T, and Nazarewicz W 2011 *Phys. Rev.* C **84** 024311.
[7] Stoitsov M, Kortelainen M, Nakatsukasa T, Losa C and Nazarewicz W 2011 *Phys. Rev.* C **84** 041305(R).
[8] Bulgac A 2007 *Phys. Rev.* A **76** 040502(R).
[9] Bulgac A and Forbes M M 2008 *arXiv*:0808.1436.
[10] Fann G I, Pei J, Harrison R J, Jia J, Hill J, Ou M, Nazarewicz W, Shelton W A and Schunck N 2009 *J. Phys. Conf. Ser.* **180** 012080.
[11] Fann G I, Harrison R J, Beylkin G, Jia J, Hartman-Baker R, Shelton W A and Sugiki S 2007 *J. Phys. Conf. Ser.* **78** 012018.
[12] http://code.google.com/p/m-a-d-n-e-s-s/
[13] Alpert B 1993 *SIAM J. of Math. Anal.* **24**(1) 246.
[14] Alpert B, Beylkin G, Gines D and Vozovoi L 2002 *J. of Comp. Phys.* **182** 149.
[15] Beylkin G and Mohlenkamp M 2002 *Proc. Natl. Acad. Sci. USA.* **99** 10246.
[16] Harrison R, Fann G, Yanai T, Gan Z and Beylkin G 2004 *J. Chem. Phys.* **23** 11587.
[17] Partridge G B et al 2006 *Science* **311** 503.
[18] Baksmaty L O, Lu H, Bolech C J and Pu H 2011 *Phys. Rev.* A **83** 023604.

Numerical study of hot strongly interacting matter

P. Petreczky
Physics Department, Brookhaven National Laboratory, Upton, NY 11973, USA

Abstract. I review recent progress in study of strongly interacting matter at high temperatures using Monte-Carlo simulations in lattice QCD.

1. Introduction

It is expected that strongly interacting matter undergoes a transition in some temperature interval from hadron gas to deconfined state also called the quark gluon plasma (QGP) [1]. Creating deconfined medium in a laboratory is the subject of the large experimental program at RHIC [2] and LHC [3]. Attempts to study QCD thermodynamics on the lattice go back to the early 80's when lattice calculations in $SU(2)$ gauge theory provided the first rigorous theoretical evidence for deconfinement [4, 5, 6]. The problem of calculating thermodynamic observables in pure gluonic theory was solved in 1996 [7], while simulations involving dynamical quarks were limited to large quark masses and had no control over the discretization errors [8, 9, 10]. Lattice calculations of QCD thermodynamics with light dynamical quarks remained challenging until recently. During the past six years calculations with light u, d quarks have been performed using improved staggered fermion actions [11, 12, 13, 14, 15, 16, 17, 18, 19, 20, 21, 22, 24]. There was also progress in lattice QCD calculations at non-zero temperature using Domain Wall and Wilson fermion formulations [25, 26, 27]. However, due to much larger computational costs the corresponding results are far less extensive.

To get reliable predictions from lattice QCD the lattice spacing a should be sufficiently small relative to the typical QCD scale, i.e. $\Lambda_{QCD}a \ll 1$. For staggered fermions, discretization errors go like $\mathcal{O}((a\Lambda_{QCD})^2)$ but discretization errors due to flavor symmetry breaking turn out to be numerically quite large and dominate the cutoff dependence of thermodynamic quantities at low temperatures. To reduce these errors one has to use improved staggered fermion actions with so-called fat links [28]. At high temperature the dominant discretization errors go like $(aT)^2$ and therefore could be very large. Thus it is mandatory to use improved discretization schemes, which improve the quark dispersion relation and eliminate these discretization errors. Lattice fermion actions used in numerical calculations typically implement some version of fat links as well as improvement of quark dispersion relation and are referred to as *p4*, *asqtad*, *HISQ/tree* and *stout*.

In this contribution I am going to discuss lattice QCD calculations on the equation of state, study of deconfinement aspects of the QCD transition, including color screening and fluctuations of conserved charges, determination of the chiral transition temperature, as well as calculations of temporal and spatial meson correlation functions.

2. Equation of State

The equation of state has been calculated using different improved staggered fermion actions *p*4, *asqtad*, *stout* and *HISQ/tree*. In the lattice calculations of the equation of state and many other quantities the temperature is varied by varying the lattice spacing at fixed value of the temporal extent N_τ. The temperature T is given by the lattice spacing and the temporal extent, $T = 1/(N_\tau a)$. Therefore taking the continuum limit corresponds to $N_\tau \to \infty$ at the fixed physical volume. The calculation of thermodynamic observables proceeds through the calculation of the trace of the energy momentum tensor, $\epsilon - 3p$, also known as trace anomaly or interaction measure. This is due to the fact that this quantity can be expressed in terms of expectation values of local gluonic and fermionic operators, (see e.g. Ref. [18]). Different thermodynamic observables can be obtained from the interaction measure through integration of the trace anomaly [1]. The pressure can be written as

$$\frac{p(T)}{T^4} - \frac{p(T_0)}{T_0^4} = \int_{T_0}^{T} \frac{dT'}{T'^5}(\epsilon - 3p). \tag{1}$$

The lower integration limit T_0 is chosen such that the pressure is exponentially small there. Furthermore, the entropy density can be written as $s = (\epsilon + p)/T$. Since the interaction measure is the basic thermodynamic observable in the lattice calculations it is worth to discuss its properties more in detail. In Fig. 1 (left panel) I show the results of calculation with *p*4 and *asqtad* actions using $N_\tau = 6$ and 8 lattices and light quark masses $m_l = m_s/10$, where m_s is the physical strange quark mass. These calculations correspond in the continuum limit to the pion mass of 220MeV and 260MeV for *p*4 and *asqtad* respectively. The interaction measure shows a rapid rise in the transition region and after reaching a peak at temperatures of about 200MeV decreases. Cutoff effects (i.e. N_τ dependence) appears to be the strongest around the peak region and decrease at high temperatures. For temperatures $T < 270$MeV calculations with *p*4 and *asqtad* actions have been extended to smaller quark masses, $m_l = m_s/20$, that correspond to the pion mass of about 160MeV in the continuum limit [17, 29]. It turns out that the quark mass dependence is negligible for $m_l < m_s/10$. Furthermore, for *astqad* action calculations have been extended to $N_\tau = 12$ lattices [29]. The trace anomaly was calculated with *HISQ/tree* action on lattices with temporal extent $N_\tau = 6$ and 8 and $m_l = m_s/20$ [29] (corresponding to $m_\pi = 160$MeV in the continuum limit). Finally, calculation of the trace anomaly and the equation of state was performed with *stout* action using $N_\tau = 4,\ 6,\ 8,\ 10$ and 12 and physical light quark masses [21]. Using the lattice data from $N_\tau = 6,\ 8,\ 10$ a continuum estimate for different quantities was given [21]. In Fig. 1 (right panel) the results of different lattice calculations of $\epsilon - 3p$ are summarized corresponding to the pion masses close to the physical value. I also compare the lattice results with the parametrization s95p-v1 of $\epsilon - 3p$. This parametrization combines lattice QCD results of Refs. [16, 18] at high temperatures with hadron resonance gas model (HRG) at low temperatures ($T < 170$MeV) [31]. At low temperatures there is a fair agreement between the results obtained with *stout* action and the results obtained with *HISQ/tree* action as well as with *asqtad* action for $N_\tau = 12$. All these lattice results are slightly above the HRG curve. There is a big difference in the height of the peak in the calculations obtained with *asqtad* and *HISQ/tree* actions and the continuum estimate for *stout* action. The *p*4 action has large cutoff effects at low temperatures and in the peak region. As the result the *p*4 results are higher in the peak region and fall below the HRG curve at low temperatures. When cutoff effects in the hadron spectrum are taken into account in the HRG model a good agreement between the *p*4 data and HRG result can be achieved [31]. Since the dominant cutoff effects of order a^2T^2 are eliminated, the N_τ-dependence is expected to be small for *p*4, *asqtad* and *HISQ/tree* action at high temperatures. We see that for $T > 250$MeV all these lattice

[1] A somewhat different approach was used in Ref. [21]

actions lead to similar results. At temperatures above 350MeV we also see a good agreement with the *stout* results.

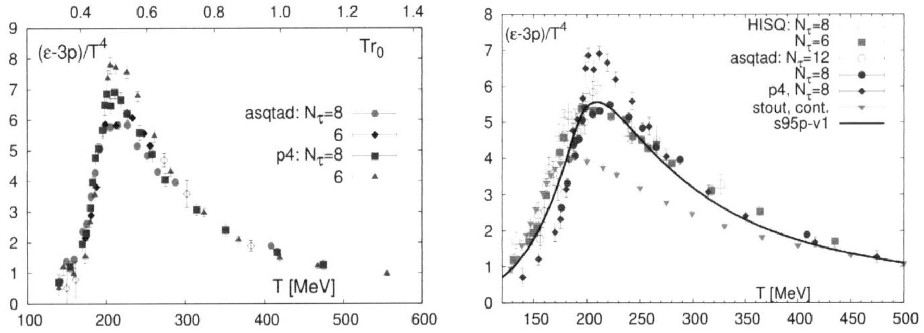

Figure 1. The interaction measure calculated with $m_l = m_s/10$ and $p4$ and *asqtad* actions [18] (left) and with $m_l = m_s/20$ for $p4$ action [17] as well as with *HISQ/tree* and *asqtad* actions [29]. Also shown in the figure are the continuum estimates obtained with *stout* action and the parametrization based on hadron resonance gas (HRG) model [31].

The pressure, the energy density and the entropy density are shown in Fig. 2. The energy density shows a rapid rise in the temperature region $(170 - 200)$ MeV and quickly approaches about 90% of the ideal gas value. The pressure rises less rapidly but at the highest temperature it is also only about 15% below the ideal gas value. In the previous calculations with the $p4$ action it was found that the pressure and the energy density are below the ideal gas value by about 25% at high temperatures [10]. Possible reason for this larger deviation could be the fact that the quark masses used in this calculation were fixed in units of temperature instead being tuned to give constant meson masses as lattice spacing is decreased. As discussed in Ref. [30] this could reduce the pressure by $10 - 15\%$ at high temperatures. In Fig. 2 I also show the entropy density divided by the corresponding ideal gas value and compare the results of lattice calculations with resummed perturbative calculation [32, 33] as well as with the predictions from AdS/CFT correspondence for the strongly coupled regime [34]. The later is considerably below the lattice results. Note that the pressure, the energy density and the trace anomaly have also been recently discussed in the framework of resummed perturbative calculations which seem to agree with lattice data quite well at high temperatures[35].

The differences between the *stout* action and the $p4$ and *asqtad* actions for the trace anomaly translates into the differences in the pressure and the energy density. In particular, the energy density is about 20% below the ideal gas limit for the *stout* action.

3. Deconfinement : Fluctuations of conserved charges

Due to the infamous sign problem lattice QCD Monte-Carlo simulations are not possible at non-zero quark chemical potentials. The pressure at non-zero chemical potentials can be evaluated using Taylor expansion. The Taylor expansion can be set up in terms of quark chemical potentials or in terms of chemical potentials corresponding to baryon number B, electric charge Q and strangeness S of hadrons. The expansion coefficients in quark chemical potential χ_{uds}^{jkl} and the hadronic ones χ_{BQS}^{jkl} are related to each other [36]. While Taylor expansion can be used to study the physics at non-zero baryon density they are interesting on their own right as they are related to the fluctuations of conserved charges. The later are sensitive probes of deconfinement. This is because fluctuation of conserved charges are sensitive to the underlying degrees of freedom which could be hadronic or partonic. Fluctuations of conserved charges have

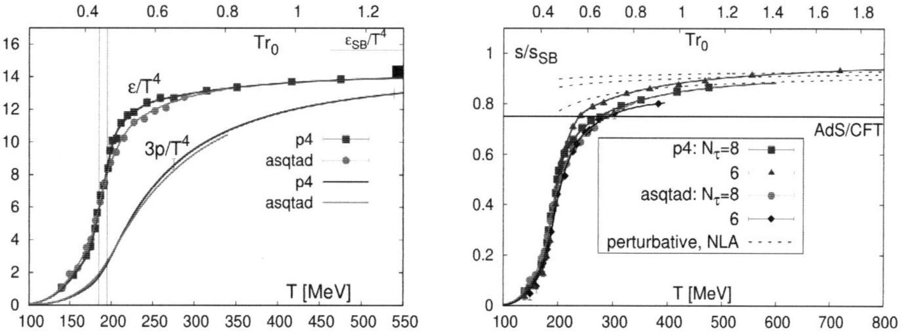

Figure 2. The energy density and the pressure as function of the temperature (left), and the entropy density divided by the corresponding ideal gas value (right). The dashed lines in the right panel correspond to the resummed perturbative calculations while the solid black line is the AdS/CFT result.

been studied using different staggered actions [11, 18, 20, 36, 37]. As an example in Fig. 3 I show the quadratic strangeness fluctuation as the function of the temperature. Fluctuations are suppressed at low temperatures because conserved charges are carried by massive strange hadrons (mostly kaons). They are well described by Hadron Resonance Gas (HRG) model at low temperatures. Strangeness fluctuations rapidly grow in the transition region of $T = (160 - 200)$ MeV as consequence of deconfinement. At temperatures $T > 250$ MeV strangeness fluctuations are close to unity that corresponds to the massless ideal quark gas. Similar picture can seen in other fluctuations, in particular for light quark number fluctuation, χ_l, which is also show in Fig. 3. The more rapid rise of χ_l in the transition region is presumably a quark mass effect.

The transition from hadronic to quark degrees of freedom can be particularly well seen in the temperature dependence of the Kurtosis, which is the ratio of quartic to quadratic fluctuations. This quantity can be also measured experimentally. In Fig. 4 I show the Kurtosis of the baryon number as a function of the temperature. At low temperatures it is temperature independent and is close to unity in agreement with the prediction of the HRG model. In the transition region it sharply drops from the hadron resonance gas value to the value corresponding to an ideal gas of quarks. Since the fluctuations of conserved charges are so well described by ideal quark gas it is interesting to compare the lattice results for quark number fluctuations with resummed perturbation theory. Quadratic quark number fluctuations are also called quark number susceptibilities and have been calculated in resummed perturbation theory [38, 39, 40]. The comparison of the resummed perturbative results with lattice calculations performed with $p4$ action [41] for the quark number susceptibility divided by the ideal gas value is shown in Fig. 4. The figure shows a very good agreement between the lattice calculations and resummed perturbative results. Let me finally note that the discretization errors in the lattice calculations of the fluctuations are small at high temperatures when calculated with the $p4$ action. The same is true for *asqtad* action [11, 18].

Fluctuations of conserved charges have also been calculated using gauge/string duality in Refs. [42, 43]. These calculations could reproduce several features of the lattice data.

4. Deconfinement : color screening

One of the most prominent feature of the quark gluon plasma is the presence of chromoelectric (Debye) screening. The easiest way to study chromoelectric screening is to calculate the Polyakov loop. The Polyakov loop is an order parameter for the deconfinement transition in pure gauge

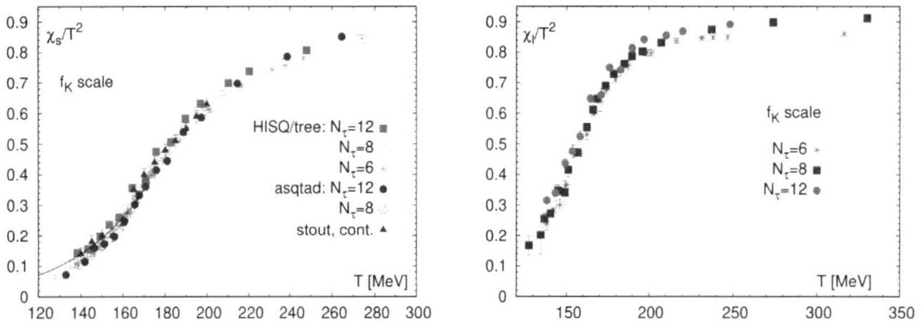

Figure 3. Strangeness fluctuation calculated with *asqtad* and *HISQ/tree* actions [24] and compared with the continuum estimate obtained with stout action [20] as well as the prediction of the HRG model shown as a black line (left panel). In the right panel the light quark number fluctuations are shown for *HISQ/tree* action [24].

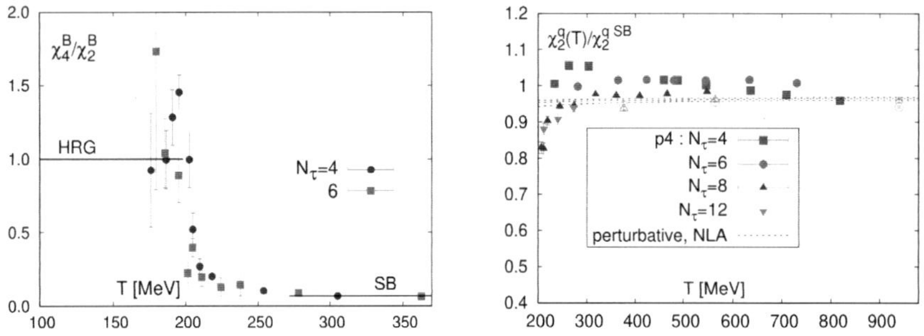

Figure 4. The Kurtosis of baryon number (left) and the quark susceptibility compared with resummed perturbative calculations (right). The open symbols on the right plot correspond to *asqtad* results for strange quark number susceptibility [11]. The resummed perturbative results were obtained in next-to-leading approximation (NLA) [39].

theory, which is governed by $Z(N)$ symmetry. For QCD this symmetry is explicitly broken by dynamical quarks. There is no obvious reason for the Polyakov loop to be sensitive to the singular behavior close to the chiral limit although speculations along these lines have been made [52]. The Polyakov loop is related to the screening properties of the medium and thus to deconfinement. After proper renormalization, the square of the Polyakov loop characterizes the long distance behavior of the static quark anti-quark free energy; it gives the excess in free energy needed to screen two well-separated color charges. The renormalized Polyakov loop has been studied in the past in pure gauge theory [51, 53] as well as in QCD with two [54], three [50] and two plus one flavors [16, 18]. The renormalized Polyakov loop, calculated on lattices with temporal extent N_τ, is obtained from the bare Polyakov

$$L_{ren}(T) = z(\beta)^{N_\tau} L_{bare}(\beta) = z(\beta)^{N_\tau} \left\langle \frac{1}{3} \mathrm{Tr} W(\vec{x}) \right\rangle, \quad W(\vec{x}) = \prod_{x_0=0}^{N_\tau - 1} U_0(x_0, \vec{x}), \tag{2}$$

where $U_0 = \exp(igaA_0)$ denotes the temporal gauge link and $z(\beta)$ is the renormalization constant determined from the $T = 0$ static potential [24]. The numerical results for the renormalized Polyakov loop for the *HISQ/tree* action are shown in Fig. 5. As one can see from the figure the cutoff (N_τ) dependence of the renormalized Polyakov loop is small. We also compare our results

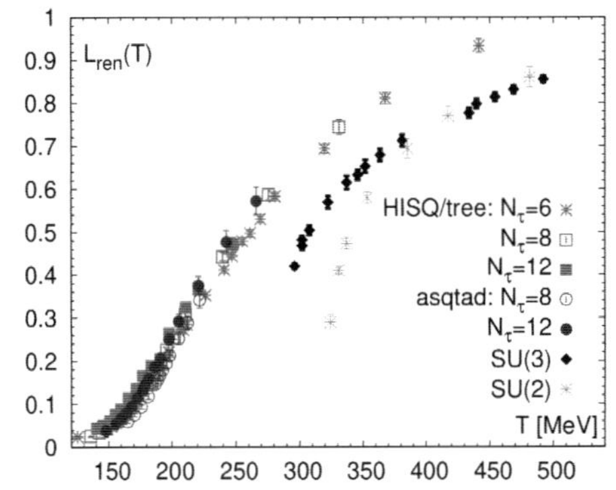

Figure 5. The Polyakov loop as function of the temperature in pure gauge theory and 2+1 flavor QCD.

with the continuum extrapolated *stout* results [20] and the corresponding results in pure gauge theory [51, 53]. We find good agreement between our results and the *stout* results. We also see that in the vicinity of the transition temperature the behavior of the renormalized Polyakov loop in QCD and in the pure gauge theory is quite different.

Further insight on chromoelectric screening can be gained by studying the singlet free energy of static quark anti-quark pair (for reviews on this see Ref. [55, 56]), which is expressed in terms of correlation function of temporal Wilson lines in Coulomb gauge

$$\exp(-F_1(r,T)/T) = \frac{1}{3}\mathrm{Tr}\langle W(r)W^\dagger(0)\rangle. \tag{3}$$

Instead of using the Coulomb gauge the singlet free energy can be defined in gauge invariant manner by inserting a spatial gauge connection between the two Wilson lines. Using such definition the singlet free energy has been calculated in $SU(2)$ gauge theory [57]. It has been found that the singlet free energy calculated this way is close to the result obtained in Coulomb gauge [57]. The singlet free energy turned out to be useful to study quarkonia binding at high temperatures in potential models (see e.g. Refs. [44, 45, 46, 47, 48]). The singlet free energy also appears naturally in the perturbative calculations of the Polyakov loop correlators at short distances [58].

The singlet free energy was recently calculated in QCD with one strange quark and two light quarks with masses corresponding to pion mass of 220MeV on $16^3 \times 4$ lattices [49]. The numerical results are shown in Fig. 6. At short distances the singlet free energy is temperature independent and coincides with the zero temperature potential. In purely gluonic theory the free energy grows linearly with the separation between the heavy quark and anti-quark in the confined phase. In presence of dynamical quarks the free energy is saturated at some finite value at distances of about 1 fm due to string breaking [55, 50, 54]. This is also seen in Fig. 6. Above the deconfinement temperature the singlet free energy is exponentially screened at sufficiently large distances [51, 53] with the screening mass proportional to the temperature , i.e.

$$F_1(r,T) = F_\infty(T) - \frac{4}{3}\frac{g^2(T)}{4\pi r}\exp(-m_D(T)r),\ m_D \sim T. \tag{4}$$

Therefore in Fig. 6 I also show the combination $F_1(r,T) - F_\infty(T)$ as a function of rT. As one can see from the figure this function shows an exponential fall-off at distances $rT > 0.8$. The fact that the slope is the same for all temperatures means that $m_D \sim T$, as expected.

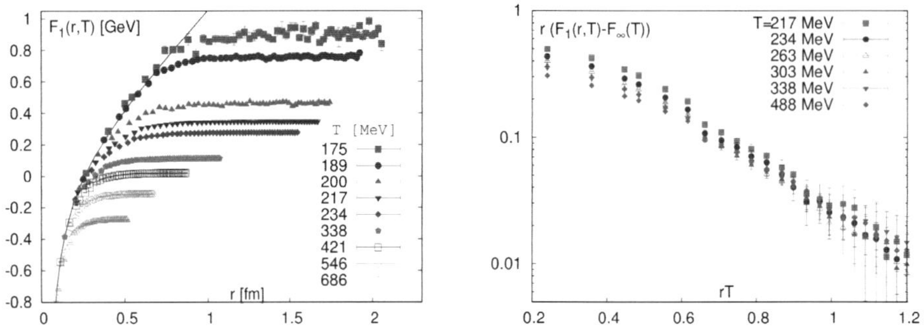

Figure 6. The singlet free energy $F_1(r,T)$ calculated in Coulomb gauge on $16^3 \times 4$ lattices (left) and the combination $F_1(r,T) - F_\infty(T)$ as function of rT (right). The solid black line is the parametrization of the zero temperature potential.

Let me finally note that contrary to the electro magnetic plasma the static chormomagnetic fields are screened in QGP. This is due to the fact that unlike photons gluons interact with each other (the stress tensor is non-linear in QCD). Magnetic screening is non-perturbative, i.e. it does not appear at any finite order of pertubation theory. In lattice calculations chromomagnetic screening is studied either in terms of spatial Wilson loops [59] or in terms of spatial gluon propagators [60, 61, 62]. The numerical results obtained so far show that the length scale related to magnetic screening is larger than the one related to electric screening.

5. Chrial transition

The Lagrangian of QCD has an approximate $SU_A(3)$ chiral symmetry. This symmetry is broken in the vacuum. The chiral symmetry breaking is signaled by non-zero expectation value of the quark or chiral condensate, $\langle \bar{\psi}\psi \rangle \neq 0$ in the massless limit. This symmetry is expected to be restored at high temperatures and the quark condensate vanishes. There is an explicit breaking of the chiral symmetry by the small value of u, d and s quark masses. While due to the relatively large strange quark mass ($m_s \simeq 100$ MeV) $SU_A(3)$ may not be a very good symmetry its subgroup $SU_A(2)$ remains a very good symmetry and is relevant for the discussion of the finite temperature transition in QCD. If the relevant symmetry is $SU_A(2)$ the chiral transition is expected to be second order for massless light (u and d) quarks belonging to the $O(4)$ universality class. Recent calculations with $p4$ action support this pictures [63]. This also means that for non-zero light quark masses the transition must be a crossover. The later fact seems to be supported by calculations in Ref. [64]. The $U_A(1)$ symmetry is explicitly broken in the vacuum by the anomaly but it is expected to be effectively restored at high temperatures as non-perturbative vacuum fluctuations responsible for its breaking are suppressed at high temperatures. If the $U_A(1)$ symmetry is restored at the same temperature as the $SU_A(2)$ symmetry the transition could be first order [65]. Recent calculations with staggered as well as with domain wall fermions suggest that $U_A(1)$ symmetry gets restored at temperature that is significantly higher than the chiral transition temperature [25, 66].

For massless quark the chiral condensate vanishes at the critical temperature T_c^0 and is the order parameter. Therefore in the lattice studies one calculates the chiral condensate and its derivative with respect to the quark mass called the chiral susceptibility. For the staggered fermion formulation most commonly used in the lattice calculations these quantities can be

Journal of Physics: Conference Series **402** (2012) 012036 doi:10.1088/1742-6596/402/1/012036

written as follows:

$$\langle\bar{\psi}\psi\rangle_{q,x} = \frac{1}{4}\frac{1}{N_\sigma^3 N_\tau}\mathrm{Tr}\langle D_q^{-1}\rangle, \tag{5}$$

$$\chi_{m,q}(T) = n_f\frac{\partial\langle\bar{\psi}\psi\rangle_{q,\tau}}{\partial m_l} = \chi_{q,disc} + \chi_{q,con} \quad q = l, s, \tag{6}$$

where the subscript $x = \tau$ and $x = 0$ will denote the expectation value at finite and zero temperature, respectively. Furthermore, $D_q = m_q \cdot 1 + D$ is the fermion matrix in the canonical normalization and $n_f = 2$ and 1 for light and strange quark. In Eq. (6 we made explicit that chiral susceptibility is the sum of connected and disconnected Feynman diagrams. The disconnected and connected contributions can be written as

$$\chi_{q,disc} = \frac{n_f^2}{16N_\sigma^3 N_\tau}\left\{\langle(\mathrm{Tr}D_q^{-1})^2\rangle - \langle\mathrm{Tr}D_q^{-1}\rangle^2\right\}, \tag{7}$$

$$\chi_{q,con} = -\frac{n_f}{4}\mathrm{Tr}\sum_x\langle\, D_q^{-1}(x,0)D_q^{-1}(0,x)\,\rangle\,, \quad q = l, s. \tag{8}$$

The disconnected part of the light quark susceptibility describes the fluctuations in the light quark condensate and is directly analogous to the fluctuations in the order parameter of an $O(N)$ spin model. The second term ($\chi_{q,con}$) arises from the explicit quark mass dependence of the chiral condensate and is the expectation value of the volume integral of the correlation function of the (isovector) scalar operator $\bar{\psi}\psi$. Let me note that in the massless linit only $\chi_{l,disc}$ diverges.

5.1. *The temperature dependence of the chiral condensate*

The chiral condensate needs a multiplicative, and also an additive renormalization if the quark mass is non-zero. Therefore the subtracted chiral condensate is considered

$$\Delta_{l,s}(T) = \frac{\langle\bar{\psi}\psi\rangle_{l,\tau} - \frac{m_l}{m_s}\langle\bar{\psi}\psi\rangle_{s,\tau}}{\langle\bar{\psi}\psi\rangle_{l,0} - \frac{m_l}{m_s}\langle\bar{\psi}\psi\rangle_{s,0}}. \tag{9}$$

In Fig. 7 I show the results for $\Delta_{l,s}$ calculated with *HISQ/tree* action and compared to the renormalized Polyakov loop and light quark number fluctuation discussed in relation to the deconfining transition. Interestingly the rapid decrease in the subtracted chiral condensate happens at temperatures that are smaller than the temperatures where the Polyakov loop rises rapidly. On the other hand the rapid change in χ_l and $\Delta_{l,s}$ happen roughly in the same temperature region.

Another way to get rid of the multiplicative and additive renormalization is to subtract the zero temperature condensate and multiply the difference by the strange quark mass, i.e. consider the following quantity

$$\Delta_q^R = d + n_f m_s r_1^4(\langle\bar{\psi}\psi\rangle_{q,\tau} - \langle\bar{\psi}\psi\rangle_{q,0}) \; q = l, s. \tag{10}$$

As before, $n_f = 2$ for light quarks and $n_f = 1$ for strange quarks, while d is a normalization constant. The factor r_1^4 was introduce to make the combination dimensionless. Here r_1 is the scale parameter defined from the zero temperature static potential [24]. It is convenient to choose the normalization constant to be the light quark condensate for $m_l = 0$ multiplied by $m_s r_1^4$. In Fig. 8 the renormalized quark condensate is shown as function of temperature for *HISQ/tree* and *stout* actions. We see a crossover behavior for temperature of $(150 - 160)$MeV, where Δ_q^R drops by 50%. The difference between the *stout* and *HISQ/tree* results is a quark

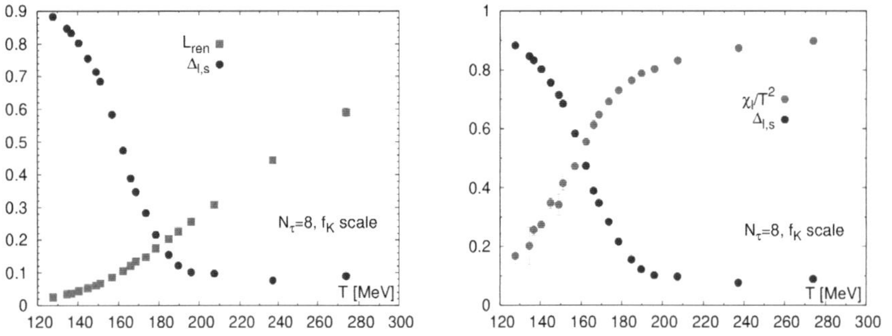

Figure 7. The subtracted chiral condensate calculated with $HISQ/tree$ action on $N_\tau = 8$ lattice and compared to the renormalized Polyakov loop (left) and light quark number fluctuation (right).

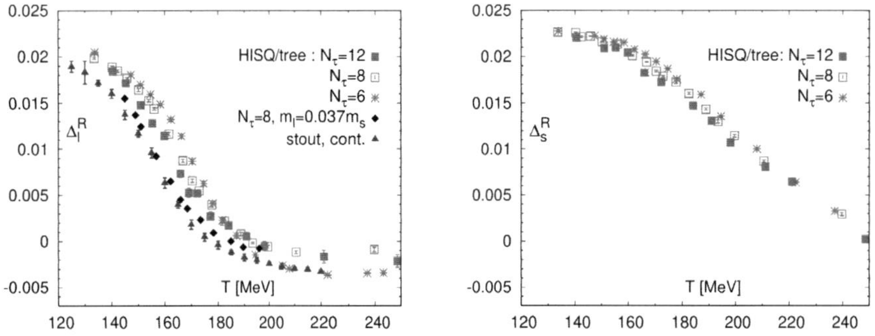

Figure 8. The renormalized chiral condensate Δ_l^R for the $HISQ/tree$ action with $m_l/m_s = 0.05$ is compared to the *stout* data. In the right panel, we show the renormalized strange quark condensate Δ_s^R for the $HISQ/tree$ action.

mass effect. Calculations for $HISQ/tree$ action were performed for $m_\pi = 160$ MeV, while the *stout* calculations were done for the physical quark mass. For a direct comparison with *stout* results, we extrapolate the $HISQ/tree$ data in the light quark mass and also take care of the residual cutoff dependence in the $HISQ/tree$ data. This was done in Ref. [24] and the results are shown in the figure as black diamonds demonstrating a good agreement between $HISQ/tree$ and *stout* results. Contrary to Δ_l^R the renormalized strange quark condensate Δ_s^R shows only a gradual decrease over a wide temperature interval dropping by 50% only at significantly higher temperatures of about 190 MeV.

5.2. The crial susceptibility

For a true chiral phase transition the chiral susceptibility diverges at the transition temperature. For physical value of the quark masses we expect to see a peak in the chiral susceptibility at certain temperature that defines the crossover temperature. The chiral susceptibility also needs a multiplicative and additive renormalization Therefore the following quantity is considered

$$\frac{\chi_R(T)}{T^4} = \frac{m_s^2}{T^4}\left(\chi_{m,l}(T) - \chi_{m,l}(T=0)\right). \tag{11}$$

The numerical results for this quantity were presented in Ref. [14] for *stout* action and in Ref. [24] for the *HISQ/tree* action. It was shown that the results of the two calculations agree quite well up to small quark mass effects [24] [2].

5.3. O(N) scaling and the transition temperature

In the vicinity of the chiral phase transition, the free energy density may be expressed as a sum of a singular and a regular part,

$$f = -\frac{T}{V}\ln Z \equiv f_{sing}(t,h) + f_{reg}(T, m_l, m_s)\ . \tag{12}$$

Here t and h are dimensionless couplings that control deviations from criticality. They are related to the temperature T and the light quark mass m_l as

$$t = \frac{1}{t_0}\frac{T-T_c^0}{T_c^0}\quad ,\quad h = \frac{1}{h_0}H \quad ,\quad H = \frac{m_l}{m_s}\ , \tag{13}$$

where T_c^0 denotes the chiral phase transition temperature, *i.e.*, the transition temperature at $H = 0$. The scaling variables t, h are normalized by two parameters t_0 and h_0, which are unique to QCD and similar to the low energy constants in the chiral Lagrangian. These need to be determined together with T_c^0. In the continuum limit, all three parameters are uniquely defined, but depend on the value of the strange quark mass.

The singular contribution to the free energy density is a homogeneous function of the two variables t and h. Its invariance under scale transformations can be used to express it in terms of a single scaling variable

$$z = t/h^{1/\beta\delta} = \frac{1}{t_0}\frac{T-T_c^0}{T_c^0}\left(\frac{h_0}{H}\right)^{1/\beta\delta} = \frac{1}{z_0}\frac{T-T_c^0}{T_c^0}\left(\frac{1}{H}\right)^{1/\beta\delta} \tag{14}$$

where β and δ are the critical exponents of the $O(N)$ universality class and $z_0 = t_0/h_0^{1/\beta\delta}$. Thus, the dimensionless free energy density $\tilde{f} \equiv f/T^4$ can be written as

$$\tilde{f}(T, m_l, m_s) = h^{1+1/\delta} f_f(z) + f_{reg}(T, H, m_s)\ , \tag{15}$$

where f_f is the universal scaling function and the regular term f_{reg} gives rise to scaling violations. This regular term can be expanded in a Taylor series around $(t,h) = (0,0)$.

It should be noted that the reduced temperature t may depend on other couplings in the QCD Lagrangian which do not explicitly break chiral symmetry. In particular, it depends on light and strange quark chemical potentials μ_q, which in leading order enter only quadratically,

$$t = \frac{1}{t_0}\left(\frac{T-T_c^0}{T_c^0} + \sum_{q=l,s}\kappa_q\left(\frac{\mu_q}{T}\right)^2 + \kappa_{ls}\frac{\mu_l}{T}\frac{\mu_s}{T}\right)\ . \tag{16}$$

The transition temperature can be defined as peaks in susceptibilities (response functions) that are second derivatives of the free energy density with respect to relevant parameters. Since there are two relevant parameters we can define three susceptibilities:

$$\chi_{m,l} = \frac{\partial^2\tilde{f}}{\partial m_l^2},\quad \chi_{t,l} = \frac{\partial^2\tilde{f}}{\partial t\partial m_l},\quad \chi_{t,t} = \frac{\partial^2\tilde{f}}{\partial t^2}. \tag{17}$$

[2] In Ref. [14] the light quark mass was used instead of m_s in Eq. (11). In the comparison with was taken into account [24].

Thus three different pseudo-critical temperatures $T_{m,l}$, $T_{t,l}$ and $T_{t,t}$ can be defined. In the vicinity of the critical point the behavior of these susceptibilities is controlled by three universal scaling function that can be derived from f_f. In the chiral limit $T_{m,l} = T_{t,l} = T_{t,t} = T_c^0$. There is, however, an additional complication for $O(N)$ universality class: while $\chi_{m,l}$ and $\chi_{t,l}$ diverge at the critical point for $m_l \to 0$

$$\chi_{m,l} \sim m_l^{1/\delta-1}, \quad \chi_{t,l} \sim m_l^{(\beta-1)/\beta\delta}, \tag{18}$$

$\chi_{t,t}$ is finite because $\alpha < 0$ for $O(N)$ models ($\chi_{t,t} \sim |t|^{-\alpha}$). Therefore, one has to consider the third derivative of $\tilde{f}$ with respect to t :

$$\chi_{t,t,t} = \frac{\partial^3 \tilde{f}}{\partial t^3}. \tag{19}$$

In the vicinity of the critical point the derivatives with respect to t can be estimated by taking the derivatives with respect to μ_l^2, i.e. the response functions $\chi_{t,l}$ and $\chi_{t,t,t}$ are identical to the first Taylor expansion coefficient of the quark condensate and the sixth order expansion coefficient to the pressure, respectively. The former controls the curvature of the transition temperature as function of the quark chemical potential μ_q and was studied for $p4$ action using $N_\tau = 4$ and 8 lattices [67]. The later corresponds to the sixth order quark number fluctuation which is related to the deconfinement aspects of the transition. The fact that this quantity is sensitive to the chiral dynamics points to a relation between deconfining and chiral aspects of the transition. In the following I discuss the determination of the transition temperature defined as peak position of $\chi_{m,l}$, i.e. $T_c = T_{m,l}$.

5.4. *Determination of the transition temperature*

The $O(N)$ scaling described in the above subsection can be used to determine the pseudo-critical temperature of the chiral transition. For the study of the $O(N)$ scaling it is convenient to consider the dimensionless order parameter

$$M_b = m_s \frac{\langle \bar{\psi}\psi \rangle_l}{T^4}. \tag{20}$$

The subscript "b" refers to the fact that this is a bare quantity since the additive UV divergence is not removed. From the point of view of the scaling analysis this divergent term is just a regular contribution. For sufficiently small quark mass and in the vicinity of the transition region we can write

$$M_b(T,H) = h^{1/\delta} f_G(t/h^{1/\beta\delta}) + f_{M,reg}(T,H). \tag{21}$$

Here $f_G(z)$ is the scaling function related to f_f and was calculated for $O(2)$ and $O(4)$ spin models [68, 69, 70]. The regular contribution can be parametrized as [24]

$$\begin{aligned} f_{M,reg}(T,H) &= a_t(T)H \\ &= \left(a_0 + a_1 \frac{T - T_c^0}{T_c^0} + a_2 \left(\frac{T - T_c^0}{T_c^0} \right)^2 \right) H. \end{aligned} \tag{22}$$

Then we have the following behavior for the light chiral susceptibility

$$\begin{aligned} \frac{\chi_{m,l}}{T^2} &= \frac{T^2}{m_s^2} \left(\frac{1}{h_0} h^{1/\delta-1} f_\chi(z) + \frac{\partial f_{M,reg}(T,H)}{\partial H} \right), \\ &\text{with } f_\chi(z) = \frac{1}{\delta}[f_G(z) - \frac{z}{\beta} f'_G(z)]. \end{aligned} \tag{23}$$

One then performs a simultaneous fit to the lattice data for M_b and $\chi_{m,l}$ treating $T_c^0, t_0, h_0, a_0, a_1$ and a_2 as fit parameters [24]. This gives a good description of the quark mass and temperature dependence of $\chi_{m,l}$ and allows to determine accurately the peak position in $\chi_{m,l}$. As an example in Fig. I show the $O(4)$ scaling fits for $N_\tau = 8$ lattice data obtained with $HISQ/tree$ action. The scaling fit works quite well. Similar results have been obtained for $N_\tau = 6$ and 12 as well as for *asqtad* action on $N_\tau = 8$ and 12 lattices [24]. Furthermore, scaling fits have been performed assuming $O(2)$ universality class. The quality of these fits were similar to the $O(4)$ ones and the resulting transition temperatures turned out to be the same within statistical errors [24]. Having determined T_c for $HISQ/tree$ and *asqtad* action for each N_τ a combined continuum extrapolation was performed using different assumption about the N_τ dependence. of T_c This analysis gave [24]:

$$T_c = (159 \pm 9)\text{MeV}. \tag{24}$$

The analysis also demonstrated that $HISQ/tree$ and *asqtad* action give consistent results in the continuum limit. The Budapest-Wuppertal collaboration found $T_c = 147(2)(3)$MeV, 157(3)(3)MeV and 155(3)(3)MeV defined as peak position in χ_R, inflection points in $\Delta_{l,s}$ and Δ_l^R respectively [20]. These agree with the above value within errors.

6. Temporal meson correlars and spectral functions

Information on hadron properties at finite temperature as well as the transport coefficients are encoded in different spectral functions. In particular the fate of different quarkonium states in QGP can studied by calculating the corresponding quarkonium spectral functions (for a recent review see Ref. [56]). On the lattice we can calculate correlation function in Euclidean time. This is related to the spectral function via integral relation

$$G(\tau, T) = \int_0^\infty d\omega \sigma(\omega, T) K(\tau, \omega, T), \quad K(\tau, \omega, T) = \frac{\cosh(\omega(\tau - 1/2T))}{\sinh(\omega/2T)}. \tag{25}$$

Given the data on the Euclidean meson correlator $G(\tau, T)$ the meson spectral function can be calculated using the Maximum Entropy Method (MEM) [72]. For charmonium this was done by using correlators calculated on isotropic lattices [73, 74] as well as anisotropic lattices [75, 76, 77] in the quenched approximation. It has been found that quarkonium correlation

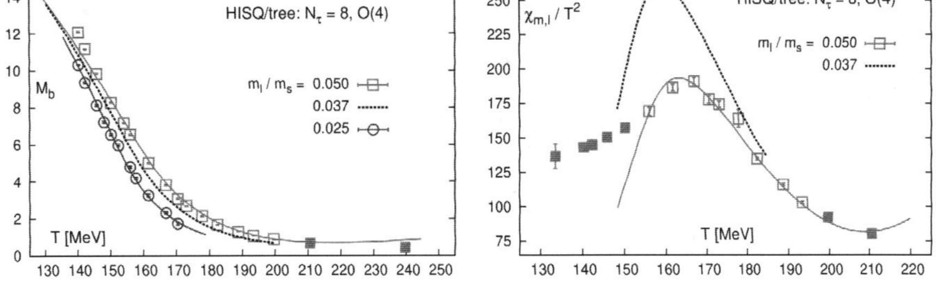

Figure 9. Scaling fits and data for the chiral condensate M_b calculated with the $HISQ/tree$ action on lattices with temporal extent $N_\tau = 8$ (left) and the chiral susceptibility $\chi_{m,l}$(right). The data for M_b at $m_l/m_s = 0.025$ and for M_b and $\chi_{m,l}$ at $m_l/m_s = 0.05$ are fit simultaneously using the $O(4)$ scaling Ansatz. The points used in the scaling fits are plotted using open symbols. The dotted lines give the data scaled to the physical quark masses.

function in Euclidean time show only very small temperature dependence [74, 77]. In other channels, namely the vector, scalar and axial-vector channels stronger temperature dependence was found [74, 77]. The spectral functions in the pseudo-scalar and vector channels reconstructed from MEM show peak structures which may be interpreted as a ground state peak [74, 75, 76]. Together with the weak temperature dependence of the correlation functions this was taken as strong indication that the 1S charmonia (η_c and J/ψ) survive in the deconfined phase to temperatures as high as $1.6T_c$ [74, 75, 76]. A detailed study of the systematic effects show, however, that the reconstruction of the charmonium spectral function is not reliable at high temperatures [77], in particular the presence of peaks corresponding to bound states cannot be reliably established. Presence of large cutoff effects at high frequencies also complicates the analysis [80]. The only statement that can be made is that the spectral function does not show significant changes within the errors of the calculations. Recently quarkonium spectral functions have been studied using potential models and lattice data for the singlet free energy of static quark anti-quark pair [46, 47, 48]. These calculations show that all charmonium states are dissolved at temperatures smaller than $1.2T_c$, but the Euclidean correlators do not show significant changes and are in fairly good agreement with available lattice data both for charmonium [74, 77] and bottomonium [77, 78]. This is due to the fact that even in absence of bound states quarkonium spectral functions show significant enhancement in the threshold region [45]. Therefore previous statements about quarkonia survival at high temperatures have to be revisited. Exploratory calculations of the charmonium correlators and spectral functions in 2-flavor QCD have been reported in Ref. [79] and the qualitative behavior of the correlation functions was found to be similar.

The large enhancement of the quarkonium correlators above deconfinement in the scalar and axial-vector channel can be understood in terms of the zero mode contribution [45, 81] and not due to the dissolution of the $1P$ states as previously thought. Similar, though smaller in magnitude, enhancement of quarkonium correlators due to zero mode is seen also in the vector channel [77]. Here it is related to heavy quark transport [44, 82]. Due to the heavy quark mass the Euclidean correlators for heavy quarkonium can be decomposed into a high and low energy part $G(\tau, T) = G_{\text{low}}(\tau, T) + G_{\text{high}}(\tau, T)$ The area under the peak in the spectral functions at zero energy $\omega \simeq 0$ giving the zero mode contribution to the Euclidean correlator is proportional to some susceptibility, $G^i_{low}(\tau, T) \simeq T\chi^i(T)$, which have been calculated on the lattice in Ref. [83]. It is natural to ask whether the generalized susceptibilities can be described by a quasi-particle model. The generalized susceptibilities have been calculated in Ref. [84] in the free theory. Replacing the bare quark mass entering in the expression of the generalized susceptibilities by an effective temperature dependent masses one can describe the zero mode contribution very well in all channels [83].

While temporal correlators are not sensitive to the change in the spectral functions spatial quarkonium correlation functions could be more sensitive to this. Recent lattice calculations show indication for significant change in spatial charmonium correlators above deconfinement[85].

The analysis of the meson spectral functions was also performed in the light quark sector [86, 87, 89, 88]. Somewhat surprisingly this analysis revealed peak structures in the deconfined phase which are difficult to interpret. Most likely the peak structures are artifacts of the MEM analysis. The new analysis of the vector spectral function performed on fine lattices and using several lattice spacings did not show any evidence for such structures [90]. Furthermore, the same analysis revealed the expected transport peak at low frequencies and estimated the electric conductivity to be [90]:

$$1/3 < \frac{\sigma}{C_{em}T} < 1, \quad C_{em} = e^2 \sum_q Q_q^2. \tag{26}$$

7. Conclusions

Recently significant progress has been made in lattice QCD calculations at non-zero temperature. Chiral and deconfining aspects of the QCD transition have been studied using improved staggered quark formulation allowing to control discretization effects. It has been shown that in the continuum limit different discretization schemes give consistent results. In particular, agreement has been reached on the value of the chiral transition temperature. There is still disagreement in the lattice calculation of the equation of state. Finally significant progress has been made in lattice calculations of meson spectral functions that encode in-medium meson properties and transport coefficients.

Acknowledgments

This work was supported by U.S. Department of Energy under Contract No. DE-AC02-98CH10886. Computations have been performed using the USQCD resources on clusters at Fermilab and JLab and the QCDOC supercomputers at BNL, as well as the BlueGene/L at the New York Center for Computational Sciences (NYCCS).

[1] D. Gross, R. Pisarski, L. Yaffe, Rev. Mod. Phys. **53**, 43 (1981)
[2] B. Müller and J. Nagle, Ann. Rev. Nucl. Part. Sci. **56**, 93 (2006)
[3] U. Wiedemann, Nucl. Phys. A, **830**, 74c (2009)
[4] J. Kuti, J. Polónyi and K. Szlachányi, Phys. Lett. B **98** (1981) 199;
[5] L. D. McLerran and B. Svetitsky, Phys. Rev. D **24** (1981) 450;
[6] J. Engels, F. Karsch, H. Satz and I. Montvay, Phys. Lett. B **101** (1981) 89.
[7] G. Boyd *et al.*, Nucl. Phys. B **469**, 419 (1996)
[8] C. W. Bernard et al., [MILC Collaboration], Phys. Rev. D **55** (1997) 6861
[9] A. Ali Khan et al., [CP-PACS Collaboration], Phys. Rev. D **64** (2001) 074510
[10] F. Karsch, E. Laermann and A. Peikert, Phys. Lett. B **478** (2000) 447;
[11] C. Bernard *et al.* [MILC Collaboration], Phys. Rev. D **71**, 034504 (2005)
[12] Y. Aoki *et al.* JHEP **0601**, 089 (2006)
[13] M. Cheng *et al.*, Phys. Rev. D **74**, 054507 (2006)
[14] Y. Aoki *et al.*, Phys. Lett. B **643** (2006) 46 [arXiv:hep-lat/0609068].
[15] C. Bernard *et al.*, Phys. Rev. D **75** (2007) 094505
[16] M. Cheng *et al.*, Phys. Rev. D **77** (2008) 014511
[17] M. Cheng *et al.*, Phys. Rev. D **81**, 054504 (2010)
[18] A. Bazavov *et al.*, Phys. Rev. D **80** (2009), 014504
[19] Y. Aoki, *et al*, JHEP **0906** (2009) 088
[20] S. Borsanyi *et al.* [Wuppertal-Budapest Collaboration], JHEP **1009**, 073 (2010)
[21] S. Borsanyi *et al*, JHEP **1011**, 077 (2010)
[22] A. Bazavov and P. Petreczky [HotQCD collaboration], J. Phys. Conf. Ser. **230**, 012014 (2010);
[23] A. Bazavov and P. Petreczky [HotQCD Collaboration], PoS **LAT2009**, 163 (2009)
[24] A. Bazavov *et al.*, arXiv:1111.1710 [hep-lat].
[25] P. Hegde, arXiv:1112.0364v1 [hep-lat]
[26] S. Borsanyi *et al.*, arXiv:1111.3500 [hep-lat].
[27] T. Umeda *et al.* [WHOT-QCD Collaboration], PoS LATTICE **2010**, 218 (2010); arXiv:1101.5582 [hep-lat].
[28] K. Orginos, D. Toussaint and R. L. Sugar [MILC Collaboration], Phys. Rev. D **60** (1999) 054503
[29] A. Bazavov *et al.* [for the HotQCD Collaboration], PoS LATTICE **2010**, 169 (2010); W. Soldner [HotQCD Collaboration], PoS LATTICE **2010**, 215 (2010)
[30] F. Csikor, *et al.*, JHEP **0405** (2004) 046
[31] P. Huovinen and P. Petreczky, Nucl. Phys. A **837**, 26 (2010)
[32] J. P. Blaizot, E. Iancu and A. Rebhan, Phys. Rev. Lett. **83** (1999) 2906;
[33] J. P. Blaizot, E. Iancu and A. Rebhan, Phys. Rev. D **63** (2001) 065003
[34] S.S. Gubser, I.R. Klebanov and I.I Tseytlin, Nucl. Phys. B **534** (1998) 202
[35] J. O. Andersen, L. E. Leganger, M. Strickland and N. Su, arXiv:1009.4644 [hep-ph].
[36] M. Cheng *et al.*, Phys. Rev. D **79** (2009) 074505
[37] S. Borsanyi *et al.* [Wuppertal-Budapest Collaboration], J. Phys. G G **38**, 124060 (2011)
[38] J. P. Blaizot, E. Iancu and A. Rebhan, Phys. Lett. B **523** (2001) 143
[39] A. Rebhan, arXiv:hep-ph/0301130.
[40] N. Haque and M. G. Mustafa, arXiv:1007.2076 [hep-ph].

[41] P. Petreczky, P. Hegde and A. Velytsky [RBC-Bielefeld Collaboration], PoS **LAT2009**, 159 (2009)
[42] K. -Y. Kim and J. Liao, Nucl. Phys. B **822**, 201 (2009)
[43] Y. Kim, Y. Matsuo, W. Sim, S. Takeuchi and T. Tsukioka, JHEP **1005**, 038 (2010)
[44] P. Petreczky, C. Miao and A. Mocsy, Nucl. Phys. A **855**, 125 (2011)
[45] Á. Mócsy and P. Petreczky, Phys. Rev. D **73**, 074007 (2006)
[46] Á. Mócsy and P. Petreczky, Phys. Rev. Lett. **99**, 211602 (2007);
[47] Á. Mócsy and P. Petreczky, Phys. Rev. D **77**, 014501 (2008);
[48] Á. Mócsy and P. Petreczky, Eur. Phys. J. ST **155**, 101 (2008)
[49] P. Petreczky, J. Phys. G **37**, 094009 (2010)
[50] P. Petreczky and K. Petrov, Phys. Rev. D **70**, 054503 (2004)
[51] O. Kaczmarek *et al.*, Phys. Lett. B **543**, 41 (2002)
[52] see e.g. Y. Hatta and K. Fukushima, arXiv:hep-ph/0311267, and references therein.
[53] S. Digal, S. Fortunato and P. Petreczky, Phys. Rev. D **68**, 034008 (2003)
[54] O. Kaczmarek and F. Zantow, Phys. Rev. D **71**, 114510 (2005)
[55] P. Petreczky, Eur. Phys. J. C **43**, 51 (2005)
[56] A. Bazavov, P. Petreczky and A. Velytsky, arXiv:0904.1748 [hep-ph].
[57] A. Bazavov, P. Petreczky and A. Velytsky, Phys. Rev. D **78**, 114026 (2008)
[58] N. Brambilla *et al.*, Phys. Rev. D **82**, 074019 (2010)
[59] M. Cheng *et al.*, Phys. Rev. D **78**, 034506 (2008)
[60] F. Karsch, M. Oevers and P. Petreczky, Phys. Lett. B **442**, 291 (1998)
[61] A. Cucchieri, F. Karsch and P. Petreczky, Phys. Lett. B **497**, 80 (2001)
[62] A. Cucchieri, F. Karsch and P. Petreczky, Phys. Rev. D **64**, 036001 (2001)
[63] S. Ejiri *et al.*, Phys. Rev. D **80**, 094505 (2009)
[64] Y. Aoki, *et al.*, Nature **443**, 675 (2006)
[65] R.D. Pisarski and F. Wilczek, Phys. Rev. D **49**, 338 (1984)
[66] M. Cheng *et al.*, Eur. Phys. J. C **71**, 1564 (2011)
[67] O. Kaczmarek *et al.*, Phys. Rev. D **83**, 014504 (2011)
[68] J. Engels, S. Holtmann, T. Mendes and T. Schulze, Phys. Lett. B **492**, 219 (2000).
[69] D. Toussaint, Phys. Rev. D **55**, 362 (1997).
[70] J. Engels and T. Mendes, Nucl. Phys. B **572**, 289 (2000).
[71] J. Engels, S. Holtmann, T. Mendes and T. Schulze, Phys. Lett. B **514**, 299 (2001).
[72] M. Asakawa, T. Hatsuda and Y. Nakahara, Prog. Part. Nucl. Phys. **46**, 459 (2001)
[73] S. Datta et al., Nucl. Phys. Proc. Suppl. **119**, 487 (2003)
[74] S. Datta, *et al.*, Phys. Rev. D **69**, 094507 (2004)
[75] T. Umeda, K. Nomura and H. Matsufuru, Eur. Phys. J. C **39S1**, 9 (2005)
[76] M. Asakawa and T. Hatsuda, Phys. Rev. Lett. **92**, 012001 (2004)
[77] A. Jakovác *et al.*, Phys. Rev. D **75**, 014506 (2007)
[78] S. Datta *et al.*, AIP Conf. Proc. **842**, 35 (2006)
[79] G. Aarts *et al.*, Phys. Rev. D **76**, 094513 (2007)
[80] F. Karsch, *et al.*, Phys. Rev. D **68**, 014504 (2003) [arXiv:hep-lat/0303017].
[81] T. Umeda, Phys. Rev. D **75**, 094502 (2007)
[82] P. Petreczky and D. Teaney, Phys. Rev. D **73**, 014508 (2006)
[83] P. Petreczky, Eur. Phys. J. C **62**, 85 (2009)
[84] G. Aarts and J. M. Martinez Resco, Nucl. Phys. B **726** (2005) 93
[85] S. Mukherjee, Nucl. Phys. A **820** (2009) 283C
[86] I. Wetzorke, F. Karsch, E. Laermann, P. Petreczky and S. Stickan, Nucl. Phys. Proc. Suppl. **106**, 510 (2002)
[87] F. Karsch *et al.*, Phys. Lett. B **530**, 147 (2002);
[88] F. Karsch *et al.*, Nucl. Phys. A **715**, 701 (2003)
[89] M. Asakawa, T. Hatsuda and Y. Nakahara, Nucl. Phys. A **715**, 863 (2003)
[90] H. -T. Ding *et al.*, Phys. Rev. D **83**, 034504 (2011)

IUPAP C20 Conference on Computational Physics (CCP 2011) IOP Publishing
Journal of Physics: Conference Series **402** (2012) 012037 doi:10.1088/1742-6596/402/1/012037

From nucleons to nuclei to fusion reactions

S. Quaglioni[1], P. Navrátil[2,1], R. Roth[3], and W. Horiuchi[4]

[1] Lawrence Livermore National Laboratory, P.O. Box 808, L-414, Livermore, CA 94551, USA
[2] TRIUMF, 4004 Wesbrook Mall, Vancouver, BC V6T 2A3, Canada
[3] Institut für Kernphysik, Technische Universität Darmstadt, 64289 Darmstadt, Germany
[4] RIKEN Nishina Center, Wako 351-0198, Japan

E-mail: quaglioni1@llnl.gov

Abstract. Nuclei are prototypes of many-body open quantum systems. Complex aggregates of protons and neutrons that interact through forces arising from quantum chromo-dynamics, nuclei exhibit both bound and unbound states, which can be strongly coupled. In this respect, one of the major challenges for computational nuclear physics, is to provide a unified description of structural and reaction properties of nuclei that is based on the fundamental underlying physics: the constituent nucleons and the realistic interactions among them. This requires a combination of innovative theoretical approaches and high-performance computing. In this contribution, we present one of such promising techniques, the *ab initio* no-core shell model/resonating-group method, and discuss applications to light nuclei scattering and fusion reactions that power stars and Earth-base fusion facilities.

1. Introduction

Low-energy nuclear reactions are the fuel of stars such as our Sun, but also of research facilities directed toward developing fusion power by either magnetic (*e.g.* ITER [1]) or inertial (*e.g.* NIF [2]) confinement. Consequently, astrophysics models as well as fusion-experiment simulations rely on various nuclear input data such as cross sections (or, equivalently, astrophysical S-factors), energy spectra, angular distributions, *etc.*, for thermonuclear reactions.

Providing the research community with accurate nuclear fusion data is one of the longstanding challenges of experimental and theoretical nuclear physics both. On one hand, due to the extreme low energies and ionized state of matter at which they take place, nuclear fusion reactions can be very challenging or even impossible to measure in beam-target experiments, leaving to theory a large role in extracting (often extrapolating) the astrophysically important information. On the other hand, a fully developed fundamental theory able to provide accurate predictions and uncertainties for a large range of relevant reactions is still missing, the main obstacle being the treatment of scattering states for many-nucleon systems based on the constituent nucleons and the realistic interactions among them.

Some of the outstanding light-nucleus uncertainty sources in astrophysics applications include: reactions leading to the nucleosynthesis of ^{8}B (and the production of the solar neutrinos measured in terrestrial experiments) such as the ^{7}Be(p,γ)^{8}B and ^{3}He(α,γ)^{7}Be radiative capture rates; the thermonuclear reaction rates of α capture on ^{8}Be and ^{12}C nuclei during the stellar helium burning; and fusion reactions that affect the predictions of Big Bang nucleosynthesis for the abundances of light elements, such as the ^{3}He(d,p)^{4}He. At the same time, large uncertainties in reactions such as the ^{3}H+$d \rightarrow ^4$He+$n+\gamma$ bremsstrahlung process or the ^{3}H(^{3}H$, 2n)^4$He fusion are limiting factors in the understanding of how the fuel is assembled in deuterium-tritium based fusion experiments. The increasingly sophisticate experimental effort devoted to these and other reactions needs to be accompanied by accurate calculations with predictive capability in the low-energy limit.

At the same time, developing a comprehensive description of nuclear properties requires also understanding exotic nuclei, short-lived nuclei that inhabit remote regions of the nuclear landscape, where the neutron-to-proton rations are larger or smaller than those naturally occurring on earth. These nuclei are difficult to study, due to their fragile nature and the often small production cross section. They present new phenomena, such as halo densities, vanishing of magic numbers, and "abnormal" spin-parity of ground states. Once again, the new experiments with exotic nuclei that will be carried out at the Facility for Rare Isotope Beams need to be supported by predictive calculations.

In both instances, a fundamental theory is needed that will enhance our ability of predicting nuclear behaviors in regions that might not be accessible to experiments. Developing such a theory requires abandoning the "traditional" separated treatment of discrete states and scattering continuum, and seeking a unified description of structural and reaction properties. Indeed, nuclei are prototypical many-body open quantum systems. They exhibit bound states, resonances, scattering states, all of which can be strongly coupled. Nuclear structure properties in the vicinity of thresholds are affected by the continuum of scattering and decay channels. At the same time, low-energy scattering and reactions, such as fusions rates, are affected by the internal structure of the interacting nuclei.

In the framework of the interacting shell model, first attempts to incorporate continuum effects in structure calculations based on the Feshbach projection formalism [3, 4] led to various formulations of the continuum shell model [5–9]. Modern versions of this are the shell model embedded in the continuum [8, 10], and the time-dependent approach to the continuum shell model [11]. A more recent attempt of extending the interacting shell model to the treatment of open quantum systems is the complex-energy or Gamow shell model, in which a symmetric description of bound, resonant and scattering single-particle states is achieved by working within a Berggren ensemble [12–15]. The cluster orbital shell model is an other variant of complex-energy shell model [16].

An alternative to the shell model picture of the nucleus was proposed in 1937 by Wheeler [17], in the form of the resonating-group method [18], a microscopic approach which explicitly takes cluster correlations into account. Here, nuclear bound states, resonances and reactions are treated within the same framework by means of expansions over fully antisymmetric cluster wave functions. A modern and physically equivalent incarnation of this approach is the generator coordinate method [19–22]. In this techniques, which make often use of central effective NN interactions and simplified cluster wave functions that are not necessarily eigenstates of the chosen Hamiltonian, the clustering of nuclei is introduced explicitly, so that the treatment of reactions becomes straightforward.

Finally, a unified description of the structural and reaction properties of light-to-medium mass nuclei is recently starting to be accomplished also within an *ab initio* framework, thanks mainly to the Green's function Monte Carlo [23], no-core shell model/resonating-group method (NCSM/RGM) [24–26] and coupled cluster technique with a Gamow-Hartree-Fock basis [27].

In this contribution, we give an overview of the NCSM/RGM approach. In Sec. 2 we briefly present the NCSM/RGM formalism, while a summary of its most important applications is given in Sec. 3. Conclusions are drawn in Sec. 4.

2. *Ab initio* NCSM/RGM

The *ab initio* nuclear reaction approach that we are developing is an extension of the *ab initio* no-core shell model (NCSM) [28]. The innovation which allows us to go beyond bound states and treat reactions is the use of cluster basis states in the spirit of the resonating-group method,

$$|\Phi^{J^\pi T}_{\nu r}\rangle = \Big[\big(|A{-}a\,\alpha_1 I_1^{\pi_1} T_1\rangle\,|a\,\alpha_2 I_2^{\pi_2} T_2\rangle\big)^{(sT)}\,Y_\ell\left(\hat r_{A-a,a}\right)\Big]^{(J^\pi T)}\,\frac{\delta(r-r_{A-a,a})}{r r_{A-a,a}}\,, \tag{1}$$

in which each nucleon cluster is described within the NCSM. The above translational invariant cluster basis states describe two nuclei (a target and a projectile composed of $A-a$ and a nucleons, respectively) whose centers of mass are separated by the relative coordinate $\vec r_{A-a,a}$ and that are traveling in a $^{2s}\ell_J$ wave or relative motion (with s the channel spin, ℓ the relative momentum, and J the total angular momentum of the system). Additional quantum numbers characterizing the basis states are parity $\pi=\pi_1\pi_2(-1)^\ell$ and total isospin T. For the intrinsic (antisymmetric) wave functions of the two nuclei we employ the eigenstates $|A{-}a\,\alpha_1 I_1^{\pi_1}T_1\rangle$ and $|a\,\alpha_2 I_2^{\pi_2}T_2\rangle$ of the $(A-a)$- and a-nucleon intrinsic Hamiltonians, respectively, as obtained within the NCSM approach. These are characterized by the spin-parity, isospin and energy labels $I_i^{\pi_i}, T_i$, and α_i, respectively, where $i=1,2$. In our notation, all these quantum numbers are grouped into a cumulative index $\nu=\{A{-}a\,\alpha_1 I_1^{\pi_1}T_1;\ a\,\alpha_2 I_2^{\pi_2}T_2;\ s\ell\}$. Finally, we note that the channel states (1) are not antisymmetric with respect to exchanges of nucleons pertaining to different clusters. Therefore, to preserve the Pauli principle one has to introduce the appropriate inter-cluster antisymmetrizer, schematically

$$\hat{\mathcal{A}}_\nu=\sqrt{\frac{(A{-}a)!a!}{A!}}\left(1+\sum_{P\neq id}(-)^p P\right)\,, \tag{2}$$

where the sum runs over all possible permutations of nucleons P different from the identical one that can be carried out between two different clusters (of $A-a$ and a nucleons, respectively), and p is the number of interchanges characterizing them. The operator (2) is labeled by the channel index ν to signify that its form depends on the mass partition, $(A-a,a)$, of the channel state to which is applied.

The channel states (1), fully antisimmetrized by the action of the antisymmetrization operator $\hat{\mathcal{A}}_\nu$, are used as a continuous basis set to expand the many-body wave function,

$$|\Psi^{J^\pi T}\rangle \;=\; \sum_\nu \int dr r^2\,\hat{\mathcal{A}}_\nu|\Phi^{J^\pi T}_{\nu r}\rangle\frac{[\mathcal{N}^{-1/2}\chi]^{J^\pi T}_\nu(r)}{r}\,, \tag{3}$$

where $\chi^{J^\pi T}_\nu(r)$ represent continuous linear variational amplitudes that are determined by solving the orthogonalized RGM equations:

$$\sum_{\nu'}\int dr' r'^2[\mathcal{N}^{-\frac12}\mathcal{H}\mathcal{N}^{-\frac12}]^{J^\pi T}_{\nu\nu'}(r,r')\frac{\chi^{J^\pi T}_{\nu'}(r')}{r'}=E\,\frac{\chi^{J^\pi T}_\nu(r)}{r}. \tag{4}$$

Here $\mathcal{N}^{J^\pi T}_{\nu\nu'}(r,r')$ and $\mathcal{H}^{J^\pi T}_{\nu\nu'}(r,r')$, commonly referred to as integration kernels, are respectively the overlap (or norm) and Hamiltonian matrix elements over the antisymmetrized basis (1), *i.e.*:

$$\mathcal{N}^{J^\pi T}_{\nu'\nu}(r',r)=\left\langle\Phi^{J^\pi T}_{\nu' r'}\right|\hat{\mathcal{A}}_{\nu'}\hat{\mathcal{A}}_\nu\left|\Phi^{J^\pi T}_{\nu r}\right\rangle\,,\qquad \mathcal{H}^{J^\pi T}_{\nu'\nu}(r',r)=\left\langle\Phi^{J^\pi T}_{\nu' r'}\right|\hat{\mathcal{A}}_{\nu'}H\hat{\mathcal{A}}_\nu\left|\Phi^{J^\pi T}_{\nu r}\right\rangle \tag{5}$$

where H is the microscopic $A-$nucleon Hamiltonian and E is the total energy in the center of mass (c.m.) frame. The calculation of the above many-body matrix elements, which contain all the nuclear structure and antisymmetrization properties of the system under consideration, represents the main task in performing RGM calculations. In the following we will review the various steps required for one of such calculations within the NCSM/RGM approach.

2.1. Input: nuclear Hamiltonian and cluster eigenstates

We start from the microscopic Hamiltonian for the $A-$nucleon system,

$$H = \frac{1}{A}\sum_{i<j=11}^{A}\frac{(\vec{p}_i-\vec{p}_j)^2}{2m} + \sum_{i<j=1}^{A} V_{ij}^{NN} + \sum_{i<j<k=1}^{A} V_{ijk}^{NNN}, \qquad (6)$$

where m is the nucleon mass, and V^{NN} and V^{NNN} the nucleon-nucleon (NN) –nuclear plus point-Coulomb– and three-nucleon (NNN) interactions, respectively. For the purpose of the RGM approach, it is convenient to separate Eq. (6) into the intrinsic Hamiltonians for the $(A-a)$- and a-nucleon systems, respectively $H_{(A-a)}$ and $H_{(a)}$, plus the relative motion Hamiltonian according to:

$$H = T_{\rm rel}(r) + \bar{V}_{\rm C}(r) + \mathcal{V}_{\rm rel} + H_{(A-a)} + H_{(a)}\,. \qquad (7)$$

Here, $T_{\rm rel}(r)$ is the relative kinetic energy, $\bar{V}_{\rm C}(r) = Z_{1\nu}Z_{2\nu}e^2/r$ ($Z_{1\nu}$ and $Z_{2\nu}$ being the charge numbers of the clusters in channel ν) the average Coulomb interaction between pairs of clusters, and $\mathcal{V}_{\rm rel}$ is localized relative (inter-cluster) potential given by:

$$\mathcal{V}_{\rm rel} \;=\; \sum_{i=1}^{A-a}\sum_{j=A-a+1}^{A} V_{ij}^{NN} + \sum_{i<j=1}^{A-a}\sum_{k=A-a+1}^{A} V_{ijk}^{NNN} + \sum_{i=1}^{A-a}\sum_{j<k=A-a+1}^{A} V_{ijk}^{NNN} - \bar{V}_{\rm C}(r)\,. \qquad (8)$$

Besides the nuclear components of the interactions between nucleons belonging to different clusters, it is important to notice that the overall contribution to the relative potential (8) coming from the Coulomb interaction,

$$\sum_{i=1}^{A-a}\sum_{j=A-a+1}^{A}\left(\frac{e^2(1+\tau_i^z)(1+\tau_j^z)}{4|\vec{r}_i-\vec{r}_j|} - \frac{1}{(A-a)a}\bar{V}_{\rm C}(r)\right), \qquad (9)$$

is also localized, presenting an r^{-2} behavior, as the distance r between the two clusters increases.

The other main input required for calculating the RGM integration kernels of Eq. (5) are the eigenstates of the projectile and target wave functions. In the NCSM/RGM approach, these are obtained by diagonalizing $H_{(A-a)}$ and $H_{(a)}$ in the model spaces spanned by the $(A-a)$- and a-nucleon NCSM bases, respectively. We adopt complete HO bases, the size of which is defined by the maximum number, $N_{\rm max}$, of HO quanta above the lowest configuration shared by the nucleons (the definition of the model-space size coincides for eigenstates of the same parity, differs by one unity for eigenstates of opposite parity). The same $N_{\rm max}$ value and HO frequency Ω are used for both clusters. Thanks to the unique properties of the HO basis, we can make use of Jacobi-coordinate wave functions [29, 30] for both nuclei or only for the lightest of the pair (typically the projectile, $a \leq 4$), and still preserve the translational invariance of the problem. In the second case we expand the eigenstates of the heavier cluster (typically the target) on a Slater-determinant (SD) basis, and remove completely the spurious c.m. components as explained in Sec. II.B.2 of Ref. [25]. Such dual approach can be used as a way of verifying our results. The use of the SD basis is computationally advantageous and allows us to explore reactions involving p-shell nuclei. In this case, the Hamiltonian matrix can reach dimensions up to 10^9, and diagonalizations are obtained by means of the Lanczos algorithm [31, 32].

Because of the complexity of the nuclear force among protons and neutrons, most nuclear interaction models generate strong short-range nucleon-nucleon correlations and the large but finite model spaces computationally achievable are not sufficient to reach the full convergence through a "bare" calculation. In these cases it is crucial to make the nuclear many-body problem computationally more tractable by means of effective interactions obtained through unitary transformations of the initial Hamiltonian. Such effective interactions can be derived, for target, projectile and compound A-nucleon system, through the Lee-Suzuki procedure in a consistent and formally exact way within the NCSM basis (see Refs. [28, 33] and Sec. II.B of Ref. [25]). Alternatively, one can perform variational calculations using bare similarity-renormalization-group (SRG) [34, 35] evolved potentials. This second choice is preferable in the NCSM/RGM approach, where, to avoid possible inconsistencies, it is desirable that the same nuclear interaction be used to obtain the structure of projectile and target, as well as the overall projectile-target potential (*i.e.*, in $H_{(A-a)}$, $H_{(a)}$, and $\mathcal{V}_{\rm rel}$). In most of the applications presented in Sec. 3 we employ SRG-evolved chiral N^{3}LO [36] NN potentials (SRG-N^{3}LO).

2.2. Calculation of norm and Hamiltonian kernels

From Eqs. (2) and (7) it follows that the norm and Hamiltonian kernels can be factorized into "full-space" and "model-space" components according to:

$$\mathcal{N}^{J^\pi T}_{\nu'\nu}(r',r) = \delta_{\nu'\nu}\frac{\delta(r'-r)}{r'r} + \mathcal{N}^{\rm ex}_{\nu'\nu}(r',r) \tag{10}$$

and,

$$\mathcal{H}^{J^\pi T}_{\nu'\nu}(r',r) = \Big[T_{\rm rel}(r') + \bar{V}_C(r') + E^{I_1'T_1'}_{\alpha_1'} + E^{I_2'T_2'}_{\alpha_2'}\Big]\,\mathcal{N}^{J^\pi T}_{\nu'\nu}(r',r) + \mathcal{V}^{J^\pi T}_{\nu'\nu}(r',r)\,, \tag{11}$$

where the exchange part of the norm, $\mathcal{N}^{\rm ex}_{\nu'\nu}(r',r)$, and the potential kernel, $\mathcal{V}^{J^\pi T}_{\nu'\nu}(r',r)$, are obtained in the truncated model space by expanding the Dirac delta function of Eq. (1) on a set of HO radial wave functions with identical frequency Ω and model-space size $N_{\rm max}$ as those used for the two clusters, more in detail:

$$\mathcal{N}^{\rm ex}_{\nu'\nu}(r',r) = \sum_{n'n} R_{n'\ell'}(r')R_{n\ell}(r) \times \begin{cases} \langle\Phi^{J^\pi T}_{\nu'n'}|\sum_{P\neq id}(-)^p P\,|\Phi^{J^\pi T}_{\nu n}\rangle & \text{if } a' = a \\ \langle\Phi^{J^\pi T}_{\nu'n'}|\sqrt{\frac{A!}{(A-a')!a'!}}\hat{\mathcal{A}}_\nu\,|\Phi^{J^\pi T}_{\nu n}\rangle & \text{if } a' \neq a \end{cases} \tag{12}$$

and

$$\mathcal{V}^{J^\pi T}_{\nu'\nu}(r',r) = \sum_{n'n} R_{n'\ell'}(r')R_{n\ell}(r)\,\langle\Phi^{J^\pi T}_{\nu' r'}|\sqrt{\frac{A!}{(A-a')!a'!}}\mathcal{V}_{\rm rel}\hat{\mathcal{A}}_\nu\,|\Phi^{J^\pi T}_{\nu r}\rangle\,. \tag{13}$$

Such a procedure is justified for matrix elements of localized operators such as those entering the exchange part of the norm and the potential kernels. To obtain the above expressions, we introduce the HO Jacobi channel states

$$|\Phi^{J^\pi T}_{\nu n}\rangle = \Big[\big(|A{-}a\,\alpha_1 I_1^{\pi_1}T_1\rangle\,|a\,\alpha_2 I_2^{\pi_2}T_2\rangle\big)^{(sT)}\,Y_\ell\,(\hat\eta_{A-a})\Big]^{(J^\pi T)} R_{n\ell}(r_{A-a,a})\,, \tag{14}$$

take advantage of the commutation between antisymmetrizers (2) and A-nucleon Hamiltonian (6), $[\hat{\mathcal{A}}_\nu, H]$=0, and use the following relationship dictated by symmetry considerations:

$$\hat{\mathcal{A}}_{\nu'}\hat{\mathcal{A}}_\nu|\Phi^{J^\pi T}_{\nu n}\rangle = \sqrt{\frac{A!}{(A-a')!a'!}}\hat{\mathcal{A}}_\nu|\Phi^{J^\pi T}_{\nu n}\rangle\,. \tag{15}$$

As pointed out in Sec. 2, the channel states (1) are not anti-symmetric with respect to the exchange of nucleons pertaining to different clusters (fully anti-symmetric states are recovered through the action of the operator $\hat{\mathcal{A}}_\nu$). As a consequence, the Hamiltonian kernel as defined

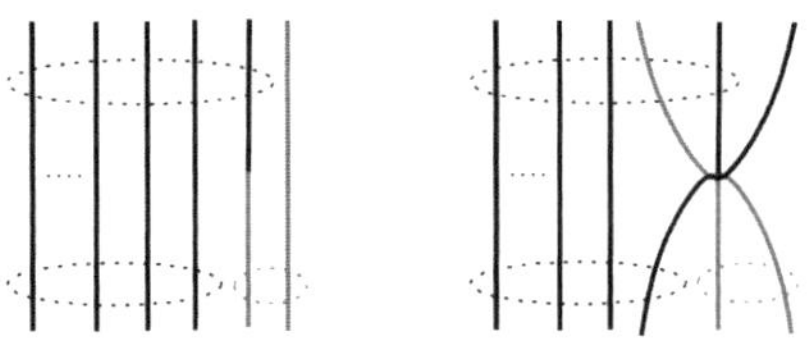

Figure 1. Diagrammatic representation of: (a) "overlap" and (b) "one-nucleon-exchange" components of the norm kernel of Eq. (18). The groups of circled lines represent the ($A-2$)-, ($A-1$)-, and two-nucleon clusters. Bottom and upper part of the diagram represent initial and final states, respectively.

in Eq. (11) is explicitly non Hermitian. Using $\hat{\mathcal{A}}_{\nu'} H \hat{\mathcal{A}}_{\nu} = \frac{1}{2}(\hat{\mathcal{A}}_{\nu'}\hat{\mathcal{A}}_{\nu} H + H\hat{\mathcal{A}}_{\nu'}\hat{\mathcal{A}}_{\nu})$, we introduce the Hermitized Hamiltonian kernel $\bar{\mathcal{H}}^{J^\pi T}_{\nu'\nu}$ in the form

$$\bar{\mathcal{H}}^{J^\pi T}_{\nu'\nu}(r',r) = \left\langle \Phi^{J^\pi T}_{\nu' r'} \right| \frac{1}{2}\left(\hat{\mathcal{A}}_{\nu'} H \sqrt{\tfrac{A!}{(A-a)!a!}} + \sqrt{\tfrac{A!}{(A-a')!a'!}} H \hat{\mathcal{A}}_{\nu}\right) \left| \Phi^{J^\pi T}_{\nu r} \right\rangle . \tag{16}$$

As a final note, we would like to point out that the exchange part of the norm kernel is explicitly hermitian, in particular:

$$\left\langle \Phi^{J^\pi T}_{\nu' n'} \right| \sqrt{\tfrac{A!}{(A-a')!a'!}} \hat{\mathcal{A}}_{\nu} \left| \Phi^{J^\pi T}_{\nu n} \right\rangle = \left\langle \Phi^{J^\pi T}_{\nu' n'} \right| \hat{\mathcal{A}}_{\nu'} \sqrt{\tfrac{A!}{(A-a)!a!}} \left| \Phi^{J^\pi T}_{\nu n} \right\rangle . \tag{17}$$

The explicit form of the inter-cluster antisymmetrizers for $a = 1$ and $a = 2$ projectiles, together with the algebraic expressions of the integration kernels for the specific cases $a' = a = 1$ and $a' = a = 2$ of equal mass partitions in initial and final states can be found in Refs. [25] and [26], respectively. For reactions involving a deuterium-nucleus entrance and nucleon-nucleus exit channels [*e.g.*, ^{3}H$(d,n)^4$He] or vice versa, and, more in general, whenever both nucleon-nucleus and deuterium-nucleus channel basis states are used in the RGM model space, one has to address the additional contributions coming from the off-diagonal matrix elements between the two mass partitions: $(A-1,1)$ and $(A-2,2)$. Here, we list these additional terms (two for the norm and five for the Hamiltonian kernel), without entering in the details of their algebraic expressions, which will be published elsewhere.

The exchange part of the norm kernel for an $(A-2,2)$ mass partition in the initial state (ν) and an $(A-1,1)$ mass partition in the final state (ν') can be cast in form:

$$\mathcal{N}^{\rm ex}_{\nu'\nu}(r',r) = \sqrt{\tfrac{A-1}{2}} \sum_{n' n} R_{n'\ell'}(r') R_{n\ell}(r) \left[2 \left\langle \Phi^{J^\pi T}_{\nu' n'} \middle| \Phi^{J^\pi T}_{\nu n} \right\rangle - (A-2) \left\langle \Phi^{J^\pi T}_{\nu' n'} \right| \hat{P}_{A-2,A} \left| \Phi^{J^\pi T}_{\nu n} \right\rangle \right] . \tag{18}$$

Two terms contribute to the above equation: the overlap between the initial $(A-2,2)$ and final (A-1,1) binary-cluster states [corresponding to diagram (a) of Fig. 1]; and a one-nucleon exchange term, corresponding to diagram (b) of Fig. 1. The corresponding Hermitized Hamiltonian kernel is obtained according to Eq. (16). In particular, the NN part of the Hermitized potential kernel

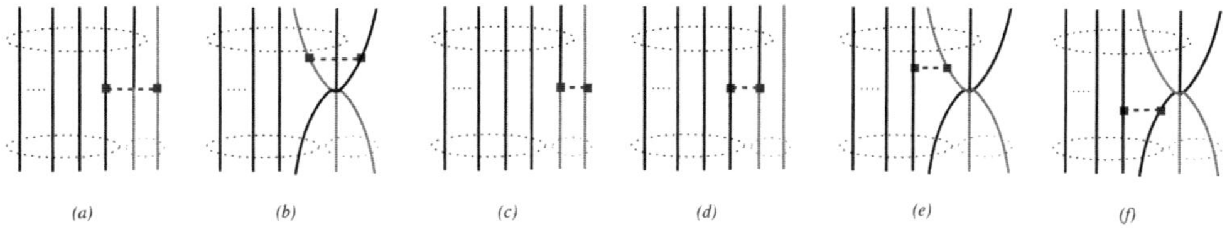

Figure 2. Diagrammatic representation of the components of the Hermitized potential kernel. The groups of circled lines represent the $(A-2)$-, $(A-1)$-, and two-nucleon clusters. Bottom and upper part of the diagram represent initial and final states, respectively.

is given by:

$$\bar{\mathcal{V}}^{NN}_{\nu'\nu}(r',r) = \sqrt{\tfrac{A-1}{2}} \sum_{n'\,n} R_{n'\ell'}(r')R_{n\ell}(r) \Big[2(A-2) \left\langle \Phi^{J^\pi T}_{\nu' n'} \right| V_{A-2,A}(1-\hat{P}_{A-2,A}) \left| \Phi^{J^\pi T}_{\nu n} \right\rangle$$
$$+ \left\langle \Phi^{J^\pi T}_{\nu' n'} \right| V_{A-1,A} \left| \Phi^{J^\pi T}_{\nu n} \right\rangle + (A-2) \left\langle \Phi^{J^\pi T}_{\nu' n'} \right| V_{A-2,A-1} \left| \Phi^{J^\pi T}_{\nu n} \right\rangle$$
$$-(A-2)(A-3) \left\langle \Phi^{J^\pi T}_{\nu' n'} \right| \tfrac{1}{2}\hat{P}_{A-2,A}V_{A-3,A-2} + V_{A-3,A-2}\hat{P}_{A-2,A} \left| \Phi^{J^\pi T}_{\nu n} \right\rangle \Big] . \quad (19)$$

In this expression, we identify four separate terms corresponding to the six diagrams of Fig. 2. The first term on the right-hand-side of Eq. (19) corresponds to diagrams (a) and (b), the second and third terms correspond to diagrams (c) and (d), respectively, while diagrams (e) and (f) represent the fourth term.

The applications presented in this contribution are obtained with the NN part only of the inter-cluster interaction. The inclusion of the three-nucleon force into the formalism, although more involved, is straightforward and is currently in progress. As an example, the potential kernel for the same $(A-1,1)$ partition in both initial and final states ($a'=a=1$) contains two additional terms due to the presence of the NNN force in the Hamiltonian. These are:

$$\mathcal{V}^{NNN}_{\nu'\nu}(r',r) = \sum_{n'\,n} R_{n'\ell'}(r')R_{n\ell}(r) \Big[\tfrac{(A-1)(A-2)}{2} \left\langle \Phi^{J^\pi T}_{\nu' n'} \right| V_{A-2,A-1,A}(1-2\hat{P}_{A-1,A}) \left| \Phi^{J^\pi T}_{\nu n} \right\rangle$$
$$- \tfrac{(A-1)(A-2)(A-3)}{2} \left\langle \Phi^{J^\pi T}_{\nu' n'} \right| \hat{P}_{A-1,A}V_{A-3,A-2,A-1} \left| \Phi^{J^\pi T}_{\nu n} \right\rangle \Big] . \quad (20)$$

As for the corresponding NN portion of the potential kernel, there are a direct and an exchange term, described by diagrams (a) and (b), and diagram (c) of Fig. 3, respectively.

The calculation of matrix elements of the type shown in Eqs. (18)-(20) represents the most computationally intensive step of the NCSM/RGM approach. Each of them requires the derivation and implementation in specialized algorithms of non trivial algebraic expressions. An advantage of the NCSM/RGM approach is that one can work with Slater-determinant target wave functions and still preserve the translational invariance of the problem (see Sec. II.B.2 of Ref. [25]). This allows the use of powerful second quantization techniques. In such a framework, the kernels are written in terms of matrix elements of the one-, two- , three-body densities, *etc.*, calculated on the target wave functions. As an example, the first two [diagrams (a) and (b) of Fig. 3] and the last [diagram (c) of Fig. 3] terms of the NNN projectile kernel (20) depend on two- and three-body density matrix elements of the target nucleus, respectively. In comparison, the NN potential kernel for an identical $(A-1,1)$ mass partition in both initial and final states depends only on matrix elements of the one and two-body densities. This

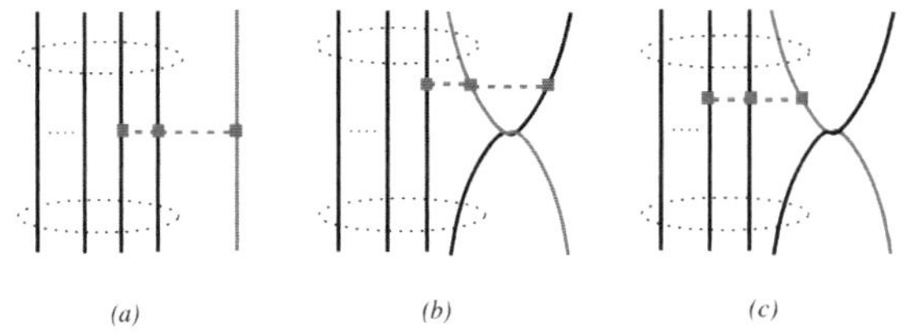

Figure 3. Diagrammatic representation of the components of the direct ((a) and (b)) and exchange components of the NNN potential kernel for the same $(A-1,1)$ partition in both initial and final states ($a' = a = 1$). The groups of circled lines represent the $(A-1)$-nucleon cluster. Bottom and upper part of the diagram represent initial and final states, respectively.

highlights the increased complexity of the NCSM/RGM approach when the three-nucleon force is included in the Hamiltonian. To conclude this section, we note that the computational demand (in terms of both CPU time and memory requirements) for the calculation of the integration kernels rapidly increases with projectile mass, number of projectile/target states, and number of channels included making the NCSM/RGM a computationally intensive approach.

2.3. Output: eigenstates, eigenenergies and Scattering matrix

Due to the Pauli exclusion principle, the integration kernels are surfaces in three dimensions, and give rise to channel-dependent non-local couplings $W^{J^\pi T}_{\nu\nu'}(r,r')$ between the unknown projectile-target relative wave functions. Indeed, by separating local diagonal and non-local terms, Eq. (4) can be cast in the form:

$$[\hat{T}_{\rm rel}(r) + \bar{V}_{\rm C}(r) - (E - E^{I_1^{\pi_1}T_1}_{\alpha_1} - E^{I_2^{\pi_2}T_2}_{\alpha_2})]\frac{\chi^{J^\pi T}_\nu(r)}{r} + \sum_{\nu'}\int dr'\, r'^2\, W^{J^\pi T}_{\nu\nu'}(r,r')\,\frac{\chi^{J^\pi T}_{\nu'}(r')}{r'} = 0. \tag{21}$$

The solution of such a set of integral-differential coupled channel equations represents a fairly standard problem in scattering theory, only slightly complicated by the presence of non local coupling potentials. A particularly efficient technique for the solution of Eq. (21) is the R-matrix method on Lagrange mesh [37, 38]. This is a method based on the microscopic R-matrix theory [39, 40] in which one assumes that $\bar{V}_{\rm C}(r)$ is the only interaction experienced by the clusters beyond a sufficiently large separation r_0, thus dividing the configuration space into an internal and an external region. In the internal region, the wave function can be written as an expansion over a set of square integrable basis functions, while in the external region it can be approximated by its asymptotic form for large r,

$$\chi^{J^\pi T}_\nu(r) = \frac{\rm i}{2}v_\nu^{-1/2}[\delta_{\nu i}H^-_\ell(\eta_\nu,\kappa_\nu r) - S^{J^\pi T}_{\nu i}H^+_\ell(\eta_\nu,\kappa_\nu r)]\,, \tag{22}$$

for scattering states, or

$$\chi^{J^\pi T}_\nu(r) = C^{J^\pi T}_\nu\, W_\ell(\eta_\nu,\kappa_\nu r)\,, \tag{23}$$

for bound states. Here, $H^\mp_\ell(\eta_\nu,\kappa_\nu r) = G_\ell(\eta_\nu,\kappa_\nu r) \mp {\rm i}F_\ell(\eta_\nu,\kappa_\nu r)$ are incoming and outgoing Coulomb functions, whereas $W_\ell(\eta_\nu,\kappa_\nu r)$ are Whittaker functions. They depend on the channel state relative angular momentum ℓ, wave number κ_ν, and Sommerfeld parameter η_ν. The corresponding velocity is denoted as v_ν. The scattering matrix $S^{J^\pi T}_{\nu i}$ (i being the initial channel) in Eq. (22) is obtained by requiring the continuity of the wave function $\chi^{J^\pi T}_\nu(r)$ and of its first

Table 1. Mean values of the relative kinetic and potential energies and of the internal ^{10}Be energy in the ^{11}Be $1/2^+$ ground state. NCSM/RGM calculation as described in the text.

NCSM/RGM	$\langle T_{\rm rel}\rangle$ (MeV)	$\langle W\rangle$ (MeV)	$E[^{10}{\rm Be(g.s., ex.)}]$ (MeV)	$E_{\rm tot}$ (MeV)
Model Space	16.65	−15.02	−56.66	−55.03
Full	6.56	−7.39	−57.02	−57.85

derivative at the matching radius r_0. The matrix elements of the scattering matrix can then be used to calculate cross sections and other reaction observables. For bound-state calculations κ_ν depends on the studied binding energy. Therefore, the determination of the bound-state energy and asymptotic normalization constant $C_\nu^{J^\pi T}$ in Eq. (23) is achieved iteratively starting from an initial guess for the value of the logarithmic derivative of the wave function at the matching radius r_0.

In the R-matrix method on Lagrange mesh, the square-integrable functions chosen to expand the wave function in the internal region are Lagrange functions [38]. This choice greatly simplifies the calculation, particularly in the presence of non local potentials. The accuracy of the R-matrix method on a Lagrange mesh is such that for a matching radius of $r_0 = 15$ fm, $N = 25$ mesh points are usually enough to determine a phase shift within the sixth significant digit. The typical matching radius and number of mesh points adopted for the calculations presented in the following section are $r_0 = 18$ fm and $N = 40$.

Finally, the solution of Eq. (21) is the least expensive of the various computing steps required for a NCSM/RGM calculation, although it can become computationally more involved as the number of coupled channels grows.

3. Applications

3.1. Parity-inversion of the ^{11}Be ground state

Developing a comprehensive description of nuclear properties requires understanding exotic nuclei, loosely bound system where the neutron-to-proton rations are larger or smaller than those naturally occurring on earth. Among light drip-line nuclei, ^{11}Be provides a convenient test of several important properties of neutron rich nuclei. The parity-inverted ground state of ^{11}Be is one of the best examples of disappearance of the $N = 8$ magic number with increasing neutron-to-proton ratio. Contrary to the shell model prediction of $\frac{1}{2}^-$ [41], the observed ground-state spin-parity of ^{11}Be is $\frac{1}{2}^+$. In 2005 Forssén *et al.* published large-scale *ab initio* NCSM calculations with several accurate NN potentials of the ^{11}Be low-lying spectrum [42]. Despite the large model space adopted, they were not able to explain the g.s. parity inversion. This result was partly attributed to the size of the HO basis, which was not large enough to reproduce the correct asymptotic of the n-^{10}Be component of the 11-body wave function. At the same time the calculations performed with the INOY (inside non-local outside Yukawa) NN potential of Doleschall *et al.* [43] suggested that the use of a realistic NNN force in a large NCSM basis might correct this discrepancy with experiment.

The correct asymptotic behavior of the n-^{10}Be wave function within ^{11}Be can be reproduced when working within the *ab initio* NCSM/RGM approach. In particular, we performed $N_{\rm max} = 6$ coupled-channel calculations [24, 25] based on n-^{10}Be channel states with four target eigenstates: ground, 2_1^+, 2_2^+, and 1_1^+ excited states. To facilitate a direct comparison with the earlier NCSM results, we used the same CD-Bonn NN interaction and HO frequency $\hbar\Omega = 13$ MeV, as in

Ref. [42]. Within this model space, we found that the NCSM and NCSM/RGM energies of the $\frac{1}{2}^-$ states are in rough agreement (-57.51 and -57.59 MeV, respectively), whereas they differ by a dramatic $\sim$ 3.5 MeV in the case of the $\frac{1}{2}^+$ state, where the $N_{\rm max} = 6$ energy is -54.39 MeV in the NCSM and -57.85 MeV in the NCSM/RGM. As a result, the $\frac{1}{2}^-$ and $\frac{1}{2}^+$ NCSM/RGM states are both bound (by 0.42 and 0.68 MeV, respectively) and the $\frac{1}{2}^+$ state is the g.s. of ^{11}Be.

To understand the reason of such a striking difference between the NCSM and NCSM/RGM results for the $\frac{1}{2}^+$ energy, we evaluated mean values of the relative kinetic and potential energies as well as the mean value of the ^{10}Be energy, and compared them to those obtained by restricting all the integration kernels to the HO model space [*i.e.* by replacing the delta function of Eq. (10) with its representation in the HO model space]. In this latter case, as in the NCSM, one loses the correct asymptotic behavior of the n-^{10}Be wave function. Indeed, it can be seen form Table 1 that the model-space-restricted calculation is similar (although not identical) to the standard NCSM calculation. In addition, due to the re-scaling of the relative wave function in the internal region when the Whittaker tail is recovered, in the full NCSM/RGM calculation we find that both average kinetic and potential energies are smaller in absolute value than those obtained within the HO model space, and this difference is larger for the relative kinetic energy. This is the origin of the dramatic decrease in energy of the $\frac{1}{2}^+$ state, which makes it bound and even leads to a g.s. parity inversion. Although the present calculations are not sufficient to exclude a role of NNN force in the inversion mechanism, it is clear that an accurate understanding of loosely-bound systems can be achieved only within a dynamic approach that encompasses the continuum.

3.2. $^4He(N,N)^4He$ scattering

The simplest system to be described in terms of binary-cluster basis states of the type described in Eq. (1) is the scattering of nucleons on ^{4}He targets. Here, energy arguments suggest that already channel states formed by a nucleon in relative motion with respect to an ^{4}He nucleus in its g.s. should provide a very good description of this process up to fairly high energies. Indeed, the ^{4}He nucleus is tightly bound and its first excited state is more than 20 MeV above its ground state. At the same time, well-determined scattering amplitudes from R-matrix fits make n- and p-^{4}He scattering calculations ideal benchmarks for our *ab initio* reaction approach.

We performed nucleon-^{4}He calculations with the SRG-N^{3}LO NN potential with $\Lambda = 2.02$ fm^{-1} in a $N_{\rm max} = 17$ NCSM/RGM model space spanned by N-^{4}He(g.s.) and N-^{4}He$^*(0_2^+)$ channel states. At $N_{\rm max} = 17$ (16, for the positive parity states) convergence of the HO expansion for the localized parts of the NCSM/RGM integration kernels and for the ^{4}He ground- and the first-excited 0^+0 states has been fully reached with this soft NN interaction.

As expected, the agreement (shown in Fig. 4) of our calculated n-^{4}He and p-^{4}He phase shifts with those obtained from an accurate R-matrix analysis of the data is quite reasonable, particularly for c.m. energies above $\sim$ 8 MeV. Correspondingly, in that energy range we can reproduce fairly well also cross-section and polarization data. As an example, Fig. 5 compares NCSM/RGM $n-$ and $p-^4$He results to the experimental data sets of Dodder *et al.* [46] and Schwandt *et al.* [45] at $E_n = 17$ and $E_p = 12$ MeV nucleon laboratory energies, respectively. The discrepancies observed in the $^2P_{3/2}$ channel, where the calculated resonance is positioned at higher energy and the phase shifts are underestimated with respect to experiment, are largely due to a reduction in spin-orbit strength caused by the omission in our calculation of the NNN interaction (chiral and SRG-induced). Efforts to include the NNN force into the NCSM/RGM formalism are currently under way. More details on these calculations can be found in Ref. [47].

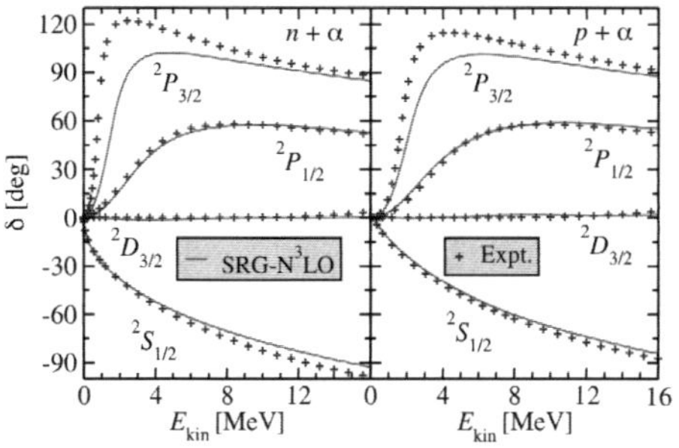

Figure 4. Adapted from Ref. [47]. Calculated $n-^4$He (left panels) and $p-^4$He (right panels) compared to the R-matrix analysis of experimental data [44]. The NCSM/RGM calculations that included the ^{4}He g.s. and the 0^+0 excited state were done using the SRG-N^{3}LO NN potential of Ref. [35] with a cutoff of 2.02 fm^{-1}. The HO frequency $\hbar\Omega$ = 20 MeV and $N_{\rm max}$ = 17 basis space were employed.

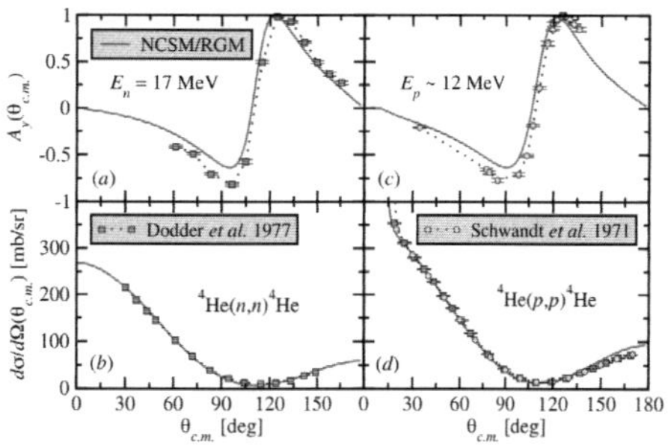

Figure 5. Calculated analyzing power (top panels) and differential cross section (bottom panels) for $n-^4$He (left panels) at a neutron laboratory energy of E_n = 17 MeV and $p-^4$He (right panels) at a proton laboratory energy of E_p = 12 MeV compared to experimental data from Schwandt *et al.* [45], and Dodder *et al.* [46]. The NCSM/RGM results as in the caption of Fig. 4.

3.3. Accurate evaluation of the $^3H(n,n)^3H$ cross section

The elastic $n-^3$H cross section induced by 14.1 MeV neutrons is important for understanding how the fuel is assembled in an inertial confinement implosion based on deuterium-tritium fuel such as those occurring at NIF [2]. To reliably infer a fuel areal density from the yield ratio between scattered and primary 14.1 MeV neutrons, this process needs to be known within about 5% accuracy. However, as shown in Fig. 6, the available data from beam-target experiments draw a rather uncertain picture for the ^{3}H$(n,n)^3$H differential cross section at this energy. In contrast, the elastic differential cross section for the mirror reaction, p-^{3}He, was measured with high accuracy at E_p=13.6 MeV [48].

In an attempt to provide an improved evaluation for the 14.1 Mev n-^{3}H elastic cross section, we performed NCSM/RGM calculations for both this and the mirror (p-^{3}H) process, using n-^{3}H(g.s.) and p-^{3}He(g.s.) channel states, respectively, in an HO model space with $N_{\rm max}$ = 17 and $\hbar\Omega$ = 20 MeV. As for the $N-^4$He case described in the previous section, we adopted the SRG-N^{3}LO NN potential with Λ = 2.02 fm^{-1}. The results of these calculations correspond to the black solid curve of Figs. 6 and 7. The agreement with the experimental ^{3}He$(p,p)^3$He differential cross section of Hutson *et al.* [48] is very good at backward angles. At forward angles, the data are underestimated by up to 15%. Such inaccuracy in the NCSM/RGM calculations is largely due to the omission of channel states with the three-nucleon system in a breakup state, which are obviously important at the laboratory energies considered here. While the inclusion of such channels was out of reach of the present calculation, we used the p-^{3}He data to quantify and correct for this inaccuracy by deducing a smooth scaling factor and applying it to the calculated n-^{3}H differential cross section. The result of this procedure, estimated to be accurate to $\sim 5\%$, are shown by red dashed curves (scaled NCSM/RGM) in Figs. 6 and 7. The n-^{3}H differential cross section at E_n=14 MeV inferred in this way, compares well with that obtained by means of an R-matrix analysis (also relying on the p-^{3}He data) [49] and with a new set of measurements obtained in a deuterium-tritium inertial confinement implosion at the OMEGA laser [50]. The integrated elastic cross section at 14 MeV obtained from the scaled NCSM/RGM calculation

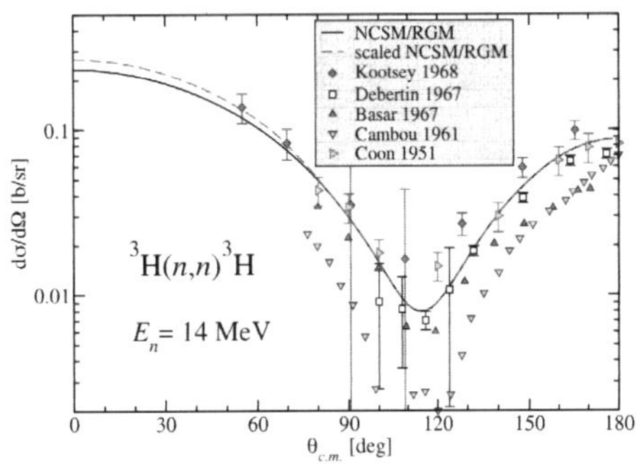

Figure 6. Neutron-^{3}H elastic differential cross section at 14 MeV. The NCSM/RGM (solid black curve) and scaled NCSM/RGM (dashed red curve) calculations described in the text are compared to available beam-target data [all].

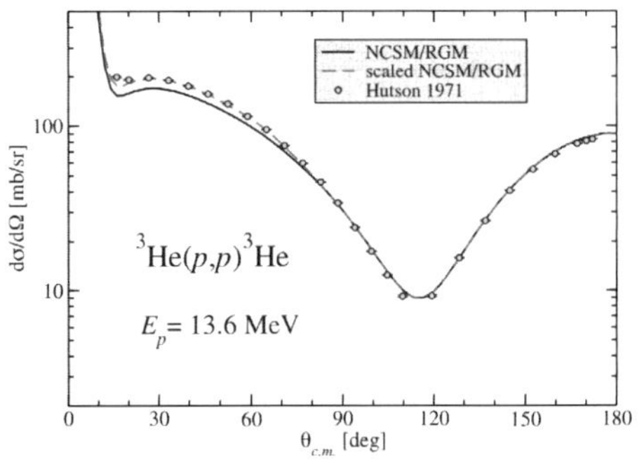

Figure 7. Proton-^{3}He elastic differential cross section at 13.6 MeV. The NCSM/RGM (solid black line) and scaled NCSM/RGM (dashed red curve) calculations described in the text are compared to the experimental data of Hutson *et al.* [48].

is 0.94 barn. Finally, we note that there is currently no *ab initio* theory capable of precisely describing the n-^{3}H reaction above the ^{3}H breakup threshold, where, in addition to n-^{3}H, one should include also three- (n-n-d) and four-body (n-n-n-p) final states. Efforts to extend the NCSM/RGM approach to include three-cluster channel states are currently under way.

3.4. The $^7Be(p,\gamma)^8B$ radiative capture

The ^{7}Be(p,γ)^{8}B radiative capture is the final step in the nucleosynthetic chain leading to ^{8}B and one of the main inputs of the standard model of solar neutrinos. Recently, we have performed the first *ab initio* many-body calculation [51], of this reaction starting from the SRG-N^{3}LO NN interaction with $\Lambda = 1.86$ fm^{-1}. Using p-^{7}Be channel states including the five lowest $N_{\rm max} = 10$ eigenstates of ^{7}Be (the $\frac{3}{2}^-$ ground and the $\frac{1}{2}^-$,$\frac{7}{2}^-$, and first and second $\frac{5}{2}^-$ excited states), we solved Eq. (4) first with bound-state boundary conditions to find the bound state of ^{8}B, and then with scattering boundary conditions to find the p-^{7}Be scattering wave functions. Former and latter wave functions were later used to calculate the capture cross section, which, at solar energies, is dominated by non-resonant $E1$ transitions from p-^{7}Be S- and D-waves into the weakly-bound ground state of ^{8}B. All stages of the calculation were based on the same HO frequency of $\hbar\Omega = 18$ MeV, which minimizes the g.s. energy of ^{7}Be. The largest model space achievable for the present calculation within the full NCSM basis is $N_{\rm max} = 10$. At this basis size, the ^{7}Be g.s. energy is very close to convergence as indicated by a fairly flat frequency dependence in the range $16 \leq \hbar\Omega \leq 20$ MeV, and the vicinity to the $N_{\rm max} = 12$ result obtained within the importance-truncated NCSM [52, 53]. The choice of $\Lambda = 1.86$ fm^{-1} in the SRG evolution of the N^{3}LO NN interaction leads to a single 2^+ bound state for ^{8}B with a separation energy of 136 keV quite close to the observed one (137 keV). This is very important for the description of the low-energy behavior of the ^{7}Be(p,γ)^{8}B astrophysical S-factor, known as S_{17}. We note that the NNN interaction induced by the SRG evolution of the NN potential is repulsive in the Λ-range $\sim$ 1.8-2.1 fm^{-1}, and, in very light nuclei, its contributions are canceled to a good extent by those of the initial attractive chiral NNN force (which is also SRG evolved) [54, 55].

The resulting S_{17} astrophysical factor is compared to several experimental data sets in Figure 8. Energy dependence and absolute magnitude follow closely the trend of the indirect Coulomb breakup measurements of Shümann et al. [56, 57], while somewhat underestimating the

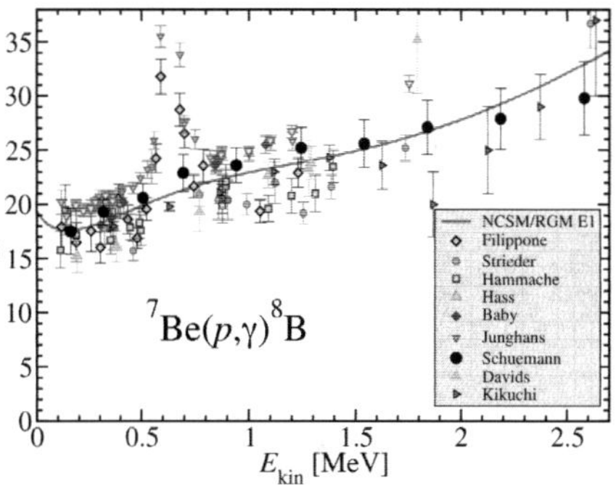

Figure 8. Calculated ^{7}Be$(p,\gamma)^{8}$B S-factor as a function of the energy in the center of mass compared to data. Only $E1$ transition were considered in the calculation.

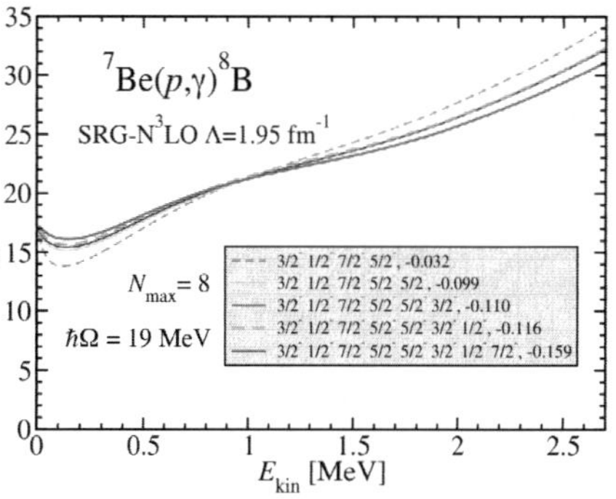

Figure 9. Convergence of the ^{7}Be$(p,\gamma)^{8}$B S-factor as a function of the number of ^{7}Be eigenstates included in the calculation (shown in the legend together with the corresponding separation energy).

direct data of Junghans *et al.* [58]. The resonance, particularly evident in these and Filippone's data, is due to the $M1$ capture, which does not contribute to a theoretical calculation outside of the narrow ^{8}B 1^{+} resonance and is negligible at astrophysical energies [59, 60]. The $M1$ operator, for which any dependence upon two-body currents needs to be included explicitly, poses more uncertainties than the Siegert's $E1$ operator. At the same time, the treatment of this operator within the NCSM/RGM approach is slightly complicated by the additional contributions coming from the core (^{7}Be) part of the wave function. Nevertheless, we plan to calculate its contribution in the future.

The convergence of our results with respect to the size of the HO model space was assessed by means of calculations up to $N_{\rm max} = 12$ within the importance-truncation NCSM scheme [52, 53]with (due to computational limitations) only the first three eigenstates of ^{7}Be. The $N_{\rm max} = 10$ and 12 S-factors are very close. As for the convergence in the number of ^{7}Be states, we explored it by means of calculations including up to 8 ^{7}Be eigenstates in a $N_{\rm max} = 8$ basis (larger $N_{\rm max}$ values are currently out of reach with more then five ^{7}Be states). This last set of calculations is presented in Fig. 9, from which it appears that, apart from the two $\frac{5}{2}^{-}$ states, the only other state to have a significant impact on the S_{17} is the second $\frac{7}{2}^{-}$, the inclusion of which affects the separation energy and contributes somewhat to the flattening of the S-factor around 1.5 MeV. We note that for these last set of calculations we used SRG-N^{3}LO interactions obtained with different Λ values with the intent to math closely the experimental separation energy in each of the largest model spaces. Based on this analysis, we conclude that the use of an $N_{\rm max} = 10$ HO model space is justified and the limitation to five ^{7}Be eigenstates is quite reasonable. Finally, our calculated $S_{17}(0) = 19.4(7)$ MeV b is on the lower side, but consistent with the latest evaluation 20.8 ± 0.7(expt)± 1.4(theory) [60].

3.5. The $^{3}H(d,n)^{4}He$ and $^{3}He(d,p)^{4}He$ fusion reactions

The ^{3}H$(d,n)^{4}$He and ^{3}He$(d,p)^{4}$He fusion reactions have important implications first and foremost for fusion energy generation, but also for nuclear astrophysics, and atomic physics. Indeed, the deuterium-tritium fusion is the easiest reaction to achieve on earth and is pursued by research facilities directed at reaching fusion power by either inertial (*e.g.*, NIF) or magnetic (*e.g.*, ITER) confinement. Both ^{3}H$(d,n)^{4}$He and ^{3}He$(d,p)^{4}$He affect the predictions of Big Bang

Journal of Physics: Conference Series **402** (2012) 012037 doi:10.1088/1742-6596/402/1/012037

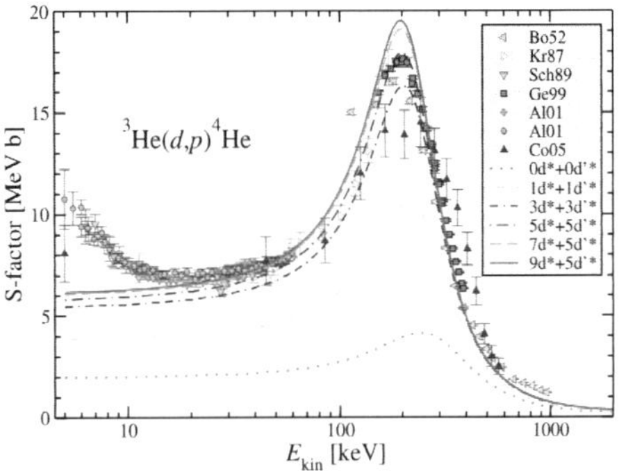

Figure 10. Calculated S-factor of the ^{3}He$(d,p)^4$He reaction compared to experimental data. Convergence with the number of deuterium pseudostates in the $^3S_1$-$^3D_1$ (d^*) and $^3D_2$ (d'^*) channels.

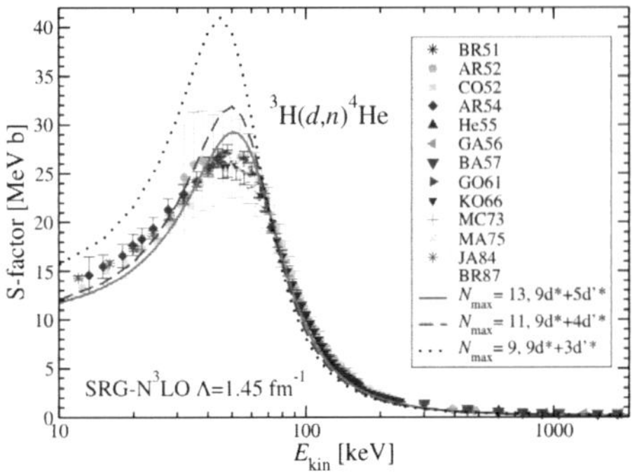

Figure 11. Calculated ^{3}H$(d,n)^4$He S-factor compared to experimental data. Convergence with $N_{\rm max}$ obtained for the SRG-N^{3}LO NN potential with $\Lambda = 1.45$ fm^{-1} at $\hbar\Omega = 14$ MeV.

nuclosynthesis for light-nucleus abundances. In addition, the deuterium-^{3}He fusion is also an object of interest for atomic physics, due to the substantial electron-screening effects presented by this reaction.

In the following we present the first *ab initio* many-body calculations [61] of these reactions starting from the SRG-N^{3}LO NN interaction with $\Lambda = 1.5$ fm^{-1}, for which we reproduce the experimental Q-value of both reactions within 1%. We adopted HO model spaces up to $N_{\rm max} = 13$ with a frequency of $\hbar\Omega = 14$ MeV. The channel basis includes n-^{4}He (p-^{4}He), d-^{3}H (d-^{3}He), d^*-^{3}H (d^*-^{3}He) and d'^*-^{3}H (d'^*-^{3}He) binary cluster states, where d^* and d'^* denote $^3S_1$-$^3D_1$ and $^3D_2$ deuterium excited pseudostates, respectively, and the ^{3}H (^{3}He) and ^{4}He nuclei are in their ground state.

Figure 10 presents the results obtained for the ^{3}He$(d,p)^4$He S-factor. The deuteron deformation and its virtual breakup, approximated by means of d pseudosates, play a crucial role. The S-factor increases dramatically with the number of pseudostates until convergence is reached for $9d^* + 5d'^*$. The dependence upon the HO basis size is illustrated by the ^{3}H$(d,n)^4$He results of Fig. 11. The convergence is satisfactory and we expect that an $N_{\rm max} = 15$ calculation, which is currently out of reach, would not yield significantly different results. The experimental position of the ^{3}He$(d,p)^4$He S-factor is reproduced within few tens of keV. Correspondingly, we find an overall fair agreement with experiment for this reaction, if we exclude the region at very low energy, where the accelerator data are enhanced by laboratory electron screening. The ^{3}H$(d,n)^4$He S-factor is not described as well with $\Lambda = 1.5$ fm^{-1}. Due to the very low activation energy of this reaction, the S-factor (particularly peak position and height) is extremely sensitive to higher-order effects in the nuclear interaction, such as three-nucleon force (not yet included in the calculation) and missing isospin-breaking effects in the integration kernels (which are obtained in the isospin formalism). To compensate for these missing higher-order effects in the interaction and reproduce the position of the ^{3}H$(d,n)^4$He S-factor, we performed additional calculations using lower Λ values. This led to the theoretical S-factor of Fig. 11 (obtained for $\Lambda = 1.45$ fm^{-1}), that is in overall better agreement with data, although it presents a slightly narrower and somewhat overestimated peak. This calculation would suggest that some electron-screening enhancement could also be present in the ^{3}H$(d,n)^4$He measured S factor below 10 keV c.m. energy. However, these results cannot be considered conclusive until more accurate calculations using a complete nuclear interaction (that includes the three-nucleon force) are

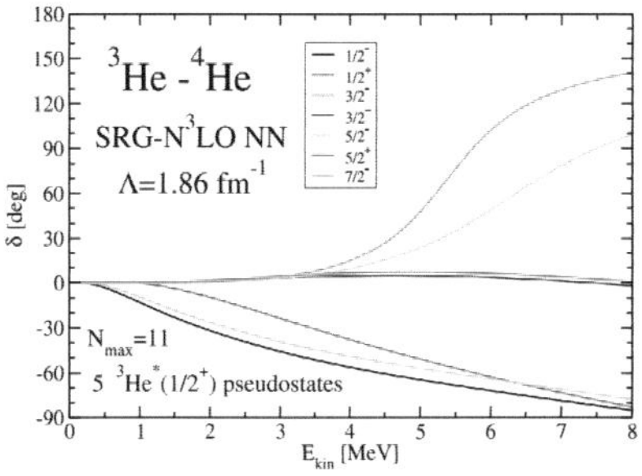

Figure 12. Preliminary results for the ^{3}He-^{4}He scattering phase shifts. The NCSM/RGM calculations, including the g.s. and first four $1/2^+$ pseudostates of ^{3}He, were obtained using the SRG-N^{3}LO NN potential with $\Lambda = 1.86$ fm^{-1}. The HO frequency $\hbar\Omega = 18$ MeV and $N_{\rm max} = 11$ basis space were employed.

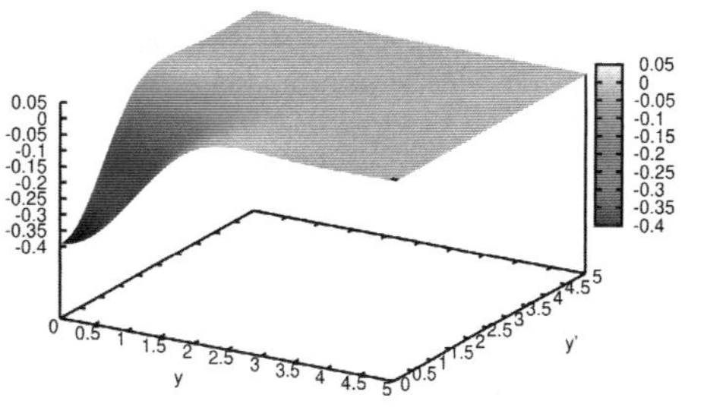

Figure 13. Preliminary results for the exchange part of the n+n+^{4}He norm kernel (in units of fm^{-6}) in the relative channel with the two neutrons (with parallel spins) in P-wave motion with respect to each other, and in overall S-wave motion with respect to the ^{4}He nucleus as a function of their distance y and y' from the target. The two neutrons in the initial and final states are at a fixed distance of 2 fm.

performed. Work in this direction is under way.

4. Conclusions and Outlook

We gave an overview of the NCSM/RGM, an *ab initio* many-body approach capable of providing a unified description of structural and reaction properties of light nuclei, by combining the RGM with the use of realistic interactions, and a microscopic and consistent description of the nucleon clusters, achieved via the *ab initio* NCSM.

Since the publication of the first results [24, 25, 47], obtained for nucleon-nucleus collisions, the NSCM/RGM has grown into a powerful approach for the description of light-ion fusion reactions. The formalism has been extended to include two-nucleon (deuteron) projectiles [26], as well as complex reactions with both nucleon-nucleus and deuteron-nucleus channels [61], based on realistic NN interactions. The treatment of three-nucleon (triton and ^{3}He) projectiles has also been included in the formalism, and will soon allow the first *ab initio* calculation of the ^{3}He(α,γ)^{7}Be radiative capture. Figure 12 shows preliminary results of our ongoing effort toward this goal. Further extensions of the approach to include the three-nucleon components of the nuclear interaction and three-cluster channel states are under way. Preliminary results for the exchange part of the n-n-^{4}He norm kernel are shown in Fig. 13.

In this contribution, we have revisited the general formalism on which the NCSM/RGM is based, and given the expressions for the terms entering the integrations kernels in the case of a deuteron-nucleus initial and nucleon-nucleus final states, as well as the additional terms appearing in the single-nucleon projectile potential kernel, when the three-nucleon force is included in the Hamiltonian.

Among the applications, we reviewed: the parity inversion of the ^{11}Be nucleus and nucleon scattering on ^{4}He, benchmark calculations that demonstrate the suitability of the NCSM/RGM approach for the description of loosely bound nuclei and low-energy scattering; the evaluation of the ^{3}H$(n,n)^3$H differential cross section for 14 MeV laboratory neutrons, an example of how the *ab initio* NCSM/RGM can provide the research community with accurate evaluations and uncertainties for less known reactions important for fusion-energy research; and the first many-body *ab initio* calculations for the astrophysically important ^{7}Be(p,γ)^{8}B radiative capture and the landmark ^{3}H(d,n)^{4}He and ^{3}He(d,p)^{4}He fusion reactions. These results are very promising and pave the way for achieving improved evaluations of reactions relevant to astrophysics and fusion research, such as the ^{3}He(α,γ)^{7}Be radiative capture or the ^{3}H+d $\rightarrow$^{4}He+n + γ bremsstrahlung process.

To conclude, we note that *ab initio* NCSM/RGM calculations such as those described in this contributions, and, more importantly, those planned for the future can rapidly grow (with projectile mass, number of projectile/target excited states, and number of channels included) into peta- and even exa-scale computing problems. As an example, the preliminary results of our ongoing investigation of ^{3}He-^{4}He scattering presented in Fig. 12 required runs with up to 64,000 cores on the Oak Ridge National Laboratory (ORNL) Jaguar [62] supercomputer.

Acknowledgments
Computing support for this work came from the LLNL institutional Computing Grand Challenge program, the Jülich supercomputer Centre and Oak Ridge Leadership Computing Facility at ORNL [62]. Prepared in part by LLNL under Contract DE-AC52-07NA27344. Support from the U. S. DOE/SC/NP (Work Proposal No. SCW1158), LLNL LDRD grant PLS-09-ERD-020, U. S. DOE/SC/NP (Work Proposal No. SCW0498), the NSERC Grant No. 401945-2011, and from the U. S. Department of Energy Grant DE-FC02-07ER41457 is acknowledged. This work is supported in part by the Deutsche Forschungsgemeinschaft through contract SFB 634 and by the Helmholtz International Center for FAIR within the framework of the LOEWE program launched by the State of Hesse. W.H. was supported by the Special Postdoctoral Researchers Program of RIKEN.

References

[1] ITER URL `http://www.iter.org/`
[2] NIF URL `https://lasers.llnl.gov/`
[3] Feshbach H 1958 *Annals of Physics* **5** 357
[4] Feshbach H 1962 *Annals of Physics* **19** 287
[5] Mahux C and Weidenmller H A 1969 *Shell Model Approach to Nuclear Reactions* (Amsterdam: North-Holland)
[6] Barz H, Rotter I and Höhn J 1977 *Nuclear Physics A* **275** 111
[7] Rotter I, Barz H and Höhn J 1978 *Nuclear Physics A* **297** 237
[8] Bennaceur K, Nowacki F, Okołowicz J and Płoszajczak M 1999 *Nuclear Physics A* **651** 289
[9] Volya A and Zelevinsky V 2006 *Phys. Rev. C* **74** 064314
[10] Rotureau J, Okołowicz J and Płoszajczak M 2006 *Nuclear Physics A* **767** 13
[11] Volya A 2009 *Phys. Rev. C* **79** 044308
[12] Michel N, Nazarewicz W, Płoszajczak M and Bennaceur K 2002 *Phys. Rev. Lett.* **89** 042502
[13] Id Betan R, Liotta R J, Sandulescu N and Vertse T 2002 *Phys. Rev. Lett.* **89** 042501
[14] Michel N, Nazarewicz W, Płoszajczak M and Okołowicz J 2003 *Phys. Rev. C* **67** 054311
[15] Papadimitriou G *et al.* 2011 *Phys. Rev. C* **84** 051304
[16] Masui H, Katō K and Ikeda K 2007 *Phys. Rev. C* **75** 034316
[17] Wheeler J A 1937 *Phys. Rev.* **52** 1083–1106
[18] Tang Y, LeMere M and Thompsom D 1978 *Physics Reports* **47** 167
[19] Horiuchi H 1977 *Prog. Theor. Phys. (Suppl.)* **90** 62
[20] Descouvemont P 2004 *Phys. Rev. C* **70** 065802
[21] Baye D, Descouvemont P and Leo F 2005 *Phys. Rev. C* **72** 024309
[22] Damman A and Descouvemont P 2009 *Phys. Rev. C* **80** 044310
[23] Nollett K M, Pieper S C, Wiringa R B, Carlson J and Hale G M 2007 *Phys. Rev. Lett.* **99** 022502
[24] Quaglioni S and Navrátil P 2008 *Phys. Rev. Lett.* **101** 092501
[25] Quaglioni S and Navrátil P 2009 *Phys. Rev. C* **79** 044606
[26] Navrátil P and Quaglioni S 2011 *Phys. Rev. C* **83** 044609

[27] Hagen G, Papenbrock T and Hjorth-Jensen M 2010 *Phys. Rev. Lett.* **104** 182501
[28] Navrátil P, Vary J P and Barrett B R 2000 *Phys. Rev. Lett.* **84** 5728
[29] Navrátil P and Barrett B R 1998 *Phys. Rev. C* **57** 562
[30] Navrátil P, Kamuntavičius G P and Barrett B R 2000 *Phys. Rev. C* **61** 044001
[31] Lanczos C 1950 *J. Res. Nat. Bur. Stand.* **45** 252
[32] Wilkinson J H 1965 *The Algebraic Eigenvalue Problem* (Oxford: Clarendon)
[33] Navrátil P, Vary J P and Barrett B R 2000 *Phys. Rev. C* **62** 054311
[34] Bogner S K, Furnstahl R J and Perry R J 2007 *Phys. Rev. C* **75** 061001
[35] Roth R, Reinhardt S and Hergert H 2008 *Phys. Rev. C* **77** 064003
[36] Entem D R and Machleidt R 2003 *Phys. Rev. C* **68** 041001
[37] Hesse M, Sparenberg J M, Raemdonck F V and Baye D 1998 *Nuclear Physics A* **640** 37
[38] Hesse M, Roland J and Baye D 2002 *Nuclear Physics A* **709** 184
[39] Lane A M and Thomas R G 1958 *Rev. Mod. Phys.* **30** 257
[40] Descouvemont P and Baye D 2010 *Rep. Prog. Phys.* **73** 036301
[41] Talmi I and Unna I 1960 *Phys. Rev. Lett.* **4** 469
[42] Forssén C, Navrátil P, Ormand W E and Caurier E 2005 *Phys. Rev. C* **71** 044312
[43] Doleschall P 2004 *Phys. Rev. C* **69** 054001
[44] Hale G M private communication
[45] Schwandt P, Clegg T and Haeberli W 1971 *Nuclear Physics A* **163** 432
[46] Dodder D C *et al.* 1977 *Phys. Rev. C* **15** 518–530
[47] Navrátil P, Roth R and Quaglioni S 2010 *Phys. Rev. C* **82** 034609
[48] Hutson R L, Jarmie N, Detch J L and Jett J H 1971 *Phys. Rev. C* **4** 17
[49] Hale G M, Dodder D C, Seagrave J D, Berman B L and Phillips T W 1990 *Phys. Rev. C* **42** 438
[50] Frenje J A *et al.* 2011 *Phys. Rev. Lett.* **107** 122502
[51] Navrátil P, Roth R and Quaglioni S 2011 *Physics Letters B* **704** 379 – 383
[52] Roth R and Navrátil P 2007 *Phys. Rev. Lett.* **99** 092501
[53] Roth R 2009 *Phys. Rev. C* **79** 064324
[54] Jurgenson E D, Navrátil P and Furnstahl R J 2009 *Phys. Rev. Lett.* **103** 082501
[55] Jurgenson E D, Navrátil P and Furnstahl R J 2011 *Phys. Rev. C* **83** 034301
[56] Schümann F *et al.* 2003 *Phys. Rev. Lett.* **73** 232501
[57] Schümann F *et al.* 2006 *Phys. Rev. C* **73** 015806
[58] Junghans A R *et al.* 2003 *Phys. Rev. C* **68** 065803
[59] Adelberger E *et al.* 2011 *Rev. Mod. Phys.* **83** 195
[60] Adelberger E *et al.* 1998 *Rev. Mod. Phys.* **70** 1265
[61] Navrátil P and Quaglioni S 2012 *Phys. Rev. Lett.* **108** 042503
[62] Information on Jaguar can be found at the URL `http://www.nccs.gov`

IUPAP C20 Conference on Computational Physics (CCP 2011) IOP Publishing
Journal of Physics: Conference Series **402** (2012) 012038 doi:10.1088/1742-6596/402/1/012038

Microscopic calculation of heavy-ion potentials based on TDHF

A S Umar and V E Oberacker

Department of Physics and Astronomy, Vanderbilt University, Nashville, TN 37235, USA

E-mail: umar@compsci.cas.vanderbilt.edu

Abstract. We present the numerical details of a new method for calculating ion-ion interaction potentials from time-dependent Hartree-Fock calculations using density as a constraint.

1. Introduction

The investigation of internuclear potentials for heavy-ion collisions is of fundamental importance for the study of fusion reactions as well as for the formation of superheavy elements and nuclei far from stability. Recently, we have developed a new method to extract ion-ion interaction potentials directly from the time-dependent Hartree-Fock (TDHF) time-evolution of the nuclear system [1]. In the density-constrained TDHF (DC-TDHF) approach the TDHF time-evolution takes place with no restrictions. At certain times during the evolution the instantaneous density is used to perform a static Hartree-Fock minimization while holding the neutron and proton densities constrained to be the corresponding instantaneous TDHF densities. In essence, this provides us with the TDHF dynamical path in relation to the multi-dimensional static energy surface of the combined nuclear system. In this approach there is no need to introduce constraining operators which assume that the collective motion is confined to the constrained phase space. In short, we have a self-organizing system which selects its evolutionary path by itself following the microscopic dynamics. Some of the effects naturally included in the DC-TDHF calculations are: neck formation, mass exchange, internal excitations, deformation effects to all order, as well as the effect of nuclear alignment for deformed systems. The DC-TDHF theory provides a comprehensive approach to calculating fusion barriers in the mean-field limit. The theory has been applied to calculate fusion cross-sections for ^{64}Ni+^{132}Sn, ^{64}Ni+^{64}Ni, ^{16}O+^{208}Pb, ^{70}Zn + ^{208}Pb, ^{48}Ca + ^{238}U, and ^{132,124}Sn+^{96}Zr systems [2, 3, 4, 5, 6, 7, 8]. In this paper we will outline the DC-TDHF method and give examples of its application to the calculation of fusion cross-sections for various systems.

2. Theory

In many branches of science, highly complex many-body systems are often described in macroscopic terms, this is particularly true in the case of relativistic and non-relativistic heavy-ion collisions. For example, the time evolution of the nuclear surface and the corresponding geometrical shape provides a very useful parameter to help organize experimental data. Using this approach numerous evolutionary models have been developed to explain particular aspects of experimental data. These methods provide a useful and productive means for quantifying multitudinous reaction data. In practice, they require a quantitative understanding of the data as well as a clear physical picture of the important aspects of the reaction dynamics. The depiction of the collision must be given at the onset, including the choice of coordinates which govern the evolution of the reaction. Guessing the correct degrees of freedom is extremely hard, without a full understanding of the dynamics, and can easily lead to misbegotten results. More importantly, it is most often not possible to connect these macroscopic classical parameters, describing nuclear matter under extreme excitation and rearrangement, with the more fundamental properties of the nuclear force. Such difficulties can only be overcome with a fully microscopic theory of the collision dynamics.

2.1. Time-dependent Hartree-Fock method

The theoretical formalism for the microscopic description of complex many-body quantum systems and the understanding of the nuclear interactions that result in self-bound, composite nuclei possessing the observed properties are the underlying challenges for studying low energy nuclear physics. The Hartree-Fock approximation and its time-dependent generalization the time-dependent Hartree-Fock theory has provided a possible means to study the diverse phenomena observed in low energy nuclear physics [9].

Given a many-body Hamiltonian containing two and three-body interactions

$$H = \sum_{i}^{N} t_i + \sum_{i<j}^{N} v_{ij} + \sum_{i<j<k}^{N} v_{ijk} \;, \tag{1}$$

the time-dependent action S can be constructed as

$$S = \int_{t_1}^{t_2} dt < \Phi(t)|H - i\hbar\partial_t|\Phi(t) > \;. \tag{2}$$

Here, Φ denotes the time-dependent, many-body wavefunction, $\Phi(\mathbf{r_1}, \mathbf{r_2}, \ldots, \mathbf{r_A}; t)$, and t_i is the one-body kinetic energy operator. General variation of S recovers the time-dependent Schrödinger equation. In TDHF approximation the many-body wavefunction is replaced by a single Slater determinant and this form is preserved at all times. The determinental form guarantees the antisymmetry required by the Pauli principle for a system of fermions. In this limit, the variation of the action yields the most probable time-dependent path between points t_1 and t_2 in the multi-dimensional space-time phase space

$$\delta S = 0 \rightarrow \Phi(t) = \Phi_0(t) \;. \tag{3}$$

In practice $\Phi_0(t)$ is chosen to be a Slater determinant

$$\Phi_0(t) = \frac{1}{\sqrt{N!}} det|\phi_\lambda(\mathbf{r}\,, t)| \;, \tag{4}$$

where $\phi_\lambda(\mathbf{r}\,, t)$ are the single-particle states with quantum numbers λ. If the variation in Eq.(3) is performed with respect to the single-particle states ϕ_λ^* we obtain a set of coupled, nonlinear, self-consistent initial value equations for the single-particle states

$$h\left(\{\phi_\mu\}\right)\phi_\lambda = i\hbar\dot{\phi}_\lambda \qquad \lambda = 1, ..., N \;. \tag{5}$$

These are the fully microscopic time-dependent Hartree-Fock equations which preserve the major conservation laws such as the particle number, total energy, total angular momentum, etc. As we see from Eq.(5), each single-particle state evolves in the mean-field generated by the concerted action of all the other single-particle states. Static equations can be obtained from Eq.(5) by taking out a trivial phase from the single-particle states

$$\begin{aligned} h(\{\chi_\mu\})\chi_\lambda &= \epsilon_\lambda \chi_\lambda \\ \phi_\lambda(\mathbf{r}, t) &= e^{-i\epsilon_\lambda t/\hbar}\chi_\lambda(\mathbf{r}) \,. \end{aligned} \tag{6}$$

In TDHF, the initial nuclei are calculated using the static Hartree-Fock (HF) theory. The resulting Slater determinants for each nucleus comprise the larger Slater determinant describing the colliding system during the TDHF evolution. Nuclei are assumed to move on a pure Coulomb trajectory until the initial separation between the nuclear centers used in TDHF evolution. Using the Coulomb trajectory we compute the relative kinetic energy at this separation and the associated translational momenta for each nucleus. The nuclei are than boosted by multiplying the HF states with

$$\Phi_j \to \exp(\imath \mathbf{k}_j \cdot \mathbf{R})\Phi_j \,, \tag{7}$$

where Φ_j is the HF state for nucleus j and $\mathbf{R}$ is the corresponding center of mass coordinate

$$\mathbf{R} = \frac{1}{A_j}\sum_{i=1}^{A_j} \mathbf{r}_i \,. \tag{8}$$

The Galilean invariance of the TDHF equations assures the evolution of the system without spreading and the conservation of the total energy for the system. In TDHF, the many-body state remains a Slater determinant at all times. The final state is a filled determinant, even in the case of two well separated fragments. This phenomenon is commonly known as the "cross-channel coupling" and indicates that it is not possible to identify the well separated fragments as distinct nuclei since each single particle state will have components distributed everywhere in the numerical box. In this sense it is only possible to extract *inclusive* (averaged over all states) information from these calculations.

2.2. DC-TDHF method

The concept of using density as a constraint for calculating collective states from TDHF time-evolution was first introduced in Ref. [10], and used in calculating collective energy surfaces in connection with nuclear molecular resonances in Ref. [11].

In this approach we assume that a collective state is characterized only by density ρ, and current $\mathbf{j}$. This state can be constructed by solving the static Hartree-Fock equations

$$< \Phi_{\rho\mathbf{j}}|a_h^\dagger a_p \hat{H}|\Phi_{\rho\mathbf{j}} >= 0 \,, \tag{9}$$

subject to constraints on density and current

$$\begin{aligned} < \Phi_{\rho\mathbf{j}}|\hat{\rho}(\mathbf{r})|\Phi_{\rho\mathbf{j}} > &= \rho(\mathbf{r}, t) \\ < \Phi_{\rho\mathbf{j}}|\hat{\jmath}(\mathbf{r})|\Phi_{\rho\mathbf{j}} > &= \mathbf{j}(\mathbf{r}, t) \,. \end{aligned}$$

Choosing $\rho(\mathbf{r}, t)$ and $\mathbf{j}(\mathbf{r}, t)$ to be the instantaneous TDHF density and current results in the lowest energy collective state corresponding to the instantaneous TDHF state $|\Phi(t)>$, with the corresponding energy

$$E_{coll}(\rho(t), \mathbf{j}(t)) =< \Phi_{\rho\mathbf{j}}|\hat{H}|\Phi_{\rho\mathbf{j}} > \,. \tag{10}$$

This collective energy differs from the conserved TDHF energy only by the amount of internal excitation present in the TDHF state, namely

$$E^*(t) = E_{TDHF} - E_{coll}(t) \,. \tag{11}$$

However, in practical calculations the constraint on the current is difficult to implement but we can define instead a static adiabatic collective state $|\Phi_\rho >$ subject to the constraints

$$\begin{aligned} < \Phi_\rho|\hat{\rho}(\mathbf{r})|\Phi_\rho > &= \rho(\mathbf{r},t) \\ < \Phi_\rho|\hat{\jmath}(\mathbf{r})|\Phi_\rho > &= 0\,. \end{aligned}$$

In terms of this state one can write the collective energy as

$$E_{coll} = E_{kin}(\rho(t),\mathbf{j}(t)) + E_{DC}(\rho(\mathbf{r},t))\,, \tag{12}$$

where the density-constrained energy E_{DC}, and the collective kinetic energy E_{kin} are defined as

$$\begin{aligned} E_{DC} &= < \Phi_\rho|\hat{H}|\Phi_\rho > \\ E_{kin} &\approx \frac{m}{2}\sum_q \int d^3r\, \mathbf{j}_q^2(t)/\rho_q(t)\,, \end{aligned}$$

where the index q is the isospin index for neutrons and protons ($q = n,p$). From Eq. 12 is is clear that the density-constrained energy E_{DC} plays the role of a collective potential. In fact this is exactly the case except for the fact that it contains the binding energies of the two colliding nuclei. One can thus define the ion-ion potential as [1]

$$V = E_{\mathrm{DC}}(\rho(\mathbf{r},t)) - E_{A_1} - E_{A_2}\,, \tag{13}$$

where E_{A_1} and E_{A_2} are the binding energies of two nuclei obtained from a static Hartree-Fock calculation with the same effective interaction. For describing a collision of two nuclei one can label the above potential with ion-ion separation distance $R(t)$ obtained during the TDHF time-evolution. This ion-ion potential $V(R)$ is asymptotically correct since at large initial separations it exactly reproduces $V_{Coulomb}(R_{max})$. In addition to the ion-ion potential it is also possible to obtain coordinate dependent mass parameters. One can compute the "effective mass" $M(R)$ using the conservation of energy

$$M(R) = \frac{2[E_{\mathrm{c.m.}} - V(R)]}{\dot{R}^2}\,, \tag{14}$$

where the collective velocity $\dot{R}$ is directly obtained from the TDHF evolution and the potential $V(R)$ from the density constraint calculations.

2.3. Skyrme interaction

Almost all TDHF calculations have been done using the Skyrme interaction. A variety of calculations have shown that the TDHF results are very sensitive to the different parametrization of the Skyrme force [12]. The Skyrme energy density functional contains terms which depend on the nuclear density, ρ, kinetic-energy density, τ, spin density, $\mathbf{s}$, spin kinetic energy density, $\mathbf{T}$, and the full spin-current pseudotensor, $\mathbf{J}$, as

$$E = \int d^3r\, \mathcal{H}(\rho,\tau,\mathbf{j},\mathbf{s},\mathbf{T},\mathbf{J};\mathbf{r})\,. \tag{15}$$

The time-odd terms ($\mathbf{j}$, $\mathbf{s}$, $\mathbf{T}$) vanish for static calculations of even-even nuclei, while they are present for odd mass nuclei, in cranking calculations, as well as in TDHF. The spin-current pseudotensor, $\mathbf{J}$, is time-even and does not vanish for static calculations of even-even nuclei. Our TDHF program includes all of the time-odd terms in the Skyrme interaction.

3. Numerical methods

In this section we discuss the numerical details of performing TDHF calculations of nuclear collisions as well as the density-constraint method, which is crucial for ion-ion potential calculations.

3.1. Discrete variation and lattice equations

The lattice solution of differential equations on a discretized mesh of independent variables may be viewed to proceed in two steps: (1) Obtain a discrete representation of the functions and operators on the lattice. (2) Solve the resulting lattice equations using iterative techniques. Step (1) is an interpolation problem for which we could take advantage of the techniques developed using the basis-spline functions [13]. The use of the basis-spline collocation method leads to a matrix-vector representation on the collocation lattice with a metric describing the transformation properties of the collocation lattice.

In order to obtain a set of lattice equations which preserve the conservation laws associated with the continuous equations it is essential to develop a modified variational approach. This goal is achieved by performing a variation to the discretized form of a conserved quantity, i.e. total energy. Consequently, the resulting equations will preserve all of the conserved quantities on the lattice. For the TDHF equations we consider a general discretized form of the action

$$S = \int dt \sum_{\alpha\beta\gamma} \Delta V_{\alpha\beta\gamma} \left\{ \mathcal{H}(\alpha\beta\gamma) - \left[i\hbar \sum_{\mu} \psi_\mu^*(\alpha\beta\gamma) \frac{\partial \psi_\mu}{\partial t}(\alpha\beta\gamma) \right] \right\} , \tag{16}$$

where indices α, β, γ denote the lattice points in three-dimensional space, and $\Delta V_{\alpha\beta\gamma}$ is the corresponding infinitesimal volume element. Due to the presence of derivative operators in the Hamiltonian the explicit form of these expressions will depend non-locally on the lattice indices. The general variation, which preserves the properties of the continuous variation, is given by

$$\frac{\delta \psi_\mu^*(\alpha\beta\gamma)}{\delta \psi_\lambda^*(\alpha'\beta'\gamma')} = \frac{1}{\Delta V_{\alpha\beta\gamma}} \delta_{\lambda\mu} \delta_{\alpha'\alpha} \delta_{\beta'\beta} \delta_{\gamma'\gamma} . \tag{17}$$

Until recently, most HF and TDHF calculations have been performed using finite-difference lattice techniques. The details of the discrete variation for the finite-difference case are given in Refs. [14, 15]. Below we outline a procedure for using the BSCM for the numerical solution of HF and TDHF equations. Further details of the BSCM is published elsewhere [13].

3.2. Basis-splines

Given a set of points or *knots* denoted by the set $\{x_i\}$ a basis-spline (B-spline denoted by B_i^M) function of order M is constructed from continuous piecewise polynomials of order $M-1$ [16]. B-splines have continuous derivatives up to $(M-2)^{nd}$ derivative and a discontinuous $(M-1)^{st}$ derivative. We only consider odd order splines or even order polynomials for reasons related to the choice of the collocation points. The ith B-spline is nonzero only in the interval (x_i, x_{i+M}). This property is commonly referred to as limited support. The knots are the points where polynomials that make up the B-spline join. In the interval containing the tail region B-splines fall off very rapidly to zero. An example of order $M = 5$ splines extending over a physical region is illustrated in Fig.1. We can also construct exact derivatives of B-splines provided the derivative order does not exceed $M - 1$.

A continuous function $f(x)$, defined in the interval (x_{min}, x_{max}), can be expanded in terms of B-spline functions as

$$f(x) = \sum_i B_i^M(x) c^i , \tag{18}$$

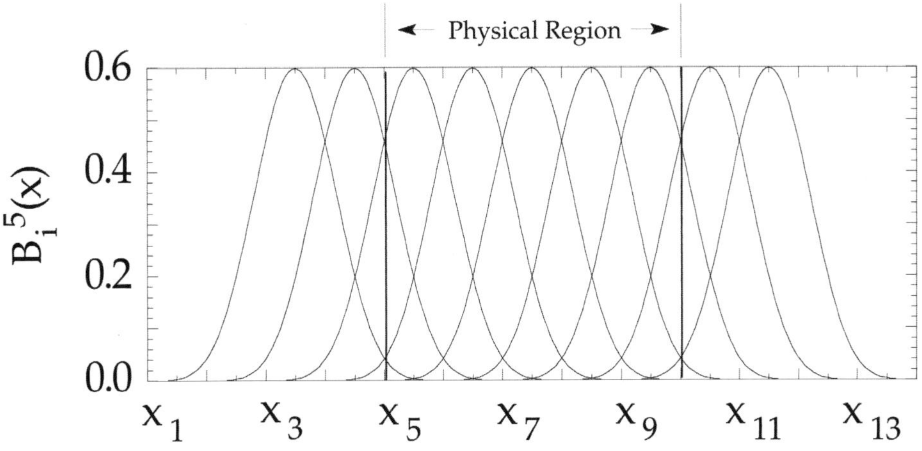

Figure 1. A region of space with physical boundaries located at knots x_M and x_{M+N} for $M = 5$ and $N = 5$. The B-spline B_1^M which begins at the first knot x_1 has its tail in the physical region. The last B-spline which begins within the physical boundaries is B_{N+M-1}^M. It extends up to the last knot x_{N+2M-1}.

where quantities c^i denote the expansion coefficients. We can solve for the expansion coefficients in terms of a given (or to be determined) set of function values evaluated at a set of data points more commonly known as *collocation points*. There are a number of ways to choose collocation points [16, 13], however, for odd order B-splines a simple choice is to place one collocation point at the center of each knot interval within the physical boundaries

$$x_\alpha = \frac{x_{\alpha+M-1} + x_{\alpha+M}}{2} \quad , \quad \alpha = 1, \ldots, N. \tag{19}$$

Here, $x_M = x_{min}$, $x_{N+M} = x_{max}$, and N is the number of collocation points. We can now write a linear system of equations by evaluating (18) at these collocation points

$$f_\alpha = \sum_i B_{\alpha i} c^i \ , \tag{20}$$

where $f_\alpha \equiv f(x_\alpha)$, and $B_{\alpha i} \equiv B_i^M(x_\alpha)$. In order to solve for the expansion coefficients the matrix $\mathbf{B}$ needs to be inverted. However, as it stands matrix $\mathbf{B}$ is not a square matrix, since the total number of B-splines with a nonzero extension in the physical region is $N + M - 1$. In order to perform the inversion we need to introduce additional linear equations which represent the boundary conditions imposed on $f(x)$ at the two boundary points, x_M and x_{M+N}. The essence of the lattice method is to eliminate the expansion coefficients c^i using this inverse matrix. The details of using the boundary conditions (or periodic boundaries) and inverting the resulting square matrix are discussed in Ref. [13]. Following the inversion the coefficients are given by

$$c^i = \sum_\alpha \left[\mathbf{B}^{-1}\right]^{i\alpha} f_\alpha \ . \tag{21}$$

One can trivially show that all local functions will have a local representation in the finite dimensional collocation space

$$f(x) \longrightarrow f_\alpha \ . \tag{22}$$

The collocation representation of the operators can be obtained by considering the action of an operator $\mathcal{O}$ onto a function $f(x)$

$$\mathcal{O}f(x) = \sum_i [\mathcal{O}B_i^M(x)]c^i \;. \tag{23}$$

If we evaluate the above expression at the collocation points x_α we can write

$$[\mathcal{O}f]_\alpha = \sum_i [\mathcal{O}B]_{\alpha i} c^i \;. \tag{24}$$

Substituting from Eq. (21) for the coefficients c^i we obtain

$$\begin{aligned} [\mathcal{O}f]_\alpha &= \sum_{i\beta} [\mathcal{O}B]_{\alpha i} \left[\mathbf{B}^{-1}\right]^{i\beta} f_\beta \\ &= \sum_\beta O_\alpha^\beta f_\beta \;, \end{aligned} \tag{25}$$

where we have defined the collocation space matrix representation of the operator $\mathcal{O}$ by

$$O_\alpha^\beta = \sum_i [\mathcal{O}B]_{\alpha i} \left[\mathbf{B}^{-1}\right]^{i\beta} \;. \tag{26}$$

Notice that the construction of the collocation space operators can be performed once and for all at the beginning of a calculation, using only the given knot sequence and collocation points. Due to the presence of the inverse in Eq. (26) the matrix O is not sparse. In practice, operator $\mathcal{O}$ is chosen to be a differential operator such as d/dx or d^2/dx^2. By a similar construction it is also possible to obtain the appropriate integration weights on the collocation lattice [13].

3.3. Discrete HF equations

Since the detailed derivation of the BSCM representation of the TDHF equations involve many terms that are present in the energy functional, we will only show few terms. The three-dimensional expansion in terms of B-splines is a simple generalization of Eq. (18)

$$\psi_\lambda(x,y,z) = \sum_{ijk} c_\lambda^{ijk} B_i(x)B_j(y)B_k(z) \;. \tag{27}$$

The knots and collocation points for each coordinate can be different. With the appropriate definition of boundary conditions all of the discretization techniques discussed in the previous section can be generalized to the three-dimensional space. The details of this procedure are given in [13].

As an example for a local term let us consider a part of the t_0 contribution to the total energy

$$\frac{t_0}{2}(1+\frac{x_0}{2})\int d^3r\rho^2 = \frac{t_0}{2}(1+\frac{x_0}{2})\sum_{\alpha\beta\gamma} w^\alpha w^\beta w^\gamma [\rho(\alpha\beta\gamma)]^2 \;, \tag{28}$$

where on the right-hand side we have written the discretized form on a collocation lattice with collocation weights denoted by w. Here, α, β, γ represent the collocation points in x, y, and z directions, respectively. In order to be able to perform the variation with respect to the single-particle states ψ_λ^* we rewrite equation (28) explicitly

$$\frac{t_0}{2}(1+\frac{x_0}{2})\sum_{\alpha\beta\gamma} w^\alpha w^\beta w^\gamma \sum_{\mu\nu} \psi_\mu^*\psi_\mu\psi_\nu^*\psi_\nu \;. \tag{29}$$

Using Eq.(17) in the variation of Eq.(29) we obtain (after replacing the primed indices with unprimed ones) the contribution

$$t_0(1+\frac{x_0}{2})\rho(\alpha\beta\gamma)\psi_\lambda(\alpha\beta\gamma)\,, \tag{30}$$

where we have rewritten a summation as the total density. The same procedure can be carried out for the nonlocal terms in the energy density. A typical term is illustrated below

$$\begin{aligned}(\nabla\psi^{\pm}_{\lambda})_{\alpha\beta\gamma} &= \sum_{\alpha'} D^{\alpha'}_{\alpha}\,\psi^{\pm}_{\lambda}(\alpha'\beta\gamma)\,\hat{\imath} + \sum_{\beta'} D^{\beta'}_{\beta}\,\psi^{\pm}_{\lambda}(\alpha\beta'\gamma)\,\hat{\jmath} \\ &+ \sum_{\gamma'} D^{\gamma'}_{\gamma}\,\psi^{\pm}_{\lambda}(\alpha\beta\gamma')\,\hat{k}\end{aligned}$$

where the matrices $\mathbf{D}$ denote the first derivative matrices in x, y, z directions (they can be different although the notation does not make this obvious), calculated as described in the previous section. Finally, the HF equations can be written as matrix-vector equations on the collocation lattice

$$h\psi^{\pm}_{\lambda} \longrightarrow \mathbf{h}\cdot\boldsymbol{\psi}^{\pm}_{\lambda}\,. \tag{31}$$

The essence of this construction is that the terms in the single-particle Hamiltonian $\mathbf{h}$ are matrices in one coordinate and diagonal in others. Therefore, $\mathbf{h}$ need not be stored as a full matrix, which allows the handling of very large systems directly in memory. The details of this procedure are discussed below.

3.4. Solution of the discrete equations

In this subsection we will outline some of the numerical methods developed for the solution of the discretized HF and TDHF equations. The subsection is divided into two parts; first part discusses the static iteration methods and the solution of the field equations and in particular the powerful damped relaxation method [17]. In addition we discuss the implementation of external constraints on the HF equations. The second part of the subsection introduces a number of time-evolution methods used in our calculations. Typical numerical accuracies are also discussed.

3.4.1. Static solutions The solution of the HF equations (31) represent the problem of finding the few lowest eigenvalues of a very large Hamiltonian matrix. Furthermore, due to the fact that we are dealing with a self-consistent problem the matrix elements must be recalculated at every iteration. However, in practice the matrix elements need not be stored. Instead, one can make use of the inherent sparsity to dynamically construct the operation of the single-particle Hamiltonian onto a statevector. The basic operation is

$$\boldsymbol{\psi}' = \mathbf{h}\cdot\boldsymbol{\psi}\,, \tag{32}$$

where the construction of the right hand side is done by explicitly programming the required linear combinations of the elements of $\boldsymbol{\psi}$ to give $\boldsymbol{\psi}'$. In this approach the only storage requirements are for the statevectors and small matrices present in the Hamiltonian.

The lattice equations are solved by using the damped relaxation method described in Refs. [11, 17]. A simple way for introducing the damped relaxation method is by pointing out its resemblance to the so-called imaginary time method. A more formal discussion is given in Ref. [17] where the generalization to the relativistic Dirac equation is also introduced. We start with the TDHF equations

$$i\hbar\frac{\partial\boldsymbol{\psi}_{\lambda}}{\partial t} = \mathbf{h}(t)\boldsymbol{\psi}_{\lambda}\,. \tag{33}$$

In terms of the discretized time $t_n = n\Delta t$ the solution at time $n+1$ can be obtained from time n by (see below)

$$\psi_\lambda^{n+1} = e^{-i\Delta t \mathbf{h}^n/\hbar}\psi_\lambda^n \,, \tag{34}$$

where $\mathbf{h}^n$ is the single-particle Hamiltonian at the nth iteration. The imaginary time-step method consists of the transformation $\Delta t \to -i\Delta t$

$$\chi_\lambda^{n+1} = e^{-x_0(\mathbf{h}^n - \epsilon_\lambda^n)}\chi_\lambda^n \,, \tag{35}$$

where $x_0 = \Delta t/\hbar$, and we have taken out a trivial phase from ψ_λ^n. The expansion of the exponential to first order in x_0 yields the imaginary time iteration scheme

$$\chi_\lambda^{n+1} = \mathcal{O}[\chi_\lambda^n - x_0(\mathbf{h}^n - \epsilon_\lambda^n)\chi_\lambda^n] \,. \tag{36}$$

where $\mathcal{O}$ stands for Gram-Schmidt orthonormalization, which is necessary to ensure the orthonormality of the single-particle states at each iteration. In Eq. (36) the index n is no longer associated with time and it simply becomes an iteration counter. It is clear from Eq. (35) that the exponential acts as a filter in selecting the lowest eigenvalues of $\mathbf{h}$ and leads to the minimization of the HF energy. The generalization of the imaginary time method, where we introduce the damping matrix $\mathbf{D}$, results in the damped relaxation method

$$\chi_\lambda^{n+1} = \mathcal{O}[\chi_\lambda^n - x_0\mathbf{D}(E_0)(\mathbf{h}^n - \epsilon_\lambda^n)\chi_\lambda^n] \,. \tag{37}$$

The damping operator $\mathbf{D}$ is chosen to be

$$\begin{aligned}\mathbf{D}(E_0) &= \left[1 + \frac{\mathbf{T}}{E_0}\right]^{-1} \\ &\approx \left[1 + \frac{\mathbf{T}_x}{E_0}\right]^{-1}\left[1 + \frac{\mathbf{T}_y}{E_0}\right]^{-1}\left[1 + \frac{\mathbf{T}_z}{E_0}\right]^{-1} ,\end{aligned}$$

where $\mathbf{T}$ denotes the kinetic energy operator. Limits can be established for the ranges of the parameters x_0 and E_0 [17], but in practice fine-tuning is necessary for optimal performance. Two convergence criteria are used in practical calculations; one being the fractional change in the HF energy

$$\Delta E^n = \frac{E^{n+1} - E^n}{E^n} \,, \tag{38}$$

and other the fluctuations in energy

$$\eta \equiv \sqrt{< H^2 > - < H >^2} \,. \tag{39}$$

The fluctuations are a more stringent condition than the simple energy difference between two iterations. In practice, we have required η to be less than 10^{-6}. For this value of the energy fluctuation the fractional change in the HF energy is about 10^{-13}.

The calculation of the HF Hamiltonian also requires the evaluation of the direct Coulomb contribution. However, since the calculation of the three-dimensional Coulomb integral is very costly, instead one could solve the corresponding differential equation

$$\nabla^2 U_C(\mathbf{r}) = -4\pi e^2 \rho_p(\mathbf{r}) \,. \tag{40}$$

Details of solving the Poisson equation using the BSCM is given in Ref. [13].

3.4.2. Constrained HF calculations It is sometimes desirable to solve the static HF equations away from the global minimum in energy. Such situations usually arise in the study of fission barriers and in the study of long-lived superdeformed states of nuclei that can be formed during low energy heavy-ion collisions. These methods have been instrumental for the understanding of the formation of nuclear molecules [11]. All of these cases require the existence of a stable minimum which does not coincide with the ground state configuration. The usual approach is to study the HF energy of a nuclear system by keeping certain macroscopic degrees of freedom at pre-specified values. This results in a multi-dimensional *energy surface* from which extremum values can be obtained. The reliability of these results depend strongly on the correct identification of the relevant macroscopic degrees of freedom. However, as we will see below a special constrained HF method, *density constrained HF*, has also been developed which allows the minimization of the energy along a TDHF trajectory.

The goal is to devise an iteration scheme such that the expectation value of an arbitrary operator $\hat{Q}$ does not change from one static iteration to next

$$\sum_\lambda < \chi_\lambda^{n+1}|\hat{Q}|\chi_\lambda^{n+1} >= \sum_\lambda < \chi_\lambda^{n}|\hat{Q}|\chi_\lambda^{n} > . \tag{41}$$

Furthermore, we require this expectation value to be a fixed number Q_0. A procedure can be developed by using Lagrange multipliers that are dynamically adjusted [11]. We start with the static HF iteration scheme modified by the addition of a constraint (we have omitted the damping matrix $\mathbf{D}$ in the equations below for simplicity)

$$\boldsymbol{\chi}_\lambda^{n+1} = \mathcal{O}[\boldsymbol{\chi}_\lambda^{n} - x_0(\mathbf{h}^n + \lambda\hat{Q} - \epsilon_\lambda^n)\boldsymbol{\chi}_\lambda^n] . \tag{42}$$

In Ref.[11] we give a set of exact equations which preserve the expectation of the constraining operator to order x_0^2. However, these equations involve the calculation of exchange terms and may become costly. Instead, one can develop a simpler iterative scheme as follows. Perform an intermediate step

$$\boldsymbol{\chi}_\lambda^{n+1/2} = \mathcal{O}[\boldsymbol{\chi}_\lambda^{n} - x_0(\mathbf{h}^n + \lambda^n\hat{Q} - \epsilon_\lambda^n)\boldsymbol{\chi}_\lambda^n] , \tag{43}$$

and calculate the difference

$$\delta Q^{n+1/2} = \sum_\lambda < \chi_\lambda^{n+1/2}|\hat{Q}|\chi_\lambda^{n+1/2} > - \sum_\lambda < \chi_\lambda^{n}|\hat{Q}|\chi_\lambda^{n} > . \tag{44}$$

In analogy with the exact case the Lagrange parameter λ is altered to reduce this difference

$$\lambda^{n+1} = \lambda^n + c_0 \frac{\delta Q^{n+1/2}}{2x_0 \sum_\lambda < \chi_\lambda^n|\hat{Q}^2|\chi_\lambda^n > + d_0} , \tag{45}$$

where c_0 and d_0 are empirical parameters replacing the exchange terms. In terms of these intermediate states the $(n+1)st$ step is given by

$$\boldsymbol{\chi}_\lambda^{n+1} = \mathcal{O}[\boldsymbol{\chi}_\lambda^{n+1/2} - x_0(\lambda^{n+1} - \lambda^n + \delta\lambda^n)\hat{Q}\boldsymbol{\chi}_\lambda^{n+1/2}] , \tag{46}$$

where

$$\delta\lambda^n = \frac{\sum_\lambda < \chi_\lambda^n|\hat{Q}|\chi_\lambda^n > - Q_0}{2x_0 \sum_\lambda < \chi_\lambda^n|\hat{Q}^2|\chi_\lambda^n > + d_0} . \tag{47}$$

The extension of the method which allows the entire density to be constrained is straightforward. In this case we would like to constrain a continuous density

$$\rho^n(\mathbf{r}) = \sum_\lambda |\chi_\lambda^n(\mathbf{r})|^2 , \tag{48}$$

to be equal to $\rho_0(\mathbf{r})$. The constraining operator $\hat{Q}$ becomes the density operator $\hat{\rho}(\mathbf{r})$ defined in the single-particle space

$$< \chi_\lambda^n|\hat{\rho}(\mathbf{r})|\chi_\lambda^n >= |\chi_\lambda^n(\mathbf{r})|^2 \tag{49}$$

and the product $\lambda\hat{Q}$ is replaced by an integral

$$\lambda\hat{Q} \longrightarrow \int d^3r\, \lambda(\mathbf{r})\hat{\rho}(\mathbf{r}) = \lambda(\mathbf{r})\,. \tag{50}$$

The last equality is due to the fact that in coordinate space $\hat{\rho}(\mathbf{r})$ is a delta function. An iterative scheme for $\lambda^n(\mathbf{r})$ is given by

$$\lambda^{n+1}(\mathbf{r}) = \lambda^n(\mathbf{r}) + c_0\frac{\delta\rho^{n+1/2}}{2x_0\rho^n(\mathbf{r}) + d_0}\,, \tag{51}$$

where

$$\delta\rho^{n+1/2}(\mathbf{r}) \equiv \rho^{n+1/2}(\mathbf{r}) - \rho_0(\mathbf{r}) \tag{52}$$

is obtained from half-time iteration step

$$\boldsymbol{\chi}_\lambda^{n+1/2} = \mathcal{O}[\boldsymbol{\chi}_\lambda^n - x_0(\mathbf{h}^n + \lambda^n(\mathbf{r}) - \epsilon_\lambda^n)\boldsymbol{\chi}_\lambda^n]\,. \tag{53}$$

Note that in these equations we require that the density remain equal to $\rho_0(\mathbf{r})$ at every iteration and not just at the final step. Using these wavefunctions the full iteration can be written as

$$\boldsymbol{\chi}_\lambda^{n+1} = \mathcal{O}[\boldsymbol{\chi}_\lambda^{n+1/2} - x_0(\lambda^{n+1}(\mathbf{r}) - \lambda^n(\mathbf{r}) + \delta\lambda^n(\mathbf{r}))\boldsymbol{\chi}_\lambda^{n+1/2}]\,, \tag{54}$$

where

$$\delta\lambda^n(\mathbf{r}) = c_0\frac{\rho^n(\mathbf{r}) - \rho_0(\mathbf{r})}{2x_0\rho_0(\mathbf{r}) + d_0}\,. \tag{55}$$

During the density constrained HF iterations the single-particle states readjust to minimize the energy while the initial density is kept fixed. In practical calculations the parameter x_0 has been replaced by the damping operator and the constants c_0 and d_0 were chosen to be 1.9 and 5×10^{-5}, respectively.

3.4.3. Time evolution The formal solution of the TDHF equations (5) is

$$\psi_\lambda(t) = U(t, t_0)\psi_\lambda(t_0)\,, \tag{56}$$

where we have omitted the spatial coordinates for simplicity and the time propagator $U(t, t_0)$ is given by

$$U(t, t_0) = \mathcal{T}\,exp\left[\frac{-i}{\hbar}\int_{t_0}^{t} dt' h(t')\right]\,. \tag{57}$$

The quantity $\mathcal{T}$ denotes time-ordering which is necessary in the general case. In practical calculations we discretize time as

$$t_n = n\Delta t \qquad\qquad n = 0, 1, 2, ...N\,, \tag{58}$$

and express the evolution operator in successive infinitesimal pieces

$$U(t, t_0) = U(t, t_{N-1})U(t_{N-1}, t_{N-2})...U(t_1, t_0)\,. \tag{59}$$

In this case the time-ordering operator can be ignored. For three-dimensional calculations the exponential operator is expanded as a Taylor series

$$U(t_{n+1}, t_n) \approx \left(1 + \sum_{k=1}^{K} \frac{(-i\Delta t\, \mathbf{h}/\hbar)^k}{k!}\right) . \tag{60}$$

The expansion of the operator requires repeated applications of **h** onto the wavefunctions. In practice, only $6-8$ terms are needed for the conservation of the norm at 1 part in 10^{-10} level during the entire time-evolution. The expansion method is also attractive due to the fact that it only involves matrix vector operations which could be easily customized for vector or parallel computers.

4. Applications

In this section we give examples of DC-TDHF calculation of potential barriers and fusion cross-sections. Calculations were done in 3-D geometry and using the full Skyrme force (SLy4) [12] without the center-of-mass correction as described in Ref. [18, 19]. We have performed density constraint calculations every 20 time steps. For the calculation of the ion-ion separation distance R we use the hybrid method, which relates the coordinate to the quadrupole moment for small R values, as described in Ref. [6]. The accuracy of the density constraint calculations is commensurate with the accuracy of the static calculations.

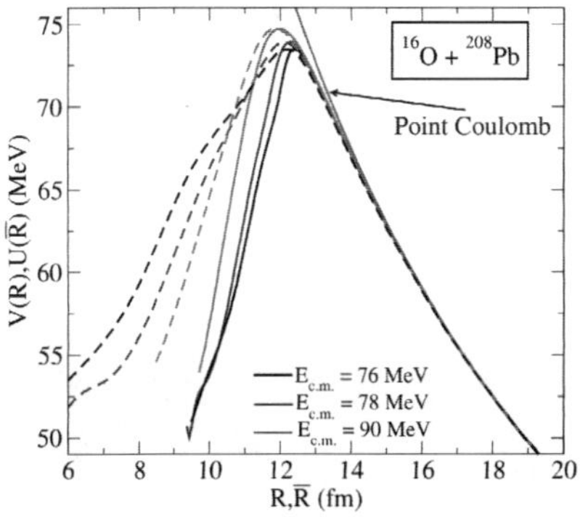

Figure 2. Potential barriers obtained from density constraint TDHF calculations at three different energies. The three dashed curves correspond to the transformed potential using coordinate dependent masses.

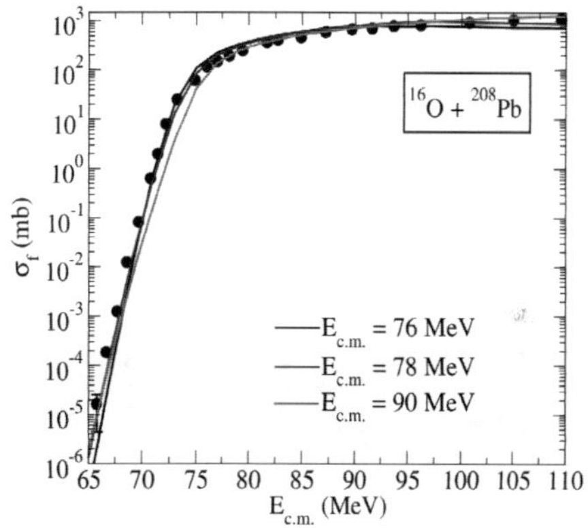

Figure 3. Total fusion cross section as a function of c.m. energy using the potentials of Fig. 2. The three curves correspond to the transformed potential using coordinate dependent masses.

4.1. Spherical system

The DC-TDHF method is expected to do best for nuclei that are well described by the Skyrme HF calculations. One such reaction is the fusion of $^{16}O+^{208}Pb$ system. In Fig. 2 we show an example of microscopic potentials for the $^{16}O+^{208}Pb$ system at three difference center-of-mass energies [6]. The dashed curves are the corresponding potentials transformed via the microscopically calculated effective mass, $M(R)$. We observe that all of the scaled barriers give a very good description of the fusion cross-section at higher energies suggesting these cross-sections are primarily determined by the barrier properties in the vicinity of the barrier peak, whereas for the extreme sub-barrier cross-sections are influenced by what happens in the inner part of the barrier and here the dynamics and consequently the coordinate dependent mass becomes very important (see Fig. 3). Specifically, we can see from Fig. 2 that as the c.m. energy is increased the ion-ion potential peak increases but the inner part of the barrier becomes narrower. This is due to the fact that for high energies the system does not have enough time for rearrangements in the density to occur and the barrier approaches the frozen-density limit. However, at lower energies substantial density rearrangements occur which modifies the inner part of the barrier. This modification is important for fusion cross-sections are deep sub-barrier energies.

4.2. Deformed systems

The collision of the ^{64}Ni+^{132}Sn system represents a good example of a collision involving a deformed (oblate) nucleus, ^{64}Ni and a neutron rich nucleus. Fusion cross-sections for this system have been experimentally measured [20] and initially a significant discrepancy was observed with standard coupled-channel calculations. We have used the DC-TDHF method to study this system [3, 4]. The ion-ion potentials corresponding to two extreme orientations of the ^{64}Ni nucleus are shown in Fig. 4 as well as an empirical barrier used in barrier penetration calculations in Ref. [20]. Two important points are observed from this plot. The first is the strong dependence of the barrier height and location on the alignment of the deformed nucleus. We also see that the empirical barrier is very close to the equatorial orientation, which is closer to the assumption of spherical nuclei. The accuracy of our result with no parameters or normalization is impressive. The second point has to do with the meaning of *sub-barrier*; as seen from Fig. 4, while the experimental energies appear to be all sub-barrier with respect to the spherical barrier, two of them are above the barrier with respect to the $\beta = 90^o$ potential barrier and the third faces a considerably narrower barrier. This explains the anomalous observation of enhanced fusion at these energies.

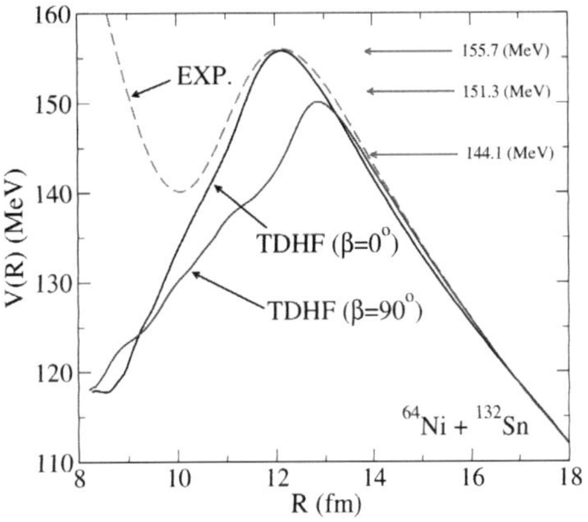

Figure 4. Potential barriers, $V(R,\beta)$, obtained from DC-TDHF calculations for the ^{64}Ni+^{132}Sn system. Angle β indicates different orientations of the deformed ^{64}Ni nucleus in $\Delta\beta = 10°$ intervals. Also shown are the experimental energies.

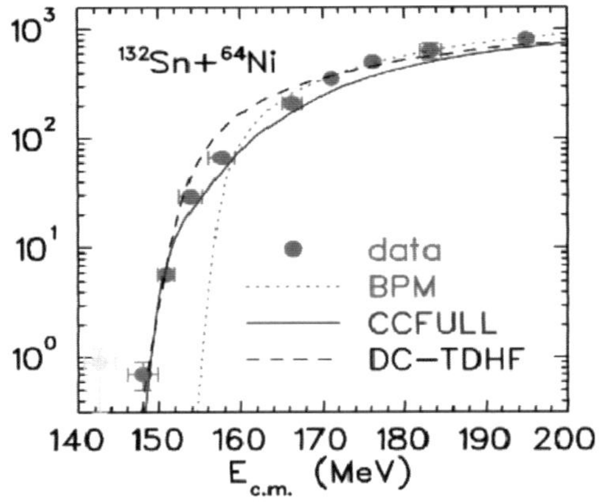

Figure 5. Total fusion cross section as a function of $E_{\mathrm{c.m.}}$. Shown are the experimental data (filled circles), the latest coupled-channel calculations [20] including neutron transfer (blue solid curve), and the DC-TDHF cross sections (dashed curve).

4.3. Superheavy systems

Ion-ion interaction potentials calculated using DC-TDHF correspond to the configuration attained during a particular TDHF collision. For light and medium mass systems as well as heavier systems for which fusion is the dominant reaction product, DC-TDHF gives the fusion barrier with an appreciable but relatively small energy dependence. On the other hand, for reactions leading to superheavy systems fusion is not the dominant channel at barrier top energies. Instead the system sticks in some dinuclear configuration with possible break-up after exchanging a few nucleons. For this reason the energy dependence of the DC-TDHF interaction barriers for these systems is not just due to the dynamical effects for the same final configuration but actually represent different final configurations. For the same reasons calculations presented here can only address the capture cross-section for these systems since the long-time evolution to complete fusion or break-up is beyond the scope of TDHF due to the absence of quantum decay processes and transitions.

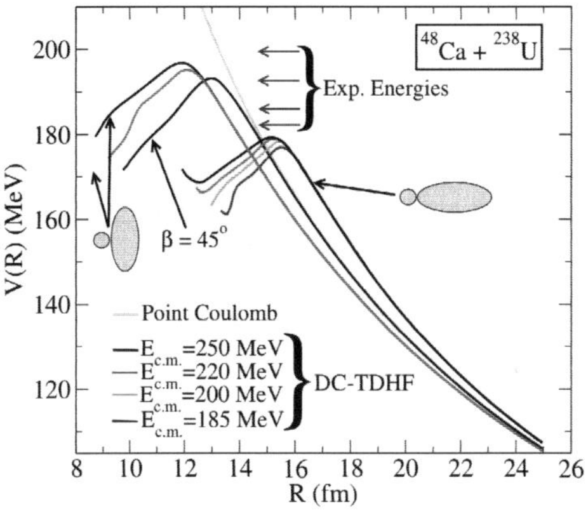

Figure 6. Potential barriers, $V(R)$, for the $^{48}Ca+^{238}U$ system obtained from DC-TDHF calculations as a function of $E_{\rm c.m.}$ energy and for selected orientation angles β of the ^{238}U nucleus. Also, shown are the experimental c.m. energies.

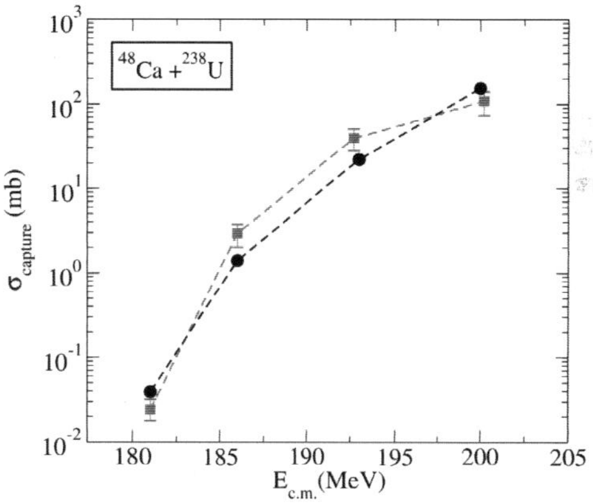

Figure 7. Capture cross-sections for the $^{48}Ca+^{238}U$ system as a function of $E_{\rm c.m.}$ energy (black circles). Also, shown are the experimental cross-sections (red squares) [21].

As an example of superheavy formation from a hot-fusion reaction we have studied the $^{48}Ca+^{238}U$ system [7]. Hartree-Fock (HF) calculations produce a spherical ^{48}Ca nucleus, whereas ^{238}U has a large axial deformation. The large deformation of ^{238}U is expected to strongly influence the interaction barriers for this system. This is shown in Fig. 6, which shows the interaction barriers, $V(R)$, calculated using the DC-TDHF method as a function of c.m. energy and for three different orientations of the ^{238}U nucleus. The alignment angle β is the angle between the symmetry axis of the ^{238}U nucleus and the collision axis. Also shown in Fig. 6 is the point Coulomb potential corresponding to this collision. The deviations from the point Coulomb potential at large R values are due to the deformation of the ^{238}U nucleus. We first notice that the barriers corresponding to the polar orientation ($\beta = 0^o$) of the ^{238}U nucleus are much lower and peak at larger ion-ion separation distance R. On the other hand, the barriers corresponding to the equatorial orientation of ^{238}U ($\beta = 90^o$) are much higher and peak at smaller R values. For the intermediate values of β the barriers rise rapidly as we increase the orientation angle from $\beta = 0^o$, as can be seen for $\beta = 45^o$. The rise in the barrier height as a function of increasing β values is not linear but seems to rise more rapidly for smaller β values.

We also see that for lower energies central collisions with polar orientation of ^{238}U are the only orientations which result in the sticking of the two nuclei, while the equatorial orientations of ^{238}U result in a deep-inelastic collision. Also, shown in Fig. 6 are the experimental energies [21] for this reaction. We observe that all of the experimental energies are above the barriers obtained for the polar alignment of the ^{238}U nucleus. For the calculation of the capture cross-section we need to average over all possible alignments of the ^{238}U nucleus. In Fig. 7 we show the capture cross-sections for the ^{48}Ca+^{238}U system as a function of $E_{\mathrm{c.m.}}$ energy (black circles). Also, shown are the experimental cross-sections (red squares) [21].

Acknowledgments

This work has been supported by the U.S. Department of Energy under grant No. DE-FG02-96ER40963 with Vanderbilt University.

References

[1] Umar A S and Oberacker V E 2006 *Phys. Rev.* C **74** 021601(R)
[2] Umar A S and Oberacker V E 2006 *Phys. Rev.* C **74** 024606
[3] Umar A S and Oberacker V E 2006 *Phys. Rev.* C **74** 061601(R)
[4] Umar A S and Oberacker V E 2007 *Phys. Rev.* C **76** 014614
[5] Umar A S and Oberacker V E 2008 *Phys. Rev.* C **77** 064605
[6] Umar A S and Oberacker V E 2009 *Eur. Phys. J.* A **39**, 243
[7] Umar A S, Oberacker V E, Maruhn J A and Reinhard P-G 2010 *Phys. Rev.* C **81** 064607
[8] Oberacker V E, Umar A S, Maruhn J A and Reinhard P-G 2010 *Phys. Rev.* C **82** 034603
[9] Negele J W 1982 *Rev. Mod. Phys.* **54** 913
[10] Cusson R Y, Reinhard P-G, Strayer M R, Maruhn J A and Greiner W 1985 *Z. Phys.* A **320** 475
[11] Umar A S, Strayer M R, Cusson R Y, Reinhard P-G and Bromley D A 1985 *Phys. Rev.* C **32** 172
[12] Chabanat E, Bonche P, Haensel P, Meyer J and Schaeffer R 1998 *Nucl. Phys.* A **635** 231; *Nucl. Phys.* A **643** 441
[13] Umar A S, Wu J , Strayer M R and Bottcher C 1991 *J. Comp. Phys* **93** 426
[14] Umar A S, Strayer M R, Reinhard P-G, Davies K T R and Lee S-J 1989 *Phys. Rev.* C **40** 706
[15] Davies K T R and Koonin S E 1981 *Phys. Rev.* C **23** 2042
[16] Boor C De 1978 *Practical Guide to Splines* (New York: Springer).
[17] Bottcher C, Strayer M R, Umar A S and Reinhard P-G 1989 *Phys. Rev.* A **40** 4182
[18] Umar A S and Oberacker V E 2006 *Phys. Rev.* C **73** 054607
[19] Umar A S, Strayer M R, Wu J-S, Dean D J and Güçlü M C 1991 *Phys. Rev.* C **44** 2512
[20] Liang J F *et al.* 2007 *Phys. Rev.* C **75** 054607
[21] Oganessian Yuri 2007 *J. Phys. G: Nucl. Part. Phys.* **34** R165

Journal of Physics: Conference Series **402** (2012) 012039 doi:10.1088/1742-6596/402/1/012039

A model for stable interfacial crack growth

Knut S Gjerden[1], **Arne Stormo**[1], **Alex Hansen**[1]

[1] Department of Physics, Norwegian University of Science and Technology, N-7491 Trondheim, Norway

E-mail:
knut.skogstrand.gjerden@gmail.com,arne.stormo@gmail.com,Alex.Hansen@ntnu.no

Abstract. We present a model for stable crack growth in a constrained geometry. The morphology of such cracks show scaling properties consistent with self affinity. Recent experiments show that there are two distinct self-affine regimes, one on small scales whereas the other at large scales. It is believed that two different physical mechanisms are responsible for this. The model we introduce aims to investigate the two mechanisms in a single system. We do find two distinct scaling regimes in the model.

1. Introduction

On a fine enough scale, fractures are rough. How to describe this roughness quantitatively? An answer to this question was first proposed by Mandelbrot et al. in 1984 [1]: Fractures are fractal and their morphology may be described quantitatively by a fractal dimension. In 1990 Bouchaud et al. [2, 3], using the more precise concept of self-affinity rather than fractality to characterize the fracture morphology of metals, proposed that the scaling properties are *universal,* i.e., they do not depend on the material that fractures. An important conceptual step forwards was taken by Bouchaud et al. [4] when they regarded the fracture surface to represent a "footprint" of a passing fluctuating line, the crack front. Schmittbuhl et al. [5] then made the suggestion that one may simplify the problem by constraining the crack growth to appear between two sintered plates and then follow the fluctuations of the crack front as it moves when the two plates are plied apart as show in Figure 1. In 1997, Schmittbuhl and Måløy [6] realized this system experimentally by sintering two sandblasted plexiglass plates together.

Schmittbuhl et al. [5] presented a numerical model for constrained crack growth based on the fluctuating line picture. Here an elastic line moves through a plane covered with pinning centers. The elastic forces that the line feels is transmitted through the media constituting the two sintered plates. Their main result was to determine the roughness exponent of the self-affine fracture front. If the front is given by $h = h(x)$, where the x-direction is orthogonal to the crack growth direction, then self affinity manifests itself through the following invariance: If $p(h, x)$ is the probability density that the crack is at height h at x when it is $h = 0$ at $x = 0$, then we have

$$\lambda^{\zeta}\, p(\lambda^{\zeta} h, \lambda x) = p(h, x)\ , \tag{1}$$

where ζ is the *Hurst* or *roughness* exponent. Schmittbuhl et al. [5] found $\zeta = 0.35$. This value was later refined to $\zeta = 0.39$ by Rosso and Krauth [7]. The experiments [6], on the other hand, gave a much larger roughness exponent, $\zeta = 0.55 \pm 0.05$. In 2003, Schmittbuhl et al. [8] proposed

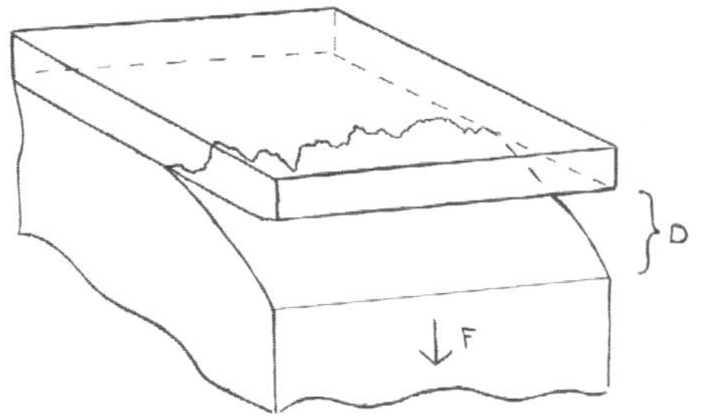

Figure 1. Sketch of the system we model. The two external control parameters are F and D, representing either force-controlled or displacement-controlled loading, respectively.

a model based on the crack front propagating due to coalescence of damage appearing in front of it. This model gave a roughness exponent $\zeta = 0.60 \pm 0.05$. Recently, Santucci et al. [9] have analyzed larger experimental systems than had been considered earlier. They find a crossover between two scaling regimes in their data; a small-scale regime where $\zeta_1 = 0.60 \pm 0.05$ and a large-scale regime where $\zeta_2 = 0.35 \pm 0.05$. Combining this result with the previous mechanism that has been suggested — a fluctuating elastic line — it is tempting to propose that the fracture coalescence mechanisms is at work on small scales, whereas on larger scales, the system effectively behaves as the fluctuating line model indicates.

It is the aim of this project to construct a single model capable of capturing both mechanisms and then to study the crossover between them. In the next section, we present the model, which is a variant of the fiber bundle model [10, 11]. The aim of this paper is to present this model in detail, including the computational side, which will be covered in Section 3. We present the treadmill technique in Section 4, which allows us to maintain a stable crack growth. In Section 5 we present our results, two different roughness exponents $\zeta_1 = 0.45$ and $\zeta_2 = 0.3$. These values are low in comparison with the experimental values, $\zeta_1 = 0.60$ and $\zeta_2 = 0.3$. As we will argue, this is probably due to the limited system sizes we so far have considered.

2. The model

Our model is based on an idea originally proposed by Batrouni et al. [10], but with some significant alterations. The basis of the model is a variation of the fiber bundle model, in which linear elastic fibers stretch according to the difference between their local displacement u_i and a global displacement D,

$$f_i = -k_i(u_i - D). \tag{2}$$

This relation contains the proportionality constant k_i, which defines the response of the fiber. $L \times L$ fibers are arranged in parallel, with each fiber connected to two elastic blocks. Each block behaves linearly elastic and can have its own elastic constant, but for simplicity and without loss of generality, one of the regions is set infinitely stiff and the other one elastic with Young's modulus E. The response of the fibers are transmitted through the elastic region via the Green's function [12],

$$u_i = \sum_j G_{ij} f_i, \tag{3a}$$

$$G_{ij} = \frac{1 - \nu^2}{\pi E a^2} \iint_{-a/2}^{a/2} \frac{dx' dy'}{|\vec{r}_i(x, y) - \vec{r}_j(x', y')|}. \tag{3b}$$

where ν is the Poisson ratio and a^2 is the area the force of each fiber acts on, thus giving the discretization of the system. $\vec{r}_i - \vec{r}_j$ gives the distance between fiber i and j. The two regions are pulled apart either by controlling an external force or the displacement between them, as shown in Figure 1. The fibers break when they are stretched beyond a given threshold distribution. In matrix notation, the problem can be re-written as

$$(\mathbb{I} + \mathbb{KG})\vec{f} = \mathbb{K}\vec{D}, \tag{4}$$

where $\mathbb{K}$ is an $L^2 \times L^2$ diagonal matrix where $\mathbb{K}_{ii} = k_i$. To simplify the problem we keep all constants equal to unity. This is to isolate the disorder in the system to the thresholds, thus avoiding the complications of simultaneously dealing with two quenched disorder distributions. Because of the $1/r$-dependence in the Green's function (3), there are long-reaching forces in the system and every fiber is connected to every other fiber through the Green's function, making $\mathbb{G}$ a dense $L^2 \times L^2$ matrix. $\mathbb{I}$ is the identity matrix and we have chosen to give the displacement vector $\vec{D}$ and solve for the forces in the L^2 vector $\vec{f}$.

We implement our model on a square lattice using bi-periodic boundary conditions with respect to the transmission of the forces. Such boundary conditions are necessary due to numerical effects that we detail in the following. We solve the matrix equation (4) and locate the fiber with the highest ratio of strain to threshold. Then we break that fiber i by setting k_i to zero, indicating that it no longer has any load-bearing capability. The process is repeated until all fibers are broken.

2.1. Loading schemes

One of the new concepts of our model is the way we load our system. Providing a uniform displacement vector amounts to pulling the system apart, keeping the two surfaces parallel. Our main goal using this model was to observe a fracture front, meaning that we need some sort of gradient in the system to emulate the loading conditions of the experiments of Schmittbuhl and Måløy [6]. There are at least three ways this can be done. The gradient can either be in the force, the displacement, or the thresholds.

Implementing a gradient in the force applied to the system would amount to solving the inverse problem since the system now would be load controlled. This is only possible by either reformulating the problem or adding a matrix inversion per broken fiber in a simulation, which would be very computationally costly. Given a choice between the displacement and the thresholds, choosing the thresholds enables us compare our results to that of gradient percolation which then would be an extreme limit of the system where it is the treshold distribution that completely dominate how the breakdown process proceeds rather than the force distribution of the fibers. Implementing a linear gradient in the thresholds models the pulling apart of the two blocks at a constant contact angle. This simplifies the problem and maintains a close approximation to realistic loading.

3. Numerical procedures

We use the conjugate gradient (CG) algorithm with Fourier acceleration [10, 14] of the matrix multiplications to solve the system, meaning that the complete system we solve is

$$(\mathbb{I} + \mathbb{KF}^{-1}\mathbb{F}^{1}\mathbb{G})\mathbb{F}^{-1}\mathbb{F}^{1}\vec{f} = \mathbb{K}\vec{D}. \tag{5}$$

The choice of an iterative solver is due to the matrix $\mathbb{KG}$ evolving from dense to sparse as the simulation progresses. We can Fourier accelerate because the Green's function, which in essence is a two-point correlation function, is only dependent on the distance between the points in question. This property makes the matrix $\mathbb{G}$ diagonal in Fourier space. Exploiting this, along with the symmetric properties of the resulting matrix on the left-hand side of (5), all matrix multiplications are performed in Fourier space and we only store matrices of size $n = L \times L$, not $L^2 \times L^2$, thereby avoiding large constraints on memory.

The higher the ratio between the Young's modulus and system size, i.e., the stiffer the system, the less impact the Green's function (3) has on the system and the more well-behaved the equations are, leading to convergence in fewer iterations. For stiff systems, the pre-conditioning scheme of Batrouni et al. [10] works very well to further reduce the calculation times. Unfortunately, for softer systems, the Green's function can lead to large differences in neighboring forces which in turn leads to a very complicated energy landscape for the CG algorithm to traverse. This drastically increases the number of iterations required for convergence. In extreme cases, it does not converge at all. This also destroys the effect of the pre-conditioner, and we have not yet found an effective replacement.

Another issue to address is the singularity in the Green's function (3). To work around this, we use the solution of Love [13], expressing the displacement at point i with coordinates (x, y) due to a force f_j acting on the area a^2 at the origin as

$$
\begin{aligned}
u_i = \frac{f_j(1-\nu^2)}{\pi E a^2} \times \Bigg\{ & (x+a)\ln\left[\frac{(y+a)+\{(y+a)^2+(x+a)^2\}^{1/2}}{(y-a)+\{(y-a)^2+(x+a)^2\}^{1/2}}\right] \\
& +(y+a)\ln\left[\frac{(x+a)+\{(y+a)^2+(x+a)^2\}^{1/2}}{(x-a)+\{(y+a)^2+(x-a)^2\}^{1/2}}\right] \\
& +(x-a)\ln\left[\frac{(y-a)+\{(y-a)^2+(x-a)^2\}^{1/2}}{(y+a)+\{(y+a)^2+(x-a)^2\}^{1/2}}\right] \\
& +(y-a)\ln\left[\frac{(x-a)+\{(y-a)^2+(x-a)^2\}^{1/2}}{(x+a)+\{(y-a)^2+(x+a)^2\}^{1/2}}\right] \Bigg\}.
\end{aligned} \tag{6}
$$

Part of the derivation behind (6) relies on the assumption that the transition from the area uniformly acted on by the force f_j to the area outside can be approximated as continuos. For stiff systems, this approximation is valid because there is little difference in the force carried by neighboring fibers. This remains valid over a large range of elastic moduli. For very soft systems, however, the difference in displacement and force carried can vary immensely across neighboring fibers, producing large local gradients eventually causing the assumption on a smooth transition to break down. We must emphasize that this is only in the case of the lower limit of the elastic modulus parameter, just as the case for the upper limit, an infinitely stiff system, equation (2) reduces to

$$f = kD,$$

the purely global load sharing fiber bundle model. The point we are making is that when considering the Young's modulus E as control parameter, the limits are one-sided in that the fundament for the equations breaks down before the lower limit of $E = 0$ is reached. Between these two limits we show in this paper the existence of two distinct regimes controlling the roughness of the crack front.

To match the experimental setup as closely as possible, we want to use boundary conditions periodic along the emerging crack front. However, for soft systems, it is necessary to implement bi-periodic boundary conditions to aid convergence in the CG solver. In the next section we explain the treadmill technique, allowing us to follow the evolution of the crack front indefinitely. This also allows us to keep the crack front as far away from the edges as possible. For stiff systems, periodicity in one direction is sufficient and the priority is to have as much of unbroken material ahead of the front as possible. For soft systems, the Green's function becomes more important, forcing us to change to bi-periodicity and keep the front in the middle of system, as the crack front now runs the risk of being impeded by the mirror image behind it.

4. The treadmill technique

The basis of the treadmill technique is 1) the data is stored in matrix, or vector representation of a matrix, form and 2) as the simulation progresses, matrix elements are irreversibly changed to a common value. In our case, these requirements are automatically met, since our simulations are driven forward by breaking links, represented by the value of the load-bearing capacity of the element in question being set to zero. When the problem includes a gradient, the matrix changes from completely dense to sparse to completely empty in a logical manner. If the gradient is very sharp, half-way through the simulation, only the top half of the matrix will have non-zero elements and the other half will be computationally "useless." Generally, only about half of the simulation can be used to gather reliable data, because the front needs time to fully develop in the beginning, and towards the end, it will eventually begin to feel the boundary conditions. To be conservative, on average, 40% of a simulation is useable with respect to gathering trustworthy statistics on the crack front. Since models of this type are computationally costly and scales badly, increasing this percentage is a priority.

To increase the yield, of a given sample, the idea is to extend the middle part of a simulation. Accomplishing this would amount to getting 100% simulation yield after a simulation length of one system size has been reached. To do this, we have a criterion that tells us when the front has completely moved a pre-specified distance into the sample, and when this occurs, we shift all the values in the matrix downwards, effectively forgetting about the lowest completely broken part of the sample, and generate new, statistically identical material at the top of the matrix. Thus, we can control at what level the average position of the front should stay at, and follow this system indefinitely: A treadmill for interfacial fracture propagation.

5. Results

Our main result is to demonstrate that our model contains a transition from a regime where the crack front is primarily driven forward by coalescence of damaged regions ahead of the crack front, to a regime where the crack front appears to behave like an elastic line moving through the medium, by variation of the Young's modulus. These two types of behavior is exemplified in Figure 2. Note that, due to the treadmill technique, the crack fronts in the figure has propagated beyond the original size of the system, in this case $L = 128$. The figure shows fronts from two simulations. In the first case, the system has a higher value for the Young's modulus, while the other example has a much lower value for E. In the stiffer system, we filter away the damage in front of and behind the crack front itself. This is illustrated in the upper part of Figure 2. As the system softens, the crack front changes behavior to not moving forward until all damaged material is part of the front, eliminating any "damage islands" as seen behind the front in the top left image in Figure 2. Tuning the system even softer, eventually only the crack front itself is seen as the interface between undamaged and damaged material, no isolated broken fibers ahead of the front is observed.

The choice of gradient in the system has an impact. In the results we present here, we have chosen a gradient that clearly demonstrates the different behaviors. A more complete study of the gradient effects will be presented elsewhere. For now, suffice to say that too steep a gradient will constrain the crack front to a smaller width, to the extreme limit where the crack front spans only one or two pixels. Setting a too low gradient will have the effect of increasing the width of the front to the point of the width of the crack front surpassing the length of the system. Zero gradient amounts to pulling the system apart, generating two parallel surfaces, but no crack front.

To analyze the front, we do a solid-on-solid (SOS) thus removing the overhangs. Then we do an averaged wavelet coefficient (AWC) analysis [15, 16], to determine the roughness exponent of the front. The AWC method consists of first doing a wavelet transform of the front $h(x)$, and then, for each length scale a, averaging the resulting wavelet coefficients $w(a, b)$ over position b

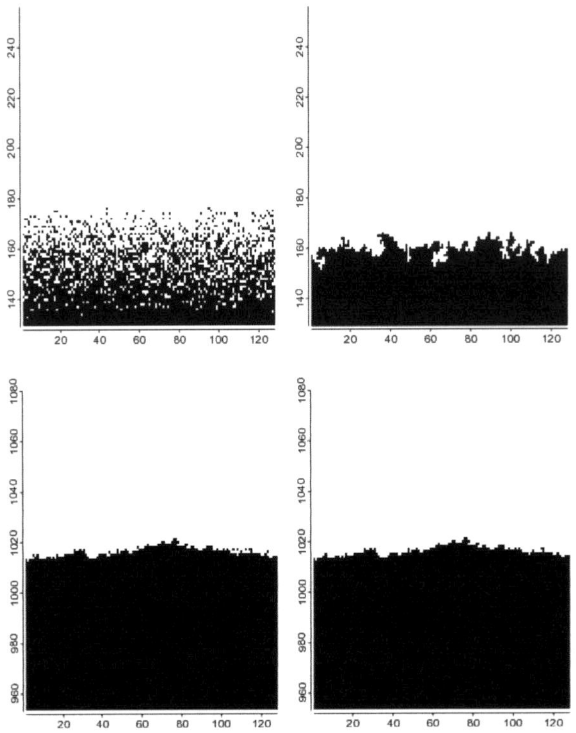

Figure 2. Snapshots from the simulations showing clear differences between two distinct regions. The two upper images are from the same system, a system with a high Young's modulus, while the two lower images are from a simulation with a low Young's modulus. The leftmost images are direct printouts, whilst the rightmost images have been filtered to find the largest connected damage cluster, yielding the crack front. These images are the basis for a solid-on-solid analysis, to remove the overhangs. Note the difference in needed filtering between the two regions.

along the front. For self-affine signals, the averaged coefficients $W(a) = \langle|w(a,b)|\rangle_b$, should scale as

$$W(a) \sim a^{\zeta+1/2}. \tag{7}$$

The result of this analysis is shown in Figure 3, giving a roughness exponent of

$$\zeta_1 = 0.45$$

for the stiff system in Figure 3(a), and

$$\zeta_2 = 0.3$$

for the soft system in Figure 3(b). We note that these values are lower than the values reported in the experiments [9] and in the numerical studies, $\zeta_c = 0.6$ seen in the coalescence model [8] and $\zeta_l = 0.39$ seen in the fluctuating line model [7]. We also show the data for a smaller system size in the same figure. We see that those data would have given lower values for the roughness exponents. Hence, we believe that the largest system sizes we have presented here are still too small for the effective roughness exponents to have attained their asymptotic values. We are currently testing several methods for running our code on massive parallel computer systems to increase the range of available system sizes.

If we do a small size analysis based on equations (2) and (3), and equation (6), using values from Schmittbuhl and Måløy [6] where they found a response of plexiglass of $F \sim 1$N at a displacement of $\delta = 0.1$mm and a Young's modulus of $E \approx 3.3$GPa, and assuming that the area covered by a single fiber is smaller than δ, we find that the system sizes we have covered so far have room for forces acting at distances of the order of $0.1 \sim 1$mm. A system of $L = 1024$ would probably reach on the order of 10mm, which would cover the range of the data used for the analysis of the experiment in both [6] and [9]. Based on the results of Santucci et al. [9], we believe that at larger system sizes we will find a crossover between the different scaling behaviors in the same simulation.

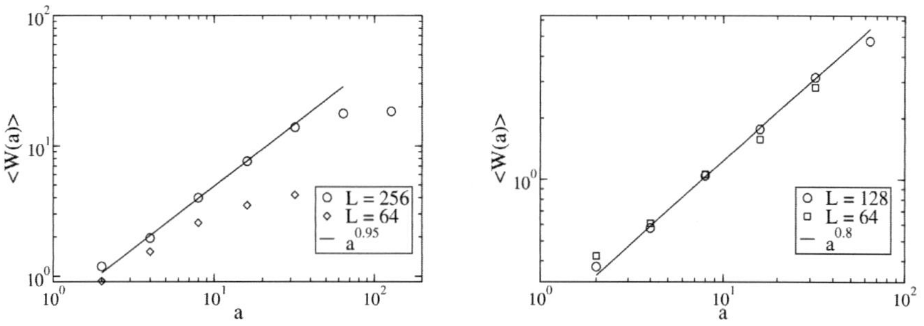

(a) Stiff system

(b) Soft system

Figure 3. Log-log plot of the average wavelet coefficients of a front extracted from the stiff region and from the soft region. Decent power laws are observed, indicating self-affine behavior of two different exponents. Two system sizes are plotted, indicating that the scaling exponents have not yet reached their asymptotic values, so that our results are underestimations.

6. Conclusions

We have shown that our model contains a transition from one regime characterized by a scaling exponent of $\zeta_1 = 0.45$, to another regime characterized by a lower exponent of $\zeta_2 = 0.3$. This transition is achieved through the variation of a single parameter, the Young's modulus. We believe that these roughness exponents are slight underestimations of the exponents found in the coalescence model and fluctuating line model, respectively, making our model the first to capture these two fracture mechanisms simultaneously — and that the idea that the two roughness regimes seen in the experiments [9] indeed is correct.

References

[1] Mandelbrot B B, Passoja D E and Paullay A J 1984 *Nature* **308** 721
[2] Bouchaud E, Lapasset G and Planès J 1990 *Europhys. Lett.* **13** 73
[3] Bonamy D and Bouchaud E 2011 *Phys. Rep.* **498** 1
[4] Bouchaud J P, Bouchaud E, Lapasset G and Planès J 1993 *Phys. Rev. Lett.* **71** 2240
[5] Schmittbuhl J, Roux S, Vilotte J-P and Måløy K J 1995 *Phys. Rev. Lett.* **74** 1787
[6] Schmittbuhl J and Måløy K J 1997 *Phys. Rev. Lett.* **78** 3888
[7] Rosso A and Krauth W 2002 *Phys. Rev.* E **65** R025101
[8] Schmittbuhl J, Hansen A and Batrouni G G 2003 *Phys. Rev. Lett.* **90** 045505

[9] Santucci S, Grob M, Toussaint R, Schmittbuhl J, Hansen A, and Måløy K J 2010 *Europhys. Lett.* **92** 44001
[10] Batrouni G G, Hansen A, and Schmittbuhl J 2002 *Phys. Rev.* E **65** 036126
[11] Pradhan S, Hansen A and Chakrabarti B K 2010 *Rev. Mod. Phys.* **82** 499
[12] Johnson K L 1985 *Contact Mechanics* (Cambridge: Cambridge University Press)
[13] Love A E H 1929 *Phil. Trans. R. Soc.* A **228** 377-420
[14] Batrouni G G and Hansen A 1988 *J. Stat. Phys.* **52** 747
[15] Mehrabi A R, Rassamdana H, and Sahimi M 1997 *Phys. Rev.* E **56** 712
[16] Simonsen I, Hansen A, and Nes O M 1998 *Phys. Rev.* E **58** 2779

Preconditioner methods applied to simulations of two-phase flow in porous media

Morten Grøva
Department of Physics, NTNU, NO-7491 Trondheim, Norway
E-mail: `Morten.Grova@ntnu.no`

Abstract.
Simulations of two-phase flow in porous media require solving a set of linear equations which can be both large and sparse, and doing so repeatedly, to obtain local pressures under conditions of viscous flow. A variation of the incomplete cholesky preconditioned conjugate gradient solver has recently been developed for this purpose, and is presented here. It is found to be both stable and efficient. The variation consists of determining the sparsity structure of the preconditioner from a neighborhood search. Also, if the viscosities of the two fluids are different, the preconditioner should be updated during the course of the simulation. How often this should be done depends on the magnitude of the viscosity ratio, on the cost of updating the preconditioner and on how quickly the simulation configuration changes. To address this, a dynamic preconditioner update criterion is formulated.

1. Introduction

The transport of two immiscible fluids through a porous medium is greatly affected by the surface tension between the fluids [1, 2]. Menisci separating the two fluids within the pore space and the capillary pressure drops across them modify and are modified by flowing conditions. Understanding the resulting dynamics is of importance to oil recovery, for example.

Numerical simulations have a long history of complementing laboratory experiments in the study of this problem. Invasion percolation [3] and diffusion-limited aggregation [4] are two popular models which have been popular in the field of computational physics and have shed light on immiscible displacement processes.

When flow is intermediate between very slow (quasi-static) and very fast (dispersive) it is of key importance to solve the internal pressure- and flow-field given the external conditions. For example, it may be desired to solve for a given configuration of meniscus positions and a given total flow-rate. This requires solving a set of linear equations, assuming that the relations between flow and pressure drops are locally linear. Each equation expresses local volume conservation for the flow into and out of each pore. There are as many equations as there are pores, and each equation depends on the pressures within the adjacent pores as well as the pore itself. This results in a sparse, symmetric and positive-definite matrix $\mathbf{A}$ which gives the general equation

$$\mathbf{A} \cdot \mathbf{x} = \mathbf{b}, \tag{1}$$

to be solved for $\mathbf{x}$, which is a vector of node pressures. The right-hand side $\mathbf{b}$ is a vector of flow-rates caused by the menisci.

Eq. 1 may be solved by Cholesky factorization, $\mathbf{A} = \mathbf{L}\mathbf{L}^{\mathrm{T}}$, followed by forward and backward substitution. For a sparse matrix **A** this direct method has worse scaling of performance than the iterative method of Conjugate Gradients (CG). This means that for some sufficiently large system, CG solvers will outperform the direct method.

The performance and scaling of CG solvers can be greatly improved by preconditioning the matrix **A**, meaning that the original problem is substituted with a new,

$$\mathbf{MA} \cdot \mathbf{x} = \mathbf{M} \cdot \mathbf{b}, \tag{2}$$

where the preconditioner **M** approximates $\mathbf{A}^{-1}$, producing a better conditioned problem with fewer iterations required by the CG solver. For ill-conditioned problems such as the one considered here, preconditioning is an essential technique.

The Incomplete Cholesky preconditioned Conjugate Gradient solver (ICCG(n), [5]) enforces the sparsity pattern of **A** upon its Cholesky factors to produce a preconditioner **M** which satisfies $\mathbf{M}^{-1} = \tilde{\mathbf{L}}\tilde{\mathbf{L}}^{\mathrm{T}}$, where $\tilde{\mathbf{L}}$ is the incomplete Cholesky factor of **A**, which is now as sparse as **A**. The matrix **M** is not explicitly calculated; multiplications with **M** are done by forward and backward substitution with the incomplete Cholesky factors. The sparsity pattern of these factors may be augmented by including n additional diagonals along the main diagonal. ICCG(n) is a popular method when applicable, combining the advantages of the Cholesky factorization with those of the iterative CG solver.

This work presents a variation of the ICCG(n) linear solver. After a brief description of the simulations which motivated its creation and implementation, the algorithm is presented. It consists of a neighborhood search which gives the sparsity pattern of the preconditioner. Some examples of solver performance are given, before a preconditioner update criterion is presented. The update criterion is used to determine when the preconditioner should be updated to provide optimal performance through the course of the simulation.

2. The network simulator

In order to study transients and steady-state properties of two-phase flow in porous media a network simulator has been developed. A brief description is given here, for more details see [6, 7].

The porous medium is represented as a transport network where pore throats are links and pores are nodes. Knowing the flow-rate through each throat requires knowing the pressures in each pore. In addition to the pressure difference between pores, flow is driven by the pressure drops across menisci. The relation between total pressure drop and flow-rate is assumed to be linear, with the permeability dependent on the cross-section of the throat. Disorder is assigned to throat radii.

Enforcing volume conservation at each pore leads to a set of linear equations which can be solved with respect to an external control parameter like total flow-rate. This problem is analogous to solving the Kirchhoff equations for a random resistor network.

Mappings of real porous media to networks are available, however for simplicity the networks considered in this work are square lattices. They are inclined at 45° relative to the main direction of flow. Boundary conditions are bi-periodic, meaning that the network may be mapped onto a torus. The reason for choosing this topology is to be able to study steady-state properties. However, this also causes the traditional method of adding extra diagonals to the incomplete preconditioner to become less efficient, as will be discussed.

The capillary pressure drops across menisci are given by the radius of the throat at their position. Starting with a set of meniscus positions the flow-rates contributed by them are calculated and placed in the right-hand side vector **b** of Eq. 1. The system is iterated forward in time by solving for the pore pressures contained in **x** to obtain the local flow-rates. Once these are known, the meniscus positions can be updated and the process is repeated.

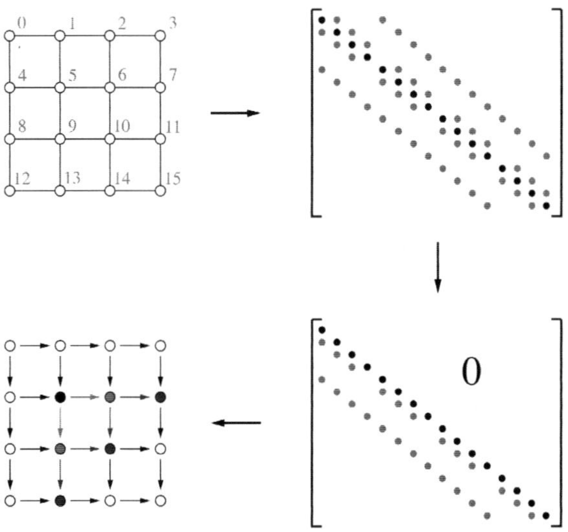

Figure 1. Schematic illustration of the neighborhood search algorithm. Taking a 4x4 square lattice (top left) as an example, the pattern of the Laplacian matrix is shown (top right) with the red dots signifying the non-zero pattern of the adjacency matrix. Removing the upper part of the matrix (bottom right) and interpreting the resulting pattern in terms of a directed graph (bottom left), the idea behind the neighborhood search can be easily explained. The filled black circle has two nearest-neighbors (red) and three next-to-nearest neighbors (blue), when the direction of the arrows is enforced. With n equal to 1, column 5 now has non-zero elements in rows 7, 10 and 13, in addition to rows 6 and 9. For n equal to 2 the next-to-next-to-nearest neighbors should also be included, and so on.

The matrix **A** contains local permeabilities and effective viscosities within each throat. If the two phases have unequal viscosities the local effective viscosity is volume-weighted in each throat. As the menisci move the local effective viscosity changes, changing **A**. If the two phases have identical viscosity, **A** does not change for the duration of the simulation.

The viscosity ratio M between the two phases is defined by

$$\mathrm{M} = \frac{\mu_{nw}}{\mu_w}, \tag{3}$$

where μ is viscosity and the subscripts refer to the non-wetting and wetting phase.

3. The linear solver

The single issue dominating computational performance is the rapid repeated solving of Eq. 1. For the simulations just described, **A** has a static sparsity pattern. The off-diagonal elements may be interpreted as an adjacency matrix, with non-zero elements where two pores share a pore throat. Given that the porous medium is static (consolidated), the sparsity pattern is therefore also static.

If the phases have equal viscosity, the entries of **A** do not change in the course of the simulation. It might then be expected that Cholesky factorization and repeated forward and backward substitution would be the most efficient solver. This is true only for small systems. Because the size of the Cholesky factors scale as $\sim N^{1.5}$ where N is the number of pores, and CG methods scale better than this, large systems are not well treated by the direct method.

Also, the size of the factors may require more memory than todays workstations have available. This means that even if the phases have equal viscosity, CG methods are preferable to Cholesky factorization.

At this point a minor point needs to be addressed. The matrix **A** has a unique solution **x** to every **b**, up to an overall absolute pressure. In other words, only pressure differences enter into the equations; the absolute pressure is an irrelevant additive constant. As it stands, **A** is therefore only semi-definite. In order to make it definite and hence invertible (stable with regards to Cholesky factorization) without destroying its symmetry a simple trick is used. One diagonal element of the matrix is multiplied by an arbitrary factor. As all but one of the equations remain unchanged, and the original matrix had exactly one redundant equation, all solutions to the new set of equations are also solutions to the old set. The only effect the arbitrary factor has is the only effect it can have, which is to fix the overall absolute pressure. As this is not important, the arbitrary factor is indeed arbitrary.

Steady-state simulations require the use of bi-periodic boundary conditions. This means that the bottom boundary of the network connects to the top boundary, and the left to the right. While ICCG(0) represents a considerable improvement over non-preconditioned CG and inverse diagonal preconditioned CG for these simulations, the traditional ICCG(n) method of adding additional diagonals does not result in any substantial additional increase of performance. This is in contrast to what is reported for several other problems, which typically deal with solving differential equations using some finite-difference method. The suspected reason for this lack of improvement is the use of bi-periodic boundary conditions. These lead to numerically large elements in the full Cholesky factors which are far from the main diagonal. Thus, the bi-periodic boundary conditions are a strong motivation for using a neighborhood search to determine the sparsity pattern.

The neighborhood search algorithm is illustrated in Fig. 1. The search proceeds in the direction of increasing node index on a directed graph obtained from the lower half of **A**. The maximum number of nodes allowed between the origin and the most distant node still considered a neighbor is n. Because the search proceeds along a directed graph the ordering of node indices is significant; this mirrors the significance of node ordering in the Cholesky factorization algorithm. An ordering which keeps near neighbors close in index value will result in a sparsity pattern which is most dense in the proximity of the main diagonal.

Proceeding only in the direction of increasing node index when mapping out the neighborhood structure is not a strictly necessary condition. If the neighborhood search is performed on the undirected graph a stable preconditioner with similar performance is obtained. There will however be more entries in the sparsity pattern which are zero-valued in the Cholesky factorization. These elements can be identified and eliminated as part of the factorization. The main reason for only searching on the directed graph is the symmetry with the factorization algorithm, which also proceeds on a directed graph.

Fig. 2 illustrates the result of the procedure. The original matrix is shown together with a full Cholesky factor (which is semi-sparse) and an $n = 4$ incomplete factor. In this example, the neighborhood search was performed on an undirected graph.

Even though the example shows a very small system, it should be clear from this that the size of the incomplete factor can be controlled using n, and that the neighborhood search produces a sparsity pattern which captures features of the full factor. For example, the entries in the lower-left corner and the outlying horizontal lines reflect the bi-periodic boundary conditions. These entries would not be present if the method of adding additional diagonals was used. It can also be noticed that the near-main diagonals are not completely regular. The slightly jagged pattern seems innocent, but is important with respect to parallelizability.

For large problems, parallelizability is an important consideration. Consider a factorization $\mathbf{A} = \mathbf{L}\mathbf{L}^{\mathrm{T}}$. Solving this requires forward backward substitution, which is not perfectly

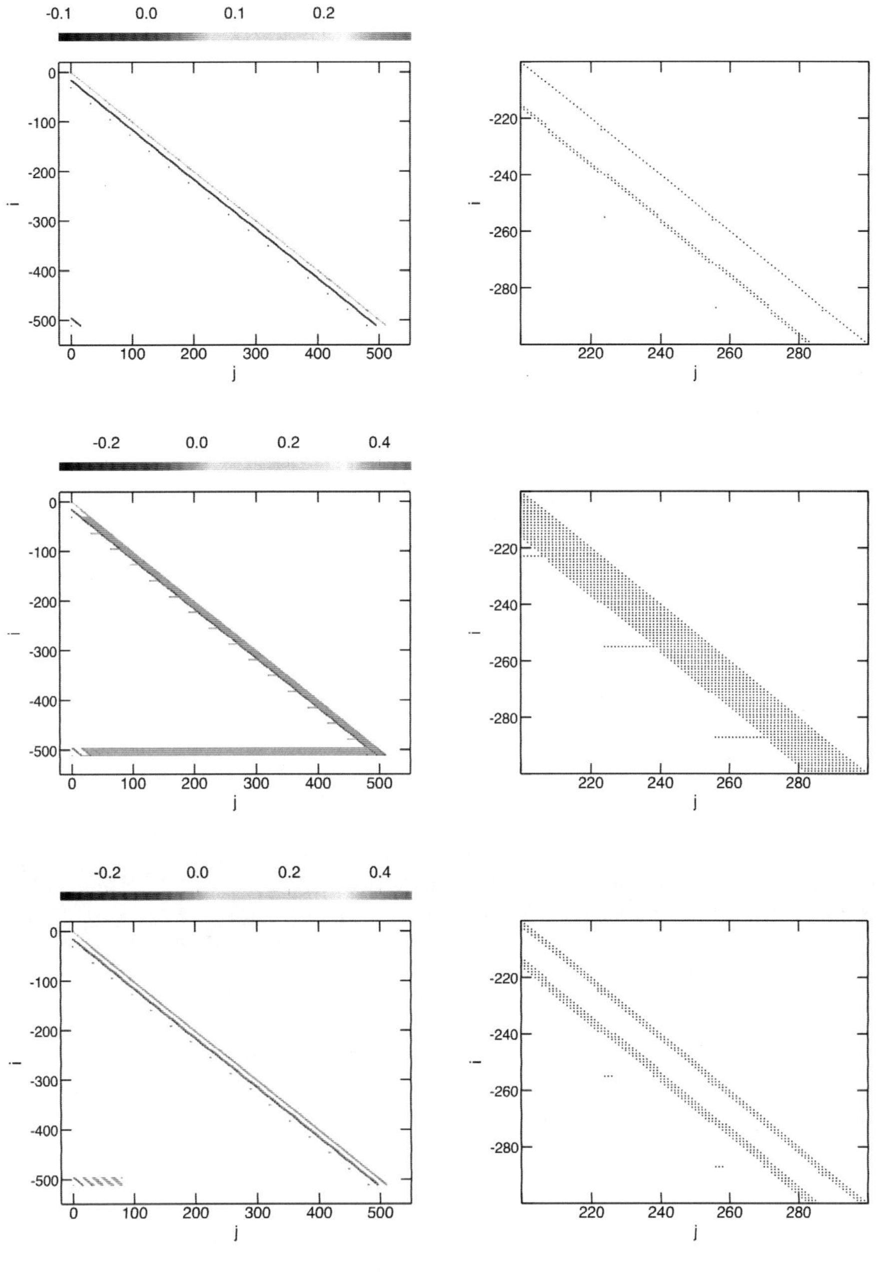

Figure 2. Visualization of matrix structure. System size is 16x32 pores. The top row is the lower half of **A**, the middle row is the complete Cholesky factor **L** and the bottom row is the incomplete Cholesky factor $\tilde{\mathbf{L}}$, where the neighborhood search algorithm has been applied with n = 4. The left column shows the full matrices with entries color coded by their values. The entries are indexed by i and j. The right column shows a zoom in. Non-zero entries are black, highlighting the sparsity pattern.

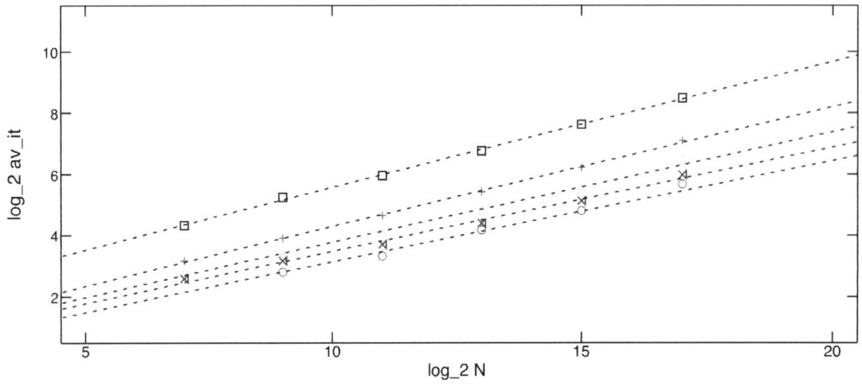

Figure 3. Scaling of solver performance measured in the number of iterations required, with system size N. Various values of n are shown. Counting from above, n is 0, 2, 4, 6 and 8. Scaling exponents are 0.41, 0.39, 0.36, 0.34 and 0.33, respectively.

parallelizable. This is the most significant challenge in parallelizing the ICCG(n) method. Consider a factor **L** which consists of only the main diagonal and the elements directly below it. This corresponds to a directed graph which is simply a one-dimensional chain. That directed graph gives the sequence in which the substitution must be carried out. It should be clear that this problem is not parallelizable at all. In general, complete diagonals prevent parallelization. Thus, the method of adding extra diagonals results in a method with poor parallelizability.

In contrast, using a directed neighborhood search results in a directed graph which can be partitioned as easily as a domain decomposition of the original problem. Because the neighborhood search preserves the original spatial layout it provides a method with excellent parallelizability.

The stability of a given sparsity pattern is not guaranteed. For example, it might be tempting to threshold the full Cholesky factor, keeping only the largest non-zero values and setting the rest to zero. This procedure results in a sparsity pattern which can then be used for subsequent calculations of incomplete Cholesky factors. However, numerical experiments have shown that this procedure results in an unstable preconditioner.

The sparsity pattern determined by a neighborhood search has resulted in a stable preconditioner for all realizations of the simulations so far, and is now a trusted method. Stability is not trivial, which makes this an important point.

The neighborhood search to determine the preconditioner sparsity pattern is expensive when implemented for general networks. However, because the sparsity pattern does not change in the course of the simulation the search only has to be done once. Because the solver is called a large number of times ($> 10^6$) the cost of the neighborhood search is negligible. In other words, the high initial cost is offset by repeated application.

Fig. 3 shows some performance results for typical simulation parameters. The number of iterations maps proportionally to overall performance for a given n, however larger n gives larger incomplete Cholesky factors and thus a larger cost per iteration. Scaling results can be found for constant n, however it is also found that the optimal n increases for larger systems. The actual scaling of the ICCG(n) linear solver with N therefore requires some scheme for varying n with system size. This scheme can not be considered independent of the architecture on which

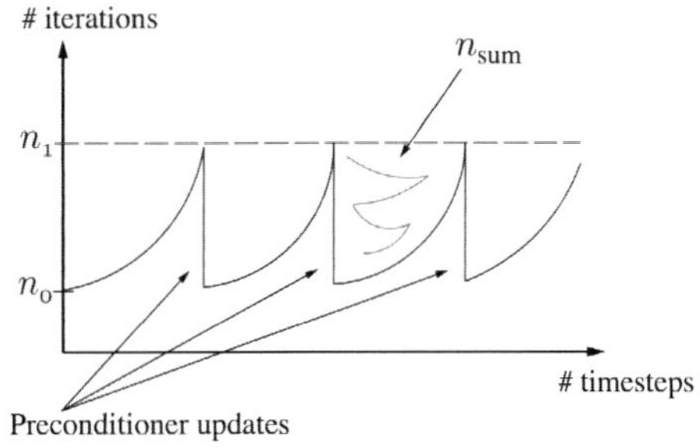

Figure 4. Visual description of preconditioner update criterion. n_0 is the number of iterations required by the CG method immediately after a preconditioner update. The area marked by a red wiggle represents the benefit of keeping the number of iterations on the black line compared to the blue line. For the interval between preconditioner updates this area should be kept equal to the cost of updating the preconditioner.

the solver is to be run.

If the scaling exponent of CG iterations with N is referred to as α and the scaling exponent of CG solver time with N as β, an idealized implementation of the CG method will scale as $\beta = \alpha + 1$. This is because every CG iteration consists of scalar-vector and sparse matrix-vector operations, all of which scale linearly with N. In any actual implementation of a CG method the situation is more complicated, as hardware issues such as memory locality become relevant and give implementation dependent scaling exponents.

If 0.33 is taken as the value for β and an idealized implementation is assumed, a value of 1.33 is obtained as an upper bound on the scaling of the neighborhood search ICCG(n) method. This is for some constant value of n. Better scaling can be achieved, because n should be increased when N increases. The exponent should be compared with 1.45, which is how these simulations scale when the inverse diagonal is used as preconditioner, and 1.5, which is for non-preconditioned CG as well as Cholesky factorization.

Perhaps more important for practical purposes is the reduction of the prefactor. For system sizes and parameters of typical interest a speed-up of a factor 4 has been achieved, when comparing additional diagonals ICCG(n) with neighborhood search ICCG(n).

Eventually, memory constraints will place a limit on n, as the size of the incomplete Cholesky factor grows rapidly with increasing n. Partitioning the factor across nodes on a distributed memory architecture will ease this constraint, but then there are additional considerations of message passing cost to take into account.

It should be clear by now that the parameter n in the ICCG(n) linear solver is a crucial parameter. Finding its optimal value is difficult to do in general, as it depends on the hardware. Of course, it can always be determined by trial and error. Tuning n allows the solver to balance various constraints of the hardware, such as available memory and message passing bandwidth.

4. The preconditioner update criterion

When **A** changes, the preconditioner based on an old **A** becomes progressively worse. The cost of re-calculating the preconditioner can be high compared to the cost of a single call to the linear solver, so this can not be done for every timestep. However, the benefit of keeping the number

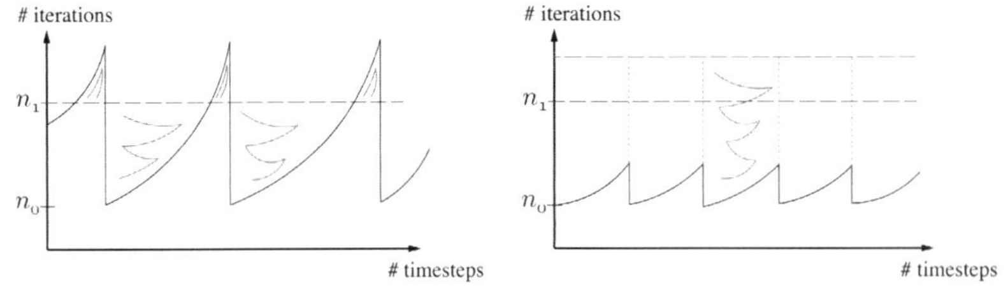

Figure 5. Visual argument for criterion optimality. The areas marked by red wiggles are meant to be identical in size to those in Fig. 4. Left: the areas marked by green wiggles represents additional cost caused by updating the preconditioner too infrequently. Right: the area between the green and blue dashed lines represents additional cost caused by updating the preconditioner too frequently.

of iterations of the conjugate gradient method low can also be large. Therfore, a criterion is needed to judge whether to upgrade the preconditioner or not.

If $n_{\text{it}}(i)$ is the number of iterations required in the CG method i timesteps after a preconditioner update, the quantity n_{sum} can be defined as

$$n_{\text{sum}} = \sum_i (n_1 - n_{\text{it}}(i)), \tag{4}$$

where the sum starts at $i = 0$ and n_1 is some value which n_{it} should not exceed. n_1 should satisfy

$$n_{\text{sum}} = n_{\text{cost}}, \tag{5}$$

where n_{cost} is the cost of updating the preconditioner, measured in units of single iteration cost. This cost can be calculated the first time the preconditioner is updated. A visual description of Eqs. 4 and 5 is given in Fig. 4. n_{sum} can be estimated without knowing n_1,

$$n_{\text{sum}} \approx \sum_i i \cdot (n_{\text{it}}(i) - n_{\text{it}}(i-1)), \tag{6}$$

where the sum starts at $i = 0$, and only runs over those values of i for which the difference within the parantheses is positive. In some rare cases n_{it} may decrease for some random reason, which will cause Eq. 6 to only be an approximation.

The application of the preconditioner update criterion in Eq. 5 can now be described as follows. The sum of Eq. 6 is updated for each timestep following a preconditioner update. Considering this to be a good approximation to n_{sum} as defined in Eq. 4, the preconditioner is updated when n_{sum} exceeds n_{cost}. Then, the sum is zeroed out and the process is repeated. Notice that n_1 never needs to be explicitly calculated.

The criterion and its application has now been described, but not yet fully motivated. The reason for considering Eq. 5 to be a criterion for the optimal preconditioner update frequency is easily understood from considering Fig. 5. If the preconditioner is updated less frequently, the CG method will require an unnecessarily large number of iterations in some interval preceding a preconditioner update. If the preconditioner is updated more frequently, the cost of the update is never fully regained by the benefit within a single interval beween updates.

Fig. 6 compares two simulations with and without applying the preconditioner update criterion. The viscosity ratio used in this example is M = 10. For viscosity ratios larger than this, the importance of updating the preconditioner increases even more. In the example, the difference is about a factor 4. It can be seen how the preconditioner update criterion enforces a constant ceiling on the number of iterations.

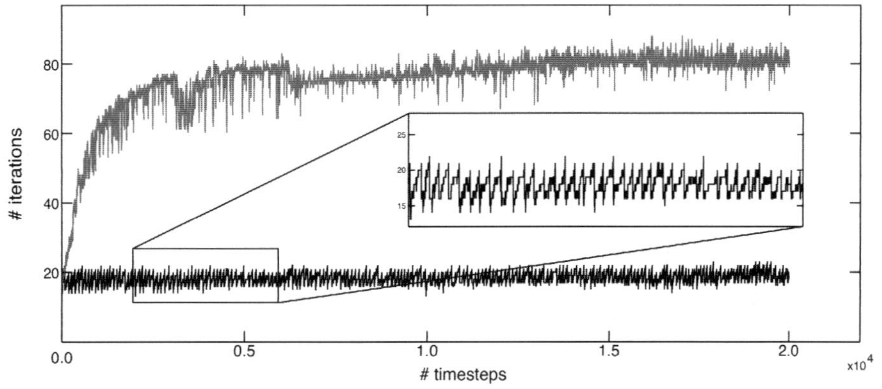

Figure 6. Number of iterations in the conjugate gradient algorithm as a function of the simulation timestep. Comparison of updating (black) and not updating (red) the preconditioner. System size is 32x64, M = 10.

5. Conclusion

The need for repeatedly solving a set of linear equations with static neighbor structure motivated this work. A neighborhood search is used to determine the sparsity structure of the incomplete cholesky preconditioner. This preconditioner has proven to be stable, and more efficient for the use considered here than previously implemented preconditioners. It is also more efficient than the method of augmenting the incomplete cholesky factors with additional diagonals.

Due to the high initial cost of the neighborhood search the method is unlikely to be generally useful. However, problems where the solver needs to be called a large number of times could benefit. In particular, the method should be beneficial to problems with complex topologies or irregular neighbor structure and problems where parallel performance is required.

A preconditioner update criterion has also been formulated, which could be of general use for problems where the solver is called repeatedly and preconditioner performance decreases gradually. Its implementation is straight-forward. It dynamically balances the cost of updates with the benefit of a reduced number of iterations.

Discussions with Knut Gjerden and Arne Stormo on the topic of preconditioners have been useful, and are greatly appreciated.

References

[1] Dullien F A L 1992 *Porous Media: Fluid Transport and Pore Structure, 2nd ed.* (Academic Press, San Diego)

[2] Sahimi M 1995 *Flow and transport in porous media and fractured rock* (VCH Verlagsgesellschaft)

[3] Wilkinson D and Willemsen J F 1983 *Journal of Physics A: Mathematical and General* **16** 3365

[4] Witten T A and Sander L M 1981 *Physical Review Letters* **47** 1400–1403

[5] Meijerink J A and van der Vorst H A 1977 *Mathematics of Computation* **31** 148–162

[6] Aker E, Måløy K J, Hansen A and Batrouni G G 1998 *Transport in Porous Media* **32** 163–186

[7] Knudsen H A, Aker E and Hansen A 2002 *Transport in Porous Media* **47** 99–121

Statistical Law of Turbulence in Granular Gas

Masaharu Isobe
Graduate School of Engineering, Nagoya Institute of Technology, 466-8555, Japan.
E-mail: isobe@nitech.ac.jp

Abstract. A novel description of granular gases from the viewpoint of "fluid turbulence" is presented. In the large-scale simulation by event-driven molecular dynamics simulations, we obtained clear evidence for the energy and enstrophy (square of vorticity) cascade, in which we found $-5/3$ (3D) and -3 (2D) exponents predicted by the Kolmogorov and Kraichnan-Leith-Bachelor energy spectrum, respectively. The growth of the Reynolds number based on the Taylor microscale confirmed that the cascade regimes were that of fully developed turbulence. During the evolving process of the regime in fully developed turbulent inhomogeneous state, we discovered a new regime arose from the collision between "clusters", in which the previous theoretical predictions of the energy decay in this regime collapse.

1. Introduction

Granular gases are fascinating systems from the viewpoint of extending classical equilibrium statistical physics. They are defined by the fact that the characteristic time of granular motion driven by external forces (*e.g.*, gravity or a vibrating plate) is small compared to the mean free time between inelastic collisions (See, review papers [1, 2]) Because granular motion is strongly affected by the energy source in general, the inelastic hard sphere (IHS) model without gravity, called the "freely evolving granular gas", has been much investigated as a simple model, using methods based on several different levels of analysis. These levels include the microscopic (molecular dynamics (MD)) [3 - 8], the mesoscopic (Enskog-Boltzmann equation) [9, 10] and the macroscopic (Navier-Stokes (NS) equations, Ginzburg-Landau theory) [11 - 14].

In the inviscid limit of three-dimensional (3D) and two-dimensional (2D) incompressible NS turbulence (*i.e.*, the Euler equation at high Reynolds number), energy and enstrophy are conserved in time. The enstrophy Z in 2D is defined by $Z = \frac{1}{2}\langle(\nabla \times \mathbf{u})^2\rangle$, where $(\nabla \times \mathbf{u})$ and $\mathbf{u} = (u_x, u_y)$ are the vorticity and the velocity, respectively. We also define the rate of volumetric dilatation as $Q = \frac{1}{2}\langle(\nabla \cdot \mathbf{u})^2\rangle$. As a result, the most prominent dynamic features of 2D turbulence are two fluxes in the k-space. By dimensional analysis, a $-5/3$ (3D) or -3 (2D) power law in energy spectrum is derived under the assumption that the energy spectrum only depends on the energy or enstrophy flux and the wave number, which is called Kolmogorov or Kraichnan-Leith-Bachelor (KLB) theory,

$$E(k) = C\epsilon^{2/3}k^{-5/3}, \quad (3D) \tag{1}$$

$$E(k) = C'\eta^{2/3}k^{-3} \quad (2D) \tag{2}$$

where C and C' are constant; ϵ and η are the energy and enstrophy transfer rate, respectively [15 - 17]. These cascade mechanisms are phenomenological, highly nontrivial and difficult to understand especially at the microscopic molecular level.

In the granular system, although these analogies have been pointed out by several authors [18, 19, 20], there are very few studies related to turbulence, because the dissipation mechanism of viscosity in the NS equation is viewed as completely different from that of inelastic collisions. First, Taguchi found numerically that the energy spectra from particle displacements at mid-depth in a 2D vibrated granular bed, which was composed of a few hundred particles, have a $-5/3$ exponent, similar to Kolmogorov scaling in 3D turbulence [18]. Then, neglecting the influence of gravity, a 2D granular gas with a Langevin-type thermostat was studied by Peng and Ohta in a system with about 10^3 particles [19]. Radjai and Roux [20] reported the spatial power-law spectrum of the particle displacements, as in Taguchi's work, in the quasi-static motions of the dense granular flow with a Parrinello-Rahman type boundary in a 2D MD model. All studies were performed in 2D and showed the power-law scaling of the energy spectrum and some relationship to NS turbulence. However, the special features of 2D NS turbulence were not shown. The main possible reasons for this are the following: (a) the statistical errors are quite large because of the small number of particles, (b) the system size is too small and is comparable to the minimal spatial correlation lengths (such as the Kolmogorov dissipative length), (c) no systematic parameter surveys exists, and (d) the unique features of 2D turbulence are not well-known in the granular physics community.

In our previous study [21], we have investigated the macroscopic statistical properties on the freely evolving quasi-elastic IHS model by performing a large-scale event-driven molecular dynamics with mainly 512^2 particles system and found that remarkably analogous to an enstrophy cascade process in the decaying 2D fluid turbulence. We found that there are four typical regimes in the freely evolving inelastic hard disk system, which are homogeneous, shearing (vortex), clustering and final state. In the shearing regime, the self-organized macroscopic coherent vortices become dominant. In the clustering regime, the energy spectra are close to the expectation of KLB theory and the squared two-particle separation strictly obeys Richardson law. In this paper, we have performed the simulation with 2048^2 for 2D and 128^3 for 3D particles system. The present system in 2D is 16 times larger than that of previous our works. We have confirmed the previous study on a clear quantitative evidence of the enstrophy cascade. To compare with the direct numerical simulation (DNS) of fluid turbulence quantitatively, we discuss the Reynolds number based on the Taylor microscale in our simulation.

2. Model and Numerical method

The freely decaying IHS model is so simple that the system is completely characterized by only three parameters: the restitution coefficient r, the total number of disks N, and the packing fraction ν. The system size L in units of disk diameter σ is $L/\sigma = \sqrt{\pi N/\nu}/2$ (2D) and $\sqrt[3]{4\pi N/(3\nu)}/2$ (3D). All disks are identical, so the system is mono-dispersive. For the dissipation, the normal restitution coefficient r for binary instantaneous collision is introduced (*i.e.*, parallel to the relative position of the two colliding disks in contact).

Even in the quasi-elastic thermodynamic limit, the system becomes unstable and several spatio-temporal scales appear. The standard evolution scenario is that spatial instabilities occur from the initial homogeneous cooling state (HCS: 1st stage) to the velocity field (shearing regime: 2nd stage), and then to the density field (clustering regime: 3rd stage). Finally, global structures in the system (final attractor regime: 4th stage) appear as a non-equilibrium steady state, or local inelastic collapse occurs.

The system consists of up to about 4.2 million hard disks (2D) or 2.1 million hard spheres (3D) placed in a square or cubic box with periodic boundaries without any external force. To perform a large scale simulation, a simple and efficient event-driven algorithm is used [22]. Initially, the system is prepared in the equilibrium state (*i.e.*, the density is uniform and the disk velocities are Maxwell-Boltzmann distributed) by a long preparation run in an elastic system (*i.e.*, $r = 1$; the so-called the Alder solid-fluid transition occurs when the packing fraction is $\nu \sim 0.70$ for 2D and

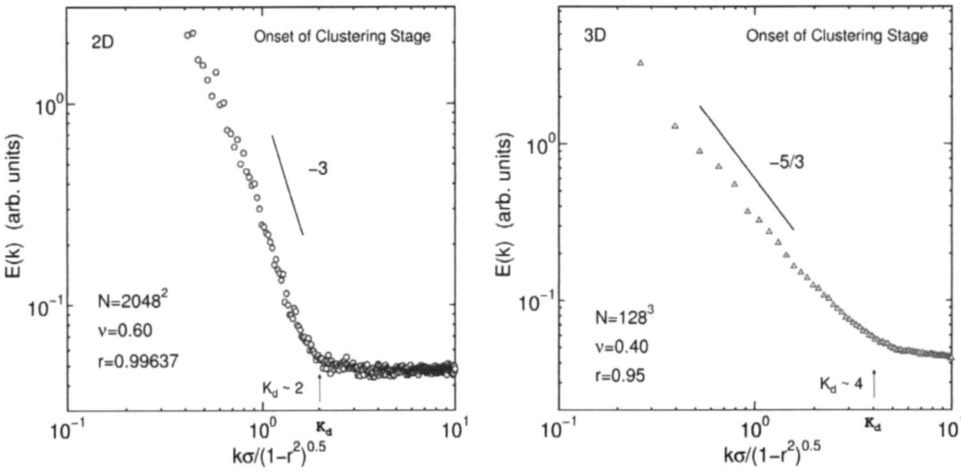

Figure 1. The energy spectra at the onset of clustering regime (3rd stage) are shown. The exponents develop a maximum close to -3 for 2D (the left) and $-5/3$ for 3D (the right). The parameters are set at (r, N, ν)=$(0.99637, 2048^2, 0.60)$ for 2D and $(0.95, 128^3, 0.40)$ for 3D, respectively.

0.48 for 3D.). Because the total energy is monotonically decreasing in a freely evolving process, constant energy in time can be realized by attaching a thermostat. There are several types of thermostats to keep the total kinetic energy constant, roughly categorized as deterministic or stochastic. The advantage of a deterministic thermostat is that the trajectory of particles does not change compared with the non-thermostat case. Here, we introduce the new-scaled time t_s, which is described by $t_s = \int_0^t \beta(t)^{-1}dt, \quad \beta(t) = \sqrt{T(0)/T(t)}$, where $T(t)$ is the average kinetic energy per particle as a function of the usual time t. Introducing a new-scaled time t_s instead of t is completely equivalent to adding a deterministic thermostat to the hard sphere system [23, 24].

3. Results

3.1. Energy Spectrum

In Fig. 1, the energy spectra $E(k)$ (power spectra of the "*kinetic temperature*" field, which are discretized by grids of size the disk diameter σ) at $(N, \nu) = (2048^2, 0.60)$ for 2D (left) and $(128^3, 0.40)$ for 3D (right) with respect to the nondimensional wave number divided by the energy dissipation for inelastic collisions $\sqrt{1-r^2}$ are shown. Although the local equilibrium assumption in a granular system is not valid, we regard the granular temperature as the "*kinetic temperature*" since the restitution coefficient is set near the elastic case between collisions and the system is still regarded as homogeneous state in density before the clustering regime. Our large-scale simulations show that $E(k)$ develops towards exponents -3 (2D) and $-5/3$ (3D) for small wavenumbers (large scale), until the onset of the clustering regime (3rd stage) [21], as is expected by the KLB and Kolmogorov theory for freely decaying 2D and 3D NS fluid turbulence, respectively.

In 2D case, we calculated the time evolution of the dissipation rate for the scaled enstrophy $-d\tilde{Z}/dt_s$ and found that the enstrophy peak is a criterion to estimate the boundary time between the shearing regime (2nd stage) and the clustering regime (3rd stage). The enstrophy dissipation is almost finished around $t_s \sim 500$, at which time the exponent of the energy spectrum reaches

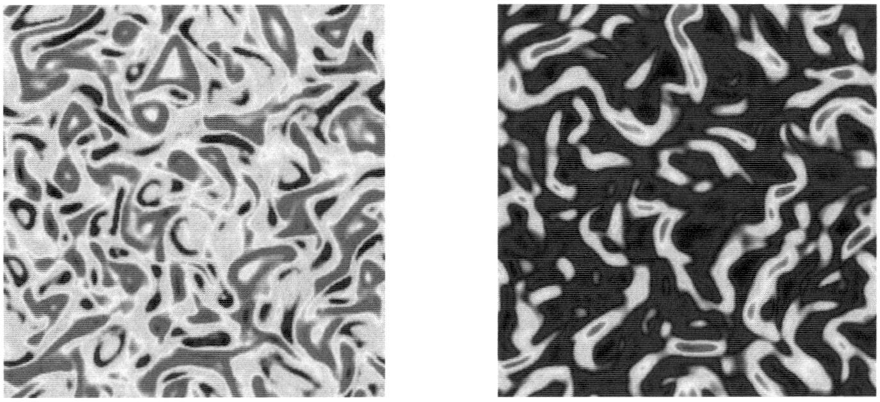

Figure 2. Developed turbulent patterns in the vorticity (left) and temperature (right) fields around $t_s \sim 500$ in 2D. The parameters are the same as in the left of Fig. 1.

a maximum. Typical vorticity and temperature fields around $t_s \sim 500$ are shown in Fig. 2.

In Fig. 1, we can also estimate the characteristic spatial scales K_d $(= k\sigma/\sqrt{1-r^2}) \sim 2$ (2D) and ~ 4 (3D) as the minimal dissipative domain, which are composed of about $\sim 10^3$ disks and $\sim 10^2$ spheres, respectively. Although the particle number in the dissipative domain depends on the packing fraction and the restitution function, we also confirmed that the values of the minimal wavenumber K_d scaled with a function of the restitution coefficient $\sqrt{1-r^2}$, as predicted by kinetic theory. Figure 3 shows various values of $E(k)$ in 2D, scaled by a function of the restitution coefficient over a wide parameter space when the packing fraction is fixed at $(N, \nu) = (2048^2, 0.60)$. During the evolution of $E(k)$ for small wavenumbers, the minimal dissipation scale K_d appears not to change. Even when the systems have already evolved to the beginning of the clustering instability, the scaling behavior for the minimal dissipation is valid. When the restitution coefficient is close to unity, we found that the inertial range grows gradually. This suggests that the Reynolds number becomes higher in the quasi-elastic limit.

3.2. Microscale Reynolds Number

In the study of fluid turbulence, the dimensionless Reynolds number Re is estimated to characterize the quantitative measurement of the turbulent state, which is given by the ratio of inertial forces to viscous forces. To analyze the small-scale statics in the DNS of NS turbulence, the Taylor's microscale Reynolds number is often used. In this subsection, we also estimated the Taylor microscale Reynolds number in the granular gas.

The Taylor microscale λ in homogeneous isotropic incompressible D $(= 2$ or $3)$ dimensional NS turbulence with zero mean velocity can be derived as,

$$\lambda = \sqrt{\frac{D(D+2)\nu_{\text{vis}} \langle u_x^2 \rangle}{\langle \epsilon \rangle}}, \tag{3}$$

where ν_{vis} and $\langle \epsilon \rangle$ are the kinematic viscosity and the energy dissipation rate, respectively. For a hard core fluid, the kinematic viscosity $\nu_{\text{vis}} = \eta_E/\rho$ can be derived by the density $\rho(= mn)$ and Enskog shear viscosity η_E, which can numerically be estimated by the expression of the Enskog theory for hard disk systems by Gass(1971) [25] with the third Sonine polynomial approximation in 2D and by Wainwright(1964) [26] with the equation of state by Ree & Hoover (1967) [27] in 3D.

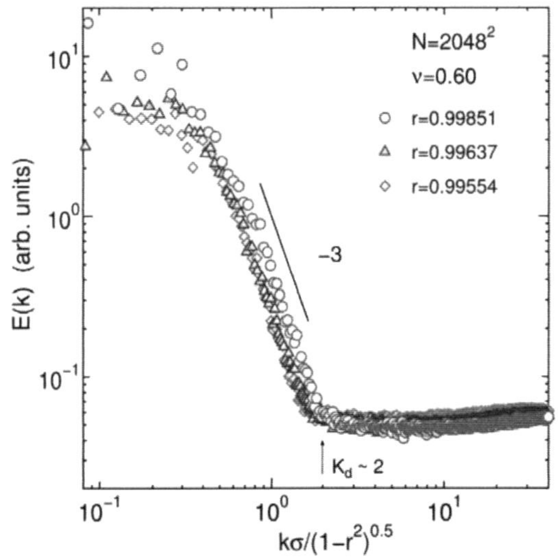

Figure 3. Scaling behavior of the energy spectra at the onset of the 3rd stage for different restitution coefficients r at a fixed $(N, \nu) = (2048^2, 0.60)$.

For example, the expression of Enskog shear viscosity η_E for 2D is,

$$\eta_E = \eta_0 \gamma\rho \left\{ \frac{1}{\gamma\rho\chi} + 1 + \gamma\rho\chi \left(\frac{1}{4} + \frac{2}{\pi\gamma_0(3)} \right) \right\}, \tag{4}$$

where $\eta_0 = (\gamma_0(3)/2\sigma)\sqrt{mkT/\pi}$, $\gamma\rho = \pi\sigma^2 n/2 = 2y$, $\chi = (1 - 7y/16)/(1 - y)^2$(Enskog scaling factor), $y = \pi\sigma^2 n/4$, and n is number density. Here we use in the third Sonine polynomial approximation $\gamma_0(3) = 1.022$.

The Reynolds number based on Taylor microscale $\lambda(t_s)$ can be estimated as,

$$R_\lambda(t_s) = \frac{\langle u_x^2 \rangle^{\frac{1}{2}} \lambda(t_s)}{\nu_{\text{vis}}}. \tag{5}$$

Because we can obtain both the energy dissipation rate $\langle \epsilon \rangle$ via Haff's law and the kinematic viscosity via Enskog theory, we can roughly estimate $R_\lambda \sim 10$ at $t_s = 0$. However, R_λ increases to 327 (2D) and 1470 (3D) at the onset of the deviation from Haff's law in the kinetic energy decaying functions. Although the granular gas system is well described as the compressible macroscopic fluid equations, we can regard that the system is still almost homogeneous in the early stage of relaxation. Therefore, before the clustering regime begins, the derivation of Taylor's microscale expression may be considered valid. This assumption is also confirmed to be fairly good approximation by the calculation of the fluctuation of density as shown in the Fig. 1 of Ref. [21].

3.3. Asymptotic Energy Decay after Haff's Law and Cluster Collision Regime

During the 1st stage, uniform granular gas remains homogeneous, however, the kinetic energy (or velocity) decreases via inelastic collisions in time. In this regime, the granular temperature $T(t)$ decays according to Haff's law,

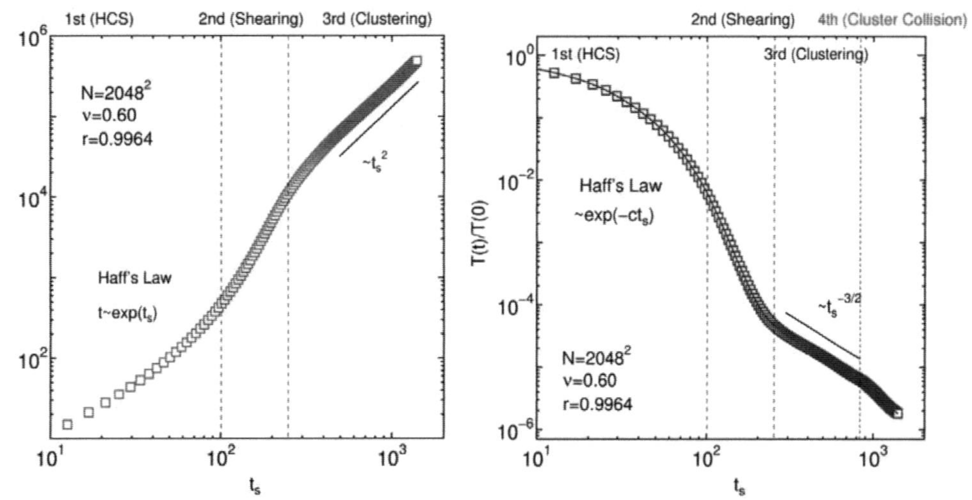

Figure 4. The relationship between time t and scaled time t_s (left) and the total kinetic energy per particle divided by its initial energy $T(t)/T(0)$ in terms of t_s in the evolving process of granular gas (right) are shown.

$$\frac{T(t)}{T(0)} = \frac{1}{(1+t/\tau_0)^2} \equiv \exp\left(-2\tau_0\tau\right) \tag{6}$$

where τ is collision number and τ_0 is a fitting parameter, which can also be estimated explicitly by the kinetic theory in dilute case.

In the clustering regime, due to inhomogeneous in density, $T(t)$ deviate from Haff's law. In our large simulation (e.g., parameters are set at $(N,\nu,r) = (2048^2, 0.60, 0.9964)$), the relationship between t and t_s are consistent with the Haff's law at 1st to 2nd stages and becomes $t \sim t_s^2$ after 3rd stages (the left of Fig. 4). We confirmed that the energy decay is also consistent with the Haff's law before clustering regime (the right of Fig. 4).

The long time (asymptotic) behavior of $T(t)/T(0)$ in a system of freely IHS model are predicted by several theories [28, 29], which are using mode coupling theory and extension of inelastic (sticky) hard rods gas. In 2D case, both theories predict $T(t)/T(0) \sim \tau^{-1}$ and $\sim t^{-1}$, respectively. In the 3rd stages, although we confirmed that the theoretical predictions are consistent with our simulations in relatively dilute system, the deviation is gradually observed in relatively dense system. In the right of Fig. 4, we can estimate that $T(t)/T(0) \sim t^{-\frac{3}{4}} \sim \tau^{-2}$, where $\tau \sim t_s^{\frac{3}{4}} \sim t^{\frac{3}{8}}$. Therefore, we found that the exponent of power decay in the 3rd stage at the dense system seems to be slower than that of dilute case, which cannot be explained by the previous theories.

Moreover, we found that the power law of the kinetic energy decay at 3rd stage is also collapsed after $t_s \sim 800$ in the right of Fig. 4. To investigate the reason of this deviation, we focus on the collision rate during the evolving process in terms of t_s, which increases rapidly at this regime. Based on the detail analysis of dynamics, we conclude that this deviation comes from collisions between "clusters". In the 3rd stage, several large clusters are created and distributed separately in space. Eventually those clusters composed of large number of inelastic particles collide each other at this regime (after $t_s \sim 800$). This regime (new 4th regime: cluster collisions) can clearly be seen more than a few million particles system, which cannot be resolved in the relatively small particles system. In the actual simulation, the spatial distribution of the rate of volumetric dilatation $Q(\mathbf{x})$ defined in Sec. 1 clearly show that the propagation of semicircular shape density shock wave in the clusters appears in this regime. The theoretical prediction of asymptotic behavior of energy decay after the clustering regime is not valid at high density and large scale system. This new finding might be considered to be important in the late stage of aggregation process in the system composed of many granular particles (such as the formation of the planetesimal (an ultimately small fraction of a planet)). We are now analyzing this regime in detail, which will appear elsewhere in the near future.

4. Concluding remarks

In this study, we showed several analogous features of both 2D and 3D NS turbulence in granular gases using the well-defined simple IHS model. By extensive event-driven molecular dynamics and systematic estimations of the spatial correlation in a wide parameter space, typical features of 2D and 3D NS turbulence (*i.e.*, the enstrophy and energy cascade) were found in the relatively dilute early stages of the granular gas. We also found that the maximum of the enstrophy dissipation rate is a clear criterion to determine the time between the shearing and clustering regimes in 2D. The energy spectra were the same as expected from KLB(2D) and Kolmogorov(3D) theory at the onset of the clustering stage. The minimal dissipation scale depended on the restitution coefficient and could be scaled by a function of the restitution coefficient. In the quasi-elastic limit, because the features of turbulence clearly appear, the system has a high Reynolds number with Taylor microscale $R_\lambda \sim 327$ (2D) and $R_\lambda \sim 1470$ (3D) in our simple estimation. In general, because the laboratory and industrial turbulent flows are characterized by $R_\lambda \sim \mathcal{O}(10^2 \sim 10^3)$, it is quite reasonable that the granular gas system shows the specific laws of turbulent behavior. These results suggest that the granular gas in the quasi-elastic and thermodynamic limit is strongly related to both 2D and 3D NS turbulence. It is indeed interesting that modern large scale molecular dynamics simulation can allow us to check the validity of the macroscopic NS equation and its numerical solution by DNS directly. These molecular dynamics simulations are hence an important bridge between the microscopic and macroscopic levels, and offer another route to approach the problem of fluid turbulence.

5. Acknowledgments

This study was supported by Grant-in-Aid for Scientific Research from the Ministry of Education, Culture, Sports, Science and Technology No. 23740293. Part of the computations for this study was performed using the facilities of the Supercomputer Center, Institute for Solid State Physics, the University of Tokyo, and Research Center for Computational Science(RCCS), Okazaki, Japan. This work is performed with the support and under the auspices of the NIFS Collaboration Research program (NIFS11KNTS010, NIFS11KNSS020).

References

[1] Kadanoff L.P. 1999 *Rev. Mod. Phys.* **71** 435.
[2] Goldhirsch I. 2003 *Annu. Rev. Fluid Mech.* **35** 267.
[3] Pöschel T. and Schwager T. 2004 *Computational Granular Dynamics*, (Springer).
[4] Chen S., Deng Y., Nie X., and Tu Y. 2000 *Phys. Lett. A* **269** 218.
[5] Miller S. and Luding S. 2004 *Phys. Rev. E* **69** 031305.
[6] Nie X., Ben-Naim E., and Chen S. 2002 *Phys. Rev. Lett.* **89** 204301.
[7] Meerson B. and Puglisi A. 2005 *Europhys. Lett.* **70** 478.
[8] Alvarez-Hamelin J. I. and Puglisi A. 2007 *Phys. Rev. E* **75** 051302.
[9] Brilliantov N. and Pöschel T. 2004 *Kinetic Theory of Granular Gases* (Oxford University Press).
[10] Brey J. J., Dufty J. W., and Santos, A. 1997 *J. Stat. Phys.* **87** 1051.
[11] van Noije T. P. C., Ernst M. H., Brito, R, and Orza, J. A. G. 1997 *Phys. Rev. Lett.* **79** 411.
[12] Wakou J., Brito R., and Ernst M.H. 2002 *J. Stat. Phys.* **107** 3.
[13] Efrati E., Livne E., and Meerson B. 2005 *Phys. Rev. Lett.* **94** 088001.

[14] Meerson B., Fouxon I., and Vilenkin A. 2008 *Phys. Rev. E* **77** 021307.
[15] Kolmogorov A. N. 1941 *Dokl. Akad. Nauk. SSSR* **30** 9.
[16] Kraichnan R. H. and Montgomery D. 1980 *Rep. Prog. Phys.* **43** 547.
[17] Tabeling P 2002 *Phys. Rep.* **362** 1.
[18] Taguchi Y. -H. 1993 *Europhys. Lett.* **24** 203.
[19] Peng G. and Ohta T. 1998 *Phys. Rev. E* **58** 4737.
[20] Radjai F. and Roux S. 2002 *Phys. Rev. Lett.* **89** 064302.
[21] Isobe M. 2003 *Phys. Rev. E* **68** 040301(R).
[22] Isobe M. 1999 *Int. J. Mod. Phys. C* **10** 1281.
[23] Komatsu T. S. 2000 *J. Phys. Soc. Jpn.* **69** 5.
[24] Soto R., Mareschal M., and Monsour M. Malek 2000 *Phys. Rev. E* **62** 3836.
[25] Gass D M 1971 *J. Chem. Phys.* **54** 1898.
[26] Wainwright T E 1964 *J. Chem. Phys.* **40** 2932.
[27] Ree F H and Hoover W G 1967 *J. Chem. Phys.* **46** 4181.
[28] Brito R and Ernst M H 1998 *Europhys. Lett.* **43** 497.
[29] Ben-Naim E, Chen S Y, Doolen G D, and Redner S 1999 *Phys. Rev. Lett.* **83** 4069.

Morphology diagrams for A_2B copolymer melts: real-space self-consistent field theory

Rajeev Kumar[1], **Yige Li**[2], **Scott W. Sides**[3], **Jimmy W. Mays**[4] **and Bobby G. Sumpter**[5]

[1]National Center for Computational Sciences, Oak Ridge National Laboratory, Oak Ridge, TN 37831

E-mail: kumarr@ornl.gov

[2]Denison University, Granville, OH 43023

[3]Tech-X Corporation, Boulder, CO 80303

[4]Department of Chemistry, University of Tennessee, Knoxville, TN 37996

[5]Center for Nanophase Materials Science, Oak Ridge National Laboratory, Oak Ridge, TN 37831

Abstract. Morphology diagrams for A_2B copolymer melts are constructed using real-space self-consistent field theory (SCFT). In particular, the effect of architectural asymmetry on the morphology diagram is studied. It is shown that asymmetry in the lengths of A arms in the A_2B copolymer melts aids in the microphase separation. As a result, the disorder-order transition boundaries for the A_2B copolymer melts are shown to shift downward in terms of $\chi N, \chi$ and N being the Flory's chi parameter and the total number of the Kuhn segments,respectively, in comparison with the A_2B copolymers containing symmetric A arms. Furthermore, perforated lamellar (PL) and a micelle-like (M) microphase segregated morphologies are found to compete with the classical morphologies namely, lamellar, cylinders, spheres and gyroid. The PL morphology is found to be stable for A_2B copolymers containing asymmetric A arms and M is found to be metastable for the parameter range explored in this work.

1. Introduction

Microphase separation in block copolymer melts[1, 2, 3, 4, 5, 6, 7] has motivated the scientific community over the last couple of decades. Delicate balance[3] of chain conformational entropy and repulsive interaction energy between unlike monomers allows block copolymers to readily self-assemble into a number of ordered morphologies. Typical morphologies[3, 8, 9] include lamellar, cylinder, sphere, gyroid and Fddd network phases, which have been proven to be equilibrium states.

Experimentally, the balance of the energy and entropy can be tuned by a number of molecular characteristics such as molecular weight[3], chain architecture[10, 11], conformational asymmetry[12, 13], polydispersity[14, 15, 16, 17, 18] and temperature. A number of studies[1, 2, 3, 4, 5, 6, 7] have been done to elucidate the effects of molecular weight, polydispersity, conformational asymmetry and temperature on the mircophase separation in linear di-block copolymer melts. However, the effects of chain architecture are not that well established. On the computational side, this is primarily due to a large parameter space for the simulations and resulting large number of morphologies, which need to be compared in terms of their free-energy to establish the morphology diagram. For the experiments, the challenge is to synthesize well-controlled macromolecules with non-linear architecture and reaching equilibrium by overcoming energy barriers.

In the 1970s, Fetters et al.[1] studied the morphologies and properties of star-block copolymers of polystyrene (PS) and polyisoprene (PI). Also in the 1970s, the synthesis of fairly well-defined multi-graft copolymers having polydiene backbones and PS side chains was reported[2, 4]. In the 1990s, systematic studies[19, 20, 21] aimed at elucidating the effect of branched architecture on the morphology and properties of block copolymers were initiated. Like the linear di-block copolymer melts, classical morphologies namely, lamellae, cylinders, spheres and gyroid, were observed in experiments[19, 20, 21] for the miktoarm star copolymers.

On the theoretical front, the morphological behavior of miktoarm star copolymers containing two kinds of blocks was described by Olvera de la Cruz[10] in the weak segregation limit and Milner[11] addressed the issue of microphase separation in the strong-segregation limit. The theory by Milner[11] accounts for the fact that the two kinds of blocks may differ in conformational characteristics, which results in a variation in the arms stretching at the junction point. Self-consistent field theory (SCFT) studies bridging the gap between the weak and strong segregation theories were done by Grason et al.[7, 22, 23]. Furthermore, dissipative particle dynamics (DPD) simulations[24] have been used to construct the morphology diagrams for A_2B copolymer melts and reasonable agreement with the SCFT predictions[7, 23] has been established with the addition of new morphologies. Quite recently, a theory addressing the effects of polydispersity[25] on the microphase separation in A_2B copolymers has been developed.

With the recent developments in the computational power and synthetic methodologies, synergetic studies on copolymers with non-linear architectures can be executed. Furthermore, theoretical studies in the weak and intermediate segregation limit can be performed. Motivated by these developments, we have studied the effects of chain architecture on the micro-phase separation in block copolymer melts and constructed morphology diagrams using the self-consistent field theory (SCFT).

In the present work, we have studied miktoarm star copolymer melts with a structure of A_2B in which the total length and length of one of the A arms is kept fixed while varying the fraction of B block. Such a variation allows us to study architecturally asymmetric miktoarm copolymers (i.e., asymmteric A arms) along with homopolymer and di-block copolymer melts. Morphology diagrams for architectural asymmetric miktoarm block copolymer melts are constructed using the real-space SCFT, without *a priori* knowledge about the morphologies. The study on the star copolymer melts is carried out in the weak and intermediate segregation limit.

This paper is organized as follows: details about the computational method (i.e., SCFT) are presented in section 2 followed by the results and conclusions in sections 3 and 4, respectively.

2. Computational method: SCFT

The SCFT is very well documented in the literature[26]. In the following, we only present the details relevant for this work. For the other details, we refer to an excellent review by Grason[7, 23].

Construction of the field theory starts from the particle description of the system in the form of a Hamiltonian having contributions from the chain stretching in Weiner form and short-range repulsive interaction energy in the Flory's χ parameter form. Representing polymer chains as continuous curves, we write the partition function, Z of the melts containing n A_2B chains as

$$Z = \frac{1}{n!}\int \prod_{\beta=1}^{n} D\mathbf{R}_\beta(s) P\left[\mathbf{R}_\beta(s)\right]\delta\left[\hat{\rho}_A(\mathbf{r})+\hat{\rho}_B(\mathbf{r})-1\right]\exp\left[-\chi\rho_0\int d\mathbf{r}\hat{\rho}_A(\mathbf{r})\hat{\rho}_B(\mathbf{r})\right] \quad (1)$$

where $P\left[\mathbf{R}_\beta(s)\right]$ is the Weiner measure for the chain connectivity and s is the contour variable specifying different locations on the curves $\mathbf{R}_\beta$. Also, $\hat{\rho}_A(\mathbf{r})$ is the microscopic density of A

defined as

$$\hat{\rho}_A(\mathbf{r}) = \frac{N}{\rho_0}\sum_{\beta=1}^{n}\int_0^1 ds\gamma_A(s)\delta\left[\mathbf{r}-\mathbf{R}_\beta(s)\right] \tag{2}$$

so that N is the total number of the Kuhn segments along each chain and ρ_0 is the reference density ($\equiv 1/l^3, l$ being the Kuhn segment length). Here, $\gamma_A(s)$ is a function, which is equal to 1 when s lies along the A blocks and 0 otherwise. A similar equation is written for $\hat{\rho}_B(\mathbf{r})$.

The chain architecture information is included in the path integral formulation written as Eq. 1. In order to switch from the particle description to the field theory, the Hubbard-Stratonovich transformation is used. In order to alleviate the complicated task of evaluating the functional integrals over the fields, these functional integrals are approximated by the value of the integrand at the optimum. This so called saddle-point approximation, leads to a set of non-linear equations, which need to be solved self-consistently using numerical methods. Furthermore, the free energy of the system at the saddle-point is readily accessible from these calculations, which, in turn, allow the construction of morphology diagrams for the co-polymer systems.

There are different ways of solving the non-linear set of equations obtained at the saddle-point. For the miktoarm copolymer melts, we have solved these set of equations in real-space starting from random guesses for the fields and densities. In particular, we have used the real-space algorithm presented in the work by Sides and Fredrickson[27], implemented in parallel. The real-space algorithm for the solution of the SCFT equations has the advantage that no *a priori* knowledge about the symmetry of morphologies is needed. This also means that the algorithm has the potential to discover new morpholgies and give us the leverage to compare the free energies of different morphologies with each other, just by starting from different initial guesses.

In this work, we have focused on star co-polymers of the form A_2B, where the two A arms can be of different lengths. For the construction of morphology diagrams, we have kept $N_{A1} + N_B = 100, N_{A2} = 25, N_j$ being the number of segments along the arm of j type. Morphology diagrams are constructed by systematically varying the fraction of B arm ($\phi_B = N_B/N$ where $N = N_{A1} + N_{A2} + N_B$) and χ parameter characterizing the repulsive interactions between the two components (A and B). Also, we have taken the Kuhn segment lengths of different arms to be the same ($= l$) and ignored the effects of conformational asymmetry. We have done this on purpose to limit the parameter space and understand the effects of architectural asymmetry on the morphology diagrams.

Morphology diagrams were constructed by starting from random guesses for the equilibrium structures and searching for the minimum free energy state. Minimum free energy state is established by comparing the free energies of the final states obtained from different random guesses. Sometimes, the relaxation scheme used for solving the non-linear equations gets trapped in metastable or weakly unstable configurations. In order to remove topological defects and achieve global free energy minimum state, we have used a "zone annealing" technique in which a zone of reduced χ is moved across the simulation grid. Other details of the technique will be presented elsewhere. This procedure allows us to sort out the lowest free energy state of the system. All of the results presented below are obtained by solving the SCFT equations on a cubic grid containing 64^3 grid points with a grid spacing of 0.1 (in units of $R_{go} = (N_{A1}/6)^{1/2}l$). For the special case of $\phi_B = 0.8$ with $N_{A2} = 25$ and $N_{A1} + N_B = 100$, $N_{A1} = 0$, (i.e., a linear di-block), we have used $R_{go} = (N_{A2}/6)^{1/2}l$ for making length dependent quantities dimensionless. Also, a time step of 0.01 is used to solve the modified diffusion equations using the pseudo-spectral method[27].

3. Results and Discussions

In Fig. 1, we present the morphology diagrams for A_2B star copolymer melts containing symmetric A arms (Fig. 1(a)) and asymmetric A arms (Fig. 1(b)). Both, the SCFT[22] and DPD[24] studies have already chalked out the morphology diagrams for A_2B copolymers with symmetric A arms. However, the SCFT study[22] has been carried out using spectral methods, where one has to presume the symmetry of morphologies. Furthermore, worm-like micelles are observed in the DPD simulations[24] in contrast to the sphere or cylindrical phase in the SCFT near the disorder-order transition around $\phi_B = 0.8$. Also, experiments by Pochan et al.[20] show the existence of microphase separated structures (micellar) without any underlying lattice.

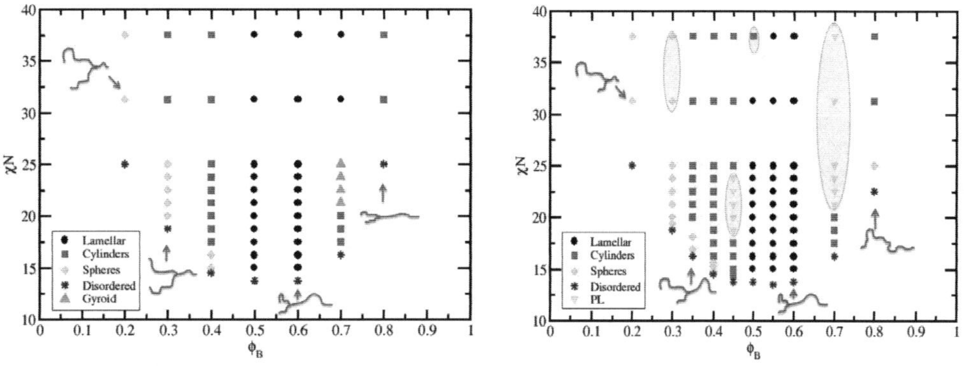

(a) (b)

Figure 1. Morphology diagrams for A_2B star co-polymer melts containing symmetric A arms (a) and asymmetric A arms (b). For Fig. 1(b), volume fraction of one of the A arms is kept fixed at $\phi_{A2} = 0.2$ and PL stands for perforated lamellar. Highlighted regions in Fig. 1(b) show the morphologies different from the case of symmetric A arms shown in Fig. 1(a).

Our real-space self-consistent field theory calculations for the A_2B copolymers with symmetric arms (Fig. 1(a)) confirms the stability of classical morphologies such as lamellar, cylinders, spheres and gyroid in different parameter ranges. Just like linear di-block copolymers, in all of the curved morphologies, the minority component forms the interior and the majority components acts like a matrix. Fig. 1(a) is in excellent agreement with the spectral SCFT[22] study. However, Fig. 1(a) is in slight disagreement with DPD[24] study. The disagreement lies in the stability of perforated lamellar (PL) and worm-like micellar morphology as observed in the DPD study near $\phi_B = 0.5$ and $\phi_B = 0.8$, respectively. Our real-space SCFT calculations starting from random numbers as initial seeds for the fields and *without the use of "zone annealing" technique* show PL for $\phi_B = 0.5$. Furthermore, these calculations reveal the existence of microphase separated but not very well-ordered micellar structures (M) for $\phi_B = 0.8$. Isosurfaces of the density profiles of A in the PL and M morphologies corresponding to $\phi_B = 0.5$ and $\phi_B = 0.8$, respectively, at $\chi N = 37.5$ are shown in Figs. 2(a) and 2(b), respectively. The M morphology has undulated cylinders and spheres coexisting with each other, without any underlying lattice. However, we have found that both the PL and M morphologies are metastable and the removal of defects using the "zone annealing" technique leads to more stable lamellar and cylindrical morphology, respectively, as shown in Fig. 1(a).

Fig. 1(b) shows the morphology diagram when we vary the volume fraction of B block (ϕ_B), while keeping volume fraction of one of the A arms fixed at 0.2. The variation allows

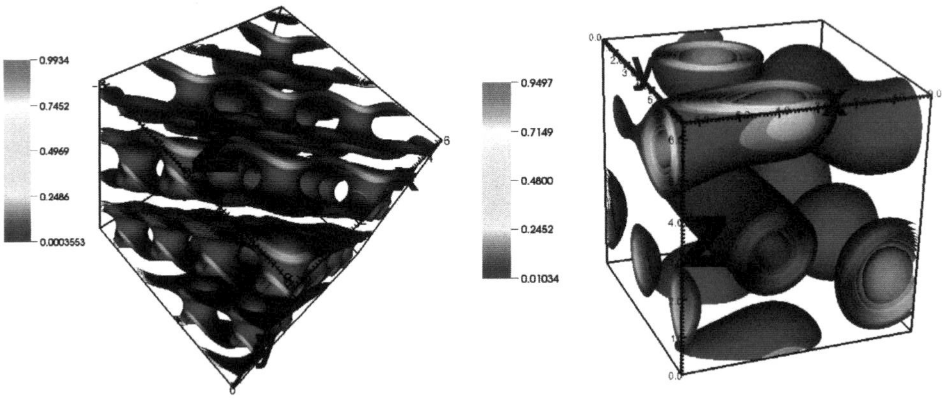

(a) (b)

Figure 2. Isosurfaces of the density distributions of A in the perforated lamellar (PL) morphology (a) and micelle-like (M) morphology (b).

us to put the morphologies of homopolymers, miktoarm copolymers with symmetric A arms and linear di-blocks on the same diagram. For example, in Fig. 1(b), the leftmost column ($\phi_B = 0.2$) represents a miktoarm star copolymer with a very small B arm. The rightmost column ($\phi_B = 0.8$) correspond to a linear di-block copolymer with A as the minority component. Morphological transitions from the disordered to the spheres to the cylinders for $\phi_B = 0.8$ are in excellent agreement with other literature studies[6]. Furthermore, the morphologies for $\phi_B = 0.6$ are the same as for the miktoarm star co-polymers containing symmetric arms as shown in Fig. 1(a).

Comparing Figs. 1(a) and 1(b), it is clear that architectural asymmetry originating from distribution of segments among different A arms has a significant effect on the morphology diagram. For example, we have found the PL morphology to be stable for miktoarm copolymers containing asymmetric A arms in contrast to Fig. 1(a). Also, cylindrical morphology (with B component forming the core) is found to be stable for $\chi N = 37.5, \phi_B = 0.5$, in qualitative agreement with Ref. [11]. Furthermore, comparison between the Figs. 1(a) and 1(b) reveals that the architectural asymmetry of A arms plays a significant role in determining the disorder-order transition boundaries. For example, comparing the values of χN at which an ordered morphology is first observed for $\phi_B = 0.8$ in Figs. 1(a) and 1(b), it is found that stronger repulsive interactions (i.e., higher χN or lower temperature values due to the inverse dependence of χ on the temperature) are required for A_2B to undergo microphase separation in comparison with the linear AB diblock containing the same fraction of B. Note that linear di-block copolymer can be imagined as the A_2B copolymer so that length of one of the A arms is zero. By this argument, A_2B copolymer in the form of a linear diblock has the highest asymmetry of A arms. So, we can infer that architectural asymmetry originating from the length of A arms helps in undergoing micro-phase separation. Comparing the χN for the disorder-order transition from Figs. 1(a) and 1(b) for the same ϕ_B's, it is clear that this is true in general and we can conclude that architectural asymmetry aids in the microphase separation. This is consistent with the spectral SCFT studies[7].

4. Conclusions

For several decades there has been considerable attention focused on block copolymer materials, due largely to their ability to self-organize into a variety of morphologies having feature sizes on the scale of nanometers. It is known that block copolymers self-assemble into a rich range of three-dimensional nanoscale structures that can provide a suitable framework for the production of new and complex materials. In contrast to the case for simple linear block copolymers, the structure-morphology relationships for nonlinear block copolymers have yet to be thoroughly understood, a problem that has largely prevented their exploitation.

Our work has demonstrated that morphologies in A_2B miktoarm star copolymer melts are sensitive to the architectural asymmetry. We have constructed the morphology diagrams for the A_2B copolymer melts containing symmetric and asymmetric A arms using the real-space self-consistent field theory without presuming the symmetry of ordered structures. In addition to the lamellar, cylinders, spheres and gyroids formed by the minority component, we have found the perforated lamellar (PL) morphology to be stable for A_2B copolymers containing asymmetric A arms. Also, we have found the PL and a micelle-like (M) morphology to be metastable for the A_2B miktoarms with symmetric A arms. This is in agreement with other literature studies carried out using spectral SCFT[22] and in disagreement with a DPD[24] study. Furthermore, comparison of the morphology diagrams for A_2B copolymers containing symmetric and asymmetric arms reveal that architectural asymmetry of A arms aids the microphase separation. While this work was under review, we have come to know about recent work by Matsen[28] on A_2B copolymers containing symmetric A arms. Our Fig. 1(a) is in agreement with Ref. [28]. Note that the PL morphology is found to be stable in Ref. [28] outside the parameter range explored in this work.

5. Acknowledgments

This work was supported by the Materials Science and Engineering Division, U.S. Department of Energy (DoE), Office of Basic Energy Sciences (BES) under Contract No. DEAC05-00OR22725 with UT-Battelle, LLC, at Oak Ridge National Laboratory (ORNL). Part of the research was done at the Center for Nanophase Materials Sciences, which is sponsored by the Scientific User Facilities Division of DOE. This research used resources of the Oak Ridge Leadership Computing Facility at the ORNL. Y-G Li received partial supports from the ORSS(HERE) program.

References

[1] Falk, J. C.; Schlott, R. J.; Hoeg, D. F.; Pendleton, J. F. Rubber Chem. Technol., 46 1044 (1973).
[2] Hadjichristidis, N.; Roovers, J. J. Plym. Sci. Polym. Phys. Ed. , 16, 851 (1978).
[3] Bates, F. S.; Fredrickson, G.H. Annual Review of Physical Chemistry 41,525-557 (1990).
[4] Pitsikalis, M.; Pispas, S.; Mays, J. W.; Hadjichristidis, N. Adv. Polym. Sci. 135, 1 (1998).
[5] Block Copolymers: Synthetic Strategies, Physical Properties, and Applications, Hadjichristidis, N.; Pispas, S.; Floudas, G.; John Wiley, Hoboken, NJ 2003.
[6] Matsen, M. W. Journal of Physics-Condensed Matter 14(2), R21-R47 (2002).
[7] Grason, G. M. Physics Reports-Review Section of Physics Letters 433(1),1-64 (2006).
[8] Matsen, M. W.; Schick, M. Phys. Rev. Lett. 72, 2660-2663 (1994).
[9] Tyler, C.A.; Morse, D.C. Phys. Rev. Lett. 94, 208302 (2005).
[10] Delacruz, M. O.; Sanchez, I.C. Macromolecules 19(10),2501-2508 (1986).
[11] Milner, S. T. Macromolecules 27, 2333 (1994)
[12] Vavasour, J. D., Whitmore, M.D. Macromolecules 26(25),7070-7075 (1993).
[13] Matsen, M. W.; Schick, M. Macromolecules 27(14): 4014-4015 (1994).
[14] Burger, C.; Ruland, W.; Semenov, A. N. Macromolecules 23, 3339-3346 (1990).
[15] Sides, S. W.; Fredrickson, G. H. Journal of Chemical Physics, 121, 4974-4986 (2004).
[16] Cooke, D. M.; Shi, A. C. Macromolecules, 39, 6661-6671 (2006).
[17] Matsen, M. W. Physical Review Letters, 99, 4 (2007).
[18] Beardsley, T. M.; Matsen, M. W. European Physical Journal E, 27, 323-333 (2008).
[19] Gido, S. P.; Lee, C. et al., Macromolecules 29(22),7022-7028 (1996).

[20] Pochan, D. J.; Gido, S.P. et al. Macromolecules 29(15): 5091-5098 (1996).
[21] Lee, C.; Gido, S.P.; Pitsikalis, M.; Mays, J.W.; Beck Tan, N.; Trevino, S. and Hadjichristidis,N. Macromolecules 30, 3732-3738 (1997).
[22] Grason, G. M.; Kamien, R.D. Macromolecules 37(19),7371-7380 (2004).
[23] Grason, G. M.; Kamien, R.D. Physical Review E 71(5) (2005).
[24] Huang, C. I. ; Yu, H.T. Polymer 48(15): 4537-4546 (2007).
[25] Aliev, M. A.; Kuzminyh, N.Y. Physica a-Statistical Mechanics and Its Applications, 390(20),3245-3261 (2011).
[26] Fredrickson, G. H.; Ganesan, V.; Drolet, F. Macrocolecules, 35, 16-39 (2002)
[27] Sides, S. W.; Fredrickson, G.H. Polymer 44(19): 5859-5866 (2003).
[28] Matsen, M.W. Macromolecules 45, 2161 (2012).

Journal of Physics: Conference Series **402** (2012) 012043 doi:10.1088/1742-6596/402/1/012043

Comparison of Phase Field Crystal and Molecular Dynamics Simulations for a Shrinking Grain

B. Radhakrishnan[1], S.B. Gorti[1], D. M. Nicholson[1] and J. Dantzig[2]

[1]Oak Ridge National Laboratory, Oak Ridge, TN 37831, USA; [2]University of Illinois at Urbana-Champagne, Urbana, IL 61801

E-mail: radhakrishnb@ornl.gov

Abstract. The Phase-Field Crystal (PFC) model represents the atomic density as a continuous function, whose spatial distribution evolves at diffusional, rather than vibrational time scales. PFC provides a tool to study defect interactions at the atomistic level but over longer time scales than in molecular dynamics (MD). We examine the behavior of the PFC model with the goal of relating the PFC parameters to physical parameters of real systems, derived from MD simulations. For this purpose we model the phenomenon of the shrinking of a spherical grain situated in a matrix. By comparing the rate of shrinking of the central grain using MD and PFC we obtain a relationship between PFC and MD time scales for processes driven by grain boundary diffusion. The morphological changes in the central grain including grain shape and grain rotation are also examined in order to assess the accuracy of the PFC in capturing the evolution path predicted by MD.

1. Introduction

The phase field crystal (PFC) method has evolved in recent years as a novel approach for investigating structural evolution at the diffusional time scale with atomic resolution. [1] The underlying feature of the method involves the representation of the atoms in a crystal as a time averaged atomic density field and evolving the density field by solving an evolution equation that drives the free energy density to a minimum. The approach eliminates the need for tracking unproductive local atomic displacements that result from energy excursions that are small compared to the activation energy required to evolve the system through atomic migration. The free energy functional that represents the three-dimensional atomic density with the required crystal symmetry is derived from the atomic density functional theory. For example, a functional that captures the periodicity of body-centred cubic (BCC) iron has recently been developed as shown in Eqn. 1.

$$F=\int_V \left[\frac{B_s R^4}{2}\rho\left(\frac{1}{R^2}+\nabla^2\right)^2\rho+\frac{B_l-B_s}{2}\rho^2+\frac{v}{3}\rho^3+\frac{w}{4}\rho^4\right]dV \qquad (1)$$

where where ρ is the atomic density. In the appendix of their paper, Jaatinen et al. [2] derive a form of the free energy functional where the cubic term is made to vanish. Setting ⍰⍰-⍰⍰=⍰ , ⍰⍰⍰4=⍰ , ⍰=1⍰0=1⍰⍰, ⍰=0, and ⍰=⍰ , and using the values based on data for BCC iron at 1772 K given in (A7)-(A11) of [1], the parameters in Eqn. (1) are obtained as B_l = 20.0417 eVÅ^3, B_s = 22.0789 eVÅ^3,

R = 0.335 ÅÅ, v = 0 and w = 8.4582 eVÅ^{9}eV Å9. The evolution equation for the free energy is based on the modified phase field crystal approach [3] given by

$$\frac{\partial^2 \rho}{\partial t^2} + \beta \frac{\partial \rho}{\partial t} = \alpha \nabla^2 \left(\frac{\delta F}{\delta \rho} \right) \tag{2}$$

where α and β are phenomenological constants. The ability of the PFC approach to capture topological defects such as dislocations and their evolution during deformation has been demonstrated in the literature [4]. The PFC approach is also known to capture the structure of grain boundaries and the variation of the grain boundary energy as a function of boundary misorientation in accordance with the Read-Shockley equation [5]. However, grain boundary motion under the influence of a driving force demonstrated by PFC has not been compared with conventional atomistic techniques such as molecular dynamics (MD) to ascertain the ability of the PFC to capture the atomistic mechanisms involved. For example, curvature driven grain boundary motion occurs by grain boundary diffusion and the grain boundary velocity is determined by the boundary energy as well as its mobility. Anisotropy in grain boundary energy or the mobility can lead to changes in the grain morphology during grain growth. It is interesting to see if such shape changes can be accurately captured by the PFC method. The objective of the present study is to compare the evolution by shrinking of a spherical grain embedded in a matrix using PFC and MD.

2. Simulations

The kinetics of shrinking and the morphological changes in the central spherical grain were investigated using PFC and MD for three different misorientations of the central grain with respect to the matrix to generate grain boundaries with different energies and mobilities. The orientation of the outer matrix grain was set to the reference state with zero rotation about the *x*, *y* or *z* - axis. The following three orientations where chosen for the central grain:

- 15 degree rotation about the *z* axis, labeled (0 0 15)
- 15 degree rotation about *x*, *y* and *z* axes, labeled (15 15 15)
- 45 degree rotation about the *z* axis, labeled (0 0 45)

The initial, two-grain geometry was created in PFC by appropriately seeding the orientations of the inner and the outer grains based using an analytical one-mode approximation for the density field outlined by Wu and Karma [5]. The input to the MD simulations was obtained by using the peak positions of the three-dimensional density profiles as the atom positions. However, the density of the PFC peaks near the grain boundary regions is lower than in the grain interior although the overall density in PFC is conserved globally. Therefore, if we assume that the probability of existence of an atom is 1.0 for peaks in the interior locations of the grain, the corresponding probability is lower at the peak positions in the grain boundary regions indicated by PFC. Therefore, in order to use the PFC data as input to MD, it is necessary to device a method that eliminates some of the PFC peaks near the grain boundary. Here, we propose three different approaches for creating bicrystals where the outer matrix grain is always in the (0 0 0) orientation and the inner grain with a spherical morphology is rotated to (0 0 15) or (15 15 15) or (0 0 45) orientation. The goal in each approach is to come up with the total number of peaks in the bicrystal that equals the number of peaks obtained in a (0 0 0) single crystal occupying the same volume, subsequently referred to as N_{000}. The first approach is defined as the baseline approach, where single crystals of (0 0 0), (0 015), (15 15 15) and (0 0 45) are first created independently by PFC. The bicrystal geometry is then created from the corresponding single crystal peak positions. Overlapping peaks in the grain boundary region are removed until the number of peaks in the bicrystal is equal to N_{000}. Another method used to create the bicrystal is defined as the "first come" method, where the outer and the inner crystals are formed directly from the liquid using the seeding approach of Wu and Karma [6]. The number of density peaks in the bicrystal is carefully monitored as the system evolves and the simulation is terminated when the number of peaks equals N_{000}. The assumption here is that the most probable atoms positions in the bicrystal especially in the grain boundary regions will first develop as PFC density peaks. The third approach involves the creation of the bicrystal geometry using PFC as in the second approach and allowing the simulation to continue until the number of PFC peaks in the system approaches the maximum value. This value is typically higher than N_{000}. The atom positions are then picked from the PFC peaks in the descending order of the peak intensity until the number of atoms becomes N_{000}.

The atomic representation of the bicrystal obtained from PFC was further evolved in MD using the Large-scale Atomic/Molecular Massively Parallel Simulator (LAMMPS) code developed at the Sandia National Laboratory [7]. The simulations were carried out for BCC iron using the embedded atom potential developed by Mendelev et al. [8] The initial atom positions from PFC were scaled to the lattice constant of BCC iron at 0K. Initially, a minimization run was performed to move the atoms in the grain boundary and interior locations to local energy minima. The next step involved an expansion of the simulation cell to a size corresponding to 873K using a linear thermal expansion coefficient of 11.8 x 10^{-6} /K while simultaneously providing a kinetic energy corresponding to 100K, followed by a minimization run. The final step involved an NVT ensemble run at 873K using a time step 0f 0.001 ps. The MD simulations were continued until the central grain completely disappeared due to curvature driven forces. The PFC simulations were also continued until the disappearance of the central grain. A system size of 360 x 360 x 360 was used in PFC to generate the bicrystal geometry that resulted in an MD domain size of 36 x 36 x 36 BCC unit cells of Fe. PFC simulations were based on a finite difference solution of Eqn. 2 based on an explicit technique with a time step of 0.005.

3. Results and Discussion

The variation of the total energy per site with PFC time is shown in Fig. 1 for three different grain boundary types. Since the excess energy is proportional to the grain boundary surface area, and since the grain boundary area decreases linearly with time for curvature-driven grain shrinkage, a linear reduction in energy with PFC time is expected. The shrinkage curve for the (15 15 15) boundary is closest to the linear behavior, while for the (0 0 45) boundary the shrinkage curve is slightly convex. The shrinkage curve for the (0 0 15) boundary appears to deviate slightly from the linear behavior.

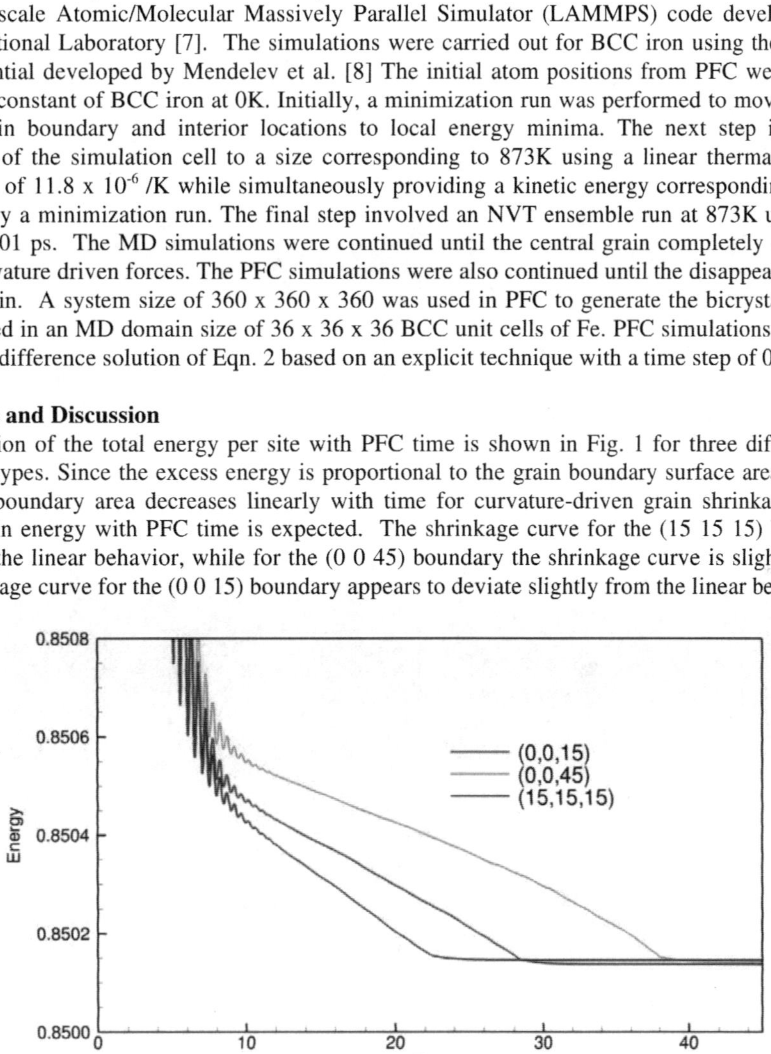

Fig 1. Variation of total energy with simulation time during PFC. The total energy decreases as the central grain shrinks.

The temporal evolution of the bicrystals is shown in Fig. 2 as contours of the density field at the mid-section perpendicular to the z-axis. Careful observation of the bicrystals indicated a small rotation of the (0 0 15) grain at the initial stages of shrinkage, with a probable change in the boundary energy and mobility that could account for the small deviation from the linear behavior shown in Fig 1. No such rotation was seen for the (15 15 15) grain that showed a linear shrinkage. The (0 0 45) grain also did not show any rotation. However, the boundary appeared to develop facets at various stages of the shrinkage process that probably resulted in energy and mobility changes. It appears that as the grain size becomes smaller, the number of facets increased, indicating that the average boundary mobility increased with decreasing grain size that results in the convex shrinkage curve shown in Fig 1. The (15 15 15) grain also showed a shape change in the *xy* plane from an initial circular to an elliptical geometry as the grain shrinks. No significant shape change was seen for (0015) and (0045) grains.

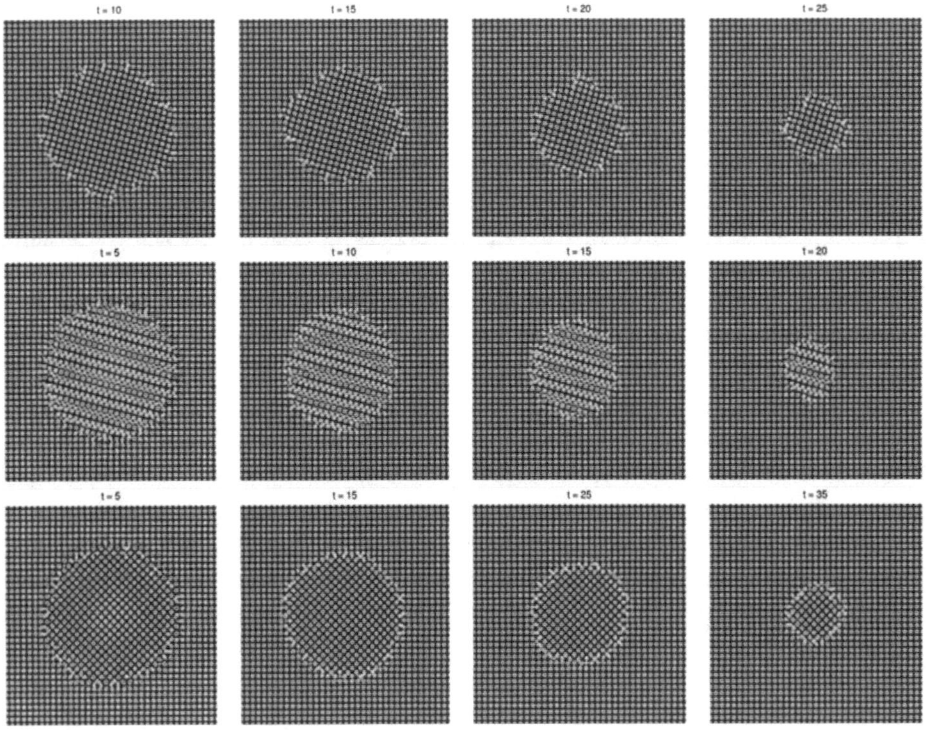

Fig 2. Two-dimensional section perpendicular to the z – axis showing the temporal evolution during PFC simulation of the (0 0 15) (top), (15 15 15) (middle) and (0 0 45) (bottom) grain situated at the center of a (0 0 0) outer grain.

Figure 3 shows the shrinkage curves obtained from MD simulations for the three different grain boundary types. Three sets of curves are shown in Fig 5 for each of the approaches described earlier for obtaining the initial atom positions from PFC. Red curves correspond to the baseline approach, the green curves correspond to the "first-come" approach, and the blue curves correspond to the sorting approach. The initial portion of the curve in which the energy increases with time is due to the increase in the thermal energy associated with heating the system to 873K. It is possible that some evolution occurred during the heating period. The shrinkage portion of the curves appears to contain many of the features exhibited by the PFC simulations. These include the almost linear behavior for

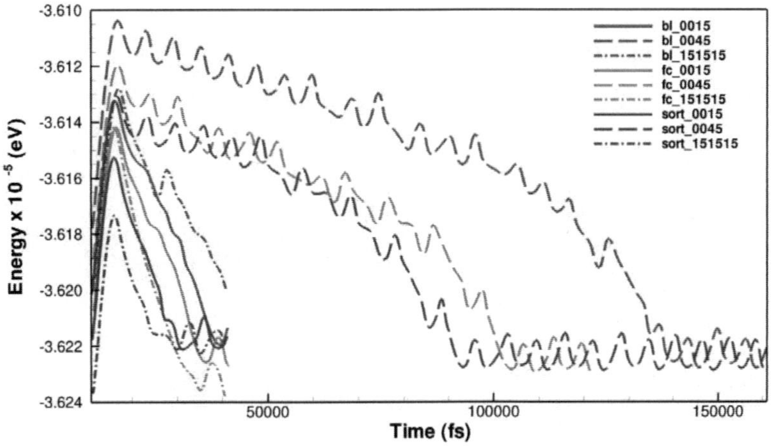

Fig 3. Variation of total energy of the bicrystal with MD simulation time.

the (15 15 15) boundary, a convex shrinkage behavior for the (0 0 45) boundary and smaller deviation from the linear behavior for the (0015) boundary. The shrinkage kinetics for the (0 0 45) boundary is significantly lower than for the other boundaries, which is the same behavior exhibited for the PFC as shown in Fig 1. The baseline approach results in the highest and the sorting approach the lowest energies for the initial bicrystal geometry. The baseline approach in which the initial single crystals are grown separately and then put together mechanically had the highest pressure during the MD runs as shown in Fig. 4. In the baseline approach, even though the total number of peaks in the bicrystal was equal to N_{000}, there was probably an overcrowding of the atom positions near the grain boundary that did not relax completely during the 873K run.

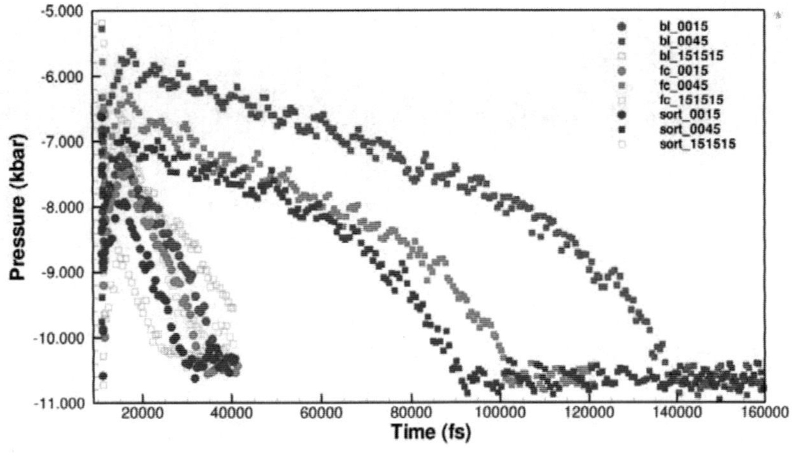

Fig 4. Variation of total pressure in the bicrystal with MD simulation time. The pressure is the highest for the (0045) case.

The "first-come" approach gives an initial energy that is between those of the baseline and the sorting approaches. The baseline approach shows a lower shrinkage kinetics for the (15 15 15) boundary compared to the (0 0 15) boundary which is not shown by either of the other two approaches, as well as by PFC. Based on these findings, it appears that the "first-come" or the sorting approach is probably better than the baseline approach for mapping atom positions from PFC for MD.

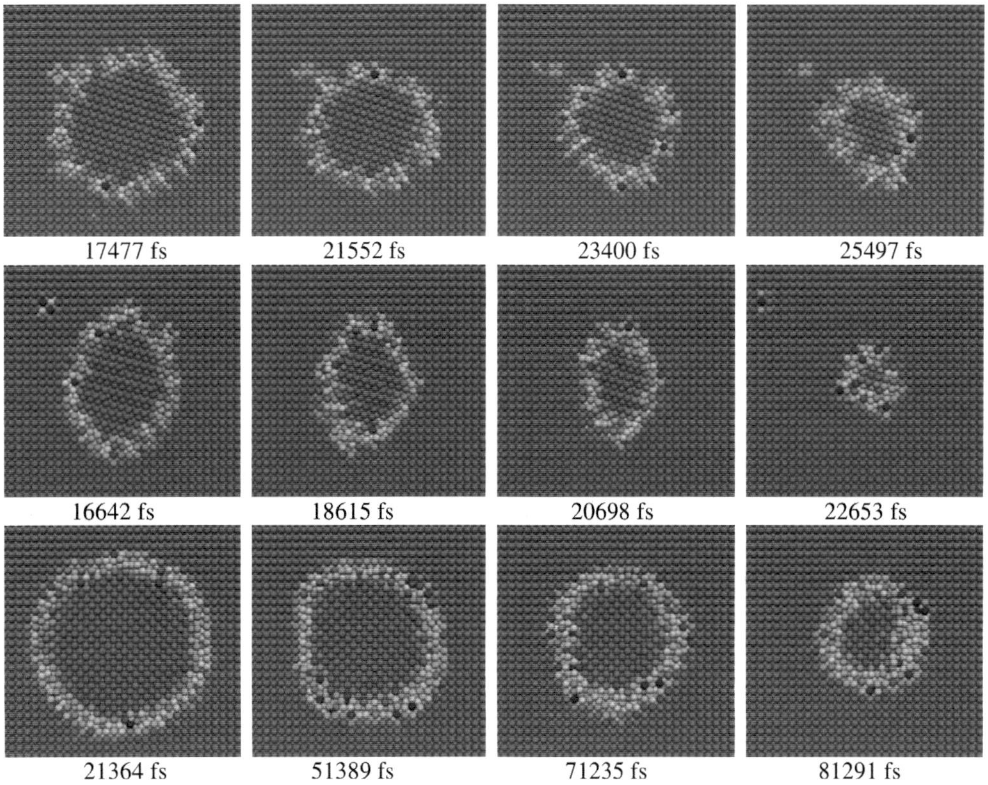

Fig 5. Two-dimensional section perpendicular to the z – axis showing the temporal evolution during MD simulations of the (0 0 15) (top), (15 15 15) (middle) and (0 0 45) (bottom) grain situated at the center of a (0 0 0) outer grain.

Figure 5 shows the temporal evolution of the grain structures obtained by the MD for the sorting approach. The atom positions shown in Fig. 5 were obtained by a minimization step carried out after the 873K NVT run for the times shown. The atoms are colored based on the potential energy of the atoms to obtain a good contrast for the grain boundary atoms. The behavior of the (0 0 15) bicrystal is very similar to that in the PFC simulations in that there is no significant faceting or shape change during shrinking. However, the (15 15 15) boundary becomes more elliptical during shrinking which is also captured by the PFC simulations. The (0 0 45) bicrystal shows evidence for faceting, which is again captured by PFC simulations. Therefore, it is safe to conclude that the PFC does capture both the relative shrinkage kinetics and the shape changes associated with changes in the grain boundary energy and mobility during shrinking of specific boundary types.

The time step in PFC simulations is a dimensionless quantity. The velocity of the grain boundary during shrinking is controlled by the phenomenological constants α and β in Eqn. 2. However, by comparing the number of time steps in PFC required to fully shrink the grain with the MD time steps we can get an estimate for the time step used in PFC. In the PFC simulations, a time step of 0.005 was used. From Fig. 1, the number of PFC time steps to shrink the (0 0 15), (15 15 15), and the (0 0 45) bicrystals were 4000, 2800 and 6000 respectively. The corresponding simulation times from the MD simulations shown in Figs. 3 and 4 are 19000 fs, 14000 fs and 87000 fs, respectively. Therefore, the equivalent time steps in the PFC simulation were 4.75 fs, 3.95 fs and 14.5 fs respectively.

4. Summary

The shrinking of a spherical grain situated in a matrix was simulated using PFC. Three different methods were developed for generating equivalent atomic positions for MD from the peak density positions obtained using PFC. The shrinking of the central grain with three different orientations was studied using PFC and MD using the different mapping procedures. PFC was able to capture the relative shrinking kinetics and the shape changes during shrinking of the central grain shown by MD. The relative speed up of PFC over MD for these grain boundary diffusion driven problems is orders of magnitude lower than for bulk diffusion driven evolution problems. Of the three methods used for mapping atom positions from PFC to MD, the method based on the sorting of PFC intensities and the method based on considering the earliest forming peaks appeared to provide reasonable comparisons between PFC and MD.

5. Acknowledgements

Research funded by the Energy Frontier Research Center (EFRC) project, Center for Defect Physics, Basic Energy Sciences, DOE at ORNL under contract DE-AC05-00OR22725. Computational resources were provided by the National Center for Computational Sciences at ORNL, which is supported by the Office of Science of the Department of Energy under Contract DE-AC05-00OR22725.

References

[1] Elder K et al 2007 Physical Review B **75** 064107
[2] Jaatinen A et al 2009 *Physical Review* E **80** 031602
[3] Stefanovic P et al 2006 *Physical Review Letters* **96** 225504
[4] Chan PY et al 2010 *Physical Review Letters* **105** 015502
[5] Provatas N et al 2007 *JOM* **59** 83
[6] Wu KA and Karma A 2007 *Physical Review* B **76** 184107
[7] lammps.sandia.gov
[8] Mendelev MI et al 2003 *Philosophical Magazine* A **83** 3977

Journal of Physics: Conference Series **402** (2012) 012044 doi:10.1088/1742-6596/402/1/012044

Molecular-Dynamics Study of Void-Formation inside Silicon Wafers in Stealth Dicing

K Shimamura[1], J Okuma[2], S Ohmura[1] and F Shimojo[1]

[1]Department of Physics, Kumamoto University, Kumamoto 860-8555, Japan

[2]Hamamatsu Photonics K.K., Shimo-Kanzo, Iwata, Shizuoka 438-0193, Japan

Email: 111d8008@st.kumamoto-u.ac.jp

Abstract. The mechanism of void formation in crystalline silicon after laser irradiation is extensively studied by molecular-dynamics simulations. When the laser-irradiated region is melted due to a rapid temperature increase, small voids are generated in that region because of large density fluctuations. Because tensile stresses are generated in the melting region, the large empty hole is formed upon cooling. When the temperature drops below the melting point, recrystallization occurs around the large void. We find that the void persists even after the recrystallization process is stopped, and that a part of the laser-irradiated region is amorphous even at room temperature. The stress distribution around the void and the amorphous region in crystalline silicon is also discussed.

1. Introduction

The process of cutting a wafer into individual IC chips, used in the semiconductor manufacturing industry, is called "dicing". An innovative dicing method, called "Stealth Dicing (SD)," was recently developed by Hamamatsu Photonics K. K. [1-4]. In this method, a permeable nanosecond laser pulse is focused inside the wafer. The accumulation of laser energy modifies the crystalline structure in the irradiated region. By applying a tensile strain, the wafer is diced into separate chips. In this process, cracks are assumed to be initiated in the irradiated region inside the wafer and they grow toward the surface. A significant advantage of this method is that the wafer surface suffers negligible damage, while only the interior is modified.

The SD method is superior to various other dicing techniques. However, the details of the irradiation-induced microscopic processes that take place inside the wafer are still unknown. Clarifying these mechanisms is essential for establishing a more reliable dicing method. So far, the void-formation mechanism in a laser-modified region has been investigated by considering the absorption coefficient of silicon [5, 6] measured experimentally [7-12]. In addition, the crack propagation process has been analyzed in terms of fracture mechanics [13].

The purpose of our study is to clarify the mechanism of void formation. Voids appear in the laser-irradiated region, as shown in Figs. 1 and 2. The reason for the creation of a void has not been clarified since the method was invented. For this purpose, we herein present the results of computer simulations based on molecular dynamics (MD).

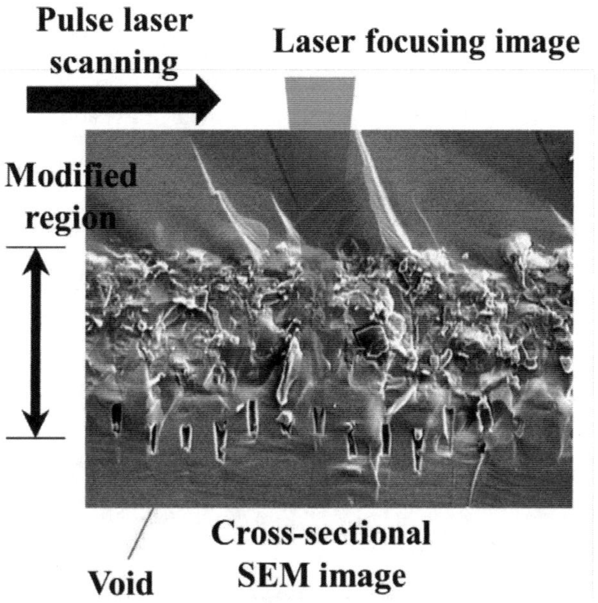

Figure 1. Cross-sectional SEM image

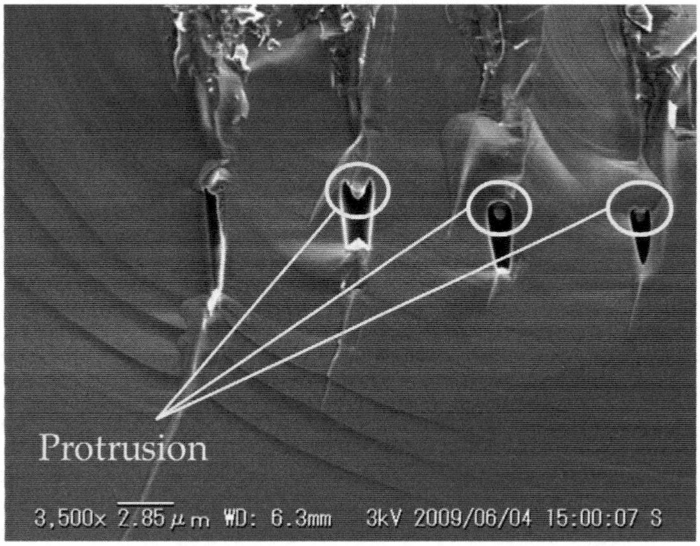

Figure 2. Magnified view of the laser-focused area in Fig. 1

2. Method of Calculation

We investigated the void-formation process by carrying out MD simulations using the micro-canonical ensemble. The Stillinger-Weber potential [14] was used to model interatomic interactions. We considered two model systems by taking into account the measured temperature dependence of the absorption coefficient of silicon [5, 6]. For the MD simulations, we set the time step Δt to 0.4 fs and controlled the temperature by velocity scaling [15]. We used an isobaric and isothermal ensemble to obtain the volume as a function of the temperature [16-18] in Fig. 3. The following section explains the details of our model systems along with the calculated results.

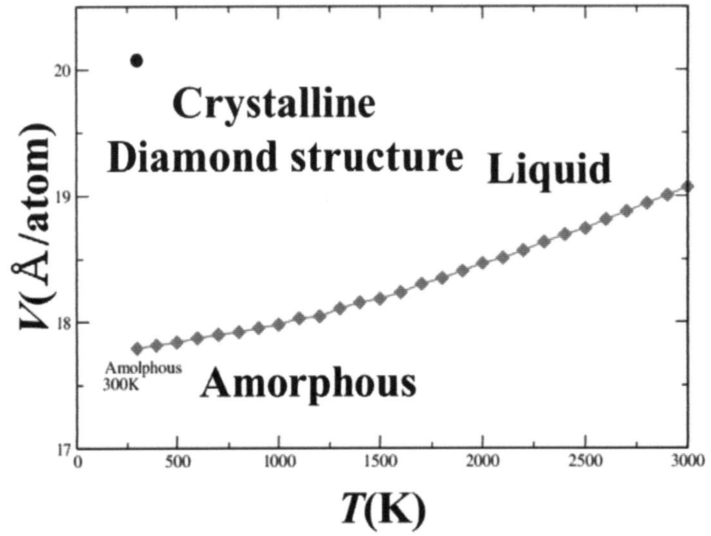

Figure 3. Variation of volume with temperature obtained by MD simulations using the isothermal-isobaric ensembles at ambient pressure.

3. Model systems and results

3.1. Void-formation process

To investigate the void-formation process, we were required to model the state of the material inside the wafer immediately after the laser irradiation. Our first model ('model I') considered the absorption coefficient, which depends on temperature. Experimental measurements showed that it increases exponentially with increasing temperature. Based on this temperature dependence, we assumed that the state of the interior of the wafer immediately after laser irradiation evolves as follows. First, the temperature rises rapidly in the laser-focused region because of the absorption coefficient and the high laser-energy density (see Fig. 4(a)). It is actually known from a thermo-elastic-plastic finite element analysis that the temperature around the laser focal point approaches 10000-40000 K due to larger absorption coefficients at higher temperatures [7]. The temperature in the surrounding region then increases as a result of thermal conduction (Fig. 4(b)). The irradiated laser energy is absorbed mainly in the region along the incident laser direction, and the temperature in this region increases rapidly (Fig. 4(c)). The region of greatest absorption shifts gradually along the incident direction because of thermal conduction and the temperature rise caused by the absorption of the laser energy (Figs. 4(d)-4(h)). It is also assumed that melting occurs in this region due to the rapid temperature increase. Based on these considerations, model I represents the shape of the melting region by a quadratic function, as shown in Fig. 5(a). The system size is $1000 \times 1000 \times 15$ Å^3, which included 844800 Si atoms. Starting with an initial diamond structure, the melting region was heated to

3000 K, while the temperature of the laser-focused region was set to 10000 K. The effects of the heating details will be studied in our future work.

Large density fluctuations occur in the laser-focused region because of the rapid temperature rise at 10000 K. Figure 3 shows that the volume of liquid silicon is smaller than that of the crystalline phase, and it decreases with decreasing temperature. Therefore, small voids are generated because of volume contraction when the system cools (see Fig. 5(b)). As shown in Fig. 5(c), the void near the laser-focused region is enlarged to relax the tensile stresses, thereby forming a large empty hole, while small voids in the melting region disappear. When the temperature of the melting region drops below the melting point, recrystallization occurs instantly. Because the melting region around the large void recrystallizes quickly, the void persists even after the recrystallization is stopped as will be discussed in the subsection 3.2. When the region directly above the void becomes recrystallized, Si atoms move in the direction of the red arrow in Fig. 5(d), because recrystallization produces a lateral compression from left and right. This displacement of Si atoms results in a gradual protrusion of the melting area into the void from above, as displayed by the red circle in Fig. 5(d). The shape of this protrusion appropriately reproduces the experimental observations of Fig. 2. We conclude that this is the mechanism for the formation of the protrusions.

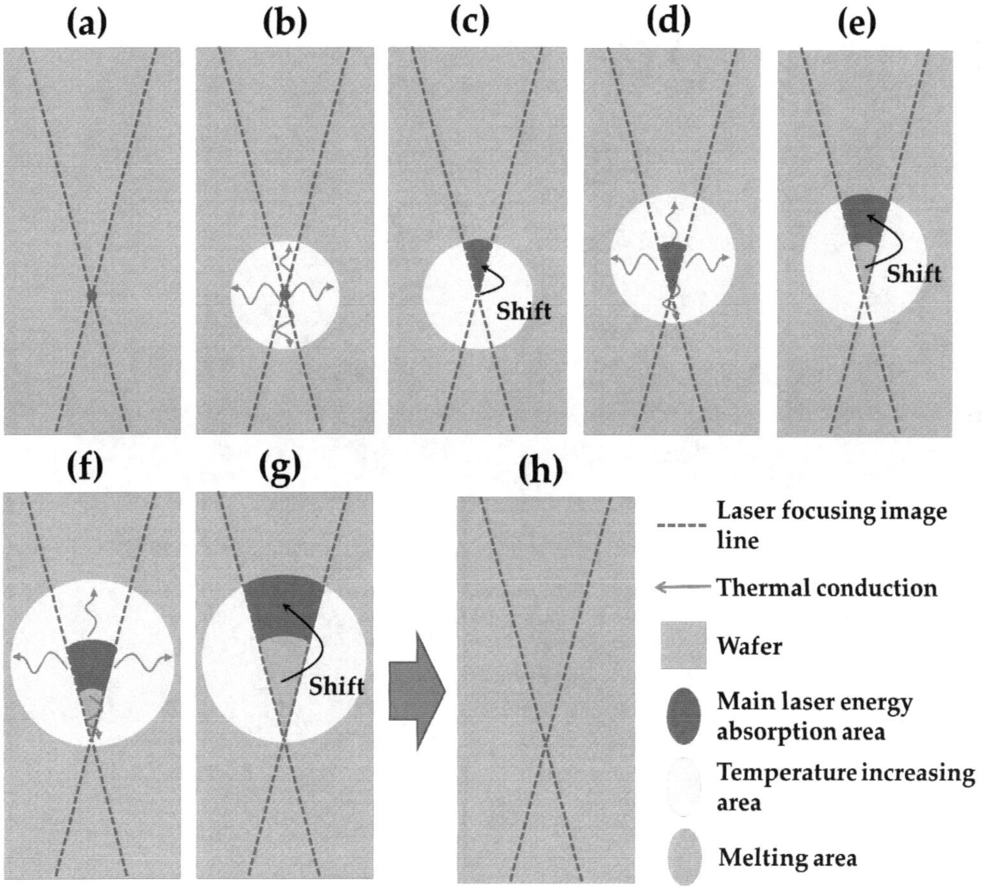

Figure 4. Schematic representation of the model I

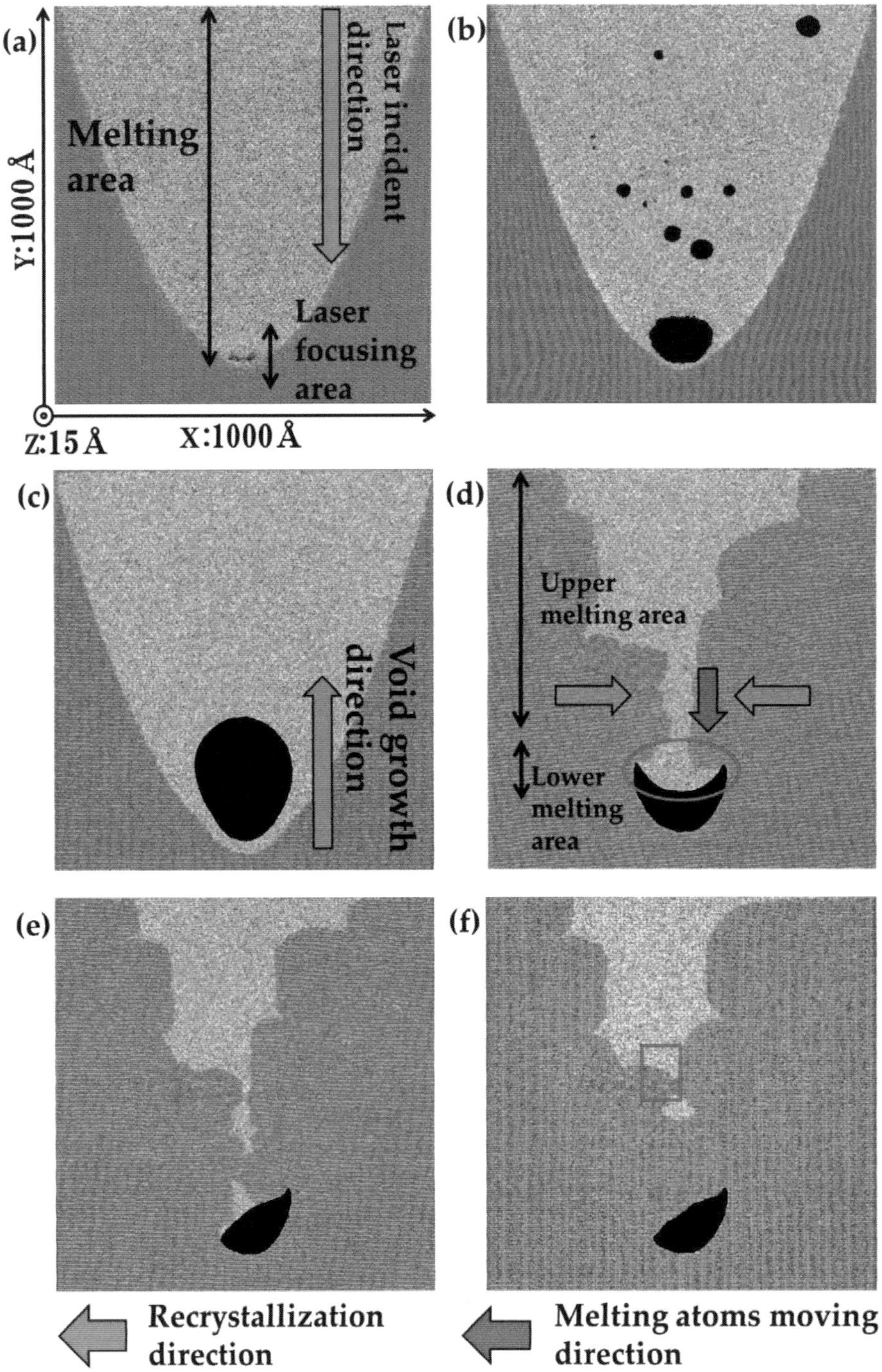

Figure 5. Time progression for model I. Approximately 15 ns passed between (a) and (f).

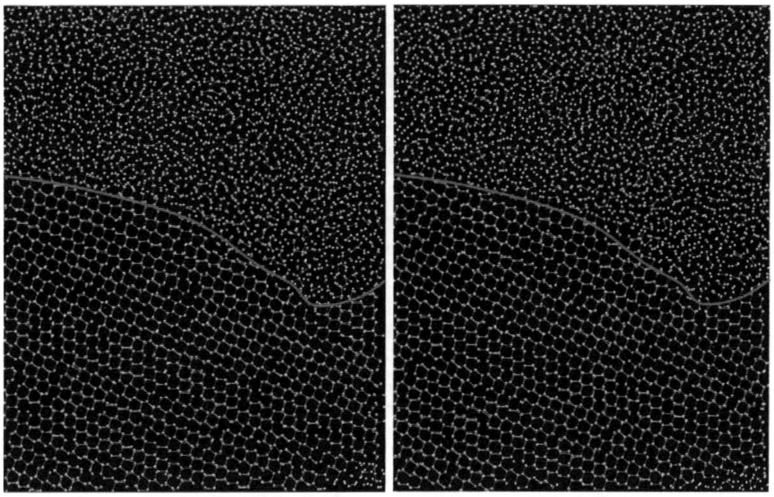

Figure 6. Magnified view of the atomic configuration enclosed in red box in Fig. 5(f)

3.2. Stress distribution around the laser-modified region inside the wafer

A few nanoseconds after the state shown in Fig. 5(d), the melting area above the void became thinner, and eventually separated from the void to give the state shown in Fig. 5(f). The atomic configuration enclosed in red box in Fig. 5(f) is magnified as shown in Fig. 6, which consists of two snapshots at 14.6 and 15.0 ns. It is clearly seen that the crystalline structure is formed below the red line, i.e. a part of the melting area was recrystallized, while the rest remained in a disordered state, namely an amorphous state, above the red line. From the time evolution of the atomic configuration from 14.6 to 15.0 ns, we see that the recrystallization was stopped because the red lines (the boundary between the crystalline and amorphous structures) are identical. For convenience, we hereafter refer to the area separated from the void, which stayed in the amorphous state when the recrystallization no longer proceeded, as the "upper melting area." Since the density of the upper melting area increased in proportion to the volume of the void, the area had a high density and could not recrystallize completely. This is the reason why a part of the melting area remained amorphous. We conclude that the upper melting area corresponds to the laser-modified region shown in Fig. 1 and that a part of the modified region remained amorphous.

To confirm this conclusion, we compared the stress distributions of the states depicted in Figs. 5(e) and 5(f) with that of the experimentally modified region [1, 2]. Figures 7(a) and 7(b) show the σ_{xx} stress distributions of the atomic configurations corresponding to Figs. 5(e) and 5(f), respectively. As shown in Fig. 7(a), tensile stresses formed in the upper melting area, whereas compressive stresses are formed in the upper melting area in Fig. 7(b), which is consistent with the experimental stress distribution.

To investigate the stress distribution in detail, we made a second model ('model II') that reproduced the behavior in the upper melting area, as shown in Fig. 8. The system size for model II was $1800\times1000\times15\text{Å}^3$, which included 1440000 Si atoms. Starting from the initial diamond structure (see Fig. 8(a)), we increased the density of an ellipsoidal region by inserting additional atoms at interstitial sites of the diamond lattice (Fig. 8(b)). The density was thus increased by approximately 8%. The ellipsoidal region was melted at 3000 K, and was recrystallized by lowering the temperature to 1200 K. The added atoms corresponded to atoms that were originally present in the void. This ellipsoidal melting area corresponded to the upper melting area (and the region modified by the laser).

Recrystallization began when the temperature decreased to 1200 K. As expected, a part of the melting region remained in an amorphous state even after recrystallization was stopped. As shown in Fig. 8(c), we created a void by extracting the same number of atoms as were inserted in the ellipsoidal region.

Figures 9(a) and 9(b) show the stress distributions for σ_{xx} before and after recrystallization, respectively. Before recrystallization, tensile stresses formed inside and round the melting region (Fig. 9(a)). However, compressive stresses appeared after recrystallization (Fig. 9(b)), and the overall distribution was in reasonable agreement with the experiment. We believe this agreement proves that a part of the laser-modified region remains in an amorphous state, and that the void-formation process is related to the formation of the laser-modified region.

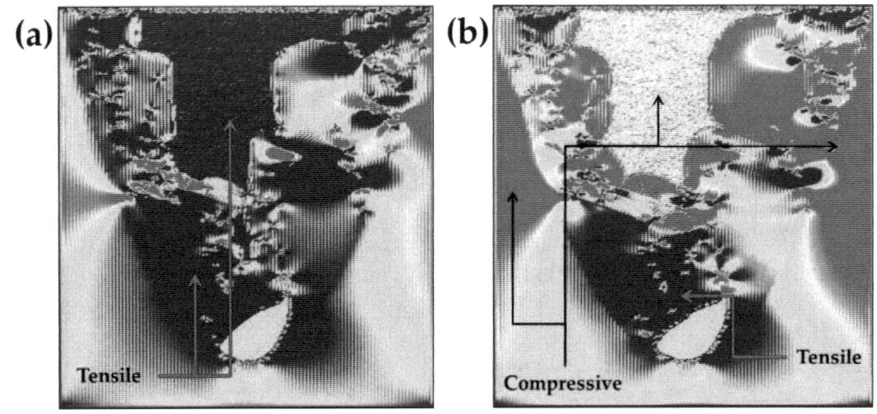

Figure 7. Stress distributions corresponding to Figs. 4(e) and 4(f). Red and blue colors show compressive and tensile stresses, respectively.

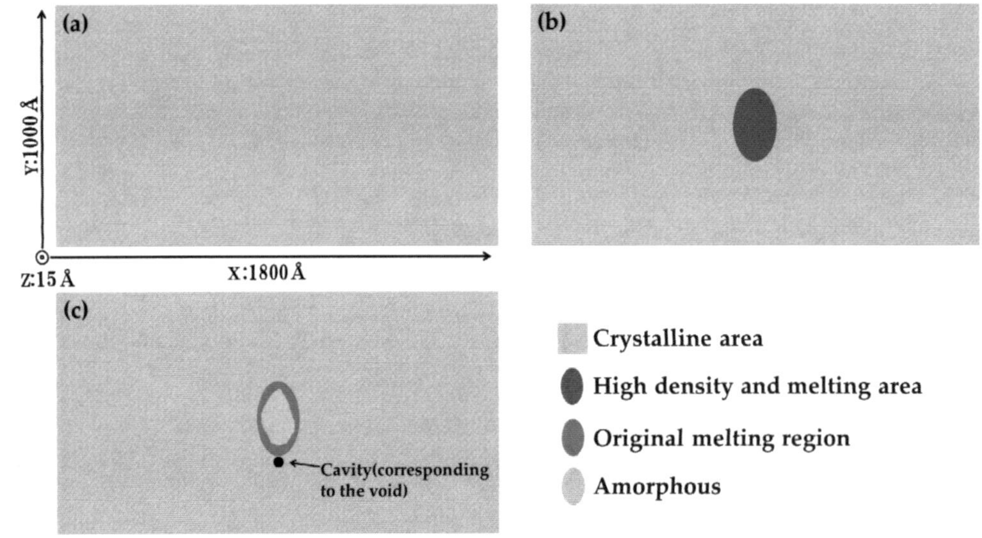

Figure 8. Schematic representation of model II

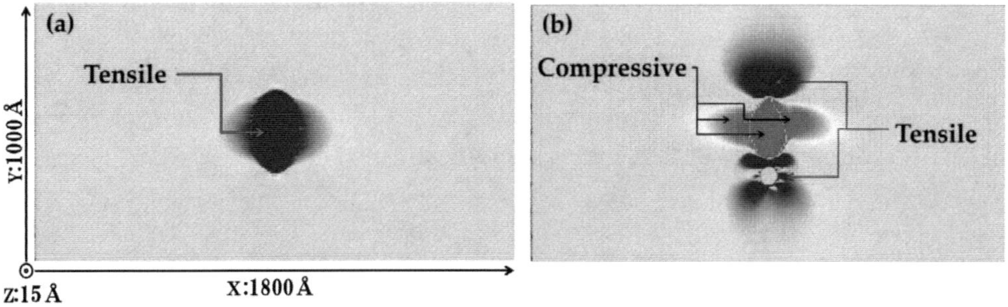

Figure 9. The stress distributions for model II. Approximately 3 ns passed between (a) and (b).

4. Summary.

We performed molecular-dynamics computer simulations for several model systems to investigate the mechanism of void formation and the stress distributions around a region modified by laser irradiation in crystalline silicon. The conclusions are summarized as follows.

(1) Melting occurs in the laser-focused region due to a rapid rise in temperature, and small voids are generated there because of large density fluctuations. The small voids are generated because of volume contraction upon cooling.

(2) These voids can lead to a relaxation of tensile stresses in the melting region, and the voids merge to form a large empty hole.

(3) When the temperature of the melting region drops below the melting point, recrystallization occurs instantly. However, the void persists even after the recrystallization process is stopped. Therefore, the density of the melting region increases in proportion to the volume of the void.

(4) The melting region near the void protrudes into the void as a result of recrystallization to form the protrusions that were observed experimentally.

(5) The melting region is not completely recrystallized because of its increasing density. We propose that a part of the region remains in an amorphous state even at room temperature and that the void-formation process is closely related to the formation of the laser-modified region.

References

[1] Fukuyo F 2005 *J. Jpn Laser Processing Soc.* **12** 17 (in Japanese)

[2] Fukuyo F, Fukumitsu K and Uchiyama N 2005 *Proc. 6th Int. Symp. on LPM 2005*

[3] Kumagai M, Uchiyama N, Ohmura E, Sugiura R, Atsumi K and Fukumitsu K 2006 *Proc IEEE Int Symp Semicond Manuf Conf 2006* 215

[4] Kumagai M, Uchiyama N, Ohmura E, Sugiura R, Atsumi K and Fukumitsu K 2007 *IEEE. Trans. Semicond. Manuf.* **20** 259

[5] Weakliem H A and Redfield D, 1979 *J. Appl. Phys.* **50** 1491

[6] Fukuyo F, Ohmura E, Fukumitsu K and Morita H 2007 *J. Jpn Laser Processing Soc.* **14** 22

[7] Monodane T, Ohmura E, Fukuyo F, Fukumitsu K, Morita H and Hirata Y 2006 *J. Laser Micro/Nanoeng.* **1** 231

[8] Ohmura E, Fukuyo F, Fukumitsu K and Morita H 2006 *J. Achiev. Mater. Manuf. Eng.* **17** 381

[9] Kumagai M, Sakamoto T and Ohmura E 2007 *Proc IEEE Int Symp Semicond Manuf Conf 2007* 521

[10] Ohmura E, Kumagai M, Nakano M, Kuno K, Fukumitsu K and Morita H 2008 *J. Adv. Mech. Des. Syst. Manuf.* **2** 540

[11] Ohmura E, Fukumitsu K, Kumagai M and Morita H 2008 *Trans. Jpn. Soc. Mech. Eng.* **74** 446

[12] Ohmura E, Kumagai M, Fukumitsu K, Nakano M, Uchiyama N and Morita H 2008 *J. Jpn. Soc. Precis. Eng.* **74** 275

[13] Ohmura E, Kawahito Y, Fukumitsu K, Okuma J and Morita H 2010 *Proc. SPIE* **7996** 799603
[14] Stillinger F H and Weber T A 1985 *Phys. Rev.* B **31** 5262
[15] Woodcock L V 1971 *Chem. Phys. Lett.* **10**, 257
[16] Nosé S 1984 *Mol. Phys.* **52** 255.
[17] Hoover W G 1985 *Phys. Rev.* A **31** 1695.
[18] Martyna G J, Tucherman M E, Tobias D J and Klein M L 1996 *Mol. Phys.* **87** 1117

IUPAP C20 Conference on Computational Physics (CCP 2011) IOP Publishing
Journal of Physics: Conference Series **402** (2012) 012045 doi:10.1088/1742-6596/402/1/012045

Influence of Heat Source Characteristics on Dimensionless Thermal Spreading Resistance

A L Wang and C P Yan

School of Astronautics, Beihang University, 37 Xueyuan Road, Haidian District, Beijing, 100191, China

E-mail: wanganliang@buaa.edu.cn

Abstract. Thermal spreading resistance (TSR) for eccentric circular, square and equilateral triangular heat sources on circular heat flux tubes are conducted in this paper. It is shown that the eccentricity has a great influence on TSR. Our numerical results concerning the TSR are in good agreement with Bairi & Laraqi's analytical data. Influence of the shapes on TSR is studied utilizing a series of isosceles triangular heat sources on circular heat flux tubes and rectangular heat sources on square heat flux tubes. The numerical results show that the TSR is strongly dependent on the ratio of inscribed circle radius to circumcircle radius of the heat source. Effect of heat source with fractal boundary on dimensionless TSR is also carried out when the heat source area of every tube is kept at a constant value. The numerical results show that TSR decreases as the construction number N increases and the variation of amplitude of TSR increases as the relative contact size ε increases.

1. Introduction

Heat is mainly transferred through the discrete contact spots between the two imperfect interfaces (figure 1)[1], which results in thermal contact resistance (TCR). TCR includes thermal constriction resistance and thermal spreading resistance (TSR). The inverse of constriction resistance is spreading resistance, and they can be depicted by one dimensionless shape factor sometimes[2]. TCR is a classical research problem in the area of heat transfer field and may play an important role in microelectronics, machinery manufacture, aerospace field and sustainable energy etc. Like many other engineering problems, the TCR research includes two aspects: engineering applications and basic theories[3]. Knowing well the mechanism of heat transfer between the complex contact surfaces is essential to predict TCR[2].

Contact spots of the interface have complex shapes and random locate in the surface as shown in figure 2[4]. TCR is a complex multi scale thermal and mechanical coupling problem. Thermo-mechanical problems of surface contact are usually modeled for engineering applications, but it is known that the physical mechanisms are the interactivity and movement of atoms (or molecules) on the contact surface. The TCR problems cover all ranges, from the atomic to the macroscopic scale. It is the common key problem to investigate the material transport, energy transfer, micro structure evolution, performance variation and active time for material science, solid mechanics and condensed state physics.

Along with the development of computer technology, people gradually approach the TCR problem in two opposite simulation models: macroscopic scale and molecular scale, also called "top-down" and "bottom-up" methods.

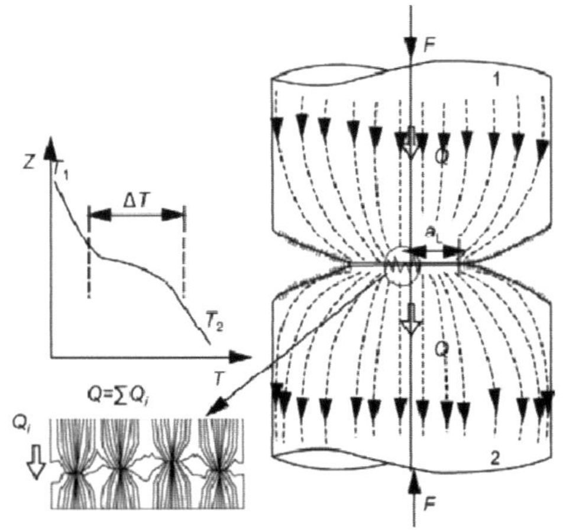

Figure 1. Contact heat transfer in two-solid interface[5].

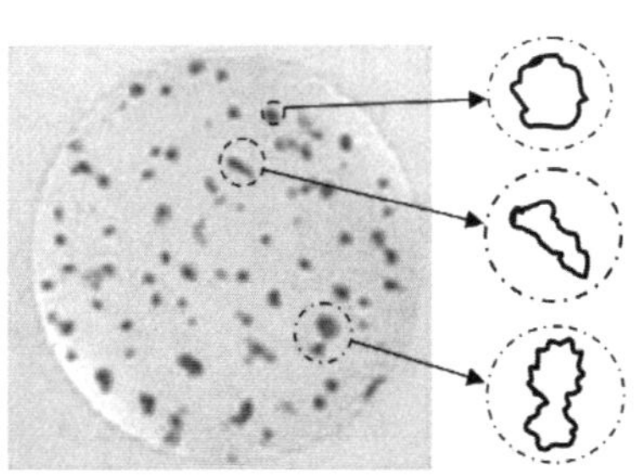

Figure 2. Imaginable shapes of micro contact points, original image of rough surface from literature [4].

On the macroscopic scale, in the 70s of the last century, Yovanovich's group has made comprehensive theoretical analysis and numerical simulation for influence of heat source shape on TCR by hypothesizing single spot as circular, rectangular, elliptical and polygonal point etc based on Fourier series expansions[2]. These results are the foundation for many TCR models of surface roughness. Muzychka et al[6] predicted TSR of rectangular flux channels using the solution for circular flux tube very well. They assessed the importance of edge cooling after comparisons with circular flux tube with edge cooling and with that adiabatic edges. Tio&Toh[7] and Vaidya&Razani[8] researched the TSR of a cylinder model between two similar materials and dissimilar materials respectively under steady state situation. They indicated that the temperature distribution on the cross section of the cylinder midway between the two solids has a relationship to solid material. Rostami[9] indicated that the gap geometry, shape of contact area and certain end surface boundary conditions strongly infect thermal constriction resistance after numerical simulation of single contact point heat transfer using CFD (Fluent version 4.3.2). Garimella's group researched influence of ratio of heat source area to contact area, gap, cone angle, conductivity of the gap and materials, coating thickness, etc on dimensionless TSR considering radiation utilizing a frustum cone model. A non-linear fitting formula and the corresponding computational domain were expressed in their paper[10],[11],[12]. Wang & Degiovanni[13],[14] researched transient TSR of a cylinder model between two similar materials, the analytical results agreed well with previous experiments. Bairi&Laraqi[15] researched TSR of an eccentric spot on a circular heat flux tube. The solutions are in good agreements with experiment results. Teertstra et al[16] researched conduction shape factor subtly utilizing hollow cylinders with nonuniform gap spacing models. The model was validated using numerical data from the literature for a variety of geometries. Recently Chaouch et al[17] studied thermal and electrical constriction resistance under the coupled thermal and electrical phenomena using finite element method. They indicated that finite element method can simulate thermal constriction resistance exactly and combined heat and electricity in the finite tube accentuates the temperature drop.

On the microscopic scale, the solid matter is made up of atoms, ions and molecules. New methods involving the quantum molecule dynamics (QMD), molecular dynamics (MD) and Boltzmann equation (BE) were used to research the non-Fourier effect, scale and boundary effect of thermal conductivity, and thin film conduction[18]. Touzelbaev & Goodson[19] investigated the thermal resistance near diamond-substrate interfaces using phonon transport theory. They found that the resistance was goverened by the number of diamond nucleation sites per unit substrate area, i.e., the nucleate density, the thickness of boundary layer. Prasher & Phelan[20] developed a scattering-mediated acoustic mismatch model (SMAMM) to describe the TCR from high temperature to low temperature. The parameters were defined at a crystal structure, while the prediction results were in good agreement with the experimental data on Rh/Mgo interfaces. Chen[21] introduced a ballistic-diffusive equation(BDE) derived from the Boltzmann equation (BE) to deal with the thin films heat conduction problems from the nano scale to the macro scale, and the contact boundary conditions were simplified. Liao & Yang[22] proposed a method coupling atomic and finite element models to simulate the TCR of dissimilar materials. The results showed that the nonuniform temperature distribution was associated with the characteristic of atoms moving along the interface.

The multi-scale thermo-mechanical problems are still great challenge to us although researchers have achieved many results and found new physical mechanism on macro scale and micro scale contact heat transfer respectively. Further research on TCR either in macro scales or micro scales should be done. In this paper, from macro perspective, we study the influence of the heat source characteristics including eccentricity, shape and fractal boundary on dimensionless TSR utilizing a series of models based on the finite element method.

2. Computational Methodology

2.1. Problem Statement

TSR arises where heat flows from a smaller to a larger area. Conduction problems may involve multiple directions and time-dependent conditions. Here, the model (figure 3) assumes a circular heat source in contact with a larger cold plate which is in turn cooled with a convective heat transfer coefficient specified over the sink surface. All edges and the region outside the heat source in the source plane are assumed to be adiabatic. This model can represent accurately numerous engineering systems.

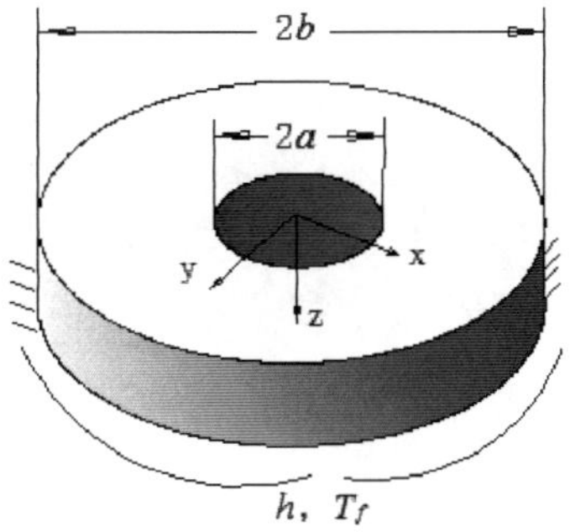

Figure 3. A circle/circle heat flux tube.

If heat is supplied to the tube for a sufficiently long time, the system will reach a steady-state situation. In the idealized system, the total thermal resistance R_t is defined as[2]:

$$R_t = (\overline{T}_{source} - \overline{T}_b) / Q \tag{1}$$

Where $\overline{T}_{source}$ and $\overline{T}_b$ are average temperature of the heat source and the bottom of the cold plate, Q is the total heat flow rate through contact.

R_t is composed of two terms: a one-dimensional resistance and a spreading resistance which vanishes as the source area approaches the substrate area, that is, $A_s \to A_b$. These two components are combined as follows[2]:

$$R_t = R_{1D} + R_s \tag{2}$$

R_{1D} is the material resistance, it can be calculated by the following formula:

$$R_{1D} = t / kA_b \tag{3}$$

where t is the length of the flux tube, k is the thermal conductivity of the cylinder.

Nondimensionalization of R_s can be made by introducing a characteristic dimension with the unit of length $\sqrt{A_s}$ [23], which can be defined as:

$$R_s^* = k\sqrt{A_s}R_s \tag{4}$$

2.2. Numerical methods

There are three classical numerical simulation methods: finite difference method, finite element method and boundary element method. They are used in different situations and have difference in computational accuracy and efficiency. The finite element technique is a good method when models have complex shapes[24],[25]. In this paper, we research TSR of heat source with complex shapes on flux tube utilizing finite element method.

Thermal spreading resistance analysis requires the solution of Laplace's equation in three dimensions.

$$\frac{\partial^2 T}{\partial x^2} + \frac{\partial^2 T}{\partial y^2} + \frac{\partial^2 T}{\partial z^2} = 0 \tag{5}$$

In most applications the following boundary conditions are applied:

$$\left.\frac{\partial T}{\partial Z}\right|_{Z=0} = 0 \qquad A_s < A < A_b \tag{6}$$

$$\left.\frac{\partial T}{\partial Z}\right|_{Z=0} = q/k \quad 0 < A < A_s \tag{7}$$

Finally, along the lower surface $Z = t$,

$$\left.\frac{\partial T}{\partial Z}\right|_{z=t} + \frac{h}{k}[T(x,y,t) - T_f] = 0 \tag{8}$$

Where h is a uniform convection heat transfer coefficient or contact conductance[6]. According to the variation principle, the key problem to solve temperature field is to solve the minimum value of the following formula:

$$J(T) = \iiint_R \frac{k}{2}[(\frac{\partial T}{\partial x})^2 + (\frac{\partial T}{\partial y})^2 + (\frac{\partial T}{\partial z})^2]dxdydz + \iint_\Gamma \alpha(\frac{1}{2}T^2 - T_f T)ds \tag{9}$$

where R is computational domain, Γ refers to Robin boundary condition.

It is transferred to get the sum of integral of all elements when computational domain are broken into many discrete elements using finite element method in three dimensional space, that is:

$$J(T) = \sum J^e \tag{10}$$

The correlation formula of J when it gains the extreme value is:

$$\frac{\partial J}{\partial T_k} = \sum_n \frac{\partial J^e}{\partial T_k} = 0 \quad (k = 1,2,\cdots,n) \tag{11}$$

where n is the number of nodes and e denotes the element in the computational domain.

It can be presented in the following form:

$$\begin{bmatrix} k_{11} & k_{12} & \cdots & k_{1n} \\ k_{21} & k_{22} & \cdots & k_{2n} \\ \vdots & \vdots & \vdots & \vdots \\ k_{n1} & k_{n2} & \cdots & k_{nn} \end{bmatrix} \begin{Bmatrix} T_1 \\ T_2 \\ \vdots \\ T_n \end{Bmatrix} = \begin{Bmatrix} p_1 \\ p_2 \\ \vdots \\ p_n \end{Bmatrix} \tag{12}$$

That is:

$$[K]\{T\} = \{P\} \tag{13}$$

where $[K]$ is temperature stiffness matrix, $\{T\}$ is temperature matrix and $\{P\}$ is load matrix.

2.3. Choice of mesh

Mesh is very important to make accurate simulation of the thermal field and acquire precise TSR. A study consisting of choosing the appropriate mesh is undertaken. Here we use one classical model from Negus[23] to make a validation of mesh rule. We take $t/D = 4$, D is the cylinder diameter. This length is sufficient for a semi-infinite cylinder assumption[12]. To best follow the spatial evolution of temperature fields, in the vicinity of the contact zone, we choose a very small mesh near the applied heat flux $\boldsymbol{q}$, and gradually increase it away from the contact area[17]. Figure 4 is part of the mesh of the circular flux tube. The commercial code ANSYS12.1 is used to solve this steady state thermal problem. The element growth rate is 0.1 and the max number of elements is about 100,000. To compare results of different cases, a dimensionless relative contact size ε is defined as the square root of the ratio of the source area to the substrate area, $\varepsilon = \sqrt{A_s / A_b}$. In figure 5 we compare the numerical results R_s^* with the analytical solutions R_p [23] and they are in good agreement. We will use the same mesh rule for the other models which are constructed in the following study.

Figure 4. Part of finite-element mesh of circle/circle flux flux.

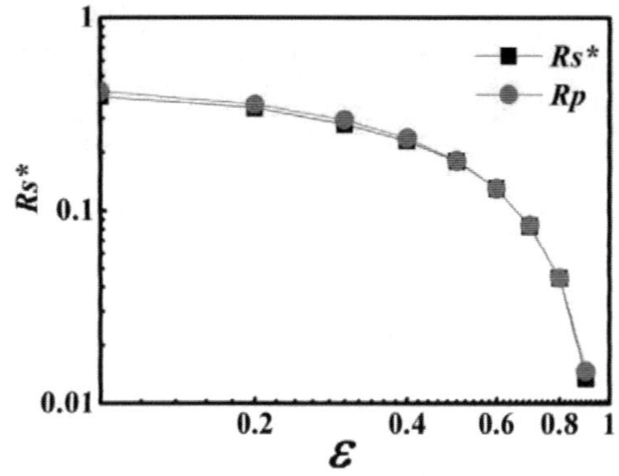

Figure 5. A comparison with ANSYS results with analytic results[23].

3. Results and Discussions

3.1. Influence of heat source eccentricity on TSR

In this part, the influence of heat source eccentricity on TSR is studied utilizing a series of eccentric circular, square, equilateral triangular spots on heat flux tubes (figure 6). The three groups of models have the same ε=0.0894. The boundary conditions are: (1) keeping the heat source suffering a constant heat flux $\boldsymbol{q}$; (2) cooling at the bottom of the cylinder; (3) the lateral and non-source top surface boundaries of the cylinder are adiabatic. To compare results of different cases, a dimensionless relative length $\xi = e/b$ is defined. Here, e is the distance between the center of the heat source and the center of the circular heat flux tube. b is the radius of the circular heat flux tube. r and R represent inscribed circle radius and circumcircle radius of the heat source respectively. Figure 7 shows the numerical results of the three groups of models when the heat source has eccentricity.

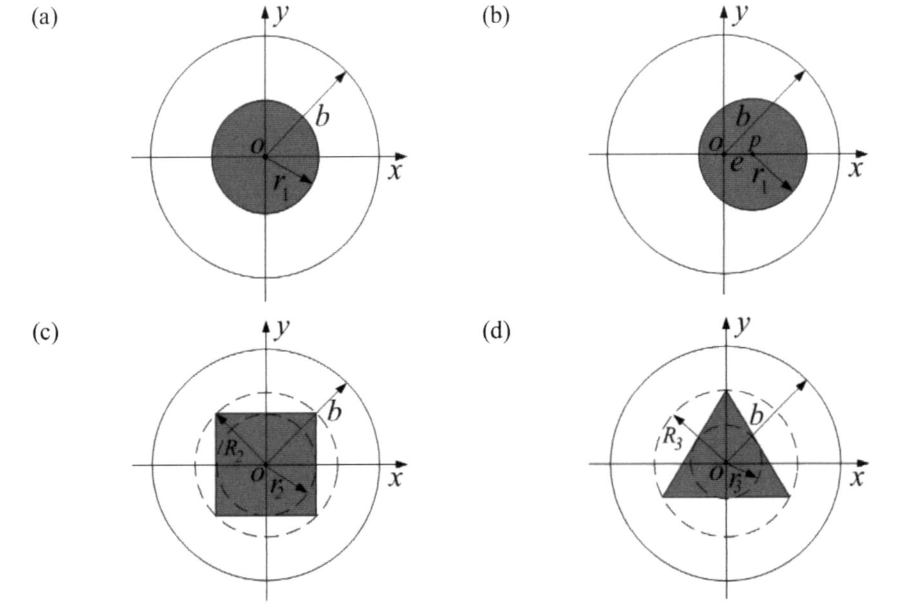

Figure 6. Contact area/cylinder cross-sectional configurations under consideration: (a) Circle/circle (ξ=0); (b) Circle/circle(ξ=0.2); (c) Square/circle (ξ=0); (d) Equilateral triangle /circle (ξ=0).

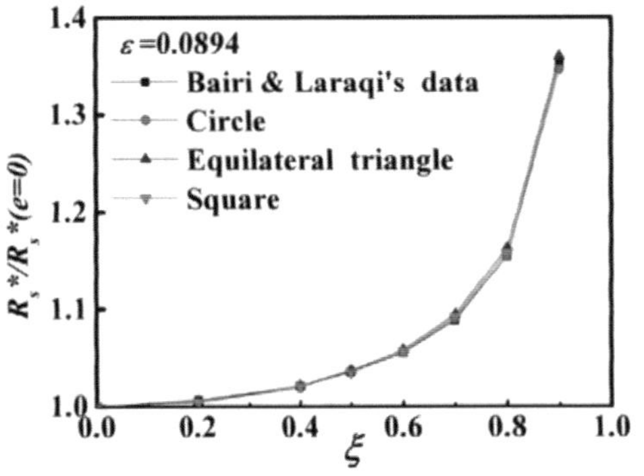

Figure 7. Influence of heat source eccentricity on TSR.

We compare the results of circle/circle flux tubes with Bairi & Laraqi's data[15]. It shows that the results are in good agreement. In figure 7 we also compare the variation of TSR of the equilateral triangular spot and square spot on circular flux tubes. It shows that the growth rates of the three group models are lower to 10 percent during $\xi = 0$ to $\xi = 0.7$ due to eccentricities of contact spots. They increase quickly after $\xi = 0.7$. This phenomenon is due to the edge effect that becomes more pronounced when the spot approaches the contour of the flux tube[15]. The maximum relative difference between an eccentred and a centred contact is about 35 percents when ξ varies from 0 to 0.9. These results are consistent with recent works for random contacts[26],[27].

3.2. Influence of heat source shape on TSR

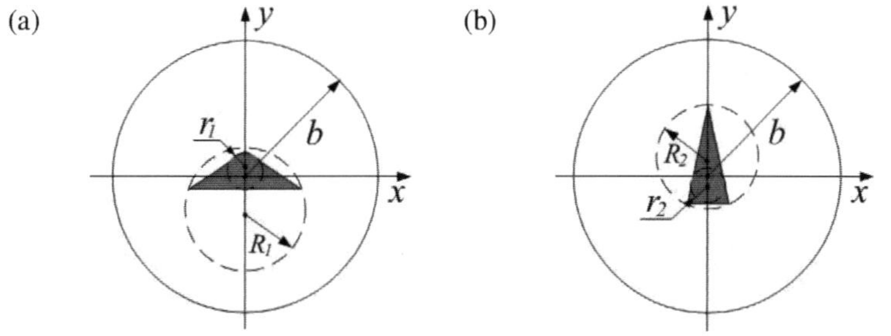

Figure 8. Isosceles triangular heat source on circular flux tubes (a) interior angle of the two sides =30° ; (b) interior angle of the two sides =80°.

3.2.1. Isosceles triangular heat source on circular flux tubes. In order to research the influence of the heat source shape on TSR, a series isosceles triangular heat source on circular flux tubes are constructed (figure 8). Here we use one shape characteristic η to depict TSR variation, and η is equal to *r/R*, where *r* and *R* represent inscribed circle radius and circumcircle radius of the isosceles triangles respectively. The models have the same heat source area and their centres of gravity are kept at the centres of the circles when the interior angles of the two sides change from 30 to 80 deg. Table 1 shows the relationship between interior angles of the two sides and η. The boundary conditions are the same with part 3.1. From figure 9 we can see TSR has the same variation trend as η. Heat will spread out when η decreases, and this phenomenon will reduce the amount of flux-line spreading between the contact surfaces, which results in a reduction of TSR.

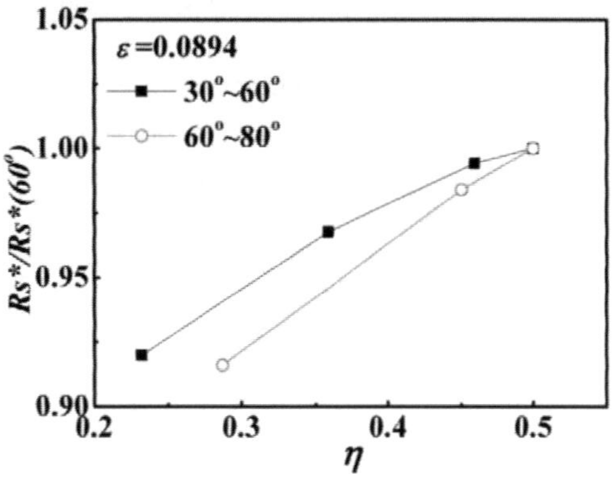

Figure 9. TSR of isosceles triangular heat source on circular flux tubes.

Table 1. Properties of isosceles triangular heat source.

interior angle of the two sides	30°	40°	50°	60°	70°	80°
$\eta=r/R$	0.232	0.358	0.459	0.500	0.450	0.287

3.2.2. *Rectangular heat source on square heat flux tubes.* Another group of models (figure 10) are constructed in this part to research TSR when heat source shape changes from a square to a rectangular. The boundary conditions are the same with 3.1. Here, we also keep the same heat source area for every model. Figure 11 shows that TSR has the same changing trend as η. From results of part 3.2.1 and that of this part, we can see that η can be a good characteristic to predict variation trend of TSR.

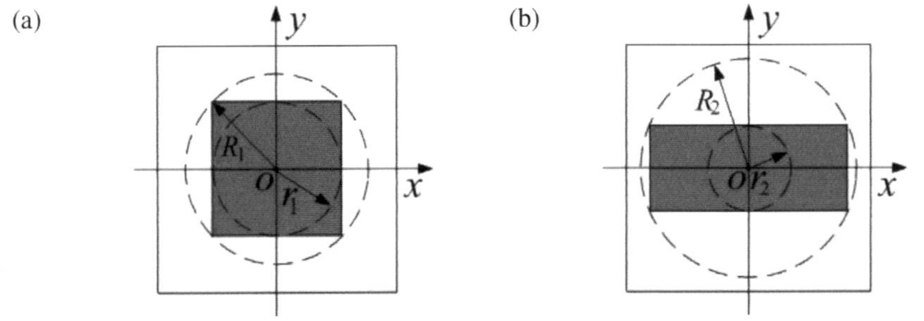

Figure 10. Contact area/cylinder cross-sectional configurations under consideration (a) Square heat source on square flux tube; (b) Rectangle heat source on square flux tube.

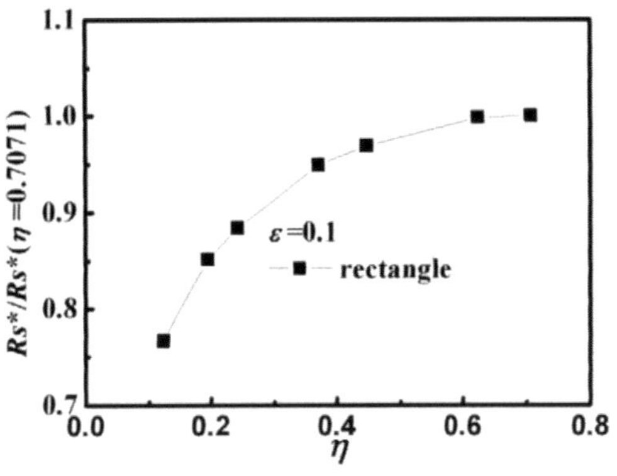

Figure 11. TSR of rectangular heat source on square heat flux tubes.

3.3. Influence of heat source with fractal boundary on TSR

Fractal geometry is a mathematical language that describes the structural disorder and chaos of a number of objects found in nature (Mandelbrot, 1982). On several occasion, however, it is crucial to know the structural details of the object at the length scales relevant to a physical phenomenon. This is where fractal geometry becomes important[28].

Koch snowflakes is one kind of famous fractal geometry. Here we develop a group of models with heat source boundaries similar to Koch snowflakes (figure 12, N represents the construction number, the heat sources are kept with the same area). The following introductions are how to develop models with boundaries similar to Koch snowflakes. Figure 12(a) is an equilateral triangle heat source which is the first model, here, N=1. The second model (the second construction, N=2) can be constructed by starting with the first model, then recursively altering each line segment as follows: (1) divide the line segment into three segments of equal length; (2) draw an equilateral triangle that has the middle segment from step (1) as its base and points outward; (3) remove the line segment that is the base of the triangle from step (2). After one iteration of this process, the result is a shape similar to the Star of David which is the second model as shown in figure 12 (b). The same iterations can be repeated to get the other models.

From table 2, we can see η just increases a little from N=1 to 2, and keeps at 0.577 after N=2 , although the circumference of the heat source keep increasing to infinite length as N goes to infinity when heat source area keep at a constant value.

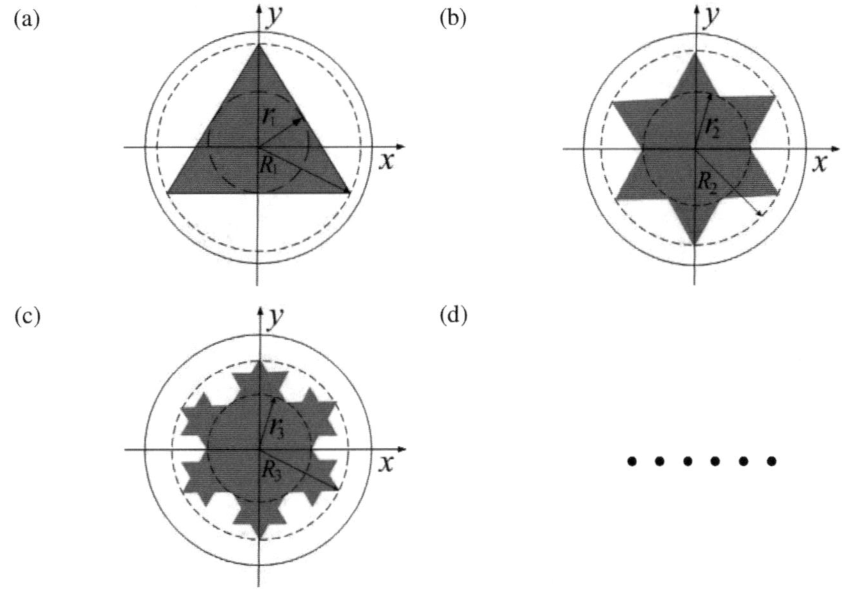

Figure 12. Contact area with fractal boundary/cylinder cross-sectional configurations under consideration (a) N=1; (b) N=2; (c) N=3; (d) N=∞.

Table 2. Properties of the heat source with boundaries similar to Koch snowflakes.

construction number	N=1	N=2	N=2	N=2	...	N=∞
Circumference(m)	0.456	0.526	0.666	0.869	...	∞
heat source area(m^2)	0.010	0.010	0.010	0.010	0.010	0.010
$\eta = r/R$	0.500	0.577	0.577	0.577	0.577	0.577

The boundary conditions are: (1) keeping the heat source at temperature 373K; (2) cooling at the bottom of the cylinder; (3) the lateral and non-source top surface boundaries of cylinder are adiabatic. We choose one-sixth model to simplify the calculation. Figure 13 and figure 14 are the mesh and the temperature field of the one-sixth model. From figure15 we can see that TSR of fractal geometries decreases as N increases. The variation of amplitude increases as ε increases. It means that fractal boundary has much more influence on TSR for the bigger ε. The related difference is about 16% at N=5 when $\varepsilon = 0.6$. We can predict that TSR will converge to a constant value as their boundaries approach Koch snowflakes from figure 15.

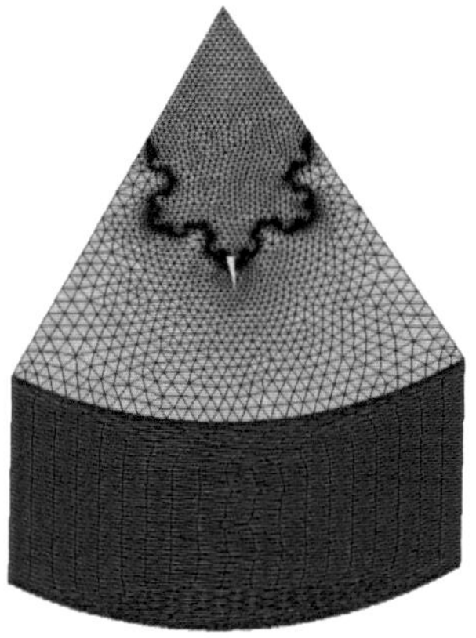

Figure 13. Mesh of one-sixth model at N=5.

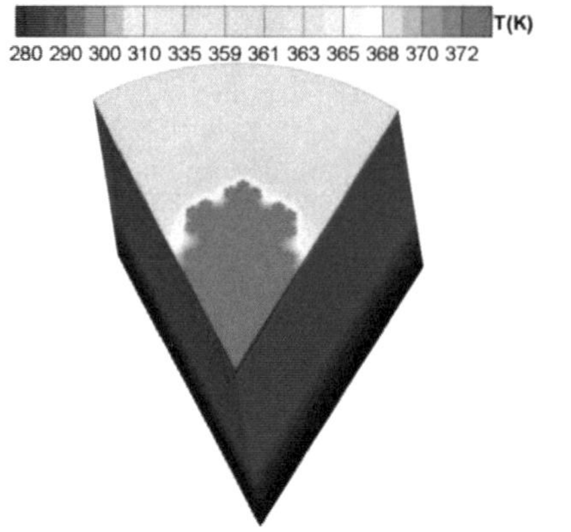

Figure 14. Temperature contour of one-sixth model at N=5.

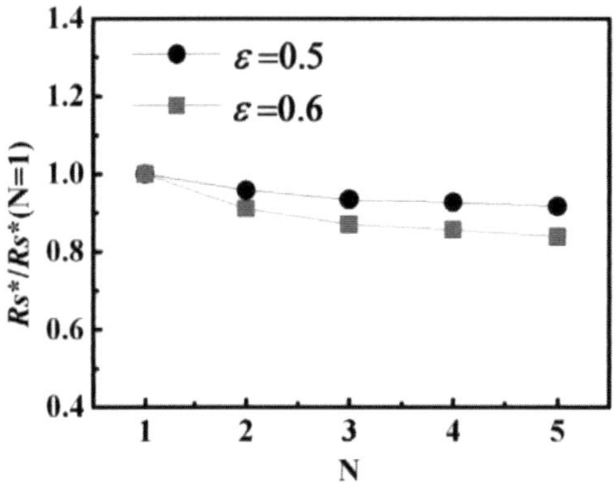

Figure 15. TSR of the heat source with fractal boundary on circular flux tubes.

At present, it is controversial that whether the actual contact surface has fractal characteristic [10]. TCR models with fractal characteristic are getting more and more attentions to researchers. Profile curve of roughness concentration may have fractal characteristic although it is hard to know whether fractal geometry can be used to predict TCR. Wavelet transformation can be used to evaluate the fractal shape curve well[29]. The results expressed in this paper may supply great reference value for predicting TCR.

4. Conclusions

From observations made with the above series of models, our numerical results demonstrate that eccentricities, shapes and fractal boundary of the heat source have significant influence on TSR. It is shown that TSR increases with the increase of eccentricity due to the heat source. Further study shows that the TSR also increase with the increase of shape characteristics factor, η. TSR decreases as construction number, N, increases, and finally converges to a constant value when the boundaries of the heat source approach the Koch snowflakes. Fractal boundary has much more influence on TSR for the bigger relative contact size ε. Hence, precise location, the shape and boundary of the heat source are important in estimating TSR.

η can be a good characteristic to predict TSR expect for the relative contact ratio ε and heat source eccentricity. The results obtained in this paper may have much reference value to the further research on three-dimensional TCR. For in-depth investigation of TCR, we will research molecular scale models with "bottom-up" methods.

5. Acknowledgements

The authors acknowledge the financial support of the Beihang University International Conference Fund for graduate. The authors also want to express their thanks to the group lead by Professor Yu Liu, School of Astronautics, Beihang University for their support and encouragement.

The authors further acknowledge the help of Professor Jianxin Zhong (Xiangtan University). Professor Arkadiusz Wojs (Wroclaw University of Technology) and Zixiang Hu (Princeton University) were very thorough in helping to find typographical errors in many equations and the authors appreciate their contributions.

References

[1] Bahrami M, Culham J R, Yovanovich M M and Schneider G E 2004 *J. Thermophys. Heat Tr.* **18** 218
[2] Yovanovich M M 1998 *Conduction and Thermal Contact Resistances (Conductances), Handbook of Heat Transf*er 3ed Rohsenow W M, Hartnett J P et al (New York: McGraw Hill)
[3] Wang A L and Zhao J F 2010 *Sci.China Tech. Sci.* **53** 1798
[4] Sadeghi E, Hsieh S and Bahrami M 2011 *Proc. Thermal Engineering Joint Conf. 8th* (Honolulu, Hawaii, USA) p T10129
[5] Bahrami M, Culham J R, Yovanovich M M and Schneider G E 2004 *J.Thermophys. Heat Tr.* **18** 209
[6] Muzychka Y S, Yovanovich M M and Culham J R 2006 *J. Thermophys. Heat Tr.* **20** 247
[7] Tio K K and Toh K C 1998 *Int. J. Heat Mass Tr.* **41** 2013
[8] Vaidya S Y and Razani A 1998 *J. Franklin. Inst.* **335** 1493
[9] Rostami A A, Hassan A Y and Lim P C 2001 *Heat Mass Transfer.* **37** 5
[10] Black A F, Singhal V, and Garimella S V 2004 *J. Thermophys. Heat Tr.* **18** 30
[11] Olsen E L, Garimella S V and Madhusudana C V 2002 *J. Thermophys. Heat Tr.* **16** 207
[12] Merrill C T and Garimella S V 2006 *J. Thermophys. Heat Tr.* **20** 346
[13] Wang H, Degiovanni A and Moyne C 2002 *Int. J. Therm. Sci.* **41** 125
[14] Wang H and Degiovanni A 2002 *Int. J. Heat Mass Tr.* **45** 2177
[15] Bairi A and Laraqi N 2004 *J. Heat Transf.* **126** 652
[16] Teertstra P M, Yovanovich M M and Culham J R 2009 *J. Thermophys. Heat Tr.* **23** 28
[17] Chaouch K T, Loulou T, Rogeon P and Benkhedda Y 2011 *Int. J. Therm. Sci.* **50** 890
[18] Liu J 2001 *Micro/nano-scale Heat Transfer(in Chinese)* (Beijing: Science Press)
[19] Touzelbaev M N and Goodson K E 1997 *J. Thermophys. Heat Tr.* **11** 507
[20] Prasher R S and Phelan P E 2001 *J. Heat Transf.* **123** 105
[21] Chen G 2002 *J. Heat Transf.* **124** 320
[22] Liao N B and Yang P 2008 *J. Thermophys. Heat Tr.* **22** 581
[23] Negus K J, Yovanovich M M and Beck J V 1989 *J. Heat Transf.* **111** 804
[24] Bhushan B and Peng W 2002 *Appl. Mech. Rev.* **55** 435
[25] Sarzala R P and Nakwaski W 1990 *J. Therm. Anal.* **36** 1171
[26] Das A K and Sadhal S S 1999 *Heat Mass Transfer.* **35** 101
[27] Laraqi N and Bairi A 2002 *Int. J. Heat Mass Tr.* **45** 4175
[28] Majumdar A and Tien C L 1991 *J. Heat Transf.* **113** 516
[29] Wang A L, Yang C X and Yuan X G 2003 *Tribol. Int.* **36** 517

Journal of Physics: Conference Series **402** (2012) 012046 doi:10.1088/1742-6596/402/1/012046

Surface adsorption of lattice HP proteins: Thermodynamics and structural transitions using Wang-Landau sampling

Ying Wai Li[1], Thomas Wüst[2] and David P Landau[1]

[1] Center for Simulational Physics, Department of Physics and Astronomy, University of Georgia, Athens, Georgia 30602, U.S.A.

[2] Swiss Federal Research Institute WSL, Zürcherstrasse 111, CH-8903 Birmensdorf, Switzerland

E-mail: ywli@physast.uga.edu, thomas.wuest@wsl.ch, dlandau@physast.uga.edu

Abstract. Wang-Landau sampling has been applied to investigate the thermodynamics and structural properties of a lattice hydrophobic-polar heteropolymer (the HP protein model) interacting with an attractive substrate. For simplicity, we consider a short HP sequence consisting of only 36 monomers interacting with a substrate which attracts all monomers in the sequence. The conformational "phase transitions" have been identified by a canonical analysis of the specific heat and suitable structural observables. Three major "transitions", namely, adsorption, hydrophobic core formation and "flattening" of adsorbed structures, are observed. Depending on the surface attractive strength relative to the intra-protein attraction among the H monomers, these processes take place in different sequences upon cooling.

1. Introduction

Protein folding and protein adsorption have long been subjects of intense research. The topics gained so much attention both because of their numerous applications in nanotechnology, biomaterials, medical and biological sciences (see e.g., [1–3] and references therein), but also because of the many interesting, yet challenging, basic scientific questions they pose. However, due to the complexity of the problems, our understanding of protein structure and folding is still incomplete and a "general theory" is lacking. The difficulties arise from the many interactions, of various strengths, among the building blocks constituting the protein organizing the molecules into primary, secondary or tertiary structures [4].

Additional protein-substrate interactions add a further level of complexity to the problem. When brought in the vicinity of a substrate, a protein rearranges its configuration to differ from its native state. Generally, the nature of surface adsorption depends on the properties of the protein, surface and interactions, but many details remain unsolved [3, 5, 6].

Simplified protein models that capture the essential features of real proteins have thus been proposed in hope of unraveling the mysteries. Such coarse-grained models have the advantage of being readily accessible to computer simulation. With the emerging computer power nowadays, numerical simulation of such protein models play therefore an ever important role to the understanding of the problems. Nevertheless, they are still computationally intensive and unexpectedly difficult despite the simplicity of the models. In this work, we will investigate the hydrophobic-polar (HP) lattice protein model [7] (one of the simplest such model), subjected to an attractive surface. Thereby, we attempt to understand the thermodynamic and *generic* structural behavior of protein adsorption from a statistical mechanics point of view. Furthermore, we will study the influence of the strength of substrate attraction on the adsorption and folding processes of an HP protein.

2. Model

In an aqueous environment, a protein's hydrophobic amino acids tend to stay away from the solvent and form an interior core, while polar amino acids form an exterior shielding shell [8, 9]. This is known as the hydrophobic effect and is believed to be the driving force for the formation of tertiary structures [10–12]. The HP model [7] is a prototypical, coarse-grained lattice protein model introduced to mimic this phenomenon. It classifies amino acids into two types of monomers according to their affinity to water: hydrophobic (H) and polar (P). Interactions are restricted to an attractive coupling, of strength ε_{HH}, between non-bonded hydrophobic monomers occupying nearest-neighbor sites.

To simulate protein adsorption, an attractive substrate is placed at $z = 0$ on a 3-dimensional cubic lattice [13]. For simplicity, in this work we consider a surface that attracts all monomers with a strength ε_S. The energy of the system is then given by:

$$E = -\varepsilon_{HH} n_{HH} - \varepsilon_S n_S, \tag{1}$$

n_{HH} being the number of H-H interacting pairs and n_S being the number of monomers adjacent to the bottom surface. In addition, a non-attractive wall is placed at $z = N+1$ to confine the HP chain from above, where N is the number of monomers in the sequence. The purpose of placing the steric upper wall is to limit the vertical translational degrees of freedom of the protein, so as to shorten the time spent on simulating desorbed conformations. Figure 1 shows a schematic diagram of the model. Here, we have used a 36mer ($P_3H_2P_2H_2P_5H_7P_2H_2P_4H_2P_2HP_2$) [14] to illustrate the effect of ε_S on the structural "phase transitions" associated with protein adsorption.

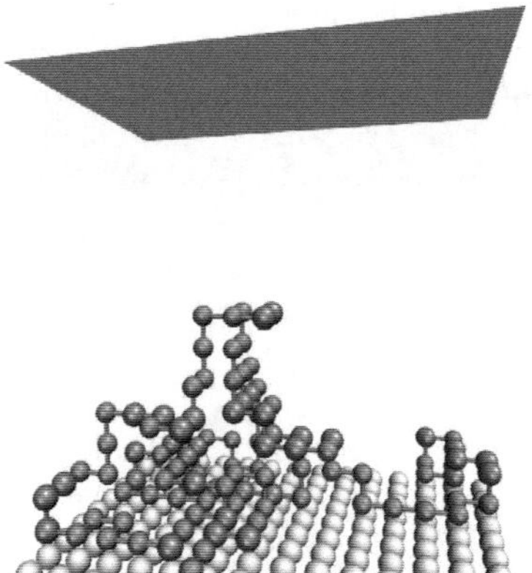

Figure 1. A schematic diagram showing the model used in this work. The gray spheres represent hydrophobic monomers, orange spheres represent polar monomers, faint spheres are the attractive molecules of the substrate and the solid top surface is non-attractive.

3. Methods

3.1. Wang-Landau sampling and production run

In order to investigate such conformational transitions, it is necessary to have a mean of sampling a protein's entire conformational/energy space efficiently. Here, in the first stage of the simulation, Wang-Landau (WL) sampling [15–18] has been employed to estimate the energy density of states, $g(E)$, which then gives access to thermodynamic quantities at any temperature. In this iterative procedure, a trial configuration of energy E_{trial} is generated with an acceptance probability inversely proportional to $g(E_{trial})$. The old configuration is retained if the trial one is rejected. A multiplicative modification factor f (with an initial value $f_{init} = e^1$ at the beginning of the simulation) is used to modify $g(E)$, i.e., $g(E) \to g(E) \times f$. A histogram in energy is also accumulated: $H(E) \to H(E) + 1$. When a "flat" histogram is attained, the simulation is brought to the next iteration: $H(E)$ is reset and f is reduced, $f \to \sqrt{f}$. A "flat" histogram is defined as when all entries in $H(E)$ are greater than $p \times H_{ave}$, where p is the flatness criterion and H_{ave} is the average of all entries in $H(E)$. All results presented in this work are obtained by using a flatness criterion $p = 0.8$ for reliable estimates of $g(E)$. The simulation is terminated when $\ln(f_{final})$ reaches a preset minimum value of 10^{-8}.

The partition function, $Z(T)$ and subsequent thermodynamics, e.g. average energy $\langle E\rangle_T$ and specific heat $C_V(T)$, then follow:

$$Z(T) = \sum_i g(E_i)e^{-E_i/kT}, \tag{2}$$

$$\langle E\rangle_T = \frac{1}{Z(T)}\sum_i E_i g(E_i)e^{-E_i/kT}, \tag{3}$$

$$C_V(T) = \frac{1}{kT^2}\left(\langle E^2\rangle - \langle E\rangle^2\right), \tag{4}$$

where k is the Boltzmann constant, T is the temperature and the sum runs over all possible energies.

The second stage of the simulation consists in a production run making use of multicanonical sampling [19,20] to generate a number of two-dimensional densities of states, $g(E, Q)$, where Q is any structural quantity of interest, such as e.g., the radius of gyration $R_g = \sqrt{\frac{1}{N}\sum_i^N (\vec{r}_i - \vec{r}_{cm})^2}$ ($\vec{r}_i$ is the position of the ith monomer and $\vec{r}_{cm}$ is the position of the center of mass), the number of surface contacts n_S, the number of surface contacts of hydrophobic monomer n_{SH}, the number of surface contacts of polar monomer n_{SP}, or the number of hydrophobic interaction pairs n_{HH}. During the production run, the inverse of $g(E)$, obtained from Wang-Landau sampling, is used as the weight of the acceptance rate, while structural quantities are calculated and two-dimensional histograms, $H(E, Q)$, are accumulated. At the end of the simulation, the $H(E, Q)$'s are reweighted by $g(E)$ in order to yield the two-dimensional densities of states, $g(E, Q)$'s. The partition function of an observable Q, $Z_Q(T)$, and its expectation value can then be calculated as

$$Z_Q(T) = \sum_{E,Q} g(E, Q)e^{-E/kT}, \tag{5}$$

$$\langle Q\rangle_T = \frac{1}{Z_Q(T)}\sum_{E,Q} Q g(E, Q)e^{-E/kT}. \tag{6}$$

Thermodynamics of the structural quantities in addition to the specific heat, C_V, are essential in identifying "transitions" between different structural "phases". In cases where the specific heat shows ambiguous signals, structural quantities help clarifying the types of transition taking place at different temperatures. In some cases distinct signals might be missing in the specific heat, whereas structural quantities are more reliable to identify structural transitions.

Journal of Physics: Conference Series **402** (2012) 012046 doi:10.1088/1742-6596/402/1/012046

3.2. Methodological pitfalls

Traditional Monte Carlo trial moves for lattice polymers either change a conformation locally (e.g. kink flip and crankshaft) or non-locally (e.g. pivot moves). Local moves generate new configurations fairly similar to the old ones as most parts of the polymer remains unchanged, inducing long correlation times in the simulation. Non-local moves do not share the same problem, but they are ineffective for dense conformations. Two inventive Monte Carlo trial moves, namely, pull moves [21] and bond-rebridging moves [22], have thus been implemented in our simulations to confront these problems. When combined with Wang-Landau sampling, they have proven to be particularly efficient in exploring conformational space [23, 24].

However, these non-traditional trial moves alone are not able to give correct low temperature thermodynamics if they are used with Metropolis sampling, which is easily trapped in metastable states. Figure 2 compares the two sampling methods, where the two transition peaks at low temperature in the specific heat are clearly missing in the Metropolis case. We thus stress that an appropriate combination of the sampling method and trial updates is crucial in obtaining correct results from a Monte Carlo simulation.

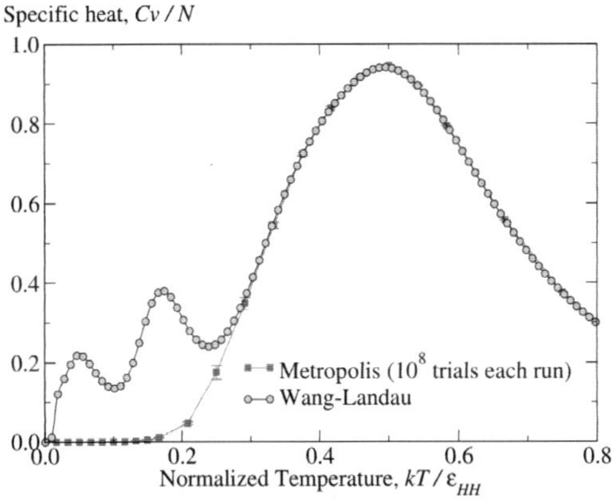

Figure 2. Comparison between Wang-Landau and Metropolis sampling in obtaining the specific heat of a 36mer interacting with a very weak attractive surface, in which $\varepsilon_{HH} = 12\varepsilon_S$. Error bars smaller than the data points are not shown.

4. Results

4.1. Identifying structural "phase transitions"

The different structural "phase transitions" are best illustrated by considering the 36mer interacting with a very weak attractive surface ($\varepsilon_{HH} = 12\varepsilon_S$). Detailed studies of this system can be found in Refs. [25, 26]. As shown in Figure 3, its specific heat has three distinct peaks, which represent three basic phase transitions, respectively (from high to low temperature): (i) hydrophobic core (H-core) formation, (ii) adsorption, and (iii) "flattening" of the adsorbed structure.

These three transitions can be identified by comparing the peak positions of the specific heat and those of the structural quantities. The thermal derivative of $\langle n_{HH}\rangle$ peaks for H-core formation, while the thermal derivatives of $\langle n_{SH}\rangle$ and $\langle n_{SP}\rangle$ peak for adsorption at a higher temperature, and flattening at a lower temperature due to the fact that the flattening process has to take place after the protein is adsorbed on cooling. In some cases, it is also possible that flattening is signaled by a shoulder, or only a peak in either $\frac{d\langle n_{SH}\rangle}{dT}$ or $\frac{d\langle n_{SP}\rangle}{dT}$ but not both.

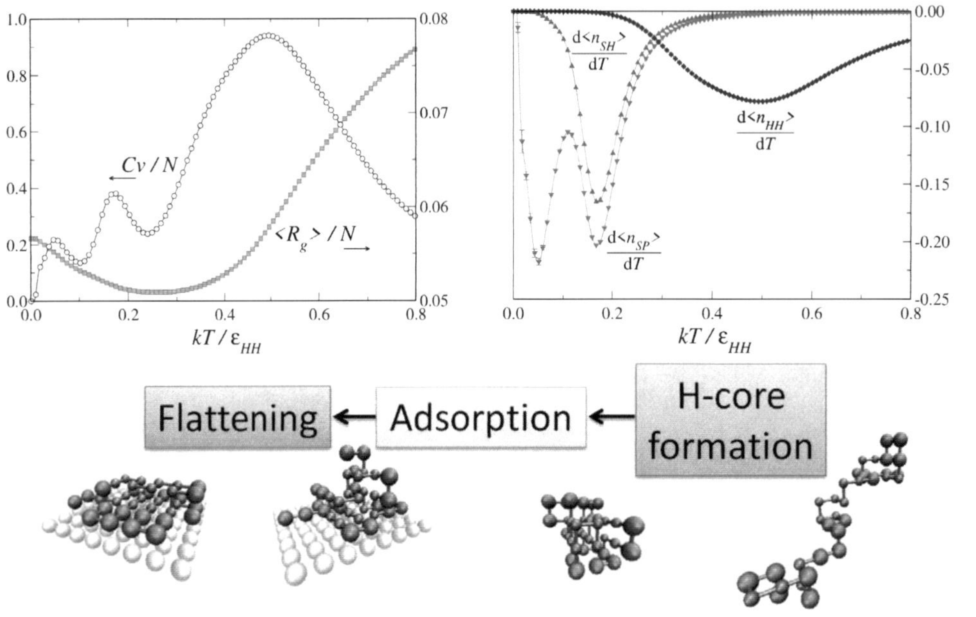

Figure 3. Specific heat and structural quantities of the 36mer interacting with a very weak attractive surface ($\varepsilon_{HH} = 12\varepsilon_S$). Error bars smaller than the data points are not shown.

4.2. Effect of surface attraction on the sequence of transitions

The three basic transitions occur at different temperatures when the surface attractive strength varies, giving rise to a *different order* in structural changes. As a consequence, an extended, desorbed protein goes through a different path in conformational space towards the acquisition of compact, adsorbed ground states. Structures of the intermediate and ground states thus vary from case to case and are completely dependent on this sequence (or order) of transitions.

When the surface attraction becomes stronger, it first affects the transitions at higher temperatures. Figure 4 shows the thermodynamics for the 36mer interacting with a surface of moderate attractive strength ($\varepsilon_{HH} = 2\varepsilon_S$). Despite the fact that only two peaks are present in the specific heat, the three basic transitions and their order of occurrence are revealed by the analysis of the structural parameters. Adsorption takes place at $kT/\varepsilon_{HH} \approx 1.0$ corresponding to the C_V peak at the higher temperature; a hydrophobic core forms at a slightly higher temperature ($kT/\varepsilon_{HH} \approx 0.35$) than that of flattening ($kT/\varepsilon_{HH} \approx 0.27$), and both processes are responsible for the C_V peak at the lower temperature.

In terms of the transition sequence, H-core formation and adsorption have swapped places compared to the former case. On cooling, a three-dimensional, adsorbed but extended structure is first formed after adsorption which takes place at the highest temperature. The lowest energy state with a two-dimensional hydrophobic core is achieved after the combined action of H-core formation and flattening. As these two processes closely overlap, no intermediate states could be singled out between them.

Further increase in surface attractive strength shifts the H-core formation to an even lower temperature as shown in Figure 5, where a strong attractive surface is used ($\varepsilon_{HH} = \frac{1}{2}\varepsilon_S$). In this case, there are also two peaks in the specific heat with a weak bump in between. A comparison with the structural properties clearly distinguishes the three basic transitions, which now occur

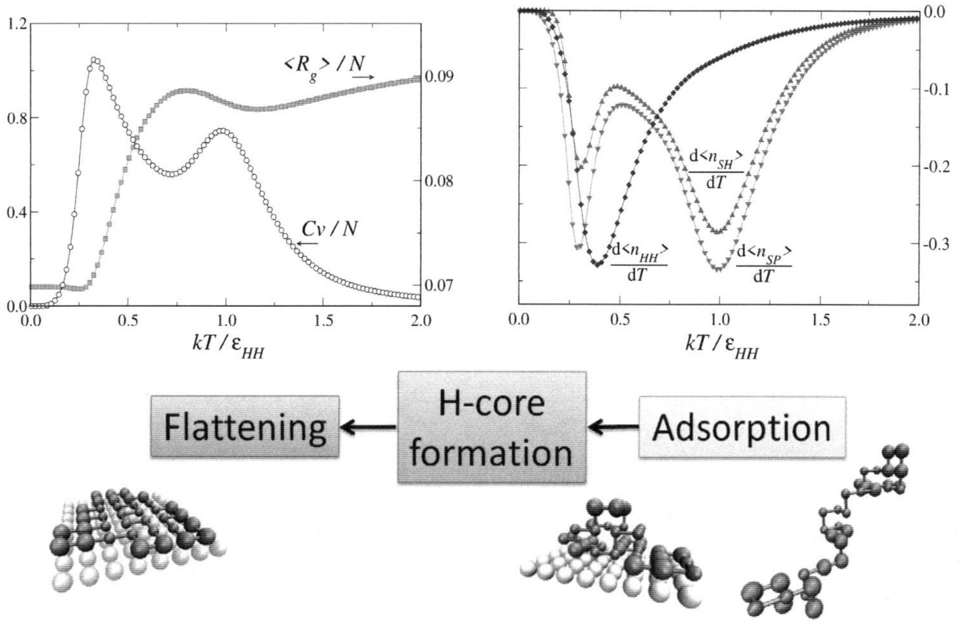

Figure 4. Specific heat and structural quantities of the 36mer interacting with a moderately attractive surface ($\varepsilon_{HH} = 2\varepsilon_S$). Error bars are smaller than the data points.

at well separated temperatures. Adsorption takes place at $kT/\varepsilon_{HH} \approx 4.0$; H-core formation at $kT/\varepsilon_{HH} \approx 0.4$; the bump occurs between $kT/\varepsilon_{HH} \approx 1.0$ and $kT/\varepsilon_{HH} \approx 3.0$ is a signal of flattening.

With this transition ordering, the desorbed, extended protein first adsorbs on the surface to form a three-dimensional, adsorbed yet extended structure. After the flattening process, most of the monomers contact with the surface but the chain is still not compact. The H-core formation finally takes place on the surface, forming a two-dimensional ground state with a hydrophobic core.

5. Conclusion

Protein adsorption has been studied using the HP lattice model and Wang-Landau sampling with efficient Monte Carlo trial updates. The combined thermodynamic signals of the specific heat and some suitable structural quantities allowed us to identify conformational "phase transitions". Three basic transitions, namely, hydrophobic core formation, adsorption and "flattening" of adsorbed structures, have been found to occur in a different sequence with varying surface attraction strength, ε_S, upon cooling. The acquisition of the adsorbed, compact, lowest energy states from a desorbed, extended coil thus goes through different paths in conformational space with different intermediates. The structures of ground state configurations may also differ from each other for different surface attractions.

Acknowledgments

This project is supported by the National Science Foundation (NSF) under grant no. DMR-0810223.

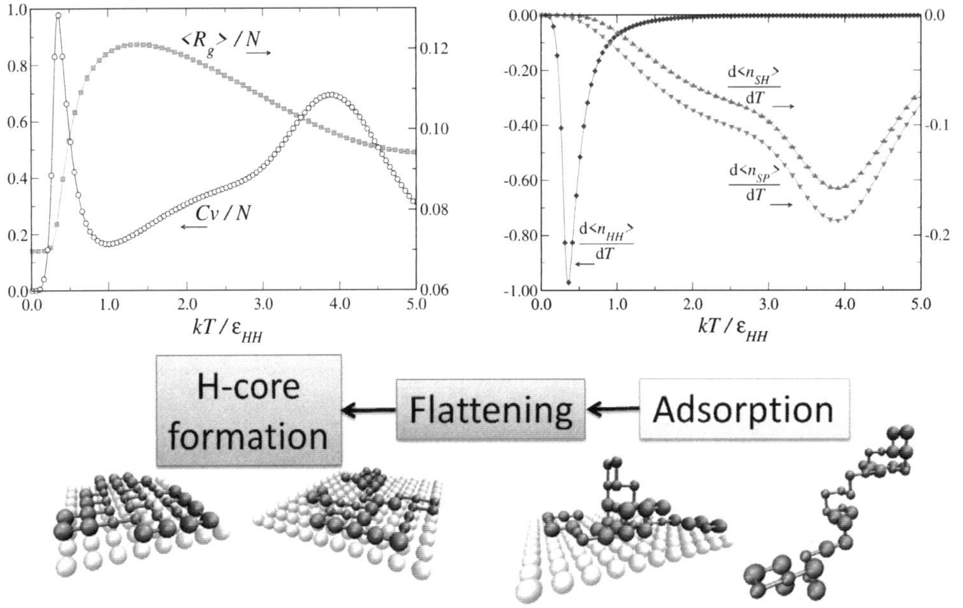

Figure 5. Specific heat and structural quantities of the 36mer interacting with a strong attractive surface, in which $\varepsilon_{HH} = \frac{1}{2}\varepsilon_S$. Error bars are smaller than the data points.

References

[1] Horbett T A and Brash J L 1995 *Proteins at Interfaces II: Fundamentals and Applications* (Washington, DC: American Chemical Society)
[2] Sarikaya M, Tamerler C, Jen A K -Y, Schulten K and Baneyx F 2003 *Nat. Mater.* **2** 577-85
[3] Hlady V and Buijs J 1996 *Curr. Opin. Biotech.* **7** 72-7
[4] Branden C and Tooze J 1999 *Introduction to Protein Structure, 2nd ed* (New York: Garland Science)
[5] Haynes C A and Norde W 1994 *Colloid Surface B* **2** 517-66
[6] Rabe M, Verdes D and Seeger S 2011 *Adv. Colloid Interface Sci.* **162** 87-106
[7] Dill K A 1985 *Biochemistry* **24** 1501
[8] Kendraw J C, Bodo G, Dintzis H M, Parrish R G, Wyckoff H and Phillips D C 1958 *Nature* **181** 662-6
[9] Kendraw J C, Dickerson R E, Strandberg B E, Hart R G, Davies D R, Phillips D C and Shore V C 1960 *Nature* **185** 422-7
[10] Sturtevant J M 1977 *Proc. Natl. Acad. Sci. USA* **74** 2236-40
[11] Baldwin R L 1986 *Proc. Natl. Acad. Sci. USA* **83** 8069-72
[12] Spolar R S, Ha J -H, Record M T Jr 1989 *Proc. Natl. Acad. Sci. USA* **86** 8382-5
[13] Bachmann M and Janke W 2006 *Phys. Rev. E* **73** 020901(R)
[14] Unger R and Moult J 1993 *J. Mol. Biol.* **231**, 75
[15] Wang F and Landau D P 2001 *Phys. Rev. Lett.* **86** 2050
[16] Wang F and Landau D P 2001 *Phys. Rev.* E **64** 056101
[17] Wang F and Landau D P 2002 *Comput. Phys. Commun.* **147** 674
[18] Landau D P, Tsai S -H and Exler M 2004 *Am. J. Phys.* **72** 1294
[19] Berg B A and Neuhaus T 1991 *Phys. Lett. B* **267** 249
[20] Berg B A and Neuhaus T 1992 *Phys. Rev. Lett.* **68** 9
[21] Lesh N, Mitzenmacher M and Whitesides S 2003 *Proc. Seventh Annual Int. Conf. on Computational Molecular Biology*, ed T Lengauer *et al.* (Berlin: RECOMB) p 188
[22] Deutsch J M 1997 *J. Chem. Phys.* **106** 8849
[23] Wüst T and Landau D P 2008 *Comput. Phys. Commun.* **179** 124-7
[24] Wüst T and Landau D P 2009 *Phys. Rev. Lett.* **102** 178101
[25] Li Y W, Wüst T and Landau D P 2011 *Comput. Phys. Commun.* **182** 1896-9
[26] Wüst T, Li Y W and Landau D P 2011 *J. Stat. Phys.* **144** 638-51

Driving Calmodulin Protein towards Conformational Shift by Changing Ionization States of Select Residues

Sunita Negi[1]**, Ali Rana Atilgan and Canan Atilgan**
Sabanci University, Faculty of Engineering & Natural Science,
Tuzla, 34956 Istanbul, Turkey

E-mail: snegi@sabanciuniv.edu

Abstract. Proteins are complex systems made up of many conformational sub-states which are mainly determined by the folded structure. External factors such as solvent type, temperature, pH and ionic strength play a very important role in the conformations sampled by proteins. Here we study the conformational multiplicity of calmodulin (CaM) which is a protein that plays an important role in calcium signaling pathways in the eukaryotic cells. CaM can bind to a variety of other proteins or small organic compounds, and mediates different physiological processes by activating various enzymes. Binding of calcium ions and proteins or small organic molecules to CaM induces large conformational changes that are distinct to each interacting partner. In particular, we discuss the effect of pH variation on the conformations of CaM. By using the pKa values of the charged residues as a basis to assign protonation states, the conformational changes induced in CaM by reducing the pH are studied by molecular dynamics simulations. Our current view suggests that at high pH, barrier crossing to the compact form is prevented by repulsive electrostatic interactions between the two lobes. At reduced pH, not only is barrier crossing facilitated by protonation of residues, but also conformations which are on average more compact are attained. The latter are in accordance with the fluorescence resonance energy transfer experiment results of other workers. The key events leading to the conformational change from the open to the compact conformation are (i) formation of a salt bridge between the N-lobe and the linker, stabilizing their relative motions, (ii) bending of the C-lobe towards the N-lobe, leading to a lowering of the interaction energy between the two-lobes, (iii) formation of a hydrophobic patch between the two lobes, further stabilizing the bent conformation by reducing the entropic cost of the compact form, (iv) sharing of a Ca^{+2} ion between the two lobes.

1. Introduction

Calmodulin (CaM) is a protein which has been extensively studied due to its pivotal role as an intracellular calcium ion receptor as well as in the calcium signaling pathways in the eukaryotic cells [1]. CaM is observed to have different conformations influenced by the external environmental factors such as temperature, ionic strength and pH (see[2] and references cited therein). Furthermore, this protein can bind to a variety of other proteins or small organic compounds, and mediates different physiological processes by activating various enzymes [3, 4]. Binding of Ca^{+2} and proteins or small organic molecules to CaM induces large conformational changes that are distinct to each interacting partner [1, 4, 5].

[1] Corresponding author.

The plethora of data on CaM conformations come from diverse sources, including (i) "static" methods such as x-ray crystallography [6], (ii) dynamic methods that are based on averaging over an ensemble of structures, such as nuclear magnetic resonance (NMR) [7], and (iii) dynamic information obtained from single molecule spectroscopy approaches, such as fluorescence resonance energy transfer (FRET) [8]. There is plethora of data that identify CaM as a protein which samples multiple conformations. They also distinguish the shifts in the distribution of their populations by changing environmental conditions such as pH, Ca^{+2} concentrations, and ionic strength. None of these, however, are able to directly identify the exact conformations that are sampled by the protein. They also fall short of directly describing the events that trigger the changes between different conformational states, which occur on millisecond time scales.

Experimentally, the conformation distributions of CaM are studied by FRET under various environmental conditions[8]. Of particular interest to us is that compaction of the protein was observed by lowering the pH from 7.4 to 5.0 [9]. The main drawback with these experimental studies is that they neither directly identify the exact conformations sampled by CaM, nor hint at the events that trigger the conformational changes[9]. Here, we aim to study the conformational multiplicity of CaM at reduced pH.

Computationally, Donnini et al. [10] present a method, implemented within molecular dynamics (MD) package GROMACS [11], for constant pH MD simulations in explicit solvent that is based on the λ-dynamics approach. Therein, the dynamics of the titration coordinate λ, which interpolates between the protonated and deprotonated states, is driven by generalized forces. Lee et al[12] introduced another method for performing constant-pH molecular dynamics (PHMD) simulations utilizing the generalized Born implicit solvent model. This approach employs an extended Hamiltonian in which continuous titration coordinates propagate simultaneously with the atomic motions of the system.

In this manuscript, we discuss the effect of pH variation on the conformations of calcium-loaded CaM by a different approach. The pKa values of the charged residues are determined; those charged residues with values shifted into the physiological range are determined. As a result, the protonation states of nine residues are changed to mimic the overall electrostatics of the protein at pH 5.0. MD simulations of length 190 ns are carried out by assigning these new protonation states. We find that CaM undergoes a conformational shift that leads to a more compact form at a lower pH value. We describe the key events that lead to the final compact conformation.

2. Computational Model: Molecular Dynamics

CaM consists of 148 amino acid residues, consisting of three domains, namely the N-lobe, the C-lobe and the central linker/helix as shown in Fig.1(a) [residues 5-68:N lobe, residues 69-91:linker, residues 92-147:C lobe]. There exist two forms of the CaM protein whereby apo-CaM is free of calcium ions, while holo-CaM is fully loaded with four calcium ions. In this study we focus on holo-CaM which is commonly observed to interact with other proteins[13]. Fully loaded CaM free of other ligands has two distinct experimentally determined conformations: In many x-ray structures, CaM has an open form with a dumbbell shape containing two domains, joined by an extended linker. It is exemplified by the protein data bank (PDB [14]) structure 3CLN. Another form is compact, whereby the central linker is bent (PDB code 1PRW). Each of the lobes holds two calcium ions (shown as grey spheres in figure 1) in their respective EF-hand motifs. Upon binding a ligand, the domains move relative to each other and the flexible linker region changes conformation accordingly.

We use the PDB structure coded 3CLN as the initial conformation. The data for the residues 1-4 and 148 are not included in this structure as the X-ray data for these residues are not available. We use the NAMD package to model the dynamics of the protein-water system[15]. The protein is soaked in a water box forming a layer of at least 10 Å measured from any residue. The protonation of the residues are determined such that we mimic the most probable state of each ionizable residue at pH 5. Details on the parameters of a similar simulation at pH 7.4 may be found in[2] where results from 120 ns long MD simulations are reported. For the reduced pH environment, we protonate nine residues as described next.

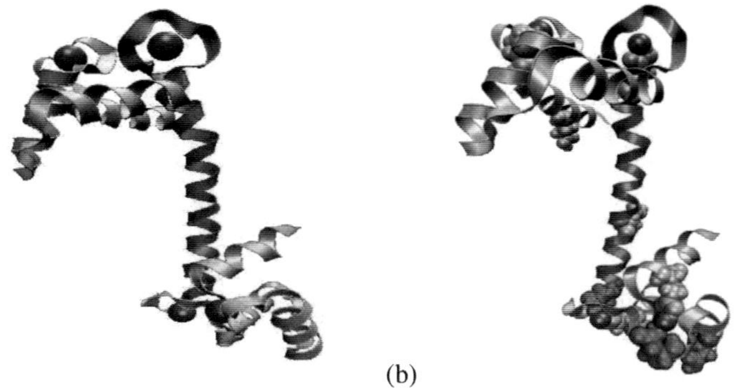

(a) (b)

Figure 1 (a) Fully loaded Calmodulin (CaM) with four calciums; N-lobe shown in green, the linker in purple and the C-lobe in cyan. (b) Calmodulin shown with the nine residues which are protonated so as to mimic the most probable ionization states at pH 5 (GLU : pink, ASP: red).

Four different servers H++ [16, 17], pKd[18, 19], propKa 3.1[20] and PHEMTO[21] are used to calculate the pKa of all the titratable residues. There are a total of 52 such sites in CaM, two of which are Tyr with high pKa values, and one His which we take to be uncharged. The standard error on the mean of the remaining Glu/Asp/Arg/Lys residues is calculated to be less than 0.5 units in 34 cases, and less than 1.3 units for 44 cases. In the remaining five, the residues are either consistently protonated above pH 5 (e.g. for residue Glu 67, we have pKa = 11.2, 5.4, 13.4 and 8.5, respectively for H++, pKd, propKa 3.1 and PHEMTO) or they are deprotonated for at least three of the four pKa predictors. This lets us assign the protonation states consistently. We therefore find that the acidic residues 11, 31, 67, 84, 93, 104, 122, 133 and 140 have upshifted pKas to 5.5 or more, compared to the standard values of ca. 4. Out of these nine residues 11, 31, 67 are on the N-lobe, 84 is on the central linker and the remaining five residues are on the C-lobe; GLU and ASP residues are shown in pink and red respectively in Fig.1(b). All other negatively charged residues are observed to be near their standard pKa values of approximately 4 in the protein environment, so that they are kept in the charged state throughout the simulation. None of the positively charged residues have downshifted values, and their protonation states are also unaltered.

The system is solvated using the VMD 1.8.7 program with solvate plug-in version 1.2[22]. The CharmM27 force field parameters are used for protein and water molecules[23]. Water molecules are described by the TIP3P model. The initial box has dimensions 72×92×72 Å containing 31660 atoms neutralized by standard addition of Na^+ and Cl^- ions. Long range electrostatic interactions are calculated by the particle mesh Ewald method[24], with a cutoff distance of 12 Å and a switching function at 10 Å. RATTLE algorithm[25] is applied to use a step size of 2 fs in the Verlet algorithm. Temperature control is carried out by Langevin dynamics with a dampening coefficient of 5/ps. Pressure control is attained by a Langevin piston. Volumetric fluctuations are preset to be isotropic. The system is run in the NPT ensemble at 1 atm and 310 K until volumetric fluctuations are stable to maintain the desired average pressure. The run in the NPT ensemble is extended to a total of 190 ns. The coordinate sets are saved at 2 ps intervals for subsequent analysis, leading to $T = 95000$ snapshots.

We note that we carried out two separate simulations. The first one is of length 190 ns, where all results reported here are calculated from. The second one is 100 ns long, and is used to confirm that the main features observed in the first trajectory are repeatable. We found that the same conclusions are reached from both sets of simulations.

3. Results and Discussion

3.1 Conformation change obtained at pH 5.0

As discussed above, we protonate nine residues so as to mimic their most probable states at pH 5 and perform extensive MD simulations. For a comparison with the FRET experiment results[9], we monitor the distance between residues 34 and 110 in CaM throughout the simulation. The cumulative distribution of separation distance between residues 34 and 110 is shown in figure 2, as the simulation proceeds. On average, the distance shrinks with time, implying that there is a change in the conformation of the protein during the simulation.

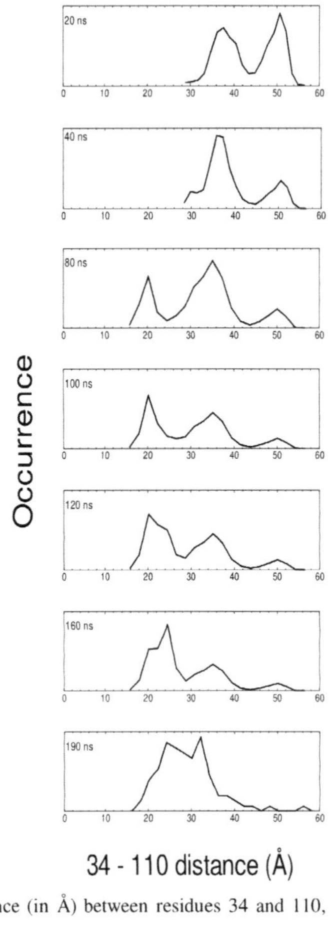

Figure 2 The distribution of the distance (in Å) between residues 34 and 110, evaluated cumulatively until the indicated instants of the simulation.

The root mean square distances (RMSDs) of the different parts of the protein are calculated with respect to the original structure and is presented in figure 3. While the overall conformation of the chain shows abrupt changes in two steps, first at ca. 15 ns, then at ca. 65 ns, the rearrangements within the different units are relatively small. Thus, the main conformational change is due to the relative arrangement of the units with respect to each other. Inspecting this figure more closely, we further find that the C-lobe motions are relatively larger, while the linker maintains its original helical structure. The latter point is important because various authors have argued that the loss of the helicity in the central linker might be the source of the conformational multiplicity of CaM [26-28].

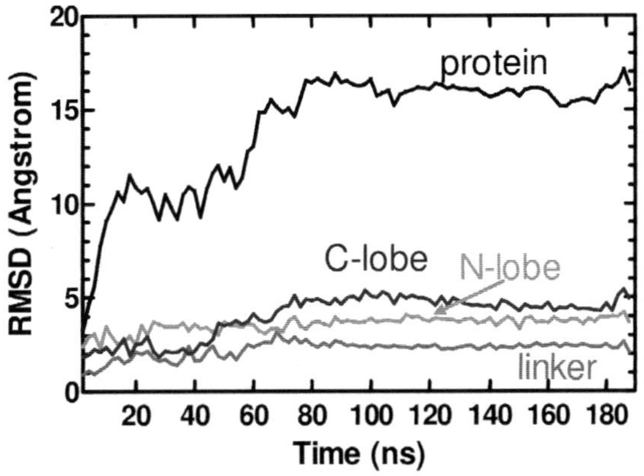

Figure 3 RMSD as a function of time at PH 5.0. RMSD of C-lobe is observed to be higher than that of N-lobe and linker

To capture the main features of the relative motions of the two lobes and the linker, we reduce the protein structure to five points in space. These points are schematically shown at the top of figure 4. Three of these points are the center of masses (COMs) of the N terminus (point 1), C terminus (point 5), and the linker (point 3). In addition, residues 69 and 91 are used to mark the beginning and end points of the linker (points 2 and 4). We then define two main degrees of freedom to capture the essence of the motions of the two lobes relative to each other. θ angle defines the bending motion observed between the N and C lobes, while φ angle defines the relative rotation of the two lobes around the linker as a virtual dihedral angle. Similar virtual dihedral angle definitions on CaM were previously made[29].

As seen in figure 4 (bottom), during the first 40 ns, the bending angle covers a relatively restricted region of values in the range 120-180°, while the torsion angle is more flexible, with values in the range 60-140°. In our previous work, where all nine acidic residues are in a deprotonated state, this range of values are observed for the whole 120 ns of the trajectory[2], while extension of those simulations further to 200 ns kept the same sampling range (data not shown). Thus, there is normally a barrier to bending of the two lobes with respect to each other. However, in this work, starting at ca. 55 ns, the bending motion is realized. At ca. 80 ns, the bending angle obtains a value near 110° implying a closer association of the N- and C-lobes. By 100 ns, the bending angle is observed to attain values as small as 90°. For the last 70 ns, the protein is observed to have come to a stable point where the (θ, φ) is (100±10°,100±20°).

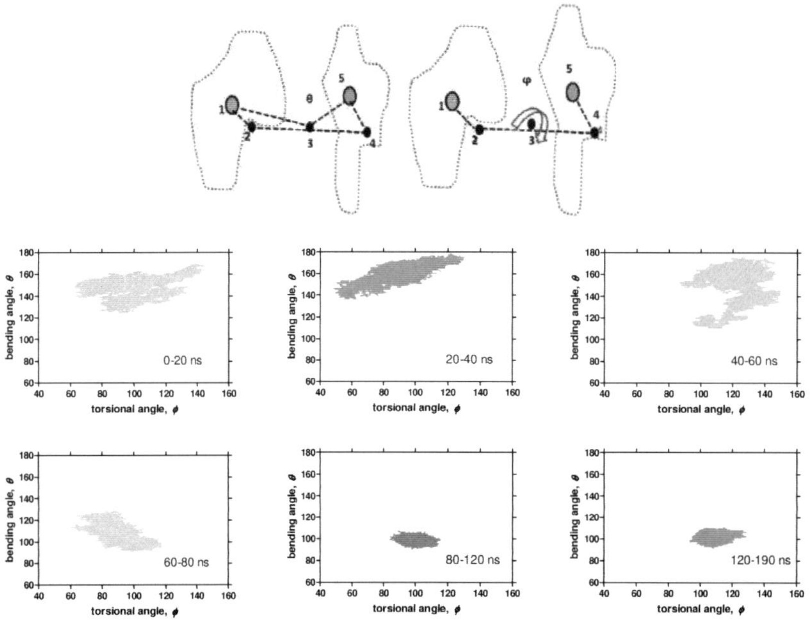

Figure 4 (top) Schematic representations of the defined bending angle θ and torsional angle φ. (bottom) (θ, φ) values covered in different parts of the simulation. A stable configuration is attained by the protein for the last 120 ns.

3.2 Shifts in the pKa values of the residues

The shift in the pKa value of a titratable residue depends on the change in its local environment, reflected as free energy differences. Therefore, large conformational shifts may substantially change the pKa values of these residues. While monitoring the key events taking place during the simulation, we also calculate the pKa values of all the charged residues. The trends for selected residues are shown in figure 5. Residues 11, 84 and 139 are observed to show a monotonic increase in their pKa values during the simulation. Note that of these, 11 and 84 are amongst the protonated residues in this MD simulation, while 139 is not. Other than these three residues, there are none which show a significant change in their pKa value during the simulation.

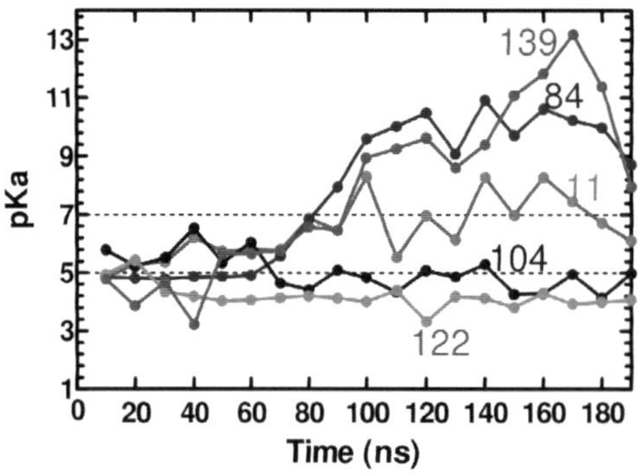

Figure 5 Changes in the pKa values of selected residues during the simulation, calculated using the propKa 3.1 server.

3.3 Interaction energy between the n and c

Using the "NAMD energy" plugin in VMD, we calculate the interaction energy between the N- and the C-lobes as shown in the figure 6. The electrostatic interactions are observed to make the major contribution to the non-bonded interaction energy. A dip is observed in the energy around 40 ns followed by two larger dips at 76 ns and 120 ns. The van der Waals part of the energy is observed to be an order of magnitude smaller. After 120 ns, the energy is observed to increase for ca. 20 ns, and stabilizes thereafter. We seek to associate the main events that take place during the conformational change with the inter-lobe interactions next. Since the overall stability of the system is determined by the free energy, we conjecture that entropic contributions must also play a role to compensate for the raise in the interaction energy along the trajectory.

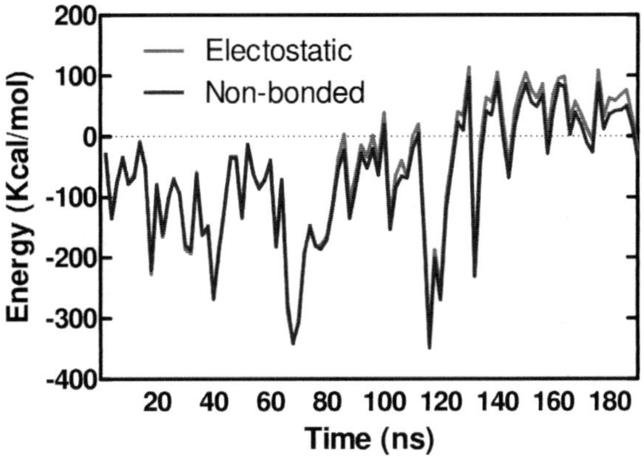

Figure 6 The non-bonded interaction energy between N and the C-lobes throughout the MD simulation; the electrostatic contribution is also shown, displaying that the van der Waals term has negligible effect on the conformational change.

3.4 Key events leading to the conformational change

In order to provide a more detailed understanding of the observed conformational changes, we study all the possible interactions between residues in the protein throughout the simulation. A salt bridge formation is observed at 40 ns between the residues 74 (ARG) of the linker and 7 (GLU) of the N-lobe as shown in figure 7a. This salt bridge restricts the motions of the N-lobe, and its formation causes a dip in the energy of the system. Once formed, it is present in the rest of the simulation. At 67 ns, the calcium ion in the third EF-hand motif, which is located in the C-lobe immediately following the linker, moves slightly away as displayed in figure 7b. This event is immediately followed by the partial unfolding of the EF-hand motif, causing a bending motion in the rest of the C-lobe. The major cause of this event is that the protonation of residue 104, which is part of this particular EF-hand motif, weakens the grip of the motif on the Ca^{+2}. This is also the reason why the C-lobe attains a higher RMSD value than both the linker and the N-lobe from this point onward (recall figure 3). The overall result of the loosened motif is the association of the C-lobe with the N-lobe which causes a dip in the energy around this point.

As shown in figure 7c, a hydrophobic patch is fully formed at 80 ns which helps stabilize the system; since hydrophobicity has entropic contribution to the free energy, this patch also compensates for the increase in the interaction energy of the two lobes, observed from 67 ns to 80 ns. Close to 120 ns, the calcium ion of the EF-hand motif 4 comes and sits in the center of the two lobes as shown in figure 7d. By 156 ns, this calcium is observed to keep this position at the center of the N and the C lobes, stabilized by two negatively charged residues 84 and 139. The linker is retains its helical structure throughout the simulation. This form of the linker is in contrast to those observed in other ligand-bound, compact forms of CaM where the linker is bent into two at position 74[29]. In order to monitor the stochasticity of our MD simulation, we repeat this simulation for 100 ns with a different random number seed and observe a similar sequence of events: A conformational change from the open to the closed form occurs at 90 ns, preceeded by a salt bridge formation between the linker and the N-lobe at 64 ns and followed by the hydrophobic patch formation at 92 ns.

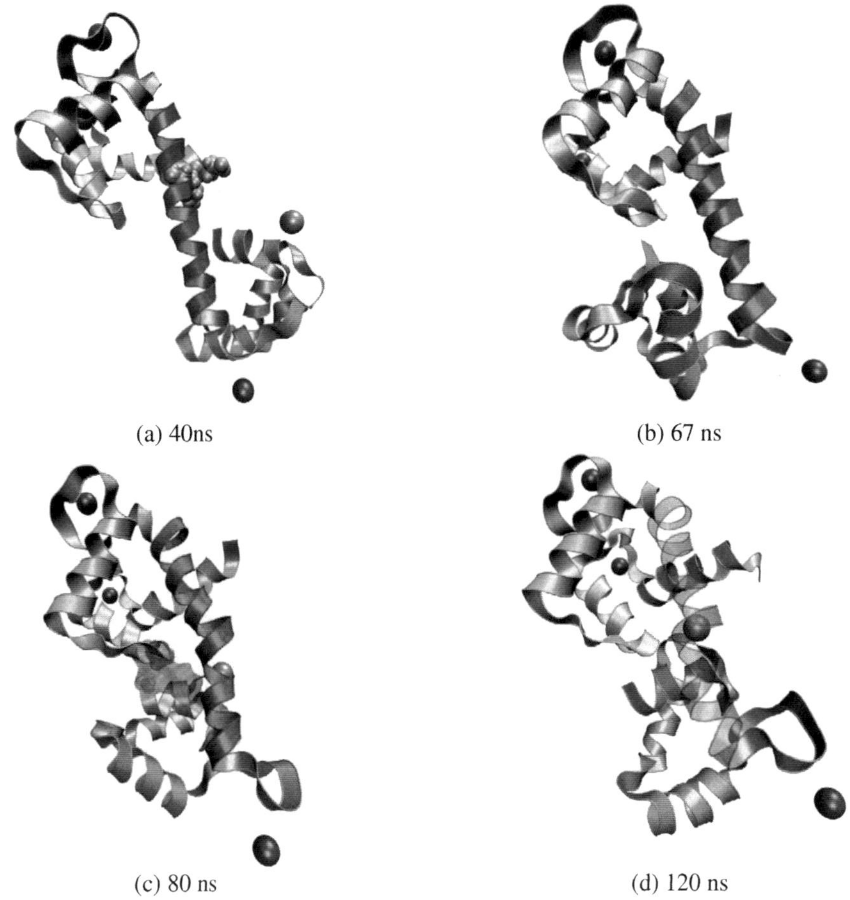

Figure 7. Snapshots from the MD simulation; N-lobe is in green, linker is in purple, C-lobe is in cyan. At (a) 40 ns, a salt bridge is formed between residues 74 (ARG) in the linker and 7 (GLU) in the N-lobe, these residues are shown in space-filling representation; (b) 67 ns, a calcium from the N-lobe is observed to move away from its EF-hand motif, this event is immediately followed by an increased mobility in the associated EF-hand motif; (c) 80 ns, a hydrophobic patch is formed on the other side of the protein shown by the new space filling atoms adjoining the two lobes; (d) 120 ns, a calcium comes and sits in the center of the two lobes, stabilized by negatively charged residues from each lobe. This form is kept in the rest of the simulation.

4. Summary

Fully loaded CaM is studied by MD simulations under conditions mimicking the most probable protonation states of residues at pH 5. By lowering the pH, local conformational changes in the protein are induced, leading to shifts in the energy landscape. These pave the way to the compact forms reminiscent of those detected by indirect experimental methods (compare figure 2 to figure 3 in ref[9]). A salt bridge and a hydrophobic patch are observed to be formed as key events leading to the conformational change by stabilizing the system. This conformation change in fully-loaded CaM is observed in two independent simulations started with different random number seeds.

Our current view suggests that at high pH, barrier crossing to the compact form is prevented by the repulsive electrostatic interactions between the two lobes. Thus, the time scale of jumps between open-closed forms of CaM is measured on the order of 100 μs. Lowering of pH significantly reduces the barrier height, while possibly also changing the overall conformational landscape.

Adaptation to different pH values in various subcellular compartments is thought to be directly related to protein stability[30] and pH of optimal binding affinity of interacting proteins[31]. Methods to determine how proteins adapt to cellular and sub-cellular pH are currently sought-after[32]. These findings lend clues as to how the conformations of proteins and therefore their associated functions may be modulated in the differentially controlled pH of the different compartments of the cell. The findings in the current study must be supported by additional MD simulations whereby the protonation states of not all but only subsets of the nine residues acted upon in this work are altered. Our work in this direction is under way.

Acknowledgments. This work was supported by the Scientific and Technological Research Council of Turkey Project (grant 110T624). SN is supported by a Turkish Academy of Sciences postdoctoral grant. We thank Ayse Ozlem Aykut for her participation in our discussions.

5. References

1. M. Ikura and J. B. Ames, *Proceedings of the National Academy of Sciences of the United States of America*, 2006, **103**, 1159-1164.
2. A. R. Atilgan, A. O. Aykut and C. Atilgan, *Journal of Chemical Physics*, 2011, **135**, 15.
3. J. L. Fallon and F. A. Quiocho, *Structure*, 2003, **11**, 1303-U1307.
4. M. Vandonselaar, R. A. Hickie, J. W. Quail and L. T. J. Delbaere, *Nature Structural Biology*, 1994, **1**, 795-801.
5. J. P. Junker and M. Rief, *Proceedings of the National Academy of Sciences of the United States of America*, 2009, **106**, 14361-14366.
6. Y. S. Babu, C. E. Bugg and W. J. Cook, *Journal of Molecular Biology*, 1988, **204**, 191-204.
7. I. Bertini, C. Del Bianco, I. Gelis, N. Katsaros, C. Luchinat, G. Parigi, M. Peana, A. Provenzani and M. A. Zoroddu, *Proceedings of the National Academy of Sciences of the United States of America*, 2004, **101**, 6841-6846.
8. B. Slaughter, M. Allen, J. Unruh, R. Urbauer and C. Johnson, *Journal of Physical Chemistry B*, 2004, **108**, 10388-10397.
9. B. D. Slaughter, J. R. Unruh, M. W. Allen, R. J. B. Urbauer and C. K. Johnson, *Biochemistry*, 2005, **44**, 3694-3707.
10. S. Donnini, F. Tegeler, G. Groenhof and H. Grubmuller, *Journal of Chemical Theory and Computation*, 2011, **7**, 1962-1978.
11. E. Lindahl, B. Hess and D. van der Spoel, *Journal of Molecular Modeling*, 2001, **7**, 306-317.
12. M. S. Lee, F. R. Salsbury and C. L. Brooks, *Proteins-Structure Function and Bioinformatics*, 2004, **56**, 738-752.
13. A. Crivici and M. Ikura, *Annuual Review of Biophysics and Biomolecular Structure*, 1995, **24**, 85-116.
14. H. M. Berman, J. Westbrook, Z. Feng, G. Gilliland, T. N. Bhat, H. Weissig, I. N. Shindyalov and P. E. Bourne, *Nucleic Acids Research*, 2000, **28**, 235-242.

15. J. C. Phillips, R. Braun, W. Wang, J. Gumbart, E. Tajkhorshid, E. Villa, C. Chipot, R. D. Skeel, L. Kale and K. Schulten, *Journal of Computational Chemistry*, 2005, **26**, 1781-1802.
16. J. C. Gordon, J. B. Myers, T. Folta, V. Shoja, L. S. Heath and A. Onufriev, *Nucleic Acids Research*, 2005, **33**, W368-W371.
17. J. Myers, G. Grothaus, S. Narayanan and A. Onufriev, *Proteins* 2006, **63**, 928-938.
18. B. M. Tynan-Connolly and J. E. Nielsen, *Nucleic Acids Research*, 2006, **34**, W48-W51.
19. T. Connolly, B.M. and J. Nielsen, *Protein Science*, 2007, **16**, 239-249.
20. M. Rostkowski, M. H. M. Olsson, C. R. Sondergaard and J. H. Jensen, *Bmc Structural Biology*, 2011, **11**, 6.
21. A. A. Kantardjiev and B. P. Atanasov, *Nucleic Acids Research*, 2009, **37**, W422-W427.
22. W. Humphrey, A. Dalke and K. Schulten, *Journal of Molecular Graphics*, 1996, **14**, 33.
23. B. R. Brooks, R. E. Bruccoleri, B. D. Olafson, D. J. States, S. Swaminathan and M. Karplus, *Journal of Computational Chemistry*, 1983, **4**, 187-217.
24. T. Darden, L. Perera, L. P. Li and L. Pedersen, *Structure with Folding & Design*, 1999, **7**, R55-R60.
25. H. C. Andersen, *Journal of Computational Physics*, 1983, **52**, 24-34.
26. G. Fiorin, R. R. Biekofsky, A. Pastore and P. Carloni, *Proteins-Structure Function and Bioinformatics*, 2005, **61**, 829-839.
27. A. Likić, R. Gooley, P. Speed and E. Strehler, *Protein Science*, 2005, **14**, 2955-2963.
28. C. M. Shepherd and H. J. Vogel, *Biophysical Journal*, 2004, **87**, 780-791.
29. W. Wriggers, E. Mehler, F. Pitici, H. Weinstein and K. Schulten, *Biophysical Journal*, 1998, **74**, 1622-1639.
30. P. Chan and J. Warwicker, *BMC Biology*, 2009, **7**, 69.
31. Z. Zhang, S. Witham and E. Alexov, *Physical Biology*, 2011, **8**, 3.
32. B. Garcia-Moreno, *Journal of Biology*, 2009, **8**, 98.91-98.94.

Conformational transitions of a confined lattice protein: A Wang-Landau study

Busara Pattanasiri[1,2,3], Ying Wai Li[1], David P. Landau[1], Thomas Wüst[4] and Wannapong Triampo[2,5,6]

[1]Center for Simulational Physics, Department of Physics and Astronomy, University of Georgia, Athens, GA 30602, USA
[2]R&D Group of Biological and Environmental Physics, Department of Physics, Faculty of Science, Mahidol University, Bangkok 10400, Thailand
[3]Department of Chemistry, Faculty of Science, Mahidol University, Bangkok 10400, Thailand
[4]Swiss Federal Research Institute WSL, Zürcherstrasse 111, CH-8903 Birmensdorf, Switzerland
[5]ThEP Center, CHE, 328 Si Ayutthaya Road, Bangkok 10400, Thailand
[6]Institute for Innovative Learning, Mahidol University, 999, Phuttamonthon 4 Road, Salaya, Nakorn Pathom 73170, Thailand

E-mail: bairng@physast.uga.edu, ywli@physast.uga.edu, dlandau@physast.uga.edu, thomas.wuest@wsl.ch, wtriampo@gmail.com

Abstract.
We use Wang-Landau sampling with suitable Monte Carlo trial moves to study a hydrophobic-polar (HP) lattice protein confined between two parallel, attractive walls. The density of states is determined iteratively by a random walk in energy space. Thermodynamic and structural properties, such as specific heat, number of surface contacts and number of H-H monomer pairs, are then calculated. When the surface attraction is comparable to the internal attraction among the hydrophobic monomers in the chain, two conformational "transitions", adsorption at higher temperature and collapse at lower temperature, have been analyzed based on these properties. This transition behavior depends on the variation of surface separation.

1. Introduction

Proteins are the most important molecules in living cells. They can fold to native structures to perform their own specific biological functions, e.g. cell structure stabilization, transporting chemical compounds, and converting chemical energy into mechanical energy [1, 2]. Misfolding to incorrect three-dimensional conformations produces inactive proteins that lead to toxicity. It can cause several diseases, such as Alzheimer's, Huntington's, Parkinson's, and cancer-related syndromes [3–5]. As a result, there is a wide interest in studying parameters that influence the conformational stability of proteins in solution or cellular environment.

Some experimental and theoretical work indicates that confinement can have a stabilizing effect on protein folding to the native state [6–10], understanding confinement effects on protein folding is thus important. Moreover, protein in confined spaces is also an interesting topic in biotechnology and food processing applications [11, 12].

A number of techniques are proposed to study proteins *in vivo* and *in vitro*, e.g. surface plasmon resonance (SPR) imaging [13], ellipsometry [14] and atomic force microscopy [15]. However, monitoring the role of proteins in the living cell is much more difficult to perform and

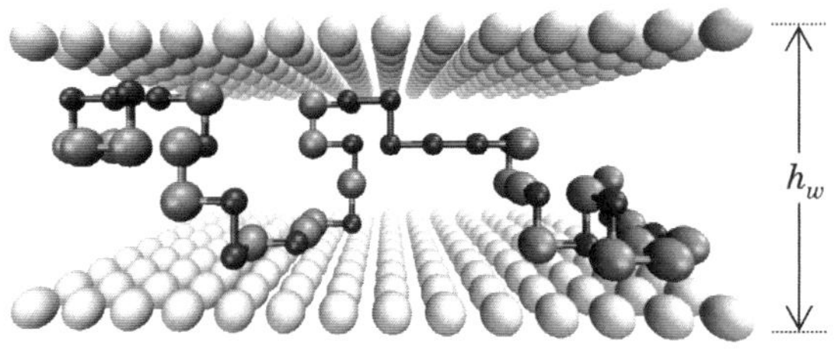

Figure 1. Schematic diagram of an HP protein between two attractive surfaces. The big orange spheres and the small black spheres represent the P-monomers and H-monomers, respectively. The light spheres are attractive surfaces molecules.

most of the methods are restricted to indirect observations. The capability of modern computer simulations offers an opportunity to overcome these limitations. A widely used subject of study is a simplified but efficient protein model, hydrophobic-polar (HP) model, introduced by Dill in 1985 [16].

In this work we focus on the thermodynamic and structural properties of a confined HP lattice protein between two parallel, attractive surfaces. We have employed Wang-Landau sampling [17–20] with suitable Monte Carlo trial updates, pull moves [21] and bond-rebridging moves [22], to obtain the density of states. The effect of volumetric spaces between two attractive surfaces is also discussed.

2. HP lattice protein model

The hydrophobic-polar (HP) model is a simplified model for investigating protein folding to a native state. In this model, an amino acid is classified as either a hydrophobic residue (H) or a polar residue (P) due to their affinity with water. The protein sequence is modeled as a self-avoiding walk on a 3-dimensional cubic lattice. To mimic the hydrophobic effect, an attraction with strength ε_{HH} is assigned between two non-bonded, nearest-neighbor hydrophobic residues so as to produce the hydrophobic core as found in the native structure of a real protein. However, it should be noted that since the protein sequence and interactions are overly simplified, the model introduces an artificial degeneracy to the system. An HP chain behaves similarly as a real protein in the sense that it folds to structures with a minimum free energy, but the "native" state structure is not unique.

In order to study a confined protein in our system, two attractive surfaces are constructed parallel to the xy-plane, at $z = 0$ and $z = h_w$. The energy function of an HP sequence in a confined space can thus be calculated as

$$E = -\varepsilon_{HH} n_{HH} - \varepsilon_S n_S, \tag{1}$$

where n_{HH} is the number of hydrophobic pairs, n_S is the number of monomers adjacent to the attractive surfaces, and ε_S is the surface attractive strength. A schematic diagram of our system is shown in Figure 1.

3. Wang-Landau sampling and trial moves

Wang-Landau (WL) sampling [17–20] is used to estimate the density of states in energy, $g(E)$. Starting from an arbitrary configuration and an initial guess of $g(E) = 1$ for all energies E, a random walk in energy space is then accomplished by generating trial configurations. The acceptance probability from the current energy E_1 to a new energy E_2 is

$$p(E_1 \to E_2) = min\left(\frac{g(E_1)}{g(E_2)}, 1\right). \tag{2}$$

Every time a new configuration is accepted, $g(E_2) \to g(E_2) \times f$; otherwise, $g(E_1) \to g(E_1) \times f$, where f is the modification factor which is initially set to $e \simeq 2.71828....$ During the random walk, the histogram of visited states is accumulated, i.e., $H(E) \to H(E) + 1$. When the histogram is sufficiently "flat", $H(E)$ is reset and a new random walk is performed with a smaller modification factor $f = \sqrt{f}$. In order to yield an accurate density of states, $H(E)$ is said to be flat when all the entries of $H(E)$ are greater than 80% of their average and the final modification factor f is less than $\exp(10^{-8})$.

During the simulation, two types of trial moves, pull moves [21] and bond-rebridging moves [22], are used for generating new configurations. These moves have been proven to be a powerful for ground state search and accurate estimation of $g(E)$ when combined with WL sampling [23]. Bond-rebridging moves are efficient to reorder the monomers in a highly dense conformation; however, its acceptance rate is rather low. To compensate for this, a higher calling probabilty (80%) is assigned to bond-rebridging moves, while 20% of the new trial configurations are generated from pull moves.

4. Thermodynamics of physical quantities

After the density of states is obtained from WL sampling, the partition function at a particular temperature, Z, with respect to the total energy of the system E can then be calculated as

$$Z = \sum_E g(E)e^{-E/k_BT}, \tag{3}$$

where T is temperature and k_B is the Boltzmann constant. From the partition function, any desired thermodynamic properties can be derived. For instance, the average energy $\langle E \rangle$ and specific heat C_V can be calculated as

$$\langle E \rangle = \frac{1}{Z}\sum_E Eg(E)e^{-E/k_BT}, \tag{4}$$

$$C_V = \frac{1}{k_BT^2}(\langle E^2 \rangle - \langle E \rangle^2). \tag{5}$$

5. Structural properties and their thermodynamics

Structural quantities are often employed to gain more understanding of the physical behavior and properties of the system. A common application of structural characterization is to identify the conformational transitions of the overall system. Examples include the number of surface contacts, n_S, and the number of H-H pairs, n_{HH}. They provide information about adsorption and hydrophobic core (H-core) formation of the polymer chain, respectively.

We can estimate the average structural quantity, $\langle Q \rangle$, at any temperature from the 2D density of states $g(E, Q)$ by performing a random walk in energy and Q space with WL sampling. However, it is time-consuming if many structural quantities are of interest. A more efficient way is proceeding with a production run from the 1D density of states $g(E)$ obtained from a previous WL sampling. In this process, the acceptance probability is proportional to $1/g(E)$ as usual. During a random walk in energy space, any desired structural quantities Q can be calculated and their corresponding 2D histograms, $H(E, Q)$, are accumulated. At the end of this process, $g(E, Q)$ is obtained by re-weighting the histogram, $g(E, Q) = g(E)H(E, Q)$.

The partition function of structural quantity Q at a particular temperature, Z, and the average of Q, $\langle Q \rangle$, can then be obtained:

$$Z = \sum_{E,Q} g(E,Q) e^{-E/k_B T}, \tag{6}$$

$$\langle Q \rangle = \frac{1}{Z} \sum_{E,Q} Q g(E,Q) e^{-E/k_B T}. \tag{7}$$

6. Results and Discussion

We performed simulations on a benchmark 48mer sequence (PHPHPPPPHPHPHPPHPHHH-HHHPPHHHPHPPHPHHPPHPHHHPPPPH), which has been designed by Yue *et al* [24] for algorithm testing purpose. In this sequence, the numbers of H-monomers and P-monomers are the same. Both surfaces interact with all monomers regardless of their type and the surface strength is set to be the same as the internal H-H interaction, i.e., $\varepsilon_S = \varepsilon_{HH} = 1$.

To understand the folding behavior of a confined protein, we simulated the case of a HP chain located between two attractive surfaces with surface separation $h_w = 49$, i.e., the chain can touch both surfaces only when it is a vertical straight chain. In this case, two peaks in specific heat, corresponding to conformational transitions, are observed as shown in Figure 2. The nature of these two peaks can also be identified by examining $d\langle n_S \rangle/dT$ and $d\langle n_{HH} \rangle/dT$ as shown in the lower panel. The peaks in C_V and in $d\langle n_S \rangle/dT$ at higher temperature, $kT/\varepsilon_{HH} \approx 2.15$, correspond to an adsorption transition, where an extended chain-like structure in free space adsorbs on a surface.

Reducing the temperature further makes the chain keep forming surface contacts until most of the monomers are on the surface. We called this process a "flattening" of the structure, and it is signaled by a small bump in $d\langle n_S \rangle/dT$ around $kT/\varepsilon_{HH} \approx 1.0$. In this case, the flattening process is not signaled in C_V. The peak at lower temperature, $kT/\varepsilon_{HH} \approx 0.4$, represents the hydrophobic core (H-core) formation in 2-dimensions as $d\langle n_{HH} \rangle/dT$ also shows a peak at the same temperature. Typical configurations are shown in the middle panel.

When the surface separation is reduced, vertical movements of the chain are restricted, resulting in a less pronounced adsorption peak due to a reduction in the number of conformations available. See the cases of $h_w = 49$ and $h_w = 25$ in Figure 3. Further reduction in the surface separation decreases the adsorption peak height until it becomes a shoulder, as in the cases of $h_w = 10$ and $h_w = 6$. However, this transition behavior is different from that for larger surface separations. In this case, the chain can have contacts with both surfaces, forming "bridges" between them even at high temperature. To maximize the number of surface contacts, the chain needs to detach from one surface and adsorb on the other until most of the monomers sit on the same surface. We called it a "debridging" process, which becomes harder to occur when the surface separation is smaller, resulting in a shift of the shoulder in C_V to a lower temperature. The debridging transition is still signaled by a peak in $d\langle n_S \rangle/dT$ as for the adsorption transition.

H-core formation temperature is not affected by surface separation. It forms a 2-dimensional H-core when the chain is already in contact with either surface, except when it is very squeezed by the two surfaces as in the $h_w = 3$ case. The chain transforms from an extended squeezed chain to a 3-dimensional squeezed globule since all monomers touch the surfaces all the time. Therefore, only a single H-core formation peak with an exceptionally large magnitude is observed in C_V and $d\langle n_{HH} \rangle/dT$, respectively.

Figure 4 shows the "adsorption" temperatures for different h_w, as obtained from the peak positions in C_V and $d\langle n_S \rangle/dT$ for the adsorption or debridging process. The difference in adsorption temperatures determined by these two measures systematically widens with the reduction of surface separation. It should be noted that there is no finite size scaling in the HP model, as each sequence is unique even with the same chain length [25]. To understand the

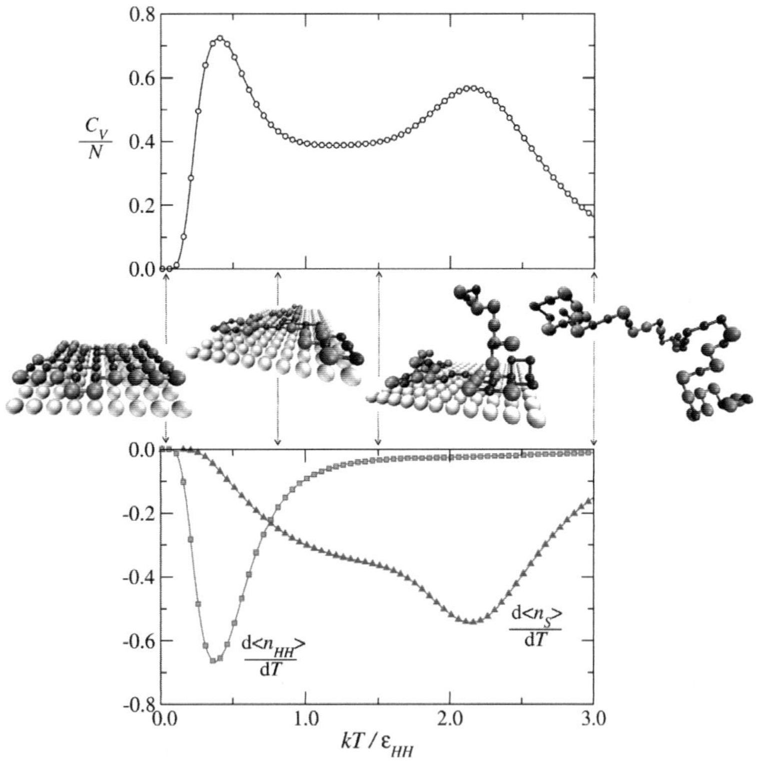

Figure 2. Thermodynamics of the 48mer confined between two attractive surfaces with $\varepsilon_S = \varepsilon_{HH} = 1$ and $h_w = 49$. Upper panel is the specific heat; middle panel is the typical configurations at each temperature; lower panel is the thermal derivative of the number of surface contacts, $d\langle n_S\rangle/dT$, and that of the number of H-H pairs, $d\langle n_{HH}\rangle/dT$. The big orange spheres and the small black spheres represent the P-monomers and H-monomers respectively. The light spheres are attractive surfaces molecules. Error bars are smaller than the data points.

origin of the difference, one may start from the energy function of the system in Eq. 1. The specific heat and its thermal derivative can be expressed as:

$$\begin{aligned} C_V &= \frac{d\langle E\rangle}{dT} \\ &= -\varepsilon_{HH}\frac{d\langle n_{HH}\rangle}{dT} - \varepsilon_S\frac{d\langle n_S\rangle}{dT}, \end{aligned} \tag{8}$$

$$\frac{dC_V}{dT} = -\varepsilon_{HH}\frac{d^2\langle n_{HH}\rangle}{dT^2} - \varepsilon_S\frac{d^2\langle n_S\rangle}{dT^2}. \tag{9}$$

By considering the lower panel in Figure 3, we see that $d^2\langle n_{HH}\rangle/dT^2 \geq 0$ at the adsorption temperature determined from the C_V peak. This implies $d^2\langle n_S\rangle/dT^2 \leq 0$, since the left hand-side in Eq. 9 is zero at the adsorption temperature. The adsorption peak in $d\langle n_S\rangle/dT$, therefore, occurs at the same or at a higher temperature than that of the peak in C_V.

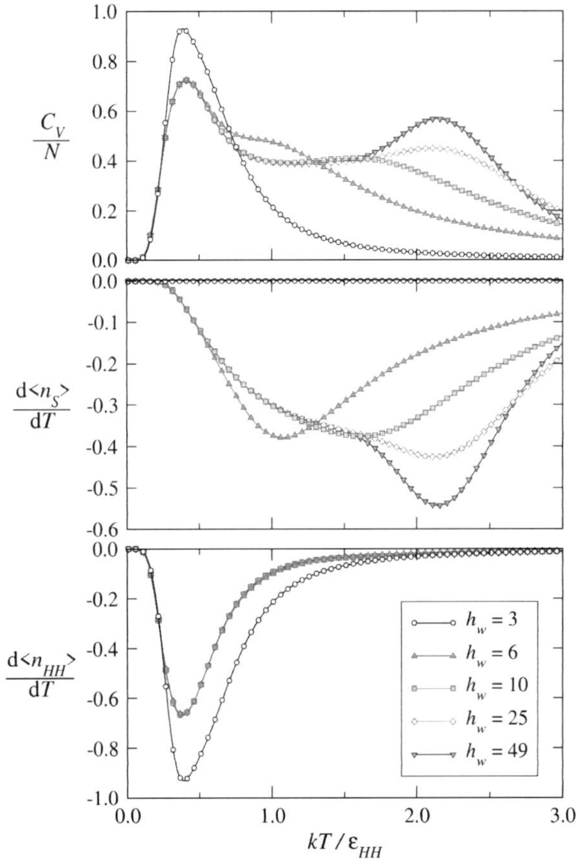

Figure 3. Thermodynamics of the 48mer confined between two attractive surfaces with different separation distances, h_w. $\varepsilon_S = \varepsilon_{HH} = 1$ in all cases. Upper panel is the specific heat; middle panel is the thermal derivative of number of surface contacts, $d\langle n_S\rangle/dT$; lower panel is the thermal derivative of number of H-H pairs, $d\langle n_{HH}\rangle/dT$. Error bars are smaller than the data points.

The systematic changes in the difference in the two adsorption temperatures determined by C_V and $d\langle n_S\rangle/dT$ is caused by the different values of $d^2\langle n_{HH}\rangle/dT^2$. Narrowing the surface separation shifts the adsorption to a lower temperature, where $d^2\langle n_{HH}\rangle/dT^2$ has a larger value. The difference between the adsorption peak positions in C_V and $d\langle n_S\rangle/dT$ are therefore enhanced.

7. Conclusions

In this work, a coarse-grained lattice model, the HP model, and Wang-Landau sampling with suitable Monte Carlo trial moves, pull moves and bond-rebridging moves, have been used to study a protein confined by two attractive surfaces. Narrowing the separation between the two attractive surfaces decreases the vertical degrees of freedom of the chain, resulting in a reduction in the magnitude and a shift to a lower temperature of the adsorption peak, while the collapse transition temperature is not affected. The difference in "adsorption" temperatures

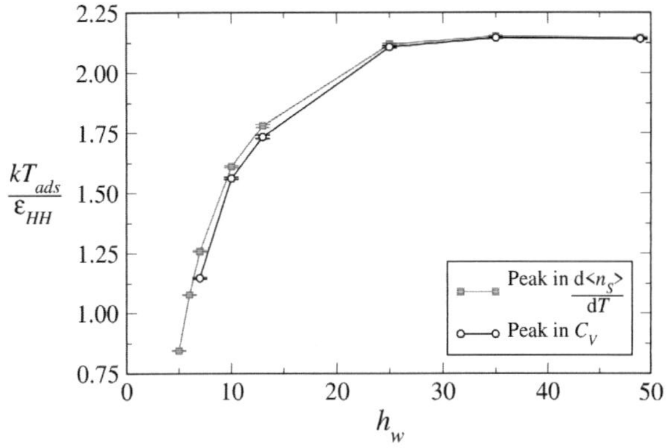

Figure 4. Adsorption temperatures for different separation distances, h_w, determined by the adsorption peaks in the specific heat, C_V, and the thermal derivative of number of surface contacts, $d\langle n_S\rangle/dT$, respectively.

determined by the peak positions in the specific heat and the derivative of the number of surface contacts is also observed, and it systematically increases with the reduction of surface separation. However, the details of this "adsorption" behavior might be different. It could simply be that a desorbed chain adsorbs on a surface, or that the bridging structure originally connecting the two surfaces gets destroyed, depending on the surface separation. Further investigation is necessary to distinguish these two conformational transition processes.

Acknowledgments

We thank M. Bachmann, S. Schnabel, and T. Vogel for fruitful discussions. This work is supported by the National Science Foundation (NSF) under grant no. DMR-0810223. B. Pattanasiri is supported by The Royal Thai Government Scholarship under the Development and Promotion of Science and Technology Talent Project (DPST), and the grant from The Institute for the Promotion of Teaching Science and Technology (IPST). W. Triampo is supported by Faculty of Science, Mahidol University, and the Thailand Center of Excellence in Physics (ThEP).

References

[1] Creighton T E 1993 *Proteins: Structures and Molecular Properties* (New York: W. H. Freeman)
[2] Collarini E J and Oxender D L 1987 *Ann. Rev. Nutr.* **7** 75–90
[3] Taubes G 1996 *Science* **271** 1493–5
[4] Selkoe D J 2004 *Nat. Cell. Biol.* **6** 1054–61
[5] Forloni G *et al* 2002 *Neurobiol Aging* **23** 957–76
[6] Javidpour L and Sahimi M 2011 *J. Chem. Phys.* **135** 125101–12
[7] Mittal J and Best R B 2008 *PNAS* **105** 20233–38
[8] Lucent D, Vishal V and Pande V S 2007 *PNAS* **104** 10430–34
[9] Tian J and Garcia A E 2011 *J. Am. Chem. Soc.* **133** 15157–64
[10] Sorin E J and Pande V S 2006 *J. Am. Chem. Soc.* **128** 6316–17
[11] Martin C R and Kohli P 2003 *Nat. Rev. Drug Discov.* **2** 29–37
[12] Foegeding E A, Luck P J and Davis J P 2006 *Food Hydrocolloids* **20** 284–92
[13] Jordan C E and Corn R M 1997 *Anal. Chem.* **69** 1449–56
[14] Seitz R, Brings R and Geiger R 2005 *Appl. Surf. Sci.* **252** 154–57
[15] Agnihotri A and Siedlecki C A 2004 *Langmuir* **20** 8846–52
[16] Dill K A 1985 *Biochemistry* **24** 1501–9
[17] Wang F and Landau D P 2001 *Phys. Rev. Lett.* **86** 2050–53

[18] Wang F and Landau D P 2001 *Phys. Rev.* E **64** 056101–16
[19] Wang F and Landau D P 2002 *Comput. Phys. Commun.* **147** 674–77
[20] Landau D P, Tsai S -H and Exler M 2004 *Am. J. Phys.* **72** 1294–302
[21] Lesh N, Mitzenmacher M and Whitesides S 2003 *Proc. Seventh Annual Int. Conf. on Computational Molecular Biology*, ed T Lengauer *et al.* (Berlin: RECOMB) p 188
[22] Deutsch J M 1997 *J. Chem. Phys.* **106** 8849–54
[23] Wüst T and Landau D P 2009 *Phys. Rev. Lett.* **102** 178101–4
[24] Yue K, Fiebig K M, Thomas P D, Chan H S, Shakhnovich E I and Dill K A 1995 *Proc. Natl. Acad. Sci. USA.* **92** 325–29
[25] Bachmann M and Janke W 2004 *J. Chem. Phys.* **120** 6779–91